Production Research:
Approaching the 21st Century

Production Research: Approaching the 21st Century

Edited by

Mark Pridham
Christopher O'Brien

Taylor & Francis

London • New York • Philadelphia

USA	Taylor & Francis Inc., 1900 Frost Road, Suite 101, Bristol, PA 19007
UK	Taylor & Francis Ltd. 4 John St. London WC1N 2ET

British Library Cataloguing in Publication Data
Computer-aided ergonomics.
1. Ergonomics
I. Karwowski, Waldemar II. Genaidy, Ashraf M. III. Asfour, Shihab S., *1952–*
620.8′2

ISBN 0-85066-753-4

British Library Cataloguing in Publication Data
International Conference on Production Research, (10th; 1989; Nottingham, England).
Proceedings of the 10th International Conference on Production Research.
1. Manufacturing industries. Technological development
I. Title II. Pridham, Mark
670.427

ISBN 0-85066-752-6

Cover design by Jordan and Jordan, Fareham, Hampshire (based on an image generated by the SAMMIE System, with permission of SAMMIE C.A.D. Ltd.)

Phototypesetting by
Graphicraft Typesetters Ltd., Hong Kong.
Printed in Great Britain by Burgess Science Press, Basingstoke.

Contents

PART I PLENARY PAPERS 1

1 Perspectives on current production research
C. O'Brien 3

2 Computer-integrated business: corporate and information system management in competitive environments
H.-J. Bullinger, J. Niemeier 9

PART II PRODUCTION PROCESSES 27

Section II.1 Grinding 29

3 A new method for grinding stability analysis
Li Li, Xu YanShen, Sun RongXi, Peng ZeMin 31

Section II.2 Cutting 37

4 An expert system for the automation of the broach design process
W.R. Terry, E.R. Sims Jr, C.W. Richards, J.A. Deno 39

5 Approaches to cutting and packing problems
H. Dyckhoff 46

6 Interactive approaches to packing problems
E.E. Bischoff 55

Section II.3 Machining 63

7 Identifying and controlling chatter in turning with a feed rate control system
Yu Junyi, Zheng Wangbin, Bao Shanfei 65

8 A study of cutting area change under chatter vibration in lathe turning
Y. Iwata 73

Section II.4 Processes 83

9 A statistical quality control design for wavesolder process
M.R. Khan, P. Rega 85

10 Small deep hole drilling by electro-discharge machining
X.K. Chen, A.W. Duffill 92

11 Review of CINPRES gas-assisted moulding process
N.J. Quick, A.P. Jarvis, K.J. Stout 98

Section II.5 Tools 105

12 The design and development of a flexible loom board system for the integrated manufacture of cable harnesses
G.P. Bertolotti, B.J. Griffiths 107

13 A graphical approach to computer-aided stress analysis in the design of bandsaw teeth
S.R. Bradbury, M. Sarwar 113

Section II.6 Measurement 121

14 Two-dimensional optical Fourier transform for surface texture assessment
V.M. Huynh, S. Kurada, W. North 123

15 Measurement of robot-cut grooves in glass
M.R. Morshedizadeh, C. Wykes, M. Howarth 132

PART III PRODUCTION AND INVENTORY MANAGEMENT 137

Section III.1 Inventory 139

16 The long-term response of inventory holdings to technological developments and economic influence
C.D.J. Waters 141

17 Inventory policies with credit-period incentives
F.J. Arcelus, G. Srinivasan 150

18 Optimal profit policy by substitutional sales promotion
T. Shimosakon, S. Kuriyama 155

Section III.2 Production Control 163

19 Automatic adaptive scheduling of multi-processor cells
F.H. Grant, S.Y. Nof 165

20 A new concept on generative bills-of-material: the generic bill-of-material
E.A. van Veen, J.C. Wortmann 173

21 Joint determination of production cycle and inspection schedules under inspection errors
H. Ohta, S. Ogawa 179

22 Job-shop evaluation using Petri nets
J.-M. Proth 189

Section III.3 Scheduling 197

23 The development of a generic bottle-neck scheduler
J. Driscoll, S. Hurley 199

24 One-machine scheduling with random tasks durations and general precedence constraints
J.-C. Hennet 208

Section III.4 Production Planning 215

25 An expansion of the virtual cell concept for scheduling computer-integrated manufacturing systems
J.R. Drolet, C.L. Moodie, B. Montreuil 217

26 Kitting systems for small size parts in manual assembly systems
M.I. Johansson 225

27 Hierarchical production planning: an approach for a two-stage production system
K. Morikawa, N. Nakamura 231

28 The Silver-Meal heuristic: worst-case bounds for finite horizons
S. Axsäter 239

29 Application of precedence and function diagrams to assembly planning
K. Thaler 244

30 The design and implementation of a factory co-ordination system
R. Bowden, J. Duggan, J. Browne 251

31 Treatment of bottle-neck machines in cellular manufacturing systems
H. Seifoddini 257

32 MRP II systems: success factors in the process of implementation
A.G. Lockett, P.R.N. Barrar, M.E. Polding 263

33 Why are production management practices different?
A. Chikán, C. Whybark 270

34 Future-orientated production structures: computer-aided methods and tools to plan and innovate integrated structures of production cells
U. Hallwachs 275

35 On the relation between batch mode, bottle-neck machines and maximum throughput
P.J. Weeda 283

36 Hierarchical integration of aggregate and detailed production planning systems
H. Tsubone, T. Tanaka, M. Sugawara 290
37 Optimal due-date setting in an assembly shop
T.C.E. Cheng 297
38 Sequencing of parts to be manufactured by flexible manufacturing systems
Zhang Bingli, Liang Xichang 305
39 Model-mix balancing: more flexibility to improve the general result
F. Fremerey 314

PART IV PRODUCTION ECONOMICS AND PRODUCTIVITY 323
Section IV.1 Productivity 325
40 Productivity practices in the United Kingdom
A.I. Stainer 327
41 Performance evaluation of unreliable workstations in flexible manufacturing systems
V. Albino, O.G. Okogbaa 332
Section IV.2 Economic Factors 337
42 The creativity curve and its measurement
Shingo Kataoka, Kozi Morooka, Jairo Hasbun, Luisa Lino 339
43 New developments in cost-management systems using the MRP data base
I. Hayakawa 353
44 Optimum surface feet in flexible manufacturing systems workstations
L. Abdel-Malek 360
45 Inventory policies with freight discounts and disposals
F.J. Arcelus, J.E. Rowcroft 366
46 Evaluation of flexible capital investments—with consequences for decisions and suppliers
I. Persson, C.-H. Nilsson, H. Nordahl 371

PART V HUMAN FACTORS ISSUES IN PRODUCTION 379
Section V.1 Ergonomics and Human Factors 381
47 Sociotechnical evaluation of assembly systems in the Swedish automobile industry
J. Lindér 383
48 Effect of learning on perceptual motor load in a combined manual and decision task
I. Ogawa, A. Raouf 392
49 Workplace design and working postures for repair and maintenance tasks
C.M. Haslegrave 401
50 A study of the occupational safety and ergonomic aspects of material flow
J. Järvinen, S. Herranen 407
51 Experimental evaluation of control selection using different robot teach pendant designs
S.V. Gray, C.S. Syan, J.R. Wilson 413
52 A multifactorial approach to the design of an industrial seat
H. Gregg, E.N. Corlett 420

PART VI MANUFACTURING SYSTEMS AND AUTOMATION 427

Section VI.1 Robotics 429

53 Precise assembly by use of reference measurement and sensory feedback
A. Mohammadion, C.N. Larsson 431

54 Image-based cell calibration in a graphical off-line programming environment
H. Aalto, J. Joensuu, I. Kauhaniemi 436

55 A distributed information-control structure for robotic systems
G. Yasuda, K. Tachibana 442

Section VI.2 Automation 451

56 Development of a programmable parts feeder for rectangular parts
P.S.K. Chua, E. Sung 453

57 Flexible porous materials handling in a CIM environment
P.R. Nicholson, M. Sarhadi 460

58 Optimal flow path design of unidirectional AGV systems
M. Kaspi, J.M.A. Tanchoco 467

59 Task planning for the flexible manufacture of sheet metal enclosures
N.W. Hird, C.F. Earl, P. Holland, R.J. Stamp 474

60 Manual versus flexible assembly: a cross comparison of performance and costs in four different countries
A. Mital 481

61 A routing control method of automated guided vehicles by the shortest path with time-windows
S. Fujii, H. Sandoh, R. Hohzaki 489

62 Assembly sequence planning using operation networks
H.-J. Bullinger, U.A. Seidel 495

63 Cause-based failure recovery approach for automated manufacturing cell components
Y.B. Moon, C.L. Moodie 504

Section VI.3 Flexible Manufacturing Systems 511

64 Determining and controlling the quality of prismatic parts machined in a flexible manufacturing cell environment
G.T. Smith 513

65 A flexible system performance analysis in probabilistic and degraded machining conditions
L. Zavanella, G.C. Maccarini, A. Bugini 525

66 Effective work organization in flexible automation
P. Seppälä, E. Tuominen 532

67 FMS in assembly and its specific demands on logistics and material supply
R. Koether 537

68 Optimal part routing in flexible manufacturing systems
C. Arbib, L. Perugini 544

69 A computer-based system for automatic cutting tool selection
C.S. Syan 551

70 A comprehensive group technology system for cellular manufacture
G. Harhalakis, I. Minis, R. Nagi, J.M. Proth 559

PART VII EXPERT SYSTEMS AND PRODUCTION SIMULATION 567

Section VII.1 Artificial Intelligence 569

71 Intelligent control in automated manufacturing environments
N. Nakamura, K. Noma, M. Isohashi 571

72 Linking shop floor and management systems in real-time
U.S. Bititci, A.S. Carrie, C.R. McMillan 578
73 The application of an AI-based vision system to robotic assembly
D.J.M. Elliott, B.J. Griffiths 589
Section VII.2 Expert Systems 595
74 Knowledge-based systems in computer-assisted production
K.-P. Fähnrich, G. Groh, M. Thines 597
75 Handling heuristic control logic in the domain of a manufacturing simulation
G. Fozzard, J. Spragg, D. Tyler, D. Sinclair, D. Parker 610
76 An expert system: tool replacement in flexible manufacturing
C. Giardini, G. Maccarini, A. Bugini 617
77 An expert system approach to the design of flexible manufacturing systems
M.C.K. Kwok, A.S. Carrie 625
78 Development of a knowledge-based system for synchronizing production in a two-stage, geographically decentralized manufacturing process
W.R. Terry, T.W. Matz 633
Section VII.3 Simulation 639
79 Obtaining resolution IV designs using Taguchi's orthogonal arrays
K.E. Bullington, J.N. Hool, S. Maghsoodloo 641
80 System performance in parallel assembly systems with sequence restrictions due to synchronized feeding of materials
M.I. Johansson 646
81 Simulation of work-time distributions for a paced assembly line
T.J. Gallwey, M. O'Muirthile 653
82 A microcomputer simulation software for scheduling production in a dynamic and flexible jobbing environment
R.S. Costa, E.G.M. Jardim 660
83 The simulation of a flexible delivery system for large scale make-to-order manufacture
B. Wu, P.H. Lowe 669
84 A simulation for scheduling in FMS
M.P. Rajith, C.E. Reddy, O.V.K. Chetty 679
85 Planning the implementation of JIT manufacturing with the aid of simulation modelling
M.D. Byrne 686
86 Simulation of the production disturbances involving human intervention at danger zones
K.K. Häkkinen, R. Booth, S. Väyrynen, J. Haijanen, M. Hirvonen 696
87 The use of simulation for short-term decision making in production—a case study at a fabric dyeworks
B.L. MacCarthy 703

PART VIII COMPUTER-AIDED MANUFACTURE 715
Section VIII.1 Computer-aided design/computer-aided manufacture 717
88 Vehicle scheduling models in automated production systems
J.P. Cyrus 719
89 Form feature recognition for manufacturing
C.S. Chen, S.P. Owusu-Ofuri 726
90 Computer-aided component selection for precision assembling
H.S. Shan, A. Satyawadi 734
91 Design and manufacture of jewellry using CAD/CAM/CAI techniques
B.S. Lim, C.K. Chua, T.T. Chong 740

92 Feature representation and extraction in CAD models for manufacturing: methods and issues
M.S. Devgun, S. Padmanabhan 749
93 A CAD/CAM system for the application of two-part adhesives using robots
R.J. Grieve, R. Herfie, C. Hitchen 764
Section VIII.2 Computer-aided process planning 771
94 Computer-aided process planning in plastics manufacture
W.E. Biles 773
95 Computer-aided process planning for electrochemical machining
Gursaran, J.L. Batra, V.K. Jain 780
Section VIII.3 Computer-integrated manufacturing 791
96 Discrete cell control: framework and architecture
J. English, I. Franks, M. Loftus, G. Williams, N. Wood 793
97 Strategies for factory automation in garment production
M. Sarhadi, P.R. Nicholson, S.J.L. Marshall 799
98 An algorithm for distributed planning and control in a computer-based manufacturing control system
A.W.M. Lung, D.M. Love 806
99 Information system design and analysis of CIM systems: the new M* approach
F. Vernadat, A. Di Leva, P. Giolito 814
Section VIII.4 Organization 827
100 Design justification in product development
J.S. Noble, J.M.A. Tanchoco 829

Preface

This volume contains selected papers from those presented at the Tenth International Conference on Production Research held at the University of Nottingham, UK, in August 1989. The conference attracted delegates from all over the world, and from the two hundred papers and poster sessions presented almost half have been selected for publication. These represent the diversity of interests and themes which characterize the way production research is progressing as we approach the 21st Century. The papers have been divided into seven sections: production processes, production and inventory management, production economics and productivity, human factors issues in production, manufacturing systems and automation, expert systems and production simulation, and computer aided manufacture. It is a reflection of the multidisciplinary nature of current research into manufacturing systems that several of the contributions could have appeared in more than one of the above categories, and we ask that both authors and readers will bear with our grouping and our allocation between headings.

The twenty years since the first International Conference on Production Research have seen substantial changes in the development of manufacturing systems. Whilst many of the basic processes have remained substantially the same, the control of these processes has developed to take advantage of the revolution which has occurred in computing and information processing. Optimization within production and inventory control has had to reflect the need to reduce stocks and throughput times to a minimum. Quality management has focused much more on control of the processes to achieve zero defects. Emphasis is placed on total factor productivity rather than labour productivity, and greater realism has emerged in the development of automated manufacturing and handling systems. All aspects of production engineering and production planning now rely on 'computer aided' approaches, and major initiatives are being made to integrate all systems, from design through to manufacture and despatch, in an attempt to reduce indirect costs and minimise product development times. Ergonomic and human factors research is having to address the changing nature of the jobs and human/computer interaction being brought about by these changes in manufacturing systems.

It is against such a background of a rapidly changing technological and economic environment that this selection of papers is presented. We hope that the reader will find much which relates to their specific interests and that the volume, taken as a whole, is a valuable addition to the series on production research around the world.

C. O'Brien
M. Pridham

Acknowledgements

Many people contributed to the success of the Tenth International Conference on Production Research, and to the production of this volume of selected papers. In particular, we would like to record our gratitude to the following.

The Chairpersons of the various technical sessions:

F.J. Arcelus, Canada
S. Axsater, Sweden
D.J. Bennett, UK
M.C. Bonney, UK
H.-J. Bullinger, Germany
J.L. Burbidge, UK
M.D. Byrne, UK
A.S. Carrie, UK
A. Chikan, Hungary
E.N. Corlett, UK
J.L. Cox, USA
B.G. Dale, UK
M. Daily, USA
S.P. Dutta, Canada
T. Gallwey, Ireland
T.J. Greene, USA
C.M. Haslegrave, UK
K.K.B. Hon, UK
B.L. MacCarthy, UK
R. Maull, UK
R. Michelini, Italy
W.A. Miller, USA
L. Ming, Peoples Republic of China
A. Mital, USA
C.L. Moodie, USA
J. Moore, USA
P. O'Grady, USA
M.S. Pridham, UK
U.E. Seidel, Germany
G.T. Smith, UK
M.L. Spengler, USA
B. Trought, UK
A. Villa, Italy
X. Wang, Peoples Republic of China
D.B. Webster, Peoples Republic of China
P.J. Weeda, Netherlands
D.J. Woodcock, UK
B.M. Worrall, Canada
P. Xu, Peoples Republic of China

Keynote Speakers:

Professor Christopher O'Brien, University of Nottingham, UK
Professor Dr Ing Hans-Jorg Bullinger, Fraunhofer Institute, Stuttgart, Germany
Dr Norman Schofield, P.A. Management Consultants, UK

The Conference Organizers:

Peter McGoldrick
Lynn McAtamney
Christopher O'Brien
Mark Pridham
Maurice Bonney
Nigel Corlett

together with all those members of the academic, technical, secretarial and research staff and students of the Department of Production Engineering and Production

Management at the University of Nottingham, UK, who ensured the smooth running of both the technical and social programmes.

To all those who organized, hosted and guided the accompanying persons' programme.

To the staff of Taylor & Francis in the preparation of this volume of selected papers and of the volume of extended summaries distributed at the conference.

To the authors and reviewers without whose efforts no conference and no volume of proceedings would be possible.

To the members of the board of the International Foundation for Production Research.

PART I
PLENARY PAPERS

1

Perspectives on current production research

Christopher O'Brien
University of Nottingham, UK

Introduction

It is appropriate at the beginning of this book reporting on the Tenth International Conference on Production Research, to consider the current state of production research and to comment on some of the directions in which research is moving, as evidenced by the contributory papers. The 20 years spanned by these conferences have seen some dramatic changes in the approach to manufacturing which have added to the variety of manufacturing environments and to the problems which they bring to the researcher. The old environments still exist, and many of the old problems still remain unsolved and continue to attract the attention of researchers, but new environments are emerging rapidly. Amongst the principal forces for change have been the explosion of information technology and the hardware and software systems which have led to developments in computer-aided design, computer-aided engineering, computer-aided process planning, flexible manufacturing systems, etc.. In addition there has been a new climate brought about by Japanese initiatives in developing efficient manufacturing organizations and highly integrated and multi-disciplinary approaches to product and process design. There has also been a recognition amongst senior management of the need to recognize manufacturing ability as a strategic weapon in their companies' battle for market share and their continued survival in increasingly competitive world markets. This recognition has led to greater interest in investment in manufacturing facilities.

The demand for manufactured goods over this 20-year period has grown considerably and continues to grow as developed nations become richer, and developing nations—despite economic difficulties—begin to acquire the capital and consumer goods of their more developed neighbours. As manufacturing output has increased, however, the number of people employed has not always increased proportionately. There is a growing awareness and acceptance that manufacturing may be following in the steps of agriculture; just as society's demand for food can be met by a small but highly mechanized workforce, so the demand for manufactured goods will be met by a smaller manufacturing workforce made infinitely more productive through modern technologies and modern organizations.

When these conferences were conceived they were intended to bring together research workers from the many component disciplines and interests which contribute to the development of manufacturing systems and to encourage an integrated approach to the design and development of production systems. They also sought to encourage research workers to consider their work in the context of improving the quality of working life and fulfilling a responsibility to the wider needs of society and the environment. A look back to the proceedings of the nine previous conferences reveals the extent to which this has been achieved. The aims of the tenth conference remained the same. The final programme contained some 290

Machine tools, tooling and processes	62
Robotics and automation	25
Flexible manufacturing systems	18
Quality, inspection and measurement	13
Expert systems	22
Computer integrated manufacturing	30
Production planning and control	41
Inventory management	12
Simulation	20
Ergonomics and organization	21
Production economics and productivity	20
Miscellaneous	9
TOTAL	293

Figure 1. Number of papers in each category accepted in conference programme

of the tenth conference remained the same. The final programme contained some 290 papers and posters spanning the usual mixture of processes, planning and control, automation, human and economic factors (Figure 1).

Several of the authors who attended the early conferences have been regular contributors over the 20 years and were present again—a little greyer, a great deal wiser, considerably more distinguished and as enthusiastic as ever.

Processes

Machining and forming processes, of course, are basic disciplines of production research; however, metal forming and manufacturing processes other than machining have never contributed to any great extent to the proceedings of these conferences, despite some distinguished papers. Perhaps researchers in these areas choose other forums for the dissemination of their findings, or perhaps this is simply that developments in these areas have been less dramatic and, thus far less influenced by developments in computer technology which have attracted the attention of so many researchers. Some 12 papers on processing were presented at the conference, including work on plastic moulding, tube drawing, rolling, electrodischarge machining and vibration drilling and turning. Machining on the other hand tended to attract a wider contribution, and some 40 papers were presented, more than half of them contributed by authors from the Peoples Republic of China. The situation for machining and forming may be changing and the future may see an upsurge in interest in processing areas as the development of better sensors, very fast signal processing, and a move towards embedded systems with intelligent distributed data processing, make the adaptive control of processes within price limits much more achievable, thereby opening up many new possibilities for research.

Robots and automation

Early conferences paid little attention to the technical aspects of developments in Robotics and Automation. These began to receive greater attention in the late 1970s with papers divided between the problems of control and specific applications.

Despite the increasing penetration of robots into all aspects of material handling, machine loading and repetitive tasks such as painting, early expectations that robots would begin to take over most assembly operations have not been realized. The failure adequately to design for automatic assembly, reduced product life cycles, the need for flexibility and the need for rapid response to change has tipped the economic balance in favour of manual assembly in most situations, to the extent that manual assembly will probably continue to dominate well into the next century and continue to employ increasing numbers of people. Robots will, however, continue to have an important role to play in improving the quality of work life by taking over many applications involving repetitive or tiring tasks or hazardous applications.

Current developments in robot technology involve the development of modular grippers, fixtures and manipulators, and the development of sophisticated intelligent control systems. Developments are proceeding towards high-speed precision robots involving lighter and less rigid structures with a much higher load to weight ratio, which obtain their positional accuracy from a better understanding of the dynamics of the structures and an embedded distributed feedback and control system. As the complexity of robot systems increases, so the effort and cost of generating the necessary task execution information increases. To cope with this there needs to be further development of application support systems, including off- and on-line programming. There is also the need for an open systems architecture approach which allows easy integration of sensor control with the robot controller and generalized dynamic cell controllers. If there is to be greater penetration by robots into the small- and medium-sized enterprises (the so-called SMEs) who do not have the integrating skills, then there is likely to be a growing market for packaged solutions addressing particular manufacturing areas, with user-friendly application software (another area needing greater attention).

Group technology

Each of the conferences has attracted particular papers on group technology. The early conferences attracted debate on coding and classification systems and on production flow analysis. In recent years group technology has come into its own with the emphasis on focused manufacturing and the move away from process or functional layouts towards cell manufacturing.

Production scheduling and control

The move towards cell manufacturing has also influenced research into production planning and control systems which have always contributed a significant number of papers. An emphasis in early years was placed upon algorithms to develop real-time shop floor schedules within a job environment. It must be remembered that in the early conferences the use of computers in production control was only just emerging and the rapid development of integrated production planning and control systems was hardly foreseen. The emergence of optimized production technology (OPT) as an approach to production scheduling resulted in renewed attention being shifted to solving the problems of bottlenecks. While both these areas continue to attract researchers, attention is increasingly being shifted to the development of real-time scheduling for use in cell manufacture operating with Just-in-Time environments. The interface between such systems and materials requirements planning (MRP) systems is also receiving attention. A new realism is emerging towards production planning and control problems with a more flexible, comprehensive and more pragmatic approach being adopted in the choice of appropriate production systems and the evaluation of the circumstances required for their success.

For many researchers interest in scheduling has shifted from the problems of job shop scheduling to the analogous problems of sequencing work through flexible manufacturing systems (FMS) and the scheduling of associated material transport systems. In such areas simulation continues to find further applications emphasizing the need for increasingly user-friendly, application-based systems of simulation.

Just-in-Time

The widespread adoption of Just-in-Time approaches to manufacturing has influenced the research being presented in scheduling and inventory control. Interest has shifted from increasingly elaborate methods of determining economic batch and order quantities, to methods of controlling and scheduling all aspects within a supply chain to minimize inventory holding costs, improve supplier service, and reduce manufacturing times. To this end better enterprise networks need to be developed addressing the planning, information and communication needs to enable close integration of the activities of different departments or sites within a single enterprise and across enterprises in multi-supplier chains. To compete successfully in international markets it is imperative that cost-effective information flow is achieved within and between enterprises to give a dynamic response to rapidly changing markets and a significant reduction in operating costs.

Computer-integrated manufacturing

The requirement for better communications lies at the core of developments towards computer-integrated manufacturing (CIM). The 20-year span of these conferences also spans the major development in the component technologies of CIM, most of which were only in embryo in 1970. The papers contributed in computer-aided design (CAD), computer-aided process planning (CAPP), computer-aided engineering (CAE), and computer aided production management (CAPM) have increased over the years and were well represented in this conference. But attempts to integrate these tools into an integrated environment have been seriously hampered by the lack of standardization of hardware and software systems. If CIM is to realize its potential, and users are not going to be dependent upon a single vendor, there needs to be greater effort made in developing the concepts of open systems architectures for all areas of an enterprise's activities from top level business planning to shop floor control systems.

In parallel with activities to provide the means for better systems integration, is the need for more research into systems for comprehensive management, planning, supervision and close loop control of manufacturing systems embracing all levels of the organization from the operator to the senior management of the enterprise. Research areas should include not only the manufacturing processes being controlled but also the entire marketing and pre-production planning activities and all post-manufacturing distribution activities. Better tools are needed to help design, plan and evaluate such systems and to improve decision making at all levels in the manufacturing business. In this context the developments in expert systems and artificial intelligence, which have been evident in recent conferences, are likely to have an increasingly important role to play.

Economic factors

Not surprisingly the developments in advanced manufacturing technology and integration are becoming increasingly expensive. As such systems have developed over the years, so researchers have turned their attention to justifying their costs. Various papers concentrated on single factor measures of productivity and performance, while others poured criticism on the short-term and short-sighted techniques of accountants in appraising investments. Current work in this area tends to be more positive with the development of multi-factor measures of productivity and better metrics for supplying the information for comprehensive cost/benefit analyses. Cost/benefit analyses can take into account estimates of the variances in costs and revenues over the lifetime of a product and manufacturing system, and the financial returns from such strategic benefits as shortening the time from concept to market which may accrue from an investment.

Ergonomics

There were fewer papers presented at this conference in the area of ergonomics, organization and participation than in many previous conferences. This may simply be the result of the increasing number of international conferences devoted to ergonomics and human factors, or may reflect a current preoccupation with understanding the technical and organizational problems and opportunities presented by the advanced technologies. Inevitably perhaps the papers submitted in recent years have shown a trend away from workplace and job design in traditional work settings to exploring the effect on job design in a CIM environment. It should be remembered that despite advances in CIM technologies a skilled and educated workforce is still the key to maintaining a competitive edge for most companies, particularly SMEs. Much more work is needed to improve the communications between engineers, planners, operators and managers in the various technical elements and systems within a CIM environment. Similarly, much greater emphasis needs to be placed on exploring the roles people should play within these complex manufacturing environments to improve decision making and to develop systems which are more flexible and robust in coping with change and in dealing with unforeseen circumstances.

Take-up of research

An early concern of these conferences was the slow rate at which industry appeared to take up the output from research. There is evidence that this situation is changing. Increasingly in manufacturing emphasis will be placed upon application systems. Orders for CAD and CAPM systems and new manufacturing equipment will be won not on which system offers the most technologically sophisticated solution, but on which combination of hardware, software and support services is easy to implement and satisfies user requirements in any given situation. Indeed, this emphasis on easy to implement solutions to particular applications may be the only way by which SMEs, which represent by far the majority of companies in all industrialized and emerging countries, can gain access to developments in CIM technologies. Otherwise they may lack the skills and expertise to develop their own solutions; trying to integrate the various technologies for themselves may involve an unacceptable degree of risk. In developing such application systems the opportunity needs to be taken for collaborative Research and Development programmes with a strong user involvement and an emphasis placed on early implementation.

However, this emphasis on applications must not be allowed to divert attention from the vital role of fundamental research in developing generic systems and technologies. Governments cannot be allowed to escape their responsibilities in ensuring monies are made available to sponsor such basic research. In the period from 1975 to 1985 Research and Development in the USA financed from public funds rose from 1·2% to 1·29% of Gross Domestic Product. Over the same period in Europe the proportion dropped from 0·81% to 0·68%. In 1986 investment in information technology on a per capita basis amounted to approximately $110 in Japan, $65 in the USA and only $43 in Europe. If further imbalance in world trade is to be avoided and healthy competition is to be encouraged in the development and supply of planning and manufacturing systems, then more money must be made more widely available for the exploration and development of the ideas upon which the next generation of production systems and products will be based. Given such appropriate support and encouragement I have no doubt that the outcome of the consequent research effort will continue to be disseminated at these International Conferences on Production Research over the next 20 years.

2

Computer-integrated business: corporate and information system management in competitive environments

Hans-Jörg Bullinger[1,2] **and Joachim Niemeier**[1]

[1]*Fraunhofer-Institut für Arbeitswirtschaft und Organisation and*
[2]*Universität Stuttgart, Stuttgart, Germany*

Abstract

Planning—the use of information technology in the office included—is carried out in every company; however, in one company it tends to be implicit and informal while in another it is explicit and formal. The key question is not whether to plan the use of information technology in the office, but how good the planning can be made. Even so, the office automation strategy must be integrated into the corporate strategy. Productivity remains just as important a consideration as always, of course, but it should not distract from a question of far more primary importance, i.e. what role is played in the strategic creation of a competitive advantage by integrated solutions for office systems? A broad field of 'scientific folklore' is opening up in this area, as is reflected by the increasing frequency with which the media are reporting on it. In the final analysis, however, only facts should count for management when deciding on an investment in information technology in the office. To this end, the following will introduce a concept for planning strategic information systems, taking international conditions of automation and representative types of competitive environment as starting points.

Information technology—a tool for raising productivity or a strategic 'weapon'?

A change of approach can be prompted by an improvement of competitiveness, by the transition from a product- and price-related type of competition to a response-related one, or by the regionalization, nationalization and internationalization of the company's offered services. These are just some of the factors spotlighting the importance of information technology for strategic creation of value. A new catchword is establishing itself increasingly for this field—information technology as a 'strategic weapon' (Bullinger, 1986; Diebold, 1987; Sommerlatte, 1987).

The more central the question as to the effect of information technology, the more clearly one begins to ask about the role played by the information. Information as a resource, i.e. potential for securing corporate success, or information as a production factor, give rise to questions concerning the conceptual and operative design of corporate divisions on the factor 'information'.

On the one hand information is produced continually in the work process as a describer of the status of the work objects and respective work situation or of their change. Hence the nature of the information is descriptive, though its function for the work process is a controlling one. These aspects—the roots of the frequently cited diversity of information—must be carefully kept apart. Minimizing the descriptive component whilst ensuring the component necessary for controlling the work process is more than just a question of system efficiency. Questions of user acceptance and aspects of the adequacy of system solutions can give rise, however, to a higher proportion of descriptive information than would otherwise be necessary for control purposes.

In the medium term a key role in companies' operative entities will be played by the interlinking of technical and administrative offices with production as part of computer-integrated manufacture (CIM) concepts. Completely new perspectives are opening up as a result of integrated systems. This approach does not preclude considering isolated solutions (office, production) in the nature of germ cells for the overall systems to be planned. It places demands not only on the company's planners and organizers, however, but also on the systems developers, who today—and increasingly in the future—have a function to perform in the field of 'integrated applications'.

The productivity effects of using information technology in the office, which have been reliably proven in the course of projects carried out in the past few years, reveal that magnitudes of 10–30% can already be deemed a success. In the production sector a factor of 2–3 is quite usual for the increase of efficiency and productivity for rationalization projects in the high technology sector. In projects of this type in the production sector, the rationalization of the organizational structures always goes hand in hand with the introduction of new technology. In the office sector technology is still frequently viewed as an exclusive solution, without regard for the principle of successful management, i.e. form the organizational structures as the first step, support them with technology as the second. 'Organization before technology' is a frequently cited maxim, but reality has shown that technology in offices is often too inflexible and too regulated in practical use, with the result that this maxim is frequently turned back to front. Furthermore, the organizers' and planners' mental approach has often been orientated to individual worksystems and individual applications and not to complete work sequences. Frequently departmental boundaries also formed the boundaries of possible design activities.

A new catchword in the discussion is 'information as a new production factor'. Though companies have learned to plan and organize with the traditional production factors and to control and supervise in quantifiable and non-quantifiable parameters, there is a distinct deficit when it comes to the production factor 'information'. Simply claiming that 'information is the most valuable resource in our company' does not in itself bring the company anything. Appropriate management concepts must first be developed to manage the production factor information. The functional aspects must be clarified along with institutional questions such as the organization of information management.

Unfortunately the necessity of strategy-directed innovations is still a seldom met corporate principle, because all too often productivity considerations conceal this deficit of action. In view of the economic environment, innovations at the right moment in the right direction are an entrepreneurial task.

Development trends are presenting themselves in possible rationalization strategies for different types of office workstation. These rationalization measures are aimed essentially at improving the operative sequences in the companies. Rationalization strategies as such are on the bottom rung of measures, which are meant in their entirety to close the company's 'strategic gap' (see later). The top levels are the object of original management decisions. Information technology has recently been discussed in this connection as a 'strategic weapon'.

An improvement of operative sequences, meaning isolated productivity increases and rationalization efforts, can close only a small part of the strategic gap, i.e. the difference between 'carrying on as before' and the desired development. They are reflected first and foremost in cost benefits when rendering the corporate services. In addition it is possible to discern four directions for expanding information and communication systems. Information management is thus responsible for:

(1) creating the focal points for information and communication systems (forming germ cells) in the administrative, technical and productive sectors,
(2) developing the general information technology infrastructure (keywords: LAN, PABX, MAP, TOP),
(3) installing company-overlapping information and communication systems,
(4) exploiting information systems services to produce original product, process and procedural innovations.

Laying down the procedural sequence for shaping information management is not only a question of quick cost benefits and productivity increases. One possible approach, for example, is to create focal points parallel to taking infrastructural measures.

Assuming that aspired growth is possible only through controlled growth, then management must have commensurate experience to master the task of this parallel strategy successfully. By comparison, creating germ cells for information and communication systems affords the advantage of being able to accumulate knowledge and experience in the company on the basis of these test type sub-solutions, thus enabling company-wide projects to be tackled with assurance.

The same applies, though with some qualification, to the installation of company-overlapping computer networks. A key aspect here is the type of link with other companies and its consequences for the development of internal and external standards that can lead to new dependencies, for example on a component supplier. The logical co-ordination of these measures is one of the rules of all-round management, i.e. information management can be understood only as a component of the corporate strategy. Information and communication systems can be utilized for market strategies such as market penetration, market and product development and diversifications. These normative requirements must be converted into effective concepts, with whose help competitive advantages can be achieved and secured. Patent recipes are available in diverse forms for this purpose. Often, however, they entail over-simplified assumptions that tend to conceal a suitable approach rather than indicate new perspectives of action (Bullinger *et al.*, 1987a). Frequently traded, over-simplified statements in this connection are:

Myth 1. We must make more intensive use of information systems in our companies in order to be able to act more successfully in the future.

Myth 2. Successful companies exploit their situational opportunities to innovate; if you don't jump on the bandwagon now, you'll cut yourself off from the future.

Myth 3. Information management is urgently required as a component of strategic planning; a strong planning arm is necessary.

Myth 4. The information manager will solve our problems; we must appoint him as quickly as possible.

Besides presenting the relationships behind these value judgments, the following sections will deal more closely with the necessity of viewing information systems as a factor of a forward-planning corporate strategy. Emphasis will be given to consideration of the competitive advantages inherent in these systems along with the company's possible approaches and lines of procedure for shaping and co-ordinating information systems.

1st myth

We must make more intensive use of information and communication systems in our companies in order to be able to act more successfully in the future

The vigorously increasing use of information is making it increasingly essential to analyse the effect of alternative corporate strategies on information equipment for planning display (Adler, 1987). This is true both for first and second generation systems (mainly data bank orientated large computers) and for systems of the third and fourth generations (interlinked computer systems and data- and symbol-processing systems). Growth is still more dynamic in the field of the classic computer generation, but newer systems are also advancing strongly. This trend is often overestimated, however, as is evidenced, for example, by the fact that the number of stable and fully tested expert systems to be found to date as 'running systems' in companies is still relatively low (Merstens, 1986), although the market for artificial intelligence (Al) workstations and expert systems shells is already substantial.

It would clearly be wrong, however, to derive criteria for assessing corporate success from this use of information. Corporate success is not conditional on the number of employed information technology investment costs per workplace, use of the most advanced technology systems such as local networks or the use of manufacturing information systems (MIS). Doubtless they make a not inconsiderable contribution, but they are not in themselves indicators of success. Parameters such as turnover and growth of earnings, increase of market shares, earning power of employed capital and factors such as a company's standing are decisive in the long run for corporate success. These are the indicators which characterize the economic success of a company.

Patent recipes like 'the use of information improves corporate success' hold too much disregard for the difficulties in developing applications and in the corporate implementation of information technology. Not to mention the difficulty of proofing the efficiency of these investments. The challenges facing a company, be they the special conditions on the labour market or in its particular branch of industry, or be they substitute products and innovative competitors, can generally be tackled with structural, personnel or technocratic adaptation strategies. All three adaptation strategies are supported by information strategies. In the past few years companies have relied increasingly on technocratic adaptation strategies, so that on this global level it is not possible to discern any difference as regards corporate success given different adaptation strategies and emphases. The intensity with which these adaptation strategies are used is practically the same in successful and unsuccessful companies. This is clearly seen in Figure 1.

To conclude that each adaptation strategy is equally successful, i.e. equivalent, would not be apt given the generalizing level of this cross-sectional view and the range of effect of information technologies. Closer differentiation of the above findings reveals quite a variety of successful strategies with corresponding strategic policies. There are certainly some environmental situations in which a company would be advised to pursue a rather more conservative policy. In other markets, on the other hand, an active and innovative approach would be of great importance for the company's success.

These markets are currently expanding, not only as part of the generally evident globalization of markets but also on account of the still accelerating rate of innovation. This produces problems for information technologies, and one of the

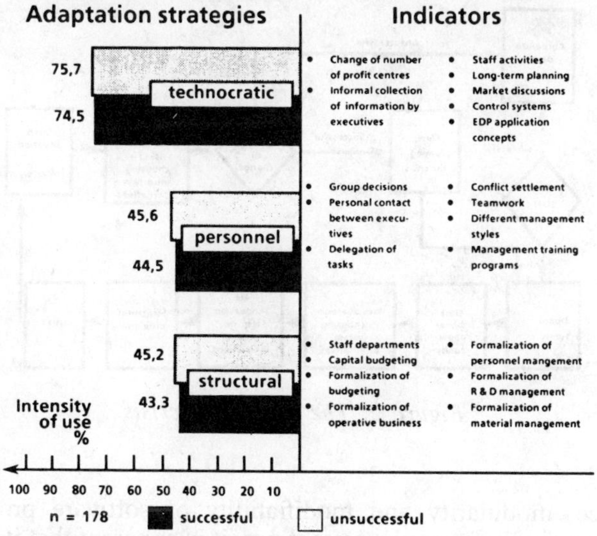

Figure 1. Effects of different adaptation strategies on corporate success in large German companies.

main aims pursued here is to highlight these problems and become aware of their inherent competitive advantages.

Interim conclusion 1st myth

As the preceding section showed, yardsticks based directly on information technology are neither an indicator for corporate success nor can this success be forced by structural, personnel or technocratic adaptation strategies. Success is only assured when all measures work together in harmony, and certainly not by applying isolated adaptation strategies (Geesink, 1987). The factors listed above are necessary for corporate success, but they are not sufficient; hence the logical requirement is for new concepts of review and representation that do justice to the problems of information technologies. The following section considers the fact that situational opportunities of utilizing strategic options are not always exploited, often directly due to a feeling of system complexity.

2nd myth

Successful companies exploit their situational opportunities to innovate; if you don't jump on the bandwagon now, you'll cut yourself off from the future

During the dominant general economic crisis of the past few years, the worst of which now seems to have been overcome through new growth impetus, information technologies have undergone further maturing. More computer power is available for lower prices, more and better single-user and LANs are being installed more and more often. Already well past the concept phase are advances toward knowledge processing (decision support systems) and artificial intelligence (AI systems) or developments in the field of new system base services in the area of distributed computer architecture, filing and retrieval of multimodal documents and the ensuing considerations of organizational structures.

A critical area—now as ever—is the development of software, which lags far behind user needs and requirements. Important aspects here are such things as the

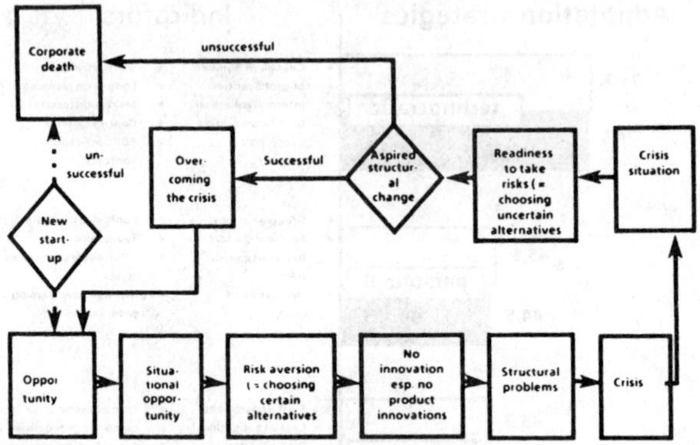

Figure 2. The success of crisis.

user convenience, modularity and modifiability of software program packages. Judging by the mass of software in current use, it is apparent that it far from reflects the present state of software engineering, although the first movements in this direction are visible.

It is also apparent that in terms of using and expanding information systems as well as introducing innovative steps, European companies lag behind the activities being undertaken elsewhere. Furthermore, there appears to be a lack of strategic concepts which structure the company's decision space and render it describable, tangible and formable. Clearly, the unused latitudes of action must provoke more than just crisis management—they should lead increasingly to a targeted management of opportunities. As shown by Perlitz (1985), however, the decision maker's behavioural spectrum often begins to take clear shape only when there is a crisis. This is illustrated by Figure 2.

It is one of the tasks of the current situation to apply future-orientated strategy concepts so as to become free from these decision pressures, or what Perlitz (1985) called the 'success of crisis'. This philosophy is sketched in the form of a decision cycle in Figure 3.

It is also obvious that decisions are often far too closely tied to operative problems and first-stage solutions. Competitive advantages are in view, albeit still not clearly discernible enough for many. This route is risky without this view, as is shown by the fact that it has not been possible to integrate any installed information system in existing companies as a pure service resource, non-selectively of value and organization. Often alterations to the organizational structure and work sequences are a consequence and are not always welcome.

Interim conclusion 2nd myth

Economic crisis is overcome by applying innovations, as is proven by companies' demonstrated actions. Economic growth is increasingly becoming a function of companies' innovative activity. What is needed, therefore, is a change of companies' purely crisis-related innovative behaviour. A study by Arthur D. Little (Little, 1985) succeeded in giving a precise picture of the significance of innovations for the corporate scene in the near future; 87% of company directors surveyed in Europe rated the importance of innovations for the future of their companies as higher or much higher than today. The question remains, however, as to whether this attitude is reflected in corporate programmes of activities and actual innovations or whether there are worlds separating the desired behaviour and corporate enforceability, which must first be bridged.

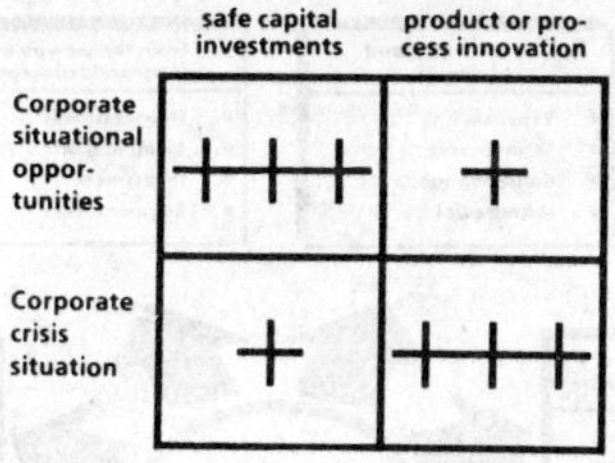

Figure 3. Risk behaviour with innovation.

3rd myth

Information management is urgently required as a component of strategic planning; a strong planning arm is essential

The situation in strategic planning is characterized by a host of different possible approaches. They are mainly market-orientated and so include individual aspects of market events in their descriptive features. Information technology usually represents one component that is viewed by a company—if at all—as having only fringe relevance for the survival of the company. Figures 4 and 5 illustrate the diversity of these strategies and strategic principles.

In the past decade information technologies have got over their development problems and now provide users with an increasingly mature technology; consequently the corporate perspectives are moving ever more centre stage. This fact is obscured by the diversity of engineering and software, by the number of equipment manufacturers, by marketing activities bordering on the dubious, and not least by the sometimes Babylonian-like multiplicity of languages and catchwords used by systems manufacturers and institutes involved in information technologies.

For information management questions of corporate perspectives—from information technologies systems beyond the use of the technological potential in the sense of job simplification—become increasingly urgent, e.g.:

(1) What effect do information technologies have on the competition, on the competitive environment, on competitive behaviour, on market events?
(2) Where are the competitive advantages that can be achieved with new technologies?
(3) Are information technologies only 'resources' or are they a factor of every serious corporate strategy?
(4) How do the tasks connected with these questions fit in with corporate management tasks?

A host of questions that produces an array of answers. At this point reference is often made to the fact that many American companies have already installed an 'information director'. All too often, however, institutional arrangements of this type promote the myth of the total plannability of the resource information.

To view the answers to these questions only as part of a restrictive company-wide planning of information systems is to misjudge both the necessity of adaptive solutions in the individual corporate divisions and the significance of these systems for the company's strategic competitive position in its markets. What is required here is an appropriate approach with the creation of development latitudes. This necessary planning approach, which is part of the essence of every strategy representing more than just coping with risks and wishing to extend beyond an operative view of problems, calls for an overall planning framework to be set up (Niemeier, 1987). Less importance must be ascribed to total plannability and more to the opportunity of learning in pilot fields, even with the risk of the final decision for a technology taking quite a different turn. Figure 6 shows the key planning interrelationships for company-wide information management.

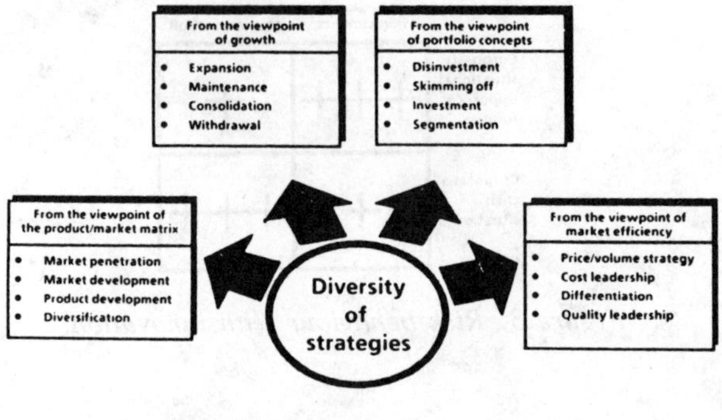

Figure 4. Diversity of strategies (I).

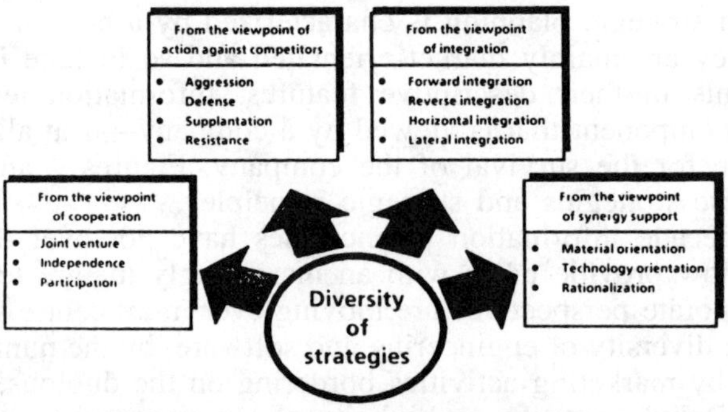

Figure 5. Diversity of strategies (II).

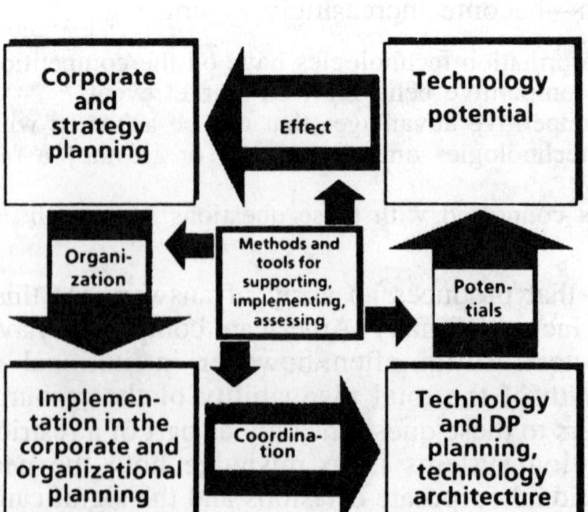

Figure 6. Company-wide information management–planning interrelationships.

Nothing would be more damaging than to embed the dynamism of corporate information systems in narrow, planning specification. Consequently it is necessary to:

(a) embed information technologies in a planning framework of corporate policy and look on investments in this area as future investments,
(b) initiate strategic search field programs for the utilization of information technologies for closing the strategic gap,
(c) move from first-phase rationalization to planning the transition,
(d) switch from individual strategies to interlinked strategies,
(e) permit development latitudes for operative problem solutions as part of the given information technologies,
(f) develop value concepts for the resource 'information', such as has already been done for the factor 'quality' in production.

Interim conclusion 3rd myth

For new strategies to replace old philosophies and behaviour patterns, their underlying measures and behaviour patterns must differ from previous procedures. This can only take place through the initiation of special programmes, changed behaviour patterns of the decision makers and new organizational focal points or through a motivating management. The result is a company as a system of divided values. Furthermore, due furnishing with resources and the regulatory control of these processes are definite requirements for the success of these efforts. The global framework for the functional tasks of information management is thus formed.

At this point it is worthwhile to recall, for example, the study on corporate management in the Federal Republic of Germany and West Berlin by Esser *et al.* (1983), according to which the yardstick of a strategically progressive corporate management was the question not just as to how far it succeeds in developing strategies and making them obligatory for the company, but also how far it succeeds in applying the chosen strategies in the operative business. On the basis of empirical evidence the authors of this study find that:

(1) Special projects are set up in strategically progressive companies for the operative application of the strategies.
(2) Budgets are agreed for applying the strategy in the operative business.
(3) Changes of the organizational structure are also taken into account so as to implement the chosen strategies.
(4) Not only are strategies planned, but the premises of the strategic planning and the progress of the strategy implementation are also checked systematically.
(5) To ensure the application of the strategies in the operative business, the information and control system continues to be adapted and staff trained accordingly. (Esser *et al.*, 1983, p. 88).

4th myth

The information manager will solve our problems; we must appoint him as quickly as possible

'Information management is a task for top management.' This at least is the message from the USA. 'Around 70% of all American companies have already installed an information director.' Nevertheless, according to a study by a major American market research institute, 80% of the personal computers in large American companies are not used (Luft, 1986).

Information management is the control of value creation processes and opens up opportunities for supporting a competitive strategy. This cannot only entail creating a new field of activity and its hierarchical arrangement (Bullinger *et al.*, 1987b). Claims like 'the information manager will solve our problems' touch only the

Period	Stage	Explanation
1900	Paperwork Management	• material control of the paperwork so as to improve efficiency
1960	Management of Automated Technology (Management of ADP)	• automation of manual techniques; automated and manual information management are considered to be fundamentally different • development of separate branches of science
1970 1980	Information Resource Management (IRM)	• integration of manual and automated methods of data processing and management • technological and functional coalesence of technologies of earlier decades • coordinated and planned attitude to all technologies
1990	Knowledge Management	• complete penetration and dependence of information technologies in the organization • emphasis on the management of information

Figure 7. Development of the tasks of information management in the USA.

surface of the problem. Closer inspection of an information manager's necessary fields of competence poses the question as to the extent to which these requirements can be met by a single person. The picture of an information manager—the subject of years of discussion—has grown so much through the continual further development and continual expansion of this field of activity that the solution can no longer be found in a single person.

Understanding information management to be a new management strategy in the sense of a new management task, a communication factor and a new controlling function not only calls for a rethink in the structuring, but also gives rise to new economic dimensions in the strategic, structural and managerial philosophy. The organizational incorporation of information management touches all spheres of activity of a reconstructional organization: functions and assignment of duties, management system and hierarchy. As is increasingly clear, it is necessary to combine the basic functions of information management with functions that are performed in specific divisions by line superiors and with dedicated backup functions.

Information has moved increasingly into the focus of a conscious organization, and the task perspectives of information management have changed accordingly. Figure 7 illustrates this development for the public administration sector in the USA.

Let us return to the catchword 'information as corporate resource'. Properly used, it is reflected in a pertinent supply of information. The often cited diversity of information is overwhelming.

Amidst the diversity of information it is frequently forgotten to consider also that the information content of communication is closely bound up with its form of presentation. Data on sales fluctuations in sterile columns of figures read differently than in neatly prepared diagrams. The visualization of information is one of the conditions of being quickly informed. These remarks touch only the surface of the problem, however. The task of information management embraces the planning, organization, regulation and control of:

the diversity of information,
the diversity of technology,
organization and personnel components.

Figure 8 illustrates this more closely. In future these tasks are to be performed in the plans of many companies by an information manager. Closer examination of

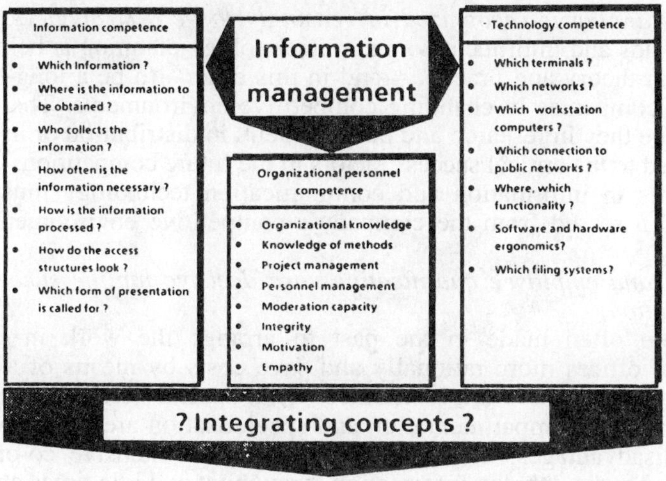

Figure 8. Fields of competence of an information manager.

the fields of competence of such an information manager poses the question as to the extent to which these requirements can be met by real persons.

Interim conclusion 4th myth

A highly qualified team of subject specialists is required to be able to pursue this route. The primary objective is more a matter of preparing channels and 'putting solutions on the tracks' and less one of initiating a 'top-heavy head' in the company. The long-term objective is service carried by the specialist divisions' own dynamism. This objective can be achieved only through a successive transfer of responsibility to these specialist divisions. At all events care must be taken to avoid separating the problem solving function completely from the operative handling factors or else, as experience shows, the problem solvers will quickly lose touch with the problem.

Office automation as a component of computer-integrated business

Versions of new computer-aided technologies are being announced in increasingly quick succession. Nevertheless, in practice there are signs at least of the beginnings of an integration of technical and organizational functions as part of a CIM concept intended to cover all corporate divisions, interlinked by information. From a realistic standpoint Scheer's remarks must be endorsed: 'Simply offering software systems from the MRP, CAD and CAM fields does not add up to CIM (...). The first prototypal solutions featuring the integration module for this are to be found in the CIM systems for implementation *en bloc* are not (yet) available for sale, however.' (Scheer, 1986, p. 51). 'CIM is not the top of the flagpole by far,' claims one electrical multinational. What he is referring to is nothing else than the merger of CIM and office automation concepts in 'computer-integrated business' (CIB).

A study of up-to-date case examples and a review of the latest literature on the subjects of CIM and office automation reveals a number of pointers calling for an integration of these rationalization systems:

Technology planning is recognized increasingly to be a strategic exercise
In the CIM fields and in office automation technology planning is being recognized increasingly in theory and practice—and in this order—to be a long-term strategic approach for companies in changing competitive environments. The call is for all investments, be they in research and development, in distribution or in organization, to be orientated to the critical success factors in the future competition. Consequently, all investments in information and communication technology must be derived, weighted and assessed from the company's competitive environment.

Organization and employee qualifications are decisive for the success of a technical solution
Attempts were often made in the past to arrange the work in technical and administrative offices more rationally and save costs by means of a rigorous and pronounced division of work combined with special (technical) support. The advantages of work compartmentation and specialization are offset, however, by a number of disadvantages. This route produced time-intensive co-ordination and communication tasks, lengthy information throughput and idle times and duplication of work. Nowadays a lot of information has to be collected, stored, processed and transferred several times. By virtue of the favourable price–performance ratio computer power no longer needs to be centralized, however. This opens up the opportunity to reintegrate work tasks and change the work content through dissolving exaggerated work sharing, enabling staff to be allocated more responsibility once again and a higher share of the overall process. In this connection it is important to ensure that isolated planning of the CIM sector and office automation does not again give rise from the outset to two worlds that can only be bridged later on with a great deal of effort.

The long-term planning of infrastructure is taking on an increasingly higher status alongside the technical and organizational planning of applications
The coalescence of technical manufacturing subsystems with organizational business subsystems and the resultant increase in the degree of integration calls for the uniform planning of the technical infrastructure in addition to the joint task of organizational structural design. It requires development of a systems architecture severed from the applications level and enabling flexible design of an integration concept on the applications level. The planning of horizontal system components within a computer level represents the smallest problem among the aspects of integration. The implications of the integrated approach and the degree of its technical complexity are not appreciated until planning the vertical integration. The first moves to a solution on the communications level are to be found in the standardization efforts of international institutions and some large companies (e.g. MAP, TOP).

Against this background we shall now define what we understand under a CIB system, i.e. an information and communication system which is:

(a) *derived from the corporate strategy*
Computer support in the production sector as well as in technical and administrative offices must be designed situationally, i.e. against the background of a host of influencing factors. Generally speaking, several directions of approach are conceivable. Their concrete design differs in respect of their operative, strategic and structural flexibility. Operative flexibility is understood to be the quantitative adaptability of the corporate base systems to volumetric fluctuations in demand. A company displays strategic flexibility when it has the capability to respond to qualitative requirements over and above making quantitative adaptions. Structural flexibility is required especially in technology-intensive companies, whose product–

service mix changes frequently and whose products and/or services are relatively short-lived. Organizational changes are implemented without disturbing the economic process.

(b) *based on a uniform infrastructure*
A role is being played here by general standardization approaches in the MAP and TOP field along with company-individual standardization efforts in the area of technical base systems (workstations, communication systems, data-bank and storage systems), fundamental software tools, application systems and man–machine communication. Implementation will be towards a distributed system. The need to observe existing standards in the communication and data-bank systems and communication between distributed systems. Joint and compatible planning of the technical infrastructure is therefore essential.

(c) *not designed as a maximal technical solution, but with optimal exploitation of synergisms*
An increase of efficiency and effectiveness is likely only when the technical solution is co-ordinated with the long-term objectives and the organizational circumstances, when the employee qualification structure is not overburdened and when a suitable line of approach is selected. The maximal technical solution is often not the appropriate one.

Starting point for the development of a company-individual CIB concept

Starting points in the strategic planning of information processing are the analysis of relevant environmental factors for determining the requirements of the competitive environment, as well as a company analysis to determine the present efficiency in the sense of strong and weak points to determine the company's support requirements. Information from both subject areas serves to derive and formulate adequate company and business field strategies and to identify concepts for operative measures. The complete framework for information processing with the appropriate organizational and technically translatable facilities can be drawn up on this basis. Individual proposals for the organizational and technical configuration can be summarized in expediently detailed projects. Through the deduction of detailed projects from the requirements of the competitive environment and the given company potential and requirement, it is possible to show the strategic significance of different projects and draw up a priority list for their realization. The mono-functional orientation in technical production target technologies to be found in many manufacturing companies is extended to cover office automation technologies under a CIB concept. As a result of the integrated technology planning of the information systems in office and production derived from competitive strategies, synergetic effects are produced which call for a review of the market, product and production technologies as a whole.

Starting points in company analysis

It is not only the competitive environment, but also a company's conditions of automation that determine the company-individual CIB strategy, which includes aspects of the application of information technologies in the manufacturing sector as well as for office automation.

The incorporation of the conditions for automation which describe the initial situation of a company from the production technology aspect, determine, apart

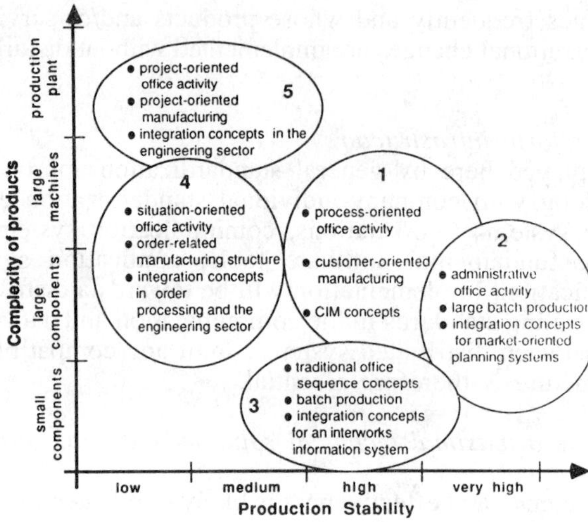

Figure 9. Automation conditions and integration types.

from the competitive environment, the development path for a CIB concept. The differing conditions for automation result in a number of types of integration (see Figure 9).

Primary companies making (large) machines and large components, e.g. manufacturers of transmissions, textile and construction equipment and agricultural machinery, can be classified under the heading integration type 1.

The production process of these companies is marked by increasing complexity in production and customer orientation. Here integration of the information systems stands particularly high on the list. Office work is seen as a procedure-oriented process. The functions of the technical office are integrated in order processing, on which other office functions are dependent. Integration type 2 companies are those with very high stability in production. In this case the integration of all technical support systems of a comprehensive CIM concept are not absolutely essential and also not economical. Information technologies have to be aimed at increasing efficiency in the production area that is directed towards CIM systems and as a result support new product planning. Office automation should be available in the form of distribution information systems, expert systems for new product planning as well as access facilities for external data banks. Integration type 3 could be made up of typical component manufacturers in the mechanical handling, fittings, pumps, tools, hydraulics and machine component field. Here the integration between information systems in production and in the office is given no great significance. Suitable technologies are information systems for controlling storage systems, numerical control (NC) machines, mechanical handling facilities, industrial robots and flexible production systems. Manufacturers of large components such as mining and testing machines, turbines, locomotives and heavy machine tools are characteristic of integration type 4. Here information systems must aim at reducing the price and technical risk in quotations, i.e. increase the conversion rate of quotations into orders.

Of strategic importance here are CAD/CAE systems with integrated office automation components (integration of CAP and project management systems, mailing systems). The technical support of situation-oriented office work is of particular significance because procedure-oriented office work is only formed in the most seldom of cases. Companies offering very complex services, which are dependent on know-how and often provided decentrally, are predestined for comprehensive use of office automation technologies. These are to be classified under integration type 5.

These are companies active in the production plant field, e.g. manufacturers of iron and steel works for complex factory plants. A considerable factor in success in this respect is the utilization of decentralized user and product know-how as well as high service in project processing. In this case simulation, planning and expert systems as well as integrated project management, planning and budgeting systems should be available.

Competitive situation and targeted measures of office automation

How does the typical environment look for our companies? The host of influencing variables can be represented in the form of five competitive environments (Niemeier, 1986). Knowledge of the current competitive environment and its dominant features can serve to identify the first firm criteria for the use of information technologies. Clearly it is not possible to derive a typical ideal strategy for the use of information systems for every competitive environment; this entails considering further factors such as the industry branches, their structure, organizational structures, corporate strategies and a company's developed connections.

The type and extent of the information technologies to be employed depends on the environment-specific framework conditions and, otherwise, mainly on the existing and potential information intensity of the products and manufacturing processes in a company (Porter and Millar, 1986, p. 34). Product complexity and product stability are dimensions which define a company's automation conditions and serve to position and categorize manufacturers (cf. Schulz, 1985, 1987). With their help it is possible to determine direct utility potentials for the use of information and communication systems and assess the effects of function-overlapping or integrated systems. In addition to pure efficiency targets through the use of scale effects and high capacity utilization, there is an increased measure of direct utility factors, such as quality and flexibility criteria in the design and production process and a stronger orientation towards customers through improved delivery fidelity, service readiness and capacity to respond to customer wishes.

From this perspective of a future integration of CIM and office information concepts we shall now illuminate the typical competitive environments.

The placid, random environment is characterized by its uniformity. This is reflected in the dimensions buyer habits, competitive behaviour, quality-related competition, novelty-related competition and change of market behaviour. The uniformity of this type of environment should be countered by efficiency raising efforts as part of rationalization strategies, e.g. in the office sector. The use of information systems mainly serves this purpose and leads to cost savings. From the production technology viewpoint, companies in the placid, random environment are usually characterized by a low product complexity.

Simple products are manufactured in large batches. For such companies the direct use of computer support in the production sector is very small. Overall cost saving potentials through CAD of between 0·7% and 1·5%, through CAP of 1·8% maximum and through CAM of between 0·9% and 2·0% have been determined for companies from the machinery and plant construction industry in this production environment (Schulz, 1987, pp. 45 f.). It must not be forgotten, however, that the value of computer support in the production sector must be orientated to the corporate and business field strategies. In a placid, random competitive environment a decisive role is played only by hard-to-quantify utility factors such as readiness to answer enquiries, observance of deadlines and readiness to deliver through CAP-systems, as this can achieve a stronger customer orientation. The operative flexibility can be increased through a higher transparency of production sequences and, in combination with the implementation of batch number advantages, can produce substantial efficiency improvements in the production sector.

The placid, clustered environment has a more distinctive look. Quality competition dominates in it. A moderate market dynamism with a heterogeneous competitive behaviour demands slight diversification trends. The lack of significant forms for the market development and the predictability of the demand point to a secure customer group. The advantages of installing company-internal and company-external computer networks for information and communication applications, support—by their interlink aspect—the required corporate consolidation and alignment to quality competition as a central parameter for corporate, success. Here the priority task is to bring about this interlinking in production, production control and regulation, so as to set up the 'quality' factor as a central success parameter within the framework of an accordingly designed information system.

The low frequency of change of market behaviour enables a largely uninterrupted material flow in a demand reduced manufacturing environment. Each activity can be controlled in such a way as to equalize the degree of capacity utilization in all stages of manufacturing. This enables a just-in-time tie-up of suppliers, because the smoother manufacturing plan makes the demand for their products predictable over the long term, thus enabling them better to plan their production (Walleigh, 1986, p. 71). Shorter feedback cycles lead to an improvement of quality. 'Office work' plays a role particularly when a batch-orientated material management has to be synchronized with a just-in-time production. In this case a form-orientated office system for individual parts and an OCR-orientated system for mass parts can replace an automated control of the material allocation.

The dynamic, reactive environment is unmistakably characterized by its pronounced quality competition, novelty-related competition and price competition. It is notable that the parameters of heterogeneity, i.e. diversification, buyer habits and competitive behaviour, as well as the market development and predictability of demand do not represent significant parameters. It is plain enough to conclude that these environmental factors can be mastered through far-sighted planning by the companies within this environment. The formation of statistics systems to describe the patterns of effect of this type of environment is an effective approach when combined with flexible manufacturing as part of CIM concepts for these industrial companies, to comply with the properties of an innovative, interestingly priced quality, as demanded by the environment. Information systems in these areas improve the response capacity considerably and so are likely to render the price-related competition less pronounced. These remarks are probably of special relevance for high-grade component manufacturers, whose competitiveness depends chiefly on production costs, quality, the capacity to respond to customer wishes and delivery times (Hellwig, 1985; Neipp, 1987). The qualitative utility effects are particularly high, as is the cost saving potential through integrating all technical support systems to form a CIM concept in the machine construction sector, given moderate production stability and product complexity (Schulz, 1987, p. 47).

The turbulent environment is the only one characterized by severe price competition. A role is played by the scarcity of raw materials. Here we rediscover suppliers with pronounced price–volume strategies. Minimization of throughput times and fast order processing times with constant capacity utilization quotas are called for. Manufacturing control systems with central computer control and direct tie-up to incoming orders are areas in which information and communication systems can exert their influence for the benefit of industrial companies in this competitive environment. Moreover, key success factors are the utilization of user know-how and the assurance of an efficient order processing and high level of service. On the systems side, it is expedient for machine construction companies to interlink design with projection, calculation, material management and order processing (Neipp, 1987, p. 234). For these companies, an essential task of technical support by specialists in the diverse office sectors is to install planning systems, especially early warning systems. Access to external data banks completes the spectrum of technical support systems.

The hyperturbulent environment is sharply characterized by a tremendous area of heterogeneity. What is meant is the pressure to diversify, strong dependence on buyer habits, i.e. trends and a high degree of dependence on the competition's behaviour, and market development. These parameters are all to be found in the companies' outer environment. Companies with a large range of products and services and great breadth and depth of assortment are likely to come in this category. Within this field of tasks are complex engineering assignments which often need to be performed decentrally using office automation technologies.

For these companies, on the other hand, the dynamic aspects do not represent significant parameters as long as they are adequately handled, i.e. areas such as product/service quality and product/service innovation and their implementation in productive structures (rate of change of production process) do not form problem areas. The targets for information and communication systems must be to use them for coping with the environmental heterogeneity and product diversity. A quickly prepared quotation, also with regard to small and very small prospective buyers, with integrated parts list breakdown and inventory optimization, together with order tracking systems, ensure the company's response capability. Also required are product development systems with computer support for securing and quickly handling the necessary product sequence and the ideas documentation.

Routes to competitive advantages—approaches for information management

The question is not one of acquiring new tools but making the best of what is available. How can organizations take hold of this tremendous resource and direct it toward securing a competitive edge? As was shown above, a host of approaches ensues from the companies' competitive environment and strategies. To answer the question about one's own company's competitive advantages, it is necessary according to Porter (Porter and Millar, 1986) to pose the general questions about:

— the intensity of information (in the product, process, etc.),
— the effect of information and communication systems on the branch structure,
— the way and manner of being able meaningfully to implement competitive advantages,
— where and how information and communication technologies can call new business branches into being.

The action plan produced on the basis of answering these central questions and covering the sequence of strategic investments, contains decisions concerning the hardware and software to be employed, new products and new processes with more information content.

References

Adler, G., 1987, Die Planung strategischer Informationssysteme. In *Strategische Waffe Informationstechnik, Internationales Management—Symposium*, Frankfurt, 1987, pp. 371–386.

Bullinger, H.-J., 1986, Wettbewerbsvorteile durch Informationsmanagement. In *Informationsmanagement für die Praxis*, edited by H.-J. Bullinger, pp. 55–119 (Berlin).

Bullinger, H.-J., Niemeier, J. and Friedrich, R., 1987a, Informationsmanagement Vom Krisenmanagement zum Chancenmanagement. In *Die Informationsverarbeitung der Zukunft Konzepte und Strategien aus der Praxis*, Vol. 1, pp. 5–29.

Bullinger, H.-J., Niemeier, J. and Schäfer, M., 1987b, Aufbauorganisation des Informationsmanagements. In *Nachrichten für Dokumentation, 1987*, pp. 193–199.

Diebold, J., 1987, Using information technology as a competitive weapon—strategies and action steps. In *Strategische Waffe Informationstechnik, Internationales Management—Symposium*, Frankfurt, 1987, pp. 33–45.

Esser, W.-M., Hofner, K., Kirsch, W. and Wieselhuber, N., 1983, Der Stand der Strategischen Unternehmensfuhrung in der Bundesrepublik Deutschland und West-Berlin, 1983.

Geesink, J.R., 1987, Organisatorische Erfolgsfaktoren zur Erzielung von Wettbewerbsvorteilen. In *Strategische Waffe Informationstechnik, Internationales Management—Symposium*, Frankfurt, 1987, pp. 357–369.

Hellwig, H.-E., 1985, CIM-Der Schritt nach CAD und CAM. *VDI-Zeitschrift*, **1985**, 135–142.

Little, A.D., 1985, *Management der Geschäfte von morgen* (Wiesbaden).

Luft, K., 1986, Strategische Investments der Nixdorf Computer AG Produkte/Innovationen. In *Nixdorf Colloquium*, 1986.

Merstens, P., 1986, Expertensysteme in der Betriebswirtschaft Stand im deutschsprachigen Raum. In *Die Zukunft der Informationssysteme. Lehren der 80er Jahre*, edited by A. Schulz, pp. 62–80 (Berlin).

Mundhenke, E., 1986, Informationsmanagement Erfolgreiche Einfuhrung in US-Verwaltungen. *Office Management*, **9**, 846–862.

Neipp, G., 1987, Rechnerintegrierte Produktion. *Zeitschrift für betriebswirtschaftliche Forschung und Praxis*, **3**, 225–238.

Niemeier, J., 1986, *Wettbewerbsumwelt und interne Konfigurationen* (Frankfurt).

Niemeier, J., 1987, Einsatzfelder für Methoden zur Analyse und Planung der Bürokommunikation. Strategische und organisatorische Perspektiven der Planungsumwelt. *Office Management*, **7–8**, 10–18.

Parker, M.M. and Benson, 1986, *Enterprise-Wide Information Management, The Available Tools and Methods* (ISETT).

Perlitz, M., 1985, Strategisches Informationsmanagement—Der zwang zur Innovation. *Innovation*, **7**, 725–731.

Porter, M.E. and Millar, V.E., 1986, Wettbewerbsvorteile durch Information. *Harvard Manager*, **1**, 26–35.

Scheer, A.-W., 1986, Strategien zur Entwicklung eines CIM-Konzeptes. *Informations Management*, **1**, 50–56.

Schulz, H., 1985, Erfolgreiche Nutzung des Potentials rechnergestützter Fabrikautomatisierung. *Werkstatt und Betrieb*, **9**, 565–568.

Schulz, H., 1987, Investitionen sind kein Glücksspiel mehr. *CWextra*, **12 Juni 1987**, 45–47.

Sommerlatte, T., 1987, Die Informationsinfrastruktur der Zukunft. In *Die Informationsverarbeitung der Zukunft. Konzepte und Strategien aus der Praxis*, Vol. 1, pp. 143–164.

PART II
PRODUCTION PROCESSES

Production processes

Research and innovation in the area of production processes is the key to the development of new products and manufacturing methods. This area of investigation is central to achieving the combined benefits of improved quality, efficiency and reliability which are so crucial to a modern manufacturing organization.

The papers in this section represent the wide range of activity currently being pursued in this area.

The CINPRES gas-assisted moulding process is described by Quick *et al.* who point out the benefits of a plastic injection moulding process, which is able to produce large mouldings with a low cooling time and up to 40% weight saving, over conventional processes. Bertolotti and Griffiths discuss the design and development of a flexible loom board system for integrated manufacture of cable harnesses.

The use of an electro-discharge machining (EDM) process to produce small deep holes in difficult-to-machine metals is reported by Chen and Duffill. They anticipate applications in industries as diverse as electronics, aerospace and medicine.

Li Li *et al.* describe a new method for analysing grinding stability and its use as a tool in optimizing processes and the design of chatter control systems. Morshedizadeh *et al.* discuss a technique for the optical measurement of robot-cut grooves in glass.

SECTION II.1. GRINDING

3

A new method for grinding stability analysis

Li Li, Xu YanShen, Sun RongXi and Peng ZeMin

Department of Mechanical Engineering, Tianjin University, Tianjin, China

Abstract

In this paper, the effect of penetration rate, which has previously been ignored, is considered in the quantitative analysis of stability for the grinding process. From the stability criteria for grinding developed by the authors, we propose a criterion index $R(w)$ which is the ratio of the distance between the centre of the Nyquist circle and any point on the characteristic straight line indicated by $D_1(w)$ to the radius of the circle denoted by $D_2(w)$, thus $R(w) = D_1(w)/D_2(w)$. The $R(w)$ stability analysis method is found to be much simpler than the traditional one by which transcendental equations must be solved in order to find the critical speeds under various grinding conditions. For the other computation methods, the method of stability analysis is more valid for the grinding system in which technological parameters have a range of choice. It has been proved that the $R(w)$ analysis method is a very useful tool in the optimizing process.

Introduction

Generally speaking, the regenerative chatter is now accepted as the main source of self-excited vibration in grinding. One of the earliest studies of regenerative chatter in grinding was conducted by Hahn (1954). Based on the proportionality of the instantaneous wheel depth of cut to the instantaneous dynamic force existing between wheel and workpiece, he deduced the unconditional stability criteria from the Nyquist diagram. Swenney studied the grinding instability by applying an algebraic theory developed by Swenney and Tobias (Swenney, 1965). He considered the regenerative effect and also the modal coupling effect which was first developed by Tlusty (Koenigsberger and Tlusty, 1970). The feature of the algebraic theory is that the dynamic characteristics of the machine structures are taken into consideration with the aid of the cross-receptance between the direction of the cutting force and the normal to the machined surface. Using this theory, Swenney (1965) determined the phase angle between the vibration and the surface wave generated on the workpiece and also the depth of the cut coefficient which describes the relationship between the variation in grinding force and the cut depth.

Gurney (1965) presented a graphical method of analysis of waviness in stability which was first used in grinding. He took further steps to derive the surface wave stability diagram from the harmonic response locus of the grinding machine. Polacek (1970) considered the regenerative effect in a close loop between the vibratory system and the grinding process. He assumed a transfer function similar to that introduced by Tobias (1958). The flexibility of the wheel was also taken into account by considering the wheel contact stiffness. Snoeys and Brown (1969) and Kaliszer (1970) analysed the regenerative effect by considering a close loop between the machine-tool elastic system, the cutting process, and two time-delay factors due to wheel waviness and workpiece waviness. They used the Nyquist stability criterion to establish stability of the system. Weck (1979) described the functional relationship between two kinds of regenerative effects and the machine-cutting process for grinding by determining characteristic parameters which describe the dynamic wear of the grinding wheel and the actual contact conditions between the machine and the workpiece.

Li *et al.* (1988) considered the effect of penetration in quantitative analysis of stability for grinding. They derived the functional relation between penetration factor and grinding stiffness from a new model of grinding force which they had developed. Further, they deduced the stability criterion for grinding which is of general significance. If the effect of penetration is neglected in the grinding process, the criterion can be changed into that developed by Snoeys and Brown (1969). However the relation between the criterion index and grinding parameters is so complicated that it is difficult for the traditional analysis method to be used in the practical grinding process, which is tedious and time consuming.

In this chapter, the authors suggest a new method for stability analysis which represents a great saving in effort as compared to graphical or explicit numerical methods for the solution of the transcendental characteristic equation. We have proposed a criterion index $R(w)$ which is the ratio of the distance between the centre of the Nyquist circle and any point on the characteristic straight line to the radius of the circle. It has been proved that the $R(w)$ analysis method, which is not restrained by any technological conditions, would be more valid than the other computation methods, e.g. regenerative spectrum (Li *et al.*, 1989; Srinivasan and Nachtigal, 1978), for the grinding system in which technological parameters have a wide range of choice. As the stability on chatter behaviour in grinding is discerned readily and quickly by calculating $R(w)$, the $R(w)$ analysis method can be used in design of grinder chatter control systems.

Model including the effect of penetration in grinding

The effect of penetration rate (which has been ignored) in the grinding process should be taken into consideration for stability analysis. Penetration rate acts as a damping factor within the system, giving a critical cut width higher than that attained by the current theory. Based on this assumption, the equation for the dynamic characteristic of grinding has been developed in combinination with the Nyquist diagram of the weakest vibrational mode of the machine structure.

Taking vertical surface grinding, for example, let us investigate the effect of penetration rate on the stability of the grinding process. The model of vertical surface grinding is shown in Figure 1.

Suppose the energy expended by the grinding wheel is equal to that consumed in cutting the workpiece, with the aid of Figure 1 the dynamic grinding force is obtained as:

$$F_n = K_w \cdot t + C_w \cdot \dot{t} \quad (1)$$

where t = instantaneous depth of cut (mm), K_w = grinding stiffness (N/mm), C_w = penetration rate coefficient in grinding (N.s/mm)

From Figure 1, we can obtain the relation between the K_w and C_w which can be expressed as:

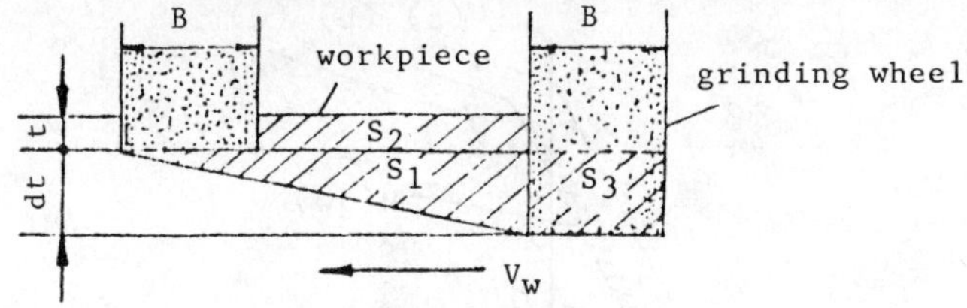

Figure 1. Model of vertical surface grinding.

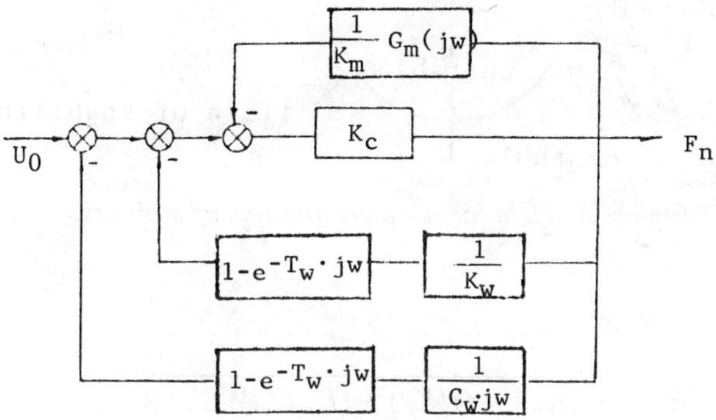

Figure 2. Block diagram of chatter loop in grinding.

$$C_w = \frac{B}{V_w} \cdot K_w \tag{2}$$

where B and V_w are the wheel width and average workpiece velocity, respectively, of the grinder.

As vertical surface grinding is mainly used in the rough-cut process, workpiece regenerative chatter holds a dominant position. So the plunge grinding system can be represented by a new block diagram, which takes the effect of penetration rate into consideration as shown in Figure 2.

The meaning of the parameters in Figure 2 is as follows: K_m = equivalent static stiffness (N/mm), $G_m(jw)$ = normalized machining system dynamic compliance, K_c = contact stiffness between wheel and workpiece (N/mm), T_w = period of workpiece rotation (s), U_o = total grinding wheel infeed (mm). Figure 2 can be reduced to the so called 'characteristic equation':

$$-\frac{1}{K_w} \cdot \frac{1}{1-\mathrm{e}^{-T_w \cdot jw}} - \frac{1}{C_w \cdot jw} \cdot \frac{1}{1-\mathrm{e}^{-T_w \cdot jw}} - \frac{1}{K_c} = \frac{1}{K_m} G_m(jw). \tag{3}$$

Stability analysis

The left-hand side of eqn (3) is composed of three elements giving the influence of the cutability of the workpiece, of the relative vibrational velocity and of the contact area deformation between wheel and workpiece. If we represent the dynamic characteristic in the grinding process $H(jw)$ for the whole left-hand side of eqn (3), we can see that $H(jw)$ represents an inclined straight line in a complex plane, which can be expressed as:

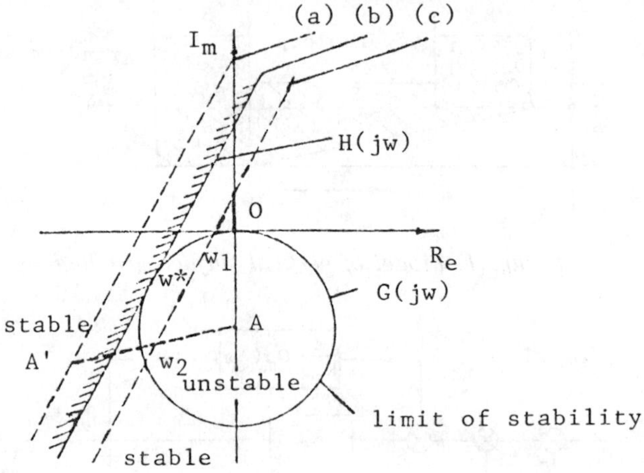

Figure 3. Graphical determination of stability.

$$\frac{X}{(1/2K_w) + (1/K_c) + [K_w/2\,(WC_w)^2]} + \frac{Y}{(1/2WC_w) + [(1/2K_w) + (1/K_c)]/4WK_wC_w} = 1. \tag{4}$$

Every point of this straight line corresponds with any combination of all technological parameters.

The right-hand side of eqn (3) represents the compliance of the grinding machine system. Usually, chatter in the grinding process corresponds with the weakest vibrational mode of the machine structure, so we can simply express the right-hand side of eqn (3) as a formal circle in the Nyquist diagram, which can be written as:

$$G(jw) = \frac{1}{K_m} G_m(jw) = \frac{K - mw^2 - jwc}{(K - mw^2)^2 + (wc)^2} \tag{5}$$

Where m, K and c are the effective mass, stiffness, and damping, respectively, of the weakest mode for the machine structure.

The traditional method for stability analysis is the one that the locating relationship for the two geometrical loci should be compared with in the Nyquist diagram. This stability criterion is illustrated by Figure 3, where the two mentioned loci are indicated.

When both loci do have coinciding points, the limit of stability has been reached. The system is stable if the point in the straight line determined by technological parameters is in the outside of the circle determined by structure. It is the same the other way round.

Since it is not easy to determine the coordinates of the points in the straight line, applying the traditional method of stability analysis to practical grinding operations is not convenient. In this paper, we have proposed a criterion index $R(w)$ which is the ratio of the distance between the centre of the Nyquist circle and any point on the characteristic straight line indicated by $D_1(w)$ to the radius of the circle denoted by $D_2(w)$. With the aid of eqns (4) and (5) for $H(jw)$ and $G(jw)$, we can express $R(w)$ as:

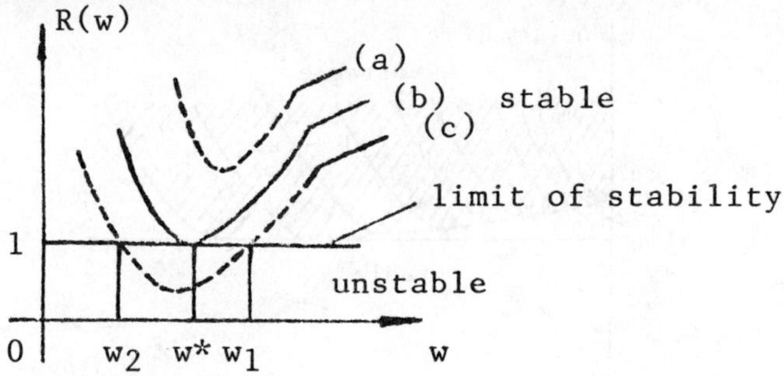

Figure 4. Characteristic curves of R(w).

$$R(w) = \frac{D_1(w)}{D_2(w)} =$$

$$\left| \frac{-(1/2K_w) - (1/K_c) + (1/2WC_w)ctg(WT_w/2)}{W_n/4KEW} + j\frac{[(1/2WC_w) + (1/2K_w)ctg(WT_w/2) + (W_n/4KEW)]}{W_n/4KEW} \right| \qquad (6)$$

where E is the ratio of relative damping, which can be written as

$$E = c/2\sqrt{Km}. \qquad (7)$$

It is clear that $R(w) > 1$ can occur if the two geometrical loci of $G(jw)$ and $H(jw)$ have no intersection, so the grinding system is always stable. The system is unstable if $R(w) < 1$, in which the loci have two intersections. In this case, if $R(w) > 1$ the system remains stable although two intersections exist. Just as the two loci are tangent or intersected each other, $R(w)$ in points of tangency or intersection is equal to zero, in which the system is at the limit of stability. The stability condition can be fully analysed as shown in Figure 4.

As indicated in Figure 4:

$$R(w) \begin{cases} > 1 & \text{system stable} \\ < 1 & \text{system unstable} \\ = 1 & \text{threshold condition} \end{cases}$$

Since $R(w)$ is a function of grinding frequency, when $R(w) = 1$, the frequency corresponds to the critical frequency at the threshold condition, and a set of critical width of cut b and work speeds N_w can be attained. So the stability curve of the grinding process is easily obtained as shown in Figure 5.

The points in Figure 5 are the average values of many experimental results in the same technological conditions. Experimental results show the same tendency as the theoretical results.

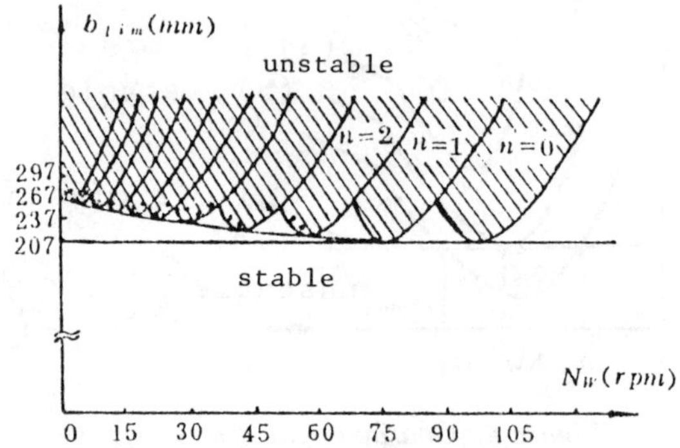

Figure 5. Stability curve of the grinding process.

Conclusions

The application of the $R(w)$ analysis method to the investigation of regenerative chatter in grinding is the primary contribution of this paper. For the conditions of any grinding system in which technological parameters have a wide range of choice so as to adapt to different needs, the stability on chatter behaviour can be discerned readily and quickly by calculating $R(w)$. Also, the $R(w)$ analysis is not restrained by any technological conditions, so it is more valid for stability analysis for the other computation methods, for example, regenerative spectrum. It has been proved that the $R(w)$ analysis method is a very useful tool in optimizing grinding processes and it can be used in the design of grinder chatter control systems.

References

Gurney, J.P., 1965, An analysis of surface wave instability in grinding. *Transactions ASME (Journal of Mechanical Engineering Science)*, **2**, 7, 198–209.

Hahn, R.S., 1954, On the theory of regenerative chatter in precision-grinding operation. *Transactions ASME*, **1**, 76, 593–597.

Kaliszer, H., 1970, Analysis of chatter vibration during grinding, Proceedings of the International 11th MTDR (Oxford: Pergamon Press), pp. 615–631.

Koenigsberger, F. and Tlusty, J., 1970, *Machine Tool Structure* (Oxford: Pergamon Press).

Li, L., Xu, Y.S., Sun, R.X. and Peng, Z.M., 1988, Regenerative stable analysis in a vertical surface grinder by taking the effect of penetration into account. *Journal of Tianjin University*, **4**, 58–66.

Li, L., Xu, Y.S., Sun, R.X. and Peng, Z.M., 1989, Dynamic analysis and design of vertical surface grinders using regenerative spectrum. *Journal of Tianjin University*, **1**, 32–39.

Polacek, M., 1970, Chatter in Grinding. *Machine Tool Structure* (Oxford: Pergamon Press), p. 1.

Snoeys, R. and Brown, D., 1969, Dominating parameters in grinding wheel and workpiece regeneration chatter. Proceedings of the International 10th MTDR (Oxford: Pergamon Press), pp. 325–348.

Srinivasan, K. and Nachtigal, C.L., 1978, Analysis and design of machine tool chatter control systems using the regeneration spectrum. *Transactions ASME (Journal of Dynamic Systems, Measurement, and Control)*, **8**, 100, 191–200.

Swenney, G., 1965, Grinding instability. Proceedings of the International 6th MTDR (Oxford: Pergamon Press), pp. 557–580.

Tobias, S.A., 1958, Theory of regenerative machine tool chatter. *Engineering*, **205**.

Weck, M., 1979, Interaction of the dynamic behaviour between machine tool and cutting process for grinding. *Annals of the CIRP*, **1**, 28, 281–285.

SECTION II.2. CUTTING

4

An expert system for the automation of the broach design process

W. Robert Terry, E. Ralph Sims, Jr, Chad W. Richards and John A. Deno

Industrial and Systems Engineering, Ohio University, Athens, OH 45701-2979, USA

Abstract

Broaching is a very efficient and precise cutting and shaping process which utilizes a linear motion, multiple tooth tool to rough-cut, size, shape and finish parts made of metal and other materials in a single stroke pass. In spite of its efficiency, the broaching process is underutilized because of the high cost of designing and manufacturing the broach tools. The application of knowledge-based, computer-aided design and the downloading of the design data to provide input to computer-aided manufacturing will significantly reduce this initial tooling cost and enhance the economics of the broaching process. This chapter develops an expert system algorithm for the knowledge-based computer design of a round broach tool.

Introduction

Broaching is a precision metal removal process whereby a multi-tooth cutting tool is pushed or pulled either through or along the surface of a workpiece. Broaching is similar to planning, and is competitive with such machining processes as milling and boring.

In its simplest form, a broach tool is '… a slightly tapering round, or flat bar with rows of cutting teeth located along the tool axis (Fennimore and Schultz, 1976). Frequently, a broach tool will consist of three distinct cutting sections—rough, semifinish and finish, with each successive tooth increasing in height from the first roughing tooth to the first finishing tooth. This unique feature of a broach tool provides its greatest strength—the ability to rough and finish machine in a single stroke.

There are basically two types of broaching: internal and external. In internal broaching, there must be a starting hole in the workpiece through which the broach is either pushed or pulled. Internal broaches are capable of producing round, oval, hexagonal, square or other irregularly shaped holes such as splines, keyways and spirals to name a few. In external broaching, the broach is usually pushed or pulled along the surface of the workpiece. In some instances, however, the broach can be held stationary while the workpiece is pushed or pulled along the broach tool surface. There are virtually no limitations to the number of contoured surface shapes which can be produced by this method.

Both internal and external broaching are used to machine parts of various sizes and composition. Applications may range from cutting small keyways to machining engine block surfaces. Almost any conceivable contoured surface can be created as long as the workpiece's machined surface remains parallel with the direction of broach travel. According to Hamm *et al.* (1983), broaching has been used on both ferrous and non-ferrous metals, plastics, hard rubber, wood, composite materials and graphite.

Broaching, if used properly, is a highly productive, economical, precise and extremely versatile process. At its best, broaching is capable of production rates as much as 25 times faster than traditional metal removal methods. This most likely stems from the ability to rough and finish cut in a single stroke. Also, with reasonably thin workpieces, broaching has the capability to process more than one part at a time by stacking a number of pieces and broaching them simultaneously. Broaching also produces superior surface finishes which, in turn, frequently eliminates the need for grinding and consistently maintains tight tolerances.

Unfortunately, despite these technical advantages, broaching has not been as well applied as other machining processes such as milling and boring. This can probably be attributed to several inherent economic factors in broaching. First, broaching requires a high capital investment in both specialized machinery and workholding devices. Also, due to their unique designs, broach tools are generally expensive and designed for job-specific tasks. They are usually not interchangeable, but rather designed for a particular operation on a particular part for a specific machine. Secondly, the lead time required to design and fabricate a broach commonly runs between 10 and 12 weeks (Savage, 1988). Unless production volumes are high, broaching can be difficult to justify economically. This point is accentuated by the availability of alternative processes, which, although usually slower, can accomplish the same machining operation at a lower cost.

The objective of this research is to lower the cost of broach tools by reducing the time required to design and fabricate them. The proposed methodology for effecting this time reduction is based on the computerization and integration of the round-hole broach design and the round-hole broach manufacturing process. Before discussing this methodology, however, a brief description of the broach design problem is needed.

In designing a round-hole broach, a number of constraints must be satisfied. First, the maximum stress a broach tool is subjected to cannot be greater than that which will cause the broach to fail in either compression (push broach) or tension (pull broach). Secondly, the deflection of the broach teeth cannot exceed the maximum allowable deflection. Thirdly, the force required to push or pull a broach cannot exceed the maximum tonnage rating of the broaching machine. The fourth constraint states that the length of a broach should not exceed 75 times the diameter of the finishing teeth for round pull-type broaches, or 25 times the finishing teeth diameter for round push-type broaches (Van DeMotter, 1988). Although there is no theoretical limit to the length of a broach, this rule of thumb can prevent much unnecessary breakage. The fifth constraint states that the length of a push of pull-type broach cannot exceed the stroke length of the broaching machine. Additionally, the length of the cutting section of the broach must be greater than the thickness of the part through which the broach is to be passed. Finally, the volume of metal to be removed must not exceed the volume of the cavity between successive teeth.

The framework for the computerization and integration of the round-hole broach design and the fabrication process consists of:

(1) *Knowledge-based system.* This system was designed to capture and use the knowledge of the expert broach designer and to represent it in a computer.
(2) *Computer-aided design interface.* This interface was developed to eliminate the process of manually transferring broach design specifications into a working drawing.

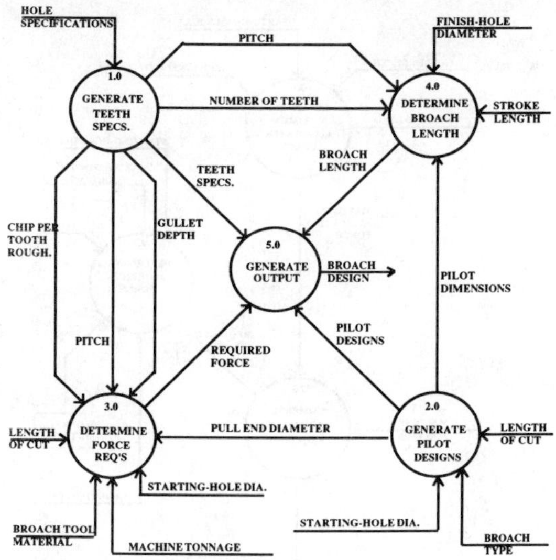

Figure 1. Zero level diagram of the round-hole broach design process.

(3) *Computer-aided manufacturing*. This step is the computer generation of the tool path based on the geometric description of the broach provided by CAD, and development of a postprocessor for translating the tool path data into computer numerical control (CNC) machine-readable code.

(4) *Physical prototyping*. This is the culmination of all previous stages—the ability to prototype physically on a real-time basis. It includes the CNC turning and grinding of the broach.

(5) *Testing and analysis*. This is the step in the broach design and fabrication methodology that provides feedback as to whether or not the knowledge-based system effectively models the expert's knowledge.

This chapter will focus on the first two process stages; the knowledge-based system and the CAD interface of the broach design and fabrication framework.

Knowledge-based system

As previously mentioned, the knowledge-based system is designed to capture the experiences, rules and heuristics of the human expert in a computer processable format.

Two experts, Christopher Van DeMotter of the Ohio Broach and Machine Company, of Willoughby, Ohio and Robert Savage of the Hassay/Savage Company, Turners Falls, Massachusetts were interviewed to develop a system model of the round-hole broach design process. Structured Systems Analysis was used to develop the system model.

Figure 1 is a zero level diagram of the round-hole broach design process. It shows that the system is composed of five processes: (1) Generate Tooth Specifications, (2) Generate Pilot Designs, (3) Define Force Requirements, (4) Determine Broach Length, and (5) Generate Output. Due to the page limitation of this chapter, only the third process—Define Force Requirements—will be discussed here. A narrative discussion of the process and associated pseudocode is provided. For further information on the remainder of the system model, please refer to Richards (1989).

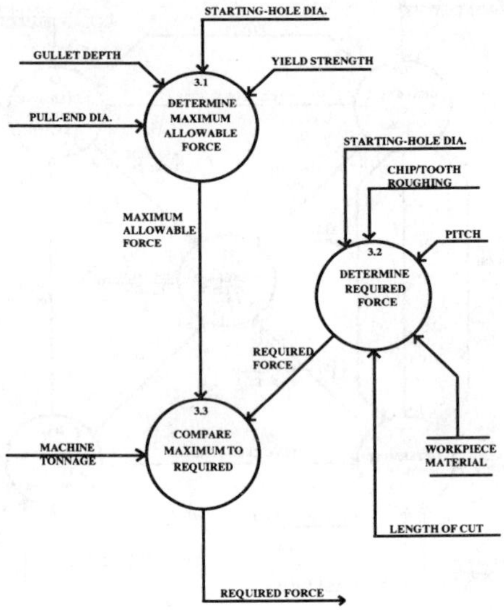

Figure 2. Subsequent transformations of process 3—Determine Force Requirements.

Process 3—determine force requirements

This process contains three subsequent transformations: (1) Determine Maximum Allowable Force, (2) Determine Required Force, and (3) Compare Maximum to Required (see Figure 2).

Transformation 3·1 calculates the theoretical maximum force that the broach can withstand without failing at the smallest cross-sectional area of the broach. The force is calculated as the product of the minimum broach cross-sectional area and the tensile yield strength of the broach tool material.

Transformation 3·2 calculates the minimum force required to pull the broach through the workpiece based on the following equation:

$$F = 3{\cdot}14 * D * N * R * C$$

where F = minimum pulling force required (lb), D = hole diameter before broaching (in), R = chip per tooth for roughing teeth (in), N = maximum number of teeth engaged in the workpiece at one time (at least three), C = empirical broaching constant representing the machinability of a metal (psi).

Transformation 3·3 compares the required pulling force (F) to the maximum allowable pulling force (P). Additionally, the required force is compared to the tonnage rating of the broaching machine (T). If F is at least 20% less than P, and F is less than T, then the broach design is deemed valid at this stage in the broach design process (Van DeMotter, 1988). However, if either condition is not met, the design must be altered to satisfy both constraints if possible. By examining the variables associated with the calculation of the required pulling force, one variable—the chip volume per tooth (R)—can be altered. The broaching constant (C) and the starting hole diameter (D) are predetermined by the user. The maximum number of teeth in contact with the workpiece material at one time (N) is usually equal to three, and the reduction of this variable could decrease the stability of the broach as it passes through the workpiece.

If it becomes necessary to reduce the required pulling force, the chip per tooth is systematically decreased. This process is continued until either the constraints are satisfied or until the chip per tooth reaches a predetermined minimum. If the chip per tooth reaches the minimum value before the force constraints are satisfied, the broach design is deemed invalid.

Pseudocode for transformation 3·0

```
VARIABLE DECLARATION
    A   = minimum cross sectional area (in ^2)
    C   = broaching constant (psi)
    D   = hole diameter before broaching (in)
    F   = required pulling force (lb)
    N   = number of teeth in contact with workpiece at one time
    P   = maximum allowable pulling force (lb)
    PED = minimum pull end diameter (in)
    PI  = 3·14159
    R   = chip per tooth (in)
    RD  = root diameter at first tooth (in)
    T   = tonage rating of the machine (lb)
    Y   = tensile yield strength of the broach tool material (psi)

END VARIABLE DECLARATION
1. Calculate the minimum cross sectional area of the broach IF the minimum is
   at the pull end
   THEN
       A = (PI/4) * PED
     ELSE
       A = (PI/4) * RD
   END IF
2. Calculate the maximum allowable pulling force
     P = A * Y
3. Calculate the required pulling force
     F = 3·14 * D * N * R * C
4. Compare F to P and F to T
     IF F < (·8*P) AND F < T
   THEN
     the design is valid
   ELSE
     DO
       A. Reduce the Chip per Tooth
       B. Re-calculate the required force
       C. Compare F to P and F to T
     LOOP until constraints are satisfied
     OR until R reaches its minimum
   END IF
   IF R reached its minimum AND the constraints are not satisfied
     THEN
       the design is invalid
     ELSE
       the design is valid up to this point in the design process
   END IF
   END
```

For information about the programming style and the knowledge representation method used to code the knowledge-based system, refer to Richards (1989).

Computer-aided design (CAD) interface

The rationale for developing a CAD interface was to eliminate the time-consuming task of translating the round-hole broach design specifications into a working drawing of the broach tool. This development represents a significant time saving and the interface provides a mechanism for generating a drawing within a matter of minutes.

The creation of the CAD interface involved two major tasks: (1) determination of the format that conventional CAD packages utilize to represent geometric data, and (2) creation of a mechanism in the knowledge-based system that is capable of reproducing this format for a particular design.

Two CAD software packages, AutoCAD and CADKEY, were analysed to determine the internal geometric representation schemes used. It was observed that both packages represent geometric data in a standard ASCII text file format. In particular, drawings are represented in terms of entity definitions. Entities represent standard geometric building blocks such as lines, arcs, circles, polygons, etc.. The definition accompanying each entity consists of its two-dimensional coordinates. For example, a line would be defined as follows:

Line
X1 coordinate
Y1 coordinate
X2 coordinate
Y2 coordinate

where X1 and Y1 represent the staring point X2 and Y2 represent the finishing point of the line.

To create the interface, a generic ASCII file structure was developed to represent the geometry of a round-hole broach. The file consisted of entity definitions structured in an order specific to round-hole broaches. A generic file structure was created for both push and pull type round-hole broaches. To generate a drawing for a specific broach design, the geometric specifications created by the knowledge-based system were read into the text file. In order to obtain a drawing of the broach, the user is required to call up the file while using CADKEY or AutoCAD.

Conclusions and recommendations

The knowledge-based system developed provides the user with direct access to expert knowledge, and affords them the ability to design a round-hole broach within minutes. The CAD interface successfully eliminates the process of translating the design specifications into a drawing of the broach tool.

It is recommended that three additional steps be taken to increase the effectiveness of the broach design system:

(1) The remaining steps in the broach design and fabrication framework should be completed if the ability physically to prototype on a real-time basis is to be realized.
(2) The knowledge-based system should be expanded to include other broach designs such as surface broaches, spline broaches, and keyway broaches to name a few.
(3) The current research in finite element analysis should be incorporated in the knowledge-based system to provide a more accurate method for analysing the rigidity of a broach tooth when it is subjected to the cutting forces.

References

Fennimore, W. and Schultz, L., 1976, Broaching. In *Tool and Manufacturing Engineers Handbook*, 3rd edn, edited by D. Dallas (New York: McGraw-Hill).

Hamm, S.J., Roseliep, R.E., Smith, L.J. and Van DeMotter, C., 1983, Broaching planning, shaping, and slotting. In *Tool and Manufacturing Engineers Handbook*, Vol. 1, Machining, edited by T.J. Drozda and C. Wick (New York: McGraw-Hill).

Richards, C., 1989, *A Knowledge-Based System for the Computerization and Integration of the Round-Hole Broach Design Process* (Athens, Ohio: Ohio University).

Savage, R., 1988, President, Hassay/Savage Company, Turners Falls, Massachusetts.

Van DeMotter, C., 1988, Manager of Engineering, The Ohio Broach and Machine Company, Willoughby, Ohio.

5

Approaches to cutting and packing problems

Harald Dyckhoff

Lehrstuhl für Industriebetriebslehre, RWTH Aachen, Templergraben 64, D-5100 Aachen, Germany

Abstract

Cutting and packing (C&P) problems as considered here are relevant for different areas of production management and appear under various names, e.g. cutting stock or trim loss problem, bin or strip packing problem, vehicle, pallet or container loading problem, nesting problem, knapsack problem, and even line balancing problem or multi-processor scheduling problem. Nevertheless, they exhibit a strong logical and structural relationship so that their solution approaches are strongly connected, too. The attempt is made to give a systematic, structured survey of these approaches based on a typology of C&P-problems. In view of the excellent existing surveys for particular kinds of solution approaches, the purpose is not to go into details but to integrate, unify and extend these surveys; hence in some way to give a certain overview of recent developments.

Common logical structure

The common logical structure of C&P-problems is characterized by two groups of basic data: the stock of '(large) objects', and the order list of '(small) items'. Both define geometric bodies of fixed shapes ('figures') in a one- or more-dimensional space of real numbers. Cutting and packing processes realize patterns as geometric combinations of small items assigned to large objects. Residual pieces, i.e. figures occuring in patterns not belonging to small items, are usually treated as 'trim loss'. In the example of Figure 1 both groups, i.e. the objects as well as the items, build rectangles of specified widths and lengths in two-dimensional space. This may be the case when cutting glass plates or loading pallets.

In a narrow sense, C&P-problems are concerned with objects and items defined by one, two or all three spatial dimensions of the Euclidean space. In cases of cutting (stock) problems the large objects are given by solid materials cut up into small items as pieces. These production processes are typical for the paper and pulp, metal, glass, wood, plastics, leather and textile industries. Packing and loading problems are characterized by large objects defined as the empty, useful space of vehicles, cars, pallets, containers, bins and so on.

Packing small material items into bins as large objects may also be looked at as cutting the empty space of bins into parts of empty spaces some of which are occupied by small items, the other being 'trim loss'. Conversely, cutting problems may be looked at as packing the space occupied by small items into the space occupied by large objects. In other words: the strong relationship between cutting

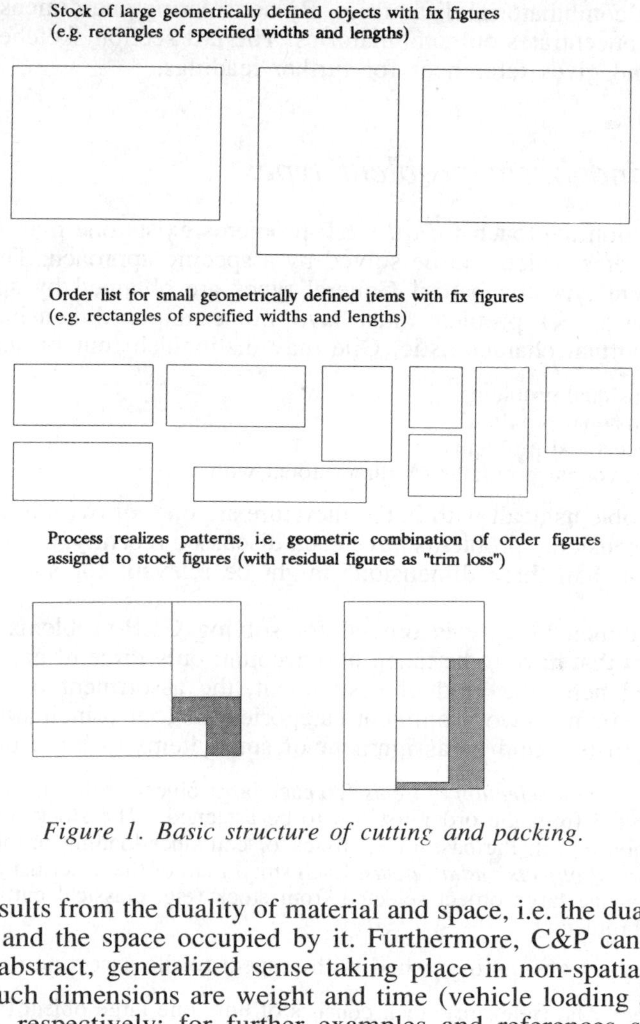

Figure 1. Basic structure of cutting and packing.

and packing results from the duality of material and space, i.e. the duality of a solid material body and the space occupied by it. Furthermore, C&P can also be considered in an abstract, generalized sense taking place in non-spatial dimensions. Examples of such dimensions are weight and time (vehicle loading and assembly line balancing, respectively; for further examples and references see Dyckhoff, 1990).

A general solution method for all C&P-problems does not exist. There are, however, some fundamental types of methods, called 'solution approaches', depending on the type of problem considered. Hence, it is an important task of (future) research to connect problem types with appropriate solution approaches, and vice versa. In this sense, a metal cutting problem and a pallet loading problem may have more in common than the cutting problems of two textile firms.

The problem type is determined by the aspects relevant to the answers requested. For example, the kind of raw material may be relevant only in so far as it implies specific conditions of the cutting process. In order to describe different types of C&P-problems one has to identify characteristics which specify them (more or less). The next section discusses a few important characteristics which allow the definition of certain general and basic types of C&P-problems.

The specific models or algorithms for the basic types are then described, in this way characterizing the concepts of the underlying solution approaches. Object- and item-oriented approaches as well as pattern-oriented approaches are distinguished.

Most important for the complexity of C&P-problems is the dimensionality of the relevant space since the geometrical difficulties grow drastically with this number. Even in the much simpler one-dimensional case, however, it is hard enough to solve the remaining combinatorial difficulties. By considering one-dimensional problems this chapter concentrates on combinatorics. The last section sketches the geometrical aspects and gives references for further readings.

General and basic problem types

Since no solution approach for *all* C&P-problems exists one may look for certain types of problems which can be solved by a specific approach. The more general such a problem type the better! General types are obtained by specifying a few properties that a C&P-problem must have. For example, dimensionality is clearly the most important characteristic. One may distinguish four principal cases:

(1) one-dimensional problems,
(2) two-dimensional problems,
(3) three-dimensional problems, and
(*N*) multi-dimensional problems (*N*-dimensional with $N > 3$).

Almost all problems dealt with in the literature are one- or two-dimensional. A very few three-dimensional problems have been discussed recently, usually for container loading. More than three dimensions might be relevant for some abstract C&P-problems.

Although dimensionality is crucial for solving C&P-problems there are more characteristics that have to be taken into account; only three of general interest will be considered here: the kind of assignment, the assortment of objects, and the assortment of items. Two prominent categories (of four principally possible ones) with respect to the kind of assignment of small items to large objects are:

(B) *All objects and a selection of items.* To each large object of the stock at least one small item selected from the order list has to be assigned. (The stock may consist of one object only, e.g. in the case of knapsack or container loading problems.)
(V) *A selection of objects and all items.* Each small item of the order list has to be assigned to exactly one large object selected from stock (e.g. classical cutting stock and bin packing problems).

Three main types are distinguished with respect to the assortment of large objects:

(O) *One object.* The (relevant) stock consists of only one large object (cf. type B above).
(I) *Identical figure.* All large objects of the stock are of the same shape (e.g. classical bin packing problems).
(D) *Different figures.* The large objects of the stock have different shapes (e.g. cutting a stock of large objects containing residual pieces of prior periods).

Similarly, four main types regarding the assortment of small items are pointed out here:

(F) *Few items.* The order list consists of only a few items, usually of different shapes (e.g. vehicle loading with about 10 items).
(M) *Many items of many different figures.* The order list contains many small items most of them having different shapes (e.g. packing hundreds and more items of lengths equally distributed between 0 and 1 into bins of length 1).
(R) *Many items of relatively few different figures.* The order list contains many small items but of relatively few different, non-congruent shapes (e.g. cutting thousands and more items of less than 50 different shapes).
(C) *Congruent figures.* All small items of the order list are congruent (e.g. pallet loading).

By combining the distinguished types of dimensionality, assignment, and assortment one obtains $4 \times 2 \times 3 \times 4 = 96$ general types of C&P-problems. They can shortly be denoted by a quartet of respective symbols $\alpha/\beta/\gamma/\delta$. For example, 3/B/O/F denotes all three-dimensional C&P-problems where one large object has to be packed with a selection out of a few small items. A missing symbol for a characteristic means that all respective properties are possible. Thus, the classical knapsack problem belongs to type 1/B/O/ , i.e. it is one-dimensional with one large object that has to be packed with a selection from the set of small items. The usual pallet loading problem differs in that it is two-dimensional with congruent items: 2/B/O/C.

The notation describes general types of C&P-problems each capturing a variety of still distinct problems. For example, nothing is stated about the shape of objects and items in two- and more-dimensional cases (e.g. regular or irregular form?). Even the classical cutting stock problems of type 1/V/I/R may be very different from each other, in particular with regard to pattern restrictions and objectives considered. Such differences result from characteristics not included in the above typology (see Dyckhoff, 1990, for a more comprehensive discussion).

Instead of adding more characteristics, one may define basic types of C&P-problems. These are very special problem types which can be uniquely described by a mathematical model and/or a corresponding data structure. Any deviation, for example because of an additional restriction or a modified objective, leads to a new, different type. For such basic problem types, definite solution methods often exist. In order to characterize some fundamental solution approaches for a general problem type, it is therefore reasonable to choose certain representative basic types—of the considered general type—and to describe appropriate solution methods for these basic types.

In the following, some particular one-dimensional basic types are chosen so that the concepts of existing solution approaches can be sufficiently illustrated as far as the combinatorial aspects of C&P-problems are concerned. A well-known type is the classical 'knapsack problem' which may be considered as a special case of the general type 1/B/O/ .

Basic 'knapsack-type' 1/B/O/ . There is (only) one one-dimensional large object of length L that has to be packed with a (non-empty) selection from a set D of one-dimensional small items d_k with lengths l_k ($k = 1, \ldots, K$) so that the sum of the individual values v_k of the packed items is maximized.

Furthermore, three special cases of the general type 1/V/I/ are of interest. They only differ in their data structures of the assortment of small items.

There are two lists S and D, called 'stock' and 'demand', consisting of an infinite number of one-dimensional large objects s_h of identical lengths L ($h = 1, 2, \ldots$) and a finite number of one-dimensional small items d_k of lengths l_k ($k = 1, \ldots, K$), respectively. The demand has to be fulfilled, i.e. all items have to be assigned to (cut from or packed into) objects of stock S. The objective is to minimize the number of large objects used.

Basic 'vehicle-loading-type' 1/V/I/F. The demand consists of only a few small items, i.e. the number K is small, e.g. $K = 10$.

Basic 'bin packing-type' 1/V/I/M. The demand contains many small items, e.g. several hundreds, most of them having different lengths.

Basic 'cutting stock-type' 1/V/I/R. The demand contains a lot of small items, e.g. thousands, but the items are of only relatively few different lengths, e.g. < 50.

Solution approaches: combinatorial aspects

For the basic knapsack-type, let the binary variables z_k describe the two cases that item d_k is packed ($z_k = 1$) or not packed ($z_k = 0$) into the large object ('knapsack'). Then the corresponding mathematical model is stated by the following relations:

Maximize $$\sum_{k=1}^{K} v_k z_k$$

such that $$\sum_{k=1}^{K} l_k z_k \leq L$$

for $$z_k \in \{0, 1\},\ k = 1, \ldots, K.$$

Solution methods to this problem are well known. Exact methods are usually based on dynamic programming or branch and bound. Because the knapsack problem is NP-complete, fast approximation algorithms may be suitable for large problem instances. For recent surveys on the knapsack problem and its generalizations see Dudzinski and Walukiewicz (1987) as well as Martello and Toth (1987).

If more than one large object exists, solution approaches to C&P-problems can in general be distinguished as (i) object- and item-oriented, or (ii) pattern-oriented. Methods of the first type immediately assign items to objects whereas pattern-oriented approaches first construct patterns and then assign large objects as well as small items to some of these patterns. It is not a sharp classification, for example, loading many pallets with identical products can be solved by considering only one (representative) pallet as a large object.

In view of the above three basic problems of general type 1/V/I/ , object- and item-oriented approaches are appropriate for the vehicle loading- and the bin packing-type, pattern-oriented ones for the cutting stock-type. Golden (1976) gives an excellent introduction to fundamental solution approaches; hence a short overview and references to more recent developments suffice here.

The three basic types can be formulated by a common mathematical model where the binary variables y_h and z_{hk} state whether object s_h of stock S is used or not and whether item d_k of demand D is assigned to object s_h or not, respectively.

Minimize $$\sum_{h=1}^{H} y_h$$

such that $$\sum_{k=1}^{K} l_k z_{hk} \leq L y_h, \qquad h = 1, \ldots, H$$

and $$\sum_{h=1}^{H} z_{hk} = 1, \qquad k = 1, \ldots, K$$

for $$y_h \in \{0,1\}, \quad z_{hk} \in \{0,1\}, \quad h = 1, \ldots, H, \quad k = 1, \ldots, K.$$

The model is essentially the same as the integer programming model of assembly line balancing (for minimizing the number of established workstations) except for the missing restrictions regarding a possibly existing partial order of small items. In general, these problems are NP-complete (Garey and Johnson, 1979). Exact solutions in a reasonable time are possible in particular cases with special data structures.

An example is the basic vehicle loading-type because of its low 'demand' for small items. It may be solved by usual methods of integer programming like, e.g. decision tree methods. Such methods were first discussed by Eilon and Christofides (1971), the objects being vehicles of standard size and the items consignments of different weights, e.g. bundles of newspaper. The one-dimensional case considered here is, however, not of particular practical importance.

Problems of the bin packing-type can, in general, be only solved approximately. In the last two decades, many fast algorithms have been developed for this purpose and impressive theoretical as well as empirical results on their behaviour have been achieved (see Coffman *et al.*, 1984, for a comprehensive survey). These 'bin packing-algorithms' are object- and item-oriented in that each object, called 'bin', as well as each item, is considered separately and usually identified by its number. The items are successively assigned to objects, one by one, without the possibility of reallocation. If the items are assigned in their natural sequence *a priori* given, it is an 'on-line algorithm'. 'Off-line algorithms' do some kind of sorting or grouping of the items before a usual on-line algorithm is applied to the changed sequence of demand. A standard sorting is that of decreasing lengths.

The simplest and fastest bin packing-algorithm is 'Next Fit' (NF). It assigns items to the actual object (bin) as long as it fits, i.e. as long as the sum of the lengths of the assigned items is not larger than the length of the bin. As soon as an item no longer fits, the bin is closed and the item is packed into the next one which now becomes the actual bin. The algorithm stops when the last item is packed. The performance of this algorithm is bad. In the worst case it needs 100% more bins than the optimal solution. The off-line version 'Next Fit Decreasing' (NFD) is not much better and is also slower because of the sorting procedure. NFD has time complexity $O(K\log K)$ instead of $O(K)$ for NF where K is the number of items.

For efficient bin packing-algorithms there is a trade-off between performance and velocity. If velocity is important an often acceptable compromise is provided by the off-line algorithm 'First-Fit-Decreasing' (FFD) with time complexity $O(K\log K)$. It arranges the items in order of decreasing length and then assigns them one by one to an appropriate bin (object) using the following rule: An item is assigned to the first bin of the given list $S = (s_1, s_2, \ldots)$ into which it (still) fits. It can be proved—although with considerable difficulty (Coffman *et al.*, 1984)—that this simple algorithm uses at most 22.3% objects more than the optimal solution, i.e. it never leads to an unnecessary trim loss larger than 18·2%. If the lengths of 200 items are uniformly distributed between 0 and L (the length of the bins), FFD has an average trim loss of only 1·9%. The average trim loss decreases to zero if the number of items, K, tends to infinity.

For large numbers of items the (average) performance of bin packing-algorithms crucially depends on the data structure of demand D. A uniform distribution is not typical in practice. Usual one-dimensional cutting stock problems in industry are characterized by a high demand for only a few different lengths. Many objects are cut down in the same manner, i.e. by the same pattern. Hence it makes sense first to construct proper patterns and then to decide how many objects are cut by a certain pattern, i.e. how often the patterns are used. In view of such pattern-oriented approaches, the above mathematical model for the three basic problems of type 1/V/I/ can be reformulated. Items of identical length l_i are no longer distinguished, but are put into the same group and are counted for their total demand b_i. (It is a sorting procedure similar to that of FFD.) A pattern j is described by the numbers a_{ij} of items of lengths $l_i\,(i = 1, \ldots, m)$ appearing in this pattern. Defining x_j as the number of objects cut by pattern $j (j = 1, \ldots, n)$ one obtains the usual linear programming formulation for cutting stock problems with cost coefficients $c_j = 1$ and with integrality conditions for the variables x_j.

Minimize $$\sum_{j=1}^{n} c_j x_j \qquad (\text{with } c_j = 1)$$

such that $$\sum_{j=1}^{n} a_{ij} x_j \geq b_i, \qquad i = 1, \ldots, m$$

for integer $$x_j \geq 0, \qquad j = 1, \ldots, n.$$

One has to keep in mind that the generation of patterns (as well as a possible necessary sorting of demand) is part of the solution procedure and therefore counts when time complexity is considered. This is of eminent importance because the number of possible patterns often is extremely large (see Dyckhoff, 1988, p. 171). Two main groups of pattern-oriented approaches can be distinguished: (i) heuristics and (ii) methods being based on linear programming relaxations.

Heuristics play an eminent role in practice (Hinxman, 1980; Dowsland, 1985; Dyckhoff *et al.*, 1988b), particularly in software packages for cutting stock management (Dyckhoff *et al.*, 1988a). They can immediately produce integer solutions for the patterns-usage. Furthermore, they are often very flexible with respect to modifications and generalizations of the considered basic problem type. Hence, various additional restrictions and objectives appearing in practice may be taken into account, e.g. the costs arising through changing the pattern on a cutting machine. An example is the often used simple heuristic of 'repeated pattern exhaustion' (Haessler, 1971; Pierce, 1964; Wäscher *et al.*, 1985):

Step 1. Construct patterns until a 'good' one is found, i.e. one which is acceptable with respect to the relevant restrictions and objectives.
Step 2. Use this pattern as often as there is demand for the items produced by the pattern, and reduce demand and stock correspondingly.
Step 3. If there is further demand go back to Step 1, otherwise stop.

The quality of this heuristic heavily depends on the problem-specific choice of 'good' patterns. When integrated into an interactive decision support system, such heuristics are able to take account of the experience of the user.

LP-based methods first solve the LP-relaxation of a pattern-oriented model and then search for an integer solution by more or less sophisticated strategies. In view of the high demands b_i for the different lengths l_i simple rounding procedures are often sufficient. Therefore, LP-based methods are heuristics or approximation algorithms too, except for the so-called '1·5-dimensional' problems where no integrality conditions on the variables x_j exist (see Dyckhoff, 1988, p. 165).

Even more important is the very large number of possible patterns, often going into hundreds of millions. In order to overcome this difficulty, Gilmore and Gomory (1961) have developed their fundamental method of 'delayed pattern generation' (Farley, 1988; Stadtler, 1988). The method starts with a feasible LP-solution by constructing a set of simple patterns (perhaps by using FFD). In each step of the revised simplex method then a new pattern corresponding to a new column improving the objective function value is generated. This is done by solving knapsack problems of the following type using the actual shadow prices v_i of the demand conditions for the different lengths l_i; the variable a_i defines how many items of length l_i appear in the pattern.

$$\text{Maximize} \qquad \sum_{i=1}^{m} v_i a_i$$

$$\text{such that} \qquad \sum_{i=1}^{m} l_i a_i \leq L$$

$$\text{for integer} \qquad a_i \geq 0, \quad i = 1, \ldots, m.$$

Besides the above, usual LP-formulation, called 'multi-cut model', there is an alternative formulation, called 'one-cut model'. The first one is based on the assumption of a single-step cutting (or packing) process with complex patterns (multi-cuts), the second one on the assumption of a multi-stage process with very simple patterns (one-cuts) in each step. Both models are equivalent in the sense that they allow for the same efficient C&P-possibilities and may pursue the same objective (see Dyckhoff, 1988, p. 172; Stadtler, 1988). In fact, it can be shown that there are many more equivalent models lying in between the two extreme ones (Dyckhoff, 1989). Because of its special structure the one-cut model can be solved by generalized network algorithms. Until recently, however, there has been no specialized algorithm faster than the Gilmore and Gomory (1961) method for the multi-cut model (Stadtler, 1988). If this method is not available, the one-cut model can be helpful because it consists of only hundreds or thousands of columns (variables) instead of millions, although the number of rows (restrictions) increases more or less. Therefore it can often be solved by standard LP-algorithms. The multi-cut model is, however, more flexible with respect to additional restrictions and objectives.

Solution approaches: geometrical aspects

The last section has given an overview of important solution approaches to one-dimensional C&P-problems. The geometrical aspects have been almost trivial. Therefore, it has been a purely combinatorial consideration. For more-dimensional C&P-problems, combinatorial and geometrical aspects are usually mixed, the combinatorial ones, however, being basically the same as in the one-dimensional case. In this sense, the previous section also gives an overview of the main combinatorial aspects of general C&P-problems. In particular, the pattern-oriented solution approaches are essentially identical except for the construction of patterns. Therefore, the main topic of a survey of more-dimensional solution approaches would be the construction of suitable patterns for single large objects, i.e. the solution of more-dimensional knapsack problems (types 2/B/O/ , 3/B/O/ , N/B/O/).

The construction of suitable more-dimensional patterns can lead to serious geometrical difficulties, particularly for nonorthogonal patterns and non-rectangular or even irregular shapes. For example, it seems at present to be unknown whether it is possible to pack 11 unit squares (lateral length 1) into a single square with an area of 15 (lateral length about 3·873). In practice, however, technological and organizational conditions have simplifying effects. Thus, a precise measurement of sizes of objects and items is not possible or may not be wanted, or only 'guillotine-cuts' may be permitted (straight 'cuts' from one end to another one dividing convex objects into two parts; see Dyckhoff, 1988, p. 152). Much recent work has been devoted to the construction of two-dimensional patterns. Surveys on particular aspects and solution methods are given by Israni and Sanders (1982), Coffman *et al.* (1984), Berkey and Wang (1985), Dowsland (1985), Terno *et al.* (1987), Farley (1988) as well as Bischoff and Marriott (1990). There exists, however, no comprehensive survey of this subject.

References

Berkey, J.O. and Wang, P.Y., 1985, Two-dimensional finite bin-packing algorithms. *Journal of the Operational Society*, **38**, 423–429.

Bischoff, E.E. and Marriott, M.D., 1990, A comparative evaluation of heuristics for container loading. *European Journal of Operational Research* (Special issue on Cutting and Packing), **44**, 267–276.

Coffman, E.G. Jr, Garey, M.R. and Johnson, D.S., 1984, Approximation algorithms for bin-packing—an updated survey. In *Approximation Algorithms for Computer System Design*, edited by G. Ausiello *et al.* (Wien: Springer), pp. 49–106.

Dowsland, W.B., 1985, Two and three dimensional packing problems and solution methods. *New Zealand Journal of Operational Research*, **13**, 1–18.

Dudzinski, K. and Walukiewicz, S., 1987, Exact methods for the knapsack problem and its generalizations. *European Journal of Operational Research*, **28**, 3–21.

Dyckhoff, H., 1988, Production theoretic foundation of cutting and related processes. In *Essays on Production Theory and Planning*, edited by G. Fandel *et al.* (Heidelberg: Springer), pp. 151–180.

Dyckhoff, H., 1991, Bridges between two principal model formulations for cutting stock processes. In *Operations Research*, edited by G. Fandel and H. Gehring (Heidelberg: Springer), (in press).

Dyckhoff, H., 1990, A typology of cutting and packing problems, *European Journal of Operational Research* (Special issue on Cutting and Packing), **44**, 145–159.

Dyckhoff, H., Finke, U. and Kruse, H.-J., 1988a, Standard software for cutting stock management. In *Essays on Production Theory and Planning*, edited by G. Fandel *et al.* (Heidelberg: Springer), pp. 209–221.

Dyckhoff, H., Kruse, H.-J., Abel, D. and Gal, T., 1988b, Classification of real world trim loss problems. In *Essays on Production Theory and Planning*, edited by G. Fandel *et al.* (Heidelberg: Springer), pp. 191–208.

Eilon, S. and Christofides, N., 1971, The loading problem. *Management Science*, **17**, 259–268.

Farley, A.A., 1988, Practical adaptations of the Gilmore–Gomory approach to cutting stock problems. *Operations Research-Spektrum*, **10**, 113–123.

Garey, M.R. and Johnson, D.S., 1979, *Computers and Intractability: A Guide to the Theory of NP-Completeness* (San Francisco).

Gilmore, P.C. and Gomory, R.E., 1961, A linear programming approach to the cutting-stock problem. *Operations Research*, **9**, 849–859.

Golden, B.L., 1976, Approaches to the cutting stock problem. *AIIE Transactions*, **8**, 265–274.

Haessler, R.W., 1971, A heuristic programming solution to a nonlinear cutting stock problem. *Management Science*, **17B**, 793–802.

Hinxman, A.I., 1980, The trim loss and assortment problems: a survey. *European Journal of Operational Research*, **5**, 8–18.

Israni, S. and Sanders, J.L., 1982, Two-dimensional cutting stock problem research: a review and a new rectangular layout algorithm. *Journal of Manufacturing Systems*, **1**, 169–182.

Martello, S. and Toth, P., 1987, Algorithms for knapsack problems. *Annals of Discrete Mathematics*, **31**, 213–258.

Pierce, J.F., 1964, *Some Large-scale Production Scheduling Problems in the Paper Industry* (Englewood Cliffs, NJ: Prentice-Hall).

Stadtler, H., 1988, A comparison of two optimization procedures for 1- and 1·5-dimensional cutting stock problems. *Operations Research-Spektrum*, **10**, 97–111.

Terno, J., Lindemann, R. and Scheithauer, G., 1987, *Zuschnittprobleme und ihre praktische Lösung* (Frankfurt: Thun).

Wäscher, G., Carow, P. and Müller, H., 1985, Entwicklung eines flexiblen Verfahrens für Zuschneideprobleme in einem Kaltwalzwerk. *Zeitschrift für Operations Research*, **29B**, 209–230.

6

Interactive approaches to packing problems

Eberhard E. Bischoff

Department of Management Science and Statistics, University College of Swansea, Singleton Park, Swansea SA2 8PP, UK

Abstract

This chapter is concerned with methodologies for tackling industrial packing problems and focuses on pallet and container loading in particular. Its central theme is the fact that much of the published algorithm-orientated work in this area has been based on the efficient utilization of the transport or storage medium as the sole objective. It is argued here that this concentration on a single objective is a serious practical shortcoming and also one which cannot be easily overcome. It is suggested that fundamentally new approaches are required to tackle packing problems involving multiple objectives and a framework for the development of such methodologies is proposed. This focuses on the user interfaces which should be provided by an algorithm. However, it is pointed out that much more research is necessary before even relatively simple multiple criteria problems can be tackled adequately.

Introduction

This chapter is concerned with methodologies for tackling industrial packing problems and focuses on approaches in the area of pallet and container loading in particular. It starts from the observation that much of the effort that has gone into the development of such approaches has been based on the efficient utilization of the transport or storage medium as the sole objective.

Many of the published approaches to constructing pallet loading patterns, for example, look upon the problem exclusively in terms of the usage of the pallet board area that can be achieved (Smith and De Cani, 1980; Bischoff and Dowsland, 1982). Although other factors (e.g. the stability of the stack), are referred to by some authors and individual techniques have been examined with regard to their performance against such criteria (Carpenter and Dowsland, 1985), there are few attempts reported in the literature to devise techniques which combine the maximization of space utilization with other objectives. Similarly, most reported techniques for producing loading arrangements for freight containers use the volume of the cargo packed as the only relevant criterion for a good solution (cf. Dowsland, 1985), although again other factors such as weight distribution or the proximity of items of the same type have received mention and are in some cases taken into account implicitly in the mechanism for constructing loading patterns (George and Robinson, 1980).

This chapter argues that this concentration on a single objective, apart from obviously limiting the applicability of the approaches put forward, is also a shortcoming which cannot be easily overcome. The arguments for this contention are discussed in the next section which also illustrates the problems encountered by means of concrete examples of multiple objectives in packing problems. The third section of the paper looks at the role of user interaction in the existing approaches and defines a framework for the development of new methodologies within which user interaction plays a central part. It is followed by some examples of how such interaction may be provided for. The final section of this paper summarizes the main points.

The inadequacy of existing approaches

It is, of course, theoretically conceivable that additional criteria can be dealt with either by grafting some meta-procedure onto an existing technique for optimizing space utilization or through otherwise using the existing technique as a building block for a multi-objective procedure. There are many examples of situations where such approaches are possible.

An initial linear programming model for investment analysis, for instance, may be based, say, on the concept of maximizing the expected net present value of the portfolio. However, if at a later stage a second objective needs to be considered—such as minimizing the payback period—this can be added to the model in several different ways. One possibility might be to substitute for the original objective a weighted sum of the two criteria, or to use a (feasible) target for one as a constraint in optimizing the other. In either case the original methodology remains essentially unchanged and the single objective model becomes an integral part of the two-objective approach.

For the type of packing problems under consideration here, however, it would appear to be very unlikely that multiple criteria methodologies can be developed simply as extensions of the existing single objective approaches. There are two basic arguments for this contention:

(a) First, the additional criteria involved are often difficult or impossible to quantify adequately.
(b) Secondly, even if some approximate quantitative measures could be found for these criteria, the method for dealing with them would need to form an integral part of the process of constructing loading patterns, i.e. necessitate changes to the packing algorithm itself.

The two points are probably best explained by reference to examples.

Pallet patterns

Stability in transport has already been mentioned above as a possible criterion in assessing the suitability of pallet patterns. How important this criterion is clearly depends on the application; if a pallet load is secured by means of strapping or stretch-wrapping, the stability of the stacking arrangement is obviously less crucial than in a situation where no such means of securing the load are used.

The objectives of stability and optimal loading area utilization are to some extent incompatible. For example, the reader will easily be able to convince himself that boxes with base dimensions of, say, 500 × 240 mm or 300 × 250 mm, while allowing a 100% utilization of a standard 1200 × 1000 mm pallet, can only be stacked with this degree of efficiency if they are placed in the same orientation (i.e. if the arrangement within layers is in straight rows and columns). The resulting stacking pattern, of course, provides no interlock between layers and is therefore highly unstable.

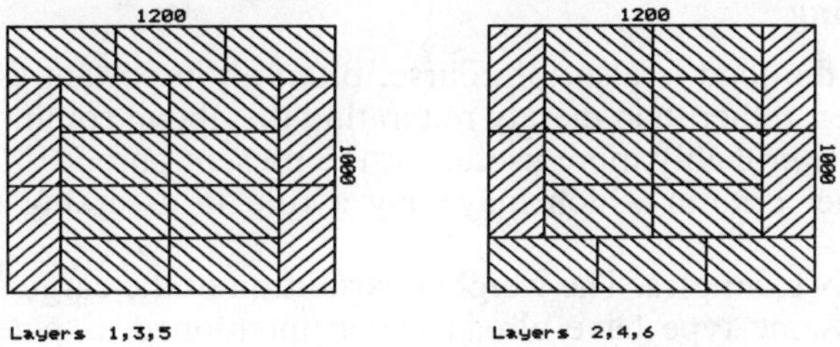

Figure 1. Loading pattern for item of dimensions 400 × 200 mm on pallet of size 1200 × 1000 mm.

On the other hand there are examples of box sizes which allow both a 100% utilization of the pallet and arrangements other than column stacking. One such example is an item with base dimensions 400 × 200 mm on a pallet of size 1200 × 1000 mm. A possible stacking pattern is shown in Figure 1.

Figure 1 also illustrates the difficulties in defining whether a stack is stable. There is a strong interlock between layers around the perimeter of the pattern, but the six boxes in the middle, of course, form a column with no interlock at all.

Carpenter and Dowsland (1985) discuss three tests for sufficient interlock, which they refer to as the 'supportive', the 'base contact' and the 'non-guillotine' criterion, respectively. Briefly put, for a stacking pattern to be classified as stable according to the 'supportive' criterion each box has to be in contact with two or more boxes on the layer below, the 'base contact' test requires a certain percentage of the base area of each box to be in contact with boxes on the layer below, and the 'non-guillotine' test concerns the length of a straight seam which runs through the stack. The stacking arrangement of Figure 1 would be considered unstable from the viewpoint of the first criterion (there are boxes which are not in contact with at least two boxes below), stable with regard to the second criterion (all boxes are completely supported by boxes below), and could be classified either way according to the third criterion, depending on the length of the seam which is considered acceptable.

There are clearly other aspects of the stacking arrangement which could be relevant from the viewpoint of stability. In the example of Figure 1 it might obviously make a difference if the column occured on the side of the pattern rather than in the middle. A classification of the pattern shown as stable or unstable, however, clearly cannot be made without reference to a specific situation. Factors such as the surface friction of the packaging material, the means of transport used and the length and the kind of journeys involved clearly have a bearing on this question. To attempt to take all these factors into account in a single measure of stability would seem to be a futile exercise.

Even if, for a given environment, a certain measure was judged to provide an acceptable definition of stability, an algorithm would still be required for constructing appropriate loading patterns. This algorithm would need to be able to produce patterns which simultaneously maximize the utilization of the pallet and fulfil the stability requirements defined—or to establish that for the box size in question no such pattern exists.

The approaches suggested in the literature cannot be easily adapted to provide such a facility. Many are only capable of producing a single two-dimensional pattern which utilizes the available loading area efficiently (e.g. Dowsland, 1987) and the few methods put forward for generating interlocking layouts provide no mechanism for ensuring that the loading area utilization of the patterns produced is optimal or even near-optimal (e.g. Kulick, 1982).

Container loading

The stability of the load is also, of course, often an important aspect in container loading. However, rather than merely reiterating the above comments in the context of containers, it is probably more useful for the present purposes to consider briefly some of the other criteria which may play a role in assessing container loading patterns.

There are many examples. The weight distribution of the cargo and the placement of items of the same type have already been mentioned as potentially important factors. A variation of the later criterion arises in multi-drop situations where items going to the same destination may need to be kept close together. The effort involved in loading a container according to the given pattern can be another major aspect. This may include not only the handling required in placing items into their respective positions within the container, but also the necessary cargo movements prior to loading an item into the container.

Considerations concerning cargo fragility to some extent will overlap with those associated with cargo stability, but there are clearly some other factors involved, too. For instance, even a slight movement during transport may lead to fragile items being damaged and they may also, of course, be crushed merely because of the weight which is placed on top of them.

It is obvious that not all of these criteria can be quantified, least of all in a generally applicable form. It is difficult, for instance, to conceive of a single measure which might adequately express the complexity of a pattern in terms of a figure for the materials handling effort required to achieve it. But even where there is a fairly straightforward measure—in the case of the weight distribution criterion, for example, it will be possible to determine the centre of gravity or the total load at given support points—a technique for dealing with the additional criterion cannot simply be superimposed onto an existing method for optimizing volume utilization. Any method combining the pursuit of several objectives would need to deal with these simultaneously and in a manner in which they have a direct bearing on the construction of loading patterns.

User interaction

As the preceding section has illustrated, the existing, single-objective, packing algorithms provide only a limited foundation for the development of methodologies for tackling multiple criteria problems. This section aims to define a conceptual framework for such methods.

The framework proposed focuses on the interfaces with the user which should be provided by an algorithm. Traditional approaches have only two such interfaces: (1) one in which the user defines the physical characteristics of the problem (e.g. the dimensions of the container and the items to be loaded) and (2) one for presenting the results obtained to the user (e.g. a diagram of the loading pattern produced). The latter type of interface in many cases takes a rather rudimentary form, but this is not the point at issue here. The issue is that these two interfaces are insufficient for tackling problems involving multiple criteria.

As was mentioned earlier, which criteria are of relevance and how important they are in relation to one another depends crucially on the situation in question. One obvious implication is that any algorithm not specifically developed for a particular, precisely defined, type of problem needs a facility for 'tuning' it according to the situation in which it is to be applied. In order for this tuning to be carried out, an interface needs to be provided through which the user can specify his requirements and thus influence the behaviour of the algorithm. This will be referred to below as 'Interface 3'.

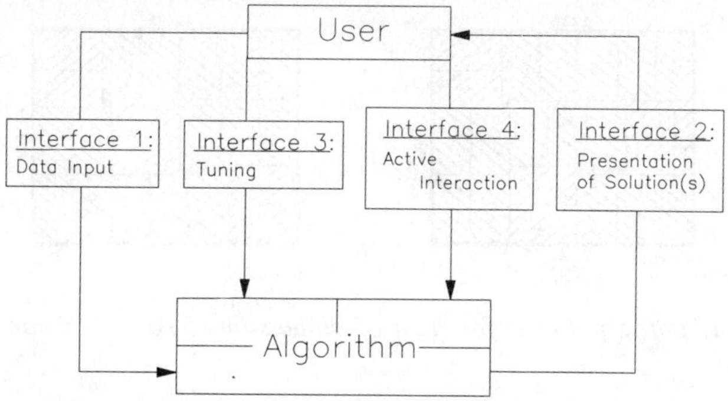

Figure 2. Structure of the proposed solution procedure.

At the simplest level this interface might take the form of a set of parameters acting as 'toggle switches' which either enable or disable certain features of the algorithm (e.g. indicate whether stability is of importance or not). A more complex interface might allow the user to assign relative degrees of importance to a set of criteria and/or to specify aspiration levels for those criteria that can be expressed numerically.

The mere addition of this third type of interface to an algorithm, however, is unlikely to overcome the difficulties arising from the existence of non-quantifiable objectives or constraints. The merits, or otherwise, of a given loading arrangement in relation to such criteria can ultimately be determined only by the user. Whether it is possible to 're-tune' the algorithm and then repeat the analysis should the solution produced prove to be unacceptable to the user, depends on the degree of sophistication of the interface provided. But even when this is theoretically possible, it may require considerable experience on the part of the user and/or a large number of iterations to steer the solution in the desired direction.

A potential alternative to repeating the whole analysis with a different set of performance parameters is to allow the user to become actively, and directly, involved in the construction of the loading arrangement. Conceptually, this can be thought of as a form of 're-tuning', but at a different level in the algorithm and using a different mode. In container loading it could, for instance, involve fixing the position of some items whose placement is acceptable prior to reapplying the algorithm for the remaining cargo. For the user to be able to influence the loading arrangement in such a way, a further interface is required.

This type of interface, referred to below as 'Interface 4', is likely to have certain elements in common with the interface used for presenting results to the user. Indeed, it will in a sense often be an 'interactive version' of the latter.

Figure 2 shows the position of the four types of interface in a structural diagram of the suggested interactive solution process. The algorithm is shown in this diagram as consisting of three sections merely to indicate the different functions within it—all three may, of course, represent integral and inseparable parts of the overall procedure.

The concept of additional user interfaces and how these may assist in the process of constructing appropriate loading patterns is illustrated in the following section by reference to methods for generating pallet loading patterns.

An illustration

While none of the existing algorithms for producing pallet patterns provide any facilities for adjustments according to the requirements of a particular situation,

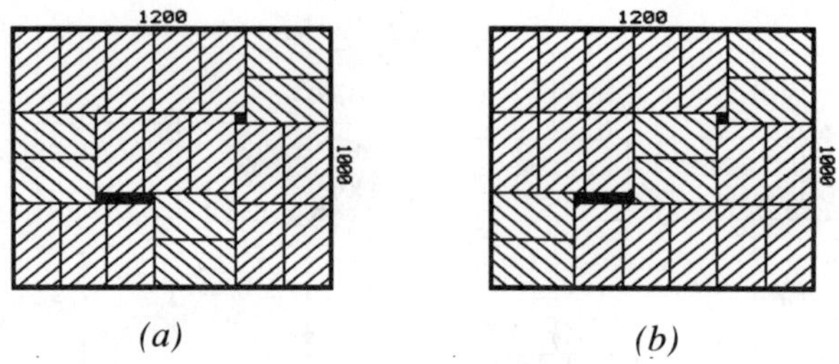

Figure 3(a), (b) Loading patterns for item of dimensions 310 × 175 mm on pallet of size 1200 × 1000 mm.

some of the procedures proposed are, at least in principle, capable of producing more than one pattern for the same problem (e.g. the methods of Smith and De Cani, 1980, and Bischoff and Dowsland, 1982). A trivial example of a 'Type 4' interface for one of these procedures would be a facility for browsing through a set of possible solutions and evaluating the resulting combinations in terms of their stacking characteristics.

A more sophisticated interface of this type could allow the user to make direct adjustments to the patterns produced. This might simply take the form of movements of boxes in order to spread gaps in the layout more evenly, or involve more extensive changes to improve the interlock in a stack.

Figure 3 illustrates how such a facility can help to produce an acceptable stacking arrangement. The diagram on the left shows a potential layout pattern for boxes with base dimensions 310 × 175 mm on a pallet of size 1200 × 1000 mm. This provides an optimal utilization of the available loading area and would also be suitable for use with a clamp truck (it can be clamped from either side). However, if the stack is to be formed by the usual method of reflecting or rotating the basic layout in alternate layers, the pattern has severe shortcomings from a stability viewpoint—whether reflections or rotations are used there are always two columns in corners of the stack which do not interlock with the rest of the load. It is easy to see that this difficulty can be overcome by replacing the spiral pattern on the lower left-hand side of Figure 3(a) by its mirror image. The result is shown in Figure 3(b). This layout pattern, when combined with reflections along the shorter axis, produces a very much more stable stacking arrangement.

Clearly, this type of interaction, in general, requires a computerized approach and its success will depend on the experience and/or the pattern recognition capabilities of the user. On the other hand, there are various simple tools which can facilitate his task and which can be provided as part of the interface, such as, for example, facilities for generating reflections and rotations of patterns and for superimposing layouts on top of each other. Also, of course, the fact that the user is actively involved in the solution process may have advantages from the viewpoint of user acceptance of the final solution.

The point that existing techniques in this area cannot be easily adapted to cater for multiple objectives has already been made and involving the user in the way described above is unlikely to lead to significant advances unless a means of influencing the behaviour of the algorithm itself is provided at the same time. This 'Type 3' interface may take a very simple form, but its existence will generally have much more far-reaching implications for the design of an algorithm than the 'Type 4' interface.

This interface may, for example, provide a facility for stipulating that the patterns must be suitable for use with clamp trucks. One possible way of dealing with such a requirement would be to apply a technique such as that of Bischoff and Dowsland (1982) with an added 'filter' for identifying patterns which satisfy the clampability requirement. However, as this algorithm has no in-built bias towards clampable layouts—and would therefore often produce no layout at all, even when several suitable ones exist—this approach will be far from satisfactory. A more sophisticated approach would require a fundamental redesign of the algorithm as such. The same applies to any other of the existing techniques.

Summary and conclusions

This chapter has suggested that fundamentally new approaches are required to tackle packing problems involving multiple objectives. It has argued that these new approaches should ideally have two additional interfaces with the user: (1) one through which the algorithm can be 'tuned' according to the requirements of the situation at hand, and (ii) one through which the user can exert a direct influence over the loading arrangement produced.

The chapter has illustrated what form these interfaces might take. However, it has also been pointed out that a great deal of research will be required before even relatively simple multiple criteria problems can be tackled adequately. Such further research is needed especially in relation to container loading algorithms where work to date has barely begun to address the issues of practical concern.

References

Bischoff, E. and Dowsland, W.B., 1982, An application of the micro to product design and distribution. *Journal of the Operational Research Society*, **33**, 271–280.

Carpenter, H. and Dowsland, W.B., 1985, Practical considerations of the pallet-loading problem. *Journal of the Operational Research Society*, **36**, 489–497.

Dowsland, K.A., 1987, An exact algorithm for the pallet loading problem. *European Journal of Operational Research*, **31**, 78–84.

Dowsland, W.B., 1985, Two and three dimensional packing problems and solution methods. *New Zealand Journal of Operational Research*, **13**, 1, 1–18.

George, J.A. and Robinson, D.F., 1980, A heuristic for packing boxes into a container. *Computers and Operations Research*, **7**, 147–156.

Kulick, A., 1982, Interlocking pallet pattern simulation program. *Industrial Engineering*, September, 22–24.

Smith, A. and De Cani, P., 1980, An algorithm to optimize the layout of boxes in pallets. *Journal of the Operational Research Society*, **31**, 573–578.

SECTION II.3. MACHINING

7

Identifying and controlling chatter in turning with a feed rate control system

Yu Junyi, Zheng Wangbin and Bao Shanfei

Department of Mechanical Engineering, Jilin University of Technology, Changchun, China

Abstract

As the basis of analysing the varying regularity of dynamic cutting force and dynamic response when the cutting process transits from the stable state to the unstable state, the peak value of function $J(\omega)$ is presented as an identifying parameter to predict chatter. $J(\omega)$ is the Fast Fourier Transform (FFT) of product multiplied by the dynamic cutting force and the dynamic response of the machining system. Once the peak value of $J(\omega)$ has exceeded a given control threshold, the control system is triggered immediately, and the chatter may be eliminated before it has time to develop.

It is confirmed by theoretical analysis and experimental results that the cutting width limit $b_{\lim}$ of the machining system will increase following the increase of feed rate. A feed rate control system was applied to control turning chatter and was shown to have good prospects for its application to production.

Introduction

In cutting conditions with a chatter predictive control system, the chatter can be suppressed for the reason that before the chatter is about to start in the cutting process, a control signal has been sent out from the identifying system to the drive system. Obviously, the predictive control of chatter is always more active and practical than the usual feedback control of chatter.

Recently, the topic of chatter predictive control in machining operations has been studied by various groups (Eman and Wu, 1980; Mei *et al.*, 1987). But it is notable that it is unknown which predictive control scheme has been applied to production until now; much yet remains to be done.

Theoretical analysis

To determine the identifying parameter

The key to studying the predictive control technique of chatter is to determine the identifying parameter which must accurately and rapidly monitor the trend of instability in the cutting process and should be measured immediately.

When the machining system is in the stable state, the cutting force is mainly

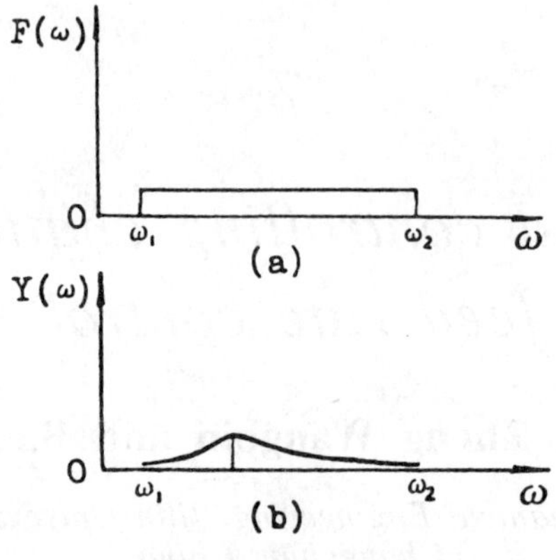

Figure 1. The spectrum drawing of cutting force and vibratory response when the machining system is in the stable state.

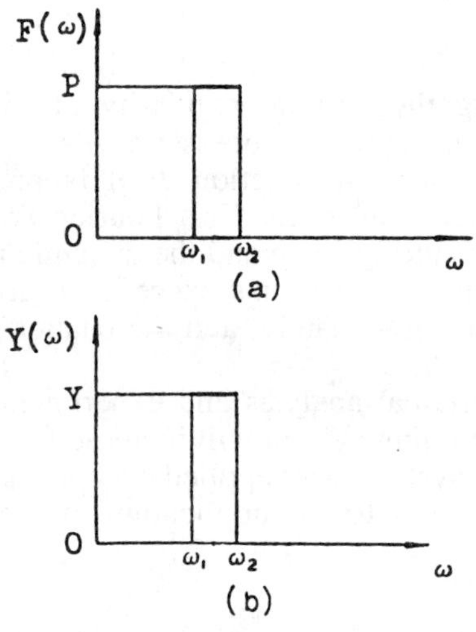

Figure 2. The spectrum drawing of cutting force and vibratory response when the machining system is beginning to enter the unstable state.

concerned with the material of the workpiece and the cutting conditions; the signal of the cutting force $f(t)$ will present a characteristic broad frequency band random signal. Figure 1 shows the spectrum $F(\omega)$ of the cutting force $f(t)$ and the spectrum $Y(\omega)$ of the vibration response, namely, $Y(\omega) = H(\omega) \cdot F(\omega)$, where $H(\omega)$ is the dynamic compliance of the vibratory system. The response is very small before the chatter occurs in the cutting process.

When the machining system begins to enter the unstable state, the chatter will be excited in the cutting process, the frequency band of the dynamic cutting force will concentrate on a narrow region and its amplitude will enlarge rapidly; so the dynamic cutting force signal can be considered as a narrow frequency band random signal as shown in Figure 2(a). Figure 2(b) shows the spectrum of vibratory response.

Based on a study of the varying regularity of dynamic parameters during the transition cutting process, we present the peak value of function $J(\omega)$ as an identifying parameter to predict chatter. $J(\omega)$ is the FFT of the product multiplied by the dynamic cutting force and the dynamic response of the machining system, i.e.

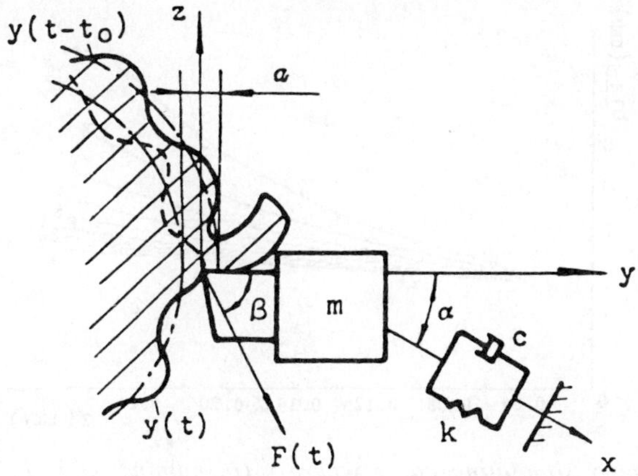

Figure 3. The dynamic model of the regenerative chatter system.

$$J(\omega) = F[f(t) \cdot y(t)] = F(\omega) * Y(\omega)/2\pi$$

where t = time, ω = angular frequency, $F(\omega)$ = the Fourier Transform of the dynamic cutting force $f(t)$,

$$F(\omega) = F[f(t)] = \begin{cases} P, & \omega_1 \le \omega \le \omega_2; \\ 0, & \text{else}. \end{cases}$$

and $Y(\omega)$ = the Fourier Transform of the dynamic response $y(t)$ of the machining system

$$Y(\omega) = F[y(t)] = \begin{cases} Y, & \omega_1 \le \omega \le \omega_2; \\ 0, & \text{else}. \end{cases}$$

Substituting $F(\omega)$ and $Y(\omega)$ into the equation of $J(\omega)$, we have

$$J(\omega) = F[f(t) \cdot y(t)] = \begin{cases} P \cdot Y \cdot (\omega - 2\omega_1), & 2\omega_1 \le \omega \le \omega_1 + \omega_2; \\ P \cdot Y \cdot (2\omega_2 - \omega), & \omega_1 + \omega_2 \le \omega \le 2\omega_2. \end{cases}$$

Clearly to adopt the peak value of $J(\omega)$ as an identifying parameter, the predicting sensitivity and anti-interference capability will be greater than that required to adopt the peak value of $F(\omega)$ or $Y(\omega)$, because the peak value of $J(\omega)$ is far greater than that of $F(\omega)$ or $Y(\omega)$.

Once the peak value of $J(\omega)$ has exceeded the given control threshold, the control system is triggered immediately and the chatter may be eliminated at its developing stage.

To determine the control parameter

Analysing the dynamic model of the regenerative chatter system as shown in Figure 3, the dynamic cutting force (in direction y $\mathrm{d}Fy$) can be expressed as follows:

$$\mathrm{d}F_y = -A\mathrm{e}^{i\omega t} \cdot [k_y \cdot b(1 - \mu \cdot \mathrm{e}^{i(\varepsilon - \omega t_0)}) + i\omega k_z ba/v] \quad (1)$$

where A = chatter amplitude, ω = angular frequency of chatter, t = time, t_0 = interval between two subsequent revolutions of cut, k_y, k_z = cutting stiffness coefficient per unit cutting width in directions y and z, respectively, ε = leading phase difference

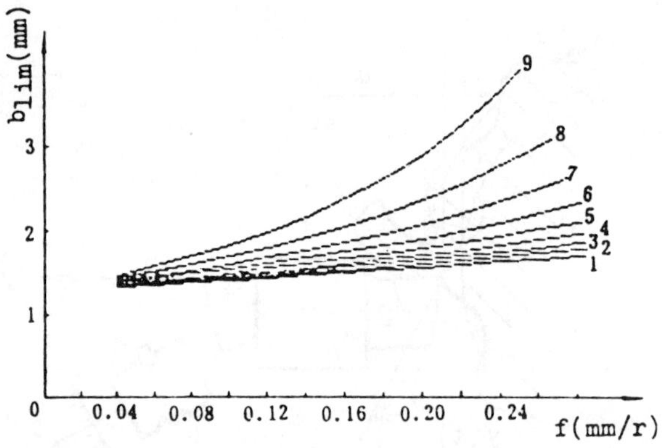

Figure 4. The computer simulating curve where v (in m/min) is: 1, 120; 2, 110; 3, 100; 4, 90; 5, 80; 6, 70; 7, 60; 8, 50; and 9, 40.

of the fluctuation in cutting force relative to vibration displacement, μ = overlap factor; a = nominal undeformed chip thickness, v = cutting speed, i = symbol of complex number.

Under the action of the dynamic cutting force, the dynamic response of the vibratory system is

$$y = W(\omega) \cdot \mathrm{d}Fy \tag{2}$$

where $W(\omega)$ = the dynamic compliance of the vibratory system, $W(\omega) = G(\omega) + iH(\omega)$; $G(\omega)$ = the real part of $W(\omega)$; $H(\omega)$ = the imaginary part of $W(\omega)$.

Substituting eqn (1) into eqn (2) and rearranging it, cutting width b may be described as

$$b \cdot [k_y \cdot \mu \cdot G(\omega) \cos(\varepsilon - \omega t_0) + k_z \omega H(\omega) \cdot a/v$$
$$- ky \cdot \mu \cdot H(\omega) \cdot \sin(\varepsilon - \omega t_0) - k_y G(\omega)] = 1.$$

In longitudinal turning, the thickness of chip a may be expressed as $a = f \cdot \sin K_r$ where f is feed rate, K_r is tool cutting edge angle. In this case, the limit cutting width can be expressed as

$$\begin{aligned} b_{\mathrm{lim}} \cdot \{ & k_y[-G(\omega)]_{\max} - k_y\mu \cdot \cos(\varepsilon - \omega t_0)\,[-G(\omega)]_{\max} - \\ & k_y\mu[H(\omega) \cdot \sin(\varepsilon - \omega t_0) + k_z\omega[H(\omega)] \cdot f \cdot \sin K_r/v\} = 1. \end{aligned} \tag{3}$$

where $[-G(\omega)]_{\max}$ = the maximum negative real part of $W(\omega)$; $[H(\omega)]$ = the imaginary part of $W(\omega)$, corresponding with $[-G(\omega)]_{\max}$; when regenerative chatter is being excited in the cutting process, in general the value of $[H(\omega)]$ is a negative number.

By analysing eqn (3) it is found that the cutting width limit increases following the increase of feed rate f, but it decreases with the increase of cutting speed v. Computer simulation of eqn (3) gives the theoretical curve of the cutting width limit following the variation of feed rate f as shown in Figure 4; the greater the cutting speed, the smaller the influence of feed rate f on the cutting width limit, b_{lim}.

It is known from the above analysis that if we choose cutting speed v as the control parameter, we pay the price of decreasing machining productivity; however, if we choose feed rate f as the control parameter, not only can chatter be suppressed, but also machining productivity can be slightly increased. Therefore, in this chapter feed rate f was chosen as the control parameter.

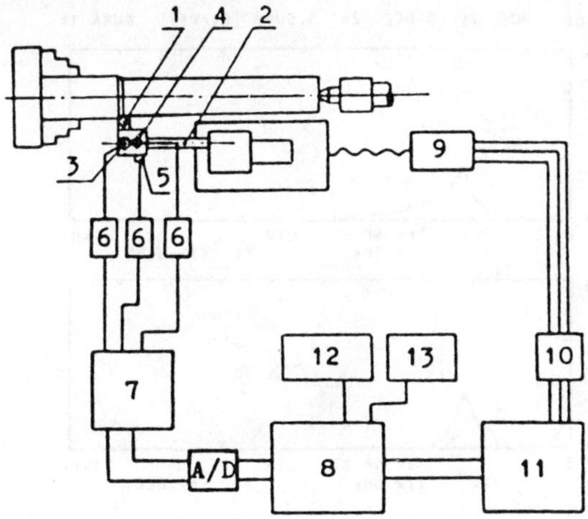

Figure 5. The scheme of experimental equipment: 1, cutting tool; 2, toolholder; 3, force sensor; 4, 5, accelerometers; 6, charge amplifiers; 7, tape recorder; 8, model TP805 microcomputer; 9, step motor; 10, amplifying circuit; 11, model TP801 single board computer; 12, display screen; 13, printer.

The experimental set-up

Experiments were performed in an engine lathe. A cutting tool was mounted on a slender toolholder (see Figure 5) which was the main vibratory system (tended closely to be a single degree of freedom) of the whole machining system. The signal of cutting force was picked up by a force sensor mounted on the left end of the toolholder. The chatter signals were picked up by two accelerometers fixed, respectively, on the directions y and z of the toolholder; they were amplified by charge amplifers and recorded synchronously with the cutting force signal by a tape recorder.

The feed movement was driven by a step motor, its export axle was jointed with the screw for longitudinal feed of the tool post. The step motor was controlled by a control case composed of a model TP801 single board computer and a power amplifying circuit.

The identification and control of chatter were performed by a model TP805 microcomputer through which the control signal was sent to the step motor, display screen and printer which gave visual information on the system.

Experimental results analysis

The varying regularity of dynamic parameters during the transitional cutting process

With the object of examining the varying regularity of dynamic parameters during the transitional cutting process, 15 turning tests of various cutting depths were performed under the following test conditions: material of sample, middle carbon steel; cutting tool material, YT14; tool geometry, $\gamma = 15°$, $\alpha = 5°$, $K_r = 60°$, $K_r' = 30°$; cutting conditions, rotational speed of workpiece $n = 160$, or 200, 250, 320 rev/min; diameter of workpiece, $D \approx 80$ mm; feed rate, $f = 0{\cdot}12$, or 0·16, 0·20, 0·26, 0·30 mm/rev; a taper work was adopted, the depth of cut a_p can be varied continuously from 0·1 to 1 mm in a working stroke.

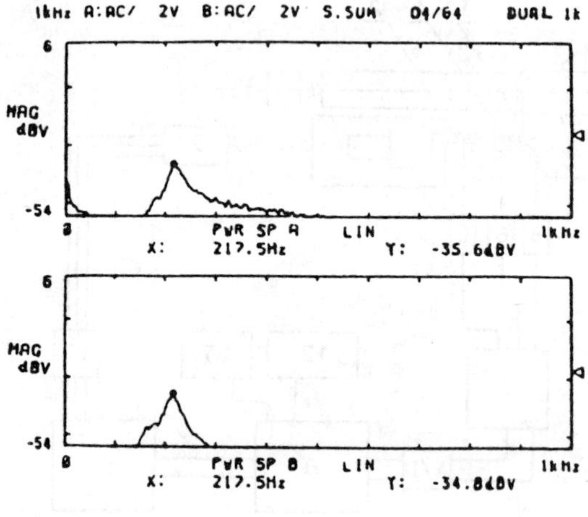

Figure 6. The spectrum drawing of cutting force $f_z(t)$ *and vibratory response* z(t) *(when the machining system is in the stable state).*

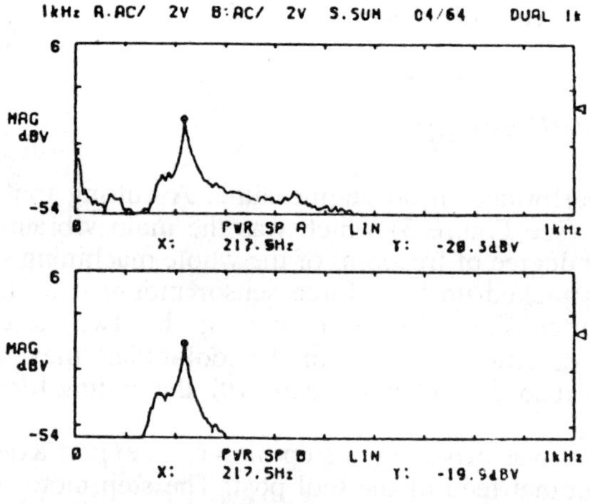

Figure 7. The spectrum drawing of cutting force $f_z(t)$ *and vibratory response* z(t) *(when the machining system is beginning to enter the unstable state).*

The experimental results of the 12th test are shown in Figures 6 and 7. During testing the cutting force signal $f(t)$ and vibration signal $z(t)$ were recorded synchronously. Through data processing of the recorded signals, we obtain a set of test data in time order. These test data revealed the characteristics of the cutting force and the response of the vibratory system at different moments during the transitional cutting process.

During testing, when the taper work is beginning to be turned, the depth of cut is very small, so the cutting process is stable. The peak values of $F_z(\omega)$ and $Z(\omega)$ are −35·6 dBV and −34·8 dBV, respectively, as shown in Figure 6.

With increasing cutting depth, the cutting process becomes unstable, the amplitude of chatter increasing rapidly. The peak values of $F_z(\omega)$ and $Z(\omega)$ shown in Figure 7 are −20·3 dBV and −19·9 dBV, respectively, which are far greater than the values for the stable cutting process.

All of the above mentioned experimental results and the other 14 similar experimental results demonstrate that when the cutting process transits from the stable to the unstable state, the dynamic cutting force and the vibratory response will show an obvious increase.

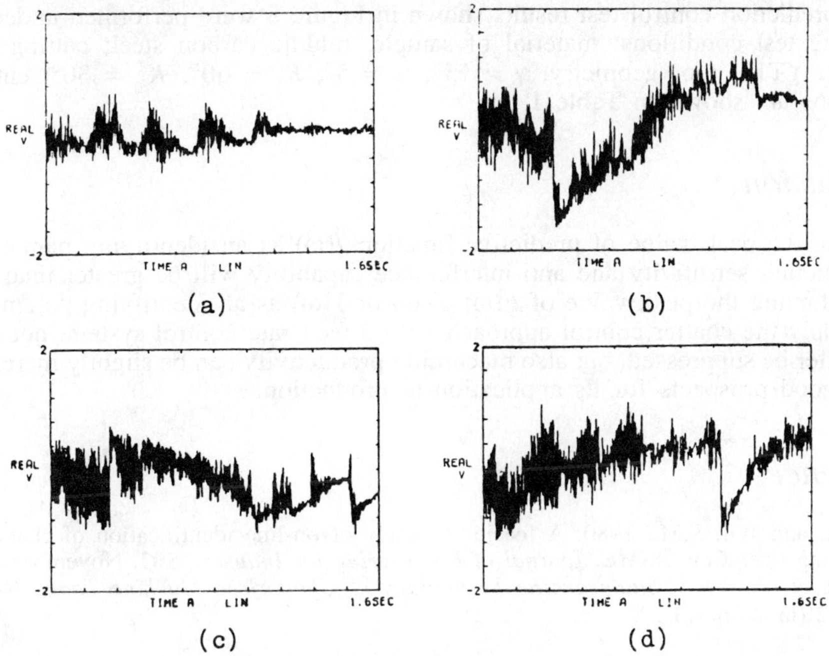

Figure 8. The wave drawing of chatter on the cutting condition with feed rate control system. For explanation of (a)–(d), see Table 1.

Table 1. Cutting conditions for the prediction control test.

		(a)	(b)	(c)	(d)
Cutting speed v (m/min)		50		70	
Depth of cut a_p (mm)		0·40	0·80	0·40	0·80
Changing range of feed rate (mm/rev)	from	0·08	0·12	0·12	0·12
	to	0·12	0·16	0·16	0·16

Predictive control of chatter

The scheme of chatter predictive control is shown in Figure 5. During testing, both signals of dynamic cutting force and acceleration are transmitted from the sensors into the charge amplifier and recorded by tape recorder, then they are transmitted to the model TP805 microcomputer to calculate the FFT of product multiplied by the dynamic cutting force and the dynamic response of the vibratory system. Then the model TP805 microcomputer takes the peak value $J(\omega)_{max}$ and compares in with the value of the given control threshold $J(\omega)_{lim}$. If $J(\omega)_{max} \geq J(\omega)_{lim}$, the micro-computer will send the control signal to the TP801 single board computer to adjust the feed rate of the tool post through the step motor. After sending out the control signal and after a short delay, the microcomputer will continue gathering the signals of the cutting force and the vibratory response. If $J(\omega)_{max} < J(\omega)_{lim}$, the cutting process will be performed continuously under the original feed rate.

Figure 8 shows the actual chatter wave drawings recording now chatter is controlled through continuously increasing the feed rate.

The prediction control test results shown in Figure 8 were performed under the following test conditions: material of sample, middle carbon steel; cutting tool material, YT14; tool geometry, $\gamma = 15°$, $\alpha = 5°$, $K_r = 60°$, $K_r' = 30°$; cutting conditions are shown in Table 1.

Conclusion

Adopting the peak value of predictive function $J(\omega)$ as an identifying parameter, the predicting sensitivity and anti-interference capability will be greater than that when adopting the peak value of $F(\omega)$, $Z(\omega)$ or $Y(\omega)$ as an identifying parameter.

To apply the chatter control approach with a feed rate control system, not only can chatter be suppressed, but also machining productivity can be slightly increased giving good prospects for its application to production.

References

Eman, K. and Wu, S.M., 1980, A feasibility study of on-line identification of chatter in turning operation. *ASME, Journal of Engineering for Industry*, **102**, November.

Mei, Z.J. *et al.*, 1987, *Study on the Technology of Chatterfree Machine Tools*, No. 9, 9–12 (in Chinese).

8

A study of cutting area change under the chatter vibration in lathe turning

Yoshiaki Iwata

Faculty of Science and Technology, Kinki University, Kowakae 3–4–1, Higashi-Osaka, Osaka 577, Japan

Abstract

Chatter vibration is influenced by alterations in the normal force during cutting of a workpiece. It is necessary to find out the change in cutting area of normal force direction during chatter vibration in lathe turning.

The external cutting of a round bar by a right-hand blade bit with a side-cutting edge angle is studied. The bit is fixed tightly in the tool post of the lathe. A round bar fitted between the centre is treated as the bending vibration. The cutting area change, altered by the bit shape and the cutting conditions, is calculated from an analysis of the phase difference before and after one rotation with chatter vibration.

The cutting area change causes a larger approach to 0·5 from 0·1 and the fraction of chatter mark on the work circumference diminishes to 0·9 from 0·5.

An increase in nose radius and side-cutting edge angle of the bit gives the greatest increase in cutting area change. Altering the bit's nose radius has a greater effect on cutting area change than altering the side-cutting edge angle.

Introduction

Chatter vibration in lathe cutting has a phase difference between the tool and the workpiece which has a direct effect upon chatter vibration. In this case, if the revolution of the workpiece is altered to N rev/min, the time of one rotation becomes $60/N$ s/rev. If the frequency of the sine wave on the workpiece is replaced by Hz, the chatter mark on the workpiece circumference after one rotation becomes:

$$\frac{60}{N} \times f = I(\text{integer}) + \varepsilon(\text{fraction})$$

The relation between fraction ε and the phase difference ϕ before and after one rotation under the chatter vibration becomes:

$$m = \frac{2\pi}{\phi}, \quad \varepsilon = \frac{1}{m}, 0 < \varepsilon < 1$$

In this study, the lip part of the right-hand blade bit goes the deepest into the workpiece by the component of vibratory amplitude X_0 in the direction of normal

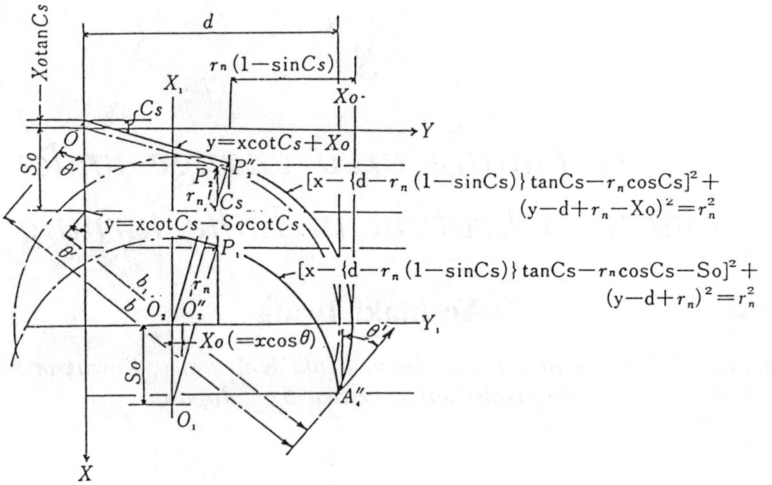

Figure 1. A cutting area after the condition of chatter vibration. (Cut in)

force from the chatter vibration before one rotation. After that, the bit is sent by only feed S_0 with one rotation of work. The lip part is treated as cutting deeply into the workpiece by the component of vibration amplitude X_0' in the direction of normal force as before one rotation. When the cutting area of work before and after one rotation is A_1'' and A_2'', respectively, the cutting area change, dA_1'', before and after is given by $A_1'' - A_2''$ under chatter vibration. The origin 0 represents the lowest lip part of the bit cutting into the workpiece. The X-axis is the reverse direction of feed and the Y-axis is the depth of cut direction. The component of vibration in work, $x\cos\theta\cos\omega t$, $x\cos\theta\cos(\omega t + \phi)$ is put in place of the component of vibratory amplitude for normal force X_0, X_0'.

From this point, the analysis advances as follows with the lip part of the cutting of the right-hand blade bit treated as follows:

(1) side-cutting edge and nose part,
(2) nose part.

The analysis of cutting area by lip part of bit

The case of the side-cutting edge and the nose part of bit

The part of the cutting of right-hand blade bit is analysed by side-cutting edge and nose part:

$$d \geqq r_n(1 - \sin Cs),\quad S_0 \leqq 2r_n \sin Ce,\quad S_0 < 2r_n \cos C_s$$

The cutting area when the bit cuts into the workpiece before one rotation is given in Figure 1, where the dotted line shows the condition before chatter vibration. In this case, the cutting area A_1'' is given by:

$$A_1'' = \frac{1}{2}\left(\tan C_s + \sqrt{\frac{4r_n^2}{S_0^2 + x^2\cos^2\theta\cos^2\omega t} - 1}\right)x^2\cos^2\theta\cos^2\omega t$$

$$+\left[\{d - r_n(1 - \sin C_s)\}\tan C_s + r_n\cos C_s + \frac{S_0}{2}\right]x\cos\theta\cos\omega t$$

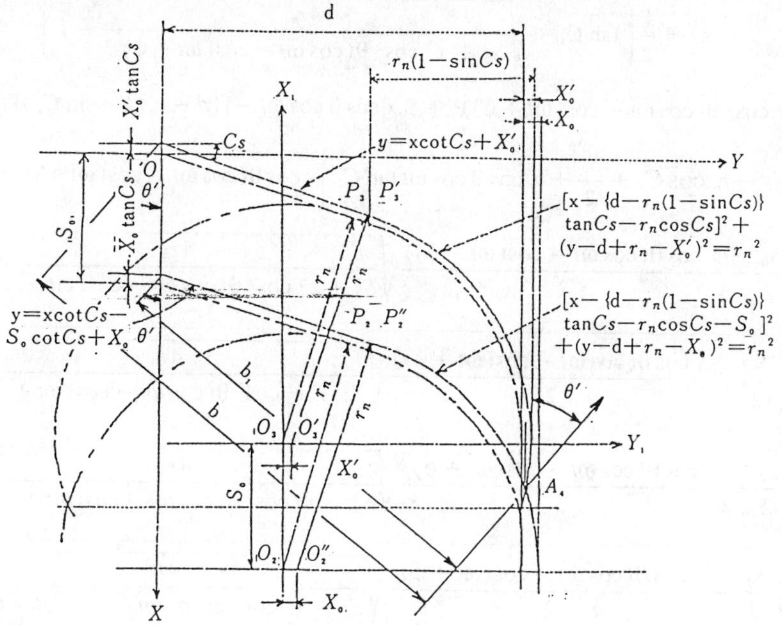

Figure 2. A cutting area under the condition of chatter vibration. (Cut in)

$$+\frac{1}{2}\left(\frac{S_0}{2}+\frac{x\cos\theta\cos\omega t}{2}\sqrt{\frac{4r_n^2}{S_0^2+x^2\cos^2\theta\cos^2\omega t}-1}\right)\cdot$$

$$\sqrt{r_n^2-\left(\frac{S_0}{2}+\frac{x\cos\theta\cos\omega t}{2}\sqrt{\frac{4r_n^2}{S_0^2+x^2\cos^2\theta\cos^2\omega t}-1}\right)^2}$$

$$+\frac{1}{2}\left(\frac{S_0}{2}-\frac{x\cos\theta\cos\omega t}{2}\sqrt{\frac{4r_n^2}{S_0^2+x^2\cos^2\theta\cos^2\omega t}-1}\right)\cdot$$

$$\sqrt{r_n^2-\left(\frac{S_0}{2}-\frac{x\cos\theta\cos\omega t}{2}\sqrt{\frac{4r_n^2}{S_0^2+x^2\cos^2\theta\cos^2\omega t}-1}\right)^2}$$

$$+\frac{r_n^2}{2}\sin^{-1}\frac{S_0+x\cos\theta\cos\omega t\sqrt{\dfrac{4r_n^2}{S_0^2+x^2\cos^2\theta\cos^2\omega t}-1}}{2r_n}$$

$$+\frac{r_n^2}{2}\sin^{-1}\frac{S_0+x\cos\theta\cos\omega t\sqrt{\dfrac{4r_n^2}{S_0^2+x^2\cos^2\theta\cos^2\omega t}-1}}{2r_n}+(d-r_n)S_0, \quad (1)$$

where X_0 is replaced by component of vibration $x\cos\theta\cos\omega t$ in the direction of normal force.

The cutting area when the bit cuts into the workpiece after one rotation is given in Figure 2 where the dotted line shows the condition before chatter vibration. In this case, the cutting area A_2'' is given by:

$$A_2'' = \frac{1}{2}\left(\tan C_s + \sqrt{\frac{4r_n^2}{S_0^2 + x^2\cos^2\theta(\cos\omega t - \cos(\omega t + \phi))^2} - 1}\right)$$

$$\cdot\, x^2\cos^2\theta(\cos\omega t - \cos(\omega t+\phi))^2 + S_0 x\cos\theta\cos\omega t - [\{d - r_n(1-\sin C_s)\}\tan C_s$$

$$\left. + r_n\cos C_s + \frac{S_0}{2} + x\cos\theta\cos\omega t\tan C_s\right] x\cos\theta(\cos\omega t - \cos(\omega t+\phi))$$

$$+\frac{1}{2}\left(\frac{S_0}{2} + \frac{x\cos\theta(\cos\omega t - \cos(\omega t+\phi))}{2}\sqrt{\frac{4r_n^2}{S_0^2 + x^2\cos^2\theta(\cos\omega t - \cos(\omega t+\phi))^2} - 1}\right)\cdot$$

$$\sqrt{r_n^2 - \left(\frac{S_0}{2} + \frac{x\cos\theta(\cos\omega t - \cos(\omega t+\phi))}{2}\sqrt{\frac{4r_n^2}{S_0^2 + x^2\cos^2\theta(\cos\omega t - \cos(\omega t+\phi))^2} - 1}\right)^2}$$

$$+\frac{1}{2}\left(\frac{S_0}{2} - \frac{x\cos\theta(\cos\omega t - \cos(\omega t+\phi))}{2}\sqrt{\frac{4r_n^2}{S_0^2 + x^2\cos^2\theta(\cos\omega t - \cos(\omega t+\phi))^2} - 1}\right)\cdot$$

$$\sqrt{r_n^2 - \left(\frac{S_0}{2} - \frac{x\cos\theta(\cos\omega t - \cos(\omega t+\phi))}{2}\sqrt{\frac{4r_n^2}{S_0^2 + x^2\cos^2\theta(\cos\omega t - \cos(\omega t+\phi))^2} - 1}\right)^2}$$

$$+\frac{r_n^2}{2}\sin^{-1}\frac{S_0 + x\cos\theta(\cos\omega t - \cos(\omega t+\phi))\sqrt{\dfrac{4r_n^2}{S_0^2 + x^2\cos^2\theta(\cos\omega t - \cos(\omega t+\phi))^2} - 1}}{2r_n}$$

$$+\frac{r_n^2}{2}\sin^{-1}\frac{S_0 - x\cos\theta(\cos\omega t - \cos(\omega t+\phi))\sqrt{\dfrac{4r_n^2}{S_0^2 + x^2\cos^2\theta(\cos\omega t - \cos(\omega t+\phi))^2} - 1}}{2r_n}$$

$$+ (d - r_n)\, S_0, \tag{2}$$

where X_0 is replaced by the component of vibration before one rotation $x\cos\theta\cos\omega t$ in the direction of normal force, and X_0' is replaced by the component of vibration, $x\cos\theta\cos(\omega t+\phi)$, after one rotation.

The case of the nose part of bit

The part of cutting of work is analysed by nose part of work

$$d < r_n\,(1 - \sin C_s), \quad S_0 \leqq 2r_n \sin C_e.$$

The cutting area when the bit cuts into work before one rotation is given in Figure 3 where the dotted line shows the condition before chatter vibration. In this case, the cutting area A_1'' is given by:

$$A_1'' = \frac{x^2\cos^2\theta\cos^2\omega t}{2}\sqrt{\frac{4r_n^2}{S_0^2 + x^2\cos^2\theta\cos^2\omega t} - 1}$$

$$+\frac{x\cos\theta\cos\omega t}{2}\left\{S_0 + \sqrt{r_n^2 - (r_n - d - x\cos\theta\cos\omega t)^2}\right\}$$

$$+\frac{1}{2}(r_n - d)\{\sqrt{d(2r_n - d)} - \sqrt{r_n^2 - (r_n - d - x\cos\theta\cos\omega t)^2}\} - (r_n - d)S_0$$

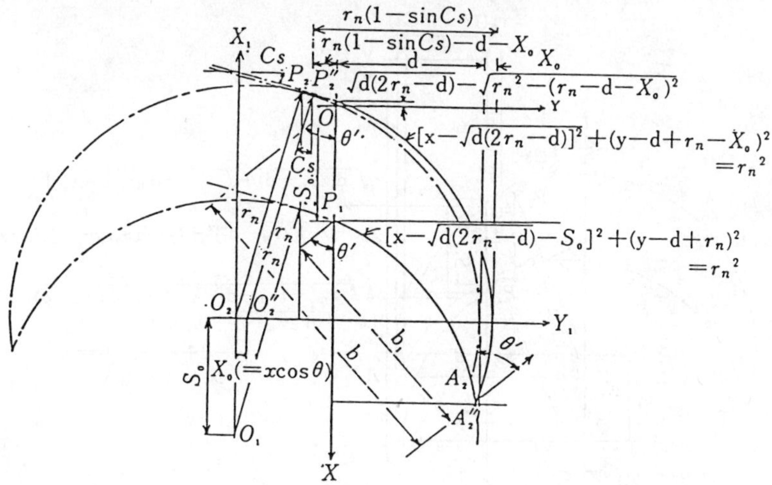

Figure 3. A cutting area after the condition of chatter vibration. (Cut in)

$$+\frac{1}{2}\left(\frac{S_0}{2}-\frac{x\cos\theta\cos\omega t}{2}\sqrt{\frac{4r_n^2}{S_0^2+x^2\cos^2\theta\cos^2\omega t}-1}\right)\cdot$$

$$\sqrt{r_n^2-\left(\frac{S_0}{2}-\frac{x\cos\theta\cos\omega t}{2}\sqrt{\frac{4r_n^2}{S_0^2+x^2\cos^2\theta\cos^2\omega t}-1}\right)^2}$$

$$+\frac{1}{2}\left(\frac{S_0}{2}+\frac{x\cos\theta\cos\omega t}{2}\sqrt{\frac{4r_n^2}{S_0^2+x^2\cos^2\theta\cos^2\omega t}-1}\right)\cdot$$

$$\sqrt{r_n^2-\left(\frac{S_0}{2}+\frac{x\cos\theta\cos\omega t}{2}\sqrt{\frac{4r_n^2}{S_0^2+x^2\cos^2\theta\cos^2\omega t}-1}\right)^2}$$

$$+\frac{1}{2}r_n^2\sin^{-1}\frac{\sqrt{r_n^2-(r_n-d-x\cos\theta\cos\omega t)^2}}{r_n}-\frac{1}{2}r_n^2\sin^{-1}\frac{\sqrt{d(2r_n-d)}}{r_n}$$

$$+\frac{1}{2}r_n^2\sin^{-1}\frac{S_0-x\cos\theta\cos\omega t\sqrt{\frac{4r_n^2}{S_0^2+x^2\cos^2\theta\cos^2\omega t}-1}}{2r_n}$$

$$+\frac{1}{2}r_n^2\sin\frac{S_0+x\cos\theta\cos\omega t\sqrt{\frac{4r_n}{S_0^2+x^2\cos^2\theta\cos^2\omega t}-1}}{2r_n} \tag{3}$$

where X_0 is replaced by the component of vibration, $x\cos\theta\cos\omega t$, in the direction of normal force.

The cutting area when the bit cuts into the workpiece after one rotation of work is given in Figure 4 where the dotted line shows the condition before chatter vibration. In this case, the cutting area A_2'' is given by:

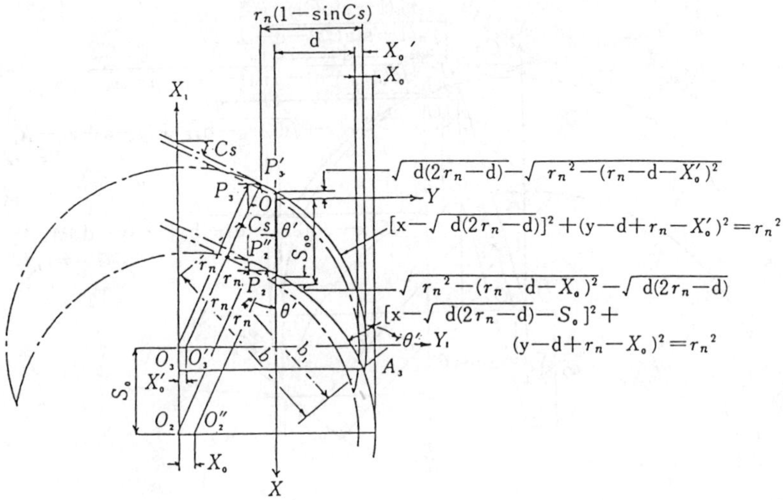

Figure 4. A cutting area under the condition of chatter vibration. (Cut in)

$$A_2'' = \frac{x^2 \cos^2\theta(\cos\omega t - \cos(\omega t + \phi))^2}{2}\sqrt{\frac{4r_n^2}{S_0^2 + x^2\cos^2\theta(\cos\omega t - \cos(\omega t+\phi))^2} - 1}$$

$$+\frac{x\cos\theta\cos\omega t}{2}\left\{S_0 - \sqrt{r_n^2 - (r_n - d - x\cos\theta\cos\omega t)^2}\right\} + \frac{x\cos\theta\cos(\omega t+\phi)}{2}$$

$$\left\{S_0 + \sqrt{r_n^2 - (r_n - d - x\cos\theta\cos(\omega t+\phi))^2}\right\} - (r_n - d)S_0$$

$$+\frac{1}{2}(r_n - d)\left\{\sqrt{r_n^2 - (r_n - d - x\cos\theta\cos\omega t)^2} - \sqrt{r_n^2 - (r_n - d - x\cos\theta\cos(\omega t+\phi))^2}\right\}$$

$$+\frac{1}{2}\left(\frac{S_0}{2} - \frac{x\cos\theta(\cos\omega t - \cos(\omega t+\phi))}{2}\sqrt{\frac{4r_n^2}{S_0^2 + x^2\cos^2\theta(\cos\omega t - \cos(\omega t+\phi))^2} - 1}\right)\cdot$$

$$\sqrt{r_n^2 - \left(\frac{S_0}{2} - \frac{x\cos\theta(\cos\omega t - \cos(\omega t+\phi))}{2}\sqrt{\frac{4r_n^2}{S_0^2 + x^2\cos^2\theta(\cos\omega t - \cos(\omega t+\phi))^2} - 1}\right)^2}$$

$$+\frac{1}{2}\left(\frac{S_0}{2} + \frac{x\cos\theta(\cos\omega t - \cos(\omega t+\phi))}{2}\sqrt{\frac{4r_n^2}{S_0^2 + x^2\cos^2\theta(\cos\omega t - \cos(\omega t+\phi))^2} - 1}\right)\cdot$$

$$\sqrt{r_n^2 - \left(\frac{S_0}{2} + \frac{x\cos\theta(\cos\omega t - \cos(\omega t+\phi))}{2}\sqrt{\frac{4r_n^2}{S_0^2 + x^2\cos^2\theta(\cos\omega t - \cos(\omega t+\phi))^2} - 1}\right)^2}$$

$$+\frac{1}{2}r_n^2\sin^{-1}\frac{\sqrt{r_n^2 - (r_n - d - x\cos\theta\cos(\omega t+\phi))^2}}{r_n}$$

$$-\frac{1}{2}r_n^2\sin^{-1}\frac{\sqrt{r_n^2 - (r_n - d - x\cos\theta\cos\omega t)^2}}{r_n}$$

Table 1. The numerical symbols of C_e, C_s, r_n, d, S_0, θ, x *and* ε.

C_e (°)	C_s (°)	r_n (mm)	d (mm)	S_0 (mm/rev)	θ	x (mm)	ε	Application
	15	1·0				0·04	0·1	d $\geqq r_n$ (1 − sin C_s)
				0·2		0·07	0·3	$S_0 \leqq 2\ r_n$ sin C_e
15	30	0·5	1·0		65°36′	0·10	0·5	$S_0 < 2\ r_n$ cos C_s
	15	1·5		0·4		0·02	0·7	d $< r_n$ (1 − sin C_s),
		2·0				0·04	0·9	$S_0 \leqq 2\ r_n$ sin C_e

$$+\frac{1}{2}r_n^2\sin^{-1}\frac{S_0 - x\cos\theta(\cos\omega t - \cos(\omega t+\phi))\sqrt{\dfrac{4r_n^2}{S_0^2 + x^2\cos^2\theta(\cos\omega t - \cos(\omega t+\phi))^2} - 1}}{2r_n}$$

$$+\frac{1}{2}r_n^2\sin^{-1}\frac{S_0 + x\cos\theta(\cos\omega t - \cos(\omega t+\phi))\sqrt{\dfrac{4r_n^2}{S_0^2 + x^2\cos^2\theta(\cos\omega t - \cos(\omega t+\phi))^2} - 1}}{2r_n} \qquad (4)$$

where X_0 is replaced by the component of vibration before one rotation, $x\cos\theta\cos\omega t$, in the direction of normal force, and X_0' is replaced by component of vibration, $x\cos\theta\cos(\omega t+\phi)$, after one rotation.

The cutting area change of a numerical example

The cutting area change is gained by cutting area A_1'' before one rotation and cutting area A_2'' after one rotation of the workpiece under the chatter vibration. The expression is shown as follows:

$$dA_1'' = A_1'' - A_2'' \qquad (5)$$

It is troublesome to calculate from expression (5). So, dA_s'' is expressed in the first term of dA_1''. The accuracy is compared in Table 1. The accuracy comparison between dA_1'' and dA_s'' is made for one cycle of vibration and the result is shown in Figure 5.

In Figure 5, the expression of cutting area change under chatter vibration is given as follows:

where $\quad d \geqq r_n\ (1 - \sin C_s),\quad S_0 \leqq 2r_n \sin C_e,\quad S_0 < 2r_n \cos C_s,$

$$dA_1'' \doteqdot dA_s'' = 2[\{d - r_n(1-\sin C_s)\}\cdot\tan C_s + r_n\cos C_s]\,x\cos\theta\cos\omega t$$

$$-\left[\{d - r_n(1-\sin C_s)\}\cdot\tan C_s + r_n\cos C_s + \frac{S_0}{2}\right]\cdot x\cos\theta\cos(\omega t+\phi). \qquad (6)$$

$$d < r_n\ (1 - \sin C_s),\quad S_0 \leqq 2r_n \sin C_e,$$

$$dA_1'' \doteqdot dA_s'' = \left\{\sqrt{d(2r_n - d)} - \frac{S_0}{2}\right\} x\cos\theta\cos\omega t + \left\{\sqrt{d(2r_n-d)} + \frac{S_0}{2}\right\}\cdot$$

$$x\cos\theta(\cos\omega t - \cos(\omega t+\phi)). \qquad (7)$$

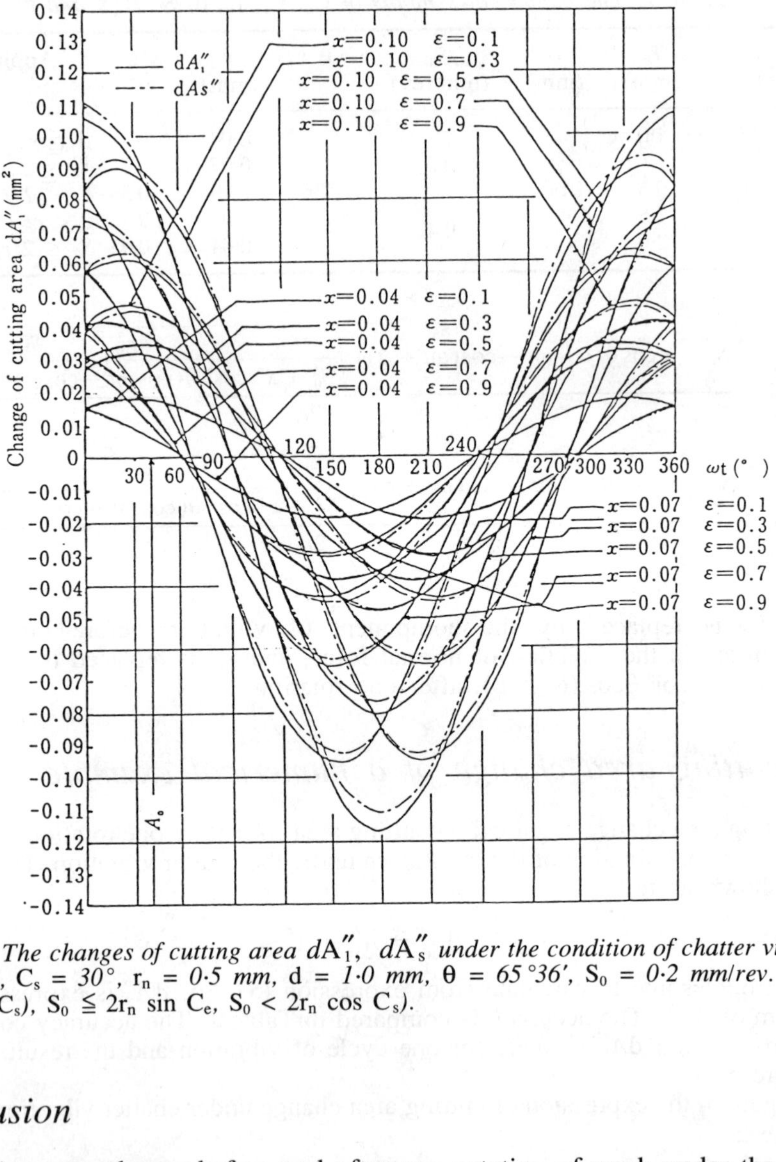

Figure 5. The changes of cutting area dA_1'', dA_s'' *under the condition of chatter vibration.* $C_e = 15°$, $C_s = 30°$, $r_n = 0{\cdot}5$ *mm*, $d = 1{\cdot}0$ *mm*, $\theta = 65°36'$, $S_0 = 0{\cdot}2$ *mm/rev*. ($d \geqq r_n (1 - \sin C_s)$, $S_0 \leqq 2r_n \sin C_e$, $S_0 < 2r_n \cos C_s$).

Conclusion

The cutting area change before and after one rotation of work under the chatter vibration is analysed. The result is as follows:

(1) The approximate expression is taken from cutting area before and after one rotation of work under the chatter vibration.
(2) The cutting area change of work under the chatter vibration causes larger approach to 0·5 from 0·1 and diminishes to 0·9 from 0·5 in the fraction.
(3) The cutting area change of work under the chatter vibration is influenced by the shape of bit. As nose radius and side-cutting edge angle of the bit increase, so the cutting area change increases. The influence of cutting area change by the change of nose radius of bit is larger than the change of side-cutting edge angle.

Acknowledgement

The author is indebted to Professor Y. Kume at Osaka Prefecture University.

Bibliography

Iwata, Y., 1983, The variation of cutting area under the condition of chatter vibration. *Journal of the Faculty of Science and Technology, Kinki University*, 267–276.

Iwata, Y., 1984, An analysis of the variation in cutting area on feed S_0 under the condition of chatter vibration. *Journal of the Faculty of Science and Technology, Kinki University*, 309–317.

Iwata, Y., 1987, A Study of Stabilizing Threshold in Lathe Turning Chatter. *Proceedings of IXth ICPR*, pp. 788–794.

SECTION II.4. PROCESSES

9

A statistical quality control design for wavesolder process

M. Riaz Khan[1] **and Peter Rega**[2]

[1]*Department of OM-MIS, University of Lowell, Lowell, MA 01854, USA and*
[2]*Raytheon Company—M.S.D., Lowell, MA 01853, USA*

Abstract

Traditional experimental designs, such as factorial design, may be employed to examine the parameters of a manufacturing process to maximize process efficiency. The effects of improving this efficiency include reduced product cost and engineering support for the firm. These tools, which can easily be applied to simple and well-defined problems, can be expanded to evaluate complex and unclear business situations. This chapter studies the process to pre-tin leadless ceramic chip carriers used on surface mount assemblies.

Introduction

Soldering with an alloy of tin/lead is by far the greatest utilized method of providing electrical interconnections in electronic assemblies. The soldering process has many variables which affect solder joint quality such as cleaning, fluxing, (conditioning) pre-heating, solder alloy type and temperature, machine parameters, and component and board solderability. An electrical component with good solderability greatly increases the probability of a good solder joint and electrical connection when soldered to printed wiring boards. The most common method of assuring component solderability is to pre-tin the component with the same solder alloy used in the assembly soldering process. This method of pre-tinning involves dipping the component leads in molten solder or passing them over a wave of molten solder.

Prior to instituting a pre-tinning process into the manufacturing cycle, a company should conduct an analysis to evaluate whether the savings anticipated as a result of the tinning operation more than offset its cost to implement and run. Obviously, the more critical and costly the process selection is to the product, the more extensive and comprehensive the analysis will be. In most cases involving pre-tinning and soldering processes, the savings will be represented by reduced rework and touch-up labour on non-conforming units. On the other hand, the cost to implement and run the new process will be in the form of new machinery, tooling, and labour cost. For firms which currently have a pre-tinning process, an analysis should be performed to determine whether the process should be continued, improved upon, or dropped. Such improvements would include increased engineering experimentation or new machinery purchases, which would provide technological advantages over present equipment.

The decision analysis can facilitate selection of technology in manufacturing processes. By utilizing the estimated probabilities for likely outcomes, Kenney *et al.* (1986) selected the most profitable manufacturing process technology for future power generation. McGardle (1985) provides a decision analysis model to evaluate the possible adoption of a new technology into a firm. As a result, management can evaluate whether a new piece of equipment or a new process would be financially advantageous for that firm to acquire. This is especially useful in technology-driven, competitive industries such as computers and electronics. Weiss and Rao (1987) developed a decision analysis technique, called Analytic Hierarchy Process, which uses scaled utility measurement for decision alternatives. This technique is especially useful when there is a large number of decision candidates in an established hierarchy.

There is a strong relationship between component solderability and joint reliability. Flattery and Arslancan (1986) use simple additive probability to predict product yields from known process defect rates. This work shows the strong dependence of yield on joint count and defect rate, especially as it relates to solder reflow parameters. Similarly, Murphy (1986) outlines process and environmental controls to improve the solderability process and presents rationale for component pre-tinning as a method of improving component solderability. Woodgate (1987) examines SPC to control the soldering process, and Roy *et al.* (1987) advocate the use of a data collection system on the soldering process, which they feel is the only practical method of control. This form of SPC was specifically applied to the flow of wavesolder process. In a study of solder joint reliability, Devore (1988) examines the relationship between surface mount component solderability and subsequent solder joint reliability. He finds mechanical failures as a direct result of poor solderability and adequate process control. Finally, Cormack (1987) outlines the requirements for process selection and control as a vehicle to improve the overall military soldering quality and reliability.

As mentioned earlier, passing components over a molten wave of solder is probably the most popular method of medium and high volume component pre-tinning. A review of soldering material pertaining to the wave soldering process is presented by Franzi (1986). He further discusses the process control systems for equipment as well as materials drastically to reduce rework and touch-up. In his study, Copeland (1987) considers the critical process parameters that must be examined in the wavesolder process. Tinning components using a wavesolder machine is advocated by Speckler and Haggis (1986) as a method of removing gold and improving solderability. Their findings show that dwell time of components in the molten solder, via conveyor speed, is a critical parameter. In an expanded study, Gulati (1988) adopts the Teguchi design of experiments to optimize component solderability parameters using the wavesolder process. This design optimizes the 'controllable factors' and minimizes the 'noise factors' to reduce variation from the intended product quality goals.

This chapter analyses a process which is followed by a Department of Defense contractor to pre-tin leadless ceramic chip carriers used on surface mount assemblies. These components consist of a ceramic package of 0·350 × 0·350 × 0·200 inches with 20 gold metallized terminations and a gold lid. During the pre-tinning process, the critical objectives to be achieved are: (1) complete removal of gold from component leads, (2) smooth, continuous solder coverage over leads (no voids, pitholes), and (3) solder 'bump' height less than 0·005 inches on component leads.

There is an extremely large number of such components which must be processed. This necessitates the use of an automated process capable of achieving the above objectives. A wavesolder machine, utilizing a conveyor which passes over a flux wave, pre-heaters, and a solderwave are currently used by the company for this purpose. A magnetic fixture is used to pass the components up the conveyor and over the wavesolder.

Prior to evaluating the specific process for possible improvement, the company performed a preliminary decision analysis to determine in which direction it would proceed with regard to the process technology. Three alternatives were considered, namely: (1) continue with the present process, (2) improve the present process with engineering manipulation of process parameters, and (3) enhance the process with new equipment and tools. Based upon known manufacturing and related rework costs and probability estimates obtained by the engineers for various outcomes, the second option was found to be most economical. The focus of this chapter, therefore, shifts to an analysis of improvement in the pre-tinning process in use at present.

Problem

Since additional engineering experimentation has been deemed the most favourable approach to the pre-tinning process, the process will be scruitinized and improved upon if possible. There are two major variables that affect the pre-tinning process and subsequently determine pre-tin quality and yields: conveyor speed and pre-heat temperature of the wavesolder machine. Other parameters, which also affect pre-tin quality but which are held at believed optimal settings, are: (1) conveyor angle, (2) flux and solder heights, (3) solder temperature, (4) flux specific gravity, (5) component positioning, and (6) component cleanliness.

It is desired to investigate the impact of variations in conveyor speed and pre-heat on pre-tin yields. It is also desired to know if the interaction of the two variables is correlated to pre-tin yields. If, for instance, conveyor speed is found to be insignificant to yields, then this parameter may be maximized to increase throughput without decreasing the yields. If pre-heat is found to be insignificant, on the other hand, then it may be eliminated, increasing throughput and saving energy. In the event that both are significant and/or that their interaction is significant, then the data might indicate optimal combinations to maximize yields. Maximum yields are critical because: (1) rework reduces component reliability, (2) rework is costly—touch labour and inspection are effectively doubled, and (3) rework uses valuable machine time for regular production and induces overtime.

Model

The most common method of evaluating the parameters is to run an experiment varying one factor, while keeping the other constant. In this case, the use of a factorial design is perhaps most appropriate. Both factors are examined at three levels, providing a total of nine treatment combinations. These nine combinations are replicated twice, providing 18 observations. The factorial design also provides levels of significance for both factors, as well as their interaction and experimental error. This information would not be extractable under the more common method of evaluating the experiment. Table 1 shows the factorial design used. $(Y_{i,j})_k$ is the observation under the ith level of Factor B and jth level of Factor A for kth replicate.

$$(Y_{i,j})_k = U + T_i + Q_j + (TQ)_{i,j} + (E_{i,j})k, \; i = 1, 2, 3; \; j = 1, 2, 3; \; k = 1, 2$$

where U = the overall mean effect, T_i = effect of ith level of Factor B, Q_j = effect of jth level of Factor A, $(TQ)_{i,j}$ = effect of interaction of T_i and Q_j, $(E_{i,j})_k$ = error component (random), and k = the kth replicate.

Table 1. Factorial design.

Factor A → B↓	Level	Conveyor Speed (ft/min) 1	3	5	$\sum_j (Y_j)$
Pre-heat temp. (°F)	70	$(Y_{1,1})_k$	$(Y_{1,2})_k$	$(Y_{1,3})_k$	
	125	$(Y_{2,1})_k$	$(Y_{2,2})_k$	$(Y_{2,3})_k$	
	250	$(Y_{3,1})_k$	$(Y_{3,2})_k$	$(Y_{3,3})_k$	
$\sum_i (Y_i)$					Y

The relevant hypotheses to be tested are:

I. H_0: $T_1 = T_2 = T_3 = 0$
H_1: At least one $T_i \neq 0$;

II. H_0: $Q_1 = Q_2 = Q_3 = 0$
H_1: At least one $Q_j \neq 0$; and

III. H_0: $(TQ)_{i,j} = 0$ for all i, j
H_1: At least one $(TQ)_{i,j} \neq 0$

Using the analysis of variance, the above hypotheses are tested, and the influence of conveyor speed and pre-heat on yields measured with some degree of confidence.

Nine combinations of conveyor speed and pre-heat temperature are run as noted in Table 1. The total number of leads out of a possible 300 per fixture that do not meet the acceptance criteria, described earlier, are totalled for all nine observations. The experiment is replicated twice in order to quantify experimental error. The sum of squares is computed for Y_i (pre-heat), Y_j (conveyor speed), $Y_{i,j}$ (interaction), and Y (total). Each sum of squares is divided by the relevant degrees of freedom (d.f.) to compute a mean square which yields an expected value.

Letting p and q represent the number of levels of Factors A and B respectively, and r the number of replicates; the d.f. for A are ($p - 1$ =) 2, for B ($q - 1$ =) 2, and for interaction A,B (d.f. A × d.f. B =) 4. The d.f. for error are given by $[pq(r - 1)]$ as 9; and the total d.f. associated with the experiment are ($pqr - 1$ =) 17.

Finally, the F-values are calculated to test the hypotheses at various levels of significance.

Results

The experiment was conducted as described for all nine treatment combinations and was replicated to provide a total of 18 treatment combinations. After the experiment, all components were thoroughly cleaned and segregated into separate packaging. Every component was then inspected, following the criteria described earlier, with the aid of vernier calipers. Data was then tabulated showing the total number of non-conforming leads per treatment. Results are shown in Table 2.

The sum of squares is now calculated from the data with SST = total sum of squares; SSA = sum of squares for conveyor speed; SSB = sum of squares for pre-heat; SSI = sum of squares for interaction of A, B; SSE = sum of squares for error. The formulas to compute these quantities are as provided below:

Table 2. Number of non-conforming leads.

Factor A → B↓	Level	Conveyor Speed (ft/min) 1	3	5	$\sum_j (Y_j)$
Pre-heat temp. (°F)	70	6,6	10,11	20,16	69
	125	6,5	11,13	17,17	69
	250	9,10	14,13	19,19	84
$\sum_i (Y_i)$		42	72	108	222

$$\text{SST} = \sum_{i=1}^{q}\sum_{j=1}^{p}\sum_{k=1}^{r} (Y_{i,j}^2)_k - Y^2/pqr;$$

$$\text{SSA} = \left(\sum_{j=1}^{p} Y_j^2\right)/qr - (Y^2/pqr);$$

$$\text{SSB} = \left(\sum_{i=1}^{q} Y_i^2\right)/pr - (Y^2/pqr);$$

$$\text{SSI} = \left[\sum_{i=1}^{q}\sum_{j=1}^{p}\left\{\sum_{k=1}^{r} (Y_{i,j})_k\right\}^2\right]/r - (Y^2/pqr) - \text{SSA} - \text{SSB};$$

$$\text{SSE} = \text{SST} - \text{SSA} - \text{SSB} - \text{SSI}.$$

In the notation used above, $(Y_{i,j})_k$ is the kth replicate for the ith level of pre-heat temperature and the jth level of conveyor speed; Y is the grand total of all the observations; Y_i is the sum of all observations for the ith level of pre-heat temperature; Y_j is the sum of all the observations for the jth level of conveyor speed. Also, there are q levels of pre-heat temperature and p conveyor speed levels, and r replicants for each of the pq combinations of speed and temperature levels.

Applying the data of Table 2 to the above equations, we obtain:

$$\text{SST} = [6^2 + 6^2 + 10^2 + 11^2 + \ldots + 19^2] - (222)^2/18 = 408;$$
$$\text{SSA} = [42^2 + 72^2 + 108^2]/6 - (222)^2/18 = 364;$$
$$\text{SSB} = [69^2 + 69^2 + 84^2]/6 - (222)^2/18 = 25;$$
$$\text{SSI} = [(6+6)^2 + (10+11)^2 + \ldots + (19+19)^2] - (222)^2/18 - 364 - 25 = 7;$$
$$\text{SSE} = 408 - 364 - 25 - 7 = 12.$$

In order to perform an analysis of variance, the mean squares are calculated by dividing sum of squares by degrees of freedom and then a comparative F-statistic is calculated by dividing that mean square by the mean square error. The analysis of variance is presented in Table 3.

A comparison of data in Table 3 with F statistics enables us to identify the important parameters. For 2 degrees of freedom (d.f.) in the numerator and 9 d.f. in the denominator, the critical value of F at 0·01% significance ($F_{0\cdot01,2,9}$) is 8·02. Since the observed F_0 values for the conveyor speed and pre-heat temperature exceed the critical value, it can be concluded that the effects of both of these factors on solder quality are significant at 99% confidence. Interestingly, variability in the

Table 3. Analysis of variance.

Source of variation	Sum of squares	Degrees of freedom	Mean square	F_0
Conveyor speed	364	2	182·00	136·84
Pre-heat temperature	25	2	12·50	9·40
Interaction	7	4	1·75	1·32
Error	12	9	1·33	
Total	408	17		

conveyor speed appears to be by far the most important parameter. Therefore, an optimal combination of these factors can be found to maximize the yield. Furthermore, since $F_{0{\cdot}05,4,9} = 3{\cdot}63$, the interaction between conveyor speed and pre-heat temperature is insignificant. Thus, the analysis concludes that null hypotheses (I) and (II) be rejected, suggesting the importance of the conveyor speed and pre-heat temperature to solder quality.

Summary

The experiment has shown that conveyor speed is the most significant parameter that affects component solder quality. Even at a confidence level of 99%, the conveyor speed is found to be significant. The interaction of pre-heat and conveyor speed is found to be of no significance to the solder quality at any level. Also, consistent with the industry belief that pre-heating the components aids in the wetting and soldering process, the analysis reveals that pre-heat temperature is significant at 99% confidence. Furthermore, the insignificance of experimental error suggests little or no variation between replicates, reinforcing the conclusions.

Based on the findings of this study, it may be recommended that, when tinning leadless chip carriers on wavesoldering equipment: (1) the conveyor is run as slowly as possible, but not so slowly that the components might be damaged by prolonged heat exposure, and (2) the components are pre-heated not only for the purpose of avoiding thermo-shock of components and cracking of large ceramic packages, but also because it contributes to the solder quality.

The use of the experimental design presented here indicates that such tools can effectively be applied to make typical manufacturing process decisions. They can facilitate engineers getting a better feel for the root relationships of the process to product quality. The obvious shortcoming of the model is that it is usually narrowly targeted towards a particular facet of a manufacturing process and quality standards.

References

Copeland, B.F., 1987, Zero defect manufacturing soldering process management. *EMPF Soldering Technology Proceedings*, **901**.

Cormack, R., 1987, The new military soldering process. *Printed Circuit Assembly*, **18**.

Devore, J.A., 1988, Solderability and surface mounting. *EMPF Soldering Technology Proceedings*, **301**.

Flattery, D.K. and Arslancan, A.N., 1986, Minimizing solder defect rates through the control of reflow parameters—Part I. *Surface Mount Technology*, **21**.

Franzi, N., 1986, Reduction of assembly costs through improvement of wave soldering process. *Circuit East Expo Proceedings*, **59**.

Gulati, B., 1988, Optimizing the hot solder coating process for component leads. *EMPF Soldering Technology Proceedings*, **7**.

Kenney, R., Lathrop, J. and Sicherman, 1986, An analysis of Baltimore Gas and Electric Company's technology choice. *Operations Research*, **34**, 18–39.

McGardle, K.F., 1985, Information acquisition and the adoption of new technology. *Management Science*, **31**, 1372–1387.

Murphy, R., 1986, Controlling the surface mount assembly process, *Circuit East Expo Proceedings*, **130**.

Roy, R., Grinovich, E. and Worthington, S., 1987, DOD 2000, the road to compliance—Part II. *Circuits Manufacturing*, **50**.

Speckler, R.G. and Haggis, D., 1986, *Soldering Techniques for Surface Mounted Leadless Chip Carriers*. RCA Technical Paper, No. 901.

Weiss, E.N. and Rao, V., 1987, AHP design issues for large scale systems. *Decision Sciences*, **18**, 43–57.

Woodgate, R.W., 1987, Solder joint criteria-cosmetics, quality, and process control. *Printed Circuit Assembly*, **23**.

10

Small deep hole drilling by electro-discharge machining

X.K. Chen[1] and A.W. Duffill[2]

[1]*Department of Mechanical Engineering, Nanjing Aeronautical Institute, China and* [2]*School of Manufacturing and Mechanical Engineering, University of Birmingham, Edgbaston, Birmingham B15 2TT, UK*

Abstract

This chapter describes a machine tool and small deep hole drilling process centred on electro-discharge machining (EDM) techniques. It illustrates and describes special apparatus which has been built into a conventional EDM machine to enable rapid drilling to occur. How the techniques are employed to result in an efficient process is also shown. For example, high metal removal rate (MRR) drilling of over 20 mm/min was achieved when producing holes of 500–1000 μm diameter to a depth of 100 mm. In addition a high degree of hole accuracy was produced.

Introduction

Over past years there have been continual developments in the field of micromachining to give holes with diameters of less than 1000 μm (the subject of this paper) and kerfs with widths of less than 250 μm.

The development of new products and the continual pressure for productivity, coupled with quality, during the production of existing products has resulted in manufacturers being faced with the need rapidly to produce small, but deep, holes of less than 1000 μm diameter.

Applications for small deep holes include electronic instrumentation, venting holes in aerospace components and accurate medical instruments. A review of small hole drilling literature (Cletis, 1970; Crookhall, 1976; Jeswani, 1979; Kagaya, 1986) indicates the limitations of using traditional drilling methods, non-conventional processes such as EDM with simple wire-shaped electrodes and even the modern approaches such as laser beam machining (LBM). These techniques are suitable for small holes within the range 100–1000 μm diameter but are limited to a depth : diameter ratio of below 10 : 1. In addition the machining speed is extremely low being less than 1 mm/min.

Some researchers (Silva and McGeough, 1986; Clifton *et al.*, 1988) have adopted other non-conventional processes to machine small deep holes. These were electrochemical machining (ECM) and a variation called electrochemical arc machining (ECAM). These authors have reported other process problems such as 'bulbing' and poor hole accuracy.

Therefore with the problems associated with the present techniques it has been

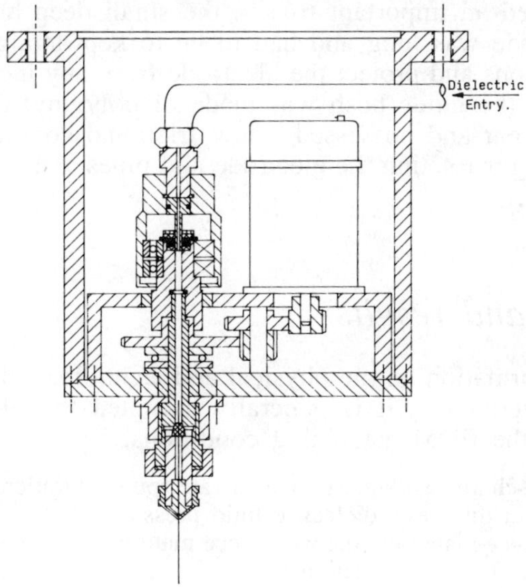

Figure 1. Fixture for tubular electrodes.

necessary for industry to continue to search for new processes which can machine small deep holes at a fast machining rate and overcome the difficulties reported to date. One such process is described below.

Principle of small deep hole drilling

The working principle for small deep hole drilling is based upon the conventional EDM process. This EDM method is to feed a small diameter wire tool into the workpiece with an oscillating motion. However, it has been noted that it is difficult to rid the eroded debris from the immediate sparking area when a small hole was machined by this process. The problem gets progressively worse as the depth: diameter ratio increases. Obviously when the hole becomes too deep and small to provide for the flow of clean dielectric fluid into the spark gap, arcing and short circuiting will occur making the EDM process unstable.

In order to improve these machining conditions some additional equipment needed to be designed so that a continual flow of clean dielectric could be provided in the spark gap. This experimental apparatus is described in the next section.

Small deep hole drilling apparatus

The overall equipment could be subdivided into three main parts: (a) an EDM machine, (b) a high pressure dielectric supply, and (c) the special fixture for guiding the small tubular electrode. The EDM machine consists of a 'C'-frame machine body, a dielectric cooling and filtering system and an 'in-house' electrical power supply. (The existing circuit was only capable of producing 80 V and had to be modified to give 200 V.) The dielectric pumping circuit consisted of a high pressure pump capable of reaching pressures of 100 kg/cm^2 and an accumulator to maintain the dielectric pressure during the cycle.

Finally, the key to the experiments was the fixture for holding and guiding the small tubular electrode (Figure 1). This consisted of a clamping head to hold the tubular electrode and a guide to allow the electrode to be fed into the workpiece.

The guide played an important role in the small deep hole drilling process as the tubular electrode was long and had to be to kept straight. The guide needed to suppress vibrations and protect the electrode from any inclination tendencies. In these experiments the guide bush was made of polyvinyl chloride (PVC), which was resistant to wear and possessed a low frictional coefficient. The fixture also contained a motor for rotating the electrode and pipes and seals to contain the high pressure dielectric.

Experiments and results

There is scant information on the physical and technological aspects of small deep hole drilling. Nevertheless, it is generally accepted that there are many factors which can affect the EDM machining conditions, e.g.:

(a) ignition and discharge voltage, tool feed rate, pulse duration and interval,
(b) dielectric and, in this case, dielectric fluid pressure,
(c) tool and workpiece materials (the workpiece materials were stainless, high carbon/high chrome and die steels and titanium).

The initial experiments were necessary to determine the optimum machining conditions for supply voltage, pulse duration and interval and feed rate. The procedure included open circuit orthogonal testing from which the following optimum machining parameters were chosen:

(a) The voltage wave form consists of a 180 V initiation voltage followed by a 70 V discharge voltage.
(b) Stainless steel machined with a pulse duration and interval of 6 and 11 μs and a tool feed rate of 150 mm/min.
(c) Titanium machined with a pulse duration and interval of 11 μs and a tool feed rate of 100 mm/min.
(d) The remaining materials had a pulse duration and interval of 6 and 11 μs and a tool feed rate of 100 mm/min.

Initially a hydrocarbon dielectric was selected but this produced a feed rate which was too slow. Hence distilled water was used, in all experiments, as this produced a much higher feed rate and had the added advantage of being readily available. The electrode material was commercially produced brass of 1000 and 450 μm external and internal diameter, respectively. This was found to be the best electrode material as copper resulted in considerable wear and steel caused short circuits and arcing. Having determined these basic conditions the present experiments were carried out. A few machining parameters were selected in turn and the workpiece metal removal rate (MRR), tool wear (relative electrode wear, REW) and surface dimensions obtained. The first experiments were to investigate the influence of water pressure on MRR and REW. The relationships between the water pressure and MRR can be seen in Figure 2 and REW in Figure 3.

In the next experiments various widths of workpiece were machined in order to investigate the effect of workpiece thickness on MRR and REW. The relationships between the workpiece depth and MRR can be seen in Figure 4 and REW in Figure 5.

The final analysis was to determine the accuracy of the hole and necessitated measuring the hole diameters at entry and exit to determine taper. The results are shown in Table 1. The surfaces were also observed, after the specimens were cut parallel to the machined axis, with an optical microscope (amplification × 100). It was only possible to obtain a vague estimate of surface finish and further work is proceeding.

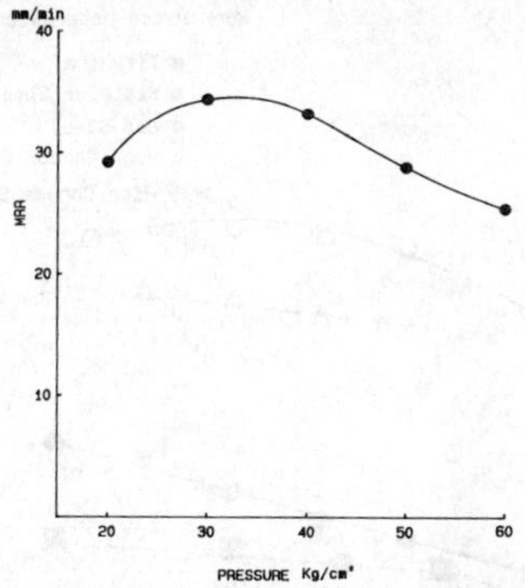

Figure 2. Relationship between pressure and the metal removal rate. Workpiece, stainless steel; thickness, 20 mm; pulse duration, 6 μs; pulse interval, 11 μs; electrode, brass tube (outside diameter 1 mm, inside diameter, 0·45 mm).

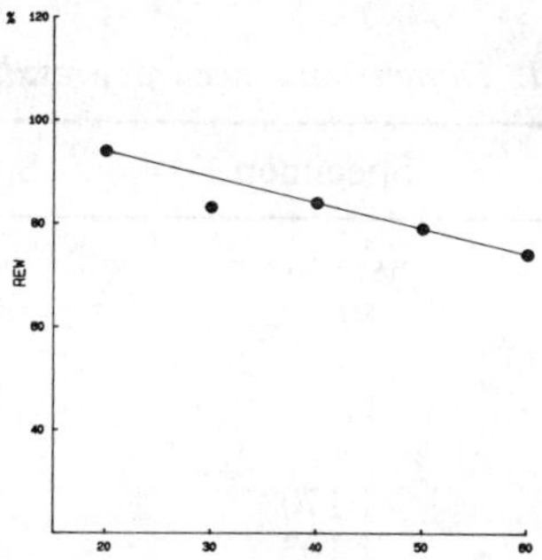

Figure 3. Relationship between pressure and the relative electrode wear. For workpiece details see Figure 2.

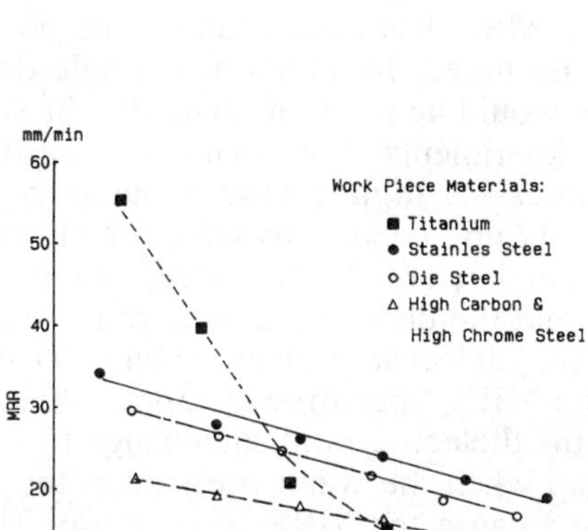

Figure 4. Relationship between depth and the metal removal rate.

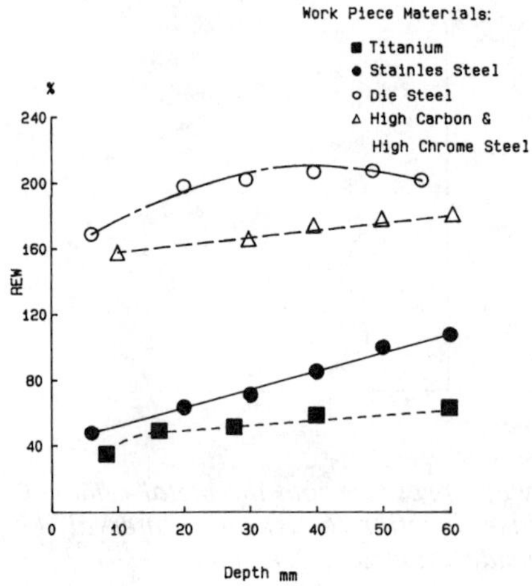

Figure 5. Relationship between depth and the relative electrode wear.

Table 1. Dimensional accuracy results.

Parameter	Specimen 1	Specimen 2	Specimen 3
Workpiece material	stainless	stainless	high C/CR
Workpiece thickness (mm)	40	50	50
Pulse duration (μs)	6	6	6
Pulse interval (μs)	11	11	11
Results			
Entry diameter (mm)	1·170	1·158	1·259
Exit diameter (mm)	1·103	1·150	1·116

Discussion

In conventional EDM the MRR has been found to be no greater than 2 mm/min when a small hole was machined. In addition, the hole depth was restricted, e.g. a 1000 μm diameter hole would be no more than 10 mm deep (or depth : diameter ratio of 10 : 1). These experiments show that following the introduction of two simple engineering features—a high pressure pumping system and a tubular electrode fixture—the machining conditions were greatly improved. It can be seen from Figure 4 that the metal removal rate exceeded the conventional small hole EDM by up to 30 times and the depth : diameter ratio could be as high as 60 : 1.

The effect of varying the dielectric pressure (Figure 2) showed that an optimum value—giving a maximum MRR—occurred at approx. 40 kg/cm^2. It is believed that the reason could be that the dielectric fluid can change from streamline to turbulent flow at the machining gap when the water pressure is too high. This would result in unstable machining and cause the MRR to decrease. The relationship between the workpiece depth and MRR is shown in Figure 4. As the machining depth increases so the MRR falls. This could be expected as the increase in machining depth will make it more difficult for the removal of the eroded particles from the machining area and a short circuit is more likely to occur. For most workpiece materials the MRR decreases with the increase of depth by a linear relationship. This however was not the case with titanium. When the thickness is less than

30 mm the MRR of titanium is linear—but with a far greater gradient and MRR than the other workpiece materials. After this point the slope changes more in line with the other materials.

A further phenomenon, found only with titanium, was a small sliver produced corresponding to the core of the tubular electrode. It indicates a much smaller spark gap or overcut, but why this should only occur with titanium requires further investigation. Perhaps the most disappointing experiment showed the high electrode wear (Figure 5). The four curves all show high wear rates. In addition there is a tendency for the electrode to wear with increase of depth. It was also found that electrode wear is almost four times greater when machining die and high carbon/ high chrome steels than stainless steel and titanium. Again further work will be required.

Finally, the small tapered hole shows good dimensional accuracy.

Conclusions

The current research has made a significant contribution to the small deep hole drilling process. The modifications made to a conventional EDM machine show considerable potential. The practical aspects of the process have been discussed and the following conclusions can be made:

(a) It has been shown that a rapid MRR can be reached by this process. An average of 30 mm/min MRR can be obtained during production of 1000 μm diameter holes in a variety of materials and to various depths.
(b) It was found that there is an optimum dielectric working pressure of about 40 kg/cm^2 and that larger or smaller pressures can result in a decrease in MRR.
(c) The applied voltage, tool material and dielectric have a most significant influence on MRR. If these parameters are incorrect then machining is unlikely to achieve any positive results.
(d) Further work is to continue which would include surface finish investigation, dielectric fluid flow theory, machining smaller and deeper holes and other tool and workpiece materials.

References

Cletis, J., 1970, An ECM process for drilling deep holes. *Metal Progress*, March, 106–111.

Clifton, D., Barker, M.B. and McGeough, J.A., 1988, An adaptive control system for fine hole drilling. *Fourth International Conference on Computer-aided Production Engineering*, Edinburgh, November, pp. 237–240.

Crookall, J.R., 1976, Performance of single electrode fine hole drilling by EDM. *Proceedings of 17th MTDR Conference*, Vol. 2, 20–24 September, pp. 313–321.

Jeswani, M.L., 1979, Small hole drilling in EDM. *International Journal of MTDR*, **19**, 3, 165–169.

Kagaya, K., 1986, Micro-electrodischarge machining using water as a working fluid micro-hole drilling'. *Precision Engineering*, **8**, 3.

Silva, A. de. and McGeough, J.A., 1986, Surface effects on alloys drilling by electrochemical arc machining. *Proceedings of the Institute of Mechanical Engineers*, **200**, B4, 237–246.

11

Review of CINPRES gas-assisted moulding process

Nick J. Quick, A. Peter Jarvis and Ken J. Stout

Department of Manufacturing Engineering, University of Birmingham, PO Box 363, Birmingham, B15 2TT, UK

Abstract

CINPRES (Controlled INternal PRESsure) gas-assisted injection moulding is arguably the most important development in plastic injection moulding this decade. It has far reaching implications for the design and manufacture of plastic mouldings. The process offers significant benefits: lighter weight mouldings with no sacrifice in strength, low moulded-in stress and low distortion, superior alignment of any chopped fibre reinforcement, faster moulding cycle times, much lower mould clamping forces, furthermore, very large mouldings can be produced more easily. This chapter describes the background to the development of the CINPRES process, the way in which the process works and the benefits it offers.

Introduction

This chapter is qualitative rather than quantative in nature. The reason for this is that whilst the process described is already proving highly successful in a wide range of practical applications, its rheology is not yet fully understood.

Background

The process was originally developed to overcome the two principal restrictions imposed on structural foam moulding. These are: (i) the poor surface finish with a 'splay' or 'swirl' caused by gas bubbles from the blowing agent which break through the surface, and (ii) the comparatively long production cycle times of between 3 and 6 min per shot. It was only as CINPRES was developing into an exploitable process that it became clear that it bridged the gap between structural foam moulding and injection moulding. Furthermore, it made possible several things which had previously been considered technically impossible. Patent applications were filed by Peerless Foam Moulding Co. Ltd of Tamworth, England in 1983 to protect the process, and in 1984 to protect products made by the process. Both patents have since been granted and after a management buy back from the UK stock market and a company restructuring, they are now held by Cinpres Ltd.

The significance of the process has been appreciated by the plastics industry world-wide. Mitsubishi were an early licencee and they now have an exclusive

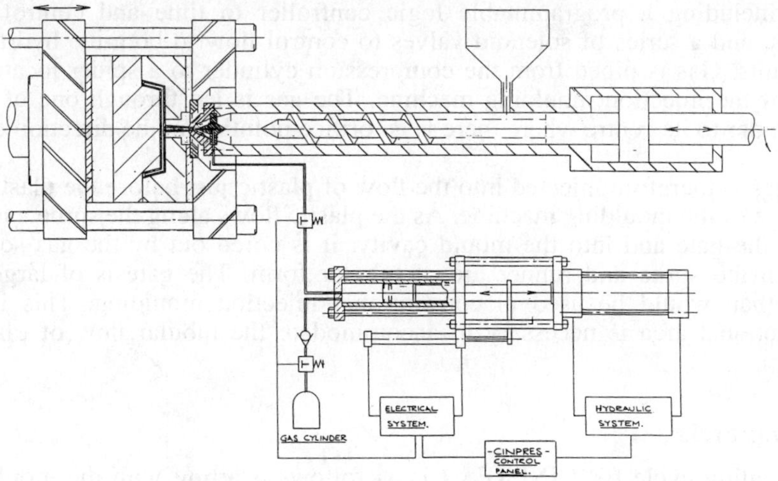

Figure 1. General arrangement of CINPRES I equipment.

licence to exploit the process in Japan. More recently they have joined with Cinpres Ltd in applying for further patents based on their own development work. Peugeot–Citröen as part of their 'CARMAT' project have been collaborating with ICI in experiments with the process for the production of exterior car body panels. Several other motor manufacturers and a wide range of trade and custom moulders are now using the process. The rate of uptake of the technology is continuously increasing as more product designers and engineers become aware of the existence of the process and the potential that it offers.

The process

The CINPRES gas-assisted moulding process should not be confused with blow moulding. The CINPRES process works by injecting inert gas (usually nitrogen) at high pressure into the moulding shortly after a flow front of molten plastic has been established. Thereafter, the gas and molten plastic are injected simultaneously. The gas remains enclosed inside the plastic and creates continuous channels inside the moulding. The gas is only exhausted after the plastic has frozen. The size and extent of the gas channels is a function of the cross-section of the mould and the volumetric ratio of the gas to plastic injected into the mould. Channels do not form in narrow sections.

CINPRES I

Equipment

The general arrangement of the basic 'CINPRES I' equipment is shown in Figure 1. The CINPRES unit is free standing. There are only two connections between the unit and the moulding machine: an electrical connection to synchronize the control of plastic and gas injection, and a small pipe to deliver the gas from the unit to the moulding machine. The unit comprises: a cylinder of nitrogen gas, a cylinder with piston to compress the gas, a hydraulic system to drive the piston, an electrical system including a programmable logic controller to time and control the gas injection, and a series of solenoid valves to control flow in both the hydraulic and gas circuits. Gas is piped from the compression cylinder to a spider located in the nozzle of the injection moulding machine. The gas is led through one of the legs of the spider to its centre where there is an orifice pointing in the direction of plastic flow.

The gas is therefore injected into the flow of plastic just before the plastic leaves the nozzle of the moulding machine. As the plastic flows along the sprue and runner, through the gate and into the mould cavity, it is cored out by the gas so that the plastic in the sprue and runner are of tubular form. The gate is of larger cross-section than would be used in conventional injection moulding. This increased cross-sectional area is necessary to accommodate the tubular flow of plastic.

Moulding cycle

The moulding cycle for CINPRES I is as follows, starting with the mould closed and the nozzle engaged against the sprue bush:

(1) Plastic injection starts, plastic injection timer starts, gas injection delay timer starts.
(2) Gas delay timer 'times out' shortly after a flow front of plastic has been established, gas injection starts.
(3) At the end of plastic injection, plastic hold timer starts.
(4) At the end of gas injection, the gas hold timer starts.
(5) When the plastic hold timer 'times out', the injection unit carriage moves back retracting the nozzle from the sprue bush. This causes sprue break which allows the gas to exhaust through the end of the sprue.
(6) When the gas hold timer 'times out' the mould opens, the moulding is ejected and the compression cylinder is recharged from the gas cylinder ready for the next cycle.

Clearly the gas must be exhausted from the moulding before the mould opens. Opening the mould whilst the moulding is still pressurized would cause the gas to blow through the moulding wall hence rendering the component useless.

CINPRES II

Process and applications

A recent development of the basic process just described is 'CINPRES II'. This alternative arrangement uses a gas injection nozzle to inject gas directly into the moulding tool at some predetermined point instead of injecting gas through the sprue and runner. This allows the process to be used with fixed nozzle moulding machines and hot runner systems as it eliminates the need for sprue break to exhaust the gas. This arrangement is shown in Figure 2(a).

Another application of CINPRES II is where it is necessary to inject the gas into the mould cavity at several discrete points. This might be necessary where there are several separate thick sections within the cavity which cannot be interconnected by gas channels, or where a multiple impression tool is being used [see Figure 2(b)].

Nozzle for CINPRES II

The gas injection nozzle for CINPRES II is shown in Figure 3. It is located in the moulding tool and is mounted on a hydraulic cylinder which is fixed to the tool.

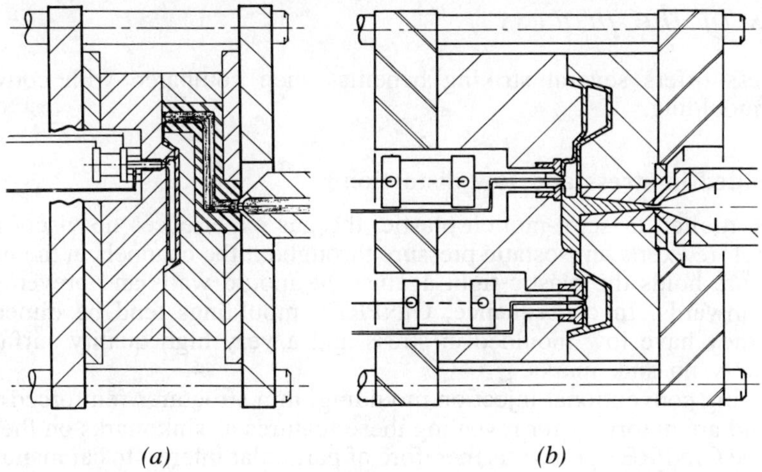

Figure 2. CINPRES II application (a) with hot runner system, (b) where moulding has two or more gas channels.

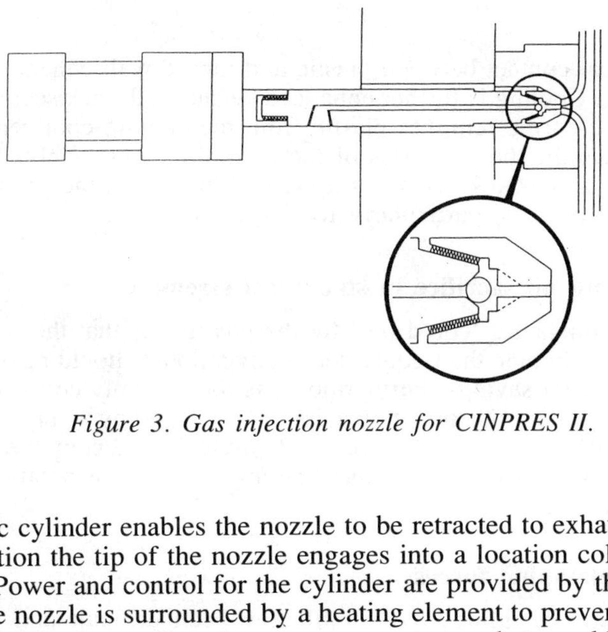

Figure 3. Gas injection nozzle for CINPRES II.

The hydraulic cylinder enables the nozzle to be retracted to exhaust the gas. In the forward position the tip of the nozzle engages into a location collar secured in the cavity wall. Power and control for the cylinder are provided by the CINPRES unit. The tip of the nozzle is surrounded by a heating element to prevent the plastic from freezing in that region so that the gas can penetrate what would otherwise be the cool outer surface of the moulding. Inside the nozzle tip is a steel ball which acts as a valve. Except when gas is being injected, this ball remains on its seat to prevent blowback and the ingress of plastic.

CINPRESS II moulding cycle

The moulding cycle is much the same as for CINPRES I except that gas injection starts when the cavity in the region of the gas injection nozzle is partly full. This means that for CINPRES II gas injection starts slightly later in the cycle than it would for CINPRES I.

Benefits of the process

The process offers several striking benefits when compared with conventional injection moulding.

Low moulded-in-stress and low distortion

Unlike the molten or semi-molten plastic, the gas experiences no viscous friction and it therefore exerts an isostatic pressure throughout the channels in the moulding. This pressure holds the plastic tight against the mould walls and prevents it from shrinking inwards. In consequence, CINPRES mouldings tend be dimensionally accurate, they have low moulded-in stress and a very high quality surface finish with virtually no sink marks.

When using conventional injection moulding, thin structures reinforced with ribs from behind are notorious for revealing these features as sink marks on the exposed surface. The CINPRES process is therefore of particular interest to car manufacturers and others wishing to mould exterior panels which must be reinforced with ribs from behind and yet show no sink marks on the exposed surface. This objective has almost been achieved by creating gas channels at the base of the ribs.

Rapid cooling

The firm mechanical contact between plastic and mould wall enhances conduction and speeds cooling. Cooling is further enhanced because all thick sections are cored out by the gas. The coring removes plastic from the slow-to-cool centres of thick sections and reduces the thermal mass of the moulding appreciably. Total cooling time is reduced by about 30%. This has a marked impact on the economics of the injection moulding process, particularly for large mouldings.

Lighter weight with no sacrifice in structural strength

The fact that mouldings are 'cored out' by the gas means that the mouldings can be as much as 40% lighter than equivalent conventional mouldings. This clearly offers considerable cost savings. Furthermore, as the gas only cores out the centre of thick sections, second moments of area are only reduced by a very small amount, and there is virtually no sacrifice in structural strength. The combined qualities of lighter weight and no sacrifice in structural strength offer considerable potential in the aerospace industry.

Reduced mould clamping forces

Very large injection mouldings have until now been difficult and expensive to produce. The reason for this is that considerable pressure must be exerted at the point of feed in order to overcome friction and push the viscous molten plastic into the extremities of the mould. This high injection pressure in turn requires a very high clamping force to hold the mould shut. These high pressures call for very robust presses and tooling, both of which are expensive. When using the CINPRES process, the isostatic pressure of the gas occupying the channels which reach into these extremities means that very much lower pressures can be used, both to fill the mould and also during the subsequent 'hold-on' period until the plastic freezes. Typically, for instance, a 1 m diameter table top moulded conventionally would require a mould clamping force of 20 000 kN. Heasman (1987) found that when using the CINPRES process this force could be reduced to 4000 kN. Furthermore, the CINPRES moulding was lighter, stress free and dimensionally stable.

Greater structural integrity

Another problem associated with large conventional injection mouldings is that it is frequently necessary to provide more than one feed point or 'gate' in order to fill the mould satisfactorily. More than one feed point implies more than one flow front of molten plastic. Where the flow fronts meet up, weld lines will occur and these weld lines of course impair the structural integrity of the moulding. With CINPRES, the isostatic pressure exerted by the gas just behind the flow front of the plastic helps the plastic to flow into the mould. This eliminates the need for multiple gating.

When using conventional injection moulding techniques for large mouldings such as car bumpers, as many as five gates have been needed to achieve a satisfactory fill, whereas one gate has proved sufficient when using CINPRES.

Fibre-reinforced mouldings

Structural mouldings are often reinforced by the addition of short-chopped glass fibre to the plastic. When using conventional injection moulding, the additional strength effected by the glass is limited for two reasons: (i) the fibres are randomly orientated in the moulding, and (ii) the fibres break the surface causing stress raisers. With CINPRES the fibres become more closely orientated to the direction of flow of the plastic and tend not to break the surface. The combination of this improved alignment of the fibres and an absence of stress raisers offers a considerable increase in strength.

Avoidance of long unsupported cores

Cores which are cantilevered from one end and are long in comparison to their width and depth often pose problems in moulding. Such cores tend to deflect under the considerable pressure of the plastic flowing into the mould. In many cases these cores are used to reduce the cross-section of the moulding and create a thin-walled, lightweight structure. Where this is the object of the core, and the inside dimensions of the cored-out section are not critical, the gas channel produced by the CINPRES process will often suffice and the core can be eliminated.

Even if dimensional accuracy of the cored hole is important and a core must be used, deflection of the core can be minimized by the use of the CINPRES process. Here the mould cavity is designed such that gas channels will form in the plastic, parallel to the core and equi-spaced about it. As CINPRES is comparatively low injection pressure process and injection velocities tend to be lower than for conventional injection moulding, there is less tendency for the core to deflect. Furthermore, as the gas channels are subject to an isostatic pressure, the pressure exerted by the gas helps to centralize the core.

Conclusions

It is very rare to find a new process which shows so many potential advantages and no obvious disadvantages. CINPRES is just such a process, and although it is still in its infancy and its rheology is not fully understood, it is proving very successful in applications which range from automotive components and garden furniture through computer housings to plastic luggage. When this exciting process comes of age it inevitably will take a major hold on the plastics industry, but in addition, it will capture markets which were previously thought inappropriate for solutions using injection moulded plastics.

References

Heasman, J., 1987, A major advance in plastic moulding technology. In *Institute of Engineering Designers Official Reference Book and Guide* 1986/87 (London: Stirling Publications), pp. 203–204.

Peerless Foam Moulding Co. Ltd and Hendry, J.W., 1983, *Injection Moulding*, UK Patent Application GB 2 139 548 A.

Peerless Foam Moulding Co. Ltd, 1984, *Injection Moulding*, UK Patent Application GB 2 158 002 A.

SECTION II.5. TOOLS

12

The design and development of a flexible loom board system for the integrated manufacture of cable harnesses

G.P. Bertolotti and B.J. Griffiths
Brunel University, Uxbridge, Middlesex, UK

Abstract

Technological advances in the cable harness industry have been substantial in recent years. New methods and new machines are continually being tried and tested. An area of interest for new hardware was identified at a major cable harness manufacturing company. This was for the manufacture and test of single cable harnesses. The approach was to build hardware to interface with a manufacturing system to make prototype cable harnesses in the same manner as production harnesses. The objective is to use the hardware as a means of verifying the manufacture of a production loom and to provide prototypes and samples of a loom which are to production methods and standards.

Based on these requirements a flexible assembly system for building cable loom boards has been developed which offers the facility of computer-assisted lay up and test. The system essentially reproduces the features that the cable harness industry in general uses and integrates these as a means of quickly generating reliable cable harnesses.

Introduction

The manufacture of a cable harness is divided between cable preparation and cable harness assembly. There has been much research in both areas and suitable systems and hardware have been developed to manufacture harnesses in moderate volumes. However, less emphasis has been placed on low volume and prototype harnesses. This is a very important area since building a prototype can provide useful information on the design of the cable harness for function and manufacture. Effectively a prototype saves valuable downtime through not having to prove the cable harness at the production stage.

The manufacture of low volume cable harnesses requires an acceptable specification within a short period of time. This requirement has led to a concentrated effort in building a loom board system which integrates the features generally used in cable harness assembly and takes into account the machines which are available to prepare cables.

The features of an integrated loom board system

The features of the loom board system can be broken down into three distinct areas. These are a flexible cable harness board which can be built automatically, computer-assisted lay up and a cable harness board which interacts with the computer.

The flexible cable harness board

The accepted method of building a cable loom board is to build a board dedicated to one particular cable harness. The advantage of a flexible cable harness board is that a loom board can be built up from a set of components to suit the requirements of a particular harness. On completion of a harness it can readily be disassembled.

The targeted area for a flexible board is in very low batch cable harness manufacture where the initial expense of a fixture kit can be justified against building a succession of dedicated boards. Examples of such fixture kits have been developed by IBM (Moore, 1986), Lansing Bagnell (Gibbons, 1984), and IPA (Warnecke *et al.*, 1985). They have one area in common—not one supports electrical testing.

Electrical testing of cable harnesses is an important feature of manufacture. It is not necessarily conducted on the same board for loom build but should be incorporated in a flexible loom board system. All of the current test systems rely on a backplane harness which is basically a second cable harness routed on the back of the loom board. This type of set-up cannot support the use of a flexible cable harness system.

Computer-assisted cable harness assembly

Routing of cables can be a very complex job for large cable harnesses. In the manufacture of large batches the time to build each successive cable harness is reduced as the operative becomes more familiar with the loom. For the manufacture of single cable harnesses this learning curve seriously reduces productivity.

There are several techniques that can be used to improve this learning curve and thus improve productivity. The most important method is to break the harness down into sub-assemblies. A complementary technique is to provide assistance at the routing stage. This assistance can be provided as an instruction sheet, a flip chart, a computer graphics routine and also by detailing the loom board. A technique which is operational and uses computer-assisted techniques is described by Mills (1984).

Interaction between the cable harness board and the computer

A prerequisite to cable harness test is to have electrical connections between the routing points on the cable harness and the processor which performs the function testing. This facility can easily be extended to provide important information in the assembly of the cable harness.

Three facilities are offered by a loom board which interacts with a computer. First, the routing path can be established by the operator earthing out the individual cable. Secondly, cables are tested immediately after they are routed, and thirdly, a complete history of the routing task can be built up if appropriate data collecting is performed.

A loom board with such capabilities enables building cable harnesses from components selected at random. Goldstein and Payne (1980) describe a technique which involves randomly picking up a connector with a bundle of flying leads and then identifying each routing path by electrically earthing each lead.

The flexible loom board system

The cable harness system was designed and developed by taking into account the facilities discussed. The system contains three separate functional components. These are a loom board, a network of slave microprocessors and a supervisory controller. An optional robot manipulator has also been developed. The loom board can readily be assembled and disassembled from a fixture kit. The components are designed so that this function can be performed automatically using a robot manipulator or by hand using computer assistance.

The salve microprocessors and the supervisory computer form the intelligence of the loom board. The microprocessors are placed at the nodal points of the cable harness and they communicate with the supervisory computer through the loom board.

The supervisory controller is responsible for the overall control of the system. It provides communication to the loom board and the robot manipulator and handles the cable routing files, the loom board build files and the graphical output which is necessary to interact with an operative.

The fixture kit

The fixturing system offers a very versatile means of building up loom boards; however, the cost considerations are very important. These were found to be very sensitive to the positional resolution of the fixtures on the board. Accuracy was not considered to be a problem as tight tolerances can be held with a fixture kit.

The resolution is dependent on the application of the final system. For example a fixturing system with a 20 mm matrix resolution would be sufficient as a cable harness test board only. However, if the fixturing system was used as a production harness lay up board the resolution is defined by the customer in which case placement within 5 mm is a typical requirement.

To prove the technology a resolution of 20 mm was chosen. It must be pointed out that defining resolutions is a very subjective matter as the requirements will depend on the routing path. Work is currently underway to establish the dimensional rules which will produce a loom of the required standard.

A series of fixture kit designs was produced to satisfy the dimensional constraints of accuracy and resolution imposed by the different cable harness criteria. The most versatile fixturing system was based on using eccentric blocks. This is demonstrated through the work of Neads (1986). However, the technology required to incorporate an electrical interconnection would make the system prohibitively expensive. To overcome this a simple system of preset components was adopted.

An electrical interconnection system was considered more appropriate than other signalling techniques on the basis of providing power to the hardware. Several systems were considered. These were a laminated interconnection system, a tensioned wire system and a printed circuit board system. They are illustrated in Figures 1, 2 and 3, respectively.

Electronics hardware

The electronics hardware provides the ability for the cable loom board to communicate with a computer. The information is simply in the form of an electrical continuity test on individual circuits. This technology is well established throughout the cable harness industry in the area of cable test.

The objective was to incorporate this facility into the fixture kit. This was achieved by building low cost individual microprocessor circuits into each node on the cable harness board. The electrical interconnection to these nodes is through the loom board and communication is by serial data transfer. This form of data

Figure 1. The laminated interconnection system.

Figure 2. The tensioned wire interconnection system.

communication common to computers and peripheral devices is also used in cable test equipment (Shelley, 1988). Figure 4 illustrates the technique.

System operation

Programs have been written to accommodate three separate functions: (1) data file handling, (2) communication with the robot manipulator and the loom board and (3) procedures for loom board build and cable harness build.

Two file handling procedures are used. The first builds up a file of loom board components from a data base of component features whilst the second contains all the routing and board layout information. Interaction between the two provides all the necessary detail for building a cable harness. The tree diagram in Figure 5 illustrates the procedures.

Figure 3. The printed circuit board interconnection system.

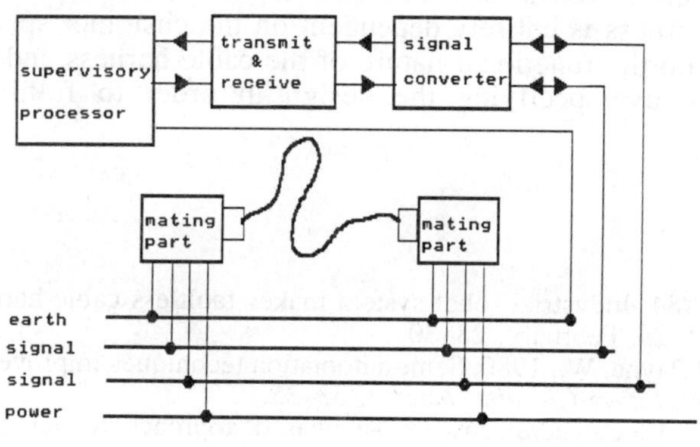

Figure 4. Through the board communication.

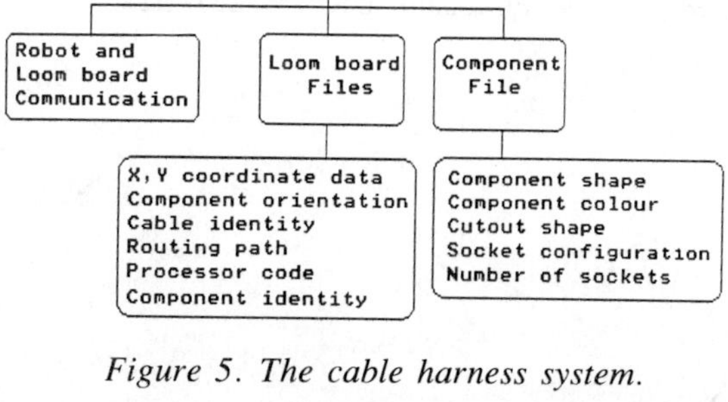

Figure 5. The cable harness system.

The communication links simply allow the transmission of serial data using ASCII codes with the loom board. Serial and parallel data transmission are used for operating the robot manipulator.

The procedure for loom board build has been developed to allow both manual build and robotic build. The manual build procedure uses a graphic display which provides information on choice and positioning of the component. This information is otherwise passed directly to the robot manipulator. On inserting a component the processor resident inside the component is interrogated by the supervisory computer. Each processor has a unique identity programmed into it and this identity must be read into the cable harness file in order to make it complete.

The procedure for cable harness build has been developed along the lines of the two separate build techniques. The first is the sequential build technique which involves a programmed plan of build. The second is the random build technique.

Conclusion

A flexible cable routing cell has been described which can form an integral part of cable harness manufacture whether it is in a machine-orientated environment or a manually intensive environment.

The combined benefit of mechanical flexibility and processing power has produced a loom board which can be used to help the operator build any cable harness from a set of routing data and component part data.

However, the effectiveness of the system to provide a fully integrated lay up and test of a cable harness is entirely dependent on the customer specification. This is dependent both on the functional nature of the cable harness and the ability of the customer in not overspecifying the design in order to fulfil these functional requirements.

References

Gibbons, R.D., 1984, Industrial robot system makes faultless cable harnesses. *Wire World International*, **26**, February, 28–30.

Goldstein, B. and Payne, W., 1980, Semi-automation techniques improve wire harness board assembly. *Insulation Circuits*, August, 33–35.

Mills, J.B., 1984, Electricad/electricam—a phased approach to automated aircraft loom manufacture. *Aerospace*, March, 19–20.

Moore, B.E., 1986, Self-storing board build system for automated harness-cable assembly. *IBM Technical Disclosure Bulletin*, **28**, no. 12.

Neads, S.J., 1986, The automatic assembly of versatile fixtures. Ph.D. Thesis. Bath University.

Shelley, T., 1988, Software untangles design and testing. *Eureka Transfers Technology*, September, 49–50.

Warnecke, H.J., Schlaich, G. and Walther, J. 1985, Application of industrial robots for flexible wiring harness manufacture. *Proceedings of the Sixth International Conference on Assembly Automation*, Birmingham, UK, pp. 221–228.

13

A graphical approach to computer-aided stress analysis in the design of bandsaw teeth

S.R. Bradbury and M. Sarwar

School of Engineering Sheffield City Polytechnic, Sheffield S1 1WB, UK

Abstract

A programme of analytical and experimental work is currently being undertaken at Sheffield City Polytechnic in association with James Neill plc, to study the performance, life and design characteristics of bandsaw blades, with an objective of developing an improved tooth form through modifying the geometry.

Previous published work by the authors has outlined the procedures and given examples of the application of the finite element method to simulate the stress behaviour in standard production bandsaw blades using data acquired through cut simulation tests (Bradbury and Sarwar, 1988; Sarwar *et al.*, 1988).

This chapter describes how analysis of the extensive output from the stress data files of the models can be simplified by adopting a graphical approach in which nomograms are developed to predict the behaviour of bandsaw teeth over the range of prevalent cutting conditions. The nomograms, referred to as yield maps, can be used to quantify the cutting forces at yielding and to identify the position within the tooth structure at which yielding commences. They can be used to compare the yielding characteristics of current production and revised bandsaw tooth profiles, thus providing the bandsaw design engineer with a technique that is simple in application, but effective in assessing the strengths and weaknesses of tooth profiles.

Introduction

The three most popular methods of sawing are hacksawing, bandsawing and circular sawing. While each of the three sawing operations have their own merits, bandsawing has rapidly gained in popularity since features of low kerf loss, high metal removal rates and flexibility in cutting have given the process a major advantage.

To obtain the output, accuracy and reliability demanded by industry, it has been essential for the bandsaw machine manufacturers to incorporate with the machine designs innovations of feed control and materials handling which have led to highly efficient cutting off operations. It is not unusual for machine and bandsaw manufacturers to team up and adopt the systems approach by having sawing centres. These have taken advantage of the reduced automation costs and have incorporated

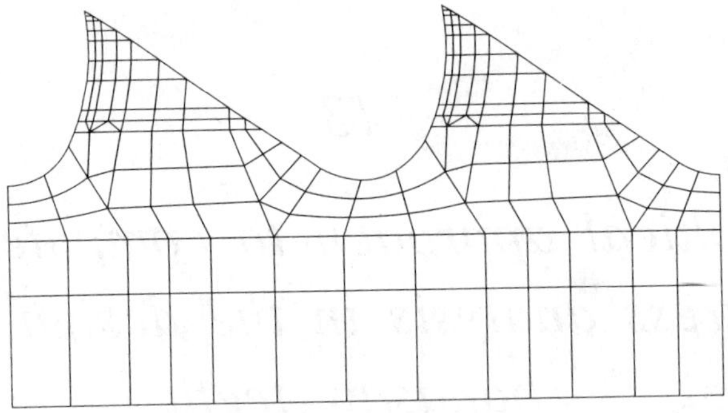

Figure 1. Twin-tooth finite element model.

material selections from the warehouse, specifying product length, cutting conditions, force per tooth, with continuous monitoring of the loads to give the best operating conditions and finally delivering the sawn length to pallets.

Whilst the saw machines have received considerable attention, very little effort has been directed into the development of the geometrical features of the blade. Although some material developments have been made from the earlier carbon steel to high speed steel bi-metal blades, geometrical features of tooth profile have not received the same attention. A three-year research programme supported by SERC/James Neill plc commenced at Sheffield City Polytechnic in 1987 to investigate the characteristics of the chip formation mechanism in multi-tooth cutting associated with bandsawing. It is hoped that the scientific approach will help the bandsaw designer to identify the geometrical features contributing to the overall performance and hence improve the blade design.

Computer-aided stress analysis of bandsaw teeth

Previous published work by the authors has discussed in detail the application of the PAFEC (Program for Automatic Finite Element Calculations) to analyse the stress behaviour developed in bandsaw teeth during cut simulation tests (Bradbury and Sarwar, 1988; Sarwar *et al.*, 1988). The work described in this chapter makes use of similar finite element models. For the reader to appreciate the current approach, a resume of the finite element model is given.

Two-dimensional, twin-tooth, finite element models of the type shown in Figure 1 were constructed using six- and eight-noded isoparametric plate elements to represent the structure of each bandsaw tooth form. The stiffness of adjacent teeth and backing strip regions were simulated in the model using restraints. Isotropic material properties were assumed and specified for each element within the model. When modelling bi-metal blades, the regions of the tip, backing strip and laminates through the weld affected zone were represented by different sets of material properties.

The models were loaded along the rake face over the tooth–chip contact length. The total cutting force (horizontal) and the total thrust force (vertical) were discretized and apportioned to all corner and mid-side nodes along the tooth–chip contact interface in accordance with Trent (1977) and the distributed loading criterion (PAFEC Ltd, 1984).

The described models have been used to simulate elastic material behaviour within the tooth structures. Analysis of the model solution revealed whether the yield strength of the constituent materials had been exceeded and if so, in which region. This was achieved by comparing the yield strengths of the tip and backing strip materials with the equivalent stress values based on von Mises yield criterion determined in the model solution.

Limitations in the finite element model solution

The model solution can be presented in either graphical or tabulated format. Interpretation of a model solution by these formats is acceptable for most finite element stress analyses; however, problems arise in analysing numerous models of similar shape, subjected to similar loading conditions. In context of the current application, the graphical format lacked sufficient resolution to distinguish between the stress behaviour within a tooth structure that resulted from slight modifications in tooth profile or loading conditions.

An extensive stress data file is created for each model solution that documents the stress behaviour at each node within each element within the finite element mesh. Consequently for a structure that contains many elements, such as the mesh shown in Figure 1, the stress data file is huge and the interpretation time consuming. The magnitude of the problem is compounded when considering the number of model solutions required to simulate a complete range of cutting conditions.

Development of a simplified approach—yield maps

Analysis of the stress data file has been simplified by considering the behaviour at a number of strategic elements within a model when subjected to a limited range of cutting conditions. The range of cutting and thrust forces was such that the yield strength of the tooth material was exceeded.

At each strategic element, the equivalent stress based on von Mises yield criterion was determined for each combination of cutting and thrust force. A nomogram was formulated to locate a contour where the equivalent stress equalled the yield strength of the material. Superimposition of the contours developed for other strategic elements formulated a nomogram, referred to as a yield map, which was used to identify potential strengths and weaknesses of a tooth design over the range of cutting conditions.

Preliminary analyses in the development of yield maps

Yield maps developed by considering the loads generated during cut simulation tests and the stress distributions predicted by the computer model. These allowed suitable ranges of cutting and thrust forces to be determined and strategic elements to be selected throughout the finite element mesh.

Determination of the load ranges

By inspection and stress analysis of teeth following cut simulation tests, a 4 teeth per inch (TPI) pitch tooth was shown to commence yielding in the range of 500–800 N for both cutting and thrust forces. In the development of yield maps, cutting and thrust forces were imposed over an extended range to cause yielding at all the strategic element locations, besides the critical location. For a 4 TPI pitch tooth form, cutting and thrust forces ranging from 400 to 1600 N were imposed in 400 N increments. Thus, 16 solutions to the finite element model were required to develop the yield map for a particular tooth form.

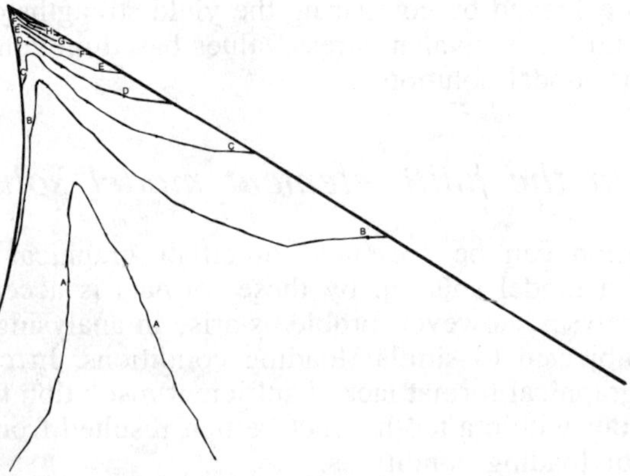

Figure 2. Distribution of equivalent stress in a bandsaw tooth tip. Loading conditions: cutting force, F_c *= 0·8 kN; thrust force,* F_t *= 1·2 kN. Equivalent stress (kN mm⁻²): A, 0·52; B, 1·58; C, 2·63; D, 3·69; E, 4·75; F, 5·80; G, 6·86; H, 7·91; I, 8·97; J, 10·02.*

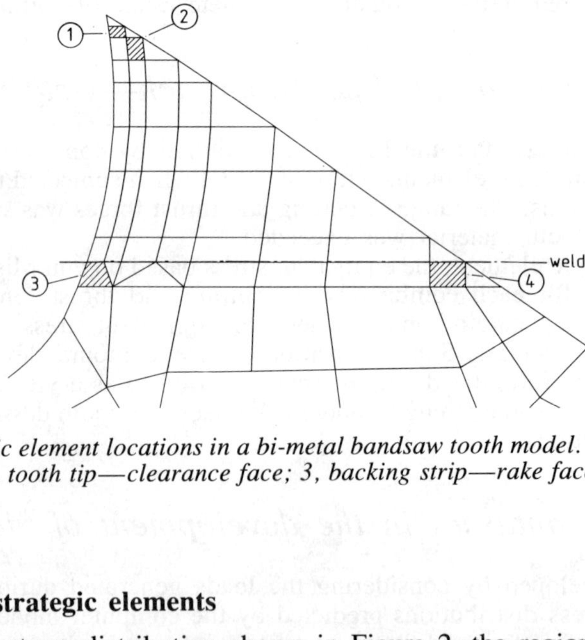

Figure 3. Strategic element locations in a bi-metal bandsaw tooth model. Elements: 1, tooth tip—rake face; 2, tooth tip—clearance face; 3, backing strip—rake face; backing strip—clearance face.

Positioning of strategic elements

Considering the stress distribution shown in Figure 2, the regions of high stress density are shown to be at the tooth surface and maximum at the tooth tip.

In the development of yield maps for bi-metal blades, strategic elements were selected at the surfaces of the rake and clearance faces in both the tip and backing strip regions. Elements near to, but not at, the tip were selected to avoid inaccuracies which arise from modelling a sharp tooth profile. Elements in the backing strip region were selected adjacent to the weld.

Development of yield maps

The stress behaviour at the strategic elements was determined for each of the 16 load cases applied to the model. From the mean principal stress values determined

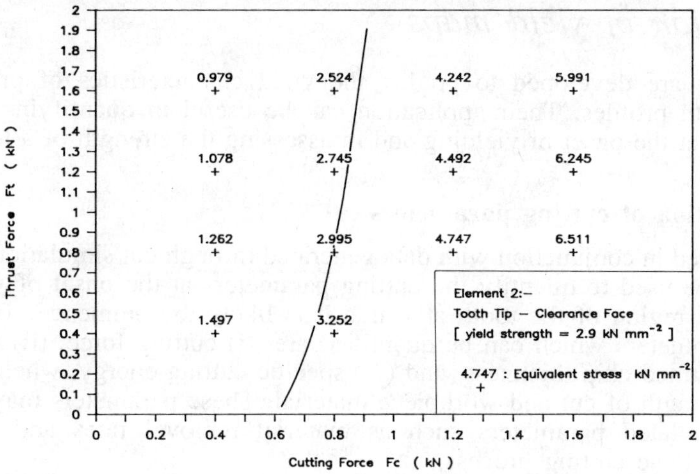

Figure 4. Yield contour developed for a strategic element.

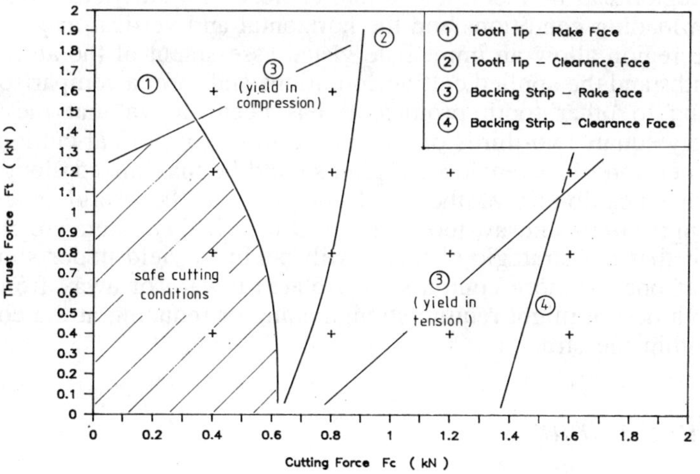

Figure 5. Yield map developed for a bi-metal bandsaw tooth form.

for each element, the equivalent stress based on von Mises yield criterion was evaluated. A comparison between these values and the yield strengths of the constituent tooth materials determined whether the stress levels had effected the onset of yielding.

Figure 4 shows the equivalent stress values for a typical strategic element related to the applied cutting and thrust forces. Points at which the equivalent stress equalled the yield strength of the element material were determined by linear interpolation, and joined to form a yield contour.

Superimposition of yield contours relating to all the strategic elements enabled the yielding characteristics of a tooth form to be determined in relation to the applied cutting and thrust forces over a full range of loading conditions, shown in Figure 5.

Application of yield maps

Yield maps are developed to predict the yield characteristics of production or revised tooth profiles. Their application can be useful in quantifying the cutting parameters at the onset of yielding and in assessing the strength of a tooth design.

Quantification of cutting parameters

When applied in conjunction with data generated through cut simulation tests, yield maps can be used to quantify the cutting parameters at the onset of yielding and identify the region of the tooth at which it is likely to commence. The principal cutting parameters which can be quantified are: (i) cutting force, (ii) thrust force, (iii) undeformed chip thickness, and (iv) specific cutting energy; when relating to a specific length of cut and workpiece material. These parameters may be used to determine related parameters such as material removal rates and the relative efficiency of the cutting process.

Assessment of strength in bandsaw tooth design

Yield maps indicate a region of 'safe cutting conditions' for the considered tooth form. This region can be identified as that enclosed by the yield contours relating to minimum loading conditions and the horizontal and vertical axes. The size and shape of the region allow an immediate visual assessment of the ability of a tooth design to withstand the applied cutting conditions and form a comparison with yield maps relating to other tooth profiles. It has been shown that the thrust force approximately equals two-thirds of the cutting force for normal cutting conditions, thus, the dimension of the enclosed region should be maximized along this vector.

The relative positioning of the yield contours can be useful in assessing the relative strengths of a bandsaw tooth design. If the yield contours are clustered then it is implicit that the strategic elements will begin to yield under similar loading conditions. If one, or more contours are isolated toward or away from the origin, then the tooth design might require strengthening or reducing at the corresponding positions within the structure.

Concluding remarks

Yield maps can be developed for any production or modified bandsaw tooth design and present the bandsaw design engineer with a unique graphical interpretation of the yield characteristics of a specific tooth form. Interpretation of a yield map allows:

(i) the cutting parameters at the onset of yielding to be quantified,
(ii) regions of the tooth structure where yielding is likely to commence to be identified,
(iii) the relative strengths and weaknesses of regions within the tooth structure to be assessed,
(iv) the size and shape of the 'safe cutting regions' associated with tooth profiles to be compared.

In the development of a revised bandsaw tooth profile, the following features make the described approach attractive:

(i) unique and easily assimilated graphical representations for each tooth profile,
(ii) no initial production of prototype teeth required,
(iii) relatively low man hours involved in developing the computer model and the yield map for a tooth profile,
(iv) adaptability of the technique to other multi-point cutting operations.

Acknowledgements

The authors thank James Neill plc and the SERC for their support in sponsoring the work, Mr Derrick Taylor (head of product development) and Mr Keith Pascoe (product development engineer, James Neill plc) for their continued encouragement throughout the work, and Mr Michael Dinsdale (research assistant) and Mr Paul Archer (senior technician, Sheffield City Polytechnic) for their assistance in the acquisition of technical data required for the preparation of this paper.

References

Bradbury, S.R. and Sarwar, M., 1988, Computer aided stress analysis in the design of bandsaw teeth. *Proceedings of the Fourth International Conference on Computer-Aided Production Engineering*, Edinburgh, November, p. 469.

PAFEC Ltd, 1984, *PAFEC Theory*, (Nottingham).

Sarwar, M., Bradbury, S.R. and Dinsdale, M., 1988, An approach to computer aided bandsaw teeth testing and design. *Proceedings of the Fourth International Conference on Production Research*, Sheffield, September, p. 494.

Trent, E.M., 1977, *Metal Cutting* (London: Butterworth).

SECTION II.6. MEASUREMENT

14

Two-dimensional optical Fourier transform for surface texture assessment

V.M. Huynh, S. Kurada and W. North

Department of Mechanical Engineering, University of Windsor, Windsor, Ontario, Canada

Abstract

An optical Fourier transform technique was developed in conjunction with a machine vision system to study the texture of machined surfaces in the roughness range up to 1·2 μm. Ground surfaces of different materials were used in the study and the spectra from the optical method were compared with those obtained by the stylus method. Correlation curves were obtained to indicate the relationship between the peak and RMS of the optical spectra and the average mechanical roughness Ra. Roughness assessment along the lay direction of the samples was also made with the optical technique and a trend was established.

Introduction

Surface texture assessment is an important part of the manufacturing process as it not only provides data for quality control of surface finish but also yields valuable information for monitoring the working condition of machine tools.

Surface texture can be studied by using either contact or non-contact methods. The stylus profilometry typifies the contact method which applies mainly to one-dimensional surfaces. This method measures the surface profile from which a Fourier spectrum is obtained to separate different spatial frequency components of the surface.

Non-contact roughness measurements involve mainly optical techniques which include laser speckle, interferometric measurement, electron microscopy, light sectioning, light scattering and optical Fourier transform methods. Among these the Optical Fourier Transform (OFT) is one of the most attractive techniques for the extraction of texture information from a surface. For relatively smooth surfaces, it was shown (Pernick, 1979) that there is a good correlation between the OFT and the power spectrum of the surface profile. It was also established that a direct interpretation of the OFT is possible when the ratio of roughness amplitude to illumination wavelength is below 0·2 (Thwaite, 1979, 1980, 1982a). In the above studies, a He–Ne laser (wavelength $\lambda = 0{\cdot}632$ μm) was used to examine the ground and lapped surfaces with roughness range up to 0·1 μm. By using longer illumination wavelengths, e.g. infrared wavelength ($\lambda = 3{\cdot}39$ μm) Thwaite (1982b) was able to extend his work to the roughness amplitude of 0·6 μm.

The work on OFT so far is strictly one dimensional and covers only a small range of roughness. This restricts the application of this technique to only a few finishing operations. The purpose of this work is to (i) increase the applicable surface roughness range for this method, (ii) extend the OFT technique to two dimensional. The aim here is to develop a versatile OFT technique to assess the surface texture of most machining processes.

The proposed method involves the use of a two-lens optical system to produce a Fraunhofer diffraction pattern of a surface which is illuminated by a He–Ne light source through a circular aperture. The pattern is then captured and digitized by a machine vision system to derive a spatial power spectrum. This optical spectrum is compared to the spectrum obtained from mechanical profilometry.

Grinding samples of different materials such as stainless steel, tool steel, brass, copper and aluminum were used for the study. Correlation between the mechanical roughness Ra and optical spectrum parameters such as root mean square (RMS) value and spectrum peak were obtained for these materials. Assessment of roughness along the lay direction was also made in terms of spectrum RMS which can be correlated to the specimen average roughness Ra.

Method and procedure

The scattering of light from a micro-rough surface which is illuminated by a monochromatic light source is determined by the diffraction phenomena of light waves. The surface in this case is modelled as gratings whose intensity distribution $g(x, y)$ on the $x - y$ plane is defined by the surface height at (x, y).

The Fraunhofer diffraction pattern of the gratings occurs at infinity and the amplitude distribution $W(f_1, f_2)$ of the pattern on the image plane is proportional to the Fourier transform of the gratings, i.e.

$$W(f_1, f_2) = C\, \mathcal{F}(g(x, y)) \tag{1}$$

where C = a proportional constant depending on the geometry of scattering,

$$\mathcal{F}(g(x,y)) = \iint_{-\infty}^{+\infty} g(x,y)e^{-2\pi i(f_1x+f_2y)}\, dx \cdot dy,$$

and f_1, f_2 = spatial frequencies of the components of $g(x, y)$ in the x and y directions.

This implies the grating function $g(x, y)$ is a combination of a series of sinusoidal waves of different amplitudes that relate to the Fourier transform by:

$$g(x,y) = \iint_{-\infty}^{+\infty} \mathcal{F}(g(x,y))e^{2\pi i(f_1x+f_2y)}\, df_1 \cdot df_2. \tag{2}$$

A lens or an optical system is used to bring the pattern at infinity to the focal plane (or Fourier plane) of the system so that the image can be digitized and processed. The experimental set-up used in this work is shown in Figure 1. It consists of a 2 mW He–Ne laser used in conjunction with a beam splitter to illuminate the surface through a 50 μm spatial filter. The transform lens is of 25 mm focal length and has a 0·25 numerical aperture. The second lens is of 100 mm focal length and is for enlarging and focusing the image of the pattern onto the image plane of a charge-coupled device (CCD) camera which has 480 × 300 line resolution. An imaging board with 512 × 512 × 8 bit resolution digitizes the image captured by the camera and stores it in one of the four frame buffers for subsequent processing by the PC-AT. The image of the pattern is displayed on a NEC Multi-sync video monitor for viewing.

To obtain the Fourier spectrum, the Fourier pattern is first digitized along the

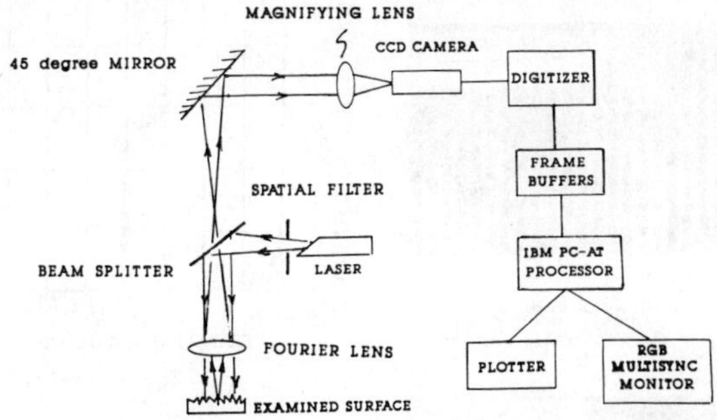

Figure 1. The experimental set-up for the optical Fourier transform method.

principal axes to derive a distribution of intensities of the pattern. The axial distance in the pattern is then transformed to a spatial frequency via the relationships:

$$f_1 = \frac{xM}{\lambda L}$$

$$f_2 = \frac{yM}{\lambda L}$$

where L = focal length of the transform lens, M = magnification of the system, and λ = wavelength of the He–Ne laser.

The present set-up gives a spatial frequency up to 200 cycles/mm.

Results and discussion

Periodic surfaces

Two standard calibrated surfaces (Ra = 0·5 and 3·0 μm) with known profiles were used to test the OFT technique and compare their spectra with those obtained by the mechanical method. The standard surfaces are of one-dimensional nature having regular and symmetric saw tooth profiles.

The Fourier pattern of the surface shown in Figure 2(a) is a series of dots aligned along an axis perpendicular to the direction of the lay [Figure 2(b)]. The intensity profile of the pattern along this axis is shown in Figure 2(c). The presence of a second order harmonic of relatively large amplitude for this sample, indicates a larger gain in the high order components. The shape of this spectrum can be related to the frequency spectrum obtained from the mechanical method; compare Figures 2(c) and 2(d).

A similar observation can be made with the standard sample having Ra = 3·0 μm (Figure 3). Again in this case, the amplitude gain is more pronounced at high order harmonics; compare Figures 3(d) and 3(e).

A series of stylus marks on the calibrated standard sample in Figure 4(a) adds a second dimension to the OFT pattern of the regular surface [see Figure 4(b)]. The pattern forms short columns of dots aligned along an axis parallel to the lay. The periodic nature of the stylus marks is revealed in the intensity distribution measured along the principal axis parallel to the lay [see Figure 4(c)].

Ground samples

The characteristics of a typical ground tool steel surface are shown in Figure 5. The spectrum has a large D.C. component superimposed on relatively small discrete

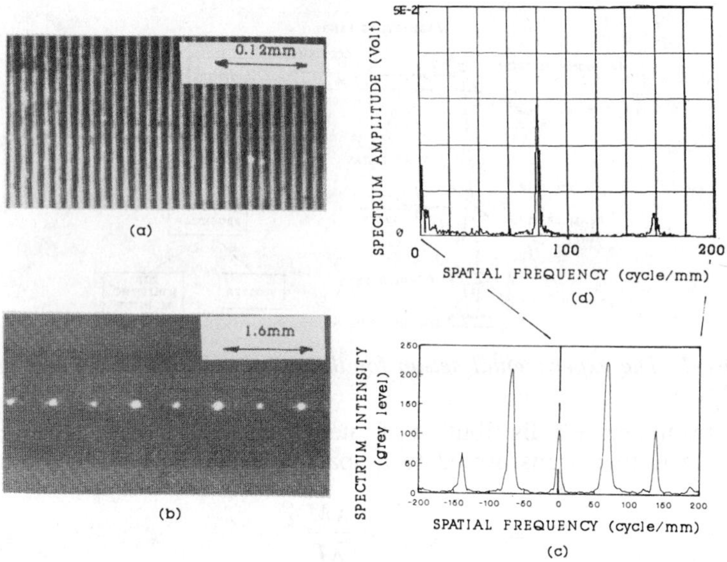

Figure 2. Typical characteristics of a standard calibrated sample (Ra = 0·5 μm): (a) the surface as seen on the microscope, (b) the Fourier pattern, (c) the power spectrum from the OFT, (d) the frequency spectrum obtained from the mechanical method.

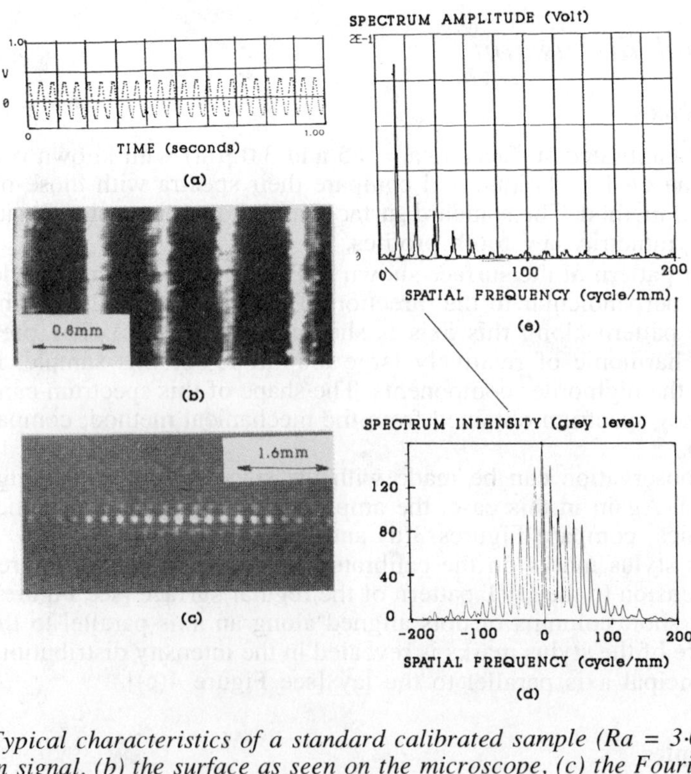

Figure 3. Typical characteristics of a standard calibrated sample (Ra = 3·0 μm): (a) the time domain signal, (b) the surface as seen on the microscope, (c) the Fourier pattern, (d) the power spectrum obtained from the OFT, (e) the frequency spectrum obtained from the mechanical method.

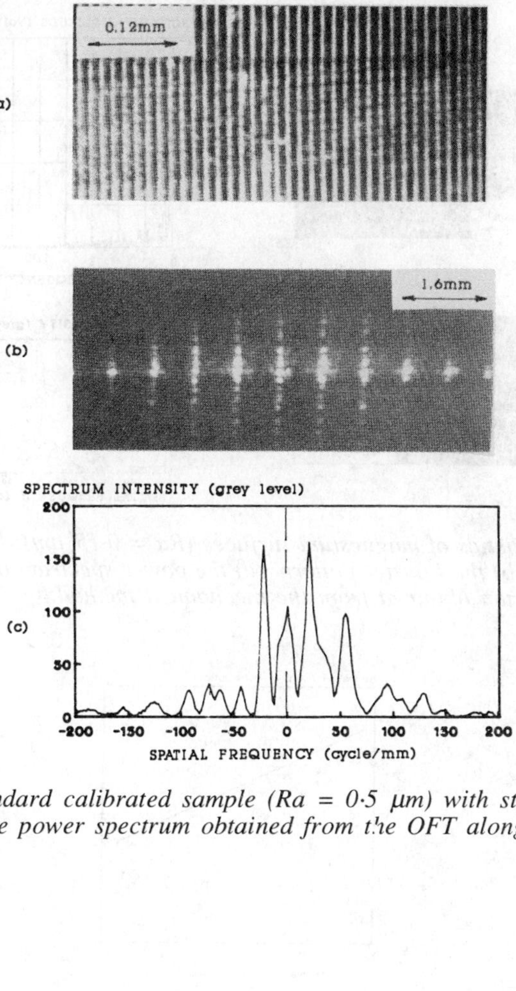

Figure 4. (a) The standard calibrated sample (Ra = 0·5 μm) with stylus marks, (b) the Fourier pattern, (c) the power spectrum obtained from the OFT along the lay direction.

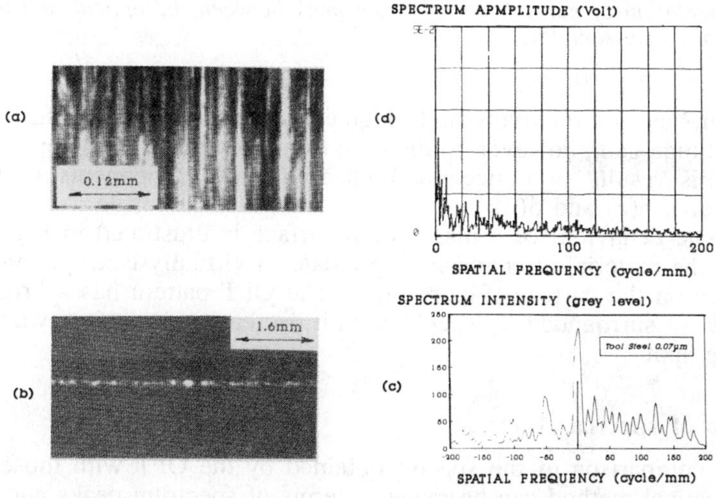

Figure 5. Typical characteristics of surface ground tool steel sample (Ra = 0·07 μm): (a) the surface as seen on the microscope, (b) the Fourier pattern, (c) the power spectrum from the OFT, (d) the frequency spectrum obtained from the mechanical method.

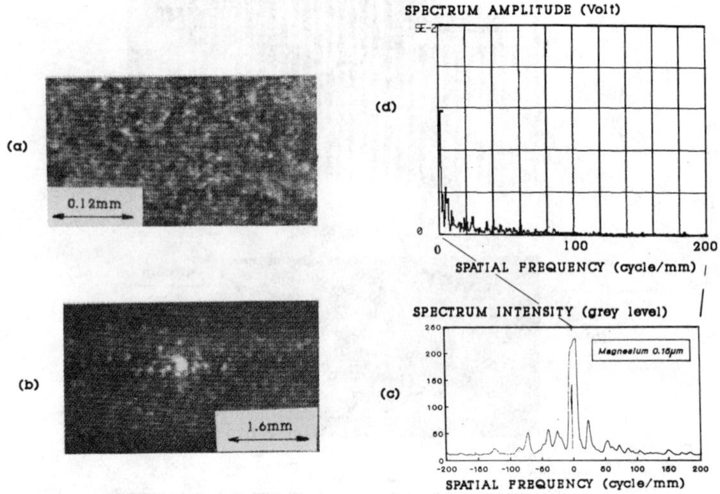

Figure 6. Characteristics of magnesium surfaces (Ra = 0·15 μm): (a) the surface as seen on the microscope, (b) the Fourier pattern, (c) the power spectrum obtained from OFT, (d) the frequency spectrum obtained from the mechanical method.

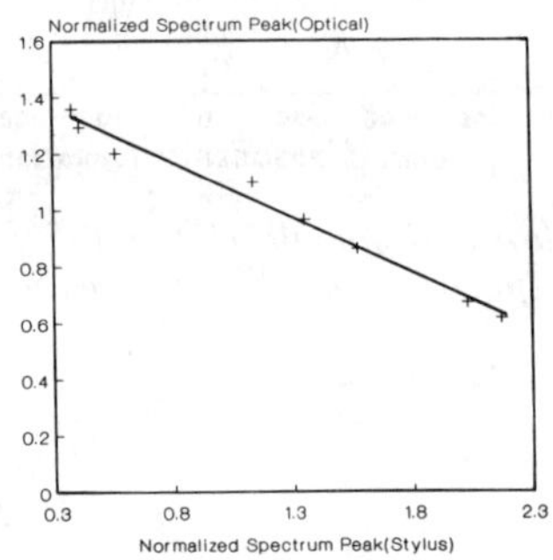

Figure 7. Correlation of normalized spectrum peak between the optical and mechanical methods for tool steel samples.

peaks. This indicates a relatively high degree of randomness in the surface. There is some amplitude gain, however, in the high frequency range of the optical Fourier spectrum. This results in noticeable high frequency components in the OFT; compare Figures 5(c) and 5(d).

The isotropic behaviour of a magnesium surface is illustrated in Figure 6. The oxidation of the material after machining creates a virtually isotropic and random grainy texture on the surface [Figure 6(b)]. The OFT pattern has a bright central D.C. component surrounded by speckles with decreasing intensity away from the centre of the spot.

Comparison of spectra

Quantitative comparison of the spectra obtained by the OFT with those obtained by the mechanical method can be made in terms of spectrum peaks and spectrum RMS. In this comparison process, the spectra are normalized with respect to the standard calibrated sample having Ra = 0·5 μm.

Figure 7 shows a linear relationship between the spectrum peaks of ground tool

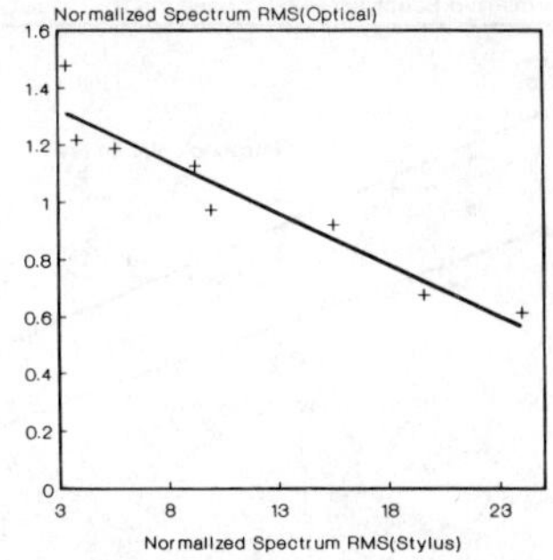

Figure 8. Correlation of the normalized spectrum RMS between the optical and mechanical methods for tool steel samples.

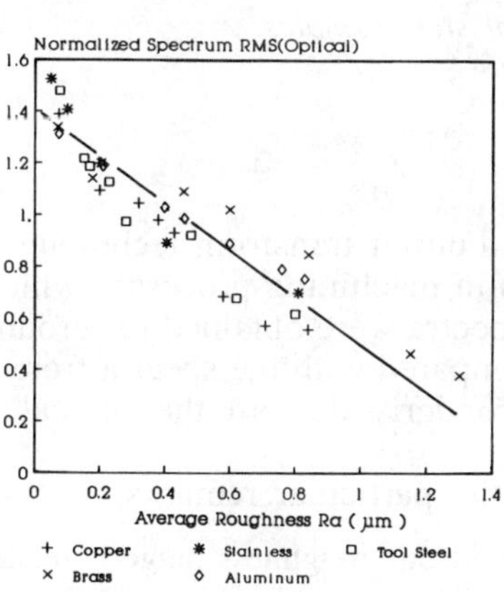

Figure 9. Correlation between spectrum RMS (optical) and surface roughness Ra for ground surfaces of five different materials.

steel samples calculated from the two methods. The slope of the curve, however, is negative in this instance.

A similar relationship is found to exist between the spectrum RMS of the same material for the two methods (Figure 8).

Optical spectra and roughness Ra

Since the OFT can be related to the frequency spectra from the mechanical method, one can use the OFT to describe the surface roughness. The RMS of the OFT power spectrum is arbitrarily selected as the optical roughness. The spectrum RMS values were obtained for machine ground samples of five different materials. They were plotted against the mechanical roughness Ra as shown in Figure 9.

An approximately linear relationship between the two variables is obtained for Ra up to 1·2 μm.

A similar optical roughness indicator can be established for the direction along the lay. Figure 10 shows the relationship between the optical spectrum RMS (along the lay) and the sample Ra value. Again a linear curve is obtained. This curve is nearly parallel to the correlation curve for the direction perpendicular to the lay (see top curve in Figure 10).

The spectrum peak can also be used as an optical roughness parameter and a similar correlation curve (as mentioned above) can be obtained.

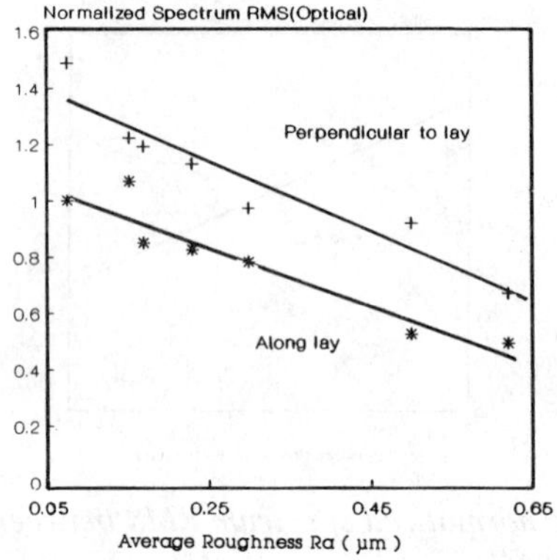

Figure 10. Correlation between the normalized spectrum RMS (optical) along lay and surface roughness Ra for tool steel samples.

Conclusions

A two-dimensional optical Fourier transform technique was developed to examine the surface texture of parts in machining processes which cover a roughness range up to 1·2 μm Ra. Fourier spectra were obtained for ground samples of five different materials and then were compared with the spectra from the stylus method. Optical roughness parameters were derived from the optical spectrum to quantify the roughness.

Following are some of the particular remarks:

(a) The intensity of the OFT in this roughness range is related to the frequency spectrum obtained by the stylus method.
(b) There is an increase in gain for the high order harmonic components in the OFT, especially at high roughness.
(c) The RMS and peak of the OFT spectrum are proportional to those obtained by the mechanical method.
(d) Linear correlation curves between the optical roughness parameters and Ra were obtained for surface ground samples of five different materials.
(e) The optical roughness correlation curves appear to be independent of the material in the range of samples tested.

The OFT technique as developed is shown to be a convenient way for assessing surface texture. It can be adopted for use in the in-line inspection of manufacturing processes where contact measuring techniques fail. Texture information is a direct measurement of the manufacturing process and thus can be used to detect tool wear, chatter, machine unbalance, etc.. Such data are essential for monitoring and controlling machining processes in a production environment.

Acknowledgement

This research project is supported by the Natural Science and Engineering Research Council of Canada under grant No. A5286.

References

Pernick, B.J., 1979, Surface roughness measurements with an optical Fourier spectrum analyzer, *Applied Optics*, **18**, 796–801.

Thwaite, E.G., 1979, The direct measurement of the power spectrum of rough surfaces by optical Fourier transformation, *Wear*, **57**, 70–80.

Thwaite, E.G., 1980, Power spectra of rough surfaces obtained by optical Fourier transformation, *Annals of CIRP*, **29**, 419–422.

Thwaite, E.G., 1982a, A quantitative comparison of the wavelength spectrum of a surface obtained by optical Fourier transformation with calculations from profile measurements, *Wear*, **83**, 181–187.

Thwaite, E.G., 1982b, The extension of optical angular scattering techniques to the measurement of intermediate scale roughness, *Annals of CIRP (Annales du Collège International pour L'Étude Scientifique des Techniques de Production Mécanique)*, **31**, 463–465.

15

Measurement of robot-cut grooves in glass

M.R. Morshedizadeh,[1] C. Wykes[1] and M. Howarth[2]

[1]*Department of Production Engineering and Production Management, University of Nottingham, University Park, Nottingham, NG7 2RD, UK and* [2]*Department of Industrial and Production Engineering, Trent Polytechnic, Nottingham, NG1 4BU, UK*

Abstract

A method of measuring the depth of grooves in robot-cut glass is reported. A fringe pattern is generated using a laser interferometer, and this is projected obliquely onto the surface of the glass, where they are viewed using a video camera. Phase-stepping is carried out on these fringes where the phase value at every pixel location on the image is measured. This phase value is directly related to the topography of the surface, and thereby the depth of the groove is found.

Introduction

The cutting of patterns on crystal glassware is a highly skilled manual operation requiring many years of training. A robot glass-cutting system has been developed to compete with Eastern European glass production and this has led to the need for a system which will inspect the finished glasses. As part of this inspection an optical system has been developed to assess the individual cuts produced by the robot system.

An image of the cut is formed on a TV camera. The length of the groove can be found directly from the length of its image. The depth of the groove is measured using a structured light method where parallel planes of light generated using a Michelson interferometer and a He–Ne laser are projected onto the groove. By applying the phase-stepping method to the projected fringes, the depth of the groove can be measured.

Contouring using fringe projection

If a grid of light is projected onto a surface, the shape of the grid which is observed on the surface is determined by the topography of the surface. When the surface is flat, the grid lines are straight, while if the surface is curved, the form of the grid is curved in accordance with the curvature of the surface. If the position of the grid lines can be measured and the illumination geometry is known, then the surface topography can be measured.

Considerable difficulties are encountered if conventional image processing techniques are used to identify grid line positions because of variations in the illumination level, variations in the brightness of the surface and optical noise in the system. Phase shifting is a technique which has been widely developed in recent years to determine the phase of sinusoidally varying light intensity fields—it is used in holographic and speckle interferometry and in Moire topography as well as in conventional interferometry.

Suppose three fringe patterns are stored in the computer where each pattern is shifted by 1/3 of the grating pitch in its own plane. Three intensity levels I_1, I_2, I_3, are stored at each pixel location. The phase, Θ, of the contour pattern at a given pixel location is given by the following equation

$$\Theta = \tan^{-1}\frac{\sqrt{3}(I_1 - I_2)}{(2I_3 - I_2 - I_1)}\,. \tag{1}$$

By evaluating this equation at every pixel location, the phase, Θ, of the fringe pattern at each point in the image and this phase value can be related directly to the topography of the surface using the equation

$$z(x) = \frac{x}{\tan\Theta} - \frac{\Theta(x)L}{2\pi\sin\phi}$$

where $z(x)$ is the position of the surface along the viewing direction, L is the width of the projected grating, ϕ is the angle at which the fringes are projected onto the surface and $\Theta(x)$ is the measured value of the phase at a point x in the image.

It can be seen that eqn (1) gives the value of the phase Θ only to within $\pm\,\pi/2$. By examining the relative value of the three intensity terms it is possible to extend this range to $\pm\,\pi$. A series of curves varying between $-\pi$ and $+\pi$ will then be obtained and a process called phase unwrapping is then required to assign the absolute values to the phases. In the absence of noise, this process would be fairly straightforward—discontinuities in the phase of about π would correspond to points where the phase changes by either π or $-\pi$; the sign of the jump could be inferred from the slope of the function. However, the results are not noise-free and software has been written to perform the phase unwrapping in the presence of noise. The following steps are performed:

(1) The difference between the current pixel intensity and the preceding one is calculated and the result stored in an array.
(2) Phase jumps will be observed in the difference data as spikes varying from zero to $+\pi$ or zero to $-\pi$, whereas noise in the data will give spikes which go from zero to π to $-\pi$ and then back to zero. This enables the jumps in phase to be distinguished from noise terms.
(3) The sign of the phase change is determined by whether the spike is positive or negative.
(4) The values of phase are thus found for the whole image.

Apparatus layout

The sinusoidal grid pattern is generated using an interference pattern which has been generated by passing He–Ne laser light through an expanding lens and then into a Michelson interferometer. The output light from the interferometer is passed through a collimating lens. By adjusting the mirrors in the interferometer, parallel planes of appropriate spacing are generated.

The fringe pattern is projected onto the glass surface and an image is formed on an analogue camera; this image is digitized and stored in a frame store linked to a computer. Each frame consists of 512 × 512 pixels with 8 bits per pixel. Each point of an image may take an intensity value ranging from 0 (usually black) to 255 (usually white) (Morshedizadeh and Wykes, 1988).

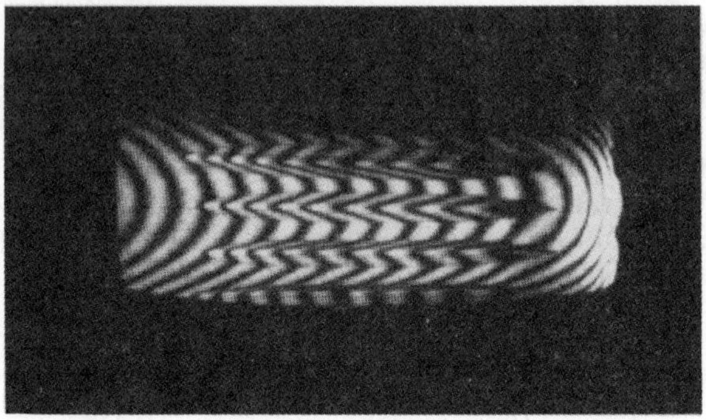

Figure 1. Fringes (100 μm) projected on the surface of a glass.

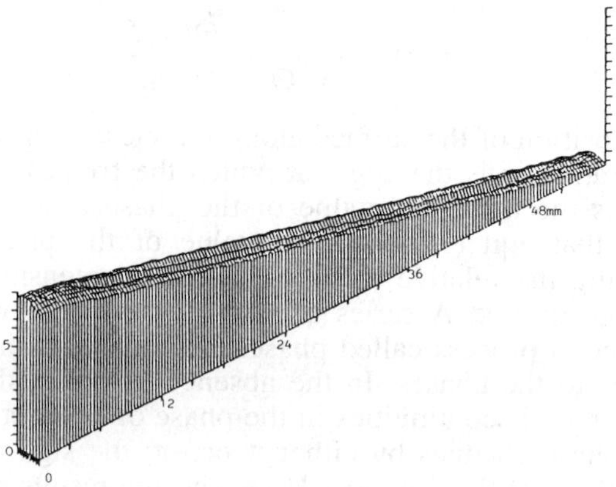

Figure 2. A phase map of one of the grooves.

Experimental method

The fringe spacing is measured by using a grating which is placed between the output of the interferometer and a screen. If the spacing of the fringe pattern and the grating are unequal, Moire fringes will be observed. The interferometer fringe spacing is adjusted by moving one of the mirrors in the interferometer until the fringe pattern on the screen is 'fluffed out', thereby giving an automatic adjustment of the fringe spacing, with the fringe spacing being equal to the grating spacing.

The magnification of the camera is found by relating the image plane coordinates to the object coordinates. This is done by replacing the object with a grid of known spacing. This grid is mounted on a flat surface and is rotated until the grid has maximum spacing in the image to ensure that it is normal to the viewing direction (Morshedizadeh and Wykes, 1989).

Experimental results

Figure 1 shows 100 μm spacing fringes projected on the surface of the glass. By windowing on the cut section of the glass and applying the phase stepping method to the projected fringes, the depth of the glass groove at different locations can be calculated. Figure 2 shows a phase map of one of the grooves in the glass.

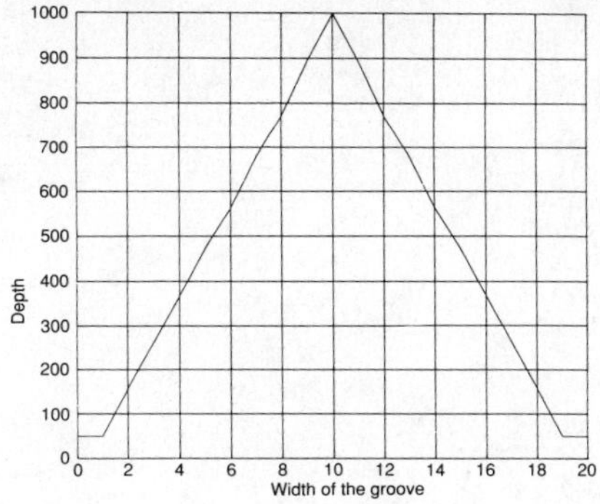

Figure 3. A cross-section of the measured cut: width (mm), depth (μm).

Figure 3 shows a cross-section of the measured cut on the glass as measured by the system.

It can be seen that the depth of the cut is 0·95 mm; the accuracy of the measurement is estimated to be 0·01 mm. The depth of the cut was measured using a depth micrometer with an accuracy of 0·01 mm and its value was also found to be 0·95 mm.

Conclusions

The proposed method has shown promise in measuring groove depth in glasses. In the current glass-cutting system, a probe is traced around the glass using the robot and this is then used as a datum to attain the correct depth of cut. The method discussed here could be used to monitor the depth of the cut as it is made and this could improve both the speed and the accuracy of the process.

Acknowledgements

The authors thank the Department of Production Engineering and Production Management for making facilities available for this work.

References

Morshedizadeh, R. and Wykes, C., 1988, Surface topography measurement using CAD generated referencing. *Proceedings of the Fourth National Conference on Production Research*, **3**, 19–29.

Morshedizadeh, R. and Wykes, C., 1989, Optical measurement of surface form using computer generated referencing, *Journal of Physics E: Scientific Instruments*, **22**, 88–92.

PART III
PRODUCTION AND INVENTORY MANAGEMENT

Production and inventory management

The papers in this section reflect the wider range of problems in production and inventory management being brought about by new approaches to the design and operation of manufacturing systems. Various methods are proposed for sequencing work flow through manufacturing cells, flexible manufacturing systems and job shops, and several papers deal with detailed and generic approaches to scheduling the bottleneck resources in such systems.

The need for better computer-aided decision support tools in planning and operating manufacturing systems is addressed by several authors, and others are concerned with methods to optimize profit policy, inventory policy and the economics of assembly operations. Two useful survey papers are included: one gives a valuable insight into the factors necessary for the successful implementation of material requirement planning (MRP) systems and the other gives an interesting comparison of differences in production practices between selected countries.

SECTION III.1. INVENTORY

16

The long term response of inventory holdings to technological developments and economic influence

C.D.J. Waters

Faculty of Management, University of Calgary, 2500 University Drive N.W., Calgary, Alberta, Canada T2N 1N4

Abstract

Views on stock holdings can vary considerably. At one extreme they can be considered valuable investments to be encouraged, while at the other extreme they are a waste of money which should be avoided. Current thinking in manufacturing industry supports the latter view.

Over many years a series of technological developments has aimed at reducing stock levels. Early work on inventory control models and demand forecasting was transferred to accounting machines and then computers. Real-time inventory control systems were introduced in the 1970s. Later on these systems were linked to manufacturing control systems which allowed Just-in-Time and zero-inventory production. The result of these developments has been a long-term decline in the proportion of output which is held as stock.

On a national scale this pattern can be observed in most industrialized countries. There are, however, other factors which affect stock holdings, particularly the prevailing economic environment. At times of rapid growth or high inflation stock levels tend to rise, while they decline during periods of recession or high real-interest rates. In this chapter changes in stock holdings of manufacturing industries in the United Kingdom are plotted over a number of years and related to technological developments and wider economic influences. Some comparisons are made with other industrialized countries.

Stock holdings in their historical context

Over recent years the idea that stock holdings are inherently wasteful of resources has become increasingly accepted. Many companies, encouraged by developments in 'zero-inventory manufacturing', believe that stocks hide problems which should be brought to light and eliminated. In this climate, it is easy to assume that minimizing stocks is the natural thing to do. In reality, this has only been attempted by a small sector of industry in the very recent past. Throughout most of history stock holdings have been considered beneficial; amassing physical goods at times when they were not needed has been a sign of personal or collective wealth.

Informal approaches to inventory control began when stocks of food were kept to cover periods of famine; those who held large stocks had the highest chance of survival. When inventories became associated with physical goods, the aim was still to maximize holdings which were seen as a measure of personal wealth. A well-known quotation from Pappilon (1677) suggests, 'The stock or riches of a kingdom doth not only consist of our money but also in our commodities and ships for trade and magazines furnished with all necessary material'.

The prevailing climate in these times was to collect as much stock as possible, and this seemed sensible when production and distribution of any commodity was uncertain. Industrialized countries are now cushioned from the need to buy goods when they are available rather than when they are needed. This change accelerated in the 19th century when production became more organized and distribution was improved. Provided a stable monetary system was in place, third parties could hold stocks and moderate the tensions between variable supply and demand.

By the turn of the 20th century supplies of goods to industrialized countries could be more or less guaranteed. This stability allowed a change of view which suggested that stock holdings were expensive and needed formal planning. This desire to plan inventories encouraged the development of relevant analyses and led to the Economic Order Quantity calculation often credited to Harris (1915). The emphasis in these systems was not the minimization of inventory, but the minimization of associated costs. Sometimes the difference was not fully appreciated and in the 1920s mistakes were made when companies tried to improve liquidity and reduced stocks to levels which made it impossible to operate effectively.

From the 1920s onwards the aim of inventory control has been to balance various factors: the cost of holding spare inventory when it is not needed is balanced by the cost of not having it when it is needed; spending too much on procurement by submitting small orders is balanced by the higher stocks with large orders; the risk of having unmet demand is balanced by the risk of keeping stock until it is obsolete, and so on. 'Scientific inventory control' is based on the observation that such balances can best be achieved by setting 'optimal' inventory levels. There are, of course, different opinions on what 'optimal' levels are, and such discussions form a notoriously difficult part of inventory control.

For some time, economists felt that a 'fixed accelerator' might be used to define an optimal stock level as some fixed proportion of sales (Abramovitz, 1950). This approach proved inadequate and a 'flexible accelerator' was suggested to allow for differences between aims and actualities, time delays, and so on (Lovell, 1961). More recently, increasingly sophisticated models and mathematical analyses have suggested flexible approaches which allow stock levels to be based on individual circumstances. Even within the past few years views have changed to suggest that optimal stock levels in some circumstances might approach zero.

This brief review has aimed at showing how views about stocks have changed over time. Until the late 19th century stocks were generally maximized. Then for most of this century inventory control has been based on a desire to minimize costs by achieving some optimal stock level. Unfortunately, this has not been easy. Considerable effort has been put into calculating or defining optimal stock levels, but it is only recently that adequate control procedures have given a reasonable chance of achieving such levels.

The means of control

Effective inventory control is based on both a desire to minimize inventory costs and an appropriate means of control. A more complete listing of requirements would include:

(a) a desire to minimize costs associated with stock holdings,
(b) the availability of models to identify optimal stock levels (and expertise in the use of these models),

(c) availability of accurate information in time to use the models and identify optimal stock levels,
(d) ability to control stocks to achieve optimal levels,
(e) knowledge of environmental influences.

The desire to control inventory costs has grown as this is recognized as a means of increasing competitiveness. The availability of models has also grown markedly since the 1920s, and these models are now so widely known that few organizations of any size will be unaware of them (this is, perhaps, one result of better management education). The next two requirements have been more elusive. Up to at least the early 1960s inventory plans were difficult to implement because of problems in collecting accurate information and effectively implementing control procedures. Despite developments in electro-mechanical accounting machines, stock control was essentially a manual process, with various systems of cards and handwritten records. Such systems had inherent deficiencies and it was certainly difficult to collect information in time to do anything about it. This problem was solved with the introduction of computers.

The advent of computers allowed information to be rapidly processed and distributed. Thus, forecast demand patterns, costs, supplier information and current positions could be monitored, with action taken when needed. Computers also did calculations which would have been previously impossible. Later developments in computer systems allowed inventory control to move from batch processing in the 1960s to on-line and interactive systems in the 1970s. In the 1980s different systems have been connected to form integrated systems of various types. The growth in computer systems has allowed the development and application of ideas which were previously of only theoretical interest. Overall, we can conclude that computers have provided a means of achieving the aims of inventory control.

Changing patterns of stock holdings

If there is uncertainty or lack of control, organizations will tend to be cautious and hold more stock than necessary. Only when an effective control mechanism is in place can they reduce stocks and remove the extra safety cushion. We could hypothesize, then, that from the 1920s to the 1960s inventory control would steadily improve and there would be a consistent reduction in stocks. Then in the 1960s, when computer systems became widely available, holdings would decline more quickly. Perhaps there would be further reductions in the 1980s with the introduction of zero-inventory manufacturing and other improved control systems.

It is difficult to discuss such changes in absolute terms, but we can relate aggregate, national inventory holdings to the Gross National Product (GNP). Then, we would expect to see inventory levels declining as a proportion of GNP. Figure 1 shows the ratio of total stock holdings to GNP for the past 40 years (Central Statistics Office, 1960–88). Reliable figures before this are more difficult to obtain, and would be distorted by wars in the 1910s and the 1940s, and by the severe economic recession between.

Figure 1 supports our hypothesis that the amount of stock held has been declining. At the end of the 1940s and into the early 1950s there was a rapid decline in stocks which can be attributed to the economy returning to normal after the second world war. From the early 1950s to the early 1970s there was a steady decline of the type suggested. Although the introduction of computers had a profound effect on inventories, the consequences would be felt over many years rather than as a short burst of change. In the early 1970s, however, there was a sudden pattern change, which was caused by the rapid increase in oil prices and the economic disruption which followed. Then the fall in stocks continued to the end of the 1970s, but the rate of decline seems to have accelerated in the 1980s.

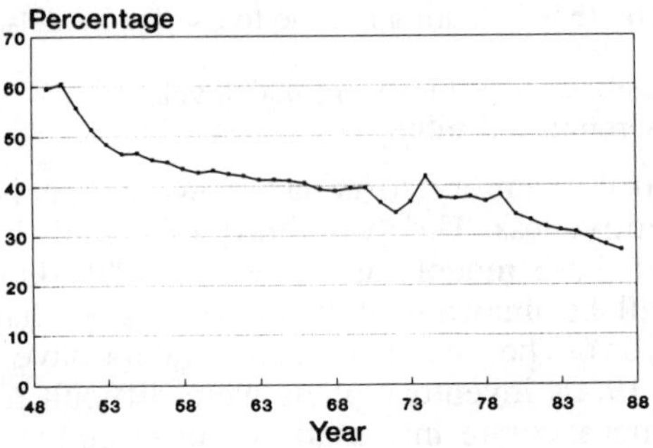

Figure 1. The percentage of Gross National Product held as stock between 1948 and 1988.

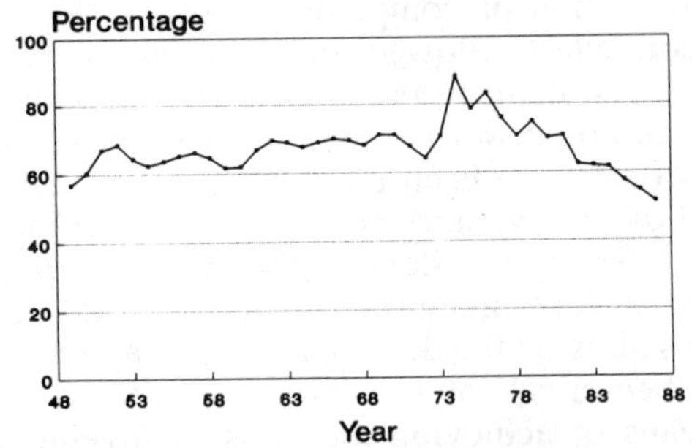

Figure 2. Percentage of output held as stock by manufacturing industry (1948–88).

Manufacturing industry has undergone major changes in the past 20 years. Although it contributes only 20% of GNP, it holds a large proportion of the nation's stocks. We would expect manufacturers, who have put a lot of effort into improving their efficiency, to be particularly careful about stock holdings. The introduction of MRP, MRP II, JIT and related systems in the 1970s and 1980s has enhanced the view that stocks in manufacturing companies can be reduced to a minimum or even eliminated. We would, therefore, expect to see manufacturing stocks falling, in contrast to, say, wholesalers and retailers who might be improving services and actually increasing their holdings. Another relevant factor here is a change of law in 1984 which stopped stock appreciation being set against corporation tax.

Figure 2 shows the proportion of output held as stock in manufacturing industry (Central Statistics Office, 1960–88). As always, some caution must be used in interpreting the figures. 'Manufacturing industry', 'stocks' and 'output' may all be interpreted in different ways, and values can be based on a number of different conventions.

It is interesting to see that manufacturing industry did not reduce its stock holdings during the 25 years from 1948 to 1973. In fact, there was a continued slight rise. It might be that stock holdings were already at optimal levels, but this seems unlikely. It would be more reasonable to suggest that inventory control systems were not being used effectively. It might be, for example, that inventory models were more appropriate for services, or that manufacturers could not risk disruption caused by shortages, or that work in progress could not be reduced, or that manufacturing was working well enough as it was. Whatever the reason, it seems that inventory control was not well implemented before the 1970s. It was only after this

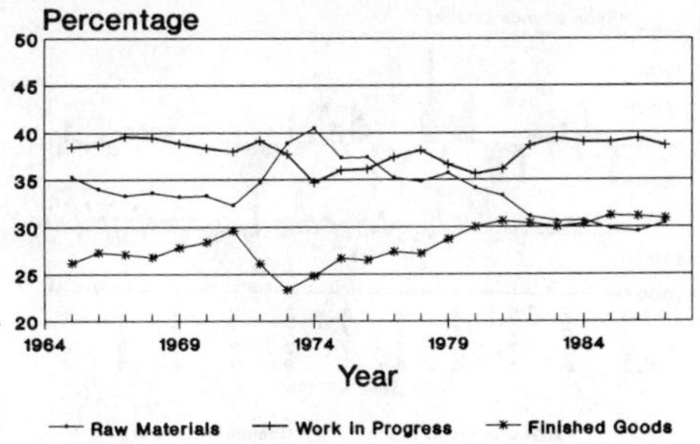

Figure 3. Changes in the proportion of raw materials, work in progress and finished goods held by manufacturers.

severe economic disruption that manufacturing joined the trend, and there has now been a steady decline in their stocks for the last 15 years.

It might be argued that recent reductions in the amount of stock held by manufacturing industry is a result of better management in general, or is a specific benefit from MRP, JIT, and other low inventory systems. Alternatively, it might be suggested that this is only an effect of short-term decisions to deal with prevailing circumstances. If the amount of stock held is classified as raw materials, work in progress or finished goods, we might be able to draw some conclusions about the effectiveness of inventory control. If, for example, changes in stock levels are brought about by long-term strategic plans or improved inventory control techniques, we would expect the amount of work in progress and raw materials to decline. Conversely, rises in these might show that short-term capacity adjustments are being made to cope with unplanned changes. Similarly, stocks of finished goods might decline with planned changes or increase if goods are being made which cannot be sold. Figure 3 shows the change in proportion of stock types over the past 20 years (Central Statistics Office, 1966–88a,b).

The patterns in Figure 3 are not very clear, but there is some suggestion that stocks of finished goods are slowly rising, while raw materials and work in progress are declining. Regression lines through the data show annual increases of 0·22% for finished goods and falls of 0·02% and 0·2% for work in progress and raw materials, respectively. Although small, these changes are consistent with manufacturers exerting better control over those stocks which are most easily handled.

In the early 1970s there was some discontinuity superimposed on the general trends, with raw material stocks rising sharply. There are two explanations for this:

(1) Shortages and rapidly increasing prices encouraged manufacturers to hoard available materials and thus ensure continued production.
(2) A rapid fall in production reduced the expected demand. Work in progress dropped suddenly, again showing industry making adjustments for reduced demand by cutting back on production rates.

Economic effects and the business cycle

Stock holdings are clearly following underlying trends, but superimposed on these are shorter-term fluctuations. Many of these are caused by wider economic influences over which individual organizations have no control. The sudden increase in oil costs in 1973 had an obvious effect on manufacturers, and no matter how good their inventory policies were they would have been disrupted by the scale of changes. There are other factors which have a less dramatic influence, often related to general business cycles, and including interest rates and inflation.

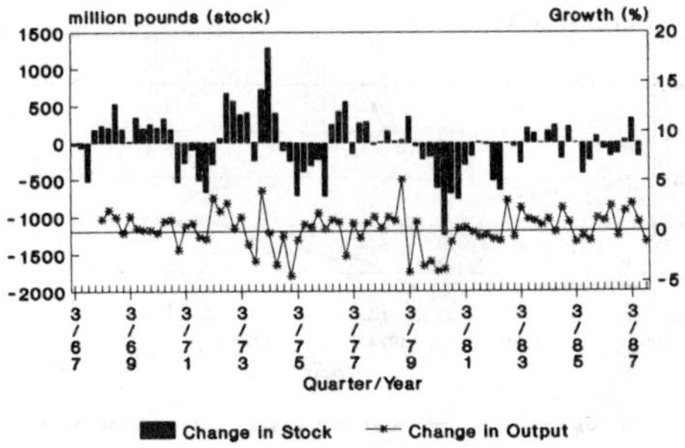

Figure 4. Changes in stocks (1980 prices) and growth of output for manufacturing industry.

A traditional view of business cycles starts with industry being optimistic about the future. Sales are expected to rise, so production is increased to match perceived future demand. Then inventories build up as current sales lag behind production, and at some point industry loses confidence and cuts back on production to deplete the excessive stocks. This causes a decline in the economy, which only picks up again when stocks are lower and production is not meeting expected demand. A precise relationship between stock levels and business cycles has not yet been found, and there is a general belief that each business cycle is in some ways unique. Nonetheless, there is a widespread belief that long-term business cycles and stock level are closely related, and in particular that inventory levels, as one of the easiest factors to change, tend to fluctuate more than the business cycle itself.

One view has fluctuations in inventories actually causing business cycles. Klein and Popkin (1961) suggested that if 75% of inventory variations had been controlled in the United States between the wars there would not have been any recessions. This kind of study led to suggestions for prohibiting wide fluctuations in inventories by taxes or other meansures, but these have never been implemented, mainly because of difficulties in defining 'excess' fluctuation (Silver and Peterson, 1985).

The general state of the economy is clearly important for stock holdings. There was an extreme illustration of this in the early 1970s when industry suddenly went through dramatic changes. Many stock items became unobtainable or the prices rose very quickly. This encouraged organizations to return to the system of grabbing items when they became available and holding on to them until they were actually needed. More generally, when the economy is in a decline we would expect inventories to change with, for example, stocks of raw materials and finished goods rising.

Figure 4 shows the changes in aggregate stock holdings of manufacturing industry (in constant 1980 prices) against the change in output over the past 20 years (Central Statistics Office, 1966–88a,b). There does seem to be a connection between these graphs, with, for example, declines in output preceding reduced stock levels in both 1974 and 1979. It is, however, very difficult to describe this connection and details of the relationship have not yet been found.

Moving to specific economic factors, it may be suggested that when interest rates are high, stock levels will be low. In one sense this is an unreasonable argument as holding stock costs 25–30% of value a year and small variations in interest rates would have no significant effect. Figure 5 shows the change in stock level against the average bank interest rate (Central Statistics Office, 1966–88a,b). Again the best we can say is that there appears to be some link between the two, but details of the relationship are hard to identify. Symptoms include the high interest rates which preceded stock reductions in both 1974 and 1979.

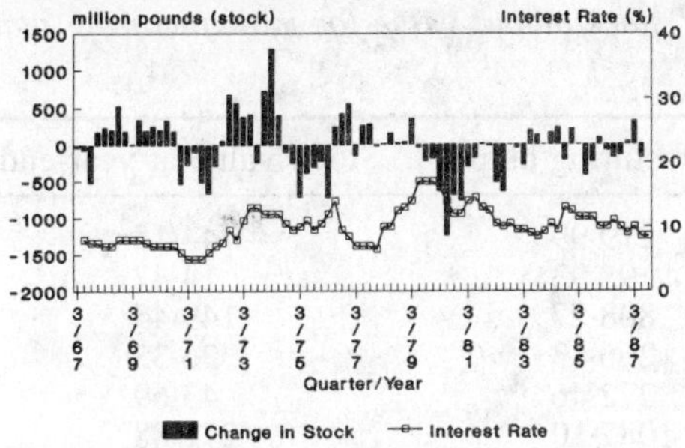

Figure 5. Changes in stocks for manufacturing industry (1980 prices) and average bank interest rates.

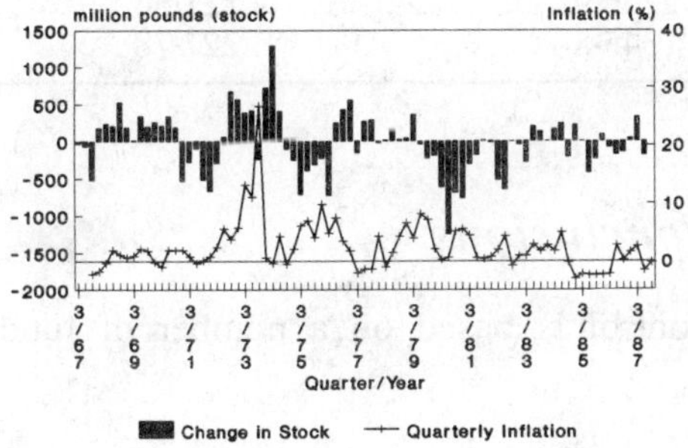

Figure 6. Changes in stocks (1980 prices) and inflation affecting manufacturing industry.

Another obvious influence on stock levels is inflation. The real cost of holding inventory can be discounted by an allowance for inflation, and during periods of high inflation we would expect stock levels to rise and decline again when inflation reduces. Figure 6 shows the changes of stock level against inflation (Central Statistics Office, 1966–88a,b). There are many measures of inflation and the one shown is the average change in price of materials and fuels purchased by manufacturing industry. Again we are forced to conclude that although there appears to be some relationship the details are not as clear as we would hope.

From these figures we can draw the general conclusion that inventory levels are certainly affected by wider economic influences, but the precise form of the relationships is very difficult to find. Simple regression and decomposition models do not give useful results. At present there are few pointers which help to describe the variations in aggregate stock holdings.

International comparisons

We have concentrated on stock holdings within the UK, but similar analyses could be done for other countries. Such data is collected by the United Nations and it is interesting that the amount of stock reported in different countries varies quite widely. Comparisons between countries depend to such an extent on the conventions used for accounting and industrial classification that they need careful interpretation. Table 1 shows some comparisons which indicate the performance of UK industry is broadly similar to other industrialized countries, but it must be emphasized that such results should be treated carefully (United Nations, 1988a, b).

Table 1. Comparison of stock/output value for manufacturing industries (values in '000m of national currency).

Country	Manufacturing output	Stock value at year-end	Stock as % of output
UK	219·99	44·15	20·1
Australia	98·93	14·47	14·6
Austria	808·27	149·45	18·5
Canada	266·83	36·33	13·6
Denmark	272·46	43·69	16·0
Hungary	1087·10	272·88	25·1
Japan	254 060	27 789	10·9
New Zealand	28·41	4·72	16·6
Norway	242·25	53·72	22·2
Portugal	2218·2	536·89	24·2
Sweden	511·0	111·88	21·9
USA	2266·9	321·56	14·2
West Germany	1345·8	227·78	16·9

Summary and conclusions

Effective inventory control is based on a number of fundamental requirements including:

(a) a desire to minimize costs,
(b) the availability of suitable models (and expertise),
(c) availability of accurate information,
(d) ability to control stocks,
(e) awareness of environmental factors.

Opinions about stock holdings have changed over time. Until the late 19th century a major aim was to maximize stocks; then for most of this century inventory control has been based on a desire to minimize costs by aiming at some optimal stock level. Considerable effort has been put into calculating or defining such optimal levels.

It is only since the introduction of computers that it has become possible to collect accurate information and implement reliable control procedures. Computers have provided the means of achieving the aims of inventory control.

From the early 1950s to the early 1970s there was a steady decline in stock levels. In the early 1970s there was a sudden pattern change, which was caused by the rapid increase in oil prices and the economic disruption which followed. For a few years at the end of the 1970s the fall in stocks continued, but the changes seem to have accelerated in the 1980s.

Manufacturing industry would be expected to play a major role in stock reduction. This was not evident, however, until after the disruptions of the early 1970s. It seems that inventory control was not well implemented and it was only after a severe economic disruption that manufacturing joined the trend. There has now been a steady decline in their stocks for the last 15 years. There is a suggestion that stocks of finished goods are slowly rising, while raw materials and work in progress are declining. These changes are consistent with manufacturers exerting better control over their stocks, and with the introduction of low inventory manufacturing systems.

Superimposed on the long-term trends in stock holdings are shorter-term fluctuations. Many of these are caused by wider economic influences which may be related to the general business cycle, and include total output, interest rates and inflation. Although there does seem to be a connection between aggregate stock holdings and change in output, it is very difficult to describe this. Details of the relationship have not yet been found. It may be suggested that when interest rates are high, stock levels will be low. Again the information seems to suggest there is some link between the two but details of the relationship are hard to identify. We are also forced to conclude that there appears to be some relationship between inflation and stock holdings, but again the details are not clear.

International comparisons of stock holdings are possible, and they show essentially similar patterns. Different conventions in collecting and analysing data mean that caution should be used when making such comparisons.

References

Abramovitz, M., 1950, *Inventories and Business Cycles* (New York: The National Bureau of Economic Research).

Central Statistics Office, 1960–88, *National Income and Expenditure* (London: HMSO).

Central Statistics Office, 1966–88a, *Economic Trends* (London: HMSO).

Central Statistics Office, 1966–88b, *United Kingdom National Accounts* (London: HMSO).

Harris, F., 1915, *Operations and Cost* (Chicago: A. Shaw & Co.).

Klein, L.R. and Popkin, J., 1961, An economic analysis of the post war relationship between inventory fluctuation and change in aggregate economic activity. In *Inventory Fluctuation and Economic Stabilization* (Washington: Joint Economic Committee).

Lovell, M.C., 1961, Manufacturers' inventories, sales expectations and the accelerator principle. *Econometrica*, **29**, 293–314.

Mack, R.P., 1967, *Information, Expectations and Inventory Fluctuations* (Columbia: Columbia University Press).

Papillon, F., 1677, *A Treatise Concerning the East India Trade* (London).

Silver, E. and Peterson, R., 1985, *Decision Systems for Inventory Management and Production Planning*, 2nd edn (New York: Wiley).

United Nations, 1988a, *Industrial Statistics Yearbook* (New York: UN).

United Nations, 1988b, *National Account Statistics* (New York: UN).

17

Inventory policies with credit-period incentives

F.J. Arcelus and G. Srinivasan

Faculty of Administration, University of New Brunswick, PO Box 4400, Fredericton, New Brunswick, Canada E3B 5A3

Abstract

This chapter evaluates the cash flow implications of a supplier's credit policy on the inventory decision. At issue here is the identification of the appropriate type of cash flow affected by the particular credit-period policy under consideration and the use of the corresponding opportunity cost to assess its impact on inventory policy. The main import of the analysis is an order quantity, independent of the length of the credit-period itself. Instead, the dependence is only upon which period, cycle or credit, is greater.

Introduction

One of the most common types of credit extended by a firm to its customers takes the form of a credit-period within which outstanding balances may be paid without incurring overdue charges. Whereas such policy will undoubtedly decrease the customer's cost of carrying inventory, its effect on the optimal order quantity remains an open question. The main reason for this is that the existence of the credit-period upsets the timing of the cash flows. The timing difference has two immediate consequences for the valuation of the inventory on hand. First, the customer does not incur any outflows associated with the purchase of the merchandise until after the credit-period is over. Second, the inflows associated with the resale of the merchandise start coming in at a uniform rate from the beginning of the planning horizon, following the usual conventions of standard inventory analysis. As a result, the traditional method of computing holding costs by averaging out the flows and then applying a charge rate on the net flows is no longer valid. The need to use a rate other than the usual financing charge for the evaluation of the holding costs has been recognized by several authors (Haley and Higgins, 1973; Kingsman, 1983; Chapman and Ward, 1984; Dave, 1985; Goyal, 1985; Daellenbach, 1986; Chand and Ward, 1987; Arcelus and Srinivasan, 1990). However these studies use only one rate on the net flows. It is the purpose of this chapter to discuss an alternate approach to the computation of the holding costs for the credit-period case and to study its effect on inventory policies. To that effect, the new approach is developed in the next section, followed by a discussion of its main implications. A conclusions section completes the chapter.

Table 1. A numerical example.

		Credit-period: t (years)	
		$t = 1/12 < T$	$t = 5/12 > T$
Parameters			
Annual Demand (units)	$D = 3600$		
Physical Holding Cost ($/unit)	$h = 1{\cdot}5$		
Unit Purchase Price ($/unit)	$p = 10$		
Outflow Interest Rate (%)	$I_c = {\cdot}24$		
Inflow Interest Rate (%)	$I_d = {\cdot}12$		
Ordering Cost ($/order)	$S = 85$		
Time Between Orders (years)	$T = 1/4$		
Ordering costs:			
Number of orders per year:	$1/T = 4$		
× ordering cost:	$S = 85$	340	340
+ *Physical holding costs*			
Average number of units on hand:	$DT/2 = 450$		
× physical holding cost rate:	$h = 1{\cdot}5$	675	675
+ *Interest outflows* (if $t < T$)			
Amount borrowed at time t:	$DTp = 9000$		
× interest rate per cycle:	$I_c\,(T - t) = 0{\cdot}04$		
× number of cycles:	$1/T = 4$	1440	
– *Interest inflows Before* T:			
Average amount invested:	$DTp/2 = 4500$		
× interest rate per cycle:	$I_d\,T = 0{\cdot}03$		
× number of cycles:	$1/T = 4$	540	540
After T: (*if* $t > T$)			
Average amount invested:	$DTp = 9000$		
× interest rate per cycle:	$I_d\,(t - T) = 0{\cdot}02$		
× number of cycles:	$1/T = 4$		720
= Average yearly inventory costs:	$Z(T)$	1915	–245

The model

For ease of exposition, our framework of analysis is the single-period, deterministic, inventory problem. Table 1 presents an example designed to illustrate the steps of the model. Demand is assumed to be constant at a rate of D units per year and no shortages are allowed. Orders are placed at regular intervals of length T (years) and no payment is due for a credit-period of length t (years). Each order consists of $Q = DT$ units, acquired at a unit cost of p. The average yearly inventory costs, Z, are computed as the sum of the ordering and of the holding costs. Ordering costs equal S/T, or the cost of placing an order, S, times the number of orders placed in a year, $1/T$. The holding costs are of two types, physical and dollar-value. The first is associated with the actual possession of the merchandise which is independent of the credit-period. Included here are the costs of storage insurance and the like, charged at a rate of \$h per unit of inventory on hand. Following the usual conventions, the merchandise is assumed to change hands at the beginning of each order cycle. Thus, the customer is expected to have an average of $DT/2$ units on hand.

The second type of costs are interest-related and the main subject of this paper. For their computation, the following three-step approach is recommended:

Step 1: Use a different charge rate for each type of flow. The traditional approach consists of using only one rate, usually that related with financing the purchase of the merchandise. Implicit in this approach is the assumption that the cost of funds to the firm, I_c, or equivalently the opportunity cost of using one's own funds for the acquisition of the orders, equals the rate of return, I_d, the firm may earn from alternate short-term investment opportunities. This is usually not correct in practice, especially when the timing differs, thus the need for the two different rates.

Step 2: Drop the accounting profit approach in favour of the cash flow approach in the computation of the costs. Even though this step is not necessary, it simplifies the process of identifying the timing of the flows. To that effect, it is useful to assume that (i) interest outflows, if any, are made as a result of interest payments on amounts borrowed at a rate of I_c per year; and that (ii) interest inflows are received from returns on amounts invested at I_d per year, a rate which may or may not be equal to I_c.

Step 3: Net the costs after using the appropriate rate for each flow. That is to say, evaluate the outflows at a rate of I_c, the inflows at a rate of I_d and then net the costs.

The formal development of the expressions for these flows may be described with the help of the example in Table 1. To compute the outflows, start with the assumption that $T > t$, say $t = 1$ month = 1/12 years. Then, an amount of DTp = \$9000 is to be borrowed at time t, at a rate of $I_c(T - t) = 4\%$ per cycle. Hence the interest paid is $DTpI_c(T - t)$ = \$360 per cycle or $DpI_c(T - t)$ = \$1440 per year. When $T < t$, inflows, rather than outflows occur as the entire \$9000 is available at time t for repayment. Formally,

$$\text{INTEREST OUTFLOWS} = DpI_c \,\text{Max}(T - t, 0). \tag{1}$$

The expression for the interest inflows may be obtained in a similar manner. When $T \geq t$, ($t = 1/12$), the average amount invested per cycle is given by $DTP/2$ = \$4500, which at a rate of $I_dT = 3\%$ yields $DT^2pI_d/2$ = \$135 of interest per cycle or $DTpI_d/2$ = \$540 per year. On the other hand, suppose that $T < t$, say $t = 5$ months = 5/12 years. Then, for the first cycle, the average amount invested and its corresponding interest are unchanged, \$4500 and \$135, respectively. However, after T, then entire DTp = \$9000 are available for investment for the remaining $t - T = 1/6$ years, which, at a rate of $I_d(t - T) = 2\%$, yields an additional $DTpI_d(t - T)$ = \$180 per cycle, or $DpI_d(t - T)$ = \$720 per year. Hence, the $T < t$ scenario yields total inflows of $DTpI_d(t - T/2)$ = \$315 per cycle or

$$DpI_d(t - T/2) = \$1260$$

per year. Formally, the yearly inflows may be given by

$$\text{INTEREST INFLOWS} = DpI_d \,\text{Max}(T/2,\, t - T/2). \tag{2}$$

The expression for the average yearly inventory costs may be obtained from eqns (1) and (2) as follows:

$$Z(T) = S/T + DTh/2 + DpI_c \,\text{Max}(T - t, 0) - DpI_d \,\text{Max}(T/2,\, t - T/2). \tag{3}$$

Using eqn (3), it can be easily shown that the optimal order quantity is given by

$$Q^* = DT^* = \sqrt{2SD/[h + p(2I_c - I_d)]} \qquad \text{if } t < T^*$$
$$= \sqrt{2SD/[h + pI_d]} \qquad \text{if } t \geq T^*. \tag{4}$$

Discussion

Several points may be made upon examination of eqns (1)–(4). First, the effect of the credit-period, t, on inventory policy can now be clearly stated. Q^* is influenced by the existence ($t < T$) or lack thereof ($t \geq T$) of interest outflows. However, once the existence issue has been resolved, it is the average yearly inventory cost, Z, and *not* the optimal order quantity, Q^*, that is dependent upon the length of the credit-period.

Second, equal interest rates for the inflows and the outflows result in the traditional economic order quantity (EOQ) policy, with $t = 0$ also yielding the EOQ value of Z. This is as it should be. If a single rate, $I_c = I_d$, is applied on the difference between the flows, then the net interest-related costs per year, $DpI_c(T/2 - t)$, yield net inflows if $t > T$, with the EOQ holding costs of $DpI_cT/2$, for $t = 0$. Moreover, it is this equality of interest rates that justifies the application of carrying costs on the average inventory, $DT/2 = Q/2$, in the standard formulation.

Third, the consignment problem is a special case of eqn (4). It occurs when payment of an order is due before the next arrives. Under these conditions, $t = T$, i.e. the cycle period equals the credit-period. Thus, no outflows take place and sales revenue may be reinvested throughout the cycle, at the end of which it will be used as payment for the current order.

Fourth, when $I_c > I_d$, the cost of borrowing, or the opportunity cost of using one's own resources, exceeds the return on alternative investments. Then, resources will tend to be shifted away from inventory and into more attractive uses. Thus, the associated order quantities will decrease in size. This result is predicated upon $t < T$; otherwise no outflows take place, which renders Q^* independent of I_c. Alternatively, when $I_c < I_d$, it is more profitable to invest in inventories than to allocate equivalent amounts to other uses. This will result in larger Q^* values, provided, as before, that $t < T$.

Fifth, Q^* increases in value as $I_d \to 0$, in the $t > T$ case. The reason for this is that lower I_d values, by decreasing the attractiveness of alternative investments, render inventory investment more profitable. This is specially true, since, due to the lack of outflows, larger orders do not convey additional carrying costs and will even decrease the ordering cost. On the other hand, when $t < T$, Q^* is an increasing function of I_d, as higher values of I_d lead to decreases in $(I_c - I_d)$ and , as shown above, to higher quantities.

Finally, when $t > T$, Q^* is independent of I_c, due to the absence of outflows. However, when $t < T$, the smaller the value of I_c, the smaller the $(I_c - I_d)$ difference and thus, as discussed above, the higher the order quantities, since the relatively cheaper borrowing rates induce the customer to hold additional inventory.

Conclusion

The traditional inventory policy has been shown to have important implications for the firm's estimation of cash flows. At issue here is the identification of the appropriate type of flow affected by the particular delay of payment policy being contemplated and the use of the corresponding opportunity cost to assess the impact of the said policy. Within this context, the framework of analysis presented here may be useful to the evaluation of any inventory policy change with cash flow implications.

Acknowledgement

The financial support to the first author from the Natural Science and Engineering Research Council of Canada for the completion of this research is gratefully acknowledged.

References

Arcelus, F.J. and Srinivasan, G., 1990, Implications of the EOQ assumptions on the valuation of the average investment. *Engineering Costs and Production Economics*, **19**, 107–114.

Chand, S. and Ward, J., 1987, A note on economic order quantity under conditions of permissible delay in payment. *Journal of the Operational Research Society*, **38**, 83–84.

Chapman, C.B., Ward, S.C., Cooper, D.F. and Page, M.J., 1984, Credit policy and inventory control. *Journal of the Operational Research Society*, **35**, 1055–1065.

Daellenbach, H.G., 1986, Inventory control and trade credit. *Journal of the Operational Research Society*, **37**, 525–528.

Dave, U., 1985, On 'Economic Order Quantity Under Conditions of Permissible Delay in Payments' by Goyal. *Journal of the Operational Research Society*, **36**, 1069–1070.

Goyal, S.K., 1985, Economic Order Quantity Under Conditions of Permissible Delay in Payments. *Journal of the Operational Research Society*, **36**, 335–338.

Haley, C.W. and Higgins, R.C., 1973, Inventory policy and trade credit financing. *Management Science*, **20**, 464–471.

Kingsman, B.G., 1983, The effect of payment rules on ordering and stockholding in purchasing. *Journal of the Operational Research Society*, **32**, 1085–1098.

18

Optimal profit policy by substitutional sales promotion

Takio Shimosakon and Sennosuke Kuriyama

Department of Industrial Management, Osaka Institute of Technology, Osaka, Japan

Abstract

Recently, some studies taking account of the substitutional characteristics among products have been made in the field of the multi-product inventory system. These studies have assumed that the seller cannot control the customers' substitutional purchasing action. Therefore, the substitutional sales promotion has not been studied for the purpose of increasing the profit and improving the service ratio. In this study, it is assumed that the seller can control the customers' substitutional purchasing action by means of price discount, and the optimal price discount rate is obtained for the customers' various substitutional purchasing actions so as to maximize the total expected profit. Consequently, the sellers' profit is increased and the service ratio is improved.

Introduction

In the case of substitutional property among products, it is possible to sell the holding product as a substitute for one in shortage by controlling substitutable products in a focused way, such as dresses of the same style and size but different colour, different brands of paint and containers of the same brand but different size. Therefore, as increase in the amount of sales and a reduction of both the holding and shortage amount can be generally expected, realizing a considerable increase in profit and improvement of the service ratio.

For these reasons, some studies taking account of substitutional property among products have been made in multi-product inventory control systems. In previous work, however, when the products the customer required are sold out, the ratio of purchases of the holding product as a substitute for the one in shortage made by the customer himself is defined as the substitution ratio; thus, the seller cannot control the substitution ratio. Therefore, the seller cannot actively increase the profit or improve the service ratio.

In this chapter, we classify the characteristics of the substitution ratio into two parts: (i) an uncontrollable part and (ii) a controllable part by the seller. The uncontrollable part relates to customer purchases of the holding product as a substitute for the product in shortage and the controllable part relates to substitutional purchasing action of the customers promoted by price discounts made by the seller. The purpose of this chapter is to obtain the optimal policy of the controllable substitution ratio by examining various situations of the uncontrollable ratio and to attempt actively to increase the profit, improve the service ratio and reduce the quantity of stock by making a precise decision.

Assumptions and notation

Assumptions

(1) Two products are treated and are mutually substitutable.
(2) The substitution ratio consists of an uncontrollable part and a controllable part by the seller.
(3) The planning horizon is one period.

Notation

The following symbols are defined for product i (i = 1, 2): c_i = purchasing price per unit, h_i = holding cost per unit, p_i = shortage cost per unit, r_i = selling price per unit, x_i = inventory amount before the purchasing, y_i = inventory amount after the purchasing, Q_i = purchasing amount ($Q_i = y_i - x_i \geqq 0$), ξ_i = demand during the period (non-negative random variable), $\phi(\xi_1, \xi_2)$ = joint probability density function of the demand, β_i = price discount rate ($0 \leqq \beta_i \leqq 1$).

Formulation of model

Substitution ratio

We defined a ratio, where a customer will purchase the product j as a substitute for the product i when the product i is sold out and j is held, as the real substitution ratio (α_{ij}). In this model, we classify this ratio into two parts; an uncontrollable part and a controllable part by the seller. The former is a ratio that a customer will purchase the substitute by himself and the latter is one that the substitutional purchasing action of the customers will be promoted by the price discount that the seller offers. Let A_{ij} and B_{ij} denote these ratios, respectively.

Now we take account of B_{ij}. It can be assumed that the real substitution ratio is 1·0 when the price discount rate is 1·0. Hence, we can obtain an equation $B_{ij} = 1{\cdot}0 - A_{ij}$ when $\beta_j = 1{\cdot}0$. If we assume, for simplification, that B_{ij} is proportional to β_j, it follows that $B_{ij} = (1{\cdot}0 - A_{ij})\beta_j$.

Therefore, summarizing these assumptions, the real substitution ratio is expressed by the following equation:

$$\begin{aligned}\alpha_{ij} &= A_{ij} + B_{ij} \\ &= A_{ij} + (1{\cdot}0 - A_{ij})\beta_j. \qquad (1)\end{aligned}$$

Purchasing cost

The purchasing cost (PC) is expressed by the following equation:

$$\begin{aligned}\text{PC} &= c_1Q_1 + c_2Q_2 \\ &= c_1(y_1 - x_1) + c_2(y_2 - x_2). \qquad (2)\end{aligned}$$

Expected sales proceeds and expected holding and shortage cost

The sales, the holding and the shortage amount can be obtained from the relation between the inventory amount after purchasing (y_1, y_2) and the demand during the period (ξ_1, ξ_2). In order to obtain these amounts, we may divide the domain into six domains with respect to y_i and ξ_i $(i = 1, 2)$; this is summarized in Table 1.

Let ESP and EHSC denote the expected sales proceeds and the expected holding and shortage cost, respectively. ESP and EHSC are expressed as follows:

$$\begin{aligned}\mathrm{ESP} = &\int_0^{y_1}\int_0^{y_2}[r_1\xi_1 + r_2\xi_2]\phi(\xi_1,\xi_2)\mathrm{d}\xi_2\mathrm{d}\xi_1\\
&+\int_0^{y_2}\int_{y_1}^{y_1+(y_2-\xi_2)/\alpha_{12}}[r_1y_1 + r_2\xi_2 + \alpha_{12}r_2(1-\beta_2)(\xi_1 - y_1)]\phi(\xi_1,\xi_2)\mathrm{d}\xi_1\mathrm{d}\xi_2\\
&+\int_0^{y_1}\int_{y_2}^{y_2+(y_1-\xi_1)/\alpha_{21}}[r_2y_2 + r_1\xi_1 + \alpha_{21}r_1(1-\beta_1)(\xi_2 - y_2)]\phi(\xi_1,\xi_2)\mathrm{d}\xi_2\mathrm{d}\xi_1\\
&+\int_0^{y_2}\int_{y_1+(y_2-\xi_2)/\alpha_{12}}^{\infty}[r_1y_1 + r_2\xi_2 + r_2(1-\beta_2)(y_2 - \xi_2)]\phi(\xi_1,\xi_2)\mathrm{d}\xi_1\mathrm{d}\xi_2\\
&+\int_0^{y_1}\int_{y_2+(y_1-\xi_1)/\alpha_{21}}^{\infty}[r_2y_2 + r_1\xi_1 + r_1(1-\beta_1)(y_1 - \xi_1)]\phi(\xi_1,\xi_2)\mathrm{d}\xi_2\mathrm{d}\xi_1\\
&+\int_{y_1}^{\infty}\int_{y_2}^{\infty}[r_1y_1 + r_2y_2]\phi(\xi_1,\xi_2)\mathrm{d}\xi_2\mathrm{d}\xi_1;\end{aligned}\tag{3}$$

$$\begin{aligned}\mathrm{EHSC} = &\int_0^{y_1}\int_0^{y_2}[h_1(y_1-\xi_1) + h_2(y_2-\xi_2)]\phi(\xi_1,\xi_2)\mathrm{d}\xi_2\mathrm{d}\xi_1\\
&+\int_0^{y_2}\int_{y_1}^{y_1+(y_2-\xi_2)/\alpha_{12}}[h_2\{(y_2-\xi_2) - \alpha_{12}(\xi_1 - y_1)\}\\
&+ p_1\{(\xi_1 - y_1) - \alpha_{12}(\xi_1 - y_1)\}]\phi(\xi_1,\xi_2)\mathrm{d}\xi_1\mathrm{d}\xi_2\\
&+\int_0^{y_1}\int_{y_2}^{y_2+(y_1-\xi_1)/\alpha_{21}}[h_1\{(y_1-\xi_1) - \alpha_{21}(\xi_2 - y_2)\}\\
&+ p_2\{(\xi_2 - y_2) - \alpha_{21}(\xi_2 - y_2)\}]\phi(\xi_1,\xi_2)\mathrm{d}\xi_2\mathrm{d}\xi_1\\
&+\int_0^{y_2}\int_{y_1+(y_2-\xi_2)/\alpha_{12}}^{\infty}[p_1\{(\xi_1 - y_1) - (y_2 - \xi_2)\}]\phi(\xi_1,\xi_2)\mathrm{d}\xi_1\mathrm{d}\xi_2\\
&+\int_0^{y_1}\int_{y_2+(y_1-\xi_1)/\alpha_{21}}^{\infty}[p_2\{(\xi_2 - y_2) - (y_1 - \xi_1)\}]\phi(\xi_1,\xi_2)\mathrm{d}\xi_2\mathrm{d}\xi_1\\
&+\int_{y_1}^{\infty}\int_{y_2}^{\infty}[p_1(\xi_1 - y_1) + p_2(\xi_2 - y_2)]\phi(\xi_1,\xi_2)\mathrm{d}\xi_2\mathrm{d}\xi_1.\end{aligned}\tag{4}$$

Total expected profit and service ratio

Let TEP denote the total expected profit. TEP is expressed by the following equation:

$$\mathrm{TEP} = \mathrm{ESP} - \mathrm{PC} - \mathrm{EHSC}.\tag{5}$$

In this model, we define the service ratio as 1·0 – (mean shortage amount)/(mean demand). Let η, Ms and $\bar{\xi}_i$ denote the service ratio, the mean shortage amount and the mean demand of product i $(i = 1, 2)$, respectively. The service ratio is expressed as follows:

$$\eta = 1{\cdot}0 - \mathrm{Ms}/(\bar{\xi}_1 + \bar{\xi}_2)\tag{6}$$

where Ms can be given by the following equation:

$$\begin{aligned}\mathrm{Ms} = &\int_0^{y_2}\int_{y_1}^{y_1+(y_2-\xi_2)/\alpha_{12}}[(\xi_1 - y_1) - \alpha_{12}(\xi_1 - y_1)]\phi(\xi_1,\xi_2)\mathrm{d}\xi_1\mathrm{d}\xi_2\\
&+\int_0^{y_1}\int_{y_2}^{y_2+(y_1-\xi_1)/\alpha_{21}}[(\xi_2 - y_2) - \alpha_{21}(\xi_2 - y_2)]\phi(\xi_1,\xi_2)\mathrm{d}\xi_2\mathrm{d}\xi_1\\
&+\int_0^{y_2}\int_{y_1+(y_2-\xi_2)/\alpha_{12}}^{\infty}[(\xi_1 - y_1) - (y_2 - \xi_2)]\phi(\xi_1,\xi_2)\mathrm{d}\xi_1\mathrm{d}\xi_2\end{aligned}$$

Table 1. Sales amount, selling price, holding amount and shortage amount in each domain.

Domain	Product	Sales amount	Selling price	Holding amount	Shortage amount
1	1	ξ_1	r_1	$y_1 - \xi_1$	0
	2	ξ_2	r_2	$y_2 - \xi_2$	0
2	1	y_1	r_1	0	$\xi_1 - y_1 - \alpha_{12}(\xi_1 - y_1)$
	2	ξ_2 $\alpha_{12}(\xi_1 - y_1)$	r_2 $r_2(1 - \beta_2)$	$y_2 - \xi_2 - \alpha_{12}(\xi_1 - y_1)$	0
3	1	ξ_1 $\alpha_{21}(\xi_2 - y_2)$	r_1 $r_1(1 - \beta_1)$	$y_1 - \xi_1 - \alpha_{21}(\xi_2 - y_2)$	0
	2	y_2	r_2	0	$\xi_2 - y_2 - \alpha_{21}(\xi_2 - y_2)$
4	1	y_1	r_1	0	$\xi_1 - y_1 - y_2 + \xi_2$
	2	ξ_2 $y_2 - \xi_2$	r_2 $r_2(1 - \beta_2)$	0	0
5	1	ξ_1 $y_1 - \xi_1$	r_1 $r_1(1 - \beta_1)$	0	0
	2	y_2	r_2	0	$\xi_2 - y_2 - y_1 + \xi_1$
6	1	y_1	r_1	0	$\xi_1 - y_1$
	2	y_2	r_2	0	$\xi_2 - y_2$

$$+\int_0^{y_1}\int_{y_2+(y_1-\xi_1)/\alpha_{21}}^{\infty}[(\xi_2-y_2)-(y_1-\xi_1)]\phi(\xi_1,\xi_2)d\xi_2 d\xi_1$$
$$+\int_{y_1}^{\infty}\int_{y_2}^{\infty}[(\xi_1-y_1)+(\xi_2-y_2)]\phi(\xi_1,\xi_2)d\xi_2 d\xi_1. \quad (7)$$

Numerical analysis

Purpose of numerical analysis

The purpose of this analysis is to calculate the optimal inventory amount, the total expected profit and the service ratio for each value of A_{ij} (= 0·0, 0·25, 0·5, 0·75, 1·0) when the price discount rate is zero and the optimal value (β_1^*, β_1^*). Then we clarify the characteristics and the effectiveness of this model.

Total expected profit and service ratio under exponential demand

We assume that each demand for product i occurs mutually independently and follows the exponential distribution, that is, $\phi(\xi_1, \xi_2) = \lambda_1\lambda_2 \exp\{-\lambda_1\xi_1 - \lambda_2\xi_2\}$. Then calculating the total expected profit and the mean shortage amount, TEP and Ms are given by the following equations.

$$\begin{aligned}\text{TEP} = {} & r_1\bar{\xi}_1 + r_2\bar{\xi}_2 - h_1(y_1-\bar{\xi}_1) - h_2(y_2-\bar{\xi}_2) - c_1(y_1-x_1) - c_2(y_2-x_2)\\ & + \bar{\xi}_1[\alpha_{12}\{r_2(1-\beta_2)+h_2+p_1\} - (r_1+h_1+p_1)]\exp\{-\lambda_1 y_1\}\\ & + \bar{\xi}_2[\alpha_{21}\{r_1(1-\beta_1)+h_1+p_2\} - (r_2+h_2+p_2)]\exp\{-\lambda_2 y_2\}\\ & + \frac{\lambda_2\alpha_{12}^2\{r_2(1-\beta_2)+h_2+p_1\}}{\lambda_1(\lambda_1-\lambda_2\alpha_{12})}\exp\{-\lambda_1 y_1 - \lambda_1 y_2/\alpha_{12}\}\\ & + \frac{\lambda_1\alpha_{21}^2\{r_1(1-\beta_1)+h_1+p_2\}}{\lambda_2(\lambda_2-\lambda_1\alpha_{21})}\exp\{-\lambda_2 y_1/\alpha_{21} - \lambda_2 y_2\}\\ & - \left[\frac{\alpha_{12}\{r_2(1-\beta_2)+h_2+p_1\}}{\lambda_1-\lambda_2\alpha_{12}} + \frac{\alpha_{21}\{r_1(1-\beta_1)+h_1+p_2\}}{\lambda_2-\lambda_1\alpha_{21}}\right]\exp\{-\lambda_1 y_1 - \lambda_2 y_2\} \quad (8)\end{aligned}$$

where $\bar{\xi}_i = \int_0^{\infty}\xi_i\lambda_i\exp\{-\lambda_i\xi_i\}d\xi_i = \frac{1}{\lambda_i} \quad (i = 1, 2)$

$$\begin{aligned}\text{Ms} = {} & \bar{\xi}_1(1-\alpha_{12})\exp\{-\lambda_1 y_1\} + \bar{\xi}_2(1-\alpha_{21})\exp\{-\lambda_2 y_2\}\\ & + \left[\frac{\alpha_{21}}{\lambda_2-\lambda_1\alpha_{21}} + \frac{\alpha_{12}}{\lambda_1-\lambda_2\alpha_{12}}\right]\exp\{-\lambda_1 y_1 - \lambda_2 y_2\}\\ & - \frac{\lambda_1\alpha_{21}^2}{\lambda_2(\lambda_2-\lambda_1\alpha_{21})}\exp\{-\lambda_2 y_1/\alpha_{21} - \lambda_2 y_2\}\\ & - \frac{\lambda_2\alpha_{12}^2}{\lambda_1(\lambda_1-\lambda_2\alpha_{12})}\exp\{-\lambda_1 y_1 - \lambda_1 y_2/\alpha_{12}\}. \quad (9)\end{aligned}$$

Data of numerical analysis

In order to carry out the above-mentioned purpose, the data in Table 2 are used.

Table 2. Data of numerical analysis.

Parameter \ Model	Model 1	Model 2	Model 3	Model 4
$\bar{\xi}_1$	10	10	5	5
$\bar{\xi}_1$	10	10	12·5	12·5
r_1	80	50	50	100
r_2	80	100	100	50
$c_1 = c_2$	20	20	20	20
$h_1 = h_2$	5	5	5	5
$p_1 = p_2$	90	90	90	90

Table 3. Results of numerical analysis.

Item		No price discount case			Optimal price discount case			
Model	A_{ij}	$y^*(y^*_1, y^*_2)$	TEP	η	(β^*_1, β^*_2)	$y^*(y^*_1, y^*_2)$	TEP	η
1	0·00	38(19,19)	226·51	85·0	(0·8,0·8)	36(18,18)	396·16	91·3
	0·25	36(18,18)	334·25	86·7	(0·6,0·6)	36(18,18)	405·83	90·7
	0·50	34(17,17)	424·17	87·8	(0·3,0·3)	34(17,17)	431·13	89·7
	0·75	34(17,17)	488·86	89·7	(0·0,0·0)	34(17,17)	488·86	89·7
	1·00	32(* ,*)	538·00	90·6	(0·0,0·0)	32(* ,*)	538·00	90·6
2	0·00	39(18,21)	146·53	85·6	(0·9,0·8)	35(14,21)	330·14	90·6
	0·25	35(15,20)	261·22	85·6	(0·7,0·6)	34(13,21)	343·51	89·1
	0·50	33(12,21)	368·98	86·1	(0·3,0·3)	33(11,22)	379·32	87·3
	0·75	31(0,31)	556·44	80·3	(0·0,0·0)	31(0,31)	556·44	80·3
	1·00	35(0,35)	902·00	91·8	(0·0,0·0)	35(0,35)	902·00	91·8
3	0·00	35(9,26)	288·14	86·4	(0·6,0·8)	32(7,25)	406·39	89·6
	0·25	33(8,25)	364·27	87·2	(0·4,0·7)	32(7,25)	413·67	89·5
	0·50	32(7,25)	422·26	87·9	(0·0,0·4)	31(6,25)	432·59	88·0
	0·75	30(0,30)	527·96	83·6	(0·0,0·0)	30(0,30)	527·96	83·6
	1·00	32(0,32)	724·53	90·8	(0·0,0·0)	32(0,32)	724·53	90·8
4	0·00	32(10,22)	−31·28	83·8	(0·6,0·8)	30(12,18)	107·76	87·7
	0·25	30(11,19)	73·73	84·8	(0·4,0·8)	29(12,17)	123·58	86·5
	0·50	28(13,15)	157·81	84·3	(0·1,0·6)	28(13,15)	160·53	85·0
	0·75	26(26, 0)	339·62	75·2	(0·0,0·0)	26(26, 0)	339·62	75·2
	1·00	32(32, 0)	724·53	90·8	(0·0,0·0)	32(32, 0)	724·53	90·8

*, *: Denotes any of (32,0), (31,0), …, (1,31), (0,32).

Results and consideration

We summarize the optimal inventory amount $y^*(y^*_1, y^*_2)$, the total expected profit (TEP) and the service ratio (η) when the price discount rate is zero and the optimal value (β^*_1, β^*_2) in Table 3.

From these results, it can be clarified that

(1) when the value of A_{ij} increases, the total expected profit increases,
(2) if the value of A_{ij} is greater than 0·5, it seems that there is no advantage of price discount,
(3) the smaller the value of A_{ij} becomes, the larger the increase of the total expected profit and the service ratio becomes,
(4) when the value of A_{ij} is smaller, the effectiveness of this model becomes more apparent.

Conclusion

(1) We constructed the model in which the seller could control the substitutional purchasing action of the customers by the price discount that the seller offers, and we formulated the total expected profit function.
(2) Both the optimal inventory level and the optimal price discount rate were calculated for each value of A_{ij}.
(3) It was clarified that the smaller the value of A_{ij} became the larger the increase of the total expected profit became.
(4) From these results, it was clarified that we could actively realize a considerable profit increase and an improvement of the service ratio for customers by controlling the substitutable products in a focused way and promoting the substitutional sale.

Bibliography

Shah, A.A., 1969, Inventory control of substitute products, Ph.D. Thesis, Case Western Reserve University.

Sivazlian, B.D. and Stanfel, L., 1975, *Analysis of System in Operations Research* (Englewood Cliffs, NJ: Prentice-Hall), pp. 370–376.

SECTION III.2. PRODUCTION CONTROL

19

Automatic adaptive scheduling of multi-processor cells

F. Hank Grant[1] and Shimon Y. Nof[2]

[1]*Pritsker Corporation, 2801 Kent Avenue, West Lafayette, IN 47906, and*
[2]*School of Industrial Engineering, Purdue University, West Lafayette, IN 47907, USA*

Abstract

Effective production scheduling can have a dramatic impact on manufacturing productivity. In modern automated manufacturing systems, scheduling can help reduce indirect costs which comprise a large portion of manufacturing costs in those systems. The objective of this research has been to study the feasibility of automatically scheduling multi-machine complexes on a real-time basis and implementing a unified computer system to provide new scheduling tools in automated systems. We are investigating real-time scheduling by separating schedule generation and regeneration tasks from schedule adaptation/recovery tasks. The article describes our general framework for automatic adaptive/predictive scheduling, and some experimental results demonstrating its effectiveness.

Introduction

Effective production scheduling can have a dramatic impact on manufacturing productivity. In modern automated manufacturing systems, scheduling can help reduce indirect costs which comprise a large portion of manufacturing costs in those systems. This research has focused on studying the feasibility of automatically scheduling multi-machine complexes on a real-time basis and the implementation of the unified computer system to provide new scheduling tools in automated systems.

The next section of this chapter will describe the research problem and the significance of the research. We will then present the research objectives which were accomplished and our feasibility conclusions and issues for future research.

Background

As the number of automated or semi-automated systems proliferate, the requirement for scheduling systems truly to support that environment becomes even more severe. Our own experience in the application of FACTOR, a shop floor scheduling system commercialized by FACTROL (now part of Pritsker Corporation) has been extremely successful in automated environments (Slutz and Munji, 1987), when used in a 'frequent batch' mode and invoked automatically by the cell controller. However, the need exists for more extensive tools to support the real-time generation of schedules to control automated systems effectively, and to provide both batch scheduling tools and adaptive scheduling tools.

Three major challenges for this research were identified: (1) automatically schedule multi-machine complexes on a real-time basis to increase their production efficiency; (2) integrate manufacturing activities at the cell-level to enable cell as well as machine automation; and (3) develop *adaptive* scheduling capability to supplement *frequent batch* scheduling. Two important research directions influenced our research: maturation and progress in interactive scheduling (Godin, 1978; Hodgson and McDonald, 1981) and decision support systems for manufacturing control (Nof and Gurecki, 1980); and emergence of expert systems for scheduling (De, 1987; Young and Rossi, 1987).

In addition, the major advances in cost-effective computer technology, e.g. speed of processing and memory availability, have already transformed some of the theoretical concepts into practicable solutions. OPT (Goldratt and Cox, 1984) was an early, significant commercial product for scheduling production operations, but has not provided a real-time solution. Our own experience with FACTOR (Robbins, 1986; Slutz and Munji, 1987) has been for interactive decision making by frequent schedule evaluation, yielding an effective timetable for computerized cells. No computer system, however, provides a generalized solution for automatic, adaptive real-time scheduling. Based on the reaction of our customers and their feedback relative to future needs, this research addresses a significant genuine problem in manufacturing.

Framework for adaptive/predictive scheduling control

The focus of this research was on adaptive, real-time scheduling in the real-world, relatively unstructured, manufacturing environment. We envision the Real-Time Adaptive Scheduling tool as receiving input from the materials resource planning (MRP) level via the cell scheduler, but further interpreting and adapting this information. The features that are required are:

(a) decide automatically how to recover from deviations and delays, and when a reschedule is required;
(b) provide a predictive measure of schedule effectiveness in terms of performance and robustness;
(c) provide on-line dispatch of schedule details, allowing for automatic change of schedule whenever this is feasible; and
(d) perform automatic what-if analysis of alternatives for schedule recovery and reschedule.

The technical approach can be described as follows. The general function of the system is to attempt continuously to get back on schedule (as much as possible), as opposed merely to rescheduling. Key elements of this approach are defined as follows:

(S_i)—A set of schedule alternatives for period i, of timed orders for each of M machines, based on 'what-if' analysis. This is the product of cell schedulers today. Alternatives are generated based on the selection of different heuristics or other modifications made to scheduling procedures. The period, i, is a scheduling interval such as a production shift.

$\{(S_i, E_i)\}$—A set of schedule alternatives, S_i, each with its relative measure of performance and schedule robustness, E_i. Performance is measured by traditional values of lateness or tardiness. Robustness reflects a schedule's sensitivity to changes and is measured by factors such as the amount of slack in a schedule, or the uncertainty in parameter estimates.

(S_i^*, E_i^*)—A given schedule S_i^* for period i that was selected based on its E_i^*, per a selection criterion.

$S_{i,M}(t)$—Real-time schedule for period i for all the components of the M machines in the

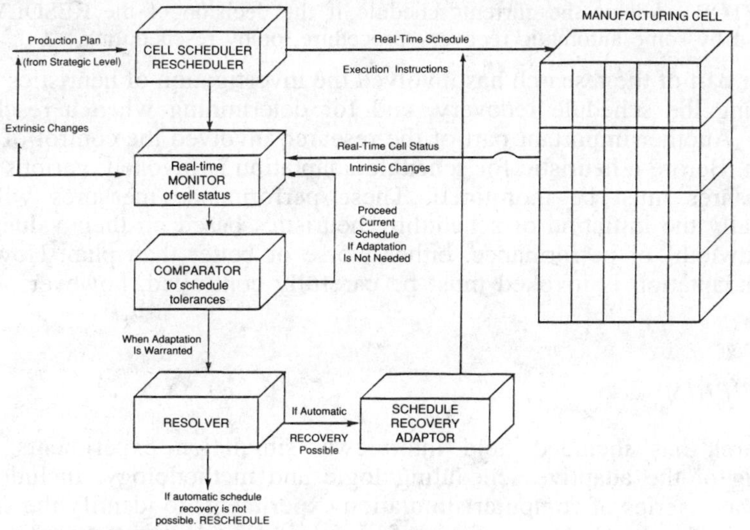

Figure 1. The automatic real-time adaptive/predictive scheduling logic.

cell that require real-time instructions. The real-time schedule may usually require greater detail than a non-real-time plan. It includes process instructions as a function of time. It can be a dynamic adaptation of S_i^*, which is revised automatically when necessary, if it is possible, or may be an adaptation of one of the previously generated schedules, S_i.

$S_{di,m}(t)$—Another real-time schedule, possibly for a shorter period, d*i*, than the complete *i*, and for just the sub-set *m* of *M* machines that are the only ones affected by a certain change.

The functions of the adaptive/predictive real-time scheduler are then:

(a) Transform a given cell schedule S_i^* (selected automatically or by the human manager) into a real-time set of work instructions, including process parameters that are a function of time for every machine in the cell.
(b) Automatically develop modified schedules for short periods (shorter relative to the period *i*), for a selected sub-set of the cell, *m*, or for the complete cell, *M*, to implement special requirements.
(c) Respond and adapt automatically to certain variations in the real-time schedule execution by certain changes in instructions, following a carefully selected recovery strategy, e.g. leave it as is or shift/delay some tasks, after evaluating that these changes do not cause major disruptions, or reroute certain operations to other equipment, or other recovery logic.

Our framework of Automatic Real-Time Adaptive/Predictive Scheduling control is summarized in Figure 1. It can be viewed as a general mechanism that includes five basic components which we define as follows:

(1) SCHEDULER—Generates a schedule of operations using factory resources for a given set of factory orders based on predicted performance and for given objectives;
(2) MONITOR—Monitors the current progress of the actual execution of the generated schedules, and the on-going new demands on the factory;
(3) COMPARATOR—Compares actual execution and demands to the predicted, planned schedule, i.e. comparing the findings of the MONITOR with the plan of the SCHEDULER;
(4) RESOLVER—Decides and selects for given criteria how to respond to the results of the COMPARATOR, what adaptation strategy (if any) to follow during the next period, or planning horizon, mainly, proceed with the current schedule, adapt it by some automatic recovery procedure, or resort to a higher level rescheduling.

(5) ADAPTOR—Adapts the current schedule if the decision of the RESOLVER is to respond by some automatic recovery procedure, or by rescheduling.

A major part of the research has involved the investigation of heuristics and logic for planning the schedule recovery, and for determining when a reschedule is warranted. Another important part of the research involved the control of schedule adaptation. Before a heuristic for schedule adaptation is invoked, various performance measures must be monitored. These performance measures will trigger automatically the initiation of scheduling heuristics based on their values relative to a 'bandwidth' of performance, either worse or better than plan. How quickly schedule adaptation is invoked must be carefully controlled, however.

Experiments

Our research has included field interviews, simulation experiments, and the prototyping of the adaptive scheduling logic and methodology. Included in our research was a series of computer simulation experiments to identify the significant challenges in the development of a unified automatic adaptive scheduling system. These experiments were useful in prototyping important concepts and exploring the feasibility of the concepts. Field interviews served in providing real-world feedback on the need for our concepts and helped us refine our prototyping design. The literature review (Nof *et al.*, 1988) has provided further attention to heuristics and the opportunity for their use in our methodology.

Two sets of experiments were performed to understand better and examine the feasibility of our approach. The first set focused on the functional structure of the control framework, and on the detailed operation of each of the framework components. The second set focused on comparing alternative real-time recovery procedures, and on alternative measures for real-time tracking of schedule execution. Both objectives of the second set were aiming at improving significantly the schedule performance despite on-going disruptions. The two experiment sets are summarized next with their main findings.

Experiment Set 1: focus on framework details and general feasibility

Two FACTOR models were used, one as a SCHEDULER to generate an initial schedule and to calculate the *expected* values of the system performance measure at specific monitoring intervals. A second FACTOR model was used to simulate the actual operation of the cell to the same schedule, and to compute, as MONITOR, the actual observed values of the performance measure at each monitoring interval. The cell model included a forklift, a mill and two grinders, producing two part types by the same process plan with total nominal process time of 147 min/part. Ten production orders were to be scheduled over a three-week horizon (nine new orders, one in-process) for the two part types, with ordered quantities varying from 22 to 50 parts each. The source of disruption in actual performance was modelled by assuming all operating times to be uniformly distributed random variables.

Three types of automatic recovery procedures were used: (i) time shifting by total performance deviation till the time of each recovery; (ii) time shifting by only the additional performance deviation (early or late) observed since the end of the previous monitoring interval till now: (iii) resequencing of all uncompleted orders. Three performance measures were used for monitoring: (i) mean order lateness to date; (ii) cumulative mean order lateness; (iii) total dynamic slack. Timing of monitoring was only at order completion, for measures (i), (ii), or at the completion of every operation, for measure (iii). Normal tolerance fences were taken at ±10% of expected value of measure at the current time (EVM). Beyond these, an automatic recovery procedure (i) or (ii) was triggered. But beyond ±20% of EVM, recovery by resequencing was applied.

Multiple tolerance fences were used in these experiments to explore resolution logic. The closer tolerance region could serve for predictive observation of trends outside normal performance. The wider tolerance could illustrate more severe cumulative disruption of expected schedule performance, providing a more powerful recovery procedure, in this case, resequencing. Deviations beyond the outside tolerance level would indicate that an external, non-automatic intervention is needed. A total of 18 simulation experiments were run for Set 1.

Experiment Set 2: focus on monitoring and adaptive functions

The focus in the second set of experiments was, first of all, to examine the feasibility of controlling execution and modifying performance positively under adverse disturbances. Different cell and production plans were studied, but the basic experimental approach of Set 1 was retained. The cell model now included a forklift, a mill and three grinders, with one grinder kept as a backup unit. The parts produced in the cell were of two types, produced by the same process plan but having different set-up times, with total nominal process time of 78 min/part. Ten new orders for the two part types, each for 60 parts in a single batch, were to be scheduled over 72 h of continuous production. Three alternative sources of disruption were (i) grinder No. 1 failure, always at the same time; (ii) unexpected order for 60 parts of one type; (iii) combination of both (i) and (ii). Three types of automatic recovery procedures were used: (i) rerouting all remaining parts of all current orders that violate tolerance to the back-up grinder; (ii) splitting orders that violate tolerance by the fraction they are off. The split order that is within tolerance is then given higher priority to expedite its production. Each order can be split only once. The back-up grinder is not used in this case; (iii) resequencing—regenerate a new schedule for all the uncompleted orders. The back-up grinder is not used in this case. Two performance measures were tried for monitoring: (i) Mean Current Fraction of Original Plan Completed (CFP) and (ii) Mean Current Critical Ratio (CCR). Timing of monitoring was every 18 min (about 1/4 of total nominal process time). Tolerance fences were defined as follows: for CFP < 0·80, recovery is activated, or for CCR > 1·20, recovery is activated. However, for each experiment, only one type of recovery procedure was applied in order to observe its specific, separate effects. A total of 18 simulation experiments were run for Set 2.

Experimental results and findings

Results of both experiment sets were analysed and sample illustrations are shown in Figures 2 and 3. Example Case la depicts the unadapted TDS (Total Dynamic Slack) with three violations of the outside tolerance limits, and two periods of violating the normal tolerance. (It should be noted that this case illustrates successful schedule completion even without adaptation and is used to describe how the approach works.) This is compared to Case 1b where effective adaptation by one time-shifting and three resequencing recoveries yielded a TDS much closer to target, and right on target from about 450 h. The same cell without adaptation as shown in Case 1a, took till the end of simulation, over 500 h to get on target. The main conclusions of Set 1 experiments are as follows.

(1) The use of several layers of recovery procedures (in this experiment, two layers: time shifting; resequencing) is useful. This approach extends the capability of the automatic adaptation by applying recovery procedures that are more complex computationally. As long as they can work effectively, human intervention is not necessary for adaptive response.

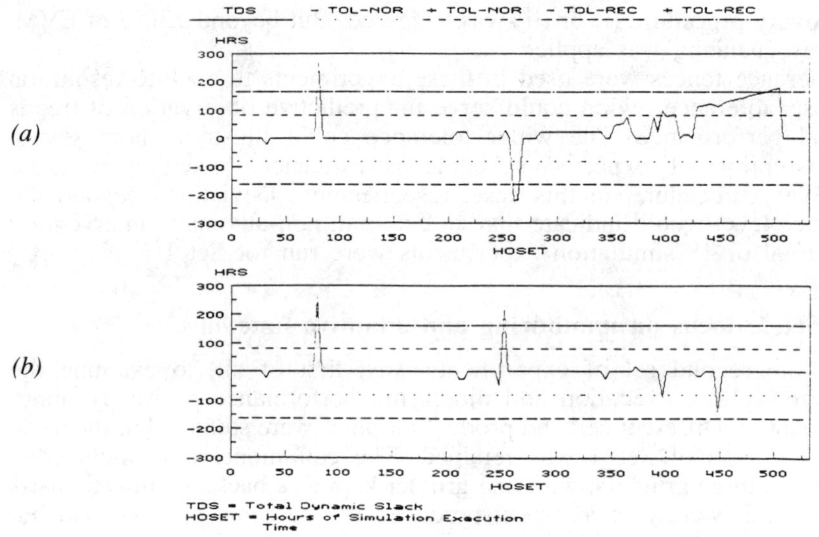

Figure 2. Sample results from experiment Set 1: (a) cell performance with no adaptation; (b) cell performance with recovery by time shifting and resequencing, based on cumulative deviations.

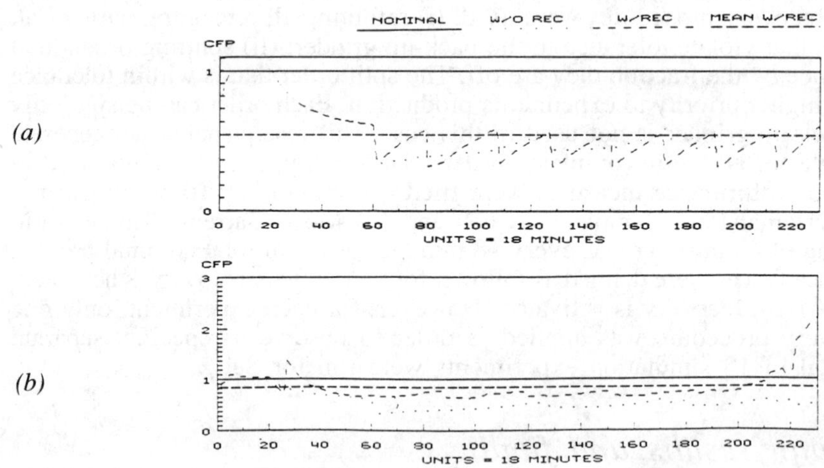

Figure 3. Sample results from experiment Set 2: disruption—both grinder failure and unexpected order; performance measure triggering recovery—CFP, mean current fraction of plan completed; recovery procedure—(a) rerouting, (b) order splitting.

(2) Since recovery by time shifting of scheduled operations adjusts and complies passively with the realistic production conditions of the cell, it always guarantees successful 'control' over the schedule. In other words, managers can tell from the current adapted schedule when production orders can realistically start and end. In our industrial survey it was found that even this type of automatic adaptation which is not commonly available today, can be highly desirable, as it will reduce the chaos often encountered on the shop floor.

(3) Comparing the three performance measures used in experiment Set 1, it was found that the Total Dynamic Slack was the best for monitoring in this case. The reason is that it is more responsive, compared to Mean Order Lateness or Cumulative Order Lateness, since it can be calculated at the end of each operation rather than wait till each order's

completion. Such a measure can be considered 'fast'. Having more responsive tracking information can certainly allow the COMPARATOR and RESOLVER to predict better the adaptation needs (as demonstrated in Case 1b). On the other hand, when using a fast monitoring measure the tolerance fences have to be specified more carefully to prevent cases of instability as a result of over-adaptation.

(4) In summary, it was demonstrated in several different ways that the adaptive framework is feasible as well as useful, and that it can significantly impact the achievement of scheduled production, if at all possible, by fine-tuning dynamically the production plan.

Analysing the second set of experiments, both the CFP and the CCR proved to be highly useful. They provided predictive monitoring capability and led to similar positive results in terms of adapted schedule performance. The original schedule, planned without any disruptions, yielded by definition a mean CFP of 100%. Without any recovery, this measure shrank down to between 53·3% when there was an unexpected 'hot' order, to 69·9% when there was a machine failure. With adaptation, the mean CFP for the unexpected order was recovered back to 100% by rerouting, or to 99·9% by resequencing. With machine failure, the mean CFP was recovered to 86% by rerouting, and to 99·8% by resequencing.

When two disruption events occurred during execution (see Figure 3), the mean CFP without any schedule adaptation went down to 55·6%. Rerouting the orders did not improve performance, but with order splitting recovery it went up to 83·5% and with resequencing it rose back up to 99·9%. Although statistical analysis is not valid here because of the relatively limited randomness in this set of experiments, it is evident that proper selection of recovery procedures could adapt cell execution to significant achievements. We consider it most impressive that in this demonstration, with all three cases of disruption, at least one of the recovery procedures applied brought performance back to 100% (99·8% with machine failure), whereas without adaptation performance went down to between about 53% and 70% of the plan. Moreover, within the adaptive framework in these experiments, even a relatively weak recovery procedure yielded better performance than no recovery at all.

Conclusions

The general approach in this research has focused on the development of an automatic control framework which will yield a general, real-time computer system for adaptive production scheduling and control. The emphasis has been on sustaining control of the manufacturing system over the production execution interval. For the purpose of feasibility analysis, we have defined 'control' to mean the generation, and subsequent possible adjustment, of achievable schedules and the communication of those achievable schedules to the manufacturing system throughout the production interval. Further, the automatic control framework is intended to provide improvements over, or at least retention of, performance objectives set by management, despite disruptions which may go on in the manufacturing process.

Overall, the technology proposed has been explored in great detail through various simulation experiments to evaluate its feasibility in the manufacturing environment. Our conclusions are that the proposed concepts and automatic control framework are feasible in real manufacturing operations. The technology has been found to be responsive to the schedule revision time demands of the factory floor and highly flexible in its ability to address the needs of various forms of manufacturing organizations. This is due in part to our integration of this new technology with the FACTOR scheduling system, which is well accepted in industry.

Acknowledgements

Research reported here was supported in part by NSF Grant No. ISI-8761000. The authors thank Doug MacFarland of Pritsker Corporation for assisting in the experimental work.

References

De, S., 1987, A knowledge-based approach to scheduling in a FMS. In *Approaches to Intelligent Decision Support*, Annals of Operations Research (Basel: Baltzer Scientific).

Godin, V.B., 1978, Interactive scheduling: historical survey and state-of-the-art. *AIIE Transactions*, **10**, 331.

Goldratt, E.M. and Cox, J., 1984, *The Goal, A Process of Ongoing Improvement* (North River Press).

Hodgson, T.J. and McDonald, G.W., 1981, Interactive scheduling of a general flowshop. *Interface*, **11**, 83.

Nof, S.Y. and Gurecki, R., 1980, MDSS: Manufacturing Decision Support System. *Proceedings of AIIE Spring Conference*, Atlanta, GA, May, p. 274.

Nof, S.Y., Rajan, V.N. and Frederick, S.W., 1988, *Dynamic, Real-time Scheduling and Rescheduling, Literature Review with some Annotated References.* Research Report, School of Industrial Engineering, Purdue University, W. Lafayette, IN 47907.

Robbins J.H., 1986, Simulation helps schedule shop. *American Machinist*, October, 106–108.

Slutz, D. and Munji, S., 1987, Development of a scheduling system for a CIM electronic component manufacturing process. *Proceedings of IXth International Conference on Production Research*, Cincinnati, OH, August.

Young, R.E. and Rossi, T., 1987, Knowledge-based approach to scheduling and control in FMS. *IIE Transactions*, **19**.

20

A new concept on generative bills-of-material: the generic bill-of-material

E.A. van Veen[†] and J.C. Wortmann

Eindhoven University of Technology, PO Box 513, 5600 MB Eindhoven, The Netherlands

Abstract

As a result of the market trends of the last two decades, many businesses have undergone a fast growth of their variety of final products. In many cases it became practically impossible to record the entire product variety by uniquely identifying each and every different product variant. However this is required by most currently available standard Production Control Information Systems (PCIS). Future PCIS should offer better means to make use of commonalities in product structures, resulting in product (structure) data to be recorded in a more effective and manageable way. In this chapter the concept of generic bills-of-material is introduced. This concept supports the possibility to define a bill-of-material (BOM) of *a set of* product variants, as opposed to being forced to define the bill-of-material for every single product variant separately. Still a specific BOM can be generated for every individual product variant the moment it is required.

Introduction

Flexibility has become a keyword in manufacturing industries. In order to remain competitive, a business must be able to manufacture products meeting specific customer requirements, short delivery lead-times, high quality, and low costs.

As a result of this market pressure, many businesses have undergone an often uncontrolled growth in their variety of products. This has led to major control problems in several organizational functions such as Engineering, Manufacturing Engineering, Forecasting, Production Control, Servicing, etc.. These control problems are largely due to both the *complexity* and the *variety* of product and process data that must be handled. There are two basically different approaches towards solving these problems of control:

(1) Getting control of the growth in complexity and variety, i.e. eliminating and preventing unnecessary complexity and variety, e.g. by defining a clear distinction between product variants that are part of a standard variety and product variants that are manufactured specifically for one customer (order) and will therefore not be part of business' standard product variety.
(2) Increasing the capability of handling massive and complex product and process data by enhancing the information systems' concepts.

[†] The author performed this research at Eindhoven University of Technology while being detached from the Institute of Information-Technology for Production Automation, Eindhoven, The Netherlands.

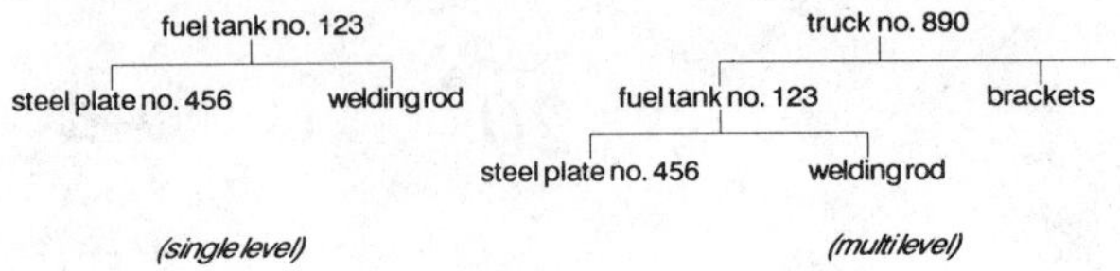

Figure 1. The single level and multi-level representation of bills-of-material.

In this chapter we will concentrate on the latter alternative. We will show that currently known concepts used in standard Production Control Information Systems (PCIS) do not support the recording of massive product (structure) data satisfactorily. We will describe a new concept that deals with the problems of recording massive product data. This concept is the generic bill-of-material.

Problem area

The research on Generic Product Descriptions in general concentrates on improving the capability of PCIS to support various descriptions of massive numbers of end-item products, sub-assemblies, etc.. In most currently available standard PCIS, product data must be defined explicitly and uniquely for each and every different product variant, disregarding any similarities in product descriptions. This approach requires large efforts for data entry and data maintenance (technical as well as organizational). In the case of an engineering change in the part of the product description that is common to many product variants, the same change must be carried into effect repeatedly for the descriptions of every single product variant.

Viewed in the light of the increasing product variety better concepts for representing product and process data in standard PCIS should be developed. Future PCIS should offer better means to make use of commonalities in product descriptions, resulting in product and process data to be recorded in a more effective and manageable way (van Veen and Wortmann, 1987).

As our primary interest lies in the area of production control, we will refer to bills-of-material as a specific type of product description.

Traditional bills-of-material

In traditional [materials requirements planning (MRP)-based] software systems the major types of product data, describing various aspects of a product are (Bertrand *et al.*, 1990):

Item related data—The attributes of the entity type 'item' describe aspects such as cost-price of the product, lead-time required to manufacture the product, etc..

Product structure data—The bill-of-material describes the 'goes-into' relationships between items. The bill-of-material (BOM) of an item (the parent-item) is a set of relationships describing all component-items of this particular parent-item. Each relationship specifies at least the quantity of the component item required to realize one parent-item. A single level BOM defines the material contents of a product. The multi-level BOM, which can be retrieved from a series of single level BOMs, additionally shows the complete material structure of the product (see Figure 1).

Routing data—The routing of a product is a set of operations describing the way a product is to be manufactured or assembled.

Figure 2 depicts the typical data structure diagram for this type of product structure representation.

PCIS which are focused on controlling stock levels of products in various stages

(a)

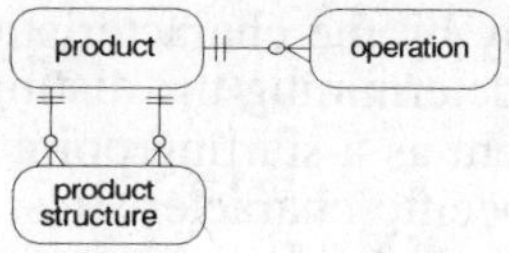

(b)

TABLE ITEM

item no.	description	unit
890	truck	each
123	fuel tank	each
456	steel plate	m^2
	welding rod	m

TABLE PRODUCT STRUCTURE

parent	component	quantity
890	123	1
123	456	2.7
	456	3.1

TABLE OPERATION

oper. no.	item	workcentre	manhrs.	tools
saw	123	burnsaw	0.5	423
bend	123	platebend	0.2	984
weld	123	welding	0.3	121

Figure 2. A typical data structure diagram (a) for representing product and process data in MRP-based PCIS. (b) An example.

of manufacture in a goods flow (MRP-based systems) rely on the separate control of different product variants. For this reason each product variant must be identified uniquely in the product data base (Orlicky *et al.*, 1972).

As the number of different product variants increases strongly it becomes more difficult—if not impossible—to record the full product variety by uniquely identifying each and every different product variant and the more difficult it will be to define all bills-of-material explicitly. This situation occurs in several types of industry, such as system-industry, all kinds of machining industry, truck industry and also in some industries for consumer goods (Wedekind and Müller, 1981; Schönsleben, 1985).

However in the customer-order specific part of the goods flow it is not necessary to identify each product variant, other than by the customer-order it belongs to. For the purpose of controlling this part of the goods flow, it will often be sufficient to generate product data describing: (i) which materials are required, and (ii) how to assemble or manufacture the product variant. In some cases additional descriptions such as drawings, testing documents, configuration data or user documentation must be generated. For this purpose the concept of generic product descriptions has been developed.

Generic bills-of-material

The basic idea of the concept of generic BOMs is that in recording product data, better use should be made of the similarities in product variants and their descriptions. BOMs should be defined for ranges of similar product variants, instead of for each individual product variant explicitly (van Veen, 1990). However if a specific product variant has been requested by a customer, the required BOM should be made available instantly. This requires mechanisms to specify specific items and generate the related BOM.

Generic items

A generic item represents a number of different product variants. These variants are represented by one generic item, because they have certain characteristics in common. Let us assume we can denote the common characteristics of all product variants related to a generic item g as $C_g : \{c_1, c_2, \ldots, c_n\}$. If an item represents only one product variant, the item will be called a specific item.

The product variants represented by a generic item differ on a number of other characteristics. These characteristics are the distinguishing characteristics, denoted by $D_i : \{s_1, s_2, \ldots, s_m\}$.

Every single product variant i is identified uniquely by the characteristics in the set $C_g \cup D_i$. That is by selecting a generic item and determining the distinguishing characteristics. In practice one will have a generic item as a starting point and will want to define a product variant that meets some specific characteristics. For this purpose, the distinguishing characteristics will be represented by *parameter values*. Sets of parameter values which are regarded as being each others' alternatives and which are therefore mutually exclusive are represented by *parameters*.

Obviously, not every permutation of related parameter values will represent an existing product variant in practice. Therefore a set of *constraints* in terms of parameter values must be defined. In order to represent an existing specific item, the distinguishing set of parameter values should meet every constraint related to the generic item.

The product variants which only exist in the customer-order related part of the goods flow do not have to be predefined explicitly within the PCIS. Once the identifying characteristics have been determined the descriptions required for material planning, assembly, testing, etc. can be generated. There are basically two ways in which a specific BOM can be generated from a generic item: *transformation* and *substitution*.

In the case of a transformation, the specific BOM is generated by specifying the distinguishing characteristics of the required product variant and computing the specific BOM. The product variant has not been defined explicitly. The way in which the computation proceeds is described later.

In the case of a substitution, a product variant is also identified by specifying the distinguishing characteristics. However the specific BOM is obtained by selecting one specific item (which only represents the required product variant) out of a set of predefined alternative specific items, associated with the generic item. For the specific item, a BOM has been defined explicitly. Of course, the substitution mechanism will in practice not be applied for end-items because of the large number of specific items and related BOMs which would then have to be explicitly predefined. The substitution mechanism often occurs for lower level items in the BOM.

The substitution mechanism: generic items

The possibility of a substitution implies that generalization–specialization relationships are to be defined. These relationships describe which predefined specific items belong to the range of variants represented by the generic item. Then, given a generic item and a number of characteristics a (more) specific item can be selected to substitute the generic item. Figure 3 shows the generic item fuel tank and the five specific variants it represents.

If a generic item is to be made specific by the substitution mechanism, a specific item must be selected on the basis of a set of distinguishing parameter values. For that purpose the specific items may have one or more 'substitution conditions' describing under which conditions in terms of parameter values the specific item must be selected to substitute the generic item it belongs to (according to a generalization relationship).

The transformation mechanism: generic bills-of-material

The bill-of-material of a product describes one specific aspect of a product, namely its material contents. The BOM of a generic item includes all BOM relationships that could be part of the BOM of one of the represented specific items. The basic element of the BOM is the entity type product structure. A specific BOM can be drafted by including or excluding specific 'BOM-relationships'.

If an item is to be made specific by transformation, a specific bill-of-material must be derived by selecting the required BOM-relationships. In order to facilitate

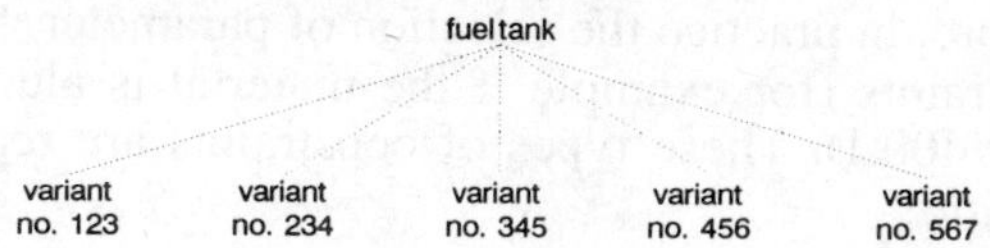

Figure 3. The generic item 'fuel tank' represents five specific items 'fuel tank'.

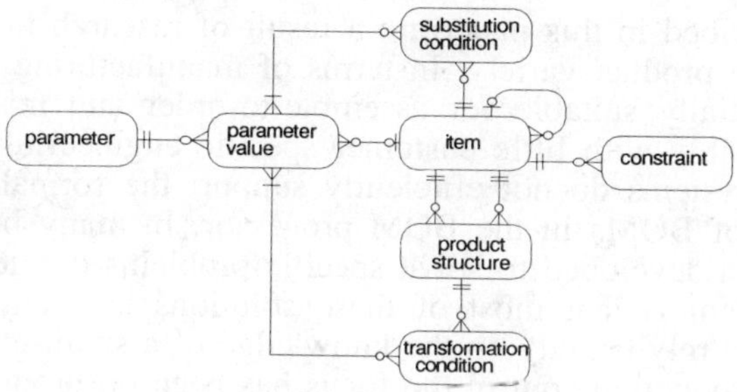

Figure 4. Extended data structure diagram for generic items and generic bills-of-material.

this selection upon a given set of parameter values, the BOM-relationships must be conditioned in terms of parameter values. These conditions are represented by entities of the type 'transformation condition'.

A generic item may have a specific BOM. In that case, all product variants represented by the generic item have a common materials content but they will differ on some other aspect such as the way it is being manufactured, viz. the routing.

Within a generic BOM structure, the following should hold:

(1) It should have a *modular character*, i.e. every item and its single level BOM can be regarded as a module and can be applied in other BOMs as a whole.
(2) If an item is dependent upon a set of distinguishing parameters P, then all items it is being applied in will be generic and will be dependent of at least the set P.
(3) Once the distinguishing parameter values of any generic item have been determined (i.e. for every parameter one value has been selected), a BOM-structure can be derived unambiguously.

Figure 4 depicts a data structure diagram by which both the substitution of generic items by specific items and the transformation of generic items by generating specific BOMs can be supported. The entity type 'item' in this model represents both generic items and specific items. The $1:n$ relationship from the entity type 'item' to itself represents the specialization–generalization relationships described earlier.

The entity type 'parameter value' represents product characteristics. For example, the fuel tank no. 123 may have as its characteristics that it is made of steel and its capacity is 150 l. Sets of characteristics which are mutually exclusive must refer to one and the same entity of the type 'parameter'. For example for the generic item 'fuel tank', a parameter 'material' can be defined with two mutually exclusive values, i.e. steel and aluminium. In fact an item indirectly refers to a set of entities of the type 'parameter'.

An item will be defined as being generic if it is associated with two or more parameter values which belong to the same parameter. In that case in order to define a specific item one value must be selected of each related parameter. On the other hand, the item is specific if it does *not* have any parameter values, or if it has exactly one value for every parameter.

As mentioned before, in practice the selection of parameter values will be subject to all kinds of constraints (for example if the material is aluminium the capacity may not be 350 l or 400 l). These types of constraints are represented by entities of the type 'constraint'.

Conclusions

The concepts described in this paper are a result of research in several industries which have a large product variety. In terms of manufacturing environments, the concepts are especially suitable for assemble-to-order and manufacture-to-order environments, possibly with little customer specific engineering.

Most standard systems do not efficiently support the formal representation of massive numbers of BOMs in the BOM processor. In many businesses, specific solutions have been developed to solve specific problems due to growing product variety. The problem is that most of these solutions lack any underlying clear overall concept and rely heavily on the knowledge of a small group of specialists.

In the area of production control the focus has been on product specification in the order entry process. The primary goal here is to improve order entry by preventing product specification errors and by shortening the order entry lead-time. The attempts in this area have been rather isolated. Relatively little attention has been paid to the generation of required BOMs and assembly instructions. Also little attention has been paid to specifying other products than end-item products, and to the relationship to engineering from a viewpoint of maintainability product data as a natural part of the engineering process.

The concept of the generic BOM and the mechanisms for substitution and transformation described here enable the formal representation of a wide variety of bills-of-material. When applying the generic BOM concept the maintenance problems will decrease, although there might be a shift in problems from dealing with massive product data redundancy to dealing with complex conditions and constraints.

The data structure diagram described here can be the basis for future BOM processors in PCIS. A first step in this direction will be the development of a prototype system of such a BOM processor.

References

Bertrand, J.W.M., Wortmann, J.C. and Wijngaard, J., 1990, *Production Control, A Structural and Design Oriented Approach* (Amsterdam: Elsevier).

Orlicky, J.A., Plossl, G.W. and Wight, O.W., 1972, Structuring the Bill-of-Material for MRP. *Production and Inventory Management*, **13**, 4, 9–17.

Schönsleben, P., 1985, *Flexibele Produktionsplanung und -Steuerung mit dem Computer* (München: CW-Publikationen).

van Veen, E.A., 1990, Modelling product structures by generic bills-of-material, Thesis, Eindhoven University of Technology.

van Veen, E.A. and Wortmann, J.C., 1987, Generic bills-of-material in assemble-to-order manufacturing. *International Journal of Production Research*, **25**, 11, 1645–1658.

Wedekind, H. and Müller, T., 1981, Stücklistenorganisation bei einer grossen Variantenzahl. *Angewandte Informatik*, **9**, 377–383.

21

Joint determination of production cycle and inspection schedules under inspection errors

Hiroshi Ohta and Sanae Ogawa

Department of Industrial Engineering, College of Engineering, University of Osaka Prefecture, Sakai, Osaka 591, Japan

Abstract

The production process tends generally to less efficiency in service because of the deterioration of machines and equipment as time elapses. Therefore, it is important to control the state of the production process by inspection and maintenance schedules. Quite recently, Lee and Rosenblatt (1987) have dealt with the problem of simultaneous determination of the economic production cycle and inspection schedules in a production system. In this chapter an extension of Lee and Rosenblatt's research is attempted considering not only Type I error but also Type II error of inspection; furthermore, quality assurance by the specified Average Outgoing Quality (AOQ) is discussed.

Introduction

The production process generally tends to lose some of its efficiency in service because of the deterioration of machines and equipment as time elapses. In consequence, the quality of products from the process decreases with time, therefore it is important to maintain the production process by inspection schedules. Recently, Lee and Rosenblatt (1987) have dealt with the problem of joint determination of the economic production cycle and inspection schedules in a production system. A simple relationship has been developed to determine whether maintenance by inspection is worthwhile or not. The problem of joint determination of the economic production cycle and inspection schedules is solved by using an approximation to the cost function.

In general, two types of inspection errors can arise: Type I error, where an 'in-control' process is diagnosed as 'out-of-control'; and Type II error, where an 'out-of-control' state is not detected. In this chapter, an extension of Lee and Rosenblatt's research is attempted with consideration of Type I and Type II errors of inspection. Although Type I error may not make the outgoing quality poorer, it has the effect of increasing the restoration cost. Type II error makes the Average Outgoing Quality (AOQ) worse. Optimal production cycle and inspection schedules under inspection errors are simultaneously determined by minimizing the expected total cost of the production system per unit time subject to assurance of the specified quality of products.

Model, assumptions and notation

The following model and assumptions are used to analyse the optimal joint determination of production cycle and inspection schedules:

(1) Consider the production of a single item on a single machine or production process.
(2) At the start of the production cycle, the production process is in an 'in-control' state, producing items of acceptable quality. In the control literature, there may still be a small percentage of defective items produced during the 'in-control' state, but in this chapter this percentage is assumed to be negligible.
(3) After a period of production time, the production process may shift to an 'out-of-control' state.
(4) The elapsed time for the process to be in the 'in-control' state, before the shift occurs, is a random variable which is exponentially distributed with mean $1/\mu$.
(5) Once a shift to the 'out-of-control' state has occurred, the production process stays in that state unless it is correctly discovered by inspection and followed by some restorative work. Inspection and restoration times are negligible.
(6) During a production of T time units, inspections of the production process are undertaken at times $T_1, T_2, \ldots, T_n$, $n \geqq 1$. Furthermore, an inspection and restorative work are always done at the end of production cycle so as to ensure that the process is in control at the beginning of each production cycle.

In this chapter, an extension of Lee and Rosenblatt's research is attempted introducing the following additional assumption:

(7) Inspections during maintenance are not perfect. Two types of inspection error, that is, Type I and Type II errors, can arise.

Notation

D : demand rate in units per year;
P : production rate in units per year;
T' : cycle time for each production lot;
T : actual production time for each cycle (production run), where $TP = T'D$; note also that $T \leqq T'$;
K : set-up cost for each production run, excluding inspection of the machines;
h : holding cost per unit per year;
s : cost incurred by producing a defective item, such as cost of rework , repair, replacement, loss of goodwill, etc.;
α : percentage of defective units produced when the process is in the 'out-of-control' state, assumed to be a function of the state of the production process and not the cumulative number of units produced;
v : cost of inspecting the production process;
r : cost of restoring the process from an 'out-of-control' state to the 'in-control' state;
n : number of inspections undertaken during each production run, $n \geqq 1$;
T_i : elapsed time from the beginning of the production run until the ith inspection takes place, $i = 1, 2, \ldots, n$;
N_i : number of defective items produced in (T_i, T_{i+1}), $i = 0, 1, 2, \ldots, n-1$;
λ : Type I error, that is, the probability that an 'in-control' state would be diagnosed as 'out-of-control' at each inspection;
β : type II error, that is, the probability that an 'out-of-control' state would be diagnosed as 'in-control' at each inspection.

Formulation of expected total cost

The total cost of the production system consists of the set-up cost, holding cost, cost of inspection, cost of restoration, and the cost of defective items.

Consider the time interval (T_i, T_{i+1}). In the case of no inspection errors, the production process begins at T_i in the 'in-control' state, because the process has been inspected at time T_i. Define τ to be the elapsed time of a shift from the 'in-control' to the 'out-of-control' state since the last inspection. Because of the memoryless nature of the exponential distribution, τ is again exponential with mean $1/\mu$. The number of defective items N_i produced in (T_i, T_{i+1}) is given by

$$N_i = \begin{cases} 0 & \text{if } \tau \geqq T_{i+1} - T_i \\ \alpha\, P(T_{i+1} - T_i - \tau) & \text{if } \tau < T_{i+1} - T_i. \end{cases} \quad (1)$$

Thus, the expected number of defective items in (T_i, T_{i+1}) is:

$$E(N_i) = \int_0^{T_{i+1}-T_i} \alpha\, P(T_{i+1} - T_i - \tau)\mu \exp(-\mu\tau)d\tau$$
$$= \alpha\, P\{T_{i+1} - T_i + (1/\mu)\exp[-\mu(T_{i+1} - T_i)] - (1/\mu)\}. \quad (2)$$

Now, in this chapter two types of inspection error, that is, Type I and Type II errors are considered. In this case the expected number of defective items in (T_i, T_{i+1}) is modified as follows:

$$E(N_0) = \int_0^{T_1-T_0} \alpha\, P(T_1 - T_0 - \tau)\mu \exp(-\mu\tau)d\tau \quad (3)$$

$$E(N_i) = \beta\, \alpha\, P(T_{i+1} - T_i)$$
$$+(1-\beta)\int_0^{T_{i+1}-T_i} \alpha\, P(T_{i+1} - T_i - \tau)\mu \exp(-\mu\tau)d\tau$$
$$(i = 1, 2, \ldots, n-1). \quad (4)$$

Note that in this case at the start of the production cycle, the production process is in control, but the production process does not necessarily begin at T_i $(i = 1, 2, \ldots, n-1)$ in the 'in-control' state, because Type II error of inspection can arise.

The total expected number of defective items in a production run is then (by defining $T_0 = 0$ and noting that $T_n = T$):

$$\sum_{i=0}^{n-1} E(N_i) = \alpha\, P\{T_1 - T_0 + (1/\mu)\exp[-\mu(T_1 - T_0)] - (1/\mu)\}$$
$$+ \beta\, \alpha\, P(T - T_1) + (1-\beta)\, \alpha\, P\{T - T_1$$
$$+ (1/\mu)\sum_{i=1}^{n-1} \exp[-\mu(T_{i+1} - T_i)] - (n-1)/\mu\}. \quad (5)$$

The cost of inspection in a production run is simply nv. For the cost of restoration, we note that, after the $(i+1)$ th inspection, the probability of the production process requiring restoration is:

$$(1-\beta)Pr\{\tau < T_{i+1} - T_i\} + \lambda\, Pr\{\tau \geqq T_{i+1} - T_i\}$$
$$= (1-\beta)\int_0^{T_{i+1}-T_i} \mu \exp(-\mu\tau)d\tau + \lambda\left\{1 - \int_0^{T_{i+1}-T_i} \mu \exp(-\mu\tau)d\tau\right\}$$
$$= (1-\beta) - (1-\lambda-\beta)\exp[-\mu(T_{i+1} - T_i)]. \quad (6)$$

Hence, the expected cost of restoration in a production run is:

$$r\sum_{i=0}^{n-1}\{(1-\beta) - (1-\lambda-\beta)\exp[-\mu(T_{i+1} - T_i)]\}.$$

As the length of a production cycle is $T' = TP/D$, then the expected total cost (TC) of the production system per unit time is given by

$$\begin{aligned}
\text{TC} &= \frac{KD}{PT} + \frac{(P-D)Th}{2} + \frac{D}{PT}[nv + r \\
&\quad \times \sum_{i=0}^{n-1}\{(1-\beta) - (1-\lambda-\beta)\exp[-\mu(T_{i+1}-T_i)]\} \\
&\quad + s\,\alpha\, P\{[T_1 - T_0 + (1/\mu)\exp[-\mu(T_1-T_0)] - (1/\mu)] + \beta(T - T_1) \\
&\quad + (1-\beta)[T - T_1 + (1/\mu)\sum_{i=1}^{n-1}\exp[-\mu(T_{i+1}-T_i)] - (n-1)/\mu]\}] \\
&= \frac{KD}{PT} + \frac{(P-D)Th}{2} + \frac{Dn}{PT}\left[v + r(1-\beta) - \frac{s\alpha P}{\mu}\left\{1 - \beta\left(\frac{n-1}{n}\right)\right\}\right] \\
&\quad + \frac{D}{PT}\left\{\frac{s\alpha P}{\mu}(1-\beta) - r(1-\lambda-\beta)\right\}\sum_{i=0}^{n-1}\exp[-\mu(T_{i+1}-T_i)] \\
&\quad + s\alpha D + \frac{s\alpha D}{\mu T}\beta\exp[-\mu(T_1 - T_0)]. \qquad (7)
\end{aligned}$$

The problem of interest here is to find the optimal production run time, T^*, the optimal number of inspections, n^*, and the associated inspection schedule, $(T_1, T_2, \ldots, T_n)$, which minimize the expected total cost TC as given by eqn (7). Once the optimal production run time, T^*, is determined, the economic production quantity, Q^*, is given by $Q^* = PT^*$.

Joint determination of optimal production cycle and inspection schedule

If n and T are given, to minimize the expected total cost TC given by eqn (7) is equivalent to minimize

$$\begin{aligned}
H &= \frac{D}{PT}\left\{\frac{s\alpha P}{\mu}(1-\beta) - r(1-\lambda-\beta)\right\}\sum_{i=0}^{n-1}\exp(-\mu x_i) + \frac{s\alpha D}{\mu T}\beta\exp(-\mu x_0) \\
&= A\sum_{i=0}^{n-1}\exp(-\mu x_i) + B\exp(-\mu x_0) \qquad (8)
\end{aligned}$$

where

$$A = \frac{D}{PT}\left\{\frac{s\alpha P}{\mu}(1-\beta) - r(1-\lambda-\beta)\right\}, \qquad (9)$$

$$B = \frac{s\alpha D}{\mu T}\beta \qquad (10)$$

$$x_i = T_{i+1} - T_i \geqq 0 \qquad (i = 0, 1, \ldots, n-1) \qquad (11)$$

and

$$\sum_{i=0}^{n-1} x_i = T. \qquad (12)$$

By using the Lagrange's method, to minimize H given by eqn (8) subject to the condition given by eqn (12) is reduced to minimize

$$U = H + \rho\left(\sum_{i=0}^{n-1} x_i - T\right)$$

$$= A\sum_{i=0}^{n-1}\exp(-\mu x_i) + B\exp(-\mu x_0) + \rho\left(\sum_{i=0}^{n-1} x_i - T\right) \tag{13}$$

where ρ is the Lagrange multiplier. Then, we have

$$\frac{\partial U}{\partial x_0} = -\mu\ A\exp(-\mu x_0) - \mu\ B\exp(-\mu x_0) + \rho$$

$$= -\mu(A + B)\exp(-\mu x_0) + \rho = 0. \tag{14}$$

$$\frac{\partial U}{\partial x_i} = -\mu\ A\exp(-\mu x_i) + \rho = 0 \quad (i = 1, 2, \ldots, n - 1). \tag{15}$$

From eqn (14) we obtain

$$\exp(-\mu x_0) = \frac{\rho}{\mu(A + B)} \tag{16}$$

that is

$$x_0 = -\frac{1}{\mu}\ln\frac{\rho}{\mu(A + B)}\ . \tag{17}$$

Similarly, from eqn (15) we have

$$x_i = -\frac{1}{\mu}\ln\frac{\rho}{\mu A} \quad (i = 1, 2, \ldots, n - 1). \tag{18}$$

Substitution of eqns (17) and (18) into eqn (12) yields

$$\ln\rho = \frac{1}{n}\{\ln\mu(A + B) + (n - 1)\ln\mu A - \mu T\}. \tag{19}$$

Accordingly, from eqns (17)–(19) we obtain

$$x_0 = (T/n) - C \tag{20}$$

$$x_i = (T/n) + C/(n - 1) \quad (i = 1, 2, \ldots, n - 1) \tag{21}$$

where

$$C = (1/\mu)(1 - 1/n)\ln\{A/(A + B)\}$$

$$= \frac{1}{\mu}\left(1 - \frac{1}{n}\right)\ln\left[\left\{\frac{s\alpha P}{\mu}(1 - \beta) - r(1 - \lambda - \beta)\right\}\Big/\left\{\frac{s\alpha P}{\mu} - r(1 - \lambda - \beta)\right\}\right] \tag{22}$$

and C does not depend on the production run time, T.

From the above results, substituting eqns (20) and (21) into eqn (7) gives

$$\mathrm{TC} = \frac{KD}{PT} + \frac{(P-D)Th}{2} + \frac{Dn}{PT}\left[v + r(1-\beta) - \frac{s\alpha P}{\mu}\left\{1 - \beta\left(\frac{n-1}{n}\right)\right\}\right]$$

$$+ \frac{D}{PT}\left\{\exp(\mu C)\exp\left(-\frac{\mu T}{n}\right) + (n-1)\exp\left(-\frac{\mu C}{n-1}\right)\exp\left(-\frac{\mu T}{n}\right)\right\}$$

$$\times\left\{\frac{s\alpha P}{\mu}(1-\beta) - r(1-\lambda-\beta)\right\}$$

$$+ s\alpha D + \frac{s\alpha D}{\mu T}\beta\exp(\mu C)\exp\left(-\frac{\mu T}{n}\right). \tag{23}$$

In general, it is not easy to solve for the optimal production run length T based on eqn (23). Thus, we will use the following approximation for the only exponential function, $\exp\{-(\mu T)/n\}$ which depends on T:

$$\exp\left(-\frac{\mu T}{n}\right) \simeq 1 - \frac{\mu T}{n} + \frac{\mu^2 T^2}{2n^2}. \tag{24}$$

The approximation, well known as Taylor's expansion, has been frequently used in the quality control literature to simplify the mathematical analysis involved (e.g. Chiu 1975a,b; Gibra 1978). Furthermore, in most cases μ is very small so that the approximation of eqn (24) is reasonable.

Using eqn (24), we can rewrite the cost function eqn (23) as

$$\mathrm{TC} = \frac{KD}{PT} + \frac{(P-D)Th}{2} + \frac{Dn}{PT}\left[v + r(1-\beta) - \frac{s\alpha P}{\mu}\left\{1 - \beta\left(\frac{n-1}{n}\right)\right\}\right]$$

$$+ s\alpha D + \left[\left\{\frac{s\alpha P}{\mu} - r(1-\lambda-\beta)\right\}\exp(\mu C)\right.$$

$$\left. + \left\{\frac{s\alpha P}{\mu}(1-\beta) - r(1-\lambda-\beta)\right\}(n-1)\exp\left(-\frac{\mu C}{n-1}\right)\right]$$

$$\times\left(\frac{D}{PT} + \frac{\mu^2 TD}{2n^2 P} - \frac{\mu D}{nP}\right). \tag{25}$$

Differentiating eqn (25) with respect to T, and setting the derivative to zero gives

$$T^*(n) = [NU/DE]^{1/2} \tag{26}$$

where

$$NU = \frac{D}{P}\left[K - n\{v + r(1-\beta) - \frac{s\alpha P}{\mu}\left\{1 - \beta\left(\frac{n-1}{n}\right)\right\}\right]$$

$$+ \left\{\frac{s\alpha P}{\mu} - r(1-\lambda-\beta)\right\}\exp(\mu C)$$

$$+\left\{\frac{s\alpha P}{\mu}(1-\beta)-r(1-\lambda-\beta)\right\}(n-1)\exp\left(-\frac{\mu C}{n-1}\right) \tag{27}$$

and

$$DE = \frac{(P-D)h}{2}+\frac{\mu^2 D}{2n^2P}+\left\{\frac{s\alpha P}{\mu}-r(1-\lambda-\beta)\right\}\exp(\mu C)$$

$$+\left\{\frac{s\alpha P}{\mu}(1-\beta)-r(1-\lambda-\beta)\right\}(n-1)\exp\left(-\frac{\mu C}{n-1}\right). \tag{28}$$

Using eqns (26)–(28), the optimal production run time $T^*(n)$ for each n can be obtained and substituting $T^*(n)$ into eqn (25) gives the minimum expected total cost TC*(n) for each n. Therefore, the optimal number of inspections n^* which minimizes $TC^*(n)$ is determined by the numerical analysis of varying n. Once the optimal production run time, T^*, and the optimal number of inspections, n^*, are determined, the associated inspection schedule, $(T_1, T_2, \ldots, T_n)$, is given by eqns (11), (20) and (21).

Quality assurance of products

It is especially important to consider the quality assurance of products in the case where Type II error of inspection can arise.

Setting $s = 1$ and $K = h = v = r = 0$ in eqn (25) gives the expected total number of defective items in production rate P in units per year as:

$$R = \frac{Dn}{PT}\left[-\frac{\alpha P}{\mu}\left\{1-\beta\left(\frac{n-1}{n}\right)\right\}\right]+\alpha D$$

$$+\left[\frac{\alpha P}{\mu}\exp(\mu C)+\frac{\alpha P}{\mu}(1-\beta)(n-1)\exp\left(-\frac{\mu C}{n-1}\right)\right]$$

$$\times\left(\frac{D}{PT}+\frac{\mu^2 TD}{2n^2P}-\frac{\mu D}{nP}\right). \tag{29}$$

Therefore, the average outgoing quality (AOQ) is given by

$$\text{AOQ} = R/P \tag{30}$$

The AOQ as given by eqn (30) depends on the production run time, T, and the number of inspections, n, so in what follows we will represent the AOQ as AOQ (T, n). Assume that a specified average outgoing quality AOQ* was given from the viewpoint of quality assurance. If AOQ (T^*, n^*) $\leqq$ AOQ* where T^* and n^* are the optimal values determined from the economic viewpoint in the previous section, then T^* and n^* are optimal from both viewpoints of economy and quality assurance. However, if AOQ (T^*, n^*) > AOQ*, we must find the optimal production run time, T^{**}, and the optimal number of inspections, n^{**}, which satisfy the specified AOQ* and minimize the expected total cost TC, simultaneously, by using eqns (26)–(28) and (30). A concrete procedure in this case will be illustrated in the following section.

Numerical examples

In order to illustrate the procedure for joint determination of optimal cycle and inspection schedule, we consider several numerical examples. Given that $K = 10$, $h = 0{\cdot}5$, $s = 1$, $\alpha = 0{\cdot}05$, $v = 4$, $r = 10$, $\lambda = 0{\cdot}01$, $D = 30\,000$ and $\mu = 50$. Table 1 shows the minimum expected total cost, TC*, the optimal production run time, T^*, the optimal number of inspections, n^*, and the optimal intervals of inspections, $T_1 - T_0$ and $T_{i+1} - T_i$ $(i = 1, 2, \ldots, n - 1)$, for various values of production rate, P, and Type II error, β. In Table 1, AOQ (T^*, n^*) is also given.

From Table 1, the following characteristics are derived:

(1) As Type II error, β, increases, both the minimum expected total cost, TC*, and AOQ (T^*, n^*) increase monotonously. Furthermore, the intervals of inspections, $T_1 - T_0$ and $T_{i+1} - T_i$ $(i = 1, 2, \ldots, n - 1)$, become longer, and the optimal number of inspections, n^*, decreases for reducing the chances of occurrence of Type II error.
(2) As the production rate, P, increases, the minimum expected total cost, TC*, increases, but both the optimal production run time, T^*, and the optimal intervals of inspections, $T_1 - T_0$ and $T_{i+1} - T_i$ $(i = 1, 2, \ldots, n - 1)$, become shorter, and AOQ (T^*, n^*) decreases.

The reason why AOQ (T^*, n^*) decreases is as follows:
From the assumptions made in the second section, at the beginning of each production cycle the process is in control by a restorative work done at the end of production cycle. Therefore, as the production run time becomes shorter, the number of restorative works increases, then the average outgoing quality is improved.

In what follows we will discuss the joint determination of the optimal production cycle and inspection schedule which satisfy the specified AOQ* and minimize the expected total cost, TC, simultaneously. Assume the AOQ (T^*, n^*) > AOQ* for a given production rate, P. In this case it is reasonable to control the production cycle and inspection schedule in order to satisfy the specified AOQ* and minimize the expected total cost, TC, simultaneously. However, it was verified from numerical analysis that the average outgoing quality is only slightly improved by controlling the production cycle and inspection schedule.

It is evident from eqn (30) that the average outgoing quality depends on the production rate, P. Therefore, we will try to improve the average outgoing quality by controlling the production rate, P. Assume that the production rate, P, is controllable by 500 units and the specified AOQ* = 0·015. From Table 1, the optimal production rate, P^*, which satisfies the specified AOQ* and minimizes the total expected cost is obtained as:

$P^* = 30\,500$ if $\beta = 0{\cdot}01$,
$P^* = 32\,000$ if $\beta = 0{\cdot}05$,
$P^* = 34\,500$ if $\beta = 0{\cdot}1$.

The minimum expected total cost, TC*, the optimal production run time, T^*, the optimal number of inspections, n^*, the optimal intervals of inspection, $T_1 - T_0$ and $T_{i+1} - T_i$ $(i = 1, 2, \ldots, n - 1)$, and AOQ (T^*, n^*) in each case are given in the block enclosed with bold lines in Table 1.

Concluding remarks

An extended procedure for joint determination of production cycle and inspection schedules has been presented with consideration of Type I and Type II errors of inspection. Especially, Type II errors make the average outgoing quality of products worse, so a procedure was proposed for joint determination of the optimal production cycle and inspection schedules which minimize the expected total cost of

Table 1. Numerical examples (K = *10*, h = *0·5*, s = *1*, α = *0·05*, D = *4*, υ = *10*, λ = *0·01*, *D* = *30 000*, μ = *50*).

β	*P*	30 500	31 000	31 500	32 000	32 500	33 000	33 500	34 000	34 500
0·01	TC*	1029·646	1049·231	1061·936	1071·165	1078·091	1083·516	11087·890	1091·191	1093·904
	T^*	0·2827	0·1912	0·1559	0·1327	0·1201	0·1084	0·0973	0·0951	0·0850
	n^*	25	17	14	12	11	10	9	9	8
	$T_1 - T_0$	0·0115	0·0115	0·0114	0·0113	0·0111	0·0110	0·0110	0·0108	0·0108
	$T_{i+1} - T_i$	0·0113	0·0112	0·0111	0·0110	0·0109	0·0108	0·0108	0·0105	0·0106
	AOQ (T^*, n^*)	0·0142	0·0139	0·0136	0·0133	0·0129	0·0126	0·0124	0·0119	0·0118
0·05	TC*	1063·585	1082·630	1095·01	1103·900	1110·748	1115·959	1119·958	1123·276	1125·734
	T^*	0·2667	0·1944	0·1584	0·1348	0·1219	0·1099	0·0986	0·0879	0·0861
	n^*	23	17	14	12	11	10	9	8	8
	$T_1 - T_0$	0·0128	0·0126	0·012	0·0123	0·0122	0·0121	0·0120	0·0120	0·0118
	$T_{i+1} - T_i$	0·0115	0·0114	0·011	0·0111	0·0110	0·0109	0·0108	0·0108	0·0106
	AOQ (T^*, n^*)	0·0159	0·0155	0·0151	0·0147	0·0143	0·0140	0·0137	0·0135	0·0131
0·1	TC*	1105·243	1123·502	1135·306	1143·745	1150·160	1155·048	1158·919	1161·941	1164·236
	T^*	0·2620	0·1779	0·1517	0·1278	0·1148	0·1028	0·0914	0·0894	0·0789
	n^*	22	15	13	11	10	9	8	8	7
	$T_1 - T_0$	0·0143	0·0142	0·0140	0·0139	0·0137	0·0136	0·0136	0·0133	0·0134
	$T_{i+1} - T_i$	0·0118	0·0117	0·0115	0·0114	0·0112	0·0111	0·0111	0·0109	0·0109
	AOQ (T^*, n^*)	0·0179	0·0175	0·0170	0·0166	0·0162	0·0158	0·0156	0·0151	0·0149

$i = 1,2, \ldots, n - 1$; unit time = year.

the production system per unit time subject to assurance of the specified quality of products. In the proposed procedure, the optimal production rate is also determined in order to satisfy the specified average outgoing quality, because it is very difficult to improve the average outgoing quality by controlling only the production cycle and inspection schedules.

There still remain, however, several problems that need additional work. For example, further extensions, such as allowing for shortages and non-exponential distribution of the elapsed time until a shift in the system's quality occurs, provide subjects for future study.

References

Chiu, W.K., 1975a, Economic design of attribute control charts. *Technometrics*, **17**, 1, 81–87.

Chiu, W.K., 1975b, Minimum cost control schemes using np charts. *International Journal of Production Research*, **13**, 4, 341–349.

Gibra, I.N., 1978, Economically optimal determination of the parameters of np-control charts. *Journal of Quality Technology*, **10**, 1, 12–19.

Lee, H.L. and Rosenblatt, M.J., 1987, Simultaneous determination of production cycle and inspection schedules in a production system. *Management Science*, **33**, 9, 1125–1136.

22

Job-shop evaluation using Petri nets

Jean-Marie Proth

INRIA-Lorraine, Cescom, Technopôle Metz 2000, 4 rue Marconi, 57070 Metz, France

Abstract

In this chapter, we study the behaviour of a job-shop manufacturing different types of parts following given ratios. The control is given by the input sequence, i.e. the order in which the part types enter the job-shop, and by the input sequence of each machine, i.e. the order in which the part types visit each machine.

The first part of the chapter is devoted to the modelling of the job-shop. In the second part, we use the previous model to analyse the job-shop. In particular, we show that it is always possible fully to utilize the bottle-neck machine(s).

The last part of the paper presents a heuristic algorithm leading to the optimal sequences, i.e. the input sequences which allow to maximize the productivity of the system with a minimal in-process inventory.

Introduction

A job-shop can be defined as a set of machines and a variety of part types which can be manufactured using the machines. Each part type is characterized by its manufacturing process, i.e. the sequence of machines each unit of this type has to visit in order to be manufactured and the time it spends on each machine. The sequence of machines visited is also known as the routing of the part type.

In this chapter, we are interested in evaluating the steady-state performance of such a job-shop assuming that the production is cyclic. The production mix is given. The control of the job-shop is the sequencing of the part types at the entrance of the system (input sequencing) and the sequencing of the part types on the machines. The sequencings comply with the production mix.

The problem is set in the next section. The Petri net model of the job-shop is developed in the third section. The fourth section is devoted to the evaluation of the job-shop system knowing the sequencing. Finally, the last section provides a heuristic algorithm leading to the maximal productivity of the job-shop with a minimum of parts in-process.

Setting the problem

We present the problem using a small example. Assume that the job-shop is composed of three machines M_1, M_2 and M_3 and is able to manufacture three part types P_1, P_2 and P_3.

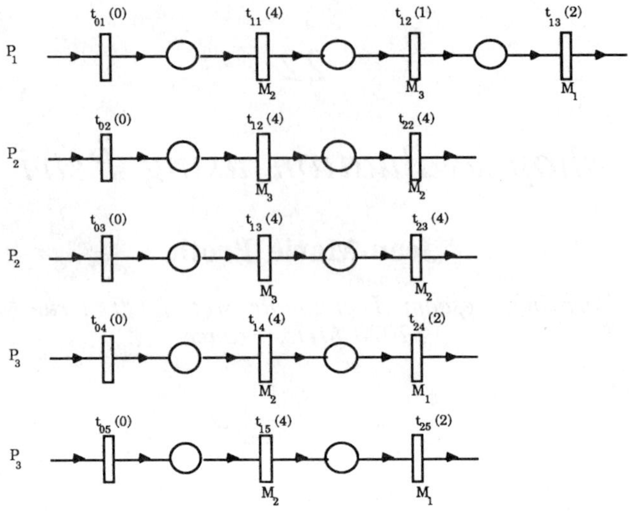

Figure 1. First step of the model design.

The manufacturing processes are the following:

P_1 : $M_2(4)$, $M_3(1)$, $M_1(2)$
P_2 : $M_3(4)$, $M_2(4)$
P_3 : $M_2(4)$, $M_1(2)$

The numbers in parentheses represent the time spent by parts on the machines. Assume that the production mix is respectively 20%, 40%, 40% between P_1, P_2 and P_3. An input sequencing could be $S_0 = (P_1, P_2, P_3, P_2, P_3)$ because it complies with the mix. M_1 being visited only by P_1 and P_3, a possible sequencing on M_1 is $S(M_1) = (P_1, P_3, P_3)$. Likewise, $S(M_2) = (P_1, P_2, P_2, P_3, P_3)$ and $S(M_3) = (P_1, P_2, P_2)$ could be chosen as sequencing on M_2 and M_3, respectively.

In this paper, we solve two problems:

(a) Knowing the sequencing, maximize the productivity using a minimal number of in-process parts.
(b) Find the sequencing allowing the maximal productivity with the number of in-process parts smaller than the one required for every other sequencing.

The Petri net model

We use the previous example to illustrate the model definition and analysis. The first step is to model the routing of each element of the input sequencing S_0, as a line of transitions (see Figure 1).

Each transition represents a machine. The values in brackets are the times spent by each part of this type on the machine. Transitions t_{0i} represent he loading of the parts into the system. The time associated with these transitions is zero, but it is possible to assign another value for this operation without changing the following results. Places correspond to storage buffers. They are called buffer places. Note that P_2 and P_3 are repeated twice.

The second step is to model the cyclic operation of the job-shop. It is obtained by changing the lines represented in Figure 1 into circuits as shown in Figure 2. These circuits will be referred to in the following as the processing circuits. The additional places are called resource places.

The final step is to model the control of the system, i.e. the input sequencing

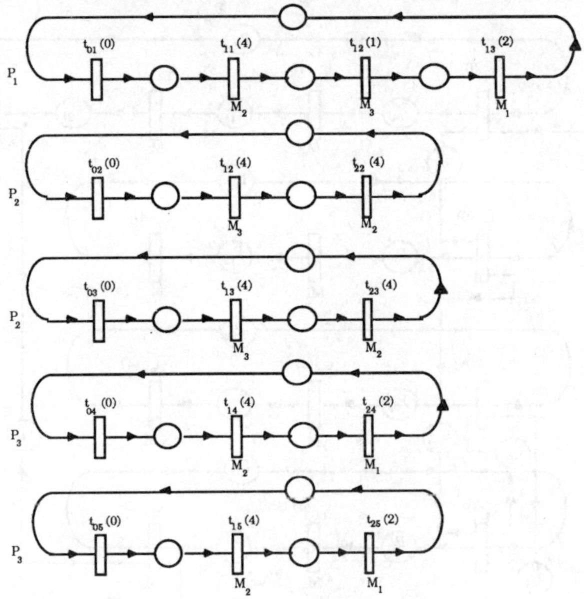

Figure 2. First step of the model.

and the sequencing of the parts on the machines. This is done by connecting all the transitions corresponding to operations performed by the same machine in a so called command circuit. This is represented in Figure 3 in which we restrict ourselves to the input sequencing and the sequencing on machine M_1.

Token circulating in process circuits represent in-process parts. The number of tokens in a process circuit is not limited. A command circuit contains only one token in order to model the fact that a given machine manufactures at most one part at a time. Command circuits represented in Figure 3 reflect S_0 and S (M_1) given in the previous section.

Analysis of the job-shop giving the control

The Petri net modelling the job-shop (see Figure 3) is a strongly connected event-graph model. An event-graph is a Petri net such that each place has exactly one input and one output transition. Commoner *et al.* (1971) showed that the total number of tokens in any elementary circuit is invariant by transition firing. There are three types of elementary circuit in the event-graph model: (i) the process circuits, (ii) the command circuits, and (iii) the mixed circuits which include transitions and places from the process circuits and the command circuits.

For instance, the elementary circuit (see Figure 3):

$$(t_{13}, q_{11}, t_{25}, p_{r5}, t_{05}, q_{05}, t_{01}, p_{01}, t_{11}, p_{11}, t_{12}, p_{21})$$

is a mixed elementary circuit.

For each elementary circuit e, we define the cycle time $C(e)$ as follows:

$$C(e) = t(e)/M(e) \tag{1}$$

where $t(e)$ is the sum of the times associated with the transitions belonging to circuit e and $M(e)$ is the number of tokens circulating in the circuit [$M(e)$ is invariant].

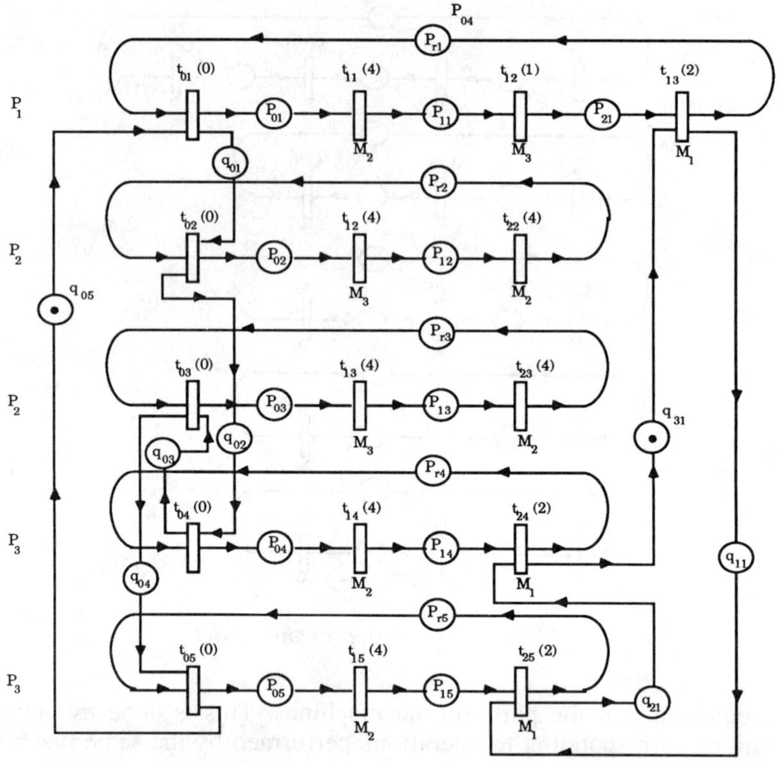

Figure 3. First step of the model design.

Let E be the set of elementary circuits associated with the model.
The maximum cycle time defined over all the elementary circuits is:

$$C^* = \max_{e \in E} C(e) \tag{2}$$

$e \in E$ such that $C(e) = C^*$ are called critical circuits.

It has been shown [see in particular Chretienne (1983)] that if the transitions fire as soon as they are enabled, the firing rate of all transitions is given by

$$r = 1/C^* \tag{3}$$

and the system works at its maximal processing time.

The following results are the consequence of the previous properties of the strongly connected event-graphs.

Result 1

It is possible to utilize the bottle-neck machine fully whatever the sequencing of the parts may be by choosing an adequate initial repartition of the parts in the job-shop.

Proof

According to the previous properties, the goal is to find an initial repartition of the tokens in the event-graph model such that the command circuit passing by the transitions corresponding to the bottle-neck machine becomes a critical circuit. Assume that the places of the graph are numbered from one to L and that the elementary circuits are numbered from one to K. If we want to find the initial repartition using the minimal number of circuits, we have to solve the following integer LP problem:

$$\text{Minimize} \quad \sum_{j=1}^{L} x_j \tag{4}$$

s.t.

$$\sum_{j=1}^{L} d_{ij}\, x_j \geq a_i \qquad \text{for } i = 1, 2, \ldots, K$$

where $d_{ij} = 1$ if the place j belongs to the elementary circuit i and 0 otherwise; x_j is the number of tokens initially in place j, a_i is the minimal integer number such that:

$$t(i)/a_i \leq t(e^*),$$

e^* being the command circuit corresponding to the bottle-neck machine. In addition, for each command circuit i:

$$\sum_{j=1}^{L} d_{ij}\, x_j = 1.$$

This problem has at least one admissible solution which is $x_j = 1$ for each j belonging to a process circuit and one token in each command circuit. Thus, problem (4) has an optimal solution. A fast heuristic algorithm has been proposed to solve this problem (see Hillion and Proth, 1988, 1989).

Note. According to the hypotheses of the problem, each token in a process circuit represents a part and a transportation resource. Thus, the solution to problem (4) provides the minimal initial repartition of the parts in the job-shop which allows full utilization of the bottle-neck machine.

Result 2

The system blocks if there exists an elementary circuit which contains no token.

Proof

Obvious from eqns (1), (2) and (3).

Optimal control

The minimal number of parts needed in the system in order fully to utilize the bottle-neck machine depends on the input sequencing S_0 and on the sequencing on the machines. The goal of this section is to find the sequencing which leads to the maximal productivity using the minimal number of tokens. The problem can be set as follows:

Find an optimal scheduling (times at which the different operations are processed) and sequencing (order in which machines process parts) on the interval $[0, t(e^*)]$.

We propose a fast heuristic algorithm to solve this problem.

Let $e(P_i)$ be a processing circuit corresponding to part-type P_i and

$$N(P_i) = \lceil t(e(P_i))/t(e^*) \rceil \tag{5}$$

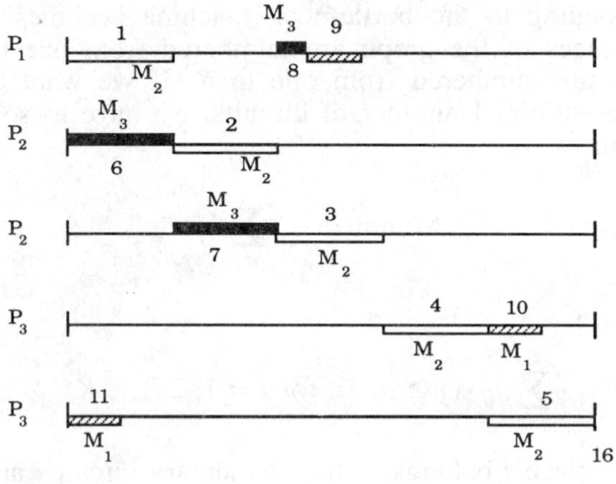

Figure 4. Sequencing and scheduling heuristic algorithm.

where $\lceil a \rceil$ is the smallest integer greater than or equal to a.

We define the degree of flexibility associated with part P_i as follows:

$$d(P_i) = [N(P_i) \cdot t(e^*) - t(e(P_i))]/[N(P_i) \cdot t(e^*)]. \quad (6)$$

We associate $N(P_i)$ elementary periods of length $t(e^*)$ with each P_i of the input sequencing.

The heuristic algorithm can be summarized as follows:

(1) Order parts in their increasing degree of flexibility.
(2) Schedule first operations of the bottle-neck machine.
(3) Schedule the remaining operations in the order of the parts, taking into account the routing and the fact that a given machine cannot be used to manufacture more than one part at the same time.

If at a given point of the algorithm, it is not possible to set the remaining operation concerning a part in the initial elementary periods, we assign a new elementary period to this part and set the operations in this new period.

Note. If we obtain a solution with just the number of elementary periods assigned at the beginning, the solution is optimal. If we have to add some elementary periods, we cannot claim that the solution is optimal.

We again consider the small example introduced in the second section. The degrees of freedom are $d(P_1) = 13$, $d(P_2) = 12$ and $d(P_3) = 14$. One elementary period is assigned to each element of the input sequencing.

The operations are placed in this period in the order indicated in Figure 4. Because no elementary period has been added to the initial ones, we can claim that the solution is optimal.

Figure 4 shows that, for the optimal solution:

$S(M_1) = (P_3, P_1, P_3)$
$S(M_2) = (P_1, P_2, P_2, P_3, P_3)$
$S(M_3) = (P_2, P_2, P_1)$.

We can also see that, with this solution, we need five tokens (i.e. resources) in the processing circuits. According to the notations these tokens are in p_{01}, p_{02}, p_{03}, p_{04}, p_{15} in Figure 3.

Conclusion

The main result of this chapter is that it is always possible fully to utilize the bottleneck machine. Another interesting point is the heuristic algorithm in the fifth section which provides a fast and near optimal (if not optimal) solution to the problem, i.e. a solution maximizing the productivity with a minimal number of transportation resources. A program running on PC/AT3 is available. Note that this algorithm encapsulates the results of the fourth section.

References

Chretienne, P., 1983, Les réseaux de Petri temporisés. Doctoral Thesis, University of Paris VI, Paris, France.

Commoner, F., Holt, A.W., Even, S. and Pnueli, A., 1971, Marked directed graphs. *Journal of Computer System Science*, **5**, 5, 511–523.

Hillion, H.P. and Proth, J.-M., 1988, Analyse de fabrications non linéaires et répétitives à l'aide de graphes d'evénements temporisés. *RAIRO/Recherche Opérationnelle*, **22**, 2, 137–176.

Hillion, H.P. and Proth, J.-M., 1989, Job-shop evaluation using Petri nets. *IEEE Transactions on Automatic Control*, **34**, 1, 3–9.

Hillion, H.P., Proth, J.-M. and Xie, X.L., 1987, A heuristic algorithm for the periodic scheduling and sequencing job-shop problem. 26th IEEE Conf. Dec. Contr., Los-Angeles, California, December.

SECTION III.3. SCHEDULING

23

The development of a generic bottleneck scheduler

J. Driscoll and S. Hurley

Mechanical Engineering, Surrey University, Guildford, UK

Abstract

This chapter concerns the preliminary development of a batch manufacturing systems modeller. The microcomputer model is an application of bottleneck scheduling with variable priority criteria. Methodologies for classifying manufacturing systems are examined in outline, leading to a physical–operational definition model, demonstrating how a manufacturing system may be defined by the operational data alone. Published works on the optimized production technology (OPT) approach, a widely publicized bottleneck scheduler, are reviewed.

Examples of operational data would include Bill-of-Materials and Job Pending file structures. Using files of this type a generic simulation model for production planning and control can be developed at an appropriate level of detail for the majority of manufacturing systems. Software models of this type are seen as having considerable potential in small and medium companies as a means of increasing effective manufacturing planning and control.

The chapter includes the modeller outline and concludes with a discussion of potential applications.

Introduction

An indication of the past limitations of scheduling software and research has been given by Woolsey and Swanson (1975), who listed as faults within scheduling software:

— tedious data refinement and preparation,
— difficult operation of software,
— time consuming and inflexible software,
— incompatible software, and
— inexperienced users.

To increase the practical utilization of scheduling and production modelling software, development should recognize the need to:

(1) *Structure software*:
  — create 'open' comprehensible procedures,
  — allow substitution, addition or withdrawal of key heuristics,
  — rationalize and limit the level of data employed, and
  — develop a core data management structure independent of operational heuristics.
(2) *Make software user-orientated*:
  — develop error limiting software,

— provide efficient expert guidance and support,
— encourage user confidence by creating professional menu, imaging and presentation procedures,
— use support documentation to shorten the learning curve, and
— increase software flexibility to suit a wide range of applications.

This chapter examines how the definition of manufacturing systems affects this process of software development, and then describes the initial specification of a microcomputer-based scheduling modeller.

Defining manufacturing systems

The development of production planning software is related to the original methods used to define the manufacturing systems. These definition methods range from general classifications to detailed functional specifications. Examples of this range can be seen from four systems approaches selected for discussion in this paper: (1) framework, (2) physical, (3) operational, and (4) generic.

Framework systems

Anthony (1965), first defined planning and control as consisting of three subsystems; strategic planning, management control and operations control. Wu (1987), further developing this framework into production systems, superimposed on Anthony's definition a dynamic analysis using the three following subsystems: (i) manufacturing, (ii) logistic, and (iii) planning and control.

Although each author was capable of qualitatively describing 'static' subsystems, the framework definitions of production from both authors remained abstract and at a level applicable to a wide range of production systems.

For modellers of manufacture, where the definition of quantitative activity is important, models in this category provide supportive background information but do not lead to a detailed understanding of operating activity.

Physical systems

Hayes and Wheelwright (1979), and Newman (1987) examined the definition of manufacturing as 'physical' systems, producing generally accepted stereotypes, for example, job shop, batch production, assembly line and continuous flow.

'Physical' descriptions, the most widely employed definition method, carry both an image of resource location and the type of manufacturing environment. Hayes and Wheelwright, and Newman, recognized 'physical' classification does not describe the method of operation. Newman proposed this deficiency be rectified by setting out basic production goals. Hayes and Wheelwright also extended the physical definition by developing a Product/Process Matrix which linked the physical production system to 'key management tasks' and 'competitive modes'. A third solution would be to describe manufacturing systems along with operational characteristics.

Operational systems

An example of an operational definition, which combines manufacturing philosophy and plant layout, is Just-In-Time (JIT), which emanated from Toyota in the 1970s. Leading papers and books explaining the JIT philosophy and application have come from Haynsworth (1984), Pegels (1984), Schonberger (1982, 1986) and Burnham (1987). Pegels (1984) gives an insight into operational definitions by stating the JIT goal as reducing work-in-progress, lead times, and finished goods inventories while pushing scrap levels down below one per million.

In plants where JIT has been fully implemented the operating philosophy has been seen to have a strong influence on the physical layout. Implementing the JIT philosophy, however, can be a lengthy and expensive process. Machines have to be moved, staff retrained, quality improved, supplier liaison increased and possible worker–manager conflict resolved.

Wilson (1985) identified problems with operational descriptions when reviewing JIT:

(1) JIT demands standardized products while the market place requests more options.
(2) Stocks cut in JIT companies pushes raw material holding costs onto suppliers.
(3) A higher shut-down probability exists as safety stocks are limited.

A second operational-generic model, of particular interest in this research, is Optimized Production Technology (OPT). The interest arises in the claimed universality (transparency), of applications. This places OPT as both an operational and a generic system and is therefore singled out for further examination.

Optimized production technology

Introduced by Goldratt and Cox (1984, 1986) and claimed to be a radical approach to manufacturing, OPT is based upon nine tenants which schedule workload around identified bottleneck machines. Non-bottleneck machines are considered 'servants' of key critical manufacturing resources. Loss of production from key resources is a loss to the whole manufacturing system.

OPT software has been applied widely in batch manufacturing environments. Operating with sales data, resource information, product routes, bill-of-materials and work patterns, production is modelled for existing factories. Both software and the OPT methodology have been reviewed widely by Everdell (1981), Jacobs (1983, 1984), Manning (1985), Meleton (1986), Plenert and Best (1986) and McManus (1987). No paper has, however, reviewed the software methods employed to identify bottle-neck resources, calculate lead times, due dates and schedules.

The OPT use of scheduling around the bottleneck machine is an example of a dominant global heuristic, where the ordering assigned to the bottleneck is applied to all connected machines. A further development within OPT is variable lead times and batch sizes. The mathematical supporting proof of variability is provided independently by David and LeFevre (1981).

Performance criteria are simplified within OPT to three 'Global Operational Measures' to judge whether a schedule is 'good'; these are:

Throughput—The rate at which the system generates money through sales.
Inventory—All the money the system invests in purchasing things the system intends to sell.
Operating expense—All the money the system spends in turning inventory into throughput.

The mathematical definition of these criteria is not available. OPT software secrecy has led to criticism. Vollmann (1986) identifies an 'A' class materials resource planning (MRP) systems with finite loading would create a package superior to OPT.

OPT can be seen therefore as a 'generic' manufacturing modeller transparent to the individual application. The software described later in this chapter initiates work on a modeller also in the 'generic' group but targeted specifically at small to medium companies.

Global scheduling, lead times and ordering rules

One of the key points from critical resource scheduling, as introduced by Baker (1974) and exploited by OPT, is the application of global scheduling, with all individual work patterns related to key machine schedules.

Global scheduling reduces the number of machine schedules required, and creates uniformity and 'flow' in the work pattern. The new modeller introduced later 'Manufacturing Simulation and Analysis' (MSA) is designed to experiment with different levels of global scheduling, with variations from total key machine dependence to individual machine sequencing.

A critical influence on sequencing is the reliability of lead times. Baker and Bertrand (1981) examined lead time variability, and the possibility of influencing lead times. They came to the conclusion that 'given' lead times are actually decision variables within the boundary of scheduling problems.

The work of Baker and Bertrand (1981), and as far as known, OPT, used a single ordering rule. MSA will allow experimentation on a key resource global model with differing sequencing heuristic rules. Simulation experimentation with the new global model should also give a feel for the accuracy of due date calculation methods typified by the Baker and Bertrand approach.

Performance criteria

When building manufacturing modellers, provision must be made for an appropriate set of performance criteria by which to judge the value of individual scheduling procedures. Most performance criteria can be split into two groups; time-based, which are generally customer related, and resource-based, which are producer related.

Time-based criteria

The most common example of time-based criteria is due date, of which many permutations exist. Two kinds of test are commonly applied: the actual delivery time variance (average time early or late—tardiness), and the frequency of failure or success in delivery (delivery reliability).

The attraction of these criteria is two-fold. First, they are closely related to the heuristics within scheduling models. Secondly, they relate to customer satisfaction, a point taken up later.

Resource-based criteria

Resource-based criteria can also be considered in two groups; usage and efficiency related criteria. Usage measures the absolute levels of resources employed, e.g. inventory levels and/or capital equipment levels. Efficiency criteria are concerned with achieving high utilization or efficiency of individual resources.

Flow time

A criterion frequently used in present manufacturing analysis is flow time, often expressed as throughput time. This criterion is popular because of a relationship to both time and resource-based groups. Low throughput times will lead to higher customer satisfaction and imply economy in the absolute levels of resources employed.

The concentration on flow time is often based on the conflict between efficient and effective manufacture. This conflict arises as time-based criteria can be

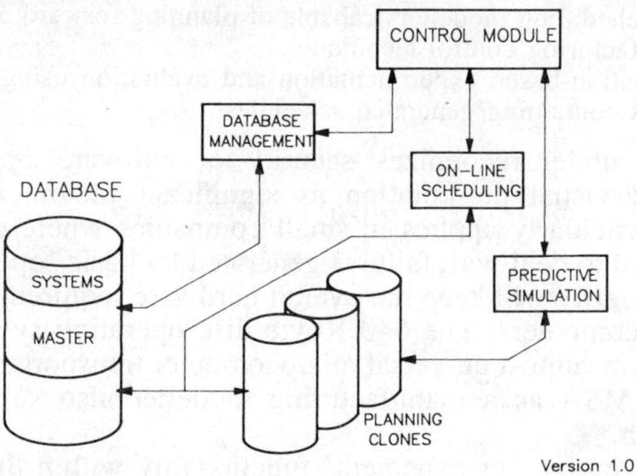

Figure 1. The conceptual model.

diametrically opposed to resource efficiency criteria. As a consequence scheduling models will seek to satisfy one of the two groups whilst measuring both. The approach supported by this work is that effective manufacture is preferable to efficient manufacture.

For this reason the generic simulator described next is based upon bottle-neck scheduling, a model concerned with satisfying customer delivery requirements and not internal resource utilization.

The 'MSA' generic modeller

Selecting the manufacturing subsystem from Anthony's (1965) framework for production as the particular area of interest within this research, work has started on a new systems modeller MSA. Developing the modeller has followed a three stage sequence:

(1) Specification of the model's conceptual framework.
(2) Definition of the structure and contents of data bases and software in line with the conceptual framework.
(3) Development and testing of the initial software version.

The conceptual framework

Two aims, application in small to medium companies and microcomputer orientation, generate a set of fundamental characteristics for a manufacturing modeller. These characteristics include a requirement for decision support, user-friendly reliable operation, flexibility, transportability in software and minimal data requirements.

Additionally, manufacturing software used in small company environments will be supporting simple production planning philosophies. This simplicity will be reflected in the planning heuristics embedded in software developed. The conceptual model eventually selected for the MSA software, which encompasses these characteristics, is shown in Figure 1.

The main features of the conceptual model are:

(1) Modular decision support and data management software, controlled through a central security and 'transaction' selection program.
(2) A two-part master data base with identified separate manufacturing and systems management information.

(3) Independent scheduling modellers capable of planning forward activity on the basis of popular manufacturing control techniques.
(4) Off-line simulation-based experimentation and evaluation using independent 'cloned' sub-data bases containing generated schedules.

The conceptual model recognizes security of software operation, transaction management and visual presentation as significant factors in present software systems. This particularly applies in small companies where computing expertise may be too limited to deal with failures generated by basic 'open access' programs.

Software modularity will keep the overall hardware requirement within the limits of present microcomputers. The 640 Kbyte disc operating system (DOS) limit, if achieved, will allow almost universal microcomputer transportability. The long-term development of MSA as a manufacturing modeller also supports an integrated modular approach.

This separate 'systems management' functionality within the conceptual model leads naturally to a two-part systems–manufacturing master data base as illustrated in Table 1.

The point has been made regarding the 'simplicity' of manufacturing approach with small companies. Alternative scheduling heuristics, progressively developed over time, can be embedded in software of this type as a series of expert planners to reflect differing popular control techniques.

Inclusion of an independent forward simulator is seen as a significant additional contribution to software of this type The process would be one of copying sets of critical manufacturing data into 'cloned' sub-data bases available for scheduling experimentation without risk of damage to the master data sets. This gives the ability to create schedules of work, forward simulate manufacture, analyse results and subsequently to implement the 'best' schedules back into the working data base at an appropriate time.

The MSA data base

The initial specification of the supporting data base for MSA consists of 11 master files and an additional 27 cloned scheduling and simulation files (Table 1). Manufacturing is represented by two product files, planning by three schedule files and resources by two files.

Four systems management files are listed. The systems control file (F-CTL) records the application specific information and also contains password control for users. Users, once given access, are controlled in the work they can undertake by file F-ACC. Systems use and performance is recorded in F-UIN and, finally, software appearance (visual images) is set by F-GRA.

Up to nine concurrent scheduling or simulation studies have been allowed for in this first software version. Three critical files, F-JOB, F-SCH and F-OUT can be copied at any time into cloned parallel files F-J**, F-S** and F-O**. This provides an up-to-date off-line sub-data base for use in developing and testing new production schedules.

MSA software

Continuing the conceptual model and data base structure, programs have been grouped into four sets; systems management, manufacturing data management, scheduling and simulation. Eighteen programs are defined in MSA Version 1·0, their logic being illustrated in Figure 2.

Complete independence for software content and removal of licensing difficulties with applications have been achieved by compiling from a high level language (Turbo Basic).

Four systems programs initialize and expand data bases, manage user access and

Table 1. MSA data file definition (Version 1·0).

Systems files definition	
F-CTL.pmt	Systems control and user vetting
F-GRA.pmt	Software graphics setting
F-UIN.pmt	User information
F-ACC.pmt	User access settings
Manufacturing	
Product	
F-BOM.inf	Bill of materials
F-PRD.inf	Product history file
Planning	
F-JOB.inf	Live jobs pending list
F-SCH.inf	Live production schedule
F-OUT.inf	Current completed work list
Resources	
F-RCH.inf	Resource characteristics
F-RST.inf	Resource performance statistics
Scheduling and simulation	
F-J01˜09.sim	Experimental clone of job list
F-S01˜09.sim	Experimental clone of current WIP locations
F-O01˜09.sim	Experimental clone of completed work

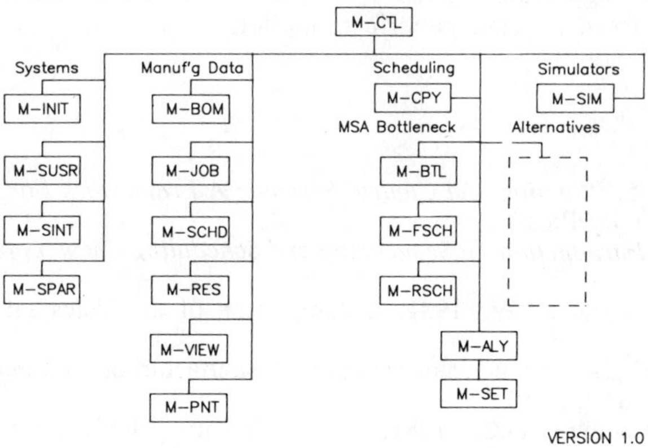

Figure 2. MSA software structure.

provide analysis of systems use. The six manufacturing programs provide expert supported access and interrogation of data base files. M-BOM, M-JOB, M-SCHD and M-RES allow the entry, editing and transfer of data. M-VIEW and M-PNT cover viewing and report printing. The three scheduling programs use bottleneck scheduling, with initial identification of bottleneck machines in program M-BTL. Generation of production schedules occurs on a forward schedule basis (M-FSCH) or alternative reverse scheduling basis (M-RSCH). Modularity will allow development of alternative scheduling approaches. Preparation of scheduling data is undertaken by M-CPY and post-scheduling analysis and re-transfer of schedules is undertaken by M-ALY and M-SET, respectively.

Forward simulation will be provided by a program set presently under the group title M-SIM.

Development progress

Development for MSA Version 1·0 has concentrated on specifying data base file contents and the creation of working programs. Data base contents are limited to scheduling model requirements and 'interfacing' to corporate data bases has been excluded.

Software for data manipulation is functional and the theoretical model for 'bottleneck' identification has been transferred to M-BTL. Present work is concentrated on completing forward and reverse scheduling programs. Experimentation, supported by general advice from an interested industrial company, is underway with the first software limits being specified as 40 products and 20 manufacturing resources.

Summary comments

This chapter has put forward a case for the development of generic manufacturing modellers based on the 'operational definition' approach. Bottleneck modellers, originally identified by Baker (1974), and brought to public attention by OPT, is an avenue worthy of further development.

Bottleneck scheduling provides a sound approach for small to medium companies, where production simplicity and/or limited size creates bottle-necks for production planning on a more frequent basis.

Within this chapter a conceptual model for a new generic manufacturing software suite has been outlined and the initial software definition discussed. Experience in developing to this stage indicates that a simple and effective manufacturing planner can be produced from a basic generic modeller.

References

Anthony, R.N., 1965, *Planning and Control Systems: A Framework For Analysis* (Boston: Harvard University Press).

Baker, K.R., 1974, *Introduction To Sequencing and Scheduling* (New York: Wiley), pp. 10–11.

Baker, K.R. and Bertrand, J.W., 1981, A comparison of due dates selection rules. *AIIE Transactions.*

Burnham, J.M., 1987, Some conclusions about JIT manufacturing. *Production and Inventory Management*, **28**, 3, 7–10.

David, H.I. and LeFevre, D.C., 1981, Finite capacity scheduling with OPT. *APICS Conference Proceedings 1981*, pp. 178–181.

Everdell, R., 1981, MRPII, JIT and OPT: not a choice but a synergy. *APICS Readings in Zero Inventory*, pp. 144–149.

Goldratt, E.M. and Cox, R.E., 1984, *The Goal: Excellence In Manufacturing* (Croton-on-Hudson, NY: North River Press).

Goldratt, E.M. and Cox, R.E., 1986, *The Race* (Croton-on-Hudson, NY: North River Press).

Hayes, R.H. and Wheelwright, S.C., (1979), Link manufacturing process and product life cycles. *Harvard Business Review*, **57**, 1, 133–140.

Haynsworth, H.C., 1984, A theoretical justification for JIT scheduling. *Production and Inventory Management*, **25**, 1, 1–3.

Jacobs, F.R., 1983, The OPT scheduling system: a review of a new production scheduling system. *Production and Inventory Management*, **24**, 3, 47–51.

Jacobs, F.R., 1984, OPT uncovered: many production planning and scheduling concepts can be applied with or without the software. *Industrial Engineering*, **16**, 10, 32–41.

Manning, E.J., 1985, OPT: a production technique case study. *BPICS Control*, **11**, 3, 17–21.

McManus, J.J., 1987, Control in the OPT environment. *Production Engineer*, **66**, 7, 20–21.

Meleton, M.P., 1986, OPT—fantasy or breakthrough? *Production and Inventory Management*, **27**, 2, 13–21.

Newman, P.A., 1987, Scheduling in CIM systems. In *Artificial Intelligence: Implications For CIM*, edited by A. Kusiak (Bedford: IFS Publications), pp. 359–402.

Pegels, C.C., 1984, The Toyota production system—lessons for american management. *International Journal of Production Management*, **4**, 1, 3–11.

Plenert, G. and Best, T.D., 1986, MRP, JIT and OPT: what's best? *Production and Inventory Management*, **27**, 2, 121–129.

Schonberger, R.J., 1982, *Japanese Manufacturing Techniques: Nine Hidden Lessons In Simplicity* (New York: Free Press).

Schonberger, R.J., 1986, *World Class Manufacturing: The Lessons Of Simplicity Applied* (New York: Free Press).

Vollmann, T.E., 1986, OPT as an enhancement to MRPII. *Production and Inventory Management*, **27**, 2, 38–47.

Wilson, G.T., 1985, Kanban scheduling—boon or bane. *Production and Inventory Management*, **26**, 3, 134–141.

Woolsey, R. and Swanson, H.S., 1975, *Operations Research For Immediate Application* (New York: Harper and Row).

Wu, C.T., 1987, A framework for production systems. In *Proceedings of the IXth ICPR*, edited by A. Mital, Cincinnati, OH, pp. 1471–1475.

24

One-machine scheduling with random tasks durations and general precedence constraints

Jean-Claude Hennet

Laboratoire d'Automatique et d'Analyse des Systèmes du CNRS, 7 avenue du Colonel Roche, 31077 Toulouse Cedex, France

Abstract

In the n-jobs, 1-machine case, the purpose of the stochastic scheduling problem (SSP) is to determine, at each decision time, what task currently has to be processed. Under the assumption of stationary probability transition matrices, the SSP is a Markov decision problem based on the maximization of the total expected discounted revenue for the whole production schedule over an infinite time-horizon. The simple case of structurally independent jobs can easily be solved as a bandit problem by the dynamic allocation index rule; but when general precedence constraints are considered, some elementary tasks can be common to several different chains in the graph. A two-level decision structure then has to be set up: simple chains of tasks can define elementary processes which are connected together to constitute 'superprocesses'. The problem analysed in this paper is how to construct superprocesses in such a way that the current optimal schedule can still be determined by index rules.

Framework of the study

The problem of optimal choice between alternative bandit processes was initially formulated by Bellmann (1956). As a discrete time Markovian stochastic control problem, it is characterized by the nature of its control variable which simply takes, at each decision time, the value of the selected process.

Each process can evolve only when it is selected. Its evolution is random and obeys a probability transition law. In the adaptive version of this problem, the transition law is not given, but has to be estimated in real-time from the available data. In this chapter, we restrict our approach to the deterministic case, when the probability transition law is a priori given.

Although the dynamic programming equations of this problem are relatively easy to establish, the combinatorial explosion puts severe limitations on their resolution.

It was only in 1972 that Gittins and Jones proposed an efficient technique for solving this problem in the case of a discounted cost function over an infinite time horizon. This technique is based on the computation of a Dynamic Allocation Index (DAI, also called Gittins index) for each process, and selection at the current time of the process having the larger index. The main advantage of this method lies in the fact that the evaluation stage is completely decomposed, even if the real-time policy to be implemented interwaves the processes.

The method was originally designed to solve a research planning problem, but it has soon been recognized as a valuable method for solving stochastic scheduling problems in the contexts of computer systems and of manufacturing industries.

In this study, we concentrate on the one-machine job scheduling problem. Some extensions of the theory are necessary to apply the index rule to relatively realistic problems.

(1) The stochastic scheduling problem should be considered in continuous time, since the completion of a task can occur at any time. Glazebrook (1976) has shown how the index rule could be applied to such *a family of alternative smuggler processes*. The optimal strategy for smuggler processes can be obtained from the discrete-time formulation of the DAI for the associated bandit problem by taking the limit of the index when the discretization period tends to 0.
(2) Most job scheduling problems impose non-pre-emption constraints on elementary tasks. These limitations can hardly be integrated in the index rule without jeopardizing the optimality of the solution. But fortunately, we can, with no major loss of generality, restrict our approach to the cases where optimal policies are naturally non-pre-emptive. A sufficient condition for a policy to be non-pre-emptive is that the index related to the job being served cannot decrease as long as the currently processed task has not been completed (Glazebrook, 1987). This property is usually verified under some mild conditions on the probability transition law when a task cannot provide any revenue as long as its achievement has not ended.
(3) Any scheduling problem is subject to precedence constraints. The structure of the problem can be described by a digraph, which is assumed to have no cycle. Elementary tasks can be represented as bandit (or smuggler) processes, and well-chosen sets of tasks as superprocesses. A condition of application of the index rule to superprocesses has been formulated by Whittle (1980). Some methods have been proposed for constructing superprocesses when the precedence graph is an out-forest in which each node has at most one predecessor (Glazebrook, 1976, 1982).

The purpose of this study is to extend the design of superprocesses to general precedence constraints while respecting Whittle's condition. In particular, the case of nodes with more than one predecessor can be solved through an appropriate definition of superprocesses. Such an extension of the methodology is quite significant in manufacturing since it concerns all the jobs with assemblage tasks.

Formulation of the index rule for superprocesses

The stochastic scheduling problem will be formulated as a problem of choice between N alternative discrete-time superprocesses, under two basic assumptions. These assumptions are satisfied by the usual data of any one-machine scheduling problem and, as will be shown in the next section, by the selected method for constructing superprocesses.

(1) The rewards generated by the production scheme can only be obtained from totally completed tasks. Any uncompleted task has a null reward. Besides, the tasks, processing time distributions are supposed to be non-decreasing, as is the case for most classical families of probability distributions. As shown by Glazebrook (1987), these conditions imply that the current optimal Gittins index cannot decrease as long as the currently processed task has not been completed. The optimal policy is then condemned to be non-pre-emptive.
(2) Whittle's condition on the validity of the index rule for superprocesses is supposed to be verified. This condition essentially stipulates that the optimal strategy internal to any superprocess i, S_i^* should not depend on the decision policy between superprocesses.

Under these two conditions, any superprocess i, denoted $(\Omega_i, P_i, R_i, \alpha)$, for $i = 1, \ldots, N$, can be represented as a completely ordered set of N_i elementary tasks. This ordering is clearly defined once the optimal strategy, S_i^*, which satisfies the non-pre-emption assumption, has been determined. The tasks which constitute the superprocess i can then be re-numbered from 1 to N_i in their processing order.

Ω_i is the state space of superprocess i. The current state $x_i(t) \in \Omega_i$ consists of the index of the current task waiting for service or currently being served and of the number of time units already spent on this task:

$$x_i(t) = (j, \theta_j) \text{ with } j = 1, \ldots, N_i + 1.$$

By convention, the state $(N_i + 1, 0)$ is obtained when the last task of superprocess i has been completed.

$P_i(\cdot)$ is a concatenation of the densities of probability $P_{ij}(\cdot)$ of the processing times required to complete tasks $j = 1, \ldots, N_i$.

α is a fixed discount factor: $0 < \alpha \leq 1$.

Let $a(t)$ be the control action, that is the number of the selected superprocess at time t.

$$P(x_i(t + \theta) = x_i(t) | a(\sigma) \neq i \quad \text{for } t \leq \sigma \leq t + \tau) = 1.$$

And by Bayes rule, we have:

$$P(x_i(t+1) = (j+1, 0) | x_i(t) = (j, \theta_j), a(t) = i) \\ = \left(\int_{\theta_j}^{\theta_{j+1}} P_{ij}(t) \cdot \mathrm{d}t\right) \Big/ \left(1 - \int_0^{\theta_j} P_{ij}(t) \cdot \mathrm{d}t\right)$$

and:

$$P(x_i(t+1) = (j, \theta_j + 1) | x_i(t) = (j, \theta_j), a(t) = i) \\ = \left(1 - \int_{\theta_j}^{\theta_{j+1}} P_{ij}(t) \cdot \mathrm{d}t\right) \Big/ \left(1 - \int_0^{\theta_j} P_{ij}(t) \cdot \mathrm{d}t\right).$$

To each task $j \in (1, \ldots, N_i)$ is associated its immediate completion revenue, denoted R_{ij}. If task j of superprocess i is completed between time t and time $t + 1$, the present value of the associated reward is $\alpha^{t+1} \cdot R_{ij}$.

The current reward function, $R_i(x_i(t), x_i(t + 1))$ can be defined as follows:

$$R_i(x_i(t), x_i(t + 1)) = R_{ij} \text{ if } x_i(t) = (j, \cdot) \text{ and } x_i(t + 1) = (j + 1, 0) \quad (1)$$

$$R_i(x_i(t), x_i(t + 1)) = 0 \text{ if } x_i(t) = (j, \cdot) \text{ and } x_i(t + 1) = (j, \cdot). \quad (2)$$

The objective function to be maximized is the total expected discounted revenue function over an infinite time horizon:

$$J = E\left(\sum_{t=1}^{\infty} \alpha^t \cdot R_{a(t-1)}(x_{a(t-1)}(t-1), x_{a(t-1)}(t))\right). \quad (3)$$

With the notation presented above, the current value at time t of the DAI for any process i, currently in state $x_i(t) = (j, \theta_j)$, is:

$$\gamma_i(x_i(t)) = \sup_{\tau^* \geq 1} \frac{E\left(\sum_{\tau=1}^{\tau^*} \alpha^\tau \cdot R_i(x_i(t + \tau - 1), x_i(t + \tau))\right)}{E(1 - \alpha^{\tau^*})} \quad (4)$$

For each superprocess i, the supremum of the fraction above is taken over all positive stopping times τ^*, assuming that before the stopping time the service is continuously assigned to superprocess i. In this way, the evaluation problem is totally decomposed and the DAIs are quite easy to compute.

Since the optimal policy is non-pre-emptive, the evaluation can be restricted to non-pre-emptive strategies. Thus, in a semi-Markov setting of the problem, decision times are only the initial time and the task completion times, and the state $x_i(t)$ of each optimally ordered superprocess, i, at decision time t, can be simply characterized by the index, j, of the next task to be processed. The partial DAI of superprocess i after completion of its $j - 1$ first tasks is, for $j \leq N_i$:

$$\gamma_i(x_i(t)) = \sup_{j \leq n^* \leq N_i} \frac{E\left(\sum_{n=j}^{n^*} \alpha^{\beta_j + \ldots \beta_n} R_{in}\right)}{1 - E(\alpha^{\beta_j + \ldots \beta_{n^*}})}. \tag{5}$$

By definition, β_n is the random processing time of task n. The optimal action at decision time t, $a(t)$ is such that:

$$\gamma_{a(t)}(x_{a(t)}(t)) = \max_{i=1,\ldots,N} \gamma_i(x_i(t)). \tag{6}$$

Under assumptions 1 and 2, the optimal scheduling policy is obtained by selecting at each time t the superprocess with the biggest current DAI, and by processing the task of this superprocess currently waiting for service. This task will be served until completion, unless the structure of the problem is changed by introducing new jobs. That case of 'arm-acquiring bandits' (Whittle, 1981) is not considered here.

Construction of superprocesses for general precedence constraints

The structure of a scheduling problem can be represented as a digraph with no cycle. The nodes are the tasks to be processed and an arc from node i to node j corresponds to the necessity of processing task i before task j. Constraints between tasks due to the common use of the same resource do not appear on the graph. The graph of the scheduling problem is determined at the beginning of the planning period. However, the real-time evolution of the processes changes the graph of active constraints, suppressing the constraints initially due to the tasks which have currently been completed, and eventually introducing in the queue some new jobs to be processed.

Application of the DAI technique to the one-machine scheduling problem implies the grouping of tasks into superprocesses. A necessary condition for the grouping process to be consistent with Whittle's condition is that superprocesses must be structurally independent. In graphical terms, this means that superprocesses must be disjoint subgraphs. Therefore, the smallest superprocesses which can be selected are the connected components of the digraph.

In order to fulfill condition 2 of the second section, any superprocess i has to be completely ordered by its internal optimal policy, S_i^*. If the ordering of this superprocess is performed according to some elementary principles, it is easy to show that it is optimal and that it will not be changed under any evolution of the system, even if some new jobs—independent from the previous ones—enter the waiting line. Only the choice between superprocesses will have to be recomputed at each decision time.

Any connected component of the precedence digraph will be transformed into a succession of evaluation graphs and ordered according to the following elementary rules:

(1) Superprocesses are analysed at the initial time. All the tasks of each superprocess have to be completely ordered by the internal optimal policy. The DAI of the superprocesses can then be compared. The first task of the selected superprocess then has to be processed. After completion of this task, the node associated with it in the precedence graph is suppressed and the index of its superprocess has to be recomputed and compared with the indices of the other superprocesses, previously computed.

(2) The ordering of a superprocess can be performed, at the initial time, by solving all the conflicts which are inherent in the structure of the superprocess. The anticipated decision times are the completion times of tasks. The precedence graph at the anticipated decision time is deducted from the initial precedence graph by suppressing all the tasks supposedly completed and all their connections to the other tasks.

The optimal order of a superprocess is such that at any stage of completion, the next task to be processed is the first task of the sequence yielding the maximal current value of the DAI for this superprocess. Clearly, only the data of the considered superprocess are involved in the task ordering scheme, and Whittle's condition is satisfied (Varaiya *et al.*, 1985). To reduce the number of feasible sequences to be evaluated, it is possible to use some properties of the precedence graph, as presented in the two following paragraphs.

(3) We call converging nodes the nodes with more than one immediate predecessor. In the initial evaluation stage, when the expected outcomes of the processing of two concurrent tasks are to be compared, any converging node following both tasks can be suppressed, as well as all the nodes following it. The tasks associated with these nodes cannot become active before complete processing of all of their preceding tasks. They do not influence the optimal ordering of the tasks preceding them since the revenues attached to them contribute by the same amount to any partial DAI associated with tasks preceding them. A converging node will be reintroduced in the evaluation graph, with the arcs connecting this node to the other nodes of this graph, as soon as n-1 of the n tasks directly preceding it have been ordered, and therefore do not appear any more in the current evaluation graph.

(4) Subgraphs with no vertex having more than one predecessor can be ordered according to a technique very similar to the one proposed by Glazebrook (1976) and called *branch contraction*. We call divergent nodes the nodes with more than one direct successor. Simple paths following a divergent node can be considered as alternative bandit processes, and ordered on the basis of a successive comparison of partial indices computed after each task completion.

The evaluation and ordering method will now be illustrated on a simple precedence graph. Partial indices for simple chains of tasks are computed by eqn (5). For instance, the index of a chain of tasks (I, J, K, L) is denoted $\gamma(I, J, K, L)$ and computed by the following equation:

$$\gamma(1, 2, 3, 4) = \sup_{1 \leq n^* \leq 4} \frac{E\left(\sum_{n=1}^{n^*} \alpha^{\beta_1 + \ldots \beta_n} R_{in}\right)}{1 - E(\alpha^{\beta_1 + \ldots \beta_{n^*}})}$$

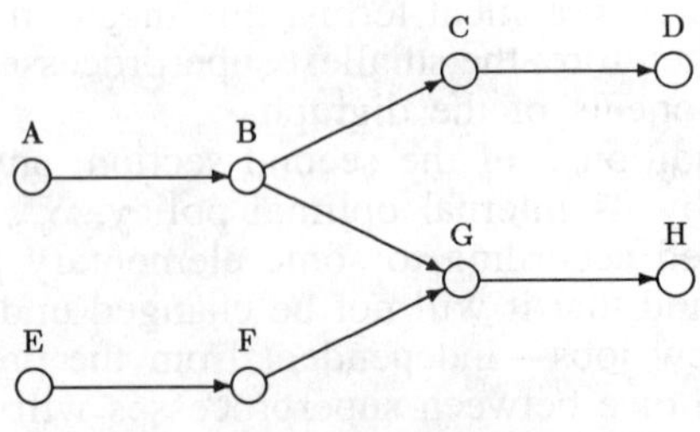

Figure 1. Example of a superprocess with eight tasks.

See Figure 1 for a superprocess with eight tasks.

First stage

This is shown in Figure 2.

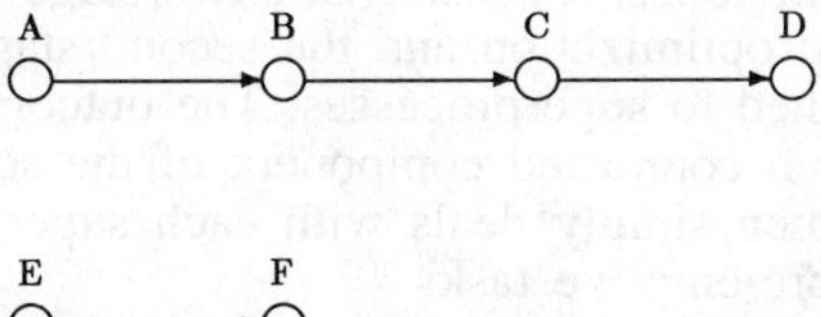

Figure 2. Suppression of the converging node, G, and of the node following it.

Second stage

Compare partial DAIs for (A, B, C, D) and (E, F); we get $\gamma(A, B, C, D) > \gamma(E, F)$. Compare partial DAIs for (B, C, D) and (E, F); we get $\gamma(E, F) > \gamma(B, C, D)$. Compare partial DAIs for (F) and (B, C, D); we get $\gamma(B, C, D) > \gamma(F)$ (see Figure 3).

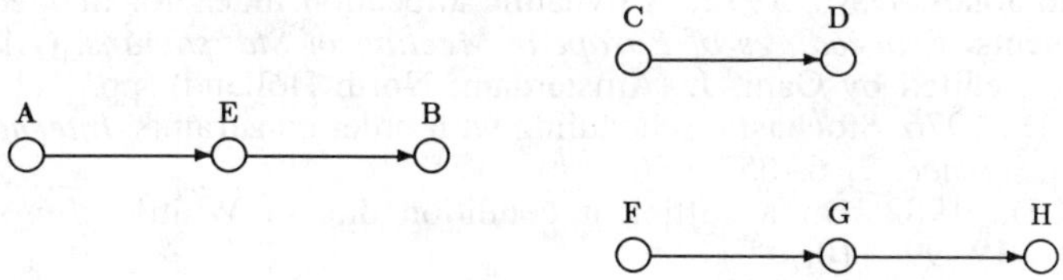

Figure 3. Task ordering for (A, E, B) and re-introduction of tasks G and H.

Third stage

Compare partial DAIs for (C, D) and (F, G, H); we get $\gamma(F, G, H) > \gamma(C, D)$. Compare partial DAIs for (C, D) and (G, H); we get $\gamma(G, H) > \gamma(C, D)$. Compare partial DAIs for (H) and (C, D); we get $\gamma(H) > \gamma(C, D)$ (see Figure 4).

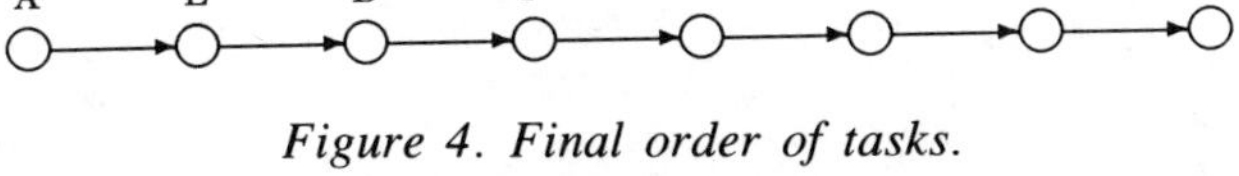

Figure 4. Final order of tasks.

The ordering algorithm is applied to all the superprocesses, that is to all the connected components of the initial precedence graph. Clearly, the ordering of each superprocess is independent and determined once and for all.

The global Gittins index of each superprocess can then be computed, at time 0, as for a completely ordered chain of tasks.

At subsequent decision times, the same order of tasks is used and the updated values of Gittins indices are computed by eqn (5). When a new job enters the queue, its tasks are optimally ordered with the rules presented before, its global Gittins index is computed and compared to the current biggest one.

Conclusion

This study presents a constructive method for solving general one-machine stochastic scheduling problems, for precedence graphs with no cycle, but with divergent and convergent nodes. It consists of a two-stage evaluation technique. The first stage is an off-line optimization and the second stage is an on-line updating of Gittins indices attached to superprocesses. The outcome of the first stage is an optimal ordering of each connected component of the scheduling graph. The on-line decision process then simply deals with each superprocess as with a totally ordered chain of non-pre-emptive tasks.

No specific assumptions other than some regularity have been put on probability laws for task completion, but they have to be known or estimated for the indices to be computed. Extension of the methodology to multi-machine scheduling problems remains to be examined, but the optimal task ordering for each job then depends on the selected machine for each task.

References

Bellmann, R., 1956, A problem in the sequential design of experiments. *Sankhya*, **16**, 221–229.

Gittins J.C. and Jones, D.M., 1972, A dynamic allocation index for the sequential design of experiments. *Proceedings of European Meeting of Statisticians*. (Also in *Progress in Statistics*, edited by Gani, J. (Amsterdam: North Holland), pp. 241–266, 1974.)

Glazebrook, K.D., 1976, Stochastic scheduling with order constraints. *International Journal of Systems Science*, **7**, 6, 657–666.

Glazebrook, K.D., 1982, On a sufficient condition due to Whittle. *Journal of Applied Probability*, **19**, 99–110.

Glazebrook, K.D., 1987, Sensitivity analysis for stochastic scheduling problems. *Mathematics of Operations Research*, **12**, 2, 205–223.

Varaiya, P.P., Valrand, J.C. and Buyukkok, C., 1985, Extensions of the multiarmed bandit problem: the discounted case. *IEEE Transactions* **AC-30**, 5, 426–439.

Whittle, P., 1980, Multi-armed bandits and the gittins index. *Journal of the Royal Statistics Society*, **B42**, 2, 143–149.

Whittle, P., 1981, Arm-acquiring bandits. *Annals of Probability*, **9**, 2, 284–292.

SECTION III.4. PRODUCTION PLANNING

25

An expansion of the virtual cell concept for scheduling computer-integrated manufacturing systems

Jocelyn R. Drolet,[1] Colin L. Moodie[2] and Benoit Montreuil[3]

[1]Industrial Engineering Department, Université du Québec à Trois-Rivières, CP 500, Trois-Rivières, Québec, Canada G9A 5H7, [2]School of Industrial Engineering, Purdue University, West Lafayette, 47907, USA and [3]Opérations et Systèmes de Décision, Université Laval, Cité Universitaire, Québec, Canada G1K 7P4

Abstract

In this chapter we present a new type of manufacturing system termed Virtual Cellular Manufacturing System (VCMS). This new manufacturing concept is an expansion of the virtual cell idea proposed by McLean *et al*. (1982). First, we present and justify such a system. Second, we present a few concepts which characterize our strategy. Third, we propose a time-phased linear programming model for scheduling these systems. Finally, we illustrate results obtained with the model.

Introductory development

The Virtual Cellular Manufacturing Concept (VCMS) is an expansion of the virtual cell concept proposed by McLean *et al*. (1982). A virtual cell is not identifiable as a fixed physical grouping of workstations, but as data files and processes in a controller. It presents the manufacturing control system with the illusion of a permanent set of assigned resources. Virtual cells extend the concept of the GT cell by allowing time sharing of workstations with other virtual cells that produce different part families but have overlapping resource requirements. As a result of workstation sharing, virtual cell based systems are believed to be superior to GT cell in terms of efficiency. In addition, these systems are expected to maintain the flexibility associated with flexible manufacturing systems (FMS).

The scheduling of jobs in FMS is known as an NP-hard problem; this means that all methods, or algorithms, currently known to solve this problem require exponentially increasing amounts of time and thus are extremely inefficient. It is unlikely that a polynomial time complexity algorithm will be discovered for scheduling FMS. Given this, we conceptualized the VCMS which is much less computation intensive in terms of scheduling.

These systems will operate in Just-in-Time mode under computer control and will have no more than 20 highly productive machines with automated material handling systems. Typically, two or three similar machines and a handling device will be put together to form a workstation. In these workstations many activities will be going on in parallel. Systems having up to 10 workstations working around the clock will not be uncommon (Solberg *et al.*, 1985). These systems although much smaller than current factories will be much more productive. The machines will be generally more versatile and products will not visit as many workstations as in today's factory (Solberg *et al.*, 1985). By the year 2000, cutting speed will most likely be up to five times faster than today's standards, hence reducing the total processing time by so much.

The reduction of lot sizing, down to a single unit will dramatically increase the demand on material handling even though the parts will generally visit fewer machines. Given this, it is clear that co-operative workstations will have to be close to each other when dealing with factories of the future. The numerical control code will not be generated until the specific routing for a certain order is known. This will permit the retention of the option of using many different machines.

A typical factory order will require only a few units up to less than a hundred. Inventory cost and the pressure for quick response time, backed up by no significant set-up times account for this. Customer orders will be produced in exact quantity Just-In-Time for shipping, hence with no inventory (Schonberger, 1986).

Virtual Cellular Manufacturing System (VCMS)

The following characteristics and operational behaviour summarize most of the specifics of a VCMS:

(1) A typical batch order contains very few up to less than a hundred units.
(2) Parts are moved between workstations, one unit, or one fixture holding parts at the time.
(3) A few parts can be accumulated at the entrance and exit of every workstation.
(4) The workstations have the ability for switching between different types of parts without significant set-up time, providing tooling is available.
(5) Most workstations have replicates distributed over the shop floor.
(6) The variety of workstation types is small, four to five at the most.
(7) The system operates in Just-In-Time mode.
(8) Shop orders come from the materials resource planning (MRP) output and from special last minute customer orders.
(9) Orders are characterized by job number, quantity of parts, precedence diagram, processing time, and tool requirement.
(10) Machines are much more productive than the ones we see in today's factories. Cutting speed is up to five times faster. Machines tend to be more versatile.
(11) Due to machine versatility, workpieces visit only a few workstations, three to five at the most.

At any given time, the VCMS is likely to contain several virtual cells, some of them overlapping over shared workstations. Figure 1 illustrates overlapping cells and some other complexities that may arise when dealing with these systems. The result is a network of routes more or less interlaced. It is clear that intelligent control of the system is required in order to avoid degenerating into anarchy.

Algorithm characteristics and implementation strategies

We developed a polynomial time complexity algorithm which can be executed in real-time for scheduling VCMS. The time-phased linear programming algorithm depicts precisely the manufacturing system under consideration. It permits the creation of virtual cells and schedules them under resource availability constraints. In addition, the algorithm responds quickly to unexpected events.

The scheduling algorithm creates the virtual cells in a way that minimizes the total distance flow. Also, when the production is behind schedule, the algorithm will

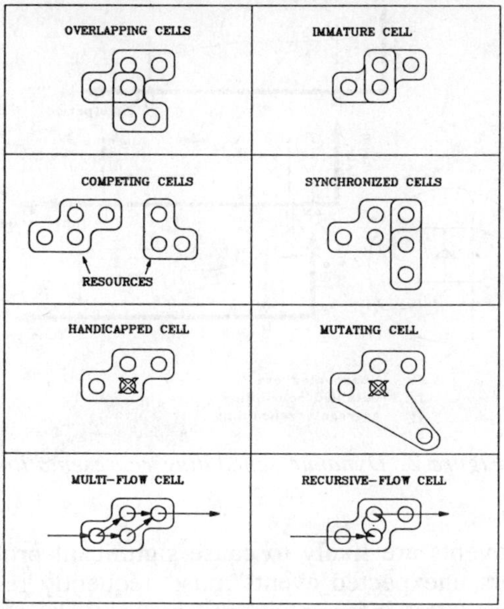

Figure 1. Cell dynamics terminology.

accept a small cost in distance flow if this permits compression of the schedule.

Strategically, a certain number of concepts have been implemented to permit the algorithm to work. Each of them are now presented.

Planning horizon

The planning horizon represents the spanning time for which a schedule will be generated and is divided into periods. The period duration, although fixed by management, should be inversely proportional to the frequency and duration of unexpected events. The planning horizon comprises two sections: the detailed horizon (h) and the aggregate horizon (H). The detailed horizon always begin at now and goes on for a few periods, say one or two hours. The aggregate horizon begins where the detailed horizon finishes and continues until all the jobs that have been considered are done. In the detailed horizon, the algorithm considers each workstation specifically, whereas in the aggregate horizon, the algorithm does not distinguish between two similar workstations. The period duration is usually longer in the aggregate is horizon.

Flow control

Flow control concerns the rate at which a set of parts is going to be sent in a virtual cell. Since these systems operate in Just-In-Time mode, one cannot have a significant accumulation of work-in-process (WIP). Thus, we balance the flow of parts to the bottle-neck of each virtual cell to determine the introduction rate of the individual parts. The most critical workstation (the bottle-neck) in a virtual cell determines the maximum introduction rate; this permits the avoidance of unnecessary WIP. The scheduling model determines the rate of introduction within these boundaries.

Quick reaction to unexpected events

VCMSs like other fast pace manufacturing systems require dynamic scheduling because their stability is continuously threatened by the occurrence of unexpected

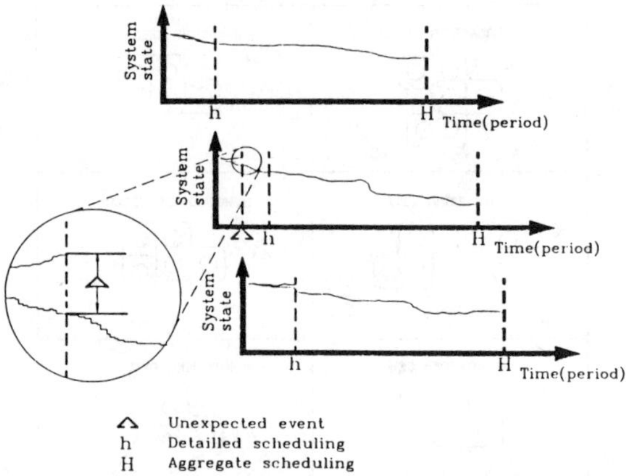

Figure 2. Dynamic scheduling representation.

events. Unexpected events are likely to cause significant problems. A scheduling strategy that considers unexpected events must frequently re-evaluate the current schedule based on the latest information and propose alternatives or adjustments if necessary. We do assume the presence of various sensors that can sense the processing advance in real-time and react accordingly.

Resource and aggregate production planning

Resource planning on a daily basis is appropriate in a VCMS. In fact, the major part of the daily production will seemingly represent orders that arrived only weeks, days, or even hours before. Given this, it is unlikely that resource and aggregate planning should be done on a long-term basis. One should be able to run the scheduling model overnight to plan the utilization of resources on an aggregate basis for the next day's production. The aggregate schedule should solely indicate when each virtual cell should be created in order not to exceed timely resource capacity. This happens to be a useful tool, particularly when concurrent operations such as off-line set-up or fixture preparation must be done.

Resource utilization profile

Each resource has a utilization profile graph attached to it. For each period of the planning horizon, the graph indicates the capacity not yet committed. This notion is used extensively when the scheduling model is operated in dynamic mode. These graphs are calculated and then passed to the scheduling model along with other information when rescheduling becomes necessary.

Figure 2 illustrates when and how the scheduling algorithm needs to be rerun. The model should be re-executed when the state of the system deviates too much from what was planned by the scheduling algorithm. The randomness of processing and travelling time, the occurrence of unexpected delays and breakdown can cause such a discrepancy.

An optimization-based scheduling model

In this section, we propose an efficient linear programming model for creating and scheduling virtual cells. The solution satisfies the capacity constraints on all considered resources; tools, workstations and fixtures.

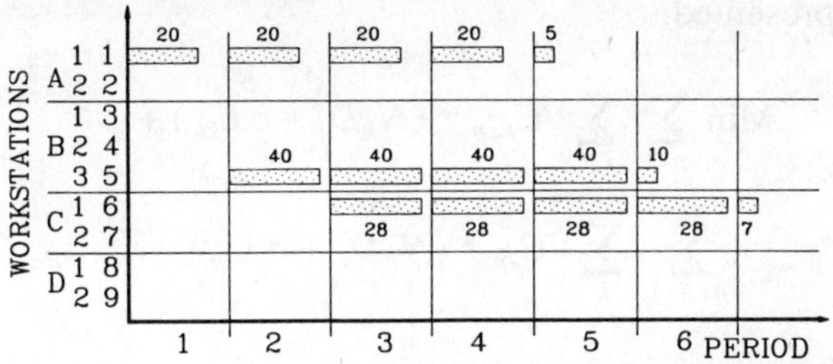

Figure 3. A gantt chart illustration of variable $VC_{3,156,1}$.

Table 1. Job orders sorted using SPT rule.

No.	Job identification	Sequence and process time (min)	Quantity	Total process time (min)
7	8476	{(D,6) (B,7)}	35	455
6	5594	{(C,4) (B,9)}	32	416
5	3847	{(B,5)}	78	390
4	5585	{(B,9) (C,4) (D,5)}	21	378
3	7674	{(A,5) (B,10) (C,7)}	17	374
2	8744	{(A,7) (B,9) (C,2)}	8	144
1	6585	{(C,5) (A,4)}	13	117

Let us now define the decisional variable VC_{k1p} which is critical to the understanding of the model. Variable VC_{k1p} represents a virtual cell created for job k using the workstations on routing 1 and starting at period p. A gantt chart which illustrates variable $VC_{3,156,1}$ is represented in Figure 3. Technological requirements and processing time for job no. 3 are shown in Table 1.

As shown, job 7674 (3) begins processing at period 1 and its routing is A1 → B3 → C1. The bottle-neck workstation determines the maximum introduction rate of parts.

The model uses the following notation:

PT_{k1pr} Parameter representing the processing time at period p for job k according with routing 1 on resource r.
VC_{k1p} A virtual cell created for job k using workstations on routing 1 and starting at period p.
VC_{kp} A virtual cell created for job k at period $p(p > h)$.
C_{pr} Capacity of resource r at period p.
L_{kp} Lateness of job k when processing begins at period p.
D_k Average travelling distance for job k.
D_1 Travelling distance associated with routing 1.
F_k Flow time for job k.
N_k Number of parts in job k.
c Any large cost.
r A resource (workstation, tool or fixture).
R Aggregate resources (a group of similar resources).
T Total planning horizon.

$$L_{kp} = \begin{cases} p - (T - F_k) & \text{if } p > T - F_k \\ 0 & \text{if } p \le T - F_k. \end{cases}$$

The model is now presented:

$$\text{Min} \sum_{p=1}^{h} \sum_{k,1} VC_{k1p} * (N_k D_1 + c\, L_{kp}) +$$

$$\sum_{p=h+1}^{H} \sum_{k} VC_{kp} * (N_k D_k + c\, L_{kp}).$$

Subject to:

$$\sum_{1}^{E} \sum_{p=1}^{h} VC_{k1p} + \sum_{p=h+1}^{H} VC_{kp} = 1 \qquad \forall(k) \tag{1}$$

$$\sum_{k,1} PT_{k1pr} * VC_{k1p} < C_{pr} \qquad \forall(r), \forall(p) \text{ where } p \leq h \tag{2}$$

$$\sum_{k} PT_{kpr} * VC_{kp} < C_{pR} \qquad \forall(R), \forall(p) \text{ where } h < p < H. \tag{3}$$

The first set of constraints makes sure that each job will be produced in its totality. The first part considers the detailed horizon in which one or more virtual cells will be created at some period in time for each job. The second part concentrates on the aggregate horizon in which specific routing 1 (by the same token specific virtual cell) are not considered. Since variables are not restricted to be 0 or 1, they sometimes happen to be between 0 and 1, thus in these circumstances the rate of introduction and the launching lot are adjusted proportionally. For example, if a variable VC_{k1p} equals 0·3 and if the bottle-neck limited the introduction rate to 20 units/period, then the rate would be adjusted to 20 * 0·3 = 6 units/period beginning at period p until an equivalent proportion of the lot has been introduced. As one can see, although the bottle-neck fixes the boundaries for the introduction rate, the rate is really fixed by the scheduling model within these boundaries.

The second set of constraints only applies to the detailed horizon. The constraints make sure that the demand will never exceed the capacity available at any workstation for any period during the detailed horizon. For this, no more than $r*p$ constraints are required.

The third set of constraints only applies to the aggregate horizon. The constraints ensure that the demand will not exceed the capacity available on any aggregate resource during any period p where $h < p \leq H$. An aggregate resource is a group of similar resources. For each group of resources R (e.g. similar workstations), there is one constraint for each period p where $h < p \leq H$. Specific workstations are not considered in the aggregate planning horizon.

The objective function creates and schedules virtual cells in a way that minimizes the total distance flow for all considered jobs. This is so as long as the lateness factor equals 0, otherwise, distance flow increases to permit the use of resources which are not as close.

Illustrative example

In this section, we continue with an example which illustrates the application of the scheduling algorithm. A set of job orders is available for processing. Table 1 shows these job orders sorted using shortest processing time (SPT) rule. The system, shown in Figure 4, has the defining parameters stated in Table 2.

Table 3 shows the results obtained with the scheduling model. This table provides

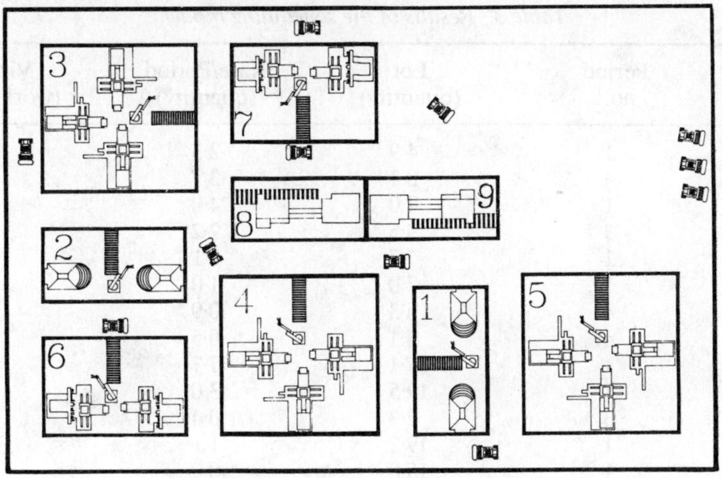

Figure 4. An experimental VCMS.

Table 2. Defining parameters for the system.

Workstation type	Workstation number	Number of machines per workstation
A	1,2	2
B	3,4,5	3
C	6,7	2
D	8,9	1

all necessary information for creating and scheduling the virtual cells for the next hour. The first column indicates the job number, the second column indicates when the virtual cells will be created, and the third column indicates the number of parts to be processed in each job or they add up to it, e.g. two lots adding up to 13 units will be launched for job 1; one of 4·9 units, and one of 8·1 units will be processed in virtual cell (6, 2). In most cases, the rate of introduction will vary and sometimes more than one virtual cell will be created for a given job. When this happens, the lot represents the portion of the job tied to the virtual cell. The rate/period indicates how many parts will be launched per period within each virtual cell. The rate stays the same as long as the lot is not completely finished, e.g. four parts/period will be launched for job 2 until all eight parts have been launched. The virtual cell controller created for this job will make the communication possible between workstations 1, 4 and 7 and will control the flow of parts between them. The reader is invited to appreciate the virtual cell's formation with the help of Figure 4. No specific virtual cell appears in the last column of Table 3 when period p in the second column is greater than four since in this example, the aggregate planning horizon begins after period four.

When the lot size is small, it is quite possible that the corresponding virtual cell will never be created. Instead, production can be increased a little in virtual cells where the rate is faster and where the lot size is larger since one has the potential for a few parts waiting before and after each workstation. In other words, the demand on any workstation can exceed the capacity a little for a few periods.

Table 3. Results of the scheduling model.

Job no.	Period no.†	Lot (quantity)	Rate/Period (quantity)	Virtual Cell (workstation no.)
1	1	4·9	2·3	6,2
1	4	8·1	3·7	6,2
2	1	8·0	4·0	1,4,7
3	3	9·5	2·2	1,4,7
4	1	7·7	1·1	3,7,8
4	1	7·0	1·0	4,7,8
4	4	6·3	0·9	4,7,8
5	4	4·1	0·5	4
5	1	67·1	7·7	5
6	1	19·5	3·0	7,3
6	3	2·3	0·4	7,4
7	1	19·2	1·1	8,3
7	1	15·8	0·9	9,5
3	7	4·3	1·0	—
3	8	0·4	0·1	—
3	10	0·5	0·1	—
3	11	2·3	0·5	—
5	8	6·9	0·8	—
6	8	4·6	0·7	—
6	9	4·9	0·8	—
6	10	0·8	0·1	—

†1 period = 15 min.

Conclusion

The size of the model grows quickly but linearly and remains reasonable for a typical size factory. A series of tests on a SUN microsystem 3/50 has shown that the model can solve any realistic problem within 2 min. Considerably less time than a human scheduler would take for diverting the flow of parts and sorting out the contentions of critical tools after the occurrence of a breakdown.

The model presented has many advantages, first, it permits consideration of critical tools and fixtures. Secondly, its dynamic mode, coupled with a reasonable computational time, provides quick reaction to unexpected events. Finally, it enables a coherent planning of resource utilization, potentially reducing possibilities of delays and deadlock, which are very common in fast pace manufacturing systems.

References

McLean, C.R., Bloom, H.M. and Hopp, T.H., 1982, The Virtual Manufacturing Cell. *Proceedings of Fourth IFAC/IFIP Conference on Information Control Problems in Manufacturing Technology*, Gaithersburg, MD, October.

Schonberger, R.J., 1986, *World Class Manufacturing: The Lessons Of Simplicity Applied.* New York, NY: The Free Press.

Solberg, J.J., Anderson, D.C., Barash, M.M. and Paul, R.P., 1985, *Factories Of The Future: Defining The Target.* A report of a research study conducted under National Science Foundation Grant MEA 8212074, Purdue University, West Lafayette, IN, January.

26

Kitting systems for small size parts in manual assembly systems

Mats I. Johansson

Department of Transportation and Logistics, Chalmers University of Technology, S-41296 Gothenburg, Sweden

Abstract

This chapter presents different kinds of materials feeding systems for small parts to manual assembly and discusses the conditions under which a certain system is favourable in relation to other solutions of materials feeding.

In the search for production systems giving a more efficient and flexible assembly, and also the possibility of better working conditions, highly parallelized assembly systems with long cycle times have been developed. One disadvantage of such systems is their complicated materials feeding. Solutions such as decentralized materials markets where the assembly operators pick their own material have been widely used. This solution is, however, not appropriate for all kinds of parallel assembly systems and in recent years storage kitting systems have been considered and developed.

Introduction

There are many demands put on material handling systems used for material feeding in assembly systems. Some of these demands emanate from the choice of assembly system, others from the production scheduling and from the characteristics of the components handled.

The assembly systems are moving towards a higher degree of parallelization at the same time as many product lines require more and more part numbers due to a growing diversity of variants (Elser, 1985). Each of these factors results in difficulties feeding materials in the traditional way, i.e. that one unit load of each part number is placed at the assembly station. Therefore, other principles of material supply systems have been developed. For many applications, however, they have been too recently introduced to allow full evaluation.

Material can be distributed for a batch of assembly objects. The batch of material consists of the necessary part numbers for building one batch of assemblies and the parts can be counted to the exact quantity needed or not. Decentralized stores, generally known as material markets, where the assemblers pick their own material have mainly been used in parallelized assembly systems. Storage kitting is another supply system sometimes used, but very seldom in relation to other systems.

This chapter discusses the conditions under which a certain material supply system for small parts is favourable in relation to other solutions. The focus is on the potential for kitting systems, since these systems are not as developed and

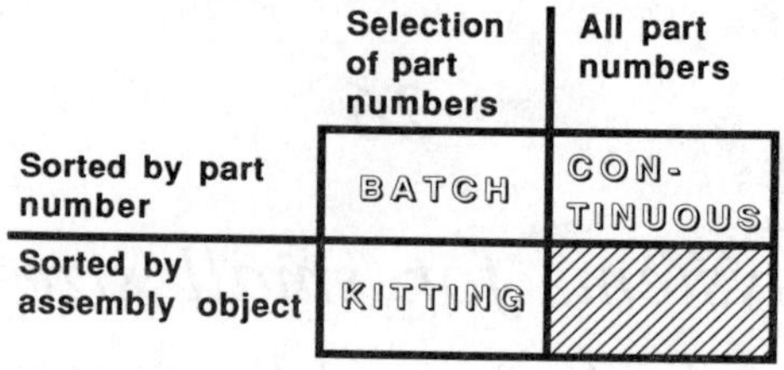

Figure 1. Categorization of material supply systems.

common as the others. After comparing the systems, one section deals with kitting systems methods and techniques for high variety and complex products in manual assembly systems. The chapter is based on experience and empirical data from several Swedish companies, collected over a long period and not one single study.

Categories of supply systems

Material supply systems can be categorized in many ways, but I here distinguish between systems in regard to the selection of part numbers exposed to the assembly stations and the way in which these part numbers are sorted at the station.

Regarding the first variable, supply systems can be divided into two types of systems. First, systems where all part numbers allowed according to existing product specifications and work division are exposed at the assembly stations and second, systems where a selection of part numbers is exposed at a time. This is a selection related to the production schedule and means that only the part numbers necessary for a fixed part of the schedule are exposed.

The second variable categorizes the supply systems into those where components with the same part number are held together and those where parts intended for one assembly object are held together.

This division gives three combinations (principles) of practical meaning: (i) continuous supply, (ii) batch supply, (iii) kitting. These three principles can exist simultaneously in one system and for different kinds of parts complement each other. Furthermore, there is a large variety of solutions within each principle and 'pure' systems can hardly be described. In this section therefore, the main characteristics of each principle will be described and these systems will then be compared in later sections.

Continuous supply

Continuous supply refers to the case where material is distributed to the assembly stations in units suitable for handling and where these units are replaced when they are empty. There is no co-ordination of replacements for different part numbers. All part numbers needed for producing every occurring product over a long period are available at the assembly station at every time.

Material is often placed at the working station in the supplier package, or if this quantity or package volume is much too big, distributed after repacking. The latter is often the case for small parts. If the parts are repacked, there can be a buffer of such repacked parts between storage and assembly. Refilling of parts at the assembly stations is often done by storemen, either in station fix bins, or by some kind of two-bin system. The assemblers then pick the parts needed for each assembly which in practice means that they make up their own kits of small parts. An alternative technique is that the assembler fills an 'assembler's box' with the small parts that he needs during the cycle. In this box he might also keep his hand tools.

Batch supply

In batch supply systems the material is supplied for a number of specific assembly objects. The batch of materials can be a batch of the necessary part numbers, or a batch of these part numbers in the requisite quantities.

The first case differs from continuous supply in the sense that fewer part numbers have to be exposed to the assembler and that different part numbers are exposed at different points in time. Repacking is only necessary if the supplier package is inappropriate (e.g. too big). The remaining material is returned to the store after completion of the batch of assemblies, unless it is to be used in the next batch. This job is eliminated in the latter case, but instead there is a need for counting the parts which requires technical and administrative systems.

Once at the assembly station, the small parts are handled in the same way as for continuous supply.

Kitting

Kitting means that the assembly is supplied with kits of components. One kit consists of a set of parts for one assembly object. This does not mean that parts (kits) for more than one object cannot be supplied at a time, but the parts for one object are held together. This differs from batch supply, where the components are sorted by part number.

The kitting process can be done in the store or in material markets. In the store, the job is generally done by the store staff, but systems where this task is part of the assembly groups' responsibility are sometimes discussed. This can be an advantage, because when kitting without sophisticated technical guidance it has proved to be advantageous if the workers are familiar with the product and its structure. On the other hand there can be a large amount of administrative work related to kitting and storage in general, which can justify a specialization. Regarding small parts, the possibility of large lot sizes when kitting and an automated kitting process are other motives for not decentralizing this job either organizationally or in the lay-out.

A materials market can be defined as a decentralized picking store common to several assembly stations. The flow of material to the material market is often organized as a continuous supply system, but the feeding of the assembly station is done by the assembler himself. This implies short distances between the stations using the market.

There are many possibilities regarding this final step in the flow. One assembler can, for example, pick material just for himself, or for a whole group of assemblers, and he can pick material for one or more objects at a time. The assemblers can also use the material market as a batch supply system and for small parts the most common case is that the workstations are fed as a continuous supply system. Note that small parts can be pre-kitted and supplied to the material market and picked as one component by the assembler.

The reason for presenting material markets as kitting systems is that they have commonly been used in Sweden for this type of feeding for medium-size parts.

Some factors affecting the choice of supply system

There are two basic factors which have a great impact on the choice of material supply principle for small parts: the assembly systems' number of parallel flows and the characteristics of the assembled product, including the single parts in the product. Many of the demands on the supply system can be derived from these two factors. Additional factors such as production scheduling will also affect the choice.

The trend towards more parallelized assembly systems has enforced development of material supply systems where the supply is planned according to the production schedule (batch and kitting systems). Problems when using continuous supply systems are related to an increase in the number of feeding points and in cycle time.

An increased cycle time usually means that more part numbers need to be exposed at each assembly station which implies a need for more space. This can be a problem in itself, but it also means greater losses through materials handling. Typical figures for this loss (as defined in Wild, 1975) on an assembly line with 2 min cycle time and continuous supply is 25%. General comparisons between the supply principles can hardly be done regarding handling time at the assembly station. Continuous and batch supply systems are efficient as long as the assembler can carry all the parts he needs, but a higher total amount of exposed part numbers and more part numbers per assembled object will make kitting more competitive. This is because kitting in general means less walking and less 'searching' for parts. On the other hand there are kits to handle and perhaps unpack.

The enlarged work content and the increased number of parts can also lead to problems with the assembly workers 'learning'. Problems in this regard can, for example, cause long learning times and quality problems through wrong components being used. Interview data from an assembly system with approximately 2 h cycle time, clearly points in the direction that kitting of small parts in comparison with other systems greatly facilitates the learning, on the assumption that the parts are grouped in kits functional to the assembly operations. One screw makes no sense, but together with a set of other parts it has an obvious application. To be efficient, batch supply systems need some assistance from the production scheduling. In a mix-model assembly system the differences between objects following each other can be very great, which brings the batch system closer to continuous supply. Batch systems require multi-model scheduling with acceptably large batch sizes. For some product lines this can be difficult to achieve and the ongoing trend towards higher diversity makes it even more difficult. This is exemplified by a company assembling a large product with high assembly time in object series and using a batch supply system for most of the parts. Over five years the number of objects in one series decreased from 80 to 40, but the number of orders generated in the store increased from 1000 to 1100 per series.

A growing number of product variants, which is a reality for many manufacturers, often results in more part numbers. This affects the system and causes the same problems as a longer cycle time. In addition there is a structural learning problem in the assembly; which part is to be assembled on this object?

Several feeding points mean a lower rate of consumption at each point which often calls for repacking of the suppliers' packaging when using continuous supply. Properly chosen packing sets can eliminate this job (Engström *et al*., 1989). Reserve and work-in-process inventories will rise as the number of feeding points increases. For many cases this capital investment is of little importance since the small parts are cheap in relation to other categories of material.

Of greater importance can be that the control of the material deteriorates. This involves, for example, costs due to poor balance control (e.g. shortage costs) and an uncertainty regarding correct part numbers being used. This latter fact is not only a direct economic question, but also a quality problem. In some cases, for example, a screw must be of a certain length and have a special kind of surface condition. Problems also arise when product changes occur. Discontinued part numbers have to be replaced with new ones at many places, which makes it difficult to match the time of introduction with the remaining number of parts. Furthermore, there is a considerable information problem. The part lists have to be changed from one specific assembly object and using the old part number after that time can cause reworking or quality problems.

Figures from one company using continuous supply and having about 2000 part numbers (one quarter of the total amount) in their small parts store shows a wastage of nearly one million dollars. Most of the parts were fasteners, small plastic and rubber parts, etc.. This figure shows that much money can be saved by having a better control of the material. Small parts are often paid little attention by the firms, which results in the assemblers not taking very good care of the parts either—another source of wastage.

Under certain circumstances kitting systems for small parts offer a number of advantages, but kitting in itself is a job to be done and the cost for this job has to be weighed against eventual savings. No generalized data on this matter can be presented, but one study was carried out to compare the costs for kitting (in-plant kitting) small parts to one station on an assembly line with short cycle time v using continuous supply (the system in use). One assembly station without any extreme characteristics was selected for the study. All physical materials handling costs between the receiving area and the assembly station (including semi-automated kitting, but excluding handling done by the assembler) and capital costs (including equipment) were compared. The result was a 5% reduction in costs if the kitting solution was chosen and this was a surprise for everyone. Note that the 'hard to calculate' kitting systems advantages expressed earlier were not included in the calculation. Neither were costs for new administrative systems, which have to be developed if kitting is used.

Empirical data from one packaging company, specializing in kitting of small components, shows that kitting speed in a semi-automated process (manual counting at a table) varies with, and can be estimated from, the number of kits in the lot, the number of parts in the kit and the number of part numbers in the kit. Very roughly, a speed of two parts per second can be used as a mean time for such components as small fasteners.

Kitting methods and techniques

Kitting of small parts as a material supply method to assembly has been commonly used only in the assembly of electronic products. These electronic applications, however, are not always of the type of kitting presented earlier in this paper. In many cases they are more related to the batch supply principle, where the parts are counted and delivered to a set of known assemblies. We can also find kitting in other applications, such as automotive spare parts and selling of unassembled furniture.

In these latter cases the 'customer' of the kits is a store and production starts at a certain inventory level. The batch size is often several thousands, which means that efficient automatic or semi-automatic kitting systems can be built up. Plastic bags are the most common kit packaging and the packing technique is well developed for such batch sizes. Plastic bags are well suited for such applications as spare parts and different kinds of consumer goods, because of high visibility and the ease of handling such bags. If the contents permit it, the bags can be handled as bulk goods. These features are also valid for kitting to assembly, where small boxes could be an alternative. High visibility helps the assembler in identifying the kit, handling and storage are facilitated through kits being handled as bulk goods, storage volume is low and the packaging is disposable which eliminates return flows.

Numerous manual and semi-automated kitting techniques exist, which can be used if the parts to be kitted are difficult to count with automated equipment, or the lot sizes are extremely low. These techniques are not discussed further in this text.

Traditional equipment for automatic counting and bagging of small parts is built up in lines. A series of feeders is connected through a synchronized conveyor ending with an automatic bagging machine. Between the feeders and the bagging machine there can be a working station for manual feeding and completion of the kits. Such systems require a high level of investment and are not very well adapted to the demands of many industrial assembly systems. Several problems concerning automated kitting of small parts face many companies producing a large number of variants and having large numbers of small parts. For example:

(a) small lot sizes due to low production volume, a variety of kits for one product function, or a production schedule fixed for a short period of time,
(b) greatly varying amounts of part numbers in the kits,
(c) a mix of automatic counting and manual counting.

Small lot sizes mean that the ratio of utilization can become too low to justify the type of automated equipment described earlier. Set-up times can even exceed the effective production time. Variety in the number of part numbers between kits and in the number of parts per part number within a kit, also decreases utilization for individual feeders. This is because the number of feeders in the kitting line will not correspond to the number of parts in many of the kits. Some feeders will then not be working. Mixed manual and automated counting in a kit means that either the man or the machine will be delayed, which is an efficiency and work-condition problem.

When constructing an automatic kitting system the following measures should, for reasons described above, be carefully considered before implementation:

(1) Computer-controlling resetting of counters, bagging machines, checking scales, etc.
(2) Group technology (kit and part level) and production scheduling systems.
(3) Disengagement between feeders and bagging machine.
(4) Division of the feeder line into smaller groups of feeders.
(5) Disengagement between manual and automatic counting.

These divisions between different tasks may demand buffers of partly complete kits and equipment and methods adapted to this principle.

When there are many small parts in one product, many kits can be required for assembling an object. There are two main principles for grouping parts together into kits. That is the principle of 'addition' and the principle of 'integration'. In the first case, one part for a certain purpose never appears in more than one kit. The latter principle allows this and means that two kits can be identical except for one single part. In a high variety production this principle can result in many kits very similar to each other. From the assembly point of view, one kit should contain a 'natural job', or part of such a job and experience shows that the integration principle is favourable in this sense. It can also be shown that this principle gives a smaller number of different kits and fewer kits to produce.

References

Elser, K., 1985, Automation of assembly—applications potential and preconditions. In *Toward the Factory of the Future*, edited by H.-J. Bullinger and H.J. Warnecke (Berlin: Springer Verlag), pp. 64–67.

Engström, T., Johansson, M.I., Lundberg, A. and Sjöstedt, L., 1989, Equipment Needs in Material Flow Systems As a Function of Integration Level. In *Transport Policy, Management and Technology Towards 2001*, (vol. II). Selected Proceedings of the 5th World Conference on Transport Research, (Ventura: Western Periodicals Co.), pp. 507–520.

Wild, R., 1975, On the selection of mass production systems. *International Journal of Production Research*, **13**, 5, 443–461.

27

Hierarchical production planning: an approach for a two-stage production system

Katsumi Morikawa and Nobuto Nakamura

Department of Industrial and Systems Engineering, Faculty of Engineering, Hiroshima University, Shitami, Saijo-cho, Higashi-Hiroshima 724, Japan

Abstract

This chapter presents a hierarchical production planning (HPP) approach for a two-stage production system which consists of part fabrication and product assembly stages. Based on the product (part) and time hierarchy, we divide the overall planning problem into four subproblems: (a) aggregate planning for the entire system, (b) disaggregation of aggregate plan to obtain product assembly schedule, (c) verification of the consistency condition, and re-planning if necessary, and (d) part lot sizing. From some relatively simple numerical examples, our HPP approach exhibited agreeable results compared with simple materials requirements planning (MRP) model when part set-up costs were lower.

Introduction

To support production planning, hierarchical planning approaches based on mathematical models have been suggested by many researchers because of their effectiveness and possibilities. The basic design of these approaches includes the partitioning of the overall planning problem, and the linkage of the resulting subproblems. Boskma (1982) distinguishes the following categories for aggregation to partition the overall planning problem:

(a) Aggregation of separate products into product categories.
(b) Aggregation of separate resources into resource categories.
(c) Aggregation of time by taking short time-slices of subperiods together for a longer time interval.

Based on the product structure, Hax and Candea (1984) have reviewed a series of thorough pioneering studies. Axsäter and Jönsson (1984) suggest a hierarchical planning procedure by grouping items and machines. But analytical studies for time aggregation are rare in comparison with these two categories.

Golovin (1975) clarifies the problem of time disaggregation and suggests several practical approaches to resolve it. By simulation experiments, Boskma (1982) shows that aggregation of time interval results in a great behavioural difference as aggregation proceeds. Saad (1988) suggests a three-level hierarchical production planning approach considering time disaggregation explicitly, but he does not discuss the problems involved in time interval disaggregation.

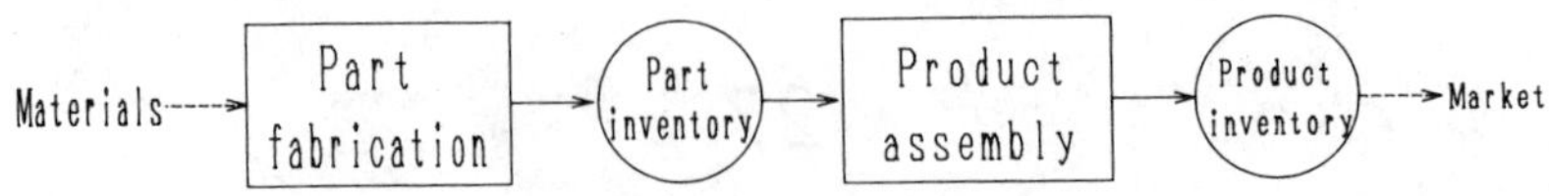

Figure 1. Two-stage production system.

Time interval aggregation and disaggregation is one of the key factors for hierarchical production planning (HPP). In this chapter, therefore, for a two-stage (i.e. product assembly stage and part fabrication stage) production system, we present a hierarchical approach for production planning involving product and time hierarchy explicitly.

Production planning model

Assumptions

We deal with the two-stage production system shown in Figure 1 subject to the following assumptions.

(1) Overtime, inventory holding and set-up costs are considered.
(2) Set-up consumes no capacity and no set-up costs are incurred at the product assembly stage.
(3) No planned back orders are allowed.
(4) Lead time for product assembly and part fabrication is negligible.
(5) Products and parts can be aggregated as follows (Bitran *et al.*, 1982):
   — *Product items* are end finished products delivered to customers.
   — *Product types* are groups of finished product items having similar direct production costs (excluding labour), holding costs per unit per period, productivities and seasonalities.
   — *Part items* are individual parts required as a component to a product item.
   — *Part types* are groups of part items having similar direct production costs, holding costs per unit per period and productivities.
(6) Planning period (e.g. months) can be divided into subperiods (e.g. weeks, days). For simplicity we assume that subperiods have the same time length and each period is made of α subperiods.

Hierarchical structure for the two-stage production system

Figure 2 shows the overview of our HPP system. The highest level of planning system determines production plans of part and product types for each period. In the next lower level, first two periods of the product types plan are scheduled for each subperiod, then the product items schedule for each subperiod is determined. At this level, the consistency condition is examined and returned to the aggregate level if re-planning is necessary. After that, required part items are calculated and fed to the part items scheduling system. Final planning is the determination of lot sizes of each part item for each subperiod using production level information obtained from the aggregate production planning level. The entire planning system consists of these procedures and is repeated under the rolling horizon manner.

Model formulation

Aggregate planning

The highest level of our planning system is the aggregate planning. Our model introduced here is formulated as a linear programming as follows:

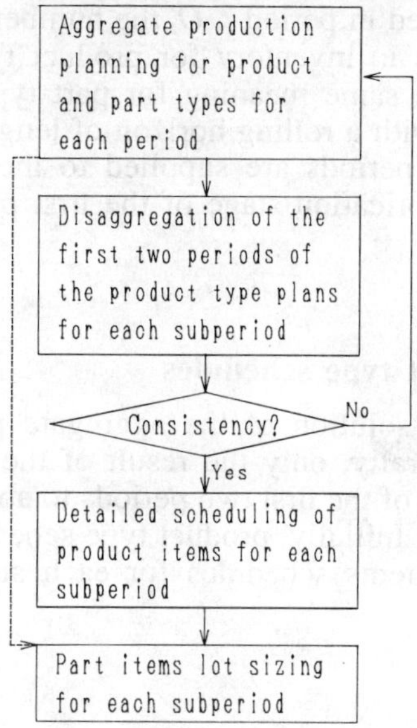

Figure 2. Overview of hierarchical production planning system.

$$\text{Minimize} \sum_{t=1}^{T}\left\{\sum_{i=1}^{I} h_i I_{it} + c\,O_t\right\} + \sum_{t=1}^{T}\left\{\sum_{j=1}^{J} \hat{h}_j \hat{I}_{jt} + \hat{c}\,\hat{O}_t\right\} \tag{1}$$

subject to:

$$I_{it-1} + X_{it} - I_{it} = \mathrm{d}_{it} \qquad i = 1, \ldots, I;\ t = 1, \ldots, T \tag{2}$$

$$\sum_{i=1}^{I} r_i X_{it} \leqq (R\max)_t + O_t \qquad t = 1, \ldots, T \tag{3}$$

$$O_t \leqq \Phi\,(R\max) \qquad t = 1, \ldots, T \tag{4}$$

$$\hat{I}_{jt-1} + \hat{X}_{jt} - \hat{I}_{jt} = \sum_{i=1}^{I} f_{ij} X_{it} \qquad j = 1, \ldots, J;\ t = 1, \ldots, T \tag{5}$$

$$\sum_{j=1}^{J} \hat{r}_j \hat{X}_{jt} \leqq (\widehat{R\max})_t + \hat{O}_t \qquad t = 1, \ldots, T \tag{6}$$

$$\hat{O}_t \leqq \hat{\Phi}(\widehat{R\max})_t \qquad t = 1, \ldots, T \tag{7}$$

$$I_{it}, X_{it}, O_t, \hat{I}_{jt}, \hat{X}_{jt}, \hat{O}_t \geqq 0 \qquad i = 1, \ldots, I;\ j = 1, \ldots, J;\ t = 1, \ldots, T \tag{8}$$

where the indices i, j, and t represent, respectively, product types, part types, and time periods, hi is the cost of holding product type i per unit for one period, c is the cost of one man-hour of overtime, d_{it} is the effective demand of units of product type i, r_i is the required man-hours to produce one unit of product type i, Φ is the allowed overtime as a fraction of the regular man-hours, and $(R\max)_t$ is the maximum regular man-hours in period t; parameters with a circumflex have the same meaning for part types.

The number of units of part type j required per unit of product type i is represented by f_{ij} which is discussed by Bitran *et al.* (1982). The variable X_{it} denotes the number of units produced in period t, O_t the number of man-hours of overtime, and I_{it} the number of units in inventory for product type i in period t; variables with a circumflex have the same meaning for part type j in period t.

This problem is solved with a rolling horizon of length T. Only the product type schedules of the first two periods are supplied to the next lower level, and the production level of part fabrication stage of the first period is also utilized in the lower level decisions.

Disaggregation of product type schedules

The disaggregation of the solution of the aggregate production problem is performed in two steps. Generally, only the result of the first period is used in this level, but we use the results of the first two periods to absorb the demand fluctuation within the first two periods. Initially, product type schedules for each subperiod are determined, then product items schedules for each subperiod are determined as follows.

Step 1: Product types schedule for each subperiod

To determine the production quantities of product types for each subperiod, the following linear programming problem needs to be solved:

$$\text{Minimize} \sum_{t=1}^{2\alpha} \left\{ \sum_{i=1}^{I} h^*_i \, I^*_{it} + c\, O^*_t \right\} \tag{9}$$

subject to:

$$I^*_{it-1} + X^*_{it} - I^*_{it} = d^*_{it} \qquad i = 1, \ldots, I;\ t = 1, \ldots, 2\alpha \tag{10}$$

$$\sum_{i=1}^{I} r_i \, X^*_{it} \leqq (R\max)^*_t + O^*_t \qquad t = 1, \ldots, 2\alpha \tag{11}$$

$$O^*_t \leqq \Phi\,(R\max)^*_t \qquad t = 1, \ldots, 2\alpha \tag{12}$$

$$\sum_{t=1}^{2\alpha} X^*_{it} = X_{i1} + X_{i2} \qquad i = 1, \ldots, I \tag{13}$$

$$I^*_{it},\ X^*_{it},\ O^*_t \geqq 0 \qquad i = 1, \ldots, I;\ t = 1, \ldots, 2\alpha. \tag{14}$$

The parameters and variables with an asterisk have the same meaning as mentioned previously except that the time-interval is subperiod instead of period. Constraints (13) request the equality between the production quantities for 2α subperiods and the amount dictated by the aggregate level for two periods.

Step 2: Product items schedule for each subperiod

In this step the number of units of each item to be assembled is determined by dividing the production quantities obtained from Step 1 above. All the product related costs have already determined in the former two levels, the following simple disaggregation is adopted for each product item $p \in i$ for each subperiod t:

$$\tilde{X}^*_{pt} = \underline{\tilde{X}}^*_{pt} + \frac{\sum_{\tau=t}^{\xi} \tilde{d}^*_{p\tau} - \tilde{I}^*_{pt-1}}{\sum_{p\in i}\left\{\sum_{\tau=t}^{\xi} \tilde{d}^*_{p\tau} - \tilde{I}^*_{pt-1}\right\}} \left\{X^*_{it} - \sum_{p\in i} \underline{\tilde{X}}^*_{pt}\right\} \quad p \in i;\, t = 1, \ldots, 2\alpha \tag{15}$$

where $\tilde{X}^*_{pt}$ is the number of units of product item p belonging to type i to be assembled at subperiod t(tilde is used to mean product item), and $\underline{\tilde{X}}^*_{pt}$ is defined by:

$$\underline{\tilde{X}}^*_{pt} = \text{Max}\,\{0,\, \tilde{d}^*_{pt} - \tilde{I}^*_{pt-1}\} \qquad p \in i;\, t = 1, \ldots, 2\alpha. \tag{16}$$

These express the minimum production quantities to satisfy the demand, and ξ is the subperiod satisfying the following equation:

$$\sum_{\tau=t}^{\xi} d^*_{i\tau} \geq \sum_{p\in i} \{\tilde{I}^*_{pt-1} - \tilde{d}^*_{pt}\} + X^*_{it} > \sum_{\tau=t}^{\xi-1} d^*_{i\tau}. \tag{17}$$

As part items lot sizing decisions may need future part items demand information, product items production quantities for two periods are calculated in spite of the fact that only the results of the first period are implemented at the product assembly stage.

Re-planning

Comparing the production quantities derived from aggregate planning problem with the result of disaggregation procedure mentioned above, eqns (18) are not always satisfied.

$$X_{i1} = \sum_{t=1}^{\alpha} X^*_{it} \qquad i = 1, \ldots, I. \tag{18}$$

If eqns (18) are not satisfied, the consistency condition of the two levels is not satisfied and disaggregated production plans may violate the capacity constraint of the part fabrication stage. To avoid this violation, after the Step 1 of disaggregation, the equality conditions (18) are checked. If not satisfied, re-planning procedures are introduced as follows:

Re-planning 1. Replace the objective function (1) with (1′) and add constraints (19) to the aggregate production problem, then solve the new problem.

$$\text{Minimize (1)} + P\sum_{i=1}^{I} \{\beta_i^+ + \beta_i^-\} \tag{1′}$$

$$X_{i1} + \beta_i^- - \beta_i^+ = \sum_{t=1}^{\alpha} X^*_{it} \qquad i = 1, \ldots, I \tag{19}$$

where β_i^+, β_i^- are deviational variables and the parameter P of (1′) is a penalty for violating constrants (18).

Re-planning 2. New constraints (20) are included in the disaggregation problem of Step 1 to keep consistency:

$$I^*_{i\alpha} = I_{i1} \qquad i = 1, \ldots, I. \tag{20}$$

Part items lot sizing

Occurrence of the set-up costs at the part fabrication stage leads to the problem of lot sizing with capacity constraints. Under the rolling horizon planning manner, future part items demand is not stable even if product items demand is deterministic, because future product assembly plans may be changed when new information updates the model. Thus lot sizing algorithms requiring future detailed demand information to generate feasible plans are not always good tools to deal with our situation. Instead of future detailed demand information, we use the production level of the part fabrication stage obtained from the aggregate production problem and employ the Eisenhut (1975) algorithm which is simple, and applicable to the capacity constrained situations.

From the aggregate production problem, the production level of the part fabrication stage for the first period is obtained. By allocating this production level to each subperiod in proportion to the regular capacity, we can get target production level for each subperiod. Setting the production level of each subperiod to target level, we expect that the feasibility will be maintained. The Eisenhut algorithm is modified to generate a production schedule up to the specified production level even if high costs may be incurred in the short-term.

Numerical examples

Compared model

Even with fairly small examples it is difficult to get optimal production schedules because of the set-up costs involved. Therefore we chose the material requirements planning (MRP) model for experimental comparison. Our simple MRP model was constructed by the following procedures.

At the top level, the master production schedule was determined by solving the linear programming problem for product items, where the first two periods of the planning horizon were divided into subperiods and relevant costs and demand information were given based on this time length. After the individual product item quantities were exploded to determine part item requirements, the Eisenhut algorithm was used to determine part item lot sizes, and the lot size adjustment was conducted if infeasibility was detected in the part fabrication stage.

Problem structures

The problem had 10 product items and 26 part items grouped into 4 product types and 10 part types, respectively. Planning horizon of the aggregate planning was set to six months and a week was selected as a unit of subperiod where one month was made of four weeks. Forecast error for product items demand was set to zero.

In order to test the robustness of the HPP and MRP methodologies, the following parameters to be varied were selected: product assembly stage capacity, part fabrication stage capacity, part items set-up costs, and part holding costs. The total combination of the values of the parameters tested results in 24 different runs. In the simulation runs covering one year, back orders were treated separately and were not included in the costs calculation.

Results and discussion

We compared the performance of the two approaches in terms of costs and back orders. Because of the separate treatment of back orders, the costs of the two approaches were not always incurred from the same production quantities, so simple

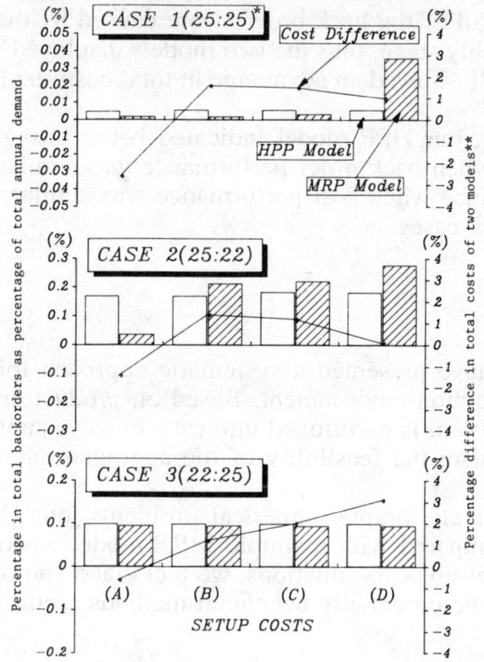

* Allowed overtime as percentage of regular capacity at (product assembly stage:part fabrication stage)

** $\frac{\text{MRP} - \text{HPP}}{\text{HPP}} \cdot 100\ (\%)$

Figure 3. A summary of the simulation results. Bars, back orders; lines, costs. Set-up costs: EOQ covering (A) 2-week, (B) 1·5 week, (C) 1 week and (D) 0·5 week demand.

comparison of these costs might not acceptable. Thus, total back orders as a per cent of annual demand, and total costs are drawn in one graph shown in Figure 3 where back orders and costs were expressed by bars and lines, respectively.

We divided the simulation results into three cases based on the production capacity of each production stage. Production capacity was expressed by the fraction of allowed overtime which was 22% or 25%, and represented by (product assembly stage: part fabrication stage) in Figure 3. Four cases of set-up costs were selected of which economic order quantity (EOQ) nearly covered 2-week (A), 1·5-week (B), l-week (C), and 0·5-week (D) demand. The results of the simulation varying part holding costs were averaged, and thus the 12 summarized results were plotted in Figure 3.

Initially, similar line trends were noticed. In the case of set-up costs (A), all three cases of simulation showed the advantage of the MRP model in terms of costs and back orders. But for the lower set-up costs cases, our model showed a cost advantage. Concerning back orders, our model was insensitive to the set-up costs change and showed relatively agreeable results compared with the MRP model, especially for the lower set-up costs cases.

In case l, enough capacity was provided in each stage, and the total back orders of the two models were small. But the MRP model incurred relatively large back orders in set-up costs (D) because of the low level of the part items inventory.

In case 2, the capacity of the part fabrication stage was restricted in comparison with the product assembly stage and almost all of the back orders occurred at the part fabrication stage. As the set-up costs declined, the total back orders were increased with the MRP model. Total costs of the two models were nearly the same when set-up costs were (D).

In case 3, almost all of the back orders were caused by the insufficient capacity of the product assembly stage, thus the two models displayed similar results in back orders. However, HPP showed an advantage in total costs excluding the set-up costs (A).

Roughly speaking, our HPP model indicated better cost performance than that of the MRP model when back order performance was similar, and indicated better back order performance when cost performance was similar with the exception of the high set-up costs cases.

Conclusions

In this chapter we have presented a systematic approach for production planning of a two-stage production environment. Based on product and time hierarchy, the overall planning problem is partitioned into easy to solve subproblems. By the kind of feedback mechanism the feasibility of the aggregate plans to detailed plans is maintained.

Using some relatively simple numerical problems, our HPP approach exhibits agreeable results compared with a simple MRP model when part set-up costs are lower. Under high set-up costs situations, we feel that a combined approach of HPP and MRP is one of the potentially beneficial methods requiring further refinement.

References

Axsäter, S. and Jönsson, H., 1984, Aggregation and disaggregation in hierarchical production planning. *European Journal of Operational Research*, **17**, 338–350.

Bitran, G.R., Haas, E.A. and Hax, A.C., 1982, Hierarchical production planning: a two-stage system. *Operations Research*, **30**, 232–251.

Boskma, K., 1982, Aggregation and the design of models for medium-term planning of production. *European Journal of Operational Research*, **10**, 244–249.

Eisenhut, P.S., 1975, A dynamic lot sizing algorithm with capacity constraints. *AIIE Transactions*, **7**, 170–176.

Golovin, J.J., 1975, *Hierarchical Integration of Planning and Control*. Technical Report No. 116, Operations Research Center (Cambridge, MA: (MIT)).

Hax, A.C. and Candea, D., 1984, *Production and Inventory Management* (New Jersey: Prentice Hall), pp. 393–464.

Saad, G.H., 1988, An optimization approach for the disaggregation of aggregate production plans. In *Expert Systems and Intelligent Manufacturing*, edited by M.D. Oliff (Amsterdam: Elsevier), pp. 305–318.

28

The Silver–Meal heuristic: worst case bounds for finite horizons

Sven Axsäter

Luleå University of Technology, S-951 87 Luleå, Sweden

Abstract

This chapter considers the classical dynamic lot size problem without backlogging over a finite horizon. One of the most popular approximate methods for solving this problem is the Silver–Meal heuristic. There is no bound for the worst case performance ratio of this heuristic in case of an infinite horizon. Since it is common in practice to apply the Silver–Meal heuristic over finite horizons, we analyse how the performance ratio of the Silver–Meal heuristic depends on the length of the horizon for extreme difficult demand sequences. A linear programming model is used for generating such extreme demand sequences.

Introduction

We consider the classical dynamic lot size problem without backlogging. It is relatively easy to derive the optimal solution of this problem by applying the well-known Wagner–Whitin algorithm (Wagner and Whitin, 1958). This method is very seldom used in practice, though. Instead it is common to use various heuristics that are easier to understand. A good heuristic will, in general, give the optimal solution or an approximate solution that is very close to optimum. In this chapter we analyse one of the most popular approximate methods, the Silver–Meal heuristic (Silver and Meal, 1973).

Although the Silver-Meal heuristic will, in most situations, provide very good solutions, there are also extreme cases when the performance is bad. It was shown in Axsäter (1982) that there is, in fact, no bound for the relative error when applying this heuristic, i.e. the worst case performance ratio is infinity. This is also the case for some other well-known heuristics; but there are also heuristics with finite worst case performance ratios. Other papers dealing with worst case analysis of lot sizing heuristics are Bitran *et al.* (1984) and Axsäter (1985).

It is easy to realize that when applying the Silver–Meal heuristic (or any other approximate method), an infinite relative error can only occur if the horizon is unbounded. Since it is quite common to apply heuristics over finite horizons, it is therefore of interest to study how the performance ratio in extreme situations depends on the length of the horizon and what the bad demand sequences look like. Kiran (1988) has previously analysed the Silver–Meal heuristic for finite horizons, and he has derived the worst case performance ratio for very short horizons (up to four periods). In this chapter we consider longer horizons (up to ten periods).

In the next section we give a problem formulation. The results of our analysis are presented in the third section and are followed by a few concluding remarks in the final section. A linear programming model that has been used for generating extreme demand sequences is described in the Appendix.

Problem formulation

The classical dynamic lot size problem can be formulated in the following way. The demand is known over a considered horizon and it must be satisfied in the period it occurs. There is a fixed set-up cost for each batch. The linear holding cost is stationary and we are penalizing the end of period inventory. There is no initial inventory. We assume that the demand in the first period is non-zero, since we can otherwise disregard this period.

Let us introduce the following notation: T = number of periods, x_i = production volume to be delivered to inventory in period i, d_i = demand in period i, I_i = inventory level at the end of period i, S = set-up cost, h = per period unit holding cost, $C(T)$ = total cost.

Our problem in mathematical terms is then to

$$\text{minimize} \qquad C(T) = \sum_{i=1}^{T} S\,\delta(x_i) + hI_i \tag{1}$$

subject to

$$\begin{aligned} I_{i-1} + x_i - I_i &= d_i \\ I_0 = 0,\ x_i,\ I_i &\geq 0 \end{aligned} \qquad i = 1, 2, \ldots, T \tag{2}$$

where

$$\begin{aligned} \delta(x_i) &= 0 \qquad \text{if } x_i = 0, \\ \delta(x_i) &= 1 \qquad \text{if } x_i > 0. \end{aligned} \tag{3}$$

It is easy to see that a production batch should always equal the demand for an integral number of periods. This is utilized by the Silver–Meal heuristic and by most other heuristics.

According to the Silver–Meal heuristic the first batch (in period one) should cover n periods if the per period cost is increasing for the first time when going from n to $n + 1$ periods, i.e.

$$\frac{S + h\sum_{j=1}^{n+1}(j-1)d_j}{n+1} > \frac{S + h\sum_{j=1}^{n}(j-1)d_j}{n}. \tag{4}$$

If no such $1 \leq n < T$ exists, the complete horizon is covered by a single set-up. If $1 \leq n < T$ exists, the same procedure is applied to the remaining problem over periods $n + 1, n + 2, \ldots, T$, etc..

By worst case for a given finite horizon T we mean a choice of parameters S, h and $d_2, d_3, \ldots, d_T$ (note that $d_1 > 0$ does not affect the solution) such that the ratio $\hat{C}(T)/C^*(T)$ between the cost $\hat{C}(T)$ when applying the Silver–Meal heuristic and the optimal cost $C^*(T)$ is as large as possible. It is trivial to see that no error can occur for $T \leq 2$.

When looking for the worst possible demand sequence it is sufficient to consider $S = 1$, $h = 1$. To see this let $S = S'$, $h = h'$. Assume that the worst case perform-

Table 1. 'Worst case' performance of the Silver–Meal heuristic.

Horizon T	Performance ratio	Demand sequence
1	1	—
2	1	—
3	1·33 (4/3)	$d_2 = 0, d_3 = S/4h + \varepsilon$
4	1·50 (3/2)	$d_2 = d_3 = 0, d_4 = S/9h + \varepsilon$
5	1·60 (8/5)	$d_2 = d_3 = d_4 = 0, d_5 = S/16h + \varepsilon$
6	1·67 (5/3)	$d_2 = d_3 = d_4 = d_5 = 0, d_6 = S/25h + \varepsilon$
7	1·71 (12/7)	$d_2 = d_3 = d_4 = d_5 = d_6 = 0, d_7 = S/36h + \varepsilon$
8	1·75 (7/4)	$d_2 = d_3 = d_4 = d_5 = d_6 = d_7 = 0, d_8 = S/49h + \varepsilon$
9	1·81 (225/124)	$d_2 = d_3 = 0, d_4 = S/9h + \varepsilon, d_5 = d_6 = d_7 = d_8 = 0, d_9 = S/25h + \varepsilon$
10	1·89 (36/19)	$d_2 = d_3 = 0, d_4 = S/9h + \varepsilon, d_5 = d_6 = d_7 = d_8 = d_9 = 0,$ $d_{10} = S/36h + \varepsilon$

ance ratio is obtained for $d_i = d_i'$ $(i = 2, \ldots, T)$. But it is then evident that $d_i = d_i' S'/h'$ will generate the same performance ratio for $S = 1$, $h = 1$. This is used in our linear programming model for derivation of difficult demand sequences (see the Appendix).

Note also that the worst case performance ratio is necessarily non-decreasing with T. If d_i $(i = 2, 3, \ldots, T)$ gives a certain performance ratio for T we obtain the same performance ratio for $T + 1$ periods with the same d_i for $i = 2, 3, \ldots, T$ and $d_{T+1} = 0$.

Worst case results

It is possible to derive a tight lower bound for the worst case performance of the Silver–Meal algorithm for a given finite horizon by solving a large number of small linear programs. We have used this method (described in the Appendix) for horizons $T \leq 6$ and it is possible to use the same technique also for longer horizons. By analysing the results of the computations for $T \leq 6$, it is easy, however, to extrapolate the bad performance to longer horizons.

We believe that the obtained results (see Table 1) are actually the worst case ratios. In any case Table 1 provides achievable very bad ratios.

The demand sequences generating the bad performance have most of the demands equal to zero and a few small non-zero demands that will cause 'unnecessary' set-ups when applying the Silver–Meal heuristic. For $T \leq 8$ there is just one non-zero demand after period one. In all cases the optimal solution is to have a single set-up in the first period. Such demand sequences have also been suggested by Kiran (1988).

In Axsäter (1982) a completely different type of demand sequence ($d_i = S/(i-1)h$) was used to prove that there is no bound for the worst case performance ratio in the infinite horizon case. This means that the Silver–Meal heuristic gives just a single set-up while the optimal solution has infinitely many. This situation is completely different from the examples in Table 1. It was shown in Axsäter (1985), though, that demand sequences like those in Table 1 may also generate arbitrarily large relative errors for infinitely long horizons.

Concluding remarks

We have analysed how the performance of the Silver–Meal heuristic in extreme difficult situations depends on the length of the horizon. Our results illustrate circumstances under which the heuristic will perform especially badly. The performance ratio is increasing relatively fast with the length of the horizon.

The linear programming methodology for derivation of 'bad' demand sequences could be applied in essentially the same way also when analysing the performance of most other common heuristics.

Appendix. A linear program for generation of difficult demand sequences

As pointed out in the second section we can without any lack of generality choose $S = h = 1$. Assume that the Silver-Meal heuristic generates a certain given cost and a certain given set-up pattern. We introduce the following notation: T = number of periods; r = number of set-ups besides the initial set-up in period 1; β = cost when applying the Silver–Meal heuristic; i_j = the period with the $(j+1)$th set-up, i.e. the jth set-up after the initial set-up, $i_0 = 1$, $i_{r+1} = T + 1$, d_j = demand in period j; V_j = optimal cost of the lot size problem over periods 1, 2, ..., j, $V_0 = 0$.

The worst performance ratio for the given cost and set-up pattern can in principle be obtained by solving:

$$\min_{d_j} \max_{V_j} V_T \tag{A1}$$

$$r + 1 + \sum_{k=0}^{r} \sum_{j=i_k+1}^{i_{k+1}-1} d_j (j - i_k) = \beta \tag{A2}$$

$$d_{i_k}(i_k - i_{k-1})^2 - \sum_{i=i_{k-1}+1}^{i_k - 1} d_i (i - i_{k-1}) = 1 \quad k = 1, 2, \ldots, r \tag{A3}$$

$$d_j (j - i_k)^2 - \sum_{i=i_k+1}^{j-1} d_i (i - i_k) \leq 1 \qquad k = 0, 1, \ldots, r; \quad j = i_k + 1, \ldots, i_{k+1} - 1 \tag{A4}$$

$$\begin{aligned} V_j &\leq V_{j-1} + 1 \\ V_j &\leq V_{j-2} + 1 + d_j \\ V_j &\leq V_{j-3} + 1 + d_{j-1} + 2d_j \qquad j = 1, 2, \ldots, T \\ &\vdots \\ V_j &\leq V_0 + 1 + \ldots + (j-1)d_j \\ d_j, V_j &\geq 0. \end{aligned} \tag{A5}$$

The constraint (A2) guarantees that the Silver–Meal solution will give the correct cost. Constraints (A2) and (A3) are the Silver–Meal constraints for the given set-ups. Note that in the worst case we can have equalities in (A3). The constraints (A5) together with the maximization of V_T imply that V_T is the optimal solution of the lot size problem for given $d_j : s$. The performance ratio is then obtained as β / V_T.

In order to obtain a linear program that is easy to solve we replace (Al) by

$$\max_{d_j V_j} V_T. \tag{A6}$$

This means evidently that we will only obtain a lower bound for the worst case performance ratio. We believe that this bound is very tight, though. First, in those cases where we know the worst case performance ratio we obtain the correct results.

Furthermore, known worst case results indicate that the worst case will occur either for $\beta = r + 1$ or when we have equalities in (A4). In both these cases the d_j:*s* are unique and we obtain the correct worst case performance ratio.

In order to consider all cases we need to evaluate all set-up patterns and for each set-up pattern all values of β. Different interesting values of β are easily evaluated by using a parametric programming approach. It is easy to see that only those values of β where we have a change of basis are of interest. There are $2^{(T-1)}$ possible set-up patterns but in each case the linear program is quite small [$2T - 1$ variables and $(T^2 + 3T)/2$ constraints]. Furthermore, it is easy to see directly that many set-up patterns cannot lead to a worst case performance. Such patterns can, of course, be excluded.

References

Axsäter, S., 1982, Worst case performance for lot sizing heuristics. *European Journal of Operational Research*, **9**, 4, 339–343.

Axsäter, S., 1985, Performance bounds for lot sizing heuristics. *Management Science*, **31**, 5, 634–640.

Bitran, G.R., Magnanti, T.L. and Yanasse, H.H., 1984, Approximation methods for the uncapacitated dynamic lot size problem. *Management Science*, **30**, 9, 1121–1140.

Kiran, A.S., 1988, A Combined Heuristic Approach to Dynamic Lot Sizing Problems, Working Paper, University of Southern California, Department of Industrial and Systems Engineering.

Silver, E.A. and Meal, H.C., 1973, A heuristic selecting lot size requirements for the case of a deterministic time-varying demand rate and discrete opportunities for replenishment. *Production and Inventory Management*, **14**, 2, 64–74.

Wagner, H.M. and Whitin, T.M., 1958, Dynamic version of the economic lot size model. *Management Science*, **5**, 9, 89–96.

29

Application of precedence and function diagrams to assembly planning

Klaus Thaler

Fraunhofer Institute for Industrial Engineering (IAO), Nobelstrasse 12, 7000 Stuttgart 80, Germany

Abstract

An approach using a combined precedence and function diagram to enable economical assembly was developed and realized within a prototype software in a research project started in 1986. Partners of this project are the Fraunhofer Institute IAO, four German companies and a software house. The research covers planning especially for flow line and flexible assembly systems within an integrated tool-box system. This chapter describes an approach to gain a better base for cost evaluation and reduction for assembly by using precedence and function diagrams in early stages of planning and design. It relates to other work described in the *ICPR Proceedings* book and highlights the aspects of modelling and using assembly functions and features to describe operations.

Introduction

Computer-integrated manufacture (CIM) proceeds on the assumption that a product, from its conception to its marketing, is accompanied and directed by information flows. According to the view that the value of an integrated system is much bigger than the total value of isolated and partial solutions, it is indispensable to think in information streams and to overcome department barriers.

Especially affected are assembly planning tasks which can only be solved successfully by the co-operation of several different departments in a company. Apart from the organizational consequences, field-overlapping work must be technically supported in a suitable manner. Informational connections between pre- and post-positioned departments have to be installed. In practice this means that, for example, communication interfaces have to be created and defined for the construction department, quality control and production planning. Concerning assembly planning, the planning process should be supported by an *integrated methodology* which is adjusted in procedure as well as in modelling the planning and design problems.

In order to meet the demand of a closer link between product planning activities and assembly planning it is helpful to remember that most of the cost influencing decisions are made during planning and design (Andreasen, 1987). Examples of problems during a planning process are:

— transformation of construction bill of materials into assembly orientated product structures,

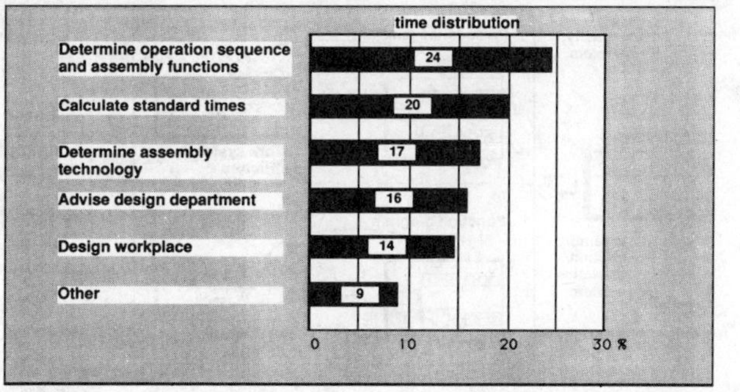

Figure 1. Planning time distribution for assembly tasks.

— analysis of component precedence,
— analysis of assembly functions,
— description of operations and technology.

An examination of relevant planning activities, which was carried out for each company involved in our project, also reflected these aspects (Figure 1).

Referring to Figure 1, it can be seen that activities concerning operation precedence, assembly functions, description of technology and design consultation are dominant problem fields.

It is worth concentrating on the precedence diagram, for it has proved to be very helpful as a common planning tool between different departments. Similar to a network plan, assembly operations which depend on each other are shown graphically as connected nodes. Unlike a network plan, each node must be directed to a row. Reading from left to right, this should be the earliest possibility of proceeding an operation. Joining backwards to a node is not allowed, which means that in a mathematical description a precedence diagram is a directed graph without cycles (Ammer, 1985). The precedence diagram can be used for (Bullinger and Ammer, 1984):

— work content planning (restrictions for assembly),
— technological planning (mechanization and automation),
— design of work system (work system structure),
— line balancing (model-mix assembly).

In order to gain the structure of a precedence diagram, all parts of the bill of material have to be considered with respect to their assembly sequence. Mostly known are comparison in pairs, disassembly models, and network approaches.

As conventional predence diagrams have many shortcomings, the OPNET model (operation network) was developed at our Institute, which enables the generation, assessment and further use of OPNETs (see Seidel, 1987). One main problem field concerning this approach is how to gain and assess such an operation structure. However, to use this structure for line balancing and capacity planning it is needed to describe which assembly restrictions exist for each operation as well as which technology and which assembly functions have to be planned. This aspect will be described in the following section.

Mapping assembly functions into a work system structure

Once a product is in the stage of design or has already been designed, it is required that costs of production will be minimal and consequently will allow an economical manufacturing process. A better basis of cost evaluation and reduction may be

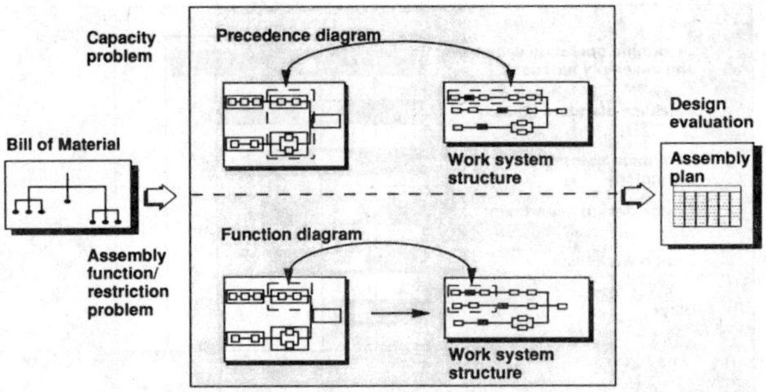

Figure 2. Mapping of assembly functions into a work system structure.

gained by drawing closer precedence structure and manufacturing process in an early stage of planning and design.

A difficulty which often arises during planning is how to deal with restrictions of the work task, e.g. simultaneous processes and other assembly features. In many cases the precedence diagram has to be structured according to corresponding and suiting operations and technology, for example:

(1) Sections of the precedence diagram should relate to sections of a given or a planned work system structure.
(2) Sections of the precedence diagram should be 'concentrated' to automated assembly devices at certain stations of a work system.

In a more theoretical view, this can be mainly seen as a problem of mapping the precedence diagram and its assembly functions into an (existing or planned) work system structure (Figure 2).

To solve this balancing problem, algorithms can be found in line balancing (see Fremerey, 1990) and capacity planning literature. In contrast to this, the focus of this paper is on the model of assembly functions and restrictions on the base of a precedence and function diagram (Thaler, 1989b). This assembly operation model is a view of the operations and their features, which are required fully to describe operations as an input to balancing methods. The main characteristics of the model will be introduced in the following section.

Assembly operation and function model

In many applications, function diagrams are used to specify assembly functions to a high degree of abstraction. In contrast to a precedence diagram a function diagram shows nodes and connecting lines for operation precedence, where relevant assembly operations are abstracted as standardized assembly function symbols. As an example, Gairola (1981) uses function diagrams and symbols to specify assembly functions as an aid for product design evaluation.

Figure 3 shows an overview of the assembly operation and function model. As an assembly operation can be described by a set of assembly functions, this possibility was one important requirement of the model. To gain further requirements, assembly operations of our industrial partners were examined.

An assembly operation is a process to join two or more parts (subassemblies) to a resulting component. For this joining process, the parts have to be placed into a certain position or device. One of these parts may be the base part (which usually follows a logistical main flow in an assembly system). The other part(s) will be called add-on parts, as they are generally moved into the direction of the base part. For the mating operation itself, joint elements such as bolts, screws etc. have to

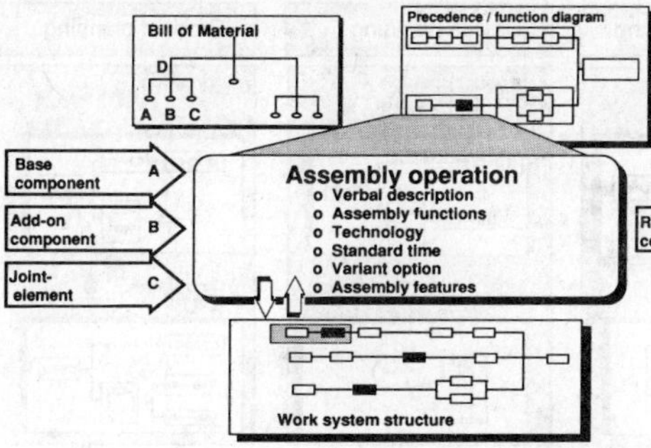

Figure 3. Assembly operation and function model.

be considered. Three main processes or types of operations can be differed: (i) process with increase of components, (ii) process with component use, and (iii) process without component use.

All direct operations which increase the number of already assembled components of a product, e.g. assemble A, B and C to D, are of the first type. The second type is usually needed to prepare or finish a direct operation, e.g. feed B into assembly device. The third type is usually needed to control or survey a direct operation. Each assembly operation can be characterized by time, technology, assembly function and a verbal description. For variant operations, a variant option can be given. For each desired operation, assembly features can be specified. The relevant features are:

— position at the product (assembly point),
— quality (of operation),
— ergonomical feature,
— accessibility at the product,
— automation feature,
— unclean operation,
— product position during assembly,
— team work operation,
— membership (to a set of operations),
— obligatory sequence (of operations),
— simultaneous operations,
— reference to work system.

The mentioned features were analysed and considered as relevant in the companies involved in our project. Although these four companies cover a wide range of manufacturing processes from automotive to small-sized assembly, the features may be only a part of an industry or company specific list.

Figure 4 shows how the precedence/function diagram and the assembly features can be used in different departments. As assembly functions are related to each operation in the assembly model, it is possible to generate an assembly function diagram automatically. Therefore, a library with standardized function symbols is used. To map the precedence and function diagram into a work system structure, an interactive procedure was developed (Thaler, 1989b).

Due to the relations between product and operation structure, all components and joint elements can be determined and allocated to operations with the help of an automatic methodology (see Seidel, 1987). In contrast to the generation of the operation structure, the determination of operation features and functions requires planner interaction, which means computer-aided description. As a planning support, the MOGRAPH system was developed at our Institute (Thaler, 1989b).

The user interacts with an assembly knowledge base, where standardized operations, their features and relevant technology can be predefined and stored. The

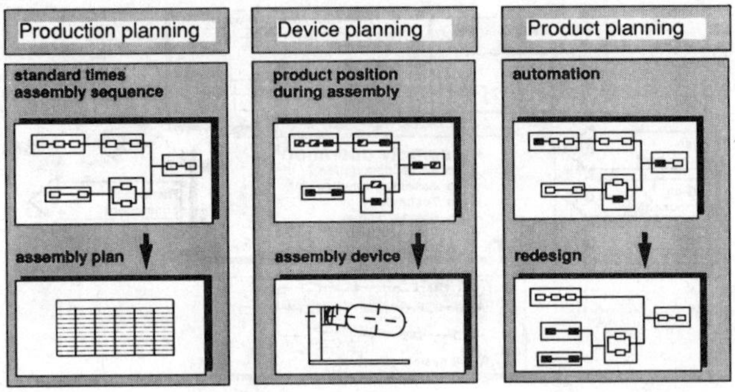

Figure 4. Examples of using the precedence/function diagram.

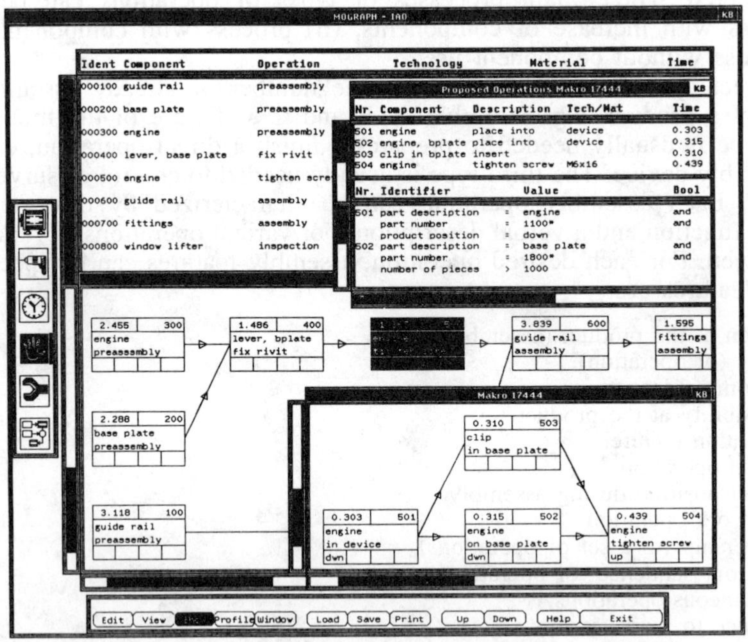

Figure 5. Application example—description of operations.

knowledge about which operation and technology is suitable is predefined as rules and can be evaluated. As a result of this, operations and technology can be proposed to the user. To enable an efficient build up or modification of the knowledge base, a rule induction algorithm is used.

An application of this approach is shown in Figure 5. The industrial example shows the operations to assemble an electric window lifter. Typical parts are base plate, engine, guide rail, lever and fittings. It can be seen that operation 500 is activated. The operations 501 to 502 were proposed as suitable and are displayed in the smaller window. For operation 504, product position 'up' is specified. Some rules which lead to this proposal are shown as well. For example, rule 502 which corresponds to operation 502 can be interpreted as: 'engine and base plate, placed into device, standard time 0·315' is valid if part description is base plate, part number starts with 1800 and the number of pieces exceeds 1000.

Figure 6 shows an example of the interactive procedure to map the result of this

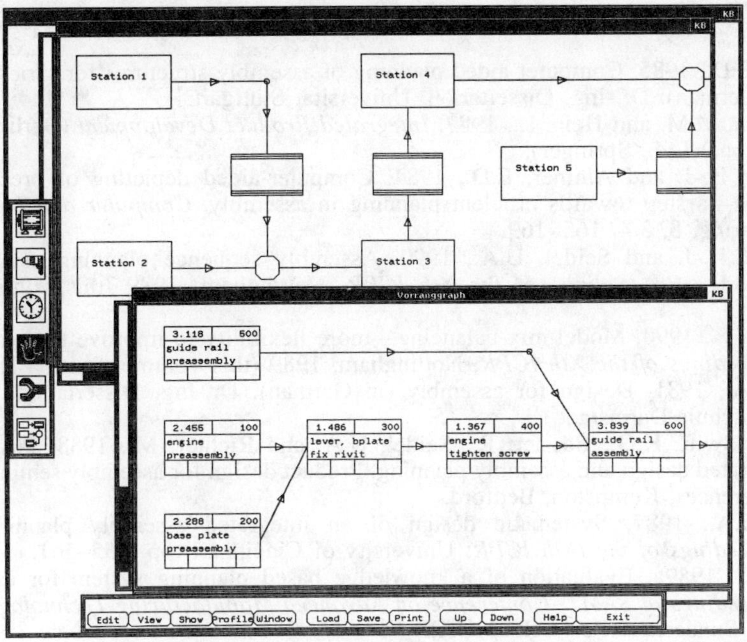

Figure 6. Application example—mapping of operations into a work system.

planning step into the work system. Work system and precedence diagram are represented as graphs.

Remarks and outlook

The methodology to use precedence and function diagrams is part of an assembly planning system described in Seidel (1987) and Thaler (1989a). A necessary basis of this planning approach is information about quantity and quality of the work task. Within the mentioned CIM process chain, the bill-of-material (BOM), relevant component joints and a production program are required as main input. The overall configuration covers different functionalities:

(1) Transformation of BOM into assembly adjusted product structure.
(2) Generation of precedence diagram (Bullinger and Seidel, 1990).
(3) Planning of work content and technology.
(4) Line balancing (Fremerey, 1990).
(5) Sequencing (production sequence).

Up to now a research prototype with focus on the further possibilities of development and an industrial prototype with attention to the relevant company interfaces have been built. The prototype is a graphically supported system running on a workstation in a local area network (LAN) environment. Window-technique, mouse, and icon-/object-oriented representation and direct manipulation technique are used. The main output of the methodology is determined and documented in an assembly plan showing all activities which are relevant to produce the described product with its specific characteristics. Current research is carried out to integrate the described model and methodology into a framework for design and engineering support (Hird *et al.*, 1988).

References

Ammer, E.D., 1985, Computer-aided planning of assembly structures for series production (in German). Dr.-Ing. Dissertation, Universität Stuttgart.

Andreasen, M.M. and Hein, L., 1987, *Integrated Product Development* (Berlin: IFS (Publications) Ltd., Springer).

Bullinger, H.-J. and Ammer, E.D., 1984, Computer-aided depicting of precedence diagrams—a step towards efficient planning in assembly. *Computer and Industrial Engineering*, **8**, 3/4, 165–169.

Bullinger, H.-J. and Seidel, U.A., 1990, Assembly sequence planning using operation networks. *Proceedings of the Xth ICPR*, Nottingham, 1989 (this volume, Section VI.2).

Fremerey, F., 1990, Model-mix-balancing—more flexibility to improve the general result. *Proceedings of the Xth ICPR*, Nottingham, 1989 (this volume, Section III.4).

Gairola, A., 1981, Design for assembly (in German). Dr.-Ing. Dissertation, Technische Hochschule Darmstadt.

Hird, G., Swift, K.G., Bässler, R., Seidel, U.A. and Richter, M., 1988, Possibilities of integrated design and assembly planning. Product design for assembly seminar note, IFS conferences, Kempston, Bedford.

Seidel, U.A., 1987, Systematic design of an integrated assembly planning toolbox. *Proceedings of the IXth ICPR*: University of Cincinnati) pp. 555–561.

Thaler, K., 1989a, Evaluation of a knowledge based planning system for assembly. In *Proceedings of SAMT, Conference on Advanced Manufacturing Technology*, (Sunderland: Sunderland Polytechnic UK).

Thaler, K., 1989b, *MOGRAPH*—a Methodology to Plan Operations and Technology (in German). Technical note, IAO.

30

The design and implementation of a factory co-ordination system

Richard Bowden, Jim Duggan and Jim Browne

CIM Research Unit, University College Galway, Ireland

Abstract

In a batch manufacturing environment, two major influences on product flow are the 'design' of the production environment in terms of group technology principles and the scheduling of operations. By placing the emphasis on the use of planning tools in these tasks, the complexity of the relevant control tasks will be reduced. This chapter contains a brief description of a tool for planning product flow throughout a factory and also discusses the need for a link between this planning tool and a 'design' tool which will structure the production environment using product based layout and product families. These two tools can be used in an electronics assembly environment to ensure that new products can be efficiently integrated into the production process and the flow of batches organized through effective planning procedures. Through the resultant stable product flow, a factory will have the flexibility to respond to fluctuations in demand.

Introduction

Factory co-ordination is a set of procedures concerned with planning the flow of products through an entire manufacturing plant over the short term (i.e. one day to one month). Factory co-ordination has close links with the 'design' of the production environment. This 'design' task involves structuring the production environment in terms of the definitions of product families and product-based layout, in order to facilitate product flow. The complexity of the planning task can be reduced if the production environment is structured efficiently. The position of factory co-ordination within the manufacturing management hierarchy is illustrated in Figure 1.

Factory co-ordination provides a link between the requirements planning stage normally carried out by a materials requirements planning (MRP) system and the control of shop floor events by a production activity control system (PAC) (Browne, 1988).

Two weaknesses in an MRP approach are (Browne *et al.*, 1988):

(1) MRP uses estimates of lead times which by definition are not accurate.
(2) The design of the production environment is ignored by the MRP approach.

PAC systems control production within each cell on the shop floor, but have no planning or control capability at a plant level.

Factory co-ordination bridges this gap by using *principles* from Just-in-Time

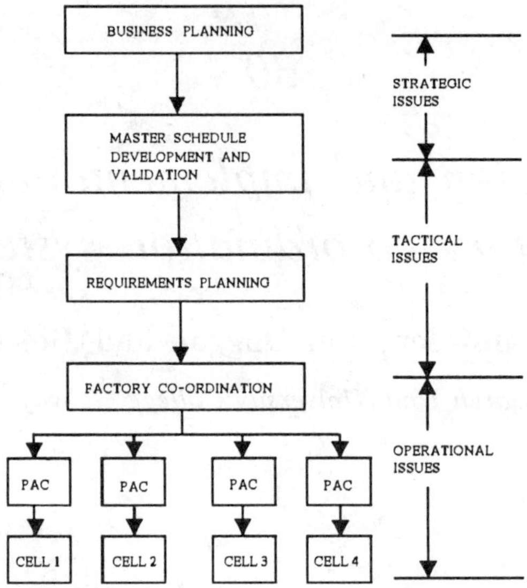

Figure 1. Manufacturing management hierarchy.

and Optimized Production Technology† (Fox, 1982). The main theme of factory co-ordination involves viewing the entire factory as one system and places great emphasis on planning in order to reduce the complexity of the control task at lower levels. The link with the 'design' of the production environment provides a facility to restructure the environment through group technology principles, but it is not used every time that the factory co-ordination system is used. This link is used when a range of new products are coming on-line for production and have to be integrated into the production environment. In electronics assembly, the introduction of new products is occurring frequently, hence the requirement for efficient integration procedures, which can increase the predictability of product flow.

The major benefits of factory co-ordination derive from the regular and stable product flow which results from the use of such a system. The introduction of new products is accomplished more efficiently and with less disruption to the process. The planning and control tasks concerning raw materials and work in progress are relatively less complex. Due to the stability of the production process, further production requirements for satisfying market demand can be determined with greater certainty. As demand fluctuates in batch manufacturing, the environment has the flexibility to handle the demand fluctuations because of its inherent stability and regularity. This chapter discusses the concept of factory co-ordination and the required architecture to fulfil the co-ordination task of structuring the production environment and planning the product flow. The chapter concludes with some observations on factory co-ordination and production environment design.

Factory co-ordination architecture

We consider a factory to consist of a number of group technology cells, each of which is controlled by a PAC system. The necessary architecture for completing the factory co-ordination task includes: a link with the production environment design module and a scheduler, dispatcher and monitor to complete the planning task, as illustrated in Figure 2.

† OPT is a registered trademark of Scheduling Technology Group Ltd.

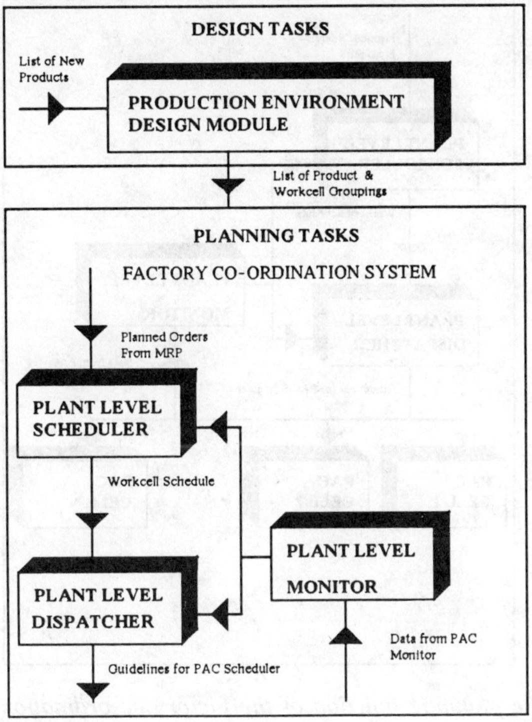

Figure 2. Links between the design and planning tasks.

Production environment design module

Product families and the resultant categorization of resources as either shared or dedicated has a major influence on organizing workflow. The main problem we are attempting to solve with this module is the proliferation of new products, which may make the workflow unpredictable, thus making the task of organizing the product flow more complex. This module will provide a concise set of procedures to help generate a list of suitable product groups and categorize the resources into workcells. The link between the 'design' of the production environment and the factory co-ordination task is illustrated in Figure 2.

Environment design procedures

There are two types of procedures used within the environment design module: variant and generative procedures. The variant approach seeks to integrate the new products within existing families according to the rules base. However through successive integration of new products into existing families the size of individual families may become excessive, and the benefits diminish. Should such a situation arise, the generative approach is used whereby existing families are disbanded, and new grouping arrangements are formed from the range of new and old products.

Environment design rules base

The rules base is derived mainly from the principles of group technology and is broken into three areas: manufacturing characteristics, design characteristics and a section to solve trade-offs between manufacturing and design considerations. The

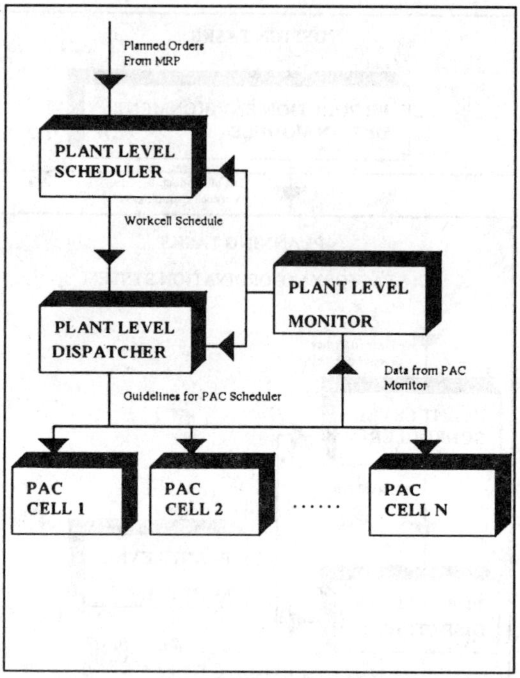

Figure 3. The planning function of the factory co-ordination architecture.

manufacturing characteristics relate to the required production processes for a product, while the design characteristics relate to the components contained in a product's bill-of-materials. The third rules base section dealing with trade-offs between the two considerations is necessary because a proposal based on manufacturing considerations may differ from a proposal based on design considerations. It is important to have to some procedure in place which can effect a compromise between two such proposals and derive an overall suitable solution.

Once a proposal is developed, it is submitted to the plant level simulator along with estimates of demand for the products. The simulator can analyse the proposal from the perspective of required capacity to produce the products and how well Just-In-Time criteria (e.g. reduced set-up and reduced batch sizes) are satisfied by the proposal.

The planning task

The exchange of data between the scheduler, dispatcher, monitor and each of the PAC systems in completing the factory-wide planning task is illustrated in Figure 3.

Plant level scheduler

The main purpose of the plant level scheduler is to produce a plan which can be used to co-ordinate the flow of batches between workcells throughout a manufacturing plant (i.e. a plant level plan). Any scheduling within the cell can be accomplished by the supervisor using the guidelines stated in the plant level plan. By using the guidelines stated in the plant level plan, any schedules which are developed for individual resources within a cell will facilitate the overall flow of batches between cells. The plant level scheduler consists of three modules: production allocation, bottle-neck management and a scheduler, which are used in determining a feasible schedule as explained below.

Production allocation
This task involves deciding on a suitable policy for determining how to spread the monthly production requirements for each product throughout the month and to decide on a suitable batch size to be assigned to each product. If there are any late batches which did not enter the process according to schedule, this task will decide how to re-allocate them and the length of the time horizon over which they are allocated. This module will determine the daily production requirements which will be used by the bottle-neck management and scheduling modules.

Bottle-neck management
The purpose of this module is to provide some guidelines to a supervisor on minimizing the effects of potential bottle-necks on product flow. Due to the effect of potential bottle-necks on product flow, they cannot be totally ignored by plant level schedule, and hence the necessity of this management module. These recommendations are separate from the plant level schedule, because the schedule is concerned with product flow between cells, and cannot be changed to facilitate one potential bottle-neck within a cell.

The potential bottle-necks are identified by a capacity check based on the daily production requirements identified in the production allocation module. The potential bottle-necks are assessed based on a finite capacity based projected utilization figure, the position of the resource in the process and the number of products using the resource. This module will act in a decision support mode, generating a range of suitable suggestions, which are based on some basic principles from Optimized Production Technology (Fox, 1982). The recommendations relating to managing the potential bottle-necks are based on three general areas: set-up reduction techniques, alternative routes and specific solutions.

Scheduler
The schedule generated by this module is intended as a plan, which initially can be used to co-ordinate the product flow between different cells in a plant.

The plant level schedule is labelled a plan because if during the course of the planning period (i.e. a day), some major problems occur within the process, the schedule may be infeasible. In this instance, there may be a requirement to generate another plant level schedule. The schedule is important because it is a common reference point for all products going through the production process.

Once a schedule is proposed, it is passed to the plant level simulator, which will analyse it from the perspective of number of batches produced daily, throughput times for each batch, work in progress levels and resource utilizations (Duggan and Browne, 1988; Duggan *et al*., 1989). The important criteria relate to the number of finished batches and batch throughput times, because through better co-ordination stable regular product flow with minimized throughput times should occur.

The factory co-ordination system should be able to function in a range of different production control environments (e.g. push or pull, etc.). Therefore the plant level scheduler contains a library of rules to create schedules for different types of environments (e.g. priority-based schedules, backward schedules, etc.).

Plant level dispatcher
The implementation of the schedule is handled by the plant level dispatcher. The plant level dispatcher communicates with each of the PAC schedulers and dispatchers. Any events which occur in a particular cell can be communicated to the other cells through the plant level dispatcher. The control capability of the plant level dispatcher is not important since it operates at a high level, most of the control

capability is handled by the PAC dispatchers. Any real-time scheduling which must be done on a plant level in relation to batches which are delayed in the process, will be handled by the plant level dispatcher. The plant level dispatcher contains different control strategies including simple heuristics such as shortest processing time.

Plant level monitor
The function of the plant level monitor is to provide relevant data on performance measurement and a real-time status of the shop floor to the dispatcher and scheduler. This data is accessed from each PAC monitor. The main data requirements will concern amount of finished daily batches, throughput times, and work in progress levels.

Conclusion

Factory co-ordination is a tool which places the emphasis on developing an efficient plan, which reduces the complexity of the shop floor production activity control task throughout a manufacturing plant. By having close links with the design task, which structures the production environment, the planning task becomes less difficult. The plan developed by the factory co-ordination system provides guidelines on starting batches at each workcell, so the detailed scheduling in each cell is developed in accordance with a strategy for the entire factory. By using one factory level plan efficiencies are gained for the overall production system, rather than at one particular set of workcells. Presently the system, described in this chapter, is being implemented in an electronics assembly environment.

References

Browne, J., 1988, Production Activity Control—a key aspect of production control. *International Journal of Production Research*, **26**, 3, 415–427.

Browne, J., Harhen J. and Shivnan, J., 1988, *Production Management Systems, A CIM Perspective* (England: Addison Wesley).

Duggan, J. and Browne, J., 1988, ESPNET: Expert system based simulator of Petri nets. *IEE Proceedings-D Control Theory and Applications*, **135**, 4, 239–247.

Duggan, J., Bowden, R. and Browne J., 1989, A simulation tool to evaluate factory level schedules. *4th International Conference on Simulation in Manufacturing*, North Carolina USA, May 1989.

Fox, R.E., 1982, OPT An answer for America Part II. *Inventories and Production Magazine*, **6**, 2.

Disclaimer: the views expressed in this paper are those of the authors and do not necessarily reflect those of Digital Equipment International.

31

Treatment of bottle-neck machines in cellular manufacturing systems

Hamid Seifoddini

Industrial and Systems Engineering Department, University of Wisconsin, Milwaukee, USA

Abstract

The bottle-neck machine problem is one of the major blocks to productivity improvement in development and implementation of cellular manufacturing systems. This chapter discusses the major approaches to the bottle-neck machine problem and highlights the advantages and disadvantages of each approach. Modification of routings is one of the simplest methods of dealing with the bottle-neck machine problem; its application, however, is limited to specific cases. Subcontracting can also be employed to deal with the bottle-neck machine problem. When neither of these two methods work, the machine duplication can be used. The duplication process, however, requires additional investment and should be based on thorough cost–benefit analysis. Finally, the formation of a hybrid cell can be employed as a simple solution to the bottle-neck machine problem. This solution, however, in some cases adversely affects the performance of the corresponding cellular manufacturing system.

Introduction

Cellular manufacturing is the application of group technology to manufacturing. In cellular manufacturing parts with similar manufacturing processes are grouped into part-families and machines are organized into machine cells each capable of processing one or more part-families. The formation of part-families and the associated machine cells is referred to as 'machine cell formation'. Ideally, all part-families assigned to a machine cell should be fully processed within that machine cell. Practically, however, there are parts that need to be processed in more than one machine cell. These are 'exceptional' parts and machines processing them are 'bottle-neck' machines (Mitrofunov, 1963; Chandrasekharan and Rajagopalan, 1986; Seifoddini, 1989a).

There are different approaches to the machine cell formation problem. Among them, production flow analysis (PFA) (Burbidge, 1975; Askin and Subramanion, 1987), rank order clustering (ROC) (Choobineh, 1988), and the similarity coefficient method (SCM) (Mitrofunov, 1963; Durie, 1970; Burbidge, 1977) are the most widely used methods. In PFA the emphasis is on the organization of manufacturing data in proper forms and identification of machine-component groups based on this data. ROC forms the machine-component groups through permutations on rows and columns of the machine-component chart (a matrix with zero/one entries containing the data on machining requirements of parts). ROC also identifies bottle-neck machines and duplicates (assigns bottle-neck machine to more than one cell) them to simplify the identification of machine-component groups. A number of researchers have employed SCM to form machine cells for cellular manufacturing applications (Mitrofunov, 1963; Durie, 1970; Sneath and Sokal, 1973; Burbidge, 1977; Kusiak, 1987). In some cases, cost factors have been considered in the selection of machine cell arrangements (Mitrofunov, 1963; Durie, 1970). Most new machine cell formation algorithms incorporate one or more relevant cost factors into the machine cell formation process (Sneath and Sokal, 1973; Chakravarty and Shtub, 1984; Owaban and Lambert, 1988).

A major problem in the implementation of cellular manufacturing systems is the existence of bottle-neck machines which tends to decrease the effectiveness of such systems due to the following reasons. First, many parts from different machine cells ought to be processed on bottle-neck machines. This creates a large number of intercellular moves which results in higher material handling costs. Secondly, the proper sequencing of similar parts, which reduces set-up times in cellular manufacturing systems, is very difficult when bottle-neck machines are present. Finally, the processing of parts outside their parent machine cells complicates the production scheduling of part-families and diminishes the major advantages of cellular manufacturing in using group scheduling. In view of these problems, the bottle-neck machine problem should be properly dealt with before an effective cellular manufacturing system can be developed.

The bottle-neck machine problem has been extensively studied and a large number of solutions to the problem have been suggested by different authors (Rajagopalan and Batra, 1982; Chandrasekharan and Rajagopalan, 1986; Seifoddini, 1989a). Among these solutions, the following are the more practical ones: the modification of routings of exceptional parts, subcontracting of exceptional parts, duplication of bottle-neck machines, and formation of a hybrid cell. While the duplication of bottle-neck machines can be applied to any cellular manufacturing situation, the other three solutions are more limited in scope.

In the following sections a brief discussion of each of these four solutions is given and their advantages and disadvantages are reviewed.

Modifications of routings

In a cellular manufacturing system, except for exceptional parts, which need to be processed on bottle-neck machines, all other parts are processed within their parent machine cell. One way of reducing the intercellular moves is carefully to examine the routings of these exceptional parts and to modify them such that they become compatible with other parts in their part-families. In some cases it is possible to eliminate some operations without significantly changing the function of the part. In other cases, some operations may be substituted by operations which are more compatible with the rest of operations in the part-family. The modification of routings, however, is limited by its feasibility and can only be applied to particular situations. This solution, therefore, at best can only partially solve the bottle-neck machine problem.

Outsourcing

When the modification of routings is not feasible, an alternative solution is to subcontract the non-compatible parts to outside producers. This eliminates these parts from the production schedule and makes it possible for the production planner to use group scheduling to reduce the scheduling effort. None the less, the subcontracting is limited by the existence of suitable outside contractors and by the disruption it may introduce into the production planning due to the creation of more uncertainty in meeting due dates.

Outsourcing can also help to reduce the indirect cost, thus making the manufacturer more competitive. For this reason outsourcing is becoming popular even for non-exceptional parts in some cases.

Machine duplication

Machine duplication is one of the most widely discussed solutions to the bottle-neck machine problem (King and Nakoranchaiv, 1982; Seifoddini, 1989a, b). In machine duplication a sufficient number of each bottle-neck machine type is assigned to the associated machine cells to eliminate the corresponding intercellular moves. The main advantage of machine duplication is its capability of totally eliminating intercellular moves and making machine cells completely independent. The major problem with machine duplication is that it usually requires additional machines and imposes additional costs. Therefore it is necessary to perform a thorough cost–benefit analysis to determine those bottle-neck machines that can be economically duplicated.

Major cost factors in the duplication of a bottle-neck machine include the duplication cost and the associated cost saving in material handling. The duplication cost is composed of the initial cost (acquisition and installation cost of the new machine) and the operating cost. The saving resulting from the duplication is mostly due to the cost reduction in material handling. To justify the duplication of a particular bottle-neck machine, the duplication cost should be compensated by the corresponding cost reduction in material handling. The duplication cost can be easily determined by considering the price, installation cost and operating cost of the new machine. The cost saving in material handling, on the other hand, is a function of the intercellular and intracellular material handling costs.

The intercellular material handling cost in a cellular manufacturing system is a function of the number of exceptional parts, production volume, travelling distances and unit transportation cost. This can be mathematically expressed as:

$$IMC = \sum_{i=1}^{MB} \sum_{j=1}^{MB} \sum_{k=1}^{N} V_{ijk} \cdot d_{ij} \cdot c_{ij} \qquad i \neq j$$

where,

IMC = intercellular material handling cost,
MB = number of machine cells containing bottle-neck machines,
N = number of exceptional parts,
V_{ijk} = number of part type K being processed in machine cells i and j,
d_{ij} = travelling distance between machine cells i and j,
c_{ij} = unit transportation cost between machine cells and i and j.

The travelling distances between machine cells are a function of the relative location of machine cells in the overall layout of the cellular manufacturing system. Therefore, it is necessary to arrange the machine cells optimally before the intercellular material handling cost is calculated. Due to the complexity of finding the optimal machine cell arrangement, most of the existing plant layout algorithms find a suboptimal solution to the problem (Seifoddini and Wolfe, 1987). There is a number of computer-aided plant layout algorithms that can be used to determine the suboptimal material handling cost for a manufacturing system. One of these algorithms can be employed to determine the suboptimal intercellular material handling costs.

Intracellular material handling costs can be calculated in a similar way. The sum of intercellular and intracellular material handling costs constitutes the total material handling cost for a cellular manufacturing system. The amount of cost saving due to the duplication of a particular bottle-neck machine can be determined by calculating the sum of intercellular and intracellular material handling costs before and after the duplication process and finding the difference between the two cases. This difference represents the cost saving due to the duplication and should be equal to or greater than the corresponding duplication cost in order for the duplication to be justified (Rajagopalan and Batra, 1982).

The duplication process improves the performance of a cellular manufacturing system in two ways. First, it replaces the intercellular moves by intracellular moves by placing the duplicated bottle-neck machine into the machine cell whose components are creating the intercellular moves. This, in turn, reduces the total material handling cost due to shorter travelling distances and less expensive material handling equipment. Secondly, the duplication eliminates the interdependence of machine cells and makes it possible to take full advantage of group scheduling. The cost saving due to better scheduling, however, is more difficult to quantify. For this reason when the duplication cost is not compensated by the corresponding cost reduction in material handling, one should not yet rule out the feasibility of the duplication process. In this case, a more detailed study of the cellular manufacturing system is warranted. Due to the complexity of the scheduling problem, simulation modelling seems to be the most promising approach to this problem. Performance measures such as set-up costs, in-process inventories, and throughput times are appropriate for such a study. This is an interesting area of research and, to the best of my knowledge, it has not been fully investigated.

Formation of hybrid cell

The formation of a hybrid cell is yet another solution to the bottle-neck machine problem. The hybrid cell is formed by collecting all bottle-neck machines from different machine cells and concentrating them into a single machine cell (hybrid cell). The formation of a hybrid cell as a solution to the bottle-neck machine problem is based on the premise that the concentration of all bottle-neck machines in one place makes them more accessible to parts from different machine cells. This simplifies the flow of material and provides an opportunity for reducing the scheduling effort. In addition, the concentration of bottle-neck machines in the hybrid cell provides an opportunity for substantial cost reduction in material handling by placing the hybrid cell close to the machine cells having the largest number of interactions with the bottle-neck machines.

Despite the promise of hybrid cells as effective solutions to the bottle-neck machine problem, a preliminary study shows that they may adversely affect the performance of the corresponding cellular manufacturing system (Seifoddini and Wolfe, 1986). This is mostly due to the increase in material handling costs as a result of increased intercellular moves. This suggests that more studies are needed before any conclusion is drawn on the effectiveness of hybrid cells as a solution to the bottle-neck machine problem. There is, however, one explanation for the deterioration in performance of cellular manufacturing systems as a result of forming hybrid cells. In the development of cellular manufacturing systems, machine cells (including bottle-neck machines) are assigned to their proper cells in the first place. Therefore, the reassignment of bottle-neck machines disturbs this process and creates a larger number of intercellular moves which translates into higher intercellular material handling costs.

Based on the above discussion the formation of hybrid cells may result in a less productive cellular manufacturing system due to higher production costs. Therefore, the decision regarding the formation of a hybrid cell as a solution to the bottle-neck machine problem should be based on a thorough economic analysis of the related factors. Some of these factors include intercellular material handling cost, intracellular material handling cost, set-up cost, and in-process inventories.

Merits of different solutions

Each of the alternative solutions discussed here has its own advantages and disadvantages. The modification of routings for exceptional parts and subcontracting of these parts to the outside contractors have minimal requirements in terms of new facilities and the required changes in the cellular manufacturing system. These solutions, however, can only be applied to special manufacturing situations and are partially effective in dealing with the bottle-neck machine problem. The nature of exceptional parts, manufacturing processes, and outside contractors significantly limits their applications.

The duplication of bottle-neck machines is the most effective solution to the bottle-neck machine problem. It can completely eliminate the intercellular moves and make machine cells independent if enough bottle-neck machines are duplicated. The duplication process, however, requires additional machines which results in under-utilization of machines in cellular manufacturing systems. This, in turn, increases the production cost and creates a major block to the acceptance of cellular manufacturing systems. For this reason, in many cases only a selected number of bottle-neck machines should be duplicated rather than all of them. The selection should be based on a thorough economic analysis of the major cost factors. The duplication of bottle-neck machines if done selectively and based on a sound economic analysis can significantly increase the productivity of the cellular manufacturing systems.

Finally, the formation of a hybrid cell is the least demanding alternative in terms of the required resources. This is true because the hybrid cell is simply a collection of the existing bottle-neck machines. It neither requires any new machine nor does it call for any changes in product mix. Its effectiveness, however, at its best is limited to special manufacturing situations. It may adversely affect the performance of the cellular manufacturing system if its implementation is not based on a careful analysis of the relevant factors such as material handling cost, set-up cost, in-process inventories cost, and so on.

In many cases, the best solution can be achieved by using a combination of the alternative solutions presented here. For example, the routings of some exceptional parts can be modified while other exceptional parts may be given to a subcontractor. The machine duplication, and formation of hybrid cell, then, can be used for these exceptional parts and bottle-neck machines which do not lend themselves to the previous two solutions.

Conclusion

Group technology classification, coding and cellular manufacturing systems are being widely used to improve productivity of batch-type manufacturing systems. They have contributed to the development of manufacturing areas such as flexible manufacturing systems, computer-aided production planning, computer-aided manufacture and Just-in-Time.

The bottle-neck machine problem is a major block to the productivity improvement in cellular manufacturing systems. Among the alternative solutions to this problem, the duplication of bottle-neck machines is the most effective in eliminating intercellular moves and making the machine cells independent. The duplication process, however, requires additional machines and imposes new costs. For this reason, any decision regarding the duplication process should be based on a sound economic analysis of the relevant cost factors.

References

Askin, R. and Subramanion, S., 1987, A cost based heuristic for group technology configuration. *International Journal of Production Research*, **25**, 639–665.

Burbidge, J.L., 1975, *The Introduction of Group Technology* (New York: Wiley).

Burbidge, J.L., 1977, A manual method of production flow analysis. *The Production Engineer*, **57**, 34–38.

Chakravarty, A. and Shtub, A., 1984, An integrated layout for group technology with inprocess costs. *International Journal of Production Research*, **22**, 431–442.

Chandrasekharan, M.P. and Rajagopalan, R., 1986, An ideal seed non-hierarchical clustering algorithm for group technology. *International Journal of Production Research*, **24**, 451–464.

Choobineh, F., 1988, A framework for the design of cellular manufacturing systems. *International Journal of Production Research*, **26**, 1161–1172.

Durie, F.R.E., 1970, A survey of group technology and its potential user's application in U.K. *The Production Engineer*, **49**, 51–61.

King, J.R. and Nakoranchaiv, V., 1982, Machine-component group formation in group technology review and extension. *International Journal of Production Research*, **20**, 117–133.

Kusiak, A., 1987, The generalized group technology concept. *International Journal of Production Research*, **25**, 961–969.

Mitrofunov, S.P., 1963, Scientific basis of group production. *Proceedings of 1st International Seminar on Group Technology*, Ilo, Turin, Italy.

Owaban, C.O.E. and Lambert, B.K., 1988, Sequence dependent machine-setup times and similarity of parts, a mathematical model. *IIE Transactions*.

Rajagopalan, R. and Batra, J.L., 1982, Design of cellular production systems: a graph-theoretic approach. *International Journal of Production Research*, **13**, 567–579.

Seifoddini, H., 1989a, Duplication process in machine cells formation in group technology. *IIE Transactions* (in press).

Seifoddini, H., 1989b, Performance evaluation of hybrid cells in cellular manufacturing systems. Working Paper, Industrial and Systems Engineering Department, University of Wisconsin, Milwaukee.

Seifoddini, H. and Wolfe, M.P., 1986, Application of the similarity coefficient method in machine cells formation in group technology. *American Institute of Industrial Engineers (AIEE) Transactions*, **18**, 271–277.

Seifoddini, H. and Wolfe, M.P., 1987, Selecting the threshold value based on material handling cost in group technology applications. *AIIE Transactions*, **19**, 266–270.

Sneath, P.H.A. and Sokal, R.R., 1973, *Numerical Taxonomy* (New York: Freeman).

32

MRP II systems: success factors in the process of implementation

A.G. Lockett, P.R.N. Barrar and M.E. Polding

Manchester Business School, University of Manchester, Booth Street West, Manchester M15 6PB, UK

Abstract

The adoption of complex manufacturing systems can be highly problematical, with many companies failing to achieve the benefits expected. This chapter reports on some early results of case research on MRP II which has been aimed at studying implementation processes from an organizational perspective. A model of these processes is presented which incorporates a number of key variables and approaches. It is argued that the philosophy behind systems such as MRP II, needs definition by top managers as explicit corporate policies and that success in implementation is dependent on getting these factors right.

Introduction

A lot of effort has gone into the application of the principles of Materials Requirements Planning (MRP) (both I and II), and it is still a very widely discussed methodology. The literature, however, suggests that most organizations have difficulties in achieving their implementation targets. Although its importance is recognized, research output on the topic is still sparse. In this chapter we make one small attempt to remedy this deficiency by presenting some early results of detailed case research. The methodologies we developed cover a range of disciplines, but we only have space to cover the main approach.

The literature

From the early days of MRP most of the literature has been of the 'how to do it kind', (e.g. Plossl and Wight, 1971). These early texts undoubtedly had an effect and led to many organizations implementing computerized MRP systems. There were some well-documented successes, Davis (1978), as well as other less favourable experiences. In parallel the growing competition between organizations led to pressures for more efficiencies and the integration of systems. Similarly, there was an increase in the complexity of data handling. With the accompanying developments in computer technology, systems such as MRP II are the logical solution to some of the above difficulties. However, the results in practice have not matched up to the original expectation. But the demand has not diminished, and clearly systems are needed. This not only applies to MRP, but also complementary methodologies such as Just-in-Time (JIT) and Optimized Production Technology (OPT), although this is not the place to discuss their respective merits. We would suggest that the results as expressed in the literature have great similarities, e.g. great expectations, Wight (1982); difficulties in implementation, e.g. Waterlow and Monniot (1986); and now a more realistic view, e.g. Tranfield *et al.* (1989).

Most of the research on this subject comes from the work on a much broader, but often simpler area of project management, especially operational research e.g. see Schultz and Slevin (1975), Lockett and Polding (1981). What is quite clear is that detailed work on MRP is in short supply. In the UK there has been some attempt to remedy this deficiency through the CAPM initiative of the ACME Directorate of the SERC and the research reported here is part of this sustained effort.

The literature on implementation stems largely from studies of Operations Research/Management Science based projects, and is a continuing theme (Markus and Robey, 1988). From the early work of Churchmann and Schainblatt (1965), to the later ideas of Schultz and Slevin (1975), there has been a gradual recognition of the importance of the non-technical aspects of the problem. Polding and Lockett's (1982) work, clearly demonstrated the importance of organizational factors and complex relationships that exist.

Since that work there has been an explosion in the capabilities of computers with the Information Technology (IT) developments, and the spotlight has been turned on to Information Systems (IS). These are becoming increasingly complex and MRP systems are part of these processes.

Work by Rockart (1982), developed the concept of 'critical success factors' (CSFs) and Slevin and Pinto (1987), have applied this together with their earlier work to produce an idealized process model of IS project implementation. This is based on a large survey of IT project leaders and is rooted in practice not theory. What is clearly shown again is that the technical aspects are only one part of the IS implementation jigsaw and factors such as 'top management' support are equally as important. The results are not based on MRP-type projects and therefore some care has to be taken. However, if we look at the characteristics of typical MRP applications it would suggest that they tend to be more complex both organizationally and technically than the ones in the survey. This would suggest that the process view is even more important in the case of MRP II systems (Barrar and Tanner (1988)).

Research methods

Given the above literature and the authors' experience it was decided to use a case study methodology. We developed a process model based on the Slevin and Pinto (1987a,b) suggestion and tested it on some small cases. This led to some refinement which allows for the particular attributes of MRP environments, but keeping the view that implementation should be seen largely as a process of organizational change.

The main research method was by longitudinal detailed case study with researchers looking at organizations over a long period of time. This work is still underway, and the results presented here are preliminary findings. The studies involve regular visits to the organizations, attending meetings, and detailed recording of interviews. A rich variety of research material results.

We have also developed a more general questionnaire which is being used on a broader group who are not necessarily part of our case sample. This includes 86 questions requesting details of the system, the organization, how the project is structured, and the implementers' perceptions of best practice. By bringing these different methodologies to bear it is hoped that stronger generalizations will emerge.

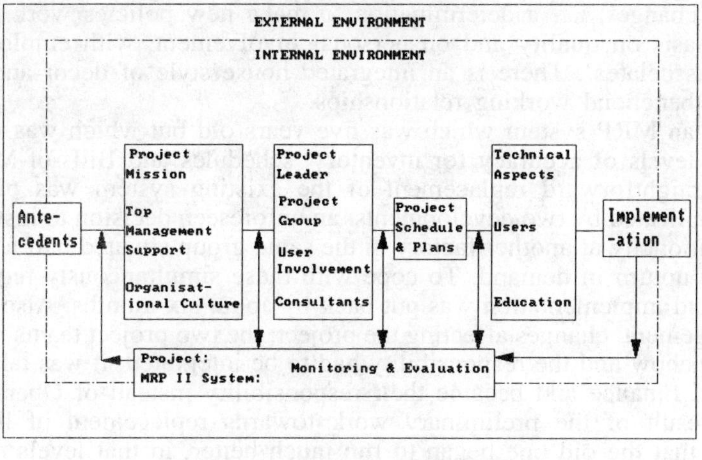

Figure 1. MRP II—process model.

In the next section we describe more detailed cases which clearly highlight the complexity of the subject and the need for a process model approach. Before that we describe a scheme that may be of value to potential implementers.

The process model

The above early results indicate that the model shown in Figure 1 would be a useful description of the generalized implementation process. It is an attempt to include the important variables that should be considered while retaining managerial validity, i.e. a tool that can be used in practice. The process model provides a prescriptive framework for analysis of systems implementation. From our case study research, we can relate factors perceived by implementers as being linked to successful projects and 'map' these onto the model. We have taken a 'process' perspective in that we are concerned not only with the strength and quality of the variables identified in the model, but also in the events and interactions which exist and which interlink these over time. Implementation strategies, like business strategies (Quinn, 1981), have an incremental logic, and at each stage in the process of implementation, the relative influence or quality of variables will alter. Hence, for example, the role of top management support may be perceived as a key element in project initiation, but at later stages, the role of top managers in clarifying goals and objectives or changing the priorities of competing projects, will influence outcomes in a different way. The process model can be likened to the frames of a moving picture—the elements remain constant, but the picture moves as the nature of the variables and their strength of interaction changes.

Some case examples

In this section we will consider three potted histories of attempts at MRP implementation which will illustrate aspects of our model.

Case A

This company is part of a well-known multinational group, which although large is still a privately owned family business. The company prides itself on having a 'culture for change', i.e. a determination to make new policies work. There is a strong emphasis on quality and on personal involvement, with employees being known as 'associates'. There is an integrated house style of decor and a climate of mutually beneficial working relationships.

They had an MRP system which was five years old but which was functioning only at low levels of accuracy for inventory, schedules and Bills-of-Materials. A relatively straightforward replacement of the existing system was planned, but became complicated by two developments: an unforeseen decision to install the new system additionally at another factory in the same group situated half a mile away, and a major upturn in demand. To cope with these simultaneously required great dedication and implementation was put back by about six months. Also there were major management changes affecting the project; the two project teams had to work together somehow and the responsibility had to be integrated; it was taken over by Services and Finance and became their responsibility instead of Operations. One noticeable result of the preliminary work towards replacement of the existing system was that the old one began to run much better, in that levels of accuracy became much improved. After a year some modules of the new system were running and it seems likely that others will be up soon. In Handy's (1976) terminology this is a primarily a 'task' culture, but the size of the operation means that there must be an element of role culture too.

Case B

This company is part of a large multinational group which consists of heterogeneous companies with little integration. This site produces small domestic appliances and demand is highly seasonal.

There is no 'culture for change' here and the systems are being replaced as a necessity, as the hardware is almost worn out and the software consists of informal systems which are largely undocumented. A reduction in storage space recently has made a new system even more urgent in order to keep the production line going.

A new system has been purchased and is being implemented, but only very slowly. There have been three different changes in a year in the overall responsibility for the system; also the company has been without an MD for most of that time. At present the FD has nominal responsibility for the system; there is a Systems Manager, but the implementation has been taken over by a new Production Manager who says that MRP is a necessity for the efficient running of the factory.

This organization seems to fit the stereotype of a 'power' culture but lacks the key individuals necessary for success. It is possible that the Production Manager will turn the system into a success.

Case C

This company is linked to a larger group but is small and autonomous. The product is fairly unusual in that the raw materials consist in part of remanufactured components: they make to order, not to stock, so their inventory needs are mainly for components. Records have in the past been kept very sketchily and it is difficult to do accurate costings.

As with the other two companies implementation has been accompanied by management changes and variations in demand. The MD and project leader are both experienced in MRP and highly motivated, and progress was good before external forces intervened. The company's major customer is withdrawing its business and the project is presently on ice. Some of the workforce will take redundancy settlements; the project leader and MD are guaranteed jobs whatever the future of the company may be. The system may remain on site or be transferred to another location.

This is a small company with a highly motivated project leader and strong support from top management; a strong 'task' culture. This looked a promising environment for implementation but external forces have intervened to prevent it.

Two of these cases, A and C, could loosely be described as successful. They both had promising antecedents: experience in MRP in the project team and a strong computing background; the project mission was very clear and the support of top management was well known. The external environment however was the feature that differentiated them so far as the outcome is concerned. Company A had a big upturn in demand and Company C had a disastrous downturn due to the decision of one large customer.

In Company B the project mission was not clearly related to business needs, although the seasonality of demand made things difficult for the project team. The company was without an MD for a period of months, so support at the top became fragmented. Unlike Company A, which is of similar size, education for the project was not formalized and programmed.

In all three companies implementation of the system was accompanied by far-reaching changes at middle and top management levels, but in the two companies that have done well in implementation these were absorbed and rationalized without too much upset; in Company B they resulted in a lack of direction for the project and a feeling of apathy in the team members.

Organizational factors

Whilst the rationale for implementing MRP II for most of the firms in our study is that of increased operational control, such MRP II systems themselves require high degrees of control (discipline) to be exercised at all levels of use of the system. For most firms implementing MRP II these disciplines do not exist, yet effective manufacturing performance and customer service is of major concern to the business. MRP II is seen by many managers as a vehicle for resolving problems of poor operational performance which, in turn, are perceived as being due to poor control stemming from inadequate current systems and procedures. In order to gain insight into these problems, data from 22 case studies have been analysed revealing that improved stock control (23%) was the primary aim for MRP II whilst increased systems integration (18%), improved manufacturing performance (18%), and flexibility (9%) were also cited.

Control is the central logic in MRP II systems, in a survey of 69 software products, every vendor claimed increased control of the business as a primary benefit of the system (Tobin, 1989). It is not surprising that the expectations of users are structured accordingly. Assumptions are inbuilt within these systems that behaviour of users will be highly predictable and will accord with the rigid procedures and disciplines required, but if these assumptions are not valid then the potential for improved control is negated. Additionally, the systems structures imply the organizational structure for the businesses within which they will operate. Functionality and structure are jointly defined by the system. Hence, users wishing to adopt MRP II need to recognize that installation of the system must also be accompanied by the implementation of what we will call the MRP II philosophy if potential benefits are to be realized. The philosophy can be described as the broad acceptance, at all levels, of the rigorous procedures and a commitment to plans and the need to resolve conflicts. In other words—a common ideology with shared values as to the way in which the business is to be managed. Success in implementation, therefore, must recognize these two complementary programmes; one based around the transfer of new 'technology' and systems installation, and the other on changes in ideology, organization and management process. Necessarily, these must be interlinked and mutually supporting, but the need to maintain the 'philosophy' is ongoing.

Project organization

Organizational factors then are clearly highly significant in the overall implementation of MRP II and our expectations of success or failure need to recognize the crucial nature of these factors.

From our case study data these characterizations largely hold. The departments reported as having responsibility for the MRP II system were: MIS (32%), production (10%), materials (27%), planning (5%) and as part of its ongoing responsibilities—the project group (10%). In Tobin's (1989) study of 69 production planning and control system software products, a total of 84 functions were claimed with an average of 17 functions per package. The maximum number claimed was 30 and the minimum was 3, with a standard deviation of 8. However, for implementation purposes modular elements typically provide a number of functions and from our case data, the mean number of system modules implemented was 4 (s.d. 2·7) with a further 3 (s.d. 2·3) planned. System development times had a mean of 3·7 years (s.d. 3·2). None of the cases studied had achieved Class 'A' status as defined by the Ollie Wight system.

Projects as a whole are guided by a steering committee (86%) composed of director level (40%) or function head managers (40%). A project manager is appointed on a full-time (64%) or part-time (36%) basis and an implementation project group composed of department managers (60%) or department representatives (10%) is appointed. Most of the project groups (77%) have a representative from each affected department and the groups (86%) work to specific targets or milestones. These targets are largely concerned with meeting implementation stage deadlines (60%), but include stage and financial targets (5%), and operational targets (10%). The majority of groups (60%) are reported as having the authority to 'get the job done'. Most of the groups report progress through progress meetings or the project manager (36%), through the steering committee (23%), or through monthly reports (14%). The major difficulties reported by groups were: insufficient user commitment (23%), lack of senior management support and understanding (23%) and lack of data accuracy or discipline. This had impacted implementation to the extent that although a number of groups reported meeting targets (23%), some were late (14%), some were only partially achieved (27%) and others (23%) were not achieved at all.

These outcomes suggest an inability on the part of the project organization to implement the MRP II philosophy, resulting from lack of 'commitment', 'support' and 'discipline'—and hence, an inability effectively to progress further with the installation of MRP II as a project.

Discussion

This short chapter illustrates the complex organizational process that is MRP II. Although only a preliminary study, the results clearly point out that many of the factors critical to successful implementation are managerial and organizational in nature. Recognition of these aspects may be vital for success and the simple process model should be helpful in avoiding their neglect in the future.

We would venture that from our research data, it is not the project perspective which is problematical but the organizational and ideological issues which follow from the adoption of the MRP II philosophy. These cannot be delegated to the project organization by top management; the new philosophy needs definition as explicit corporate policies, these need to be communicated by top managers, and, most importantly, they need to be enacted by them. Success in implementation is dependent on getting these factors right.

References

Barrar, P. and Tanner, I., 1988, Organising for Systems Integration. *BPICS 23rd European Conference on Production and Inventory Control*, Birmingham, 1988.

Churchman, C.W. and Schainblatt, A.H., 1965, The researcher and the manager: a dialectic of implementation. *Management Science*, **11**, B69–B87.

Davis, E.W., 1978, *Case Studies in Material Requirements Planning* (Falls Church, VA: American Production and Inventory Control Society).

Handy, C.B., 1976, *Understanding Organisations* (London: Penguin).

Lockett, A.G. and Polding, M.E., 1981, Organisational linkages and OR projects. *European Journal of Operational Research*, **7**, 14–21.

Markus, M.L. and Robey, D., 1988, IT and organisational change: causal structure in theory and research. *Management Science*, **4**, 5, May.

Pinto, J.K. and Slevin, D.P., 1987, Critical factors in successful project implementation. *IEEE Transactions in Engineering Management*, **EM-34**, 1, 22–27.

Plossl, G.W. and Wight, O.W., 1971, *MRP Planning by Computer* (Falls Church, VA: American Production and Inventory Control Society).

Polding, M.E. and Lockett, A.G., 1982, Attitudes and perceptions relating to implementation and success in operational research. *Journal of the Operational Research Society*, **33**, 733–744.

Quinn, J.B., 1981, Formulating strategy one step at a time. *Journal of Business Strategy*, **l**, 3, 42–63.

Rockart, J.F., 1982, The changing role of the IS executive: A C.S.F. perspective. *Sloan Management Review*, **24**, 1, 3–13.

Schultz, R.L. and Slevin, D.P. (Eds), 1975, *Implementing OR/MS* (New York: Elsevier).

Slevin, D.P. and Pinto, J.K., 1987, Balancing strategy and tactics in project implementation. *Sloan Management Review*, Fall, 33–44.

Tobin, S., 1989, Production and inventory control systems: a review of software and the claims of suppliers. Unpublished MBA Dissertation, Manchester Business School, University of Manchester.

Tranfield, D., Smith, S. and Kirkwood, R., 1989, The implementation cube for advanced manufacturing systems. *IJOPM*, **9**, 8.

Waterlow, J.G. and Monniot, J.P., 1986, *A Study of the State of the Art in Computer Aided Production Management in UK Industry*. ACME Directorate, Science and Engineering Research Council.

Wight, O.W., 1982, *The Executive's Guide to Successful MPR II*, 2nd edn (Oliver Wight Ltd).

33

Why are production management practices different?

Attila Chikán[1] and Clay Whybark[2]

[1]*University of Economics, Budapest, Hungary and* [2]*Indiana University, Bloomington, IN, USA*

Abstract

This chapter presents some results of a survey conducted in four regions of the world: South Korea, China, Western Europe and Hungary. Components of manufacturing management such as forecasting, production planning, shop floor control and purchasing and materials management are examined and the practice of firms in these fields are compared. The main conclusion is that manufacturing systems are basically influenced by two major groups of factors, related to the market situation and the cultural background.

Introduction

This chapter presents the results of the evaluation of a set of surveys which have been carried out in South Korea, China, Western Europe and Hungary. The first survey was directed by Clay Whybark in China in two industries: the non-fashion textile and the small machine tool industries. The survey was also later conducted in the other three regions. The surveys provided interesting results themselves and the cross-national comparison promises further important lessons. The methodology of the survey is described in Whybark and Rho (1988), the first results of the analysis in Rho and Whybark (1989a,b) and Chikán (1988).

The four regions which have taken part in the analysis are all very different; this means that the international comparison provides a many-sided picture which can help in the understanding of both the general characteristics of production systems which have universal effects and the special national features.

Method of the survey

The method of the survey was similar in the cases of Korea and China, namely, questionnaires were first distributed by mail, but after the disappointing results of the returns interviewers were sent to companies. Both in Korea and China about 50 companies were questioned in each industry. In Western Europe the survey started with a mail distribution (also with poor results) and this was followed up with telephone solicitation involving 35 companies in the machine-tool industry and 34 textile firms. Since the survey was carried out in Hungary after the other three places, we made use of the experiences of the previous ones. Interviewers were sent to the companies, and collected results from 79 companies. In the Hungarian survey there were not enough companies in either of the two industries included in the other surveys, so other companies in the manufacturing industry were also questioned. This sampling difference does not disturb the evaluation because in the comparison of the other three places there were only a few examples where the difference between industries was greater than between countries.

The limits of this chapter do not allow a detailed evaluation of the survey results; furthermore the questionnaire itself, which is quite lengthy is not presented. We provide a few results in each area where the survey was carried out and finally draw some general conclusions. The figures show the data of the machine-tool industry for the first three regions because in the Hungarian survey the majority of firms were from the machine industry.

The questionnaire consisted of 65 questions in the following structure:

(I) Company profile.
(II) Sales forecasting.
(III) Production planning and scheduling.
(IV) Shop floor control.
(V) Purchasing and materials management.

The following summarizes the main results of the questionnaire.

Forecasting

Forecasting methodology used in Hungary was the closest to that of the European companies and the furthest from China, regardless of the fact that Hungary and China have planned economies with a high ratio of state-owned companies. This indicated that cultural differences can sometimes (and especially in actual operations) be more important than the forms of ownership.

The level of executive attention to forecasting (namely, the highest level executive who is in charge of producing the forecast) is much higher in Hungary and in Western Europe than in China and Korea. In the use of computers we see the surprising fact that Korean firms are at the lowest level of all four regions. Importantly, the forecasting period is similar in all four regions in that there is an annual forecast which is updated quarterly. This is especially interesting because we tended to think that this time-phasing is a reminder of the 'classical' planned economy when quarterly planning was mandatory for the companies. One can now draw the conclusion that it is natural to think in these time periods. Further information about the methodology of forecasting can be obtained from Figure 1.

Production planning

Production to order is dominant in all four regions; however, there are also large differences: in Hungary over 90% of production is to order, while in Korea it is about 70%, in China only 55% in the machine-tool industry (although another 25% in China is produced according to the state plan which basically has the same role as a consumer's order). Regarding the use of computers in production planning (see Figure 2) there is a great difference again between the European and the Asian nations in both industries. The Western European and Hungarian use of computers is at a much higher level than in the Asian countries. Interestingly, however, there is not much difference between the elaborateness of the plan, namely, there is not much variance in the proportion of plans made for individual products (as opposed to groups of products). The situation is the same in the proportion of planning in physical units (as opposed to planning in value). However, one can see again that the two European regions are closer to each other than to the Asian ones.

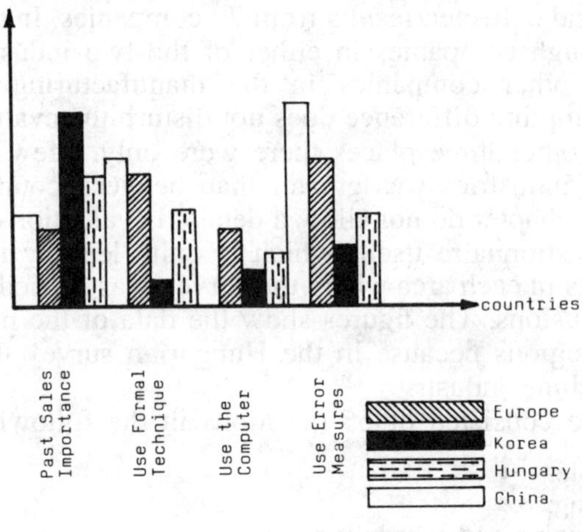

Figure 1. Forecasting.

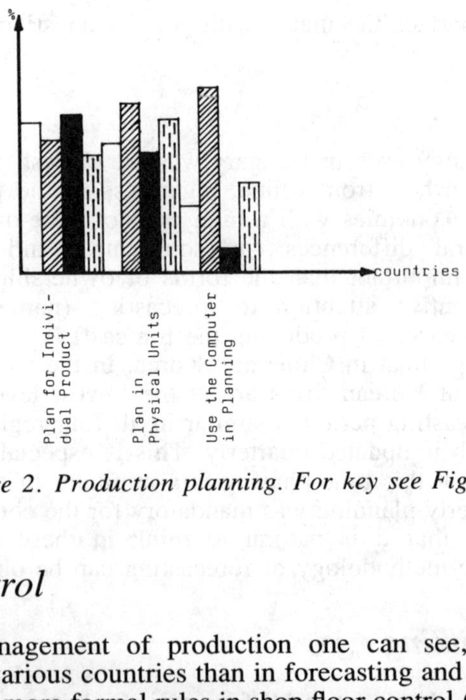

Figure 2. Production planning. For key see Figure 1.

Shop floor control

In the operative management of production one can see, in general, smaller differences between various countries than in forecasting and planning. Companies in Hungary use many more formal rules in shop floor control (if it is measured with the proportion of written priority orders or priority changes as opposed to verbal directions; see Figure 3). Initiation of changes in Europe mostly comes from marketing or administration while in Korea and China it comes from production. In Hungary, we note a very mixed situation.

Capacity limitations play an outstanding role in Korea and Western Europe as reasons for late delivery, while materials shortages are major factors in China and Hungary. It seems that not only here but in other activities as well, market shortages play an important role in internal company operations. Probably the largest differences can be experienced in the connection between market performance and production scheduling. In the Hungarian survey we made distinctions between where the company sells its goods; namely whether they are sold on the domestic, Western or Eastern markets. There is a great difference between these three

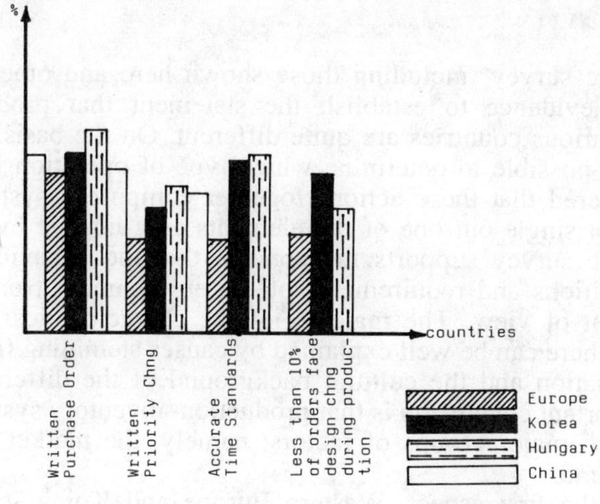

Figure 3. Shop floor control.

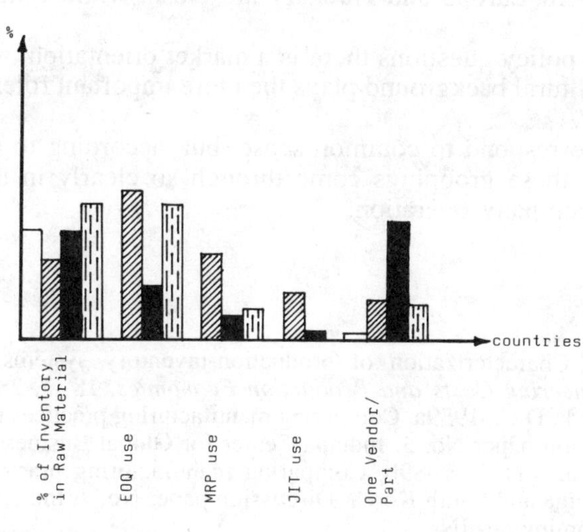

Figure 4. Purchasing and materials management. For key see Figure 3.

destinations; the Western market delivery performance was the highest of all while performance was less reliable when examined in the case of domestic and Eastern markets. The situation is, of course, not the consequence of the subjective preference of the companies, but the different requirements of the various markets.

Purchasing and materials management

The proportion of raw material (or rather, input) inventories was characteristically higher in Hungary than in any of the other countries (Figure 4). Korea and China were at about the same level, while the lowest ratio of input inventories was experienced in the Western European companies. This shows the relative uncertainty of the input markets which is the highest in Hungary and the lowest in Western Europe.

As for the use of up-to-date production–inventory techniques (like MRP or Just-in-Time) one can see that these techniques have spread basically only in Western Europe and play a very small role in the case of the other countries.

Conclusion

The results of the survey (including those shown here and others not discussed) provide enough evidence to establish the statement that production–inventory systems of the various countries are quite different. On the basis of the individual findings it is not possible to determine which type of operation is superior, but it has to be considered that these actions together compose a system of operations where you cannot single out one of the elements and judge it by itself. However, we think that the survey supports the concept that the behaviour of companies reflects the conditions and requirements of the environment from both the social and cultural point of view. The majority of the characteristics which were only briefly discussed here can be well explained by causes stemming from the economic and political situation and the cultural background of the different countries.

The most important conclusion is that production–inventory systems are basically influenced by two major groups of factors; namely, the market situation and the cultural background.

According to the first aspect, Western Europe and Korea seem to be in one cluster, while Hungary and China are also close to each other; as for the cultural background, Western Europe and Hungary are close, while China and Korea are also similar.

It seems that in policy questions there is a market orientation, while in organizational decisions cultural background plays the more important role; in methodology, they seem to mix.

These results correspond to common sense, but, according to our view, it is far from evident that these groupings come through so clearly in the analysis of an internal field of company operation.

References

Chikán, A., 1988, Characterization of production-inventory systems in the Hungarian industry. *Engineering Costs and Production Economics*, **18**, 3, 285–292.

Rho, B. and Whybark, D.C., 1989a, Comparing manufacturing practices in Europe and South Korea. Discussion paper No. 3, Indiana Center for Global Business, Bloomington, IN.

Rho, B. and Whybark, D.C., 1989b, Comparing manufacturing practices in the People's Republic of China and South Korea. Discussion paper No. 4, Indiana Center for Global Business, Bloomington, IN.

Whybark, D.C. and Rho, B., 1988, A worldwide survey of manufacturing practices. Discussion paper, No. 2, Indiana Center for Global Business, Bloomington, IN.

34

Future-orientated production structures: computer-aided methods and tools to plan and innovate integrated structures of production cells

Ulrich Hallwachs

Fraunhofer-Institut für Arbeitswirtschaft und Organization, Stuttgart, Germany

Abstract

An approach is presented, which allows planning and introduction of cost-optimized production structures with decentral responsibility units. With these small units, which can be considered as 'factories within the factory', both requirements of marketing and production will be regarded in different production strategies. This means to form part-families and machine groups, not only in point of view of similarity of production as to be done in group technology but in point of view of combining product and production related criteria. Production structuring in this context is aiming at the realization of mixed structures, which are optimal in costs by combining product-orientated and production-orientated organizational units.

Starting situation

Changing external and internal requirements necessitates that many entrepreneurs will have to reconsider their production philosophy. This situation calls for small units where you can see what's what—so-called production cells—which, as 'a factory within the factory', assume all the functions required for the manufacturing of a part or a part group with direct responsibility. The impact on the situation of the business will be heavy. Functional workshops based on process specialization will be dissolved and the machines required for complete machining of a part group will be combined to make up new organizational units. The management pyramid will be flattened out, i.e. the organizational units will be given (again) more authority and responsibility. The staff of near-production departments, such as quality assurance, will become fewer or their functions will be integrated into the production cells.

Under such radical structural changes, detailed planning is inevitable. Conventional planning strategies oriented at the individual workplace cannot achieve this (Burbridge, 1982; Hyer, 1985; Auch, 1989). On the contrary, a uniform and systematic approach oriented at the production system is required, assisted by reasonable use of EDP equipment. In this context, reasonable means that the planner will continue to produce his planning ideas as before. The use of EDP permits that considerably more and better information is available, the ideas can be calculated

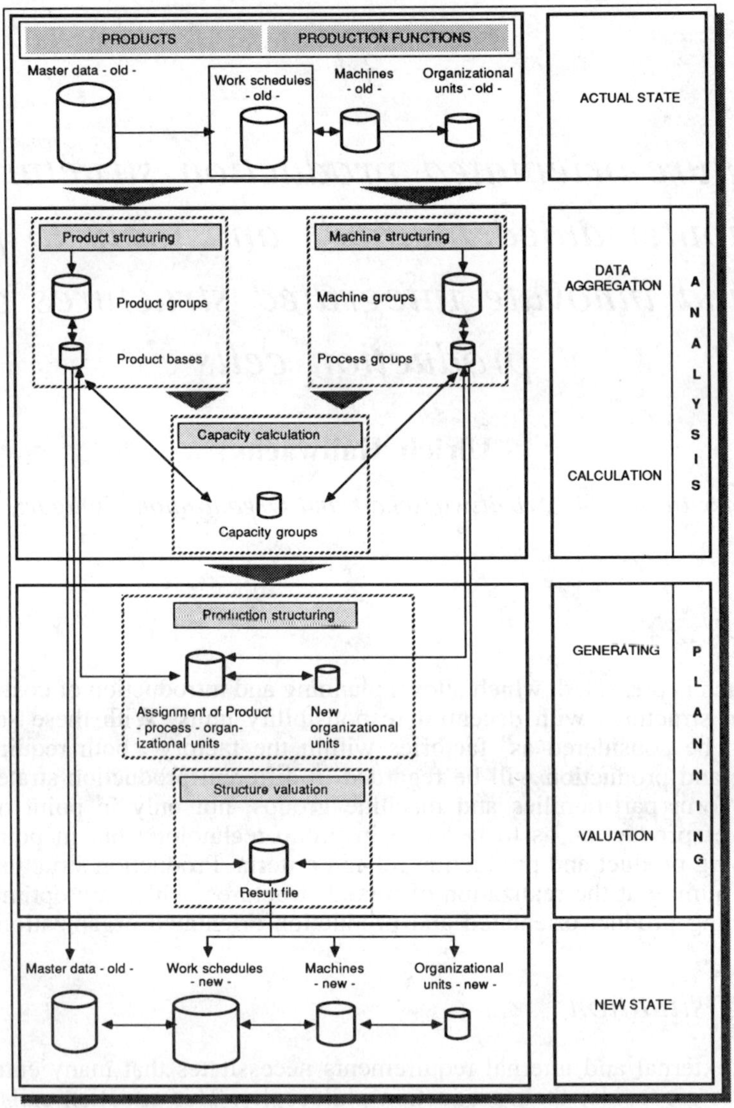

Figure 1. Production cell planning system FIPS.

and valued on the basis of the total production range—no representative planning—and many planning alternatives can be tried as a result of the support given in routine work (computations) (Hallwachs, 1988).

Production cell planning system

Based on this requirements profile, the production cell planning system (Figure 1) is being developed at the Fraunhofer Institute for Industrial Engineering under the research project 'integrated manufacturing of parts families' (Hallwachs and Schaal, 1989). Methods modules are built up for production structure planning, which support a restructuring of the entire production or parts of it from different aspects of structuring. In this connection, 'dynamic classification' is of fundamental importance; it permits variable classification codes with product related criteria

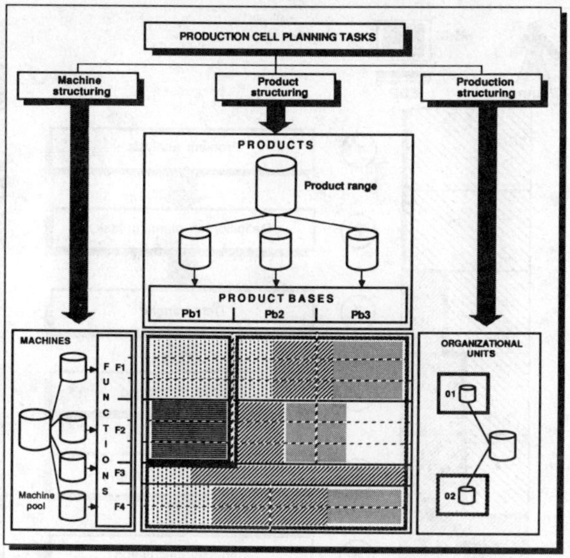

Figure 2. Planning sequence.

(size, weight), production engineering criteria (machines concerned) and/or production structure oriented criteria (associated part assemblies) to be defined automatically.

The planning system is based on a planning control loop. A division of functions between planner and EDP system is realized in such a way that the generation and decision-making phase will be assigned to the planner and the calculation and evaluation phase to the EDP system.

In this division of functions, emphasis is placed on the interaction between planner and EDP system. A graphic operating area for the planner is intended to facilitate the production of ideas of structuring and the interpretation of the results obtained. Because PCs are generally used in the planning department, the results will be implemented on PCs.

Systematic planning

The planning sequence (Figure 2) reflects the need for a target-oriented finding of problem solutions.

Starting from a problem/requirements analysis, selective approaches to product, machine and organization structure will be analysed at several planning levels (total structure, system, subsystem) in a top-down strategy (Figure 3). The necessary data will be generated at the relevant data levels through EDP-assisted qualitative and quantitative data aggregation of the existing base information—part and machine related data from master files and work schedules being transformed into product group related and production process related data.

Problem analysis

At the beginning of such a planning object, it is important to know the internal factory problems. This requires an in-depth problem analysis. Weak points present in the factory and the future requirements made on it must be known. The results

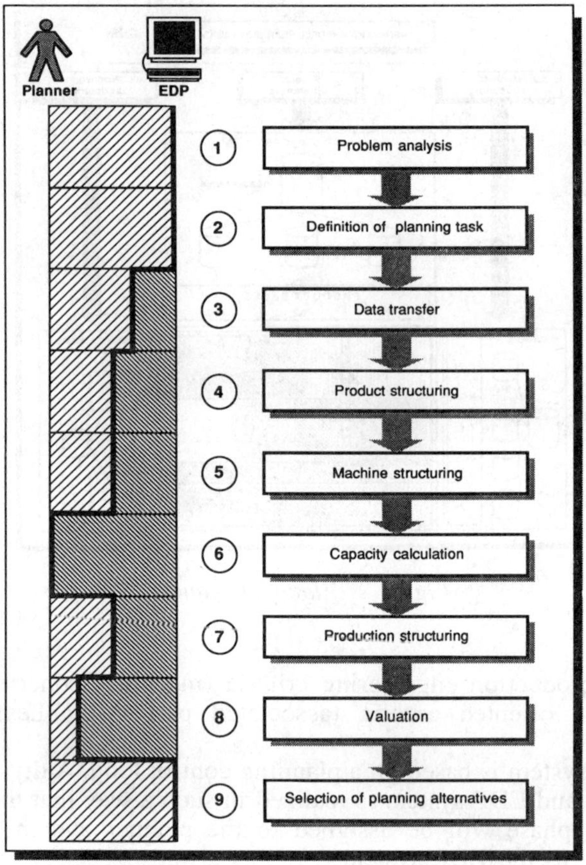

Figure 3. Production cell planning tasks.

of this analysis will then be used to define meaningful approaches to structuring the next planning steps.

Planning task definition

The planning task is defined by specifying the range and by target criteria. The planning range should be defined in two ways: size, e.g. by shop or product, on the one hand, by the scope of activities, as e.g. planning, performing and checking, or only parts thereof, on the other hand.

Based on the results of the analysis, the targets are defined by individual criteria, as e.g. responsiveness on the shop floor, lead times in line with market requirements or effective assignment of labour, and sorted according to their importance. Planning work should be oriented at these targets and the results measured.

Data transfer

For the generation/valuation of structuring alternatives, a minimum number of specific data will either have to be transferred from the central EDP system or generated. At the first stage, these are parts related quantities and work schedules. For further approaches to structuring, parts related criteria (size, weight, quality

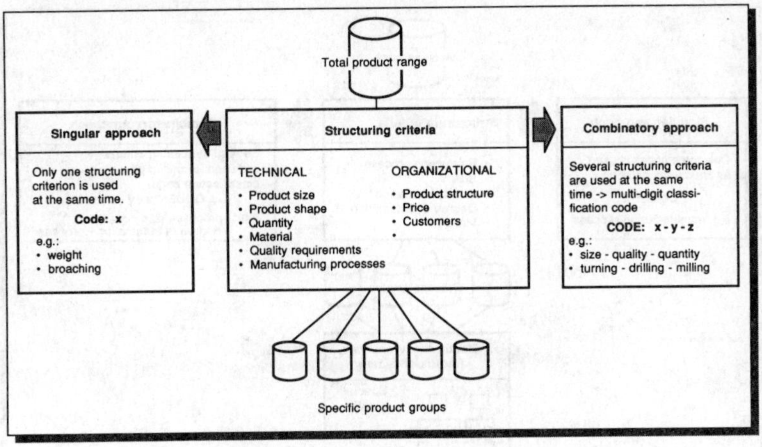

Figure 4. Product structuring.

standard) and machine related criteria (process, automation) will be required in addition. The data will be transferred from the central EDP system in two steps. First, files with selected data will be transferred from the central EDP system to the PC. The data structures will then be defined interactively and the data available will be adopted in the internal data structure via the data interface. Initialization permits a relational database to be created automatically.

Product structuring

The objective of this planning step is to divide the product range into elementary product cores. Parts groups which are homogeneous as such will be formed in a first step by combining the parts according to the defined structuring criteria. In grouping steps based on these, main product groups will then be formed according to further aspects of structuring.

Various product/product structure related and production engineering related structuring variables are available, so that the known geometry/technology and production sequence related approaches to structuring oriented at the individual part can be realized separately and in combination in the one hand, and new ideas of structuring can be applied on the other hand (Figure 4).

Regarding the production sequence related structuring, a modification has been made to the effect that so-called 'process combinations' are the subject of observation instead of flow sequences. The sequence of machining is deliberately disregarded here, as the availability of the technical resource (machine) in the organizational unit is exclusively relevant to structure planning.

Machine structuring

In this planning step, machines will be combined to make up groups according to different aspects. The objective is a classification of the machines according to technical viewpoints, such as production process, quality standard, overall size (Figure 5).

This permits an aggregated capability profile to be produced at little expense. In addition, parts related information from the work schedules will be transformed into a machining requirements profile without any time-consuming manual re-classification

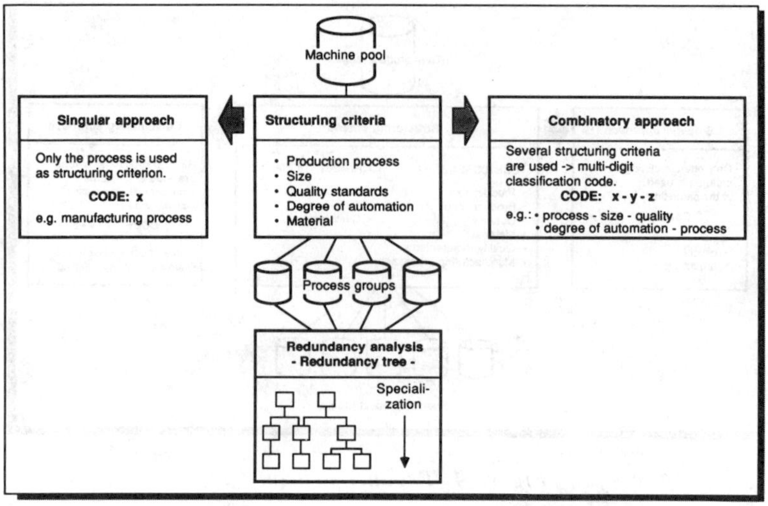

Figure 5. Machine structuring.

of the parts. The loss of information suffered is deliberately disregarded, as it does not play a decisive role in product structuring.

Capacity calculation

The parts/group related capacity requirements necessary for designing production are calculated in this planning step on the basis of the production program and the work schedules. At each planning stage, the relationship between production functions (machines or groups) and products (parts, groups) can thus be represented in a 'capacity list' in the form of the number of parts or operations concerned and the capacity requirements necessary for machining.

Production structuring

In this planning step, the future organizational units will be defined interactively with the aid of the 'capacity board' (Figure 6). This requires an allocation of functional resources (machines) and production functions (operations on the products). The effects of this allocation will be indicated on the capacity board.

A major problem of this planning step is the so-called 'resource sharing'. This means that the available resources (machines) will have to be shared, as they will be integrated into different organizational units. The difficulties which frequently arise manifest themselves in the fact that some resources cannot be split.

Capacity-sharing strategies are used for solving this problem. These strategies, for example, include the splitting of product groups, products and operations (excluding a total quantity/subquantity or specific operations on products).

Valuation

The approaches to structuring will be valued with respect to the effects they have on the fixed and current assets. In particular, additional capital expenditure incurred for machines and machine conversion will be compared with expected profits

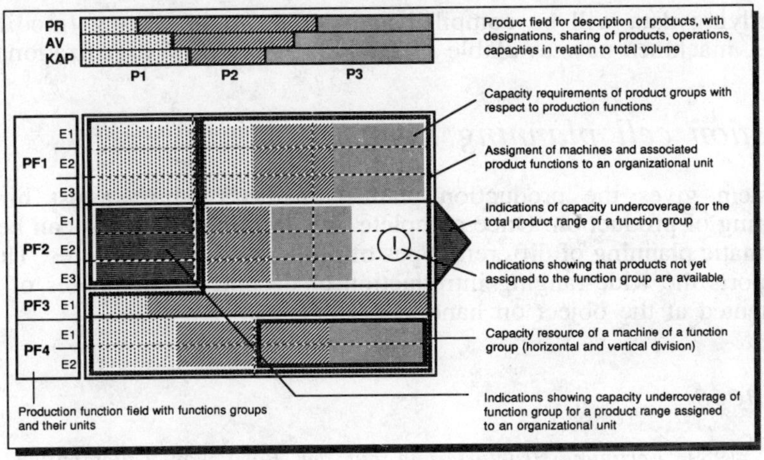

Figure 6. Production structuring.

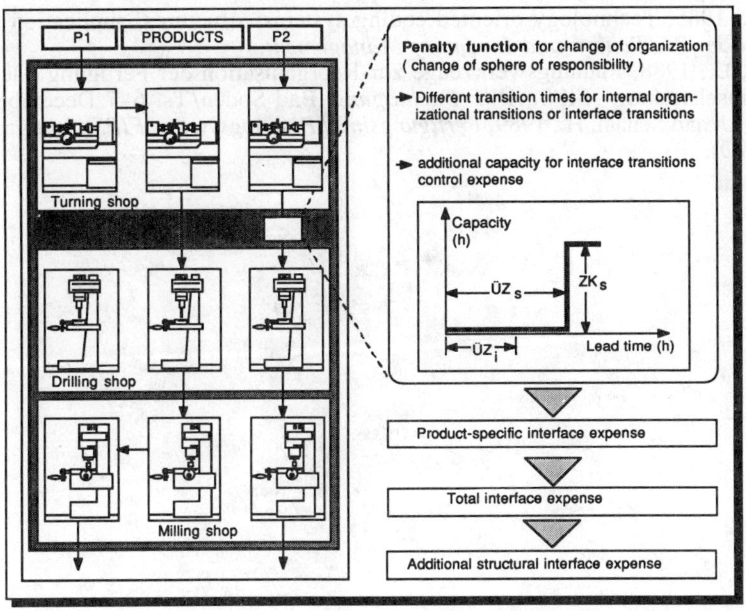

Figure 7. Interface penalty function.

by reducing the expenditure for interfaces. For calculating the expenditure for interfaces, an 'interface penalty function' is used, which provides for additional transition times for the products at the interface transitions and for additional capacities for control (Figure 7).

At the same time, valuation criteria difficult to quantify in terms of money will be considered on the basis of the target criteria.

Selecting planning alternatives

After comparison of several planning alternatives, the production structure will be defined on the basis of the above valuation data. For further planning activities, the underlying data will be compiled and put down in writing (modified work schedules, machines now available or new machines, new organizational units).

Production cell planning system

This system gives the production planner methods which assist him in the restructuring of production. Once complete, the described modules can be used for the systematic planning of different types of production cell structures. This system thus supports the wide-ranging introduction of decentralized spheres of responsibility oriented at the object on hand.

References

Auch, N., 1989, *Fertigungsstrukturierung auf der Basis von Teilefamilien* (Stuttgart: Springer).

Burbridge, J.L., 1982, The simplification of material flow systems. *International Journal of Production Research*, **20**, 3.

Hyer, N.L., 1985, Technology oriented coding systems: structures, application and implementation. *Production and Inventory Management*.

Hallwachs, U., 1988, Planungswerkzeuge zur Reorganisation der Fertigung nach der Ferti gungsinselstruktur—FIPS. *AWF-Fachtagung*, Bad Soden/Ts, 6–7 December.

Hallwachs, U. and Schaal, H., 1989, *Fertigungsinsel Planungssystem FIPS* (Stuttgart: Prospekt des IAO).

35

On the relation between batch mode, bottle-neck machines and maximum throughput

P.J. Weeda

Department of Mechanical Engineering, University of Twente, PO Box 217, 7500 AE Enschede, The Netherlands

Abstract

The relation between batch mode, bottle-neck machines and maximum throughput within a finite time allowance, is analysed for serial, convergent and divergent process configurations consisting of two machines and three processes. A key role in the analysis is played by the different types of idle times that may occur. Four types of idle times are established, which result from the succession of processes and machines, set-up times, unit processing times and the time allowance. Based on the idle times, the appropriate batch mode (or batch structure) is chosen to maximize throughput. A batch mode may be single or multiple. A single batch mode means one cycle of one batch per process. A multiple batch mode implies multiple cycles. The results specify the appropriate batch mode and bottle-neck machine for each of the process configurations.

Introduction

This chapter concentrates on the relation between batch sizes, utilization of capacities and throughput. It tries to arrive at a better understanding of this relation by analysing a number of process configurations. A process configuration is defined by a number of processes, which are run through by products in accordance with a specific routing. To each process a unique machine is assigned. Machines may be assigned to several processes. Each process and machine pair implies a set-up time and a unit processing time. Transfer between machines is instantaneous and transfer batches have one size. There is a given time allowance during which throughput should be maximized.

A batch mode indicates the batch structure. Throughput may be maximized by one single batch or by multiple batches during the time allowance. A multiple batch mode may be repetitive (equal batch sizes), or non-repetitive. For example, batch sizes may be strictly decreasing or increasing in a non-repetitive batch mode.

The method of analysis adopted here starts by establishing idle times which are intrinsic due to the succession of processes and machines, set-up times, unit processing times and the time allowance. Then it decides upon the appropriate batch mode in order to maximize throughput within the time allowance. The scheduling of the product units follows as a consequence of the batch mode.

In a previous paper (Huisman *et al.*, 1990), a particular process configuration has been analysed in order to demonstrate some aspects of Optimized Production Technology (OPT), a philosophy and a software package for production control, developed by Creative Output Inc. For a recent article on OPT by its inventor, the reader is referred to Goldratt (1988).

However, it has been recognized in Weeda (1990) that the method used to analyse this process configuration in fact deviated from OPT. OPT starts with the identification of critical capacities (bottle-necks) based on an materials resource planning (MRP)-like utilization analysis, followed by an unknown scheduling procedure. Instead of starting with the identification of critical capacities as OPT does, the intimate relationship between batch mode and utilization is analysed in advance. Once the appropriate batch mode is established, the bottle-neck machines and the actual scheduling of the product units are implied by it. The approach is explained in three serial process configurations consisting of two machines and three processes.

The results of a detailed analysis for divergent and convergent process configurations have also been obtained and can be provided (Weeda, 1989). The present chapter is restricted to an overview of all results, focusing on similarities and dissimilarities between the various configurations. Moreover, some preliminary effort will be spent on the extrapolation of the exhibited line of thinking to more complex configurations.

Notation

This chapter is restricted to process configurations involving three manufacturing processes on two machines, A and B. On machine A, two of the three processes have to be executed. On machine B only one process is executed. The two processes on machine A are denoted by A1 and A2, respectively. The single process on machine B is denoted simply by B.

The three processes and the two machines may be composed exhaustively into three serial, two convergent and two divergent process configurations. The three serial process configurations are denoted by (A1, A2, B), (A1, B, A2) and (B, A1, A2), reflecting the order of execution of the processes A1, A2 and B.

The two convergent process configurations distinguished are denoted by (A1/A2, B) and (A1/B, A2) respectively. For example, configuration (A1/A2, B) symbolizes assembly of the two supplying processes A1 and A2 by process B.

The two divergent process configurations distinguished are denoted by (A1, A2/B) and (B, A1/A2) respectively. For example, configuration (A1, A2/B) symbolizes process A1 supplying the two final processes A2 and B. The thus defined process configurations are exhaustive, since the two processes on machine A are inter-substitutable.

Set-up times are denoted by $S(A1)$, $S(A2)$ and $S(B)$. Throughout the paper the set-up time on the machine with only one process is assumed to be zero, i.e. $S(B) = 0$. The set-up times on the machine with two processes are assumed to be positive, i.e. $S(A1), S(A2) > 0$, unless stated otherwise. Unit processing times are denoted by $P(A1)$, $P(A2)$ and $P(B)$. Frequently the aggregate notations $S(A)$ for $S(A1) + S(A2)$ and $P(A)$ for $P(A1) + P(A2)$ are used. The time allowance is denoted by T.

Outline of the analysis

The analysis of the process configurations focuses on the batch mode (or batch structure). The appropriate batch mode is established by analysing idle times on the machines. Four different types of idle times may occur. They are:

(1) Inter-unit idle times between successive units of a batch, due to the larger size of the unit processing time of the preceding process.
(2) Inter-batch idle times, between successive batches under a multiple batch mode on the same machine.
(3) Inter-process idle times, between batches of successive processes on the same machine, due to an intermediate process on another machine.
(4) Initial or final idle times due to the time restriction, the succession of processes and zero initial stock.

A machine is termed a bottle-neck machine if it is uninterruptedly utilized during the time allowance, with the exception of either an initial or a final idle time. A detailed analysis of serial process configurations (Weeda, 1990) has led to the following reasoning regarding the choice of batch mode.

If machine B experiences inter-unit idle times then machine A governs throughput. In order to save set-up time a single batch is produced. Hence inter-unit idle times on machine B always imply a single batch mode. A multiple batch mode does not maximize throughput in these cases.

A multiple batch mode comes into the picture if it leads to a reduction of either the initial or the final idle time on machine B compared with a single batch mode. This is particularly true if either of these idle times depends on the batch size. Reduction of the initial idle time on machine B is appropriate if machine B is utilized after machine A. Reduction of the final idle time on machine B is appropriate if machine B is utilized ahead of machine A. However, if the use of a multiple batch mode implies inter-batch idle times on one of the machines or inter-process idle times on machine A, then a multiple batch mode does not maximize throughput and a single batch mode should be applied.

Serial process configurations

In this section the main characteristics of the three serial process configurations are summarized. They are denoted by (A1, A2, B), (A1, B, A2) and (B, A1, A2), respectively. If necessary a subdivision is made with respect to relevant relations between unit processing times.

Configuration (A1, A2, B) with $P(B) < P(A2)$

In this case machine B experiences an inter-unit idle time equal to $P(A2) - P(B)$. Hence a multiple batch mode is not appropriate. In order to minimize total set-up time on machine A, a single batch mode is appropriate. Machine A is bottle-neck. Since the last production is on machine B, the time available for pure production on machine A equals $T - S(A) - P(B)$. Dividing by $P(A)$ yields the maximum throughput.

Configuration (A1, A2, B) with $P(B) > P(A)$

In this case there are no inter-unit idle times on machine B. Under a single batch mode with batch size Q, the initial idle time on machine B is given by $S(A) + Q*P(A1) + P(A2)$. This initial idle time can be reduced by decreasing the single batch size Q. On the other hand additional set-ups on machine A are possible since $P(B) > P(A)$. These two facts are properly balanced by a repetitive batch size Q' such that $Q'*P(B) = Q'*P(A) + S(A)$ under a multiple batch mode. It guarantees the absence of the inter-batch idle times on machine B as well as on machine A. Note that the batch processing times on both machines are equal for batch size Q. Both machines are equally utilized per batch. Both machines are bottle-neck machines. It is easily proven that the thus obtained maximum throughput is in favour of a single batch maximum throughput.

Configuration (A1, A2, B) with P(A2) < *P*(B) < *P*(A1)

Also in this case there are no inter-unit idle times on machine B. Moreover a repetitive batch size Q' is infeasible. Inter-batch idle times on machine B can be prevented by a multiple batch mode with strictly decreasing batch sizes. The batch sizes $Q(1)$, $Q(2)$, ... follow from a set of recurrence relations given by

$$Q(n+1) = \frac{Q(n) * [P(\mathrm{B}) - P(\mathrm{A2})] - S(\mathrm{A})}{P(\mathrm{A1})}. \tag{1}$$

However, a computational procedure based on eqn (1) does not guarantee the maximum throughput in all cases, as a counter example reveals (cf. Huisman *et al.*, 1990). An additional search procedure in the neighbourhood of the solution obtained by applying eqn (1) may be necessary. Occasionally inter-batch idle times on both machines may occur depending in a complicated way on the numerical data. In relation to these phenomena either machine A or machine B or both machines may be bottle-neck machine(s).

Configuration (A1, B, A2)

If $P(\mathrm{B}) < P(\mathrm{A1})$, machine B experiences inter-unit idle times equal to $P(\mathrm{A1}) - P(\mathrm{B})$. In any case there exists an inevitable inter-batch idle time on machine B between the starting time of the last unit of a batch of process A2 and the finishing time of the first unit of the next batch of process A1, if a multiple batch mode is considered. Since this inter-batch idle time is independent of the batch size, a single batch mode is appropriate with machine A as a bottle-neck machine. If for a single batch of size Q the expression $Q*[P(\mathrm{A}) - P(\mathrm{B})] + S(\mathrm{A2}) - P(\mathrm{A})$ is positive then machine A is a bottle-neck machine and machine B otherwise. In the latter case there is a positive inter-process idle time on machine A equal to the opposite of this quantity.

Configuration (B, A1, A2)

This configuration is in many respects the reflection of configuration (A1, A2, B) and the implied symmetry holds very much for its analysis. If $P(\mathrm{B}) < P(\mathrm{A1})$ machine B experiences an inevitable inter-batch idle time under a multiple batch mode. A single batch mode is appropriate in this case and machine A is the bottle-neck machine. For $P(\mathrm{B}) > P(\mathrm{A})$ again a multiple batch mode with repetitive batch sizes is appropriate. Also the case $P(\mathrm{A1}) < P(\mathrm{B}) < P(\mathrm{A})$ is comparable with the similar case for configuration (A1, A2, B), however with strictly increasing batch sizes. The results for serial process configurations are summarized in Table 1.

Convergent process configurations

In this section the main characteristics of the two convergent process configurations are treated. According to configuration (A1/A2, B), two types of parts are processed on machine A and subsequently assembled to final products on machine B. The main difference with serial configuration (A1, A2, B) is the fact that the processes on machine A can be executed in arbitrary order, but not simultaneously.

According to configuration (A1/B, A2) the two types of parts are produced on different machines and assembled on one of these. The two types of parts can be produced simultaneously.

Table 1. Serial process configurations.

Configuration	Cases	Batch mode	Bottle-neck machine(s)
(A1,A2,B)	$P(B) < P(A2)$	Single	A
	$P(A2) < P(B) < P(A)$	Multiple decreasing	A or B or both
	$P(A) < P(B)$	Multiple repetitive	A and B
(A1,B,A2)	All	Single	A or B
(B,A1,A2)	$P(B) < P(A1)$	Single	A
	$P(A1) < P(B) < P(A)$	Multiple increasing	A or B or both
	$P(A) < P(B)$	Multiple repetitive	A and B

Configuration (A1/A2, B) with $P(B)$ < MAX[$P(A1)$, $P(A2)$]

If $P(B)$ < MIN[$P(A1)$, $P(A2)$] then machine B has inter-unit idle times. Hence a single batch mode is appropriate and the maximum throughput is given by $T - S(A) - P(B)$ divided by $P(A)$. If MIN[$P(A1)$, $P(A2)$] < $P(B)$ < MAX[$P(A1)$, $P(A2)$] then machine B has inter-unit idle times in case the process on machine A with the shortest unit processing time is chosen as the initial process. If the process on machine A with the longest unit processing time is chosen as the initial process then a multiple batch mode is also excluded.

Configuration (A1/A2, B) with $P(B) > P(A)$

The analysis of this case is very similar to the analysis of serial configuration (A1, A2, B) with $P(B) > P(A)$. A multiple batch mode with repetitive batch size Q' maximizes throughput. Both machines are equally utilized and one bottle-neck machines. There is one remark to make. The throughput depends on the choice of initial process on machine A. The maximum throughput is obtained if for the initial process on machine A the process with the shortest unit processing time is chosen.

Configuration (A1/A2, B) with $P(A) > P(B)$ > MAX[$P(A1)$, $P(A2)$]

The analysis of this case is quite similar to the analysis of serial configuration (A1, A2, B) with $P(A) > P(B) > P(A2)$. A multiple batch mode with non-repetitive batch sizes maximizes throughput. The computation of the batch sizes is even more complex then in the corresponding serial case. This is caused by the fact that the sequence of batch sizes may be different for the two processes on machine A. However, their sum should be equal.

Configuration (A1/B, A2)

Machine B can be utilized from the start of the time allowance. An initial batch on machine B should be ready when the last unit of the same batch of process A2 is going to be processed. Hence the initial idle time on machine B is equal to $S(A) + Q*P(A) - P(A2) - Q*P(B)$. If this idle time is positive (and it will be for any Q if $P(A) > P(B)$) then a single batch mode is appropriate in order to save set-up time on machine A. If it is negative then machine B should be processing uninterruptedly from the start, independently of the batch mode adopted.

Table 2. Convergent and divergent process configurations.

Configuration	Cases	Batch mode	Bottle-neck machine(s)
(A1/A2,B)	$P(B) <$ MAX $[P(A1), P(A2)]$	Single	A
	MAX $[P(A1), P(A2)] < P(B) < P(A)$	Multiple non-repetitive	A or B or both
	$P(A) < P(B)$	Multiple repetitive	A and B
(A1/B,A2)	All	Single	A or B
(A1,A2/B)	All	Single	A or B
(B,A1/A2)	All	Single	A or B

Divergent process configurations

In this section the main characteristics of divergent process configurations (A1, A2/B) and (B, A1/A2) are summarized.

Configuration (A1, A2/B)

Each batch of process A1 consists of two sub-batches, each feeding one of the final processes A2 and B. A set-up of process A2 has to wait until the preceding batch supplying process A1 is finished. Hence it is advantageous to start with the sub-batch for process B. The initial idle time on machine B is equal to $S(A1) + P(A1)$, independently of the batch size. If $P(B) < P(A1)$ then process B experiences inter-unit idle times, a single batch mode is appropriate and machine A is bottle-neck. If $P(B) > P(A1)$ then, independently of the batch mode, machine B will be processing uninterruptedly and the initial idle time cannot be reduced. A multiple batch mode would reduce the throughput of process A2. Hence a single batch mode is appropriate in all cases. The optimal mix of processes A2 and B can be determined by means of linear programming.

Configuration (B, A1/A2)

In this case process B supplies both final processes on machine A. Assuming no commitments on final products, throughput is maximized by selecting one of the two final processes since set-up time is saved. A single batch mode is appropriate in all cases. Let Ai denote the selected final process. If $P(Ai) < P(B)$ then B is bottle-neck machine, otherwise machine A.

The results for convergent and divergent process configurations are summarized in Table 2.

Conclusions

An interesting relationship between batch mode, bottle-neck machines and maximum throughput emerges from the results of both tables. A single machine is a bottle-neck machine in the case where throughput is maximized by a single batch mode. Both machines are bottle-neck machines in the case where throughput is maximized by a multiple repetitive batch mode. Also there are intermediate cases where, under a multiple batch mode, there are occasionally relatively small inter-batch idle times on one of the machines. These cases are characterized by three possibilities: either machine A or machine B or both machines may be bottle-neck.

The intriguing question is whether or not this picture can be extended to more complex structures with a larger number of machines and processes. In this respect it is noted that some results of this paper are immediately extendable to an arbitrary finite number of processes on machine A. For example, in serial configuration (A1, A2, B) the formulas remain valid for n processes on machine A, if P(A2) is replaced by P(An) and P(A1) by the sum of the first $n - 1$ unit processing times on machine A. P(A) and S(A) have to be interpreted, respectively, as the sum of the n unit processing times and the sum of the n set-up times of the n processes on machine A.

References

Goldratt, E.M., 1988, Computerized shop floor scheduling. *International Journal of Production Research*, **26**, 443–455.

Huisman, H.H., Polderman, G.L. and Weeda, P.J., 1990, Maximizing throughput in some simple time-constrained scheduling situations. *Engineering Costs and Production Economics*, **18**, 293–299.

Weeda, P.J., 1989, *Maximizing Throughput in Convergent and Divergent Process Configurations Consisting of Three Processes and Two Machines*. Internal report, Department of Mechanical Engineering, University of Twente, Enschede, The Netherlands.

Weeda, P.J., 1990, On the choice of batch mode in order to maximize throughput. *Engineering Costs and Production Economics*, **19**, 241–247.

36

Hierarchical integration of aggregate and detailed production planning systems

Hitoshi Tsubone,[1] Takao Tanaka,[2] and Mitsumasa Sugawara[3]

[1]*Tokyo Metropolitan Institute of Technology, 6-6 Asahigaoka, Hino-city, Tokyo, 191, Japan,* [2]*Takachiho University, Japan, and* [3]*Miyako College, Japan*

Abstract

This chapter deals with the problem of designing a hierarchical production planning system where higher level decisions impose constraints on lower level decisions. Two-level decision problems are modelled as a stochastic one-stage production process.

An analytical model can clarify the interaction mechanisms between the two different levels. Based on the results of analysis, this chapter proposes a method for designing an efficient production planning system concerning multi-purpose operations.

Introduction

The production planning problem generally involves a hierarchical decisions structure. The decisions are made from a higher level to a lower level by adding the constraints or guide, which have been determined at the higher level, to the decisions at the lower level. There are at least two distinct decision levels, depending upon whether they are aggregate decisions or detailed decisions. For example, aggregate decisions are made concerning capacity planning based on fairly aggregate information. The decision variable may be production rate or total production quantities as a suitable common denominator for aggregate groups of products. On the other hand, detailed decisions concern lot-sizing or scheduling for each product, based on detailed information.

In such a multi-level decisions system, the aggregate decisions made at a higher planning level lead to some restrictions or guide on detailed decisions at a lower level. Thus, it is necessary to integrate aggregate and detailed decisions in order to design an efficient production planning system.

General discussions on such hierarchical production planning systems can be found in reports by Bitran and Hax (1977), Bitran *et al.* (1981, 1982) and Hax and Candea (1984). Most of them have proposed algorithms which minimize the total costs relative to changes in production level, set-up costs and inventory carrying costs. However, set-up time is not considered at all in these algorithms, or set-up costs are constant and independent of the available working time.

First, the proposed analytical model will clarify how the planning rules for production will affect a shortage of products required to meet market demands, the total set-up time, the inventory level for products, and the variation of production rate concerned. We then propose a method for designing an efficient production planning system under multi-purpose operations.

Hierarchical production planning model

Aggregate model

An aggregate model is used to plan the production capacity over the planning horizon by using the aggregate forecast demand for products. The decision variable is the working time required over the planning horizon (one month). It is formulated in eqn (1) in order to generate proposals for production planning by trading off the production rate and the inventory level for products.

$$H_p = \gamma X + (1 - \gamma)R_p \tag{1}$$

where H_p is the working time required during month p ($p = 1, 2, \ldots, 12$); γ is the control parameter for smoothing production, by which the working time on month p can be determined ($0 \leq \gamma \leq 1$); X is the desired working time over the planning horizon, expressed as follows:

$$X = \bar{b} \cdot \bar{D} + L\sum_{j=1}^{J} a_j$$

where the model parameters are: L, the number of weeks in one month; $\bar{b}$, the weighted average time required to produce one unit of product; $\bar{D}$, the expected aggregate demand per month; a_j, the set-up time required for the product j; J, the set of product j; R_p, the required working time for both net processing and set-up on month p, expressed as follows:

$$R_p = \bar{b}(\hat{D}_p + SS - AI_{p-1}) + L\sum_{j=1}^{J} a_j$$

where the model parameters are: $\hat{D}_p$, aggregate forecast demand on month p; SS, aggregate safety stock level; AI_{p-1}, aggregate inventory level at the end of month $p - 1$. It is assumed that no change can be made in working time during the planning horizon, once the available working time has been set up, in order to maintain stable production activities. Accordingly, aggregate planning takes place once every month.

Detailed model

The detailed model is used to plan and schedule for individual products by using the forecast demand for each product. Since the available working time, which has been determined at the aggregate model, is divided into one week intervals, detailed production planning takes place every week for production in the immediate week over the planning horizon. The decision variable is the quantity of individual products to be produced in the immediate week.

The product quantities are determined so that the total time required for production equals the available working time, which has been determined at the aggregate planning level. They are formulated in eqn (2).

$$\sum_{j \in J^0} u(y_{j,t})(a_j + b_j \cdot y_{j,t}) = h_t. \tag{2}$$

$y_{j,t}$ is the product j quantity to be produced on week t, $u(y_{j,t})$ is the binary variable, if $y_{j,t} > 0$, $u(y_{j,t}) = 1$, otherwise $u(y_{j,t}) = 0$, J^0 is the set of products produced in the planning week, that is, J^0 contains only those products which trigger during the planning week, b_j is the net time required to produce one unit of product j, h_t is the available working time on week t, $h_t = H_p/L$ $(t = 1, 2, \ldots, L)$.

Two planning rules for production are adopted at the detailed planning level in consideration of set-up time and a shortage of products required to meet market demands.

Rule 1. This rule aims to minimize the total set-up time for switching from one kind of product to another. Available working time is allocated to manufacturing each product in proportion to the product of the expected demand and the unit set-up time for each product.

Rule 2. This rule aims to minimize any shortage of products required to meet market demands. Available working time is allocated for each product in such a way that the shortage of each product coincides with the shortage of the aggregate products.

Performance critertia

Criteria used to evaluate the manufacturing performance are as follows:

(1) shortage of products required to meet market demands ... (α),
(2) average inventory level for products ... (β),
(3) total set-up time ... (η)
(4) variation of production rate ... (V).

They are expressed in eqns (3), (4), (5), and (6), respectively.

$$\alpha = \sum_{t=1}^{T}\sum_{j=1}^{J} -\min(I_{j,t}, 0) \Big/ \sum_{t=1}^{T}\sum_{j=1}^{J} d_{j,t} \tag{3}$$

$$\beta = \sum_{t=1}^{T}\sum_{j=1}^{J} \max(I_{j,t}, 0) \Big/ \sum_{t=1}^{T}\sum_{j=1}^{J} d_{j,t} \tag{4}$$

$$\eta = \sum_{t=1}^{T}\sum_{j=1}^{J} a_j \cdot u(y_{j,t}) \Big/ \sum_{t=1}^{T}\sum_{j=1}^{J} b_j \cdot y_{j,t} \tag{5}$$

$$V = \sum_{p=1}^{12} (H_p - \overline{H})^2/(12X) \tag{6}$$

where, $I_{j,t}$ and $d_{j,t}$ are, respectively, the inventory level for product j at the end of week t and the actual demand for product j in week t; T is the number of weeks throughout one year; $\overline{H}$ is the average value of required working time.

Experimental conditions

(1) Four kinds of product are produced. Table 1 shows the relationship among the expected demands for each product in week ($\overline{d}_j$), the unit set-up time (a_j), and the unit processing time (b_j).
(2) The fluctuated demand value (e_d) is a normal distributed random variable with zero mean and known variance (σ_d^2), where, $\rho_d = \sigma_d/\overline{d}_j$ (ρ is the coefficient of variation).
(3) The forecast demand is set equal to the expected demand, based on either aggregate planning or detailed planning. Therefore, the standard deviation of demand is equivalent to the standard deviation of the forecast error.

Table 1. Experimental conditions.

Product j	Expected demand $\bar{d}_j$	Unit set-up time a_j	Unit processing time b_j
1	500	50	1
2	1000	50	1
3	500	100	1
4	1000	100	1

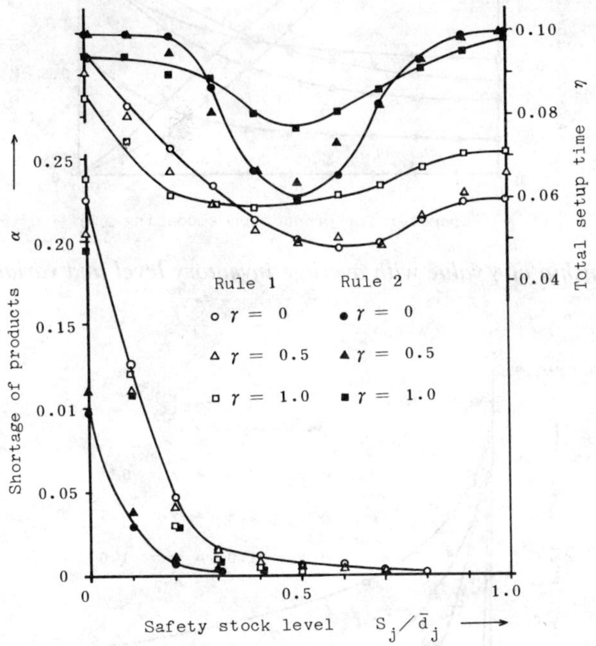

Figure 1. Relationship of safety stock level with shortage of products and total set-up time.

(4) The safety stock level for each product (S_j) is set in proportion to the expected demand for each product.
(5) One year is divided into 48 available working weeks.

Experimental results

Through a series of experiments, the following points have been clarified:

(1) The higher the safety stock level is set, the smaller the shortage of the products decreases under either *Rule 1* or *Rule 2*. However, the total set-up time increases gradually after it decreases as the safety stock level becomes higher. This is the reason that all the products are produced on the same week in order to use up the available working time, even if there is sufficient stock at the beginning of the planning week (Figure 1).
(2) The larger the value of γ, the more the average inventory level keeps increasing. However, the variation of production rate adversely keeps decreasing (Figure 2).
(3) The total set-up time is almost independent of the value of γ as far as the value of γ does not come to 1·0 (Figure 1).
(4) The average product inventory can be kept at a lower level under *Rule 2* than under *Rule 1*, considering the given shortage of product value (Figure 3).

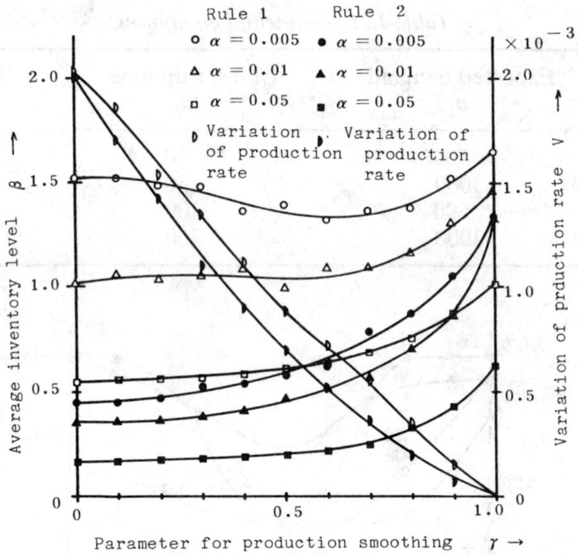

Figure 2. Relationship of γ value with average inventory level and variation of production rate.

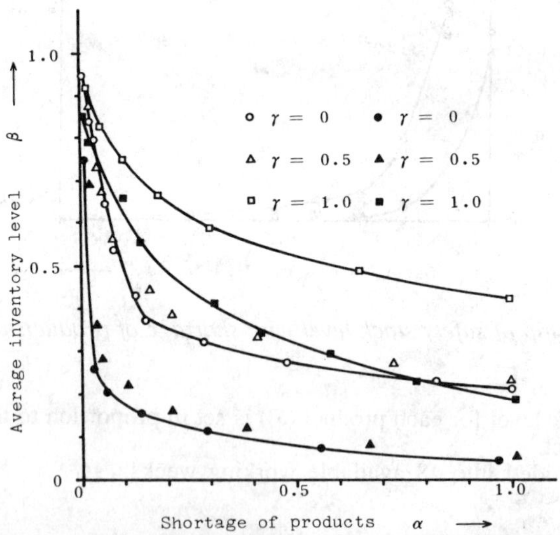

Figure 3. Relationship between shortage of products and average inventory level.

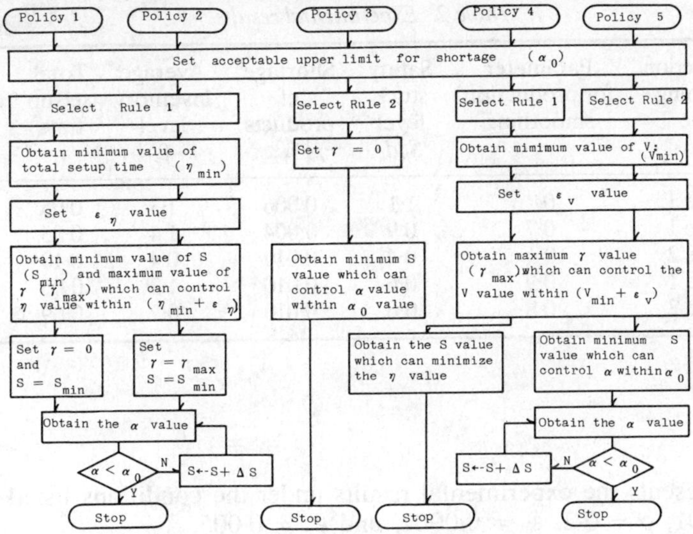

Figure 4. Flowchart for designing an efficient production planning system.

A method to design an efficient production planning system

A method is proposed to design an efficient production planning system under a selected policy.

Step 1—Give specific values to the following parameters:

$$T, \overline{D}, J, a_j, b_j, \overline{d}_j, \overline{b} \qquad (\overline{b} = \Sigma\, b_j \cdot \overline{d}_j / \Sigma\, \overline{d}_j)$$

Step 2—Set a shortage of products value which can be controlled within an acceptable upper limit (α_0), no matter which policy mentioned below will be adopted.

Step 3—Select a policy from among the following policies:

Policy 1. (1) Set the acceptable upper limit for the total set-up time (η_0) from its minimum value (η_{min}) · $\eta_0 = \eta_{min} + \varepsilon_\eta$, where ε_η is the acceptable increment of set-up time.
(2) Minimize the average inventory level for products (β) under condition (1).

Policy 2. (1) Set the acceptable upper limit for the total set-up time from its minimum value.
(2) Minimize the variation of production rate (V) throughout the year under condition (1).

Policy 3. Minimize the average products inventory level (β).

Policy 4. (1) Set the acceptable upper limit for variation of production rate (V_0) from its minimum value (V_{min}) · $V_0 = V_{min} + \varepsilon_V$, where ε_V is the acceptable increment of variation of production rate.
(2) Minimize the total set-up time under condition (1).

Policy 5. (1) Set the acceptable limit for variation of production rate from its minimum value.
(2) Minimize the average inventory level of products under condition (1).

Step 4—Select either planning rule for production, following the flowchart in Figure 4, which is able to optimize each policy.

Step 5—Obtain the optimum value (γ) of parameter for smoothing production and the optimum value (S) of safety stock level of each product.

Table 2. Experimental results.

Policy	Production planning rule	Parameter for production smoothing γ	Safety stock level $S_j/\bar{d}_j$	Shortage of products α	Average inventory level β	Total set-up time η	Variation of production rate V
1	Rule 1	0	1·1	0·006	1·3	0·05	$2{\cdot}1 \times 10^{-3}$
2	Rule 1	0·7	0·9	0·004	1·4	0·05	$0{\cdot}6 \times 10^{-3}$
3	Rule 2	0	0·4	0·010	0·4	0·10	$2{\cdot}0 \times 10^{-3}$
4	Rule 1	0·9	0·6	0·010	1·3	0·05	$0{\cdot}2 \times 10^{-3}$
5	Rule 2	0·8	0·6	0·010	0·7	0·09	$0{\cdot}2 \times 10^{-3}$

An example

Table 2 represents the experimental results under the conditions listed in Table 1 and $\alpha_0 = 0{\cdot}01$, $\rho = 0{\cdot}2$, $\varepsilon_V = 0{\cdot}0002$, and $\varepsilon_\eta = 0{\cdot}005$.

Conclusion

Our proposed analytical model has clarified the interaction mechanisms between the two different levels, by analysing how the production planning rules would affect the shortage of products, the total set-up time, the average inventory level, and the variation of production rate. Based on the results of analysis, we proposed a method to design an efficient production planning system under multi-purpose operations.

References

Bitran, G.R. and Hax A.C., 1977, On the design of hierarchical production planning systems. *Decision Science*, **8**, 28–54.

Bitran, G.R., Haas, E.A. and Hax, A.C., 1981, Hierarchical production planning: a single stage system. *Operations Research*, **29**, 717–742.

Bitran, G.R., Haas, E.A. and Hax, A.C., 1982, Hierarchical production planning: a two-stage system. *Operations Research*, **30**, 232–251.

Hax, A.C. and Candea, D., 1984, *Production and Inventory Management* (Englewood Cliffs, NJ: Prentice Hall).

37

Optimal due-date setting in an assembly shop

T.C.E. Cheng

Department of Actuarial and Management Sciences, University of Manitoba, Winnipeg, Manitoba, Canada R3T 2N2

Abstract

Under intensified pressure from both domestic and foreign competition, today's manufacturers must strive to keep delivery promises and reduce waste. This depends largely on a manufacturer's ability to assign accurate and attainable due-dates. This chapter presents an analytical method of assigning optimal total-work-content (TWK) due-dates that minimize the average amount of missed due-dates in a dynamic job shop with assembly operations. The accuracy of the method is assessed through computer simulation of hypothetical assembly shops with different job structures under various operating conditions. While it is generally more accurate for jobs with simpler structures, the simulation results indicate that, over a range of test conditions, the method is both simple and effective in assigning optimal TWK due-dates for assembly shops.

Introduction

Scheduling is an important topic in production research. Over the past three decades an abundance of research results on job shop scheduling using computer simulation has been accumulated, e.g. Conway (1965), Jones (1973), Eilon and Chowdhury (1976), Weeks and Fryer (1976), Panwalkar and Iskandar (1977), Weeks (1979), Baker and Kanet (1983, 1984) and Cheng (1988).

A fundamental assumption in most of these studies of the dynamic job shop is that the job consists of a single component routed randomly through the shop to various processing facilities on which operations are performed. In a realistic situation, however, the job essentially consists of several parts or components which must go through a number of assembly operations to become a final product. Thus, the job has a unique job structure determined by the precedence relationships of the constituent parts. Since a part cannot normally be processed until all its predecessor parts are ready, there is the problem of delays caused by staging that adds an extra dimension of difficulty to the analysis.

To date only a handful of research has been done on job shops with assembly operations, examples are: Conway *et al.* (1967), Maxwell (1969), Rochette and Sadowski (1976), Sculli (1980) and Russell and Taylor (1985). All of this research is concerned with assessing the relative performance of various sequencing rules in relation to some predefined performance measure. The results of this research confirm that rules taking account of progress information on other parts required for the same job usually perform better than those that fail to do so by synchronizing the completion times of related parts in other work centres with respect to both flow-time and due-date oriented performance criteria.

With increasing competitive pressure in the marketplace, the keys to success for a manufacturing firm competing in today's business climate lie in consistently meeting delivery promises to win satisfied customers and in holding down capital tied up in inventory to improve the cash flow performance and reduce costs. Thus the ability of a manufacturer to assign accurate and reliable order due-dates is of utmost importance. While Ragatz and Mabert (1984) have discussed the importance of due-date management in job shop production systems, Cheng (1986) has presented an analytical approach to assigning optimal due-dates in dynamic job shops.

This chapter proposes an analytical method to assign optimal total-work-content (TWK) due-dates in a dynamic job shop with assembly operations. TWK is a reasonable method of due-date assignment for job shops, see, for example, Conway *et al.* (1967). Some examples of scheduling research involving the TWK due-dates are Conway (1965) and Cheng (1984, 1988). The primary objective of this research is to assess the accuracy of the analytical method of optimal TWK due-date assignment for assembly shops with jobs with various structural complexities. Specifically, we are interested in testing the following null hypothesis:

$H_0$1: There is no significant difference in the accuracy of the analytical method of assigning optimal TWK due-dates to jobs with varying structural complexities.

Analytical method for assigning TWK due-dates

Sculli (1980) has pointed out that the difficulty of dealing with jobs with assembly operations in the scheduling environment is that there are not only delays caused by the limited resources (i.e. processing facilities), but also delays in processing a part until all the predecessor parts are available. He has proposed using the network method for project scheduling as an aid for visualization and analysis of scheduling problems involving assembly operations.

For each job we construct an operation sequence network in accordance with the precedence relationships of the various parts making up the final product. Evidently, it is difficult to perform a thorough analysis of an assembly shop due to the complexity of the operation sequence networks of the jobs. However, to make the analysis more tractable and tenable we propose an approximate analytical method which deals only with the critical path of a job without regard to the other parts of the job. It is the purpose of this study to evaluate the effectiveness of the analytical method of assigning optimal TWK due-dates under conditions of varying job structures.

The theory of the analytical method is largely based on Jackson's decomposition principle (Jackson, 1955) for queuing networks and Laplace transforms. The assumptions of the production model are:

(1) There are m single non-identical machines in the shop providing the various types of service required by the jobs.
(2) Job arrivals follow the Poisson process with a mean arrival rate λ.
(3) The processing times of the machines p_i ($i = 1, 2, \ldots, m$) are independent and identical random variables, negative-exponentially distributed with mean μ_p.
(4) Each job consists of a fixed number of parts, n, and each component part consists of a single operation only.
(5) All machines are equally likely to be selected for processing the operation of a part, but parts with immediate precedence relationships are not allowed to be processed on the same machine.

(6) The queue discipline at each machine is first-come-first-served (FCFS).
(7) The individual machine utilization factor has a common value $\rho = \lambda\mu_p$, which is less than unity.
(8) The shop load ratio ρ_s, representing the ratio of the total work load to the total production capacity available in the shop, is a system parameter and can be set at different levels to provide various shop congestion conditions.
(9) The machine set-up times are included in the processing times and are sequence-independent; transportation times between machines are neglected.
(10) Job pre-emption is not allowed.

It is also assumed that the critical path of a job is the path in which the expected total processing time of the parts is longest. In addition, we have assumed in our model that each part consists of a single operation and that the part processing times are independent and identically distributed random variables. It follows that the critical path of a job is the path with the largest number of serial parts.

Let n denote the number of parts in the critical path of a job. The second moments of the waiting time W and flow-time F of the critical path can be derived as follows:

$$E(W^2) = n^2\{\lambda E(p^2)/[2(1-\rho)]\}^2 + n\{\lambda E(p^3)/[3(1-\rho)] + \{\lambda E(p^2)/[2(1-\rho)]\}^2\}; \quad (1)$$

$$E(F^2) = n^2\{\mu_p + \lambda E(p^2)/[2(1-\rho)]\}^2 + n\{E(p^2) - \mu_p^2 + \lambda E(p^3)/[3(1-\rho)] + \{\lambda E(p^2)/[2(1-\rho)]\}^2\}; \quad (2)$$

where $E(\cdot)$ is the expected value operator.

Let a and P denote, respectively, the arrival time and total processing time of the critical path of the job and k the due-date multiple factor. Thus, with the TWK method of due-date assignment, each job is assigned a due-date D as follows:

$$D = a + kP. \quad (3)$$

Let A denote the actual job flow-time. Since a job will be penalized if it misses its due-date, be it early or late, we wish to minimize the expected value of the square of the lateness, given by:

$$E(L^2) = E(A^2) + k^2E(P^2) - 2kE(AP). \quad (4)$$

It follows that the optimal due-date multiple k^* that minimizes $E(L^2)$ can be obtained by differentiating (4) once and setting the result to zero, yielding

$$k^* = E(AP)/E(P^2). \quad (5)$$

Since

$$E(AP) = \{E(A^2) + E(P^2) - E(W^2)\}/2 \quad (6)$$

and

$$E(P^2) = n^2\mu_p^2 + n\{E(p^2) - \mu_p^2\} \quad (7)$$

we have

$$k^* = \{E(A^2) + E(P^2) - E(W^2)\}/(2\{n^2\mu_p^2 + n[E(p^2) - \mu_p^2]\}). \quad (8)$$

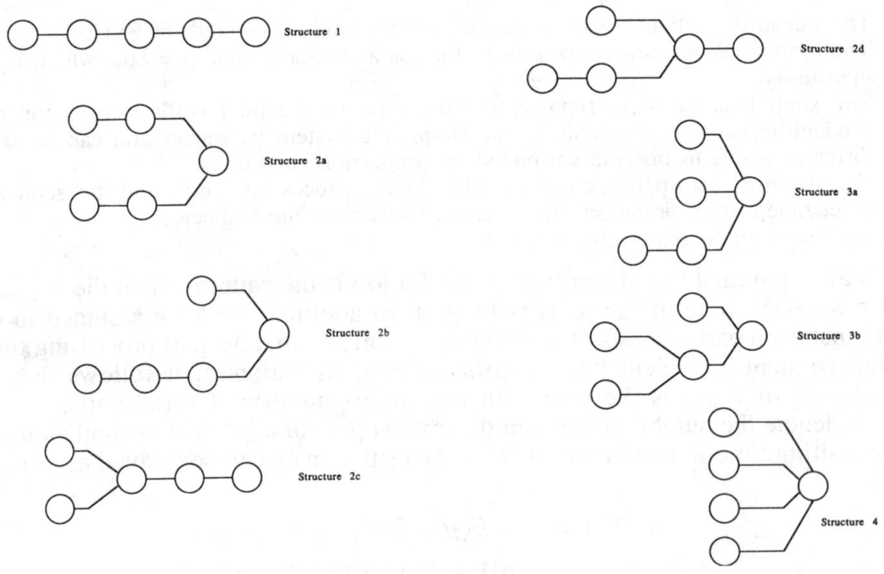

Figure 1. Eight different job structures resulting from different combinations of five parts.

Finally, approximating $E(A^2)$ by $E(F^2)$ and substituting eqns (1) and (2) into eqn (8), we obtain an approximate optimal due-date multiple factor k_c* as follows:

$$k_c* = \{n^2[2(1-\rho)\mu_p^2 + \rho E(p^2)] + 2(1-\rho)n[E(p^2) - \mu_p^2]\}/ \{2(1-\rho)[n^2\mu_p^2 + n[E(p^2) - \mu_p^2]\}. \quad (9)$$

Job structures and performance measure

In this study we assume that all jobs entering the assembly shop consist of five parts, each requiring a single operation. There are in total eight ways of combining the parts into different job structures which are studied in this research. The various job structures with their associated reference numbers are depicted in Figure 1. In the figure, the job structures are arranged in the order of increasing degree of structural complexity.

For each job structure there are two characteristics that determine its degree of structural complexity: (1) the number of branches per job N_b, and (2) the number of levels of branching, L_b. In this study these two characteristics are program parameters which are manipulated to generate the desired operating conditions for investigation. With one branch per job ($N_b = 1$), the simple job shop process will be produced. If there are more branches per job ($N_b > 1$), this generates an assembly shop where all jobs have the same number of branches. All but one of the structures studied have a single level of branching ($L_b = 1$) which can take place at any operation of a job. The exception is structure number 3b in which there are two levels of branching.

In this study it is held that the complexity of a job is largely determined by the number of branches in the job because this number determines the amount of parallel processing of component parts. Thus all structures having the same number of branches are treated as similar and grouped together for the purpose of the statistical analysis to be discussed later. Although we fully appreciate the fact that the complexity of a job is also dependent on the number of levels of branching in the job, a study of its effect on the accuracy of the approximation method has not been included in this research. In addition, as mentioned earlier, there is only one structure with more than one level of branching.

The accuracy of using the analytical method for assigning optimal TWK due-dates under various job structure conditions is evaluated by the following performance measure:

$$\alpha = \{|k^* - k_c^*|/k^*\} \times 100\%. \tag{10}$$

Simulation experiments

A hypothetical job shop provides the setting for this research. Experiments within this setting are conducted using a simulation model which has the following characteristics:

(1) The shop consists of $m = 5$ single machines.
(2) The number of parts per job, $c = 5$, is fixed and each part requires a single operation to be performed on one of the m machines. The actual machine on which a part will be processed is determined by sampling from a discrete uniform distribution with a range from 1 to 5 corresponding to the machine numbers.
(3) The processing time distributions of the machines are independent and identical, all having an exponential density function with a mean processing time $\mu_p = 1{\cdot}0$. However, due to rounding off continuous values to integral values in the simulation program, the first three moments of the processing time distribution have the actual values of $\mu_p = E(p) = 1{\cdot}582$, $E(p^2) = 3{\cdot}423$ and $E(p^3) = 10{\cdot}320$, respectively.
(4) Job arrivals follow the Poisson process with a mean arrival rate λ which has the following relationship with the shop load ratio ρ_s:

$$\rho_s = \lambda\mu_p.$$

The above result follows from the fact that $m = n = 5$ in the present case. Setting $1/\lambda$, the mean inter-arrival time of jobs, equal to 3·638 and 2·220, respectively, produces shop load ratios of 0·5 and 0·9 which are the shop load conditions used for the experiments in this study. These two load ratios, representing a lightly and a heavily loaded shop, respectively, are commonly used in earlier job shop simulation studies (Conway, 1965; Eilon and Chowdhury, 1976; Cheng, 1988).
(5) The release time of a part is equal to the longest completion time of its immediate predecessor parts.
(6) Processing of a part on a machine cannot begin until it is released to the shop and all its predecessor parts are available.
(7) A job is completed when all its component parts have been processed.
(8) A run-in period for the shop to reach steady state is allowed; this is the time from the start of the simulation to the completion of the 250th part. The statistics of the next 500 jobs (2500 parts) leaving the shop are collected for analysis. The simulation stops upon the completion of the last job in the 500 job-set. Since the shop is still in steady state when the simulation stops, adjustment of the collected data to take account of the run-out period is unnecessary.

For each shop load ratio, one replication is made using random numbers antithetic to the first experiment for variance reduction for the eight job structures; altogether 16 experiments are performed.

Determination of the actual optimal due-date multiple factor k^* is accomplished through the use of a direct searching technique to locate the optimal value of a unimodal function. The optimal value obtained from the searching method is accurate within +0·01 of the true optimal value.

Table 1. Comparison of actual and approximate optimal due-date multiple factors.

Job Structure	Actual k^*		Approximated k_c^*
	Rep. 1	Rep. 2	
Shop load ratio $\rho_s = 0{\cdot}5$			
1	1·69	1·68	1·64
2a	1·52	1·52	1·61
2b	1·76	1·75	1·63
2c	1·76	1·75	1·63
2d	1·76	1·76	1·63
3a	1·76	1·76	1·61
3b	1·76	1·76	1·61
4	1·75	2·05	1·58
Shop load ratio $\rho_s = 0{\cdot}9$			
1	7·25	6·63	6·73
2a	6·82	6·00	6·48
2b	7·52	6·84	6·64
2c	7·52	6·61	6·64
2d	7·52	6·75	6·64
3a	7·50	6·77	6·48
3b	7·76	7·03	6·48
4	9·06	8·53	6·20

Experimental results

The simulation results are shown in Table 1. We use two non-parametric tests: (1) Kruskal–Wallis one-way analysis of variance, used to test the significance of differences in the sample averages, and (2) Jonckheere's test, applied to compare the sample averages. Use of these distribution-free tests is pertinent in this situation since we do not have prior knowledge about the forms of the distributions from which the performance measures are sampled for analysis. Detailed treatment of the theories and applications of these tests can be found in the papers by Jonckheere (1954) and Kruskal and Wallis (1952).

Table 2 shows the results of the tests of the hypothesis, with a 95% level of significance, which yield the following important inferences:

Shop load ratio $\rho_s = 0{\cdot}5$—There is significant difference in accuracy of the analytical method to estimate the optimal due-date multiple factors for various job structures. The estimates are very accurate for jobs with simple structures and become rough for complex jobs.

Shop load ratio $\rho_s = 0{\cdot}9$—The test of difference in accuracy of the analytical method to estimate the optimal due-date multiple factor for different job structures is marginally not significant. However, the comparison test indicated that the accuracy of the method is inversely proportional to job structure complexity.

Conclusions

In this study an analytical method for assigning optimal TWK due-dates is proposed for use in the job shop environment with assembly operations. The method is based on the premise that reasonable estimates of the job flow-times can be derived from an analysis of the critical path of the job operation sequence network, without regard to the operations of the related parts in the network. The aim of the research

Table 2. Comparison of the accuracy of the analytical method for different job structures by the Kruskal–Wallis and Jonckheere's tests.

Job structure	α%		Mean	Rank		Rank sum R_j
	Rep. 1	Rep. 2		Rep. 1	Rep. 2	
Shop load ratio $\rho_s = 0{\cdot}5$						
1	3·02	2·56	2·79	2	1	3
2a	6·15	6·15		3·5	3·5	
2b	7·49	7·07		8·5	8·5	
2c	7·49	7·07	7·05	8·5	5·5	52
2d	7·49	7·49		8·5	8·5	
3a	8·46	8·46		12·5	12·5	
3b	8·46	8·46	8·46	12·5	12·5	50
4	9·85	23·56	16·71	15	16	31
$H = 12{\cdot}88, \chi^2_{0{\cdot}05,3} = 7{\cdot}81$; $T = 4{\cdot}11, z_{0{\cdot}05} = 1{\cdot}69$						
Shop load ratio $\rho_s = 0{\cdot}9$						
1	7·13	1·51	4·32	7	2	9
2a	4·95	7·45		6	8	
2b	11·70	3·03		11	4	
2c	11·70	0·42	6·58	11	1	55
2d	11·70	1·68		11	3	
3a	13·56	4·19		13	5	
3b	16·44	7·80	10·50	14	9	41
4	31·60	27·34	29·47	16	15	31
$H = 7{\cdot}20, \chi^2_{0{\cdot}05,3} = 7{\cdot}81$; $T = 2{\cdot}74, z_{0{\cdot}05} = 1{\cdot}69$						

is to study the effectiveness of the method applied to jobs with varying structural complexities. The appropriate statistical tests conducted on the simulation results support the fact that there are significant differences in the accuracy of the method for varying job structures. The major merit of this research is that a simple method based on the critical path analysis is proposed, which has been shown to be an effective means, capable of yielding reasonably accurate results under a variety of test conditions, of assigning optimal TWK due-dates in assembly shops.

Acknowledgements

The author thanks Professor H.J. Boom for many helpful discussions. This research was supported in part by a grant from the Natural Sciences and Engineering Research Council of Canada.

References

Baker, K.R. and Kanet, J.J., 1983, Job shop scheduling with modified due-dates. *Journal of Operations Management*, **4**, 11–19.

Baker, K.R. and Kanet, J.J., 1984, Improved decision rules in a combined system for minimizing job tardiness. *International Journal of Production Research*, **22**, 917–921.

Cheng, T.C.E., 1984, Optimal due-date determination and sequencing of n jobs on a single machine. *Journal of the Operational Research Society*, **35**, 433–437.

Cheng, T.C.E., 1986, Optimal due-date assignment in a job shop. *International Journal of Production Research*, **24**, 503–515.

Cheng, T.C.E., 1988, Simulation study of job shop scheduling with due-dates. *International Journal of Systems Science*, **19**, 383–390.

Conway, R.W., 1965, Priority dispatching and job lateness in a job shop. *Journal of Industrial Engineering*, **16**, 228–237.

Conway, R.W., Maxwell, W.L. and Miller, L.M., 1967, *Theory of Scheduling* (Reading: Addison-Wesley).

Eilon, S. and Chowdhury, I.G., 1976, Due-date in job shop scheduling. *International Journal of Production Research*, **14**, 223–238.

Jackson, J.R., 1955, *Scheduling a production line to minimize maximum tardiness*. Management Science Research Report, UCLA, California.

Jonckheere, A.R., 1954, A distribution-free *k*-sample test against ordered alternatives. *Biometrika*, **41**, 133–145.

Jones, C.H., 1973, An economic evaluation of job shop dispatching rules. *Management Science*, **20**, 293–299.

Kruskal, W.H. and Wallis, W.A., 1952, Use of ranks in one-criterion variance analysis. *Journal of the American Statistical Association*, **47**, 583–621.

Maxwell, W.L., 1969, *Priority dispatching and assembly operations via job shop*. Report Rm-5370-PR, Rand Corporation, California.

Panwalkar, S.S. and Iskandar, W., 1977, A survey of scheduling rules. *Operations Research*, **25**, 45–61.

Ragatz, G.L. and Mabert, V.A., 1984, A framework for the study of due-date management in job shops. *International Journal of Production Research*, **22**, 685–695.

Rochette, R. and Sadowski, R.P., 1976, A statistical comparison of the performance of some simple dispatching rules for a particular set of job shops. *International Journal of Production Research*, **14**, 63–75.

Russell, R.S. and Taylor, B.W., III, 1985, An evaluation of sequencing rules for an assembly shop. *Decision Sciences*, **16**, 196–212.

Sculli, D., 1980, Priority dispatching rules in job shops with assembly operations and random delays. *Omega*, **8**, 227–234.

Weeks, J.K., 1979, A simulation study of predictable due-dates. *Management Science*, **25**, 363–373.

Weeks, J.K. and Fryer, J.S., 1977, A methodology of assigning minimum cost due-dates. *Management Science*, **23**, 1362–1371.

38

Sequencing of parts to be manufactured by flexible manufacturing systems

Zhang Bingli and Liang Xichang

1st Department of Mechanical Engineering, Chongqin University, Chongqin Sichuan Province, China

Abstract

This chapter is concerned with the part-sequencing problem of flexible manufacturing systems (FMS). A new and efficient algorithm is developed in this chapter. The difference between the algorithm and other existing algorithms is that any parts-sequencing decision is based on a detailed analysis of the parts and the status of the system instead of several simple heuristic rules. With the advantages of lower computing complexity and higher optimal results the algorithm has much promise for practical application.

Introduction

How to sequence optimally the parts to be manufactured by FMS is one of the most important problems in its production planning and scheduling. Just after solving the part-batching and workload balancing problems of FMS, which decides uniquely the route of the every part in the system, the production planning and scheduling system of FMS has to schedule the part order in which the parts are processed by every workstation. The first research of part-sequencing problems can be traced back to 1950 and now a formal production scheduling theory has formed. The part-sequencing problem of FMS can be generally thought of as a scheduling problem of shop flow with the optimal objective of minimizing the maximum flow-time of a set of parts in the system in the production scheduling theory.

The problem has been proved to be an NP-complete problem by many researchers; and so many optimal algorithms based on the bound and branch searching principle or maths-programming have little use in practice except in theoretical research because of their very high computing complexity. Other approaches which have been widely been used in practice are heuristic approaches which just find nearly optimal results which can be accepted in practice instead of the optimal ones in theory; in this way the computing complexity of the algorithms would probably be reduced. However many algorithms which exist at present are based on several simple heuristic rules or on combinations of them; they make decisions by experience without rigorous theoretical analysis. Therefore most of them are too simple to get satisfactory results especially for a high capital involvement production system such as FMS.

The aim of this chapter is to develop a very simple but efficient heuristic algorithm. In this algorithm the decision of sequencing the parts is based on detailed analysis of the parts and the status of the system.

In the next section the basic concepts and theorem related to the algorithm will be given. The algorithm is built in the third section and conclusions can be found in the last section.

Basic concepts and related theorem

Let $PS = \{P_1, P_2, \ldots, P_N\}$ be a set of parts and the workstations in the system are $MS = \{M_1, M_2, \ldots, M_M\}$; the following matrix

$$A_{MN} = \begin{pmatrix} a_{11} & a_{12} & \ldots & a_{1N} \\ a_{21} & a_{22} & \ldots & a_{2N} \\ & & \ldots & \ldots \\ \ldots & & \ldots & \\ a_{M\cdot 1} & a_{M\cdot 2} & \ldots & a_{M\cdot N} \end{pmatrix}$$

is called as the workload assignment matrix of PS about the system. In the matrix a_{ij} is the time for the workstation M_i to process the related operation of part P_j. If $a_{ij} = 0$, the part P_j has no operation to be processed by M_i.

Because the type of the production system in discussion is one kind of flow shop, there is a strictly deterministic sequence of the workstations in the system, say

$$M_1, M_2, \ldots, M_M,$$

related to the parts in PS. Any part in PS, which has been processed by M_j, is no longer processed by the workstations which stand before M_j in the sequence.

Example 1: $PS = \{P_1, P_2, P_3, P_4\}$, $MS = \{M_1, M_2, M_3\}$. The workload assignment matrix is

$$A_{3\times 4} = \begin{pmatrix} 8 & 6 & 5 & 9 \\ 2 & 3 & 12 & 7 \\ 10 & 5 & 4 & 3 \end{pmatrix}.$$

The flow diagrams for the working processes of the system with part sequences $P_1\ P_2\ P_3\ P_4$ and P P P P.

Figure 1 (a) and (b) demonstrates that there is a different maximum flow-time for a same part set in the system with different part sequence to be manufactured by FMS. However what relation exists between the maximum flow-time of a set of parts and the idle time of the workstations? The following theorem can be easily proved:

THEOREM 1—Suppose B is the sum of idle time on every workstation in the system during whole process for the system to manufacture all parts in PS and T is the maximum flow time of PS, then

$$T_{\text{MAX}} = \left(B + \sum_{i=1}^{M} \sum_{j=1}^{M} a_{ij} \right) \Big/ M$$

where M is the number of workstations in the system and a is an element of the workload assignment matrix of PS about the system. □

Following theorem 1, we know that the part-sequencing problem with an optimal objective of minimizing the maximum flow-time is completely equal to the part-sequencing problem with the object of reducing the idle-time of every workstation

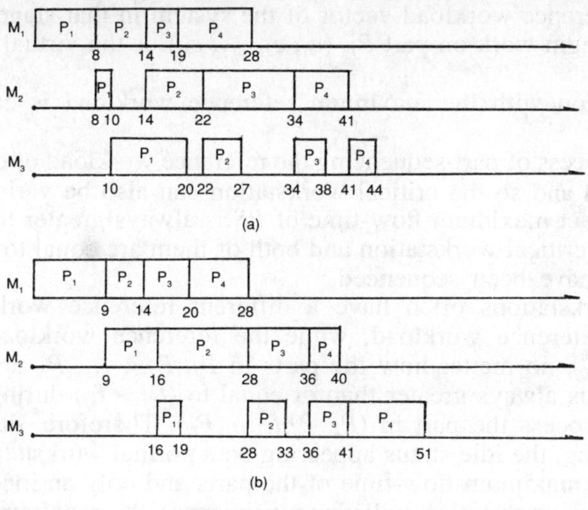

Figure 1. The working flow diagram for Example 1.

as much as possible. Therefore we will pay attention to reducing the idle-time of workstations instead of minimizing the maximum flow-time.

Workstations are idle during the working process of the system for two reasons: unbalanced workload of the workstations in the system (this idle-time cannot be reduced); and improper sequence of the parts (this can be reduced by optimally sequencing the parts to be manufactured). However, the problem arises of how to distinguish between the idle-time produced by an unbalanced workload and the idle-time associated with improper part sequence.

Definition 1—Suppose P_N is the last part in PS to be manufactured by the system MS. The following vector is called the virtual workload vector of the system:

$$E_N = \left(\sum_{j=2}^{M} a_{jN}, \sum_{j=3}^{M} a_{jN}, \ldots, a_{M-1N}, 0 \right).$$

From Figure 1 we find that when workstation M_j has finished all operations on part P_N, it has to stand idle for a time greater than $\sum_{j=j+1}^{M} a_{jN}$ before the last workstation M_M ends its work on part P_N. In other words the workstation M_j has to get an extra workload compared with M_M. This extra workload is decided only by the last part to be manufactured by FMS.

In this chapter we take the process of sequencing the parts in PS as the process of selecting parts one by one from PS.

Definition 2—Suppose that the parts in P_j, P_{j+1}, ..., P_N have not been selected out at any stage in the process of sequencing the parts. The following vector:

$$T_j = \begin{pmatrix} t_{1j} \\ t_{2j} \\ \cdots \\ t_{Mj} \end{pmatrix} = \begin{pmatrix} f_{1j} \\ f_{2j} \\ \cdots \\ f_{Mj} \end{pmatrix} + \begin{pmatrix} \sum_{k=j}^{N} a_{1k} \\ \sum_{k=j}^{n} a_{2k} \\ \cdots \\ \sum_{k=j}^{N} a_{Mk} \end{pmatrix} + \begin{pmatrix} e_1 \\ e_2 \\ \cdots \\ e_M \end{pmatrix}$$

is called the reference workload vector of the system in that stage, where f_{ij} is the time for M_i to begin work on part P_j, $(e_1, e_2, \ldots, e_M)'$ is the virtual workload vector of the system.

The workstation with the maximum reference workload is called the critical workstation.

During the process of part-sequencing the reference workload of every workstation may be changed and so the critical workstation can also be varied. At any stage, however, the exact maximum flow-time of PS is always greater than the reference workload of the critical workstation and both of them are equal to each other when all parts in PS have been sequenced.

Different workstations often have a different reference workload. When M_k has t_{kj} as its reference workload, while the reference workload of the critical workstation is t_{Dj}, no matter how the parts in P_j, P_{j+1}, …, P_N are sequenced, the idle-time of M_k is always greater than or equal to $(t_{Dj} - t_{kj})$ during the process for the system to process the part in $(P_j, P_{j+1}, \ldots, P_N)$. Therefore, during the process of part-sequencing, the idle status appearing on a normal workstation has not much influence on the maximum flow-time of the parts and only an increase of the idle-time on a critical workstation will directly increase the maximum flow-time.

From the discussion above it can be seen that an efficient algorithm of sequencing the parts should pay attention only to deducing the idle-time on critical workstations instead of on normal workstations; that is the primary base for developing our algorithm and also the difference between our algorithm and other existing algorithms.

To make it convenient to describe the algorithm, we give some matrices, vectors and special operators on the vectors.

Definition 3—$A_{M \times N}$ is the workload assignment matrix of PS for MS, the matrix B^i is deduced from $A_{M \times N}$ as follows:

$$B^i = \begin{pmatrix} b^i_{11} & b^i_{12} & \ldots & b^i_{1N} \\ b^i_{21} & b^i_{22} & \ldots & b^i_{2N} \\ \ldots & & \ldots & \\ b^i_{M1} & b^i_{M2} & \ldots & b^i_{MN} \end{pmatrix}$$

which is known as the adopted matrix of part P_i, where:

$$b^i_{kj} = \begin{cases} 0 & \text{if } k = 1 \\ a_{k \cdot i} - a_{k-1 \cdot j} & \text{if } k > 1 \,. \end{cases}$$

The vector composed of the elements in jth column in the above matrix is known as the adopted vector of part P_i about P_j, denoted as B^i_j.

Definition 4—Suppose P_i is ith part in PS to be manufactured by the system, the vector

$$V^i = (0, V^i_2, V^i_3, \ldots, V^i_M)'$$

is called the started vector of P_i, where V^i_j = the time for M_j to begin work on P_i—the time when M_{j-1} finishes work on P_i.

For example, in Figure 1(a) the started vector of P_3 is

$$V^3 = (0, 3, 0)'.$$

Definition 5—F^* is a special vector operator. If

$$E_M = (e_1, e_2, \ldots, e_M)' = F^*(d_1, d_2, \ldots, d_M)'$$

the elements in E_M are formed by the following procedure:

(1) $e_1 = d_1;\ k = 1$
(2) if $e_k < 0$, then $e_{k+1} = e_k + d_{k+1}$
else $e_{k+1} = d_{k+1}$
(3) $k = k + 1$
if $k < M$ then goto 2
else stop.

For example: $F^*\ (3, -4, 2, -1)' = (3, -4, -2, -3)'$.

Definition 6—F^+, F^- are two special vector operators. If

$$G = F^+(c_1, c_2, \ldots, c_M)' = (g_1, g_2, \ldots, g_M)'$$
$$D = F^-(c_1, c_2, \ldots, c_M)' = (d_1, d_2, \ldots, d_M)'$$

the elements in G are produced by the following rules:

$$g = \begin{cases} 0 & \text{if } c_k <= 0 \\ c_k & \text{if } c_k > 0 \end{cases} \quad k \in (1, 2, \ldots, M);$$

the elements in D are produced by the following rule:

$$d = \begin{cases} 0 & \text{if } c_k >= 0 \\ c_k & \text{if } c_k < 0 \end{cases} \quad k \in (1, 2, \ldots, M).$$

Definition 7—If P_j is processed by the system just after P_i, the vector

$$I(P_i, P_j) = (t_1, t_2, \ldots, t_M)'$$

is known as the idle-time vector of the system about P_i and P_j, where t_k is the time workstation M_k stands idle waiting for P_j's coming just after ending the work on P_i.

The important problem is how to calculate all idle-time vectors between parts during the process of part-sequencing only by operating on related vectors and matrices.

If we define a special adopted matrix and a started vector V' as follows for the beginning of the sequencing process:

$$B^0 = \begin{pmatrix} 0 & 0 & \cdots & 0 \\ -a_{11} & -a_{12} & \cdots & -a_{1N} \\ -a_{21} & -a_{22} & \cdots & -a_{2N} \\ \cdots & \cdots & \cdots & \cdots \\ -a_{M-1 \cdot 1} & -a_{M-1 \cdot 2} & \cdots & -a_{M-1 \cdot N} \end{pmatrix}$$
$$V' = (0, 0, \ldots, 0)'$$

we can calculate out every idle-time vector and started vector for every pair of parts in the part-sequence by the recursive equations below:

If P_j is processed just after P_i then

$$I(P_i, P_j) = F^-(F^*(V^j + B^i_j)) \tag{1}$$

where B is the adopted vector related P and P and V is the started vector of P.

The started vector of P which is processed just after P is :

$$V^{j+1} = F^+(F^*(V^j + B^i_j)). \tag{2}$$

The correctness of the eqns (1), (2) can be easily verified by examing Example 1.

Suppose $P_1, P_2, \ldots, P_i$ is an arranged sequence of parts and P_j is the part to be selected out. The reference workload vector of the system at that stage is

$$T_j = (t_{1j}, t_{2j}, \ldots, t_{mj})'.$$

When P_j is to be processed after P_i, the new reference workload vector of the system can be modified as:

$$T_{j+1} = T_j + I(P_i, P_j) \tag{3}$$

where $I(P_i, P_j)$ is the idle-time vector of the system about P_i, P_j.

If the maximum reference workload in T_j is d_j and the one in T_{j+1} is d_{j+1} then

$$y = \begin{cases} d_{j+1} - d_j & \text{if } d_{j+1} > d_j \\ 0 & \text{if } d_{j+1} < d_j \end{cases} \tag{4}$$

which is known as the overcost time produced by arranging for P_j to be processed after P_i.

Algorithm of sequencing the parts

In the algorithm below *DS* is the set part to be sequenced and *GS* is the series of parts having been sequenced.

ALGORITHM

Input : *PS* and *A*
Output: *GS*

Step 1—(a) Produce the adopted matrix for every part in *PS* from *A*.

$$B^0, B^1, \ldots, B^N$$

where *P* is a virtual part whose adopted matrix is:

$$B^0 = \begin{pmatrix} 0 & 0 & \ldots & 0 \\ -a_{11} & -a_{12} & \ldots & -a_{1N} \\ -a_{21} & -a_{22} & \ldots & -a_{2N} \\ \ldots & \ldots & \ldots & \ldots \\ -a_{M-1\cdot 1} & -a_{M-1\cdot 2} & \ldots & -a_{M-1\cdot N} \end{pmatrix}.$$

(b) For every part P_x in *PS*, suppose P_x is the last one to be manufactured by the system and calculate the sum of the virtual workload of workstations:

$$g(P_x) = \sum_{k=2}^{M} a_{kx} + \sum_{k=3}^{M} a_{kx} + \ldots + a_{Mx}\,.$$

(c) Select the part in *PS* with minimum *g* value (denoted as P_N) and calculate the reference workload vector of the system:

$$T = \left(\sum_{k=1}^{N} a_{1k}, \sum_{k=1}^{N} a_{2k}, \ldots, \sum_{k=1}^{N} a_{M-1k}, \sum_{k=1}^{N} a_{Mk} \right)' + \left(\sum_{k=2}^{M} a_{kN}, \sum_{k=3}^{M} a_{kN}, \ldots, \sum_{k=M}^{M} a_{kN}, \sum 0 \right)'.$$

(d) Let $DS \leftarrow PS - \{P_N\}$
$GS \leftarrow \{P_0\}$
$V^1 = (0, 0, \ldots, 0)'$
$k \leftarrow 1.$

Step 2—Suppose P_i is the $(k-1)$th part in *PS* being selected out into series *GS* by the algorithm. For every part (denoted as P_j) in *DS* do the following calculations:

(a) Suppose P_j is processed just after P_i, calculate the overcost time of the system: $y(P_j)$.

(b) For every part P_k in $Ds - \{P_j\}$ suppose that P_j has been arranged to be processed just after P_i and P_k is worked just after P_j, calculate the overcost time of the system. Denote this overcost time as $e(P_j, P_k)$ and then calculate:

$$X(P_j) = \frac{1}{D} \sum_{P_k \in DS - \{P_j\}} e(P_j, P_k)$$

where D is the number of parts in $DS - \{P_j\}$.

(c) Suppose P_j is worked after P_i, calculate the new reference workload vector of the system:

$$T^* = (t_1, t_2, \ldots, t_M)'$$

and the started vector for next part to be sequenced

$$V^* = (0, v_2, v_3, \ldots, v_M)'.$$

The start value of the workstation which is the critical one according to T^*, is denoted as $Z(P_j)$.

(d) Suppose P_j is processed just after P_i the new critical workstation is M_e calculate:

$$Z^*(P_j) = \frac{1}{D} \sum_{P_k \in DS - \{P_j\} \wedge b_{ek}^j > 0} b_{ek}^j$$

where b_{ek}^j is the value of the adopted vector related to P_k and P_j associated with M_e. $P_k \in DS - \{P_j\} \wedge b_{ek}^j > 0$ is the condition of the summary and D is the total number of the parts in $DS - \{P_j\}$.

Step 3—Select one of the parts from *DS* which satisfies the following three conditions (this part is denoted as P_k without losing generality)

$$y(P_k) - X(P_k) = \text{Min}\ \{y(P_j) - X(P_j)/P_j \in DS\} \qquad (5)$$

$$Z(P_k) = \text{Max } \{Z(P_j)/P_j \in ES\} \tag{6}$$

$$Z(P_k) = \text{Max } \{Z^*(P_j)/P_j \in FS\} \tag{7}$$

where ES is the set of parts which are selected from PS and satisfy condition (1); FS is the set of parts which are selected from ES and satisfy condition (2).

Step 4—Let $DS \leftarrow DS - \{P_k\}$
$GS \leftarrow GS + \{P_k\}$
and calculate the new reference workload vector T_{k+1} and new start vector V^{k+1}.

Step 5—If DS is not empty $k = k + 1$ go to Step 2
else: $GS\ GS + \{P_N\}$.

END Stop algorithm.

Suppose the part sequence $P_1, P_2, \ldots, P_i$ has been scheduled out by the algorithm from PS. When the algorithm is going to select the kth part from PS, there are two situations to be considered by the algorithm based on the concept of critical workstation; one is how much overcost time of the system would be produced by arranging P_j just after P_i (known as the present loss of selecting part P_j, that is the $y(P_j)$ calculated in the Step 2 of the algorithm), and the other is what influence would be paid to the following process to select parts by arranging P_j just after P_i (known as the risk loss of selecting P_j). Obviously when we select one part from PS, we should consider not only its present loss but also the risk loss. The obstacle to solving the problem is how to calculate the risk loss of the part, while the present loss of part can be exactly calculated. If we want to calculate the risk loss exactly, we have to enumerate all possible permutations of the parts in PS; this would be impossible if the number of parts in PS is very large.

In the algorithm, the X value of every part is calculated. The meaning of the X value can be explained as follows.

If P_j is not selected out at that stage, of course the present loss of P_j could be avoided at that stage. However the P_j will be selected out last. In that way, how much loss would be connected with P_j. The $X(P_j)$ can be generally thought of as that average value, i.e. the expected value of the risk loss of P_j. For part P_j, if $y(P_j) - X(P_j)$ is less, it is proper to select P_j at that stage.

$Z(P_j)$ and $Z^*(P_j)$ calculated in the algorithm are also a reflection of the risk loss of part P_j in other side. The greater $Z(P_j)$ or $Z^*(P_j)$, the less probability that the critical workstation will become idle because the workload of the critical workstation has been arranged densely at that stage.

Applying the algorithm to solve the problem in Example 1, the part sequence obtained is P, P, P, P and the maximum flow-time is 41, which can be verified to be the optimal result of the problem.

Most of the calculation in the algorithm is in Step 2. Considering the worst situation, if there are N parts to be sequenced the process between Step 2 and Step 4 should be repeated N times. For every recursion the algorithm should calculate $y(P_j)$, $Z(P_j)$, $Z^*(P_j)$ and $X(P_j)$ N times. And for every $X(P_j)$ the related $e(P_k)$ should be calculated N times. Therefore the total of calculation would not be greater than

$$N * (N + N + N + N * N) = N^3 + 3N^2,$$

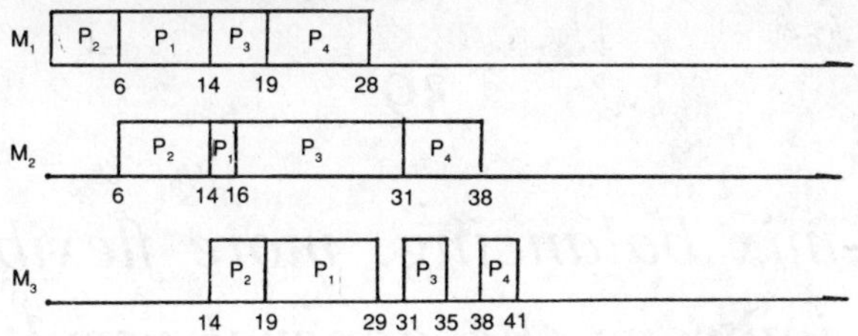

Figure 2. The working flow diagram obtained by the algorithm compared with Example 1.

which clearly shows the computing complexity of the algorithm is $0(N^3)$, which makes the algorithm have much promise for practical application.

Conclusion

How to sequence the parts to be manufactured by FMS is a very important problem. A new and very efficient algorithm for the problem is developed in this chapter. Compared with other approaches the following characterics are used in the theoretical base to construct the algorithm:

(1) In many papers the critical workstation is just the one with the heaviest workload and will not change during the whole working process of the system, while in this study the critical workstation is defined as the one with the heaviest reference workload and may change to other workstations during the process, which is obviously more consistent with practice.
(2) Based on the correct definition of the critical workstation the optimal direction of sequencing the parts becomes clear and simple, that is, how to make the critical workstation work as continuously as possible. So the algorithm gets rid of the serious trouble met in other algorithms that several optimal rules should be considered, some of which are contradictory.
(3) The concept of the risk loss and the present loss of parts is defined in the study. The algorithm should consider both the present loss and the risk loss of the part before the decision to select one part out is made. On the other side the equations to calculate exactly the present loss of parts and the equation to calculate the expected value of the risk loss are given, which makes the computing complexity of the algorithm very low.

The results of simulation experiments show that to solve the part sequencing problem with 20 parts, the maximum flow-time of the parts obtained by our algorithm is generally 15–20% less than that obtained by several heuristic scheduling rules found in Ponwalhar (1987).

References

Carrie, A.S. and Defsopoulos, A.C., 1986, Operational sequencing in FMS. *ROBOTICA*, **13**.

Ponwalhar, S.S. and Iskander, W., 1987, A survey of scheduling rules. *Operation Research*, **25**, 1.

Gravew, S.G., 1981, A review of production scheduling. *Operational Research*, **129**, 4, July–August.

Rajia Suri and Whifney, C.K., 1984, Decision supported requirement in FMS. *Journal of Manufacturing System*, **3**, 1.

Yasuda, K. and Iwata, K., 1982, Production scheduling of FMS. *Annals of the CIRP*, **31**, 1.

39

Model-mix balancing: more flexibility to improve the general result

F. Fremerey

Fraunhofer Institute for Industrial Engineering, Nobelstrasse 12c, 7000 Stuttgart 80, Germany

Abstract

In this chapter special emphasis is placed on the problem caused by standard time differences in the assembly. It is shown how the influence of balancing considerably affects the costs of model-mix and how these costs can be reduced. Influences causing these problems depend on the variant product itself as well as on the production program. For balancing, therefore, differences between the operation time of variants should be minimized in each station.

Another strong influence on balancing is caused by the structure of the assembly system. Nowadays there are less assembly lines or assembly lines with parallel stations for equal tasks. There are more and more flexible assembly systems. In these systems structural switch elements can bring different variants into different branches of the flexible assembly system. In an assembly system with automated guided vehicles (AGVS) these branches could be assembly modules (boxes).

In co-operation with four industrial companies and a software house we are designing a computer-aided assembly planning tool to meet these features. The main output result is determined and documented in an assembly sequence plan showing all activities which are relevant to produce the described product with its specific characters.

Introduction

Today, in practical operation, assembly planning is not always done with computer-aided tools. Reasons for the low use of computer-aided methods may be on one hand the complicated operation and the inferior adaptability of the existing systems. On the other hand the methods themselves may not present an acceptable decision support for the assembly planner.

Within a systematic planning methodology balancing is the most important step in generating assembly system specific sequence plans. The aim of balancing is basically to assign all assembly operations to elements (workstations, workers) of an assembly system. Hitherto this assignment has almost exclusively been considered under the capacity-oriented aspect of achieving maximum loading or a minimum number of workstations (stations, hereafter). Due to the enormous number of variants and the layout of flexible automated assembly system structures, the requirements of balancing have changed tremendously. In the near future it will therefore no longer be sufficient to maximize only capacity load but it will be essential to improve efficiency by increasing flexibility.

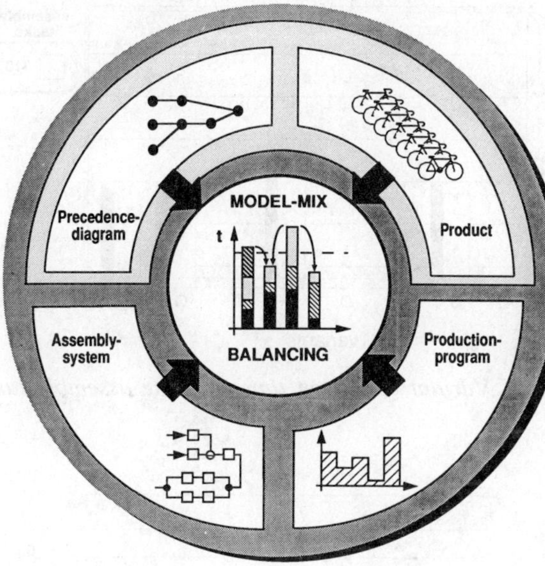

Figure 1. Influences of model-mix balancing.

The main influences of model-mix balancing which must be taken into consideration by a decision support system (DSS) are shown in Figure 1.

Some authors concentrate in detail on very special problems. They make several restrictions and give excellent solutions by optimizing a single part of the entire problem. (Görke and Lentes, 1976; Baybars, 1986; Auwärter *et al.*, 1987; Dar-El and Rabinovitch, 1988). Due to the influence of numerous parameters in practical operation, some authors try to reflect the knowledge of assembly planners in expert systems (Chu and Wang, 1987; Chu *et al.*, 1988; Köhler *et al.*, 1988; Wang and Wysk, 1988). In this ongoing research project a verified heuristic (Schad, 1986) is extended by additional components to be more suitable for the changed situation. Therefore it is necessary that the DSS is on one hand a system which 'only' offers suggestions giving the planner the choice to agree or not. On the other hand the system must offer substantial information about his problem, so that he has the best conditions to make his decision.

The computer-aided tool should present among other things a special support for assigning model-mixed assembly systems. Therefore it is essential to know how flexible stations could react on changes in the product program (program, hereafter) after the assignment. In this chapter special emphasis is placed on the problem caused by standard time differences in the assembly. It is shown how the influence of balancing considerably effects the cost of model-mix and how these can be reduced. Influences causing these problems depend on the variant product itself as well as on the production program.

Usually the operation time of an assembly operation or task (task, hereafter) is represented by a single time. In a real model-mix assembly system one task may have different operation times, e.g. the operation time for the task 'fit the engine into car' depends on the variant in which the engine is mounted. To blend a six-cylinder engine takes a longer time than for a four-cylinder engine. Or, for example, the task 'fit air conditioning' is only necessary in variants which have this extra. On the other hand the models with these options do not need the normal heating. A task may have as many different operation times as there are variants. An example of this fact is shown in Figure 2.

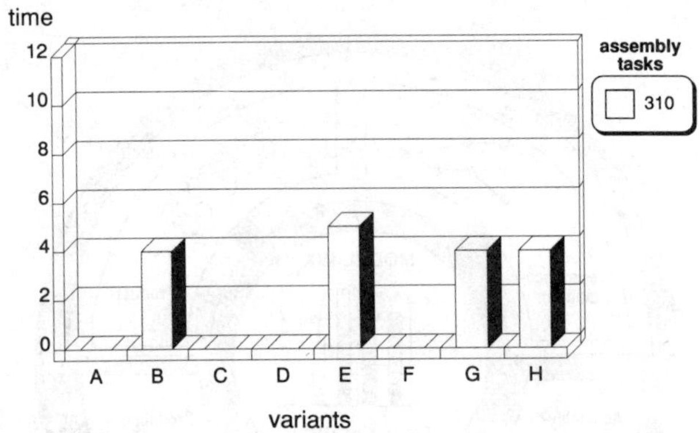

Figure 2. Variant operation times for one assembly task.

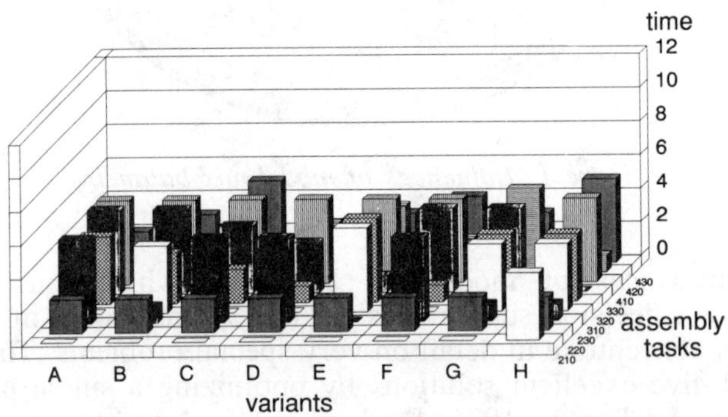

Figure 3. Variant operation times for every assembly task.

There is a separate operation time for every task and every variant (model) (see Figure 3). Due to the assignment from tasks to stations these time differences are transmitted to the stations. They could increase if an unfavourable assignment of several tasks is chosen such as in Figure 4.

Most of the model-mix problems which scheduling has to face are caused by these operation time differences between the variants. In this context the production program starts to be of vital importance. It depends on the quantity of a certain variant you want to produce. This influences the combination of variants sequenced in the assembly system, e.g. there are two variants (A and B). Let us assume that we want to produce 99% of variant A and only 1% of B. In this case the variant A determines the operation time in the station. The operation time of variant B could be ignored, or rather has no influence on the operation time in the station. If we assume, however, that we want to produce 50% of each, then both variants determine the operation time of the station equally.

For balancing, therefore, differences between the operation time of variants should be minimized in each station.

Notation

v = total number of variants,
P_i = production rate of variant i,
X = total number of stations,

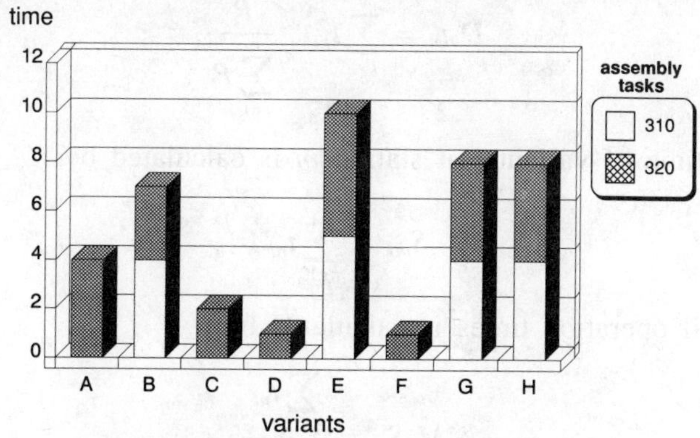

Figure 4. Strong varying operation times in a workstation.

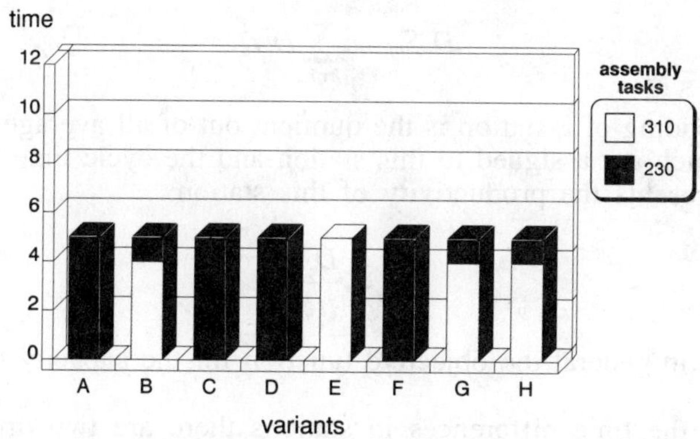

Figure 5. Non-varying operation times in a workstation.

tz_m = cycle time in station m,
Q_m = set of tasks which had been assigned to station m,
z_m = number of tasks which had been assigned to station m,
t_{ni} = operation time task n for variant i,
D_t_n = average operation time of task n,
S_{mi} = operation time of variant i in station m,
M_S_m = mean of all operation times in workstation m,
D_S_m = average operation time of all variants i in station m,
A_m = capacity loading in station m,
Q_u_m = *unweighted* coefficient of variation on time differences in station m,
Q_g_m = *weighted* (with the production program) coefficient of variation in station m.

The model

The capacity load is calculated according to a mathematical average model, which does not really exist. The average operation time of a task n represents the operation time for this model. This time is calculated by the operation time of the task multiplied by the production rate of all variants for which the task must be carried out.

$$D\_t_n = \sum_{i=1}^{v} t_{ni} * \frac{P_i}{\sum_{i=1}^{v} P_i}. \tag{1}$$

The operation time of variant i at station m is calculated by:

$$S_{mi} = \sum_{n \varepsilon Y_m} t_{ni}. \tag{2}$$

The mean of all operation times is calculated by:

$$M\_S_m = \frac{\sum_{n \varepsilon Y_m} t_{ni}}{Z_m}. \tag{3}$$

The average operation time at station m is calculated by:

$$D\_S_m = \sum_{n \varepsilon Y_m} D\_t_n. \tag{4}$$

The capacity loading of a station is the quotient out of all average operation times of the tasks which are assigned to this station and the cycle time. In this case the cycle time represents the productivity of this station.

$$A_m = \frac{D\_S_m}{tz_m}. \tag{5}$$

Equation (5) is in general the objective function for the capacity load which must be maximized.

To decrease the time differences in stations there are two different objective functions. You could either minimize the absolute or the weighted time differences. Equation (6) takes variants into account without their production rate. Equation (7) represents an estimated value for flexibility of a station concerning the actual program. The difference between these equations lies in the fact that in the second statement the operation times of the variants are weighted by their production rate.

With eqns (2), (3) and (4) the equations for the coefficient of variation of time differences are:

$$Q\_u_m = \frac{\sqrt{\left(\sum_{i=1}^{v} (S_{mi} - M\_S_m)^2\right) \Big/ V}}{(D\_S_m)} \tag{6}$$

$$Q\_g_m = \frac{\sqrt{\left(\sum \left(\left(S_{mi} * \frac{P_i}{\sum_{i=1}^{v} P_i}\right) - D\_S_m\right)^2\right) \Big/ V}}{(D\_S_m)} \tag{7}$$

For the assignment to decrease time differences another step is necessary to reflect the real behaviour in stations. The worst case in a station would be if 50% of the variants got an operation time t and the other 50% have none. In this case the coefficients have got the result 1. Every change of operation time, variants or production rate would increase the behaviour in this station. To take this into account there are two ranges:

I) $Q\_u_m < = 1 = Q\_u_m = Q\_u_m$
$Q\_g_m < = 1 = Q\_g_m = Q\_g_m$

II) $Q\_u_m > = Q\_u_m = 1/Q\_u_m$
$Q\_g_m > 1 = Q\_g_m = 1/Q\_g_m.$

The objective functions to minimize the time differences are:

$$Z1 := \text{Min} \sum_{m=1}^{X} Q\_u_m \tag{8}$$

$$Z2 := \text{Min} \sum_{m=1}^{X} Q\_g_m \, . \tag{9}$$

If the first objective function is chosen you will get the optimal assignment with the best balanced operation time differences between variants. Such a balanced assembly system can cope with changes within the program, without having big time slacks or overload between variants. The second objective function only makes sense if the program is fixed or if there is always a new assignment for every change in the program.

Additionally, time differences may increase or decrease according to the program. Therefore it makes sense to compare the weighted coefficient with the unweighted. Thus getting an estimation about the problems which may arise in production while sequencing the variants of the existing program.

In order to get a practicable interactive support system with a satisfactory runtime the optimization should not be done at all stations of the complete assembly system, but only on a selected part. For every station the precedence diagram gradually determines the task which can be assigned. At the moment we develop a modified branch-and-bound heuristic which will minimize one of the objective functions (8) or (9).

Integration into a comprehensive system for balancing assembly systems

As mentioned earlier there are many different influences which must be faced in practical operation. In order to integrate this method into a computer-aided planning tool every boundary condition should be taken into consideration.

The presented equations which can level the time differences in stations only represent a possible solution for the model-mix problem in balancing. This optimization is integrated in a heuristic for an interactive balancing which is based on a methodology from Bullinger *et al.* (1986). This DSS had been developed at Fraunhofer Institute for Industrial Engineering and is still in use in several industrial companies.

Furthermore other structural conditions (e.g. simultaneous tasks or obligatory sequence of tasks) which could not be represented by the precedence diagram can be fulfilled in the new DSS. The planner can define several technical or ergonomic restrictions (e.g. try to assign all tasks with the same technology or assembly location in one station or zone) which are taken into consideration during the balancing.

Due to methodological and data-based joining of the precedence diagram and the list of parts (Seidel, 1987) it is possible to support the material staging and to calculate the surface density (material loading). Another strong influence on balancing is caused by the structure of the assembly system. Nowadays there are less assembly lines or assembly lines with parallel stations for equal tasks. There are more and more flexible assembly systems. In these systems structural switch elements can bring different variants into different branches. In an assembly system with automated guided vehicles (AGVS) these branches could be assembly modules (boxes). Taking all these aspects into consideration a special validity language for tasks, product variants and structural switch elements was built in the system.

Every task has got a boolean string in the data base, which tells if the task must be carried out for an option. This boolean string deduces from the where-used list of the part list and is transmitted to the task which mounts this part. In the program every variant is represented by a listing of all its options. By comparing the boolean string of the task with the program, the system is able to determine if a task must be carried out for a variant.

The heuristic accomplishes that the tasks will be assigned in the branches in which they are absolutely necessary, e.g. there are two branches (the main line divides into two lines). Let us assume that a structural switch element brings in the first branch only USA-cars (all cars with the option USA-model). In the second branch it will bring in all other variants which have not got this option. Tasks, exclusively carried out for USA-models are only necessary in the first branch, but tasks which must be carried out for all models (USA and others) must be assigned in every branch.

Outlook

In co-operation with four industrial companies and a software house we are designing a computer-aided assembly planning tool to meet these features. In addition to the presented balancing this system will include other functions for assembly planning like 'transformation of construction BOM into assembly orientated product structure', 'generation of the precedence diagram', 'planning of work content (tasks) and technology' and after the balancing 'simulation module for sequencing' (Seidel, 1987; Fremerey *et al.*, 1988; H.-J. Bullinger and U.A. Seidel, this volume; K. Thaler, this volume).

The main output result is determined and documented in an assembly sequence plan showing all activities which are relevant to produce the described product with its specific characters.

References

Auwärter, A., Gröchtemeier, M. and Schad, G., 1987, Einführung eines Programmsystems zur dialogorientierten Leistungsabstimmung von Montagelinien in der Praxis. wt Werkstattstechnik **77**, 1, 215–218.

Baybars, I., 1986, A survey of exact algorithms for the simple assembly line balancing problem. *Management Science*, **32**, 909–932.

Bullinger, H.-J., Schad, G. and Lentes, H.-P., 1986, Intakt—interactive assembly line balancing. *Towards the Factory of the Future*.

Chu, C.H. and Wang, H.P., 1987, The use of artificial intelligence in process planning. IJOPM, **8**, 1, 5–18.

Chu, G.H., Wang H.P. and Chu, Y.C., 1988, On the design of an intelligent micro-based DSS for line balancing. *Computer and Industrial Engineering*, **14**, 4, 419–427.

Dar-El, E.M. and Rabinovitch, M., 1988, Optimal planning and scheduling of assembly systems. *International Journal of Production Research*, **26**, 9, 1433–1450.

Fremerey, F., Korge, A., Seidel, U.A. and Thaler, K., 1988, Methodenintegration—Voraussetzung einer CIM-gerechten Montageplanung, FhG-Bericht 2–88 1988, S.54–59.

Görke, M. and Lentes, H.-P., 1976, Leistungsabstimmung von Montagelinien. AV **13**, 3–5.

Köhler, F. and König, D., 1988, wt Werkstatttechnik 78 (1988), 'PIA, ein wissensgestütztes Programmsystem zur automatischen Erstellung von Montagearbeitsplänen', pp. 251–255.

Seidel, U.A., 1987, Systematic design of an integrated assembly planning toolbox. *Proceedings of the IX, International Conference on Production Research*, Cincinnati, 555–561.

Schad, G., 1986, *Entwicklung und Einsatz eines interaktiven Verfahrens zur Leistungsabtsimmung von Montagesystemen* (Berlin: Springer).

Wang, H.P. and Wysk, R.A., 1988, A knowledge-based approach for automated process planning. *International Journal of Production Research*, **26**, 6, 999–1014.

PART IV
PRODUCTION ECONOMICS AND PRODUCTIVITY

Production economics and productivity

Productivity improvement programmes are an important means to making organizations more efficient and competitive. In a survey by Stainer, practices in the United Kingdom are compared with the USA, and an evaluation is made of the perceptions and standing of managers involved in productivity improvement programmes.

In two studies of flexible manufacturing systems (FMS), Abdel-Malek shows that metal cutting economics is a dominating factor in minimizing cost per part, while Albino and Okogbaa evaluate the effects of reliability and maintainability of resources on workstation performance.

In today's dynamic manufacturing environment new approaches are constantly being sought to production accounting and investment appraisal. Persson *et al.* describe a method for evaluating capital investments which implies that every investment is divided into both a flexible and an inflexible part. Hayakawa reports on new developments in cost management within the Toyota system, using the material requirement planning (MRP) database. Arcelus and Rowcroft attempt to fill a gap in standard inventory theory by considering the effect of trade discounts and disposals on the inventory policies of profit maximizing firms.

SECTION IV.1. PRODUCTIVITY

40

Productivity practices in the United Kingdom

Alan Ivor Stainer

Head of Engineering Management, Middlesex Business School, Middlesex Polytechnic, The Burroughs, Hendon, London NW4 4BT, UK

Abstract

This chapter provides a critique of the present state of the art and shows the results of a survey into productivity measures, indicators and their management in various UK organizations. The survey is divided into industrial (manufacturing and extractive) and non-industrial (service) industries. Comparison is made with practices in the United States. Productivity programmes are analysed and discussed at the level of the company, plant/division and department. Also investigated are the characteristics and standing of managers involved in productivity programmes at both operational and strategic levels.

On the basis of the survey material, evaluation is made of managers' perception of the potential of productivity improvement programmes as a means of making their organization more efficient and competitive.

Productivity in perspective

Productivity is something that has been spoken about in vague terms for decades by politicians, economists and the media, usually trying to justify a point of view. In management discussions, productivity is too commonly used in a narrow sense as synonymous with labour productivity. More properly, productivity should be regarded as the relationship between output and all its inputs. The majority of these input components are under the control of the organization's management, either directly or indirectly.

In the UK particularly, much lip service has been given to the raising of productivity to meet foreign competition, but little analysis has been carried out into productivity measures and indicators currently being used by organizations. To counteract this phenomenon and to place the British experience into context, a survey was carried out in 1987–8 embracing 500 of the largest organizations in the UK.

The survey

Productivity measures were categorized into four groups:

(1) The partial measures, the ratio of output to one type of input, such as output per man-year or return on capital employed.
(2) Total factor productivity which is based on net output, that is the total output less bought-in goods, services and materials, referred to generally in the UK as added-value.

Table 1. Productivity measures and indicators—Formal measures of productivity in programmes.

	Company		Plant/division		Department	
	Industrial (%)	Non-industrial (%)	Industrial (%)	Non-industrial (%)	Industrial (%)	Non-industrial (%)
USA	[71·1]	[62·5]	[86·7]	[41·7]	[44·4]	[50·0]
UK	64·2	54·7	72·4	38·2	35·8	35·2

Table 2. Productivity measures (%).

	Organization	Plant/division	Department
Industrial			
Partial measures			
Related to labour input	38·6	42·4	50·5
Related to capital	18·7	13·6	5·2
	57·3 [51·6]	56·0 [60·0]	55·7 [50·0]
Added-value (total factor)	9·2 [11·7]	9·2 [11·7]	3·5 [3·3]
Total productivity measures	7·5 [13·3]	10·0 [20·0]	6·8 [10·0]
Other measures	26·0 [23·4]	24·8 [8·3]	34·0 [36·7]
Non-industrial (service)			
Partial measures			
Related to labour input	31·2	40·3	52·5
Related to capital	16·6	12·5	4·0
	47·8 [41·4]	52·8 [48·3]	56·5 [58·6]
Added-value (total factor)	10·5 [13·8]	8·2 [10·3]	10·5 [13·8]
Total productivity measures	10·0 [31·0]	10·0 [20·7]	4·2 [6·9]
Other measures	31·7 [13·8]	29·0 [20·7]	28·8 [20·7]

(3) Total productivity which is the ratio of total output to the sum of all input factors.
(4) All other measures, ranging from sales per employee to management by objectives achieved.

The survey was divided into industrial (manufacturing and extractive) and non-industrial (service) industries. The results are outlined in Tables 1 and 2. For comparative purposes with the United States, these results are shown against those published by Sumanth and Einspruch (1980) (bracketed). Although the English language is spoken on either side of the Atlantic, the terminology is marginally different.

It can be clearly seen that, seven years on, the UK still lags in formal measures of productivity in programmes. However, at the company (corporation) level for both industrial and non-industrial, and plant/division for industrial only, productivity programmes with formal measures were in the majority.

Productivity measures in themselves were further analysed at the level of the organization, plant/division and department, and compared with the American survey. Partial measures were in line with the USA experience and were related to both labour and capital inputs. The same is true with productivity measures based on added-value (total factor).

Table 3. Applying total productivity to various industries/services.

Industry/service	Measure of output
Hospital	Patients treated
University	High calibre graduates
Departmental store	Inventory adjusted sales
Underground coal-mine	Gigajoules of saleable coal
Airline	Tonne-kilometres
Farming	Tonnes of saleable crop
Refuse collection	Tonnes of waste
Catering	Meals served

Total productivity

In the UK, total productivity measures were employed far less than in the USA, being about half for industrial companies as well as non-industrial companies at the level of the plant/division and department. The most interesting difference is that, at the organization level, only 10% of non-industrial companies used total productivity measures compared with 31% in the USA. The category of other measures was far higher in the UK especially at the plant/division level of industrial organizations and the organizational level in non-industrial organizations.

The most widely used total productivity measure was real unit cost or its inverse, that is taking into account inflation. In some instances the capital input was based on replacement costs or using a leasing charge as an opportunity cost. The only practical way of aggregating inputs was, in economic terms, expressed in currency. For international comparisons, the economic ills of exchange rates and inflation are isolated, expressing prices and exchange rates at a base-year value. It is important to choose a relatively stable base-year as the anchor to avoid distortion of information.

The survey results show that there is a foundation of total productivity measurement in the UK within organizations. The seeds of total productivity measurement have been sown and the advantages acknowledged. Hence, a more accurate economic picture of the organization has evolved in real terms.

The measurement of outputs, and hence the basis of real unit costs, are relatively straightforward in certain industries, especially those with tangible output, such as tonnes of waste or crop. Where the output is not of a tangible nature, such as in hospitals and universities, the output has to be more carefully examined. In programmes in the UK, the output of hospitals has been taken as the number of patients treated while, in universities, it is the high calibre graduates (those above a certain level of attainment). The question of quality must never be overlooked and the measure of output must be integrated with the quality level required.

The total productivity approach has been successfully utilized in underground coal mining for nearly eight years, and, more recently, has been introduced into other industries, such as illustrated in Table 3. It has been found that, in all cases, labour productivity peaked at different times to the major objective of the total productivity measure. Strategic decisions are now being made and tested, utilizing sensitivity analysis based on target ratios and their effect on the total productivity objective—the inverse of real unit costs. Total productivity measures, along with other senior management's goal priorities, have been utilized as corporate organizational targets.

Managers' perception of productivity

One of the major problems in the UK in promoting productivity has been the lack of drive in national programmes and a lack of understanding by managers. Unlike many other countries, ranging from the USA to Israel, there are no productivity centres in the UK and the perceptions of productivity echo relatively low-level management services' tasks.

In very few instances are productivity responsibilities taken on by top management. When senior managers were asked their expectation of productivity programmes, 45% mentioned improved profitability and cost effectiveness, 21% stated the more effective running of the organization, 15% said technological change, with the remainder giving a variety of answers. When asked if productivity was one of the organization's strategic objectives, only 18% replied positively; however, 35% applied it only in the short-term operational sense.

Conclusion

The survey highlights the fact that, although the awareness and understanding of productivity in the UK still has to be inculcated into management's mind, there are presently more formal programmes and managerial appreciation than has previously been considered. It also reveals that productivity is generally taken into account in the lower hierarchy of organizations while, in the USA, many leading enterprises have vice-presidents responsible for this function. Therefore, it would appear that the UK needs to adopt a high profile strategy to enhance the cause of productivity and add a competitive edge.

As illustrated, the total productivity concept can be applied to a variety of industries, from manufacturing and extractive to public and private services. It disciplines management clearly to identify organizational objectives in productivity terms. The role of total productivity management, therefore, would be much enhanced by effective training in the total productivity approach. Hence, it is encouraging to notice that productivity management courses at undergraduate, postgraduate and post-experience levels are appearing in engineering and management programmes.

Bibliography

Adam, E.E., Hershauer, J.C. and Ruch, W.A., 1981, *Productivity and Quality: Measurement as a Basis for Improvement* (Englewood Cliffs, NJ: Prentice-Hall).

Bailey, D. and Hubert, T., 1980, *Productivity Measurement* (Aldershot, England: Gower).

Bain, D., 1982, *The Productivity Prescription* (New York: McGraw-Hill).

Cave, M., Hanney, S., Kogan, M. and Trevett, G., 1988, *The Use of Performance Indicators in Higher Education* (London: Jessica Kingsley).

Chew, W.B., 1988, No-nonsense guide to measuring productivity. *Harvard Business Review*, January–February 1988.

Craig, C.E. and Clark Harris, R., 1973, Total productivity at the firm level. *Sloan Management Review*, Spring 1973.

Delmar, D., 1985, *Operations and Industrial Management—Designing and Managing for Productivity* (New York: McGraw-Hill).

Edosomwan, J.A., 1987, *Integrating Productivity and Quality Management* (New York: Marcel Dekker).

Eilon, S., 1982, Use and misuse of productivity ratios. *Omega*, **10**, 6.

Eilon, S. and Teague, J., 1973, On measures of productivity. *Omega*, **1**, 5.

Eilon, S., Gold, B. and Soesan, J., 1976, *Applied Productivity Analysis for Industry* (Oxford, England: Pergamon Press).

Faraday, J.E., 1971, *The Management of Productivity* (London: Management Publications).
Husband, T. and Ghobadian, A., 1980, Total productivity and its link with pay. *Management Decision*, **18**, 6.
Lawlor, A., 1985, *Productivity Improvement Manual* (Aldershot, England: Gower).
Mali, P., 1978, *Improving Total Productivity* (New York: Wiley).
Miller, D.M., 1984, Profitability = productivity + price recovery. *Harvard Business Review*, May–June 1984.
Potts, G.W., 1988, Raising productivity in customer services. *Long Range Planning*, **21/2**, 108.
Scott Sink, D., 1985, *Productivity Management: Planning, Measurement and Evaluation, Control and Improvement* (New York: Wiley).
Stainer, A.I., 1979, Productivity in longwall mining of hard coal. PhD Thesis Loughborough University in Technology.
Sumanth, D.J., 1984, *Productivity Engineering in Management* (New York: McGraw-Hill).
Sumanth, D.J. (ed.), 1987, *Productivity Management Frontiers—I* (Amsterdam: Elsevier).
Sumanth, D.J., Edosomwan, J.A., Scott Sink, D. and Werther, W.B. (eds), 1988, *Productivity Management Frontiers—II* (Geneva: Inderscience Enterprises).
Sumanth, D.J. and Einspruch, N.G., 1980, Productivity awareness in the US: a survey of some major corporations. *Industrial Engineering*, October 1980.

41

Performance evaluation of unreliable workstations in flexible manufacturing systems

Vito Albino and O. Geoffrey Okogbaa†

University of Basilicata, Potenza, Italy, and University of Cincinnati, Cincinnati, USA.

Abstract

In order to enhance flexible manufacturing system performance, considerable attention should be paid to workstation performance evaluation and reliability apportionment. In this chapter workstation performance is computed and the effects of reliability and maintainability of resources are evaluated. A workstation is considered as a multi-server queuing system with limited buffer capacity. An approximation model is provided to reduce computation requirements. Some case examples are shown and the validation of the approximation model is discussed.

Introduction

Due to the extensive use of automation, flexible manufacturing systems (FMSs) usually require high capital investments. Consequently, they are only considered to operate economically if a high level of system performance is obtained. FMS consists of a set of workstations (resources) linked by a material handling system, all under computer control. The performance measures of FMS are strongly related to the characteristics of its main components. In particular, workstations are responsible for the major manufacturing activities and value-adding. Therefore, considerable attention should be paid in modelling and evaluating workstation as well as overall system performance.

Most analytical models describe FMS as a queuing network in which the customers are the jobs to be processed and the servers, the workstations. In general, each workstation of an FMS can be modelled as a multi-server queuing system with finite buffer capacity and reliable servers (Buzacott and Yao, 1986). The assumption that the machines are reliable can greatly affect system performance evaluation. The effects of machine failures on performance should be carefully analysed to ensure correct system design and management (Draper Laboratory, 1984). Co and Wysk (1986) consider the effect of a server failure as included in the service time randomness if this random variable is exponentially distributed. Vinod and Solberg (1985) address the optimal design evaluation of both reliable and unreliable flexible

† Present address: Department of Industrial and Management Systems Engineering, College of Engineering, University of South Florida, Tampa, FL 33620, USA.

manufacturing systems. Using a closed network of multi-server queues with servers prone to failure, they evaluated the FMS throughput by exponentially approximating the service completion time for jobs at each resource in the network.

In this chapter the typical workstation of an FMS is modelled as a multi-server queuing system with finite buffer capacity and two types of customers, jobs and failures, with different priority and pre-emptive discipline. An approximation model is proposed and some case examples are provided.

The model

The typical workstation in an FMS is modelled as a multi-server queuing system with two classes of customers namely, jobs and failures. Job arrival to the system follows a Poisson process of rate λ. The service time of each job is exponentially distributed with parameter μ. As jobs arrive they wait for service in a finite queue, i.e. a limited buffer capacity queue with first-in first-out discipline. Server failures can only occur if the server is engaged. The interruption of service caused by a server failure is modelled as the service of a phantom customer. Failures occur according to a Poisson process with rates dependent on the number of working servers. Also, failures have pre-emptive priority over jobs. The phantom customer source is bounded so as to equal the number of servers in the system, since a queue of failures is not allowable. As a result, the arrival rate of failures is $i\ \lambda g$ if i servers are working. Upon server failure, the job in process is pre-empted until the repair is completed, after which the job continues processing. The repair time of each server is exponentially distributed with parameter μg. The number of maintenance facilities is equal to s and on-line maintenance is always carried out except, of course, when all servers are failed.

The state of the system is described by the pair $(X(t), Y(t))$ where $X(t)$ and $Y(t)$ are the number of phantom customers and the number of jobs, respectively, in the system at time t. By letting s be the number of servers and k the maximum (finite) number of jobs allowed in the system (in the buffer and in service), then the system state can assume all values in the pair (i, j) for $0 \leq i \leq s$ and $i \leq j \leq k$. Only one transition (failure, repair, arrival or service) is assumed to occur at each time. The transition rates between the system states are estimated from the failure, repair, arrival and service rates. As the stochastic process which describes the system behaviour can be modelled as a discrete-state continuous-time Markov chain, the steady-state probabilities, Pij, are obtained by solving the resulting linear equation system.

The *THS* throughput of the system, i.e. the expected number of jobs leaving the system in a time unit, is equal to:

$$THS = \lambda \left[1 - \sum_{i=0}^{s} Pik \right]. \tag{1}$$

The expected number L of jobs in the system is given by:

$$L = \sum_{i=0}^{s} \sum_{j=i}^{k} j\ Pij. \tag{2}$$

Since there is no closed form solution for obtaining the values of Pij, an approximation method which uses the decomposition technique is employed. This method is based on the observation that the arrival and service rates of jobs in a workstation are generally considerably higher than the failure and repair rates. The technique consists of separating the total system into a two-level hierarchy of 'aggregated systems'. At the lower level, the subsystem is considered to have

$s - i$ reliable active servers and $k - i$ jobs are allowed in the subsystem for each number i of failed servers. Therefore, these subsystems are $M/M/s - i/k - i$ queuing systems. In the total system, i jobs are assumed waiting for service on i failed servers. For $i = s$, the subsystem $M/M/0/k - s$ is atypical and its stochastic behaviour is related to the $M/M/1/k - s + 1$ subsystem. At the higher level, the system is considered lumped with s servers subject to failure, where failures are phantom customers with finite source size equal to s. So, this system is an $M/M/s/s$ queuing system with finite source size, serving only phantom customers. Each possible state of this lumped system is characterized by the number i of busy (failed) servers and, as such, it can be viewed as a 'macrostate' of the total system, representing the aggregation of all states having the same number i of failed servers.

Using the above approach, the *THS* throughput is calculated by averaging the *THi* throughput of the $M/M/s - i/k - i$ subsystems. In the average computation, the probability of each subsystem is equal to the probability of the related macrostate in the lumped system. Let Pi be the probability of the macrostate with i failed servers and $Pj(i)$ the probability of j jobs in the $M/M/s - i/k - i$ subsystem, then:

$$THS = \sum_{i=0}^{s} THi\ Pi = \lambda\left[1 - \sum_{i=0}^{s} Pk - i(i)\ Pi\right]. \tag{3}$$

The expected number L of jobs in the system is given by:

$$L = \sum_{i=0}^{s} \overline{L}i\ Pi = \sum_{i=0}^{s}\sum_{j=0}^{k-i} (i + j)\ Pj(i)\ Pi. \tag{4}$$

$\overline{L}i$ is the expected number of jobs in the $M/M/s - i/k - i$ subsystem, corrected by the fact that i jobs must always be added to the number of jobs considered in the subsystem. The only approximation here is in the evaluation of the probability Pij with the product of the probabilities $Pj(i)$ and Pi. It is worthwhile observing that Pij values are obtained solving a linear equation system whereas $Pj(i)$ and Pi are given by closed formulas well known in queuing theory (Gross and Harris, 1985). For $i = s$, $Pj(s)$ cannot be computed because no servers are available for the related subsystem that, of course, only admits a trivial steady-state solution. However, $Pj(s)$ can be evaluated using $Pj(s - 1)$, $Ps - 1$ and Ps values and stochastic balance equations.

Case examples and approximation model validation

Some case examples have been developed to show the effect of failures on workstation performance. For each case, performances of the reliable and unreliable workstation are computed and compared. Note that the rates λ and μ are measured in [jobs/time unit], the rates λg and μg are measured in [failures/time unit] and [repairs/time unit], respectively. With perfectly reliable workstations, the performance values for different buffer sizes ($b = 0, 1, 2, 3, 4, 5, 10$) and numbers of servers ($s = 1, 2, 3, 4$), and for a fixed system load ($\lambda = 0{\cdot}08$ s, $\mu = 0{\cdot}1$) are shown in Table 1.

The effect of failures and repairs has been considered by evaluating the relative (percentage) variation (RV) of the workstation performance measures from the corresponding measures of the perfectly reliable workstation. Considering the same workstations ($s = 1, 2, 3, 4$ and $b = 0, 1, 2, 3, 4, 5, 10$) and the same system load ($\lambda/s = 0{\cdot}08$, $\mu = 0{\cdot}1$), the relative variation, RV, of *THS* and L is shown in Table 1 for $\lambda g = 0{\cdot}0005$ and $\mu g = 0{\cdot}01$. Note that the relative variation of *THS* depends on the buffer size and number of servers. For small buffer sizes and/or server numbers breakdowns increase the probability the system is full and arriving jobs

Table 1. Performances of reliable workstations and corresponding RV values (λ/s = 0·08, μ = 0·1, λg = 0·0005, μg = 0·01).

	b	0	1	2	3	4	5	10
s = 1	*THS*†	4·444	5·902	6·612	7·025	7·289	7·469	7·852
	RV	−2·160	−2·779	−2·995	−3·075	−3·073	−3·026	−2·483
	L	0·444	0·852	1·225	1·563	1·868	2·142	3·114
	RV	2·723	3·836	4·702	5·527	6·418	7·394	13·948
s = 2	*THS*†	10·722	12·659	13·710	14·356	14·785	15·084	15·740
	RV	−2·248	−2·259	−2·327	−2·354	−2·361	−2·334	−1·938
	L	1·072	1·475	1·836	2·161	2·453	2·713	3·628
	RV	2·639	3·465	4·428	5·483	6·605	7·788	14·580
s = 3	*THS*†	17·558	19·758	21·027	21·836	22·385	22·774	23·647
	RV	−2·250	−2·131	−2·083	−2·056	−2·019	−1·972	−1·594
	L	1·756	2·152	2·505	2·820	3·102	3·352	4·223
	RV	2·637	3·354	4·191	5·141	6·171	7·276	13·621
s = 4	*THS*†	24·699	27·061	28·483	29·414	30·057	30·517	31·568
	RV	−2·239	−2·055	−1·963	−1·894	−1·830	−1·760	−1·375
	L	2·470	2·860	3·205	3·512	3·785	4·026	4·861
	RV	2·648	3·297	4·043	4·883	5·805	6·798	12·486

†*THS* values are to be multiplied by 10^{-2}.

Table 2. Relative error (RE) and absolute error (AE) of THS *and* L *(λ/s = 0·08, μ = 0·1, λg = 0·0005, μg = 0·01).*

	b		0	1	2	3	4	5	10
s = 1	*THS*	RE	2·208	0·924	0·670	0·646	0·665	0·704	0·836
		AE	0·096	0·053	0·043	0·044	0·047	0·051	0·064
	L	RE	−2·651	−2·135	−2·082	−2·292	−2·686	−3·234	−7·709
		AE	−0·012	−0·020	−0·027	−0·038	−0·053	−0·074	−0·274
s = 2	*THS*	RE	−1·755	−1·220	−1·045	−1·006	−1·025	−1·073	−1·607
		AE	−0·184	−0·151	−0·140	−0·141	−0·148	−0·158	−0·248
	L	RE	1·935	1·704	1·711	1·803	1·916	2·041	2·437
		AE	0·021	0·026	0·033	0·041	0·050	0·060	0·101
s = 3	*THS*	RE	−1·544	−1·065	−0·879	−0·799	−0·775	−0·779	−1·057
		AE	−0·265	−0·206	−0·181	−0·171	−0·170	−0·174	−0·246
	L	RE	1·681	1·533	1·544	1·622	1·749	1·908	3·176
		AE	0·030	0·034	0·040	0·048	0·058	0·069	0·152
s = 4	*THS*	RE	−1·346	−0·947	−0·773	−0·686	−0·640	−0·627	−0·739
		AE	−0·325	−0·251	−0·216	−0·198	−0·189	−0·188	−0·230
	L	RE	1·471	1·377	1·397	1·471	1·586	1·728	2·893
		AE	0·037	0·041	0·047	0·054	0·063	0·074	0·158

are rejected. However, in systems with few servers and small buffer size utilization is greatly reduced and there are few failures. It is worthwhile considering that even if the relative variation of *L* is high, the absolute variation is always less than one unit.

For purposes of validating the approximation model, the exact solution to the above case examples is compared with the approximate solution. Two error measures have been adopted as accuracy indices: the absolute error, AE, and the relative error, RE (expressed as percentage). Table 2 shows the RE and AE of *THS* and L, respectively, for different buffer sizes and numbers of servers. Note that the RE of *THS* and L is generally low, as is the AE, especially for high numbers of servers. More in-depth analysis has shown that the approximation model is also sufficiently accurate in the practical range of values of λ, μ, λg and μg.

Conclusions

The approach proposed here has considered the effect of failures and repairs on the main performance measures of an FMS workstation modelled as a multi-server queuing system with finite buffer size. For the cases examined, consisting of most practical applications in an FMS, the capacity value of the buffer strongly affects the variation in the performance caused by failures. For large buffers, L is more influenced by failures than *THS*. Instead, for small buffers, *THS* is more dependent on failures than L. A large relative variation in the mean number of jobs in the system is generally found for large buffers but absolute variations are slight.

The approximation model is fairly easy to develop and considerably reduces computing time and storage requirements, particularly for systems with many servers and large buffers. The proposed approach is useful for performance analysis, fault tolerance evaluation and reliability apportionment of a workstation in an FMS.

References

Buzacott, J.A. and Yao, D.D., 1986, Flexible manufacturing systems: a review of analytical models. *Management Science*, **32**, 7, 890–905.

Co, H.C. and Wysk, R.A., 1986, The robustness of CAN-Q in modelling automated manufacturing systems. *International Journal of Production Research*, **24**, 6, 1485–1503.

Draper Laboratory, 1984, *Flexible Manufacturing Systems Handbook* (Park Ridge, NJ: Noyes Publications).

Gross, D. and Harris, C.M., 1985, Fundamentals of queuing theory (New York: Wiley).

Vinod, B. and Solberg, J.J., 1985, The optimal design of flexible manufacturing systems. *International Journal of Production Research*, **23**, 6, 1141–1151.

SECTION IV.2. ECONOMIC FACTORS

42

The creativity curve and its measurement

Shingo Kataoka, Kozi Morooka, Jairo Hasbun, and Luisa Lino

Department of Management Engineering, Tokai University, 1117 Kitakaname, Hiratsuka, Kanagawa, 259-12, Japan

Abstract

It is our object to understand to what extent the investment in research and development (R. & D.) influences the capacity to produce new things, or to what extent the investment in R. & D. influences the creative output of a nation. We also intend to understand the differences in the curves of each studied nation. In the IXth ICPR Morooka and Kataoka (1987) proposed a creativity curve explaining the relationship between the cumulative real R. & D. expenditure and the payments/receipts ratio of technological trade to evaluate the creativity of major Japanese industry groups and selected industrial nations over the long range. In this chapter the creativity curve analysis and results with the addition of new data expanded world-wide is described to assess various aspects of creativity of 12 selected nations' R. & D. activity.

Introduction

Human creativity determines to a great extent the pace of technological advance in many industries. In fact, not only our productivity but also our standard of living today has been largely determined by past discoveries of 'new' materials and inventions of 'new' technologies, and our future prosperity will in large part depend on the fruits of contemporary research. Internal and external creativity of companies can be used in R. & D. strategies to enhance their capabilities: the means of developing or marketing new products, the acquisition of highly or potentially highly creative companies, and the co-operation and/or partnership agreements with creative world competitors. A way must be found to measure the creativity of nations, industries and companies.

In the IXth ICPR, Morooka and Kataoka (1987) proposed a creativity curve (CC) to evaluate the creativity of major Japanese industry groups and selected industrial nations. A new quantitative analytical tool for R. & D. strategies capable of making comparisons of creativity between competitors has been developed. It is our object to understand to what extent the investment in R. & D. influences the capacity to produce new things, or to what extent the investment in R. & D. influences the creative output of a nation. We also intend to understand the differences in the curves of each studied nation. Getting the R. & D. data is not difficult, but how can we get the creativity data; that is, how can we get a measure of creativity?

Creativity measurements and R. & D. activity data

Performance of R. & D. activity on Science and Technology (S. & T.) can be quantified in our study using available data, obtained from Science and Technology Agency, Japan (1988) and OECD (1982–8).

The Gross Domestic Expenditure on Research and Development (GERD) covers all R. & D. projects carried out on national territory in the year concerned. For international comparisons, Real Gross Domestic Expenditure (R-GERD) will be adopted, i.e. GERD national currency is inflation-adjusted by using a GNP deflator (1980 = 100) and changed into Japanese Yen in Tables 1 and 3, or inflation adjusted by using implicit GDP price indices (1980 = 1·00) and exchanged by purchasing power parities obtained by OECD in Tables 2 and 5. They indicate Japan had the second largest real expenditure on R. & D. of all the nations after the USA in 1986.

As yet, there is no direct measure for the output of scientific and technological activities. For the purposes of this study, the measure of S. & T. is determinedly two types of data: technological balance of payments (TBP) and patents.

TBP registers the flow of proprietary knowledge and know-how into or out of a nation. It consists of money paid (PAY) or received (REC) for the use of patents, licences, trademarks, designs, inventions, know-how and closely-associated technical services, etc. These data can be considered as partial measures of international technology flows shown in Tables 2, 3 and 5.

Patent data can be considered to be a partial measure of the output of R. & D. in the form of inventions. Patent data is the number of domestic patent applications [DPAT] applied for in each nation by residents of the nation concerned.

Five selected nations' creativity measurements (1965–86) are shown in Figure 1 and Table 1. Japan has increased in [DPAT] from 1980, but other nations are flatter. Real Gross Domestic Expenditures on R. & D. divided by Number of Domestic Patent Applications [R-GERD/DPAT] will be considered as unit patent costs in a nation. The costs of all nations except the USA decrease from 1980. In population per domestic patent applications [POP/DPAT}, Japan has 400 people per DPAT, less than a quarter of other nations; and Japan has a negative trend in [POP/DPAT]. This may be an efficiency of human resources devoted to the creativity outputs.

Figure 2 shows a positive correlation of R-GERD and DPAT. Hence a nation having large-scale R-GERD as well as large-scale DPAT can get a good position in its technological dependence rate, as obtained by payments divided by receipts (PAY/REC), in Figures 3 and 4. However we could not recognize a relationship between Patents costs (R-GERD/DPAT) and the technological dependence rate (PAY/REC), shown in Figure 5.

Concepts of nations' creativity in R. & D. on science and technology

The definition of creativity differs among authors, making finding a measure for it even more difficult (Holt, 1983). Finding a direct measure of creativity to be too difficult we decided to try and find an indirect measure. We arrived at a creativity index using factors such as Real R. & D. expenditure and Technological dependence rate, i.e. technology import and export rate.

We will define a nation's creativity as a measurable performance of R. & D. activity and of sciences and technologies created: an example is the relationship between R-GERD on R. & D. and the rate of technological dependence on foreign nations.

In our study, the creativity rate of a nation can be defined in terms of the efficiency of improvement or reduction of technological dependence on foreign

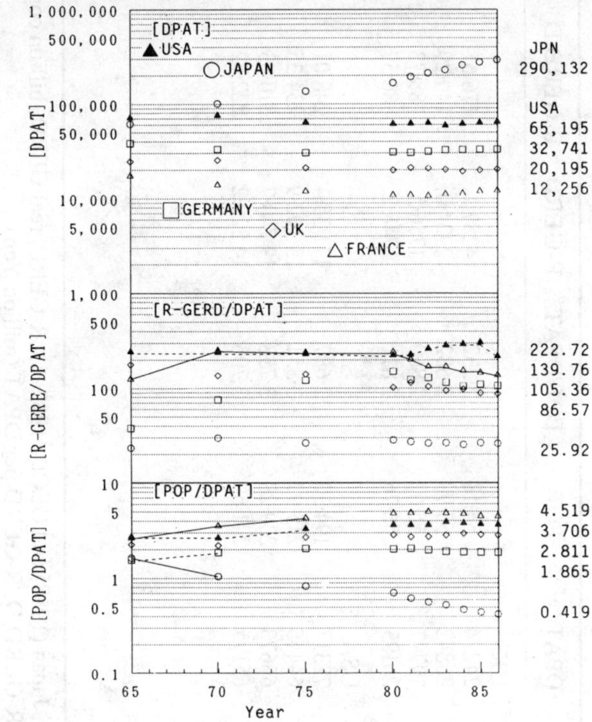

Figure 1. Historical trend of creativity measurements (1965–86). [DPAT], No. of Domestic Patents; [R-GERD/DPAT], million constant Yen (on the base 1980)/DPAT; [POP/DPAT], population/DPAT (thousand people). From Science and Technology Agency, Japan (1988) and OECD (1982–8).

technology with cumulative R. & D. activity increasing in the period of data analysis. The creativity rate can be decided by the regression nation's creativity curve of the technological dependence rate on the cumulative R. & D. activity.

Cumulative R. & D. activity

Total resources devoted to a nation's R. & D. activity can be fixed and measured by the relative quantity of R-GERD on R. & D., inflation-adjusted current expenditure by deflator. R-GERD is exchanged by Japanese constant Yen on the base 1980 for five selected nations and by constant ppp$ on the base 1980 for 12 selected nations, for an international comparison. The cumulative R. & D. activity can be obtained by cumulated R-GERD (CR-GERD). CR-GERD can also indicate as cumulated knowledge and experience through R. & D. projects. Table 3 shows five nations' long range period data, and Table 5 shows twelve nations' in the recent years. Both tables present indicators of the level and trends in total national R. & D. efforts.

Technological dependence rate

We consider technological balance of payments to indicate R. & D. outputs evaluated in international market economics. The number of patents is a quantitative index of R. & D. output without evaluation of market needs. They only have quality level of patentship.

Table. 1 Five nations' number of domestic patent applications, population and R-GERD.

Year	DPAT	POP	POP/DPAT	R-GERD	R-GERD/D	DPAT	POP	POP/DPAT	R-GERD	R-GERD/D
France						UK				
1965	17 509	48·76	2·785	2228·9	127·30	24274	54·18	2·232	4287·1	176·61
1975	12 110	52·79	4·359	2945·3	243·21	20842	55·90	2·682	2924·0	140·29
1985	12 158	55·17	4·538	1809·6	148·84	19797	56·62	2·860	1756·5	88·73
1986	12 256	55·39	4·519	1712·9	139·76	20195	56·76	2·811	1748·2	86·57
Germany						USA				
1965	38 148	59·01	1·547	1431·9	37·53	72317	194·30	2·687	17 935·4	248·01
1975	30 198	61·83	2·047	3634·4	120·35	64445	215·97	3·351	15 102·9	234·35
1985	32 708	61·02	1·866	3579·6	109·44	63673	239·28	3·758	19 740·7	310·03
1986	32 741	61·05	1·865	3449·7	105·36	65195	241·60	3·706	14 520·2	222·72
Japan										
1965	60 796	98·88	1·626	1410·0	23·19					
1975	135 118	111·57	0·826	3547·8	26·26					
1985	274 348	120·75	0·440	7195·4	26·23					
1986	290 132	121·49	0·419	7520·1	25·92					

DPAT, No. of domestic patents applications (from Science and Technology Agency, Japan (1988) and OECD (1982–8); R-GERD, real GERD (billion constant Yen); POP, 42/7 population (million); POP/DPAT, POP per DPAT thousand, R-GERD/D, R-GERD per DPAT (million Yen).

Table 2. Twelve countries' R-GERD/DPAT and PA/RE data (1986).

Nation	DPAT	R-GERD	R-GERD/D	PA/RE	Nation	DPAT	R-GERD	R-GERD/D	PA/RE
Austria	2300	1532·7	0·666	4·8085	Japan	290 132	37 702·4	0·130	1·1629
Canada	2161	4249·2	0·625	1·2724	Norway†	903	712·2	0·789	2·7078*
Denmark†	856	534·6	0·624	0·8771	Portugal†	77	81·8	1·063	9·1645
France	12 256	9327·9	0·761	1·2026	Spain	1652	933·1	0·565	4·0220
Germany	32 741	17 850·8	0·545	2·0994	UK	20 195	10 985·0	0·544	0·9266
Italy†	2063	3682·2	1·785	3·7867	US	65 195	87 161·9	1·337	0·1570

DPAT, number of domestic patent applications; R-GERD, real gross domestic expenditure (million constant ppp$ 1980); R-GERD/D = R-GERD/DPAT (million constant ppp$ per DPAT); PA/RE = payments/receipts on technology trade (technological dependence rate). †1985 data. From OECD (1982–8).

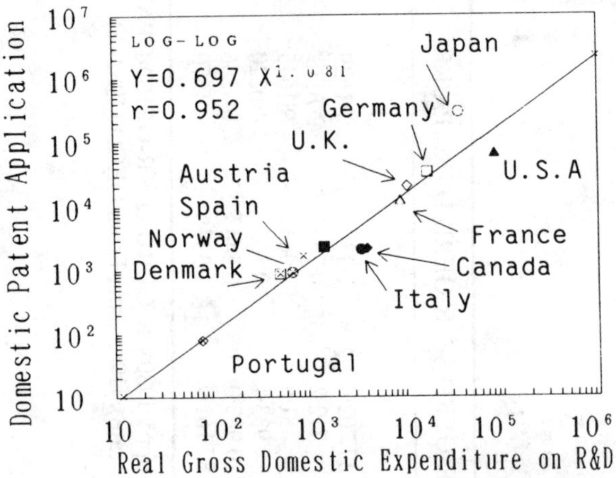

Figure 2. Relation between R-GERD and DPAT in 1986 (data for 1985 for Denmark, Italy, Norway and Portugal).

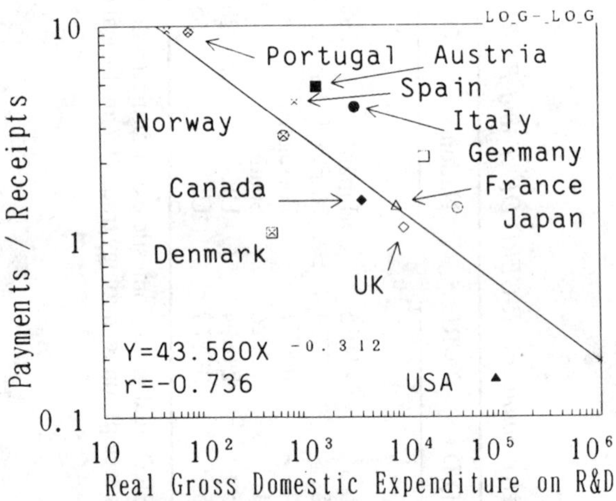

Figure 3. Relation between R-GERD and PAY/REC in 1986 (data for 1985 for Denmark, Italy, Norway and Portugal).

Thus we can define the performance of a nation's R. & D. activity output, science and technology as the rate of technological dependence on foreign nations as the coefficient obtained by dividing payments by receipts on international technological trade.

The creativity curve of a nation

A nations' creativity curve can be defined as the value of the cumulative average technological dependence rate on the value of cumulative R. & D. activity in a nation over the period of obtained historical data. The formula is:

$$Y_i = AX_i^B$$

i is the time period in historical data (year), X_i is the cumulative R. & D. activity to a year (real R. & D. expenditure cumulated), Y_i is the cumulative average

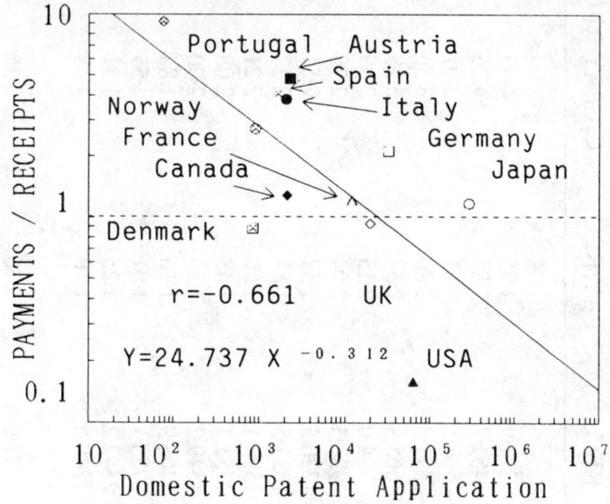

Figure 4. Relation between DPAT and PAY/REC in 1986 (data for 1985 for Denmark, Italy, Norway and Portugal).

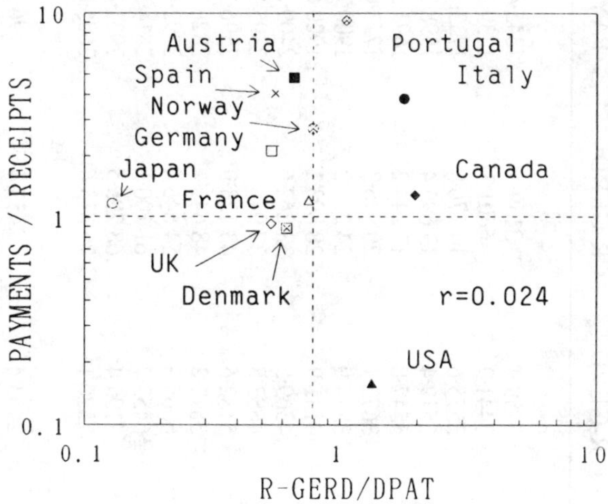

Figure 5. Relation between R-GERD/DPAT and PA/RE in 1986 (data for 1985 for Denmark, Italy, Norway and Portugal).

technological dependence to a year or technological dependence in a year, A is the potential dependence on foreign technology in the first period, B is the decline coefficient or regression slope, $C = 2^B \times 100$ is the creativity rate (per cent of CC slope).

(C) is a constant for a given CC of a nation. It is the percentage of reduction in technological dependence on foreign nations' technology when the cumulative R. & D. activity is doubled.

Results of creativity curve analysis

A regression CC is estimated for five selected nations since 1971 and twelve selected nations since 1982. The results are summarized in Figure 6 and Table 3, and Figure 7 and Table 4. There are differences in sources and periods of data.

Table 3. Example data for five selected nations' creativity curve.

Year	GERD	DEF %	R-GERD	CR-GERD	PAY	REC	PAY/REC	CA-PAY/REC
Japan (all data)								
1971	1345·9	42·6	3159·4	3159·4	1345·4	271·9	4·95	4·95
1972	1586·7	46·1	3441·9	6601·3	1739·2	451·7	3·85	4·26
1973	1980·9	55·4	3575·6	10 176·9	1733·1	508·5	3·41	3·91
1974	2421·4	68·8	3519·4	13 696·4	1598·3	571·0	2·80	3·56
1975	2621·8	73·9	3547·8	17 244·2	1691·3	665·9	2·54	3·28
1976	2941·4	80·1	3672·1	20 916·3	1773·0	834·0	2·13	2·99
1977	3233·5	84·1	3844·9	24 761·2	1900·7	933·3	2·04	2·78
1978	3570·0	86·4	4131·9	28 893·1	1920·6	1220·5	1·57	2·51
1979	4063·6	93·0	4369·5	33 262·6	2409·8	1331·5	1·81	2·37
1980	4683·8	100·0	4683·8	37 946·3	2395·3	1596·1	1·50	2·21
1981	5364·0	103·5	5182·6	43 128·9	2596·3	1751·1	1·48	2·08
1982	5881·5	106·7	5512·2	48 641·1	2826·1	1849·2	1·53	2·00
1983	6503·7	108·3	6005·3	54 646·4	2792·8	2408·9	1·16	1·86
1984	7176·5	111·2	6453·7	61 100·1	2814·5	2775·1	1·01	1·55
1985	8116·4	112·8	7195·4	68 295·5	2931·7	2342·2	1·25	1·51
1986	8415·0	111·9	7520·1	75 815·6	2605·8	2240·8	1·16	1·48
France (sample data)								
1971	1047·6	42·7	2453·4	2453·4	796·1	237·0	3·36	3·36
1982	2312·4	124·4	1858·8	29 598·9	2250·5	938·4	2·40	2·29
1985	2812·1	155·4	1809·6	35 122·3	2334·8	1214·7	1·92	2·06
1986	2795·5	163·2	1712·9	36 835·2	2075·8	1151·8	1·80	2·03

Germany (sample data)								
1971	1801·3	64·4	2797·0	27 97·0	1312·9	518·4	2·53	2·53
1982	4325·2	108·6	3982·7	42 672·3	2258·9	1225·4	1·84	2·20
1985	4181·0	116·8	3579·6	53 260·0	2373·4	1300·5	1·82	2·10
1986	4153·4	120·4	3449·7	56 709·7	2621·7	1311·8	2·00	2·09
1987	4575·5	122·9	3722·9	60 432·6	2718·6	1343·7	2·02	2·09
UK (sample data)								
1971	1086·4†	29·4	3695·2	3695·2	938·0	1001·8	0·936	0·936
1982	2557·8†	120·0	2131·5	31 593·6	1870·2	2187·0	0·855	0·860
1985	2448·6	139·4	1756·5	37 093·4	1651·1	2022·2	0·816	0·848
1986	2524·4†	144·4	1748·2	38 841·6	1530·2	1651·3	0·927	0·854
USA (sample data)								
1971	9317·9	52·8	17 647·5	17 647·5	842·0	8890·0	0·095	0·095
1982	19 757·6	116·7	16 930·2	179 397·3	1539·0	12 893·0	0·119	0·108
1985	25 623·4	129·8	19 740·7	234 674·2	2127·0	14 298·0	0·149	0·119
1986	19 326·4	133·1	14 520·2	249 194·4	1751·0	12 321·0	0·142	0·121
1987	17 793·0	137·1	12 978·1	262 172·5	1337·0	13 070·0	0·102	0·119

GERD (billion current Yen); R-GERD = Real GERD (billion constant Yen on the base 1980); collected in GERD devided by GNP deflator (by the R. & D. cost deflator in Japan): DEF %; CR-GERD = cumulative R-GERD; PAY = payments (hundred million Yen); REC = Receipts (hundred million Yen); PAY/REC (technological dependence rate); CA-PAY/REC = cumulative average PAY/REC (cumulative average technological dependence rate). †, Estimated GERD in our study. From Science and Technology Agency, Japan (1988).

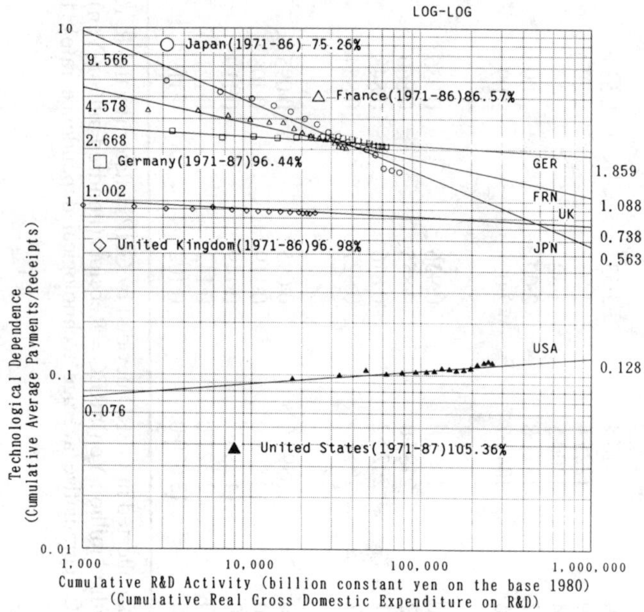

Figure 6. Creativity curves for five selected nations (log–log scale).

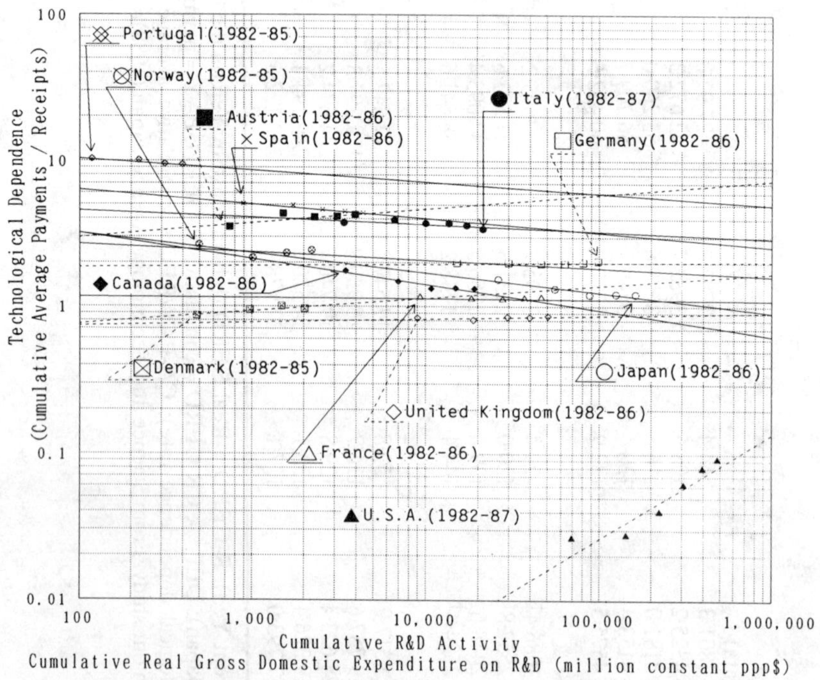

Figure 7. Creativity curves of 12 selected nations since 1982.

Table 4. Regression creativity curves (Y = AX^B) of five selected nations.

	Creativity Curve	C. rate (C, %)	R (correlation coefficient)
France	$Y = 19{\cdot}2685X^{-0{\cdot}2081}$	86·57	−0·946 (fitting period 1971–86)
Germany	$Y = 3{\cdot}8273X^{-0{\cdot}0523}$	96·44	−0·856 (fitting period 1971–87)
Japan	$Y = 162{\cdot}5934X^{-0{\cdot}4101}$	75·26	−0·973 (fitting period 1971–86)
UK	$Y = 1{\cdot}3608X^{-0{\cdot}0443}$	96·98	−0·943 (fitting period 1971–86)
USA	$Y = 0{\cdot}0450X^{0{\cdot}0754}$	105·36	0·884 (fitting period 1971–87)

The regression of CA-DR on CR-GERD, Y = CA-DR X = CR-GERD.
CA-DR = cumulative average dependence rate; CR-GERD = cumulative real gross domestic expenditure on R. & D. (billion constant Yen on the base 1980); C. rate = creativity rate (= Per cent slope); R = coefficient of log linear correlation. Data are shown in Table 4.

Five selected nations' creativity curves since 1971

Table 4 shows the five cases for large economic nations. In each, a CC is estimated by log linear regression over the period of obtained data by a straight line after plotting all data on log–log paper, as shown in Figure 6. It is clear that the fit is good. The value of the log linear correlation coefficient is significant in all reported regressions. The term (C) shown in Table 4 is the most useful parameter. Figure 6 shows Japan's creativity curve at 75%, that is, each time the cumulative R. & D. activity is doubled, the technological dependence rate is 75% of what it used to be.

Twelve nations' creativity curves since 1982

Tables 5 and 6 show twelve nations. In Table 6, for each nation a regression CC has a C. rate (CC slope per cent) ranging from 89–163%. It is clear that the fit is good. The correlation coefficient has values of 0·60 or more in two out of the 12 reported regressions, and lower values of 0·60 in only two cases: France and Germany. Figure 7 shows four nations which have positive lines, indicating that their C. rate is over 100%.

The twelve nations' curves in Figure 7 will be classified in terms of technological dependence rate (CA − PA/RE) $\leqq$ 1·00 or > 1·00; potential dependence (A) $\leqq$ 1·00, or >1·00, and creativity rate (C) < 100%, or $\geqq$100%. We can further classify these nations by their scale of R. & D. expenditure. There is a large R. & D. expenditure group (USA, Japan, Germany, UK, France); medium R. & D. expenditure group (Canada, Italy); and small R. & D. expenditure group (Austria, Norway, Denmark, Spain, Portugal).

Conclusion

In this study, creativity measurements can be defined by using the concepts of performance on R. & D. activity in science and technology. The ratio of total resources to R. & D. activity and the performance of a nation's R. & D. activity outputs are examined. These are carefully measured and quantified. Cumulative annual R. & D. resources can be considered to be cumulative knowledge or

Table 5. Data for 12 nations' creativity curves since 1982.

Nation	Year	R-GERD	CR-GERD	Payments	Receipts	PA/RE	CA-PA/RE
Austria	1982	750·7	750·7	1860	533	3·4897	3·4897
	1983	775·0	1525·6	2461	470	5·2362	4·3081
	1984	798·0	2323·7	1924	530	3·6302	4·0737
	1985	817·6	3141·3	2059	483	4·2629	4·1190
	1986	845·6	3986·9	2260	470	4·8085	4·2494
Canada	1982	3521·7	3521·7	523	299	1·7492	1·7492
	1983	3523·0	7044·7	570	444	1·2838	1·4711
	1984	3840·4	10 885·0	623	561	1·1105	1·3160
	1985	4142·2	15 027·2	729	537	1·3575	1·3281
	1986	4236·4	19 263·6	752	591	1·2724	1·3146
Denmark	1982	483·6	483·6	1112	1293	0·8600	0·8600
	1983	497·0	980·6	1247	1219	1·0230	0·9391
	1984	516·8	1497·4	1471	1330	1·1060	0·9969
	1985	534·6	2031·9	1706	1945	0·8771	0·9566
France	1982	9384·9	9384·9	6542	5609	1·1663	1·1663
	1983	9110·2	18 495·1	6920	6247	1·1077	1·1355
	1984	9290·3	27 785·4	8838	7918	1·1162	1·1277
	1985	9376·5	37 161·9	9556	8031	1·1899	1·1457
	1986	9323·2	46 485·1	9486	7888	1·2026	1·1583
Germany	1982†	15 275·0	15 275·0	2701	1360	1·9860	1·9860
	1983	14 830·8	30 105·7	3010	1508	1·9960	1·9913
	1984†	16 176·7	46 282·5	3165	1651	1·9170	1·9642
	1985	17 109·2	63 391·7	3551	1809	1·9630	1·9638
	1986†	17 850·8	81 242·5	4159	1981	2·0994	1·9961
	1987	17 918·3	99 160·8	4391	1943	2·2599	2·0461
Italy	1982	3444·1	3444·1	808 426	217 043	3·7247	3·7247
	1983	3282·9	6727·0	911 499	224 788	4·0549	3·8927
	1984	3401·4	10 128·4	989 757	289 646	3·4171	3·7044
	1985	3682·2	13 810·5	1042 441	275 288	3·7867	3·7269
	1986	3611·0	17 421·6	1056 350	326 438	3·2360	3·6067
	1987	4093·6	21 515·2	1021 082	389 893	2·6189	3·3832
Japan	1982	26 296·5	26 296·5	282 613	184 921	1·5283	1·5283
	1983	29 449·2	55 745·7	279 280	240 887	1·1594	1·3196
	1984	32 722·5	88 468·2	281 447	277 512	1·0142	1·1991
	1985	36 857·8	125 325·9	293 173	234 220	1·2517	1·2122
	1986	37 633·8	162 959·8	260 577	224 078	1·1629	1·2027
Norway	1982	499·2	499·2	487	185	2·6324	2·6324
	1983	521·8	1021·0	638	342	1·8655	2·1347
	1984	581·8	1602·8	639	235	2·7191	2·3150
	1985	633·6	2236·4	658	243	2·7078	2·4100
Portugal	1982	118·7	118·7	3147	308	10·218	10·218
	1983†	103·7	222·4	4174	422	9·8910	10·029
	1984	92·0	314·5	4201	490	8·5735	9·4443
	1985†	84·2	398·7	5682	620	9·1645	9·3500
Spain	1982	902·7	902·7	78 984	15 707	5·0286	5·0286
	1983	840·7	1743·5	88 338	18 265	4·8365	4·9253
	1984	835·4	2578·8	84 742	20 781	4·0779	4·6037
	1985	905·8	3484·7	102 957	24 536	4·1962	4·4776
	1986	933·1	4417·8	105 331	26 189	4·0220	4·3644
UK	1982†	9060·6	9060·6	415	502	0·8267	0·8267
	1983	9870·2	18 930·8	482	615	0·7837	0·8030
	1984†	10 709·8	29 640·6	591	646	0·9149	0·8440
	1985	10 247·2	39 887·8	534	654	0·8165	0·8366
	1986	10 929·0	50 816·8	619	668	0·9266	0·8561

Table 5. (Cont.)

Nation	Year	R-GERD	CR-GERD	Payments	Receipts	PA/RE	CA-PA/RE
USA	1982	69 586·3	69 586·3	189	7133	0·0265	0·0265
	1983	73 933·6	143 519·9	228	7855	0·0290	0·0278
	1984	80 104·2	223 624·0	516	8115	0·0636	0·0404
	1985	85 810·6	309 434·6	892	6166	0·1447	0·0624
	1986	87 161·9	396 596·5	1077	6862	0·1570	0·0803
	1987	88 845·9	485 442·4	1220	8101	0·1506	0·0932

R-GERD: real gross domestic expenditure on R. & D. (constant million ppp$ based on 1980 US$): inflation adjusted by implicit GDP price indices (1980 = 1·00). †Estimated GERD. CR-GERD, Cumulative R-GERD from 1982 to date; ppp$, purchasing power parities obtained by OECD. National currency GERD DATA have been coverted to US$ using ppp$. Technological balance of payments: payments = PA and receipts = RE (million national currency). CA-PA/RE: cumulative average PA/RE from 1982 to date. From OECD (1982–8).

Table 6. Regression creativity curves ($Y = AX^B$) of twelve nations since 1982.

Nation'	A	B	C. rate (C%)	R (correlation coefficient)
(1) Austria	1·9075	0·0985	107·07	0·768 (fitting period 1982–86)
(2) Canada	7·0038	−0·1737	88·66	−0·946 (fitting period 1982–86)
(3) Denmark	0·5063	0·0879	106·28	0·879 (fitting period 1982–85)
(4) France	1·2180	−0·0060	99·59	−0·274 (fitting period 1982–86)
(5) Germany	1·8156	0·0086	100·60	0·395 (fitting period 1982–87)
(6) Italy	5·6337	−0·0464	96·84	−0·671 (fitting period 1982–87)
(7) Japan	5·9387	−0·1360	91·00	−0·942 (fitting period 1982–86)
(8) Norway	3·5352	−0·0569	96·13	−0·423 (fitting period 1982–85)
(9) Portugal	14·9646	−0·0781	94·74	−0·939 (fitting period 1982–85)
(10) Spain	9·6820	−0·0942	93·68	−0·966 (fitting period 1982–86)
(11) UK	0·6642	0·0224	101·56	0·630 (fitting period 1982–86)
(12) US	8·24E–6	0·7050	163·01	0·942 (fitting period 1982–87)

The regression CC of CA-PA/RE on CR-GERD (million constant ppp$); Y = CA-PA/RE, X = CR-GERD; CA-PA/RE = cumulative average dependence rate on foreign technology; CR-GERD = cumulative real gross domestic expenditure on R. & D. (million constant ppp$ 1980); C. rate = creativity rate (= per cent slope); R = coefficient of log linear correlation. From OECD (1982–8).

experience of R. & D. in S. & T. obtained by real R. & D. expenditures over the range of historical data in a nation, as shown in Table 7.

In five selected nations, the creativity curves for R. & D. in S. & T. can explain the relationship between technology dependence and cumulative R. & D. activity over long data periods, but in 12 selected nations (OECD members) CCs can also explain it for the most recent four to six years. We find that an improvement phenomenon of technological dependence rate depends on an increase of the cumulative R. & D. activity in 7 out of 12 nations.

The scale of R. & D. expenditure is one of the primary determinants of a good position in the technological dependence rate, as explained by dividing the technology trade balance of payments by receipts. An important reason for the small

Table 7. Classification of creativity into types for 12 countries' creativity curves.

Potential technological dependence							
(A) ≦ 1·00, creative potential				(A) > 1·00, dependent potential			
Cumulative Average Technological Dependence Rate in 1986							
creative position (CA-PAY/REC) ≦ 1·00		dependent position (CA-PAY/REC) > 1·00		creative position (CA-PAY/REC) ≦ 1·00		dependent position (CA-PAY/REC) > 1·00	
Creativity Rate (C)							
Improve $C < 100\%$	Depend $C \geqq 100\%$	Improve $C < 100\%$	Depend $C \geqq 100\%$	Improve $C < 100\%$	Depend $C \geqq 100\%$	Improve $C < 100\%$	Depend $C \geqq 100\%$
	UK					France	Germany
	USA					Japan	
						Canada	
						Italy	
	Denmark			Norway	Austria	Spain	
						Portugal	

rate of technological dependence on foreign technology is that large scale R. & D. expenditure nations such as the USA and the UK have lower dependence on foreign technology. The improving trend toward a small rate of dependence on foreign technology is due in part to the so-called creativity rate (C), whereby the technological dependence of a nation declines 4–25% each time their cumulative real R. & D. expenditure doubles.

References

Holt, K., 1983, *Product Innovation Management* (London: Butterworth).

Morooka, K. and Kataoka, S., 1987, Creativity curve. *IXth ICPR* **1**, 1044–1050.

OECD (Organization for Economic Co-operation and Development), 1982–8, *Main Science and Technology Indicators*, 2.

Science and Technology Agency Japan, 1988, White Paper on S. & T.

43

New developments in cost management systems using the MRP data base

Iwao Hayawaka

Aichi Institute of Technology

Abstract

Since the introduction of computer systems, the Decision Support System (DSS) has made rapid development in the field of production management accounting. Material Requirement Planning (MRP), developed in the USA is one method of production management accounting and some standard cost computing systems utilize this method. Toyota's production system, developed in Japan, is not a contradictory system to MRP, but rather a complementary system for the formulation of comprehensive business plans. The MRP data base is integrated with a production management accounting data base and the new production system is integrated through the data base with the production management accounting system, that is, the cost and profit management systems.

In this paper I will deal with the following topics:

(1) Toyota's production system which 'smooths' daily production,
(2) incorporation of MRP with Toyota's system,
(3) methods of cost management using the MRP data base,
(4) evaluation of a company's profit growth management using the Just-in-Time (JIT) Business and MRP models.

Development of cost and profit management systems using the Toyota system

Let us consider the application of the Toyota system to cost and profit management systems from the point of view of a manager in the company. There are three types of production control accounting systems in use today: the Kanban system, MRP and the production instruction number system. The Kanban system uses a single card, known as a Kanban, to give withdrawal and production instructions. This system ensures that parts needed for later processes are delivered no earlier than necessary, and is therefore also sometimes referred to as the 'Just-in-Time' system.

The Kanban system is simply described below. For example, assume a part Z is used in one of the later processes of assembly of a product. These parts are kept in containers near the position where the assembly is carried out and a card is attached to each of the containers. This card is the 'Kanban', which strictly means 'sign' in Japanese. When one of the containers becomes empty, a carrier arrives within a specified time and takes the empty container and the card away to where parts Z completed in preceding manufacturing stages are stored. He leaves the empty container and card at the store and takes another container full of parts Z

Table 1. Comparison between traditional and smoothed production systems.

Type	Item	Number of Production/Day 1	2	3	4	5
Traditional Production System						
Z	1600	4000				4000
Y	1200		4000			
X	800			4000		
W	400				4000	
Total	4000	4000	4000	4000	4000	4000
Smoothed Production System (per day)						
Z	1600	1600	1600	1600		
Y	1200	1200	1200	1200		
X	800	800	800	800		
W	400	400	400	400		
Total	4000	4000	4000	4000		

and a new attached Kanban back to the point where they are assembled. The Kanban from the empty container is used as the production-ordering card for the preceding processes where part Z is manufactured.

This system can lead to problems with the inventories processes. The number of parts in the store must be equal to the average demand for parts each replenishment cycle added to a safety factor of a certain number of shelf items. In order to minimize the number of parts in the inventory (output of preceding processes and the parts used in subsequent processes), the length of the cycle is kept short, to one or two hours, so that parts are withdrawn from stock several times each day and the number withdrawn each day can be kept to a minimum. In other words, the lot sizes ordered by the Kanbans from previous processes can be kept as small as possible. Therefore the final products cannot be manufactured in large lots, but need to be produced in smaller lots, averaged on a daily, or even hourly, basis. An example is shown in Table 1. Four types of finished product, Z, Y, X, and W used to be produced in large lots, with 4000 of type Z only produced one day, 4000 of type Y only the next day, then 4000 of types X and W, respectively, on the following two days. However, under the smoothed production system, 1600 type Z, 1200 type Y, 800 type X and 400 type W are produced every day. That is, the production of each of the products is averaged for each day, so that the number of component parts used each day is held constant. This is one of the features of the Kanban system.

The features of the production instruction number system are that parts and materials are procured according to the orders received from the customer and then a production instruction number is attached to the products being manufactured. For example, if the production number for one order of a certain product X is #110, the same number 110 is also entered on the order forms for parts A, B and C used to manufacture X.

This system has a major weakness, however, if there is a change in the basic manufacturing plan for product X. For example, problems occur when using 50 parts A, originally purchased for the manufacture of production instruction number #110, in products with production instruction number #111. In this case, the parts are the same part A but the production instruction number attached the parts is different, so that it becomes impossible to manage the delivery deadlines for part A.

The MRP system also employs a production instruction number (for example #110) related to the order of the product X in the same way as the production

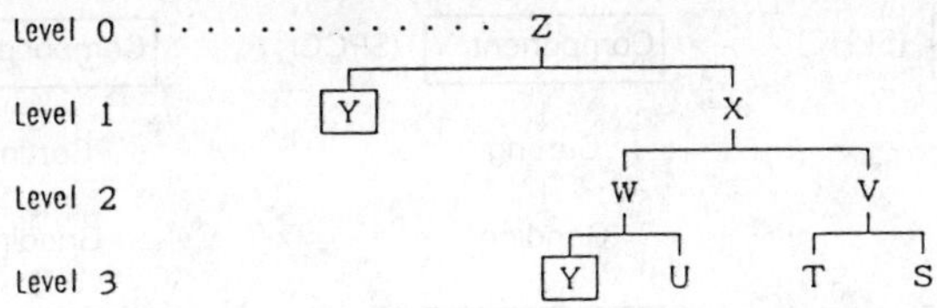

Figure 1. Low-level code of component Z.

instruction number system. However, unlike the production instruction number system, MRP does not employ a simple number display, but uses parts tables for each production stage to calculate the total number of parts required for each stage.

In this case, the parts at each level are marked with a number (e.g. #1000), which differs from the production instruction number of product X. This eliminates the effort associated with the production instruction number system of converting differing production instruction numbers for various orders of the same product to the same part number. Consequently, this system is able to respond flexibly to changes in the basic production plan to handle the delivery deadline management of all parts at every stage of manufacture. When using the production number system, cost management is carried out by adding the cost of all the component parts as these are also marked with production instruction number of the finished product. However, when using the MRP system, the production instruction number is separated from the component parts used at each level, so that the costs are calculated at each stage of manufacture. This method allows the determination of the standard prices from a standard bill-of-materials and gives a grasp of the predicted costs.

New developments in cost management systems using the MRP data base

I will now consider how the MRP data base can be applied to cost management and describe the new types of cost management system.

Cost piled-up system

This is a data base for the piling-up calculation of the standard costs and latest costs of parts or a product. It uses a 'master file for components' and 'master file for product composition' (parts table) for the direct materials costs and outsourced manufacturing costs, and a 'master file for (manufacturing) processes' and 'master file for working sections' to determine the direct labour costs and indirect manufacturing costs. The cost pile-up calculation starts from the part at the extreme bottom level of the parts table and works level by level up the table. When using multi-level parts tables, a single part may appear on several of the levels of the table, but the unit cost for the part can be computed in a single calculation using the part's low-level code, known as LLC for short (see Figure 1).

The master file

The master file for working sections contains wage rate data and indirect manufacturing cost rate distribution data. The master file for processes contains the standard operation time for each process and code data for the operations associated with the process. The labour and indirect manufacturing costs of part Z are

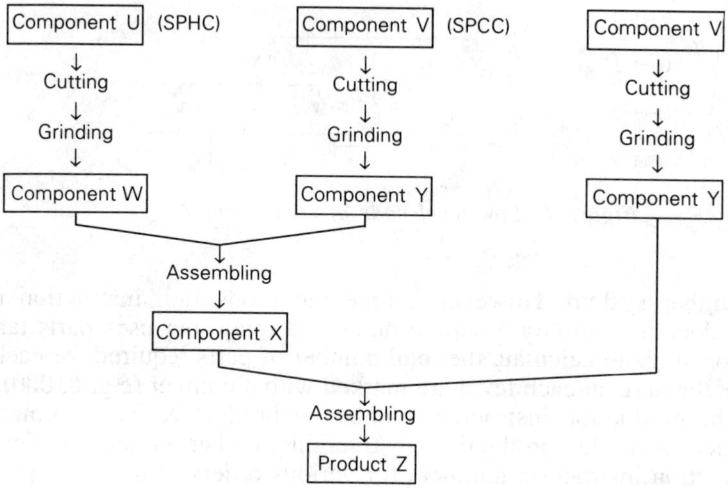

Figure 2. Flow of production processes for product Z.

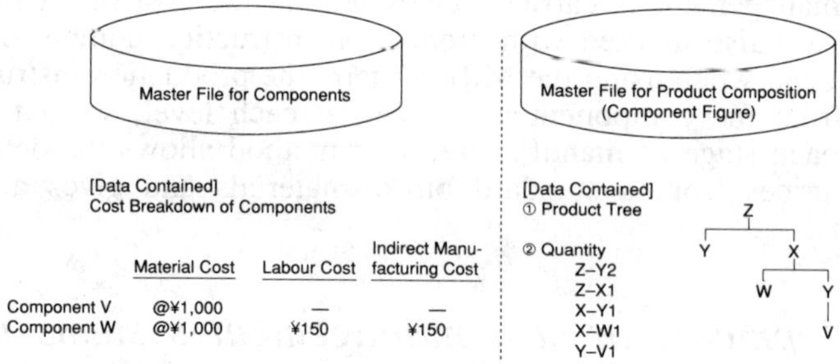

Figure 3. Data on design engineering.

calculated from this data for each process which the part undergoes, as shown below:

$$\text{Labour costs} = \sum_{i=1}^{n} (\text{standard time} \times \text{wage rate for the operation})$$

$$\text{Indirect costs} = \sum_{i=1}^{n} (\text{standard time} \times \text{indirect cost rate of the operation}).$$

Product cost calculation

The production flow for the completion of product Z is shown in Figure 2. Part V is a steel part which is cut and ground to produce part Y. Part U is a steel part which is cut and ground to produce the finished part *W*. Part X is produced by assembling parts Y and W together and then the finished product Z is made by assembling parts X and Y together.

Figure 3 shows the technical design data, which comprises parts cost data and parts composition data (product tree). The costs of part W have already been completed from the materials, labour and indirect manufacturing costs. Figure 4 shows process data for the machined part Y and assemblies X and Z. The costs of the product Z calculated from the data base shown in Figures 3 and 4 are presented in the calculated costs table (Table 2).

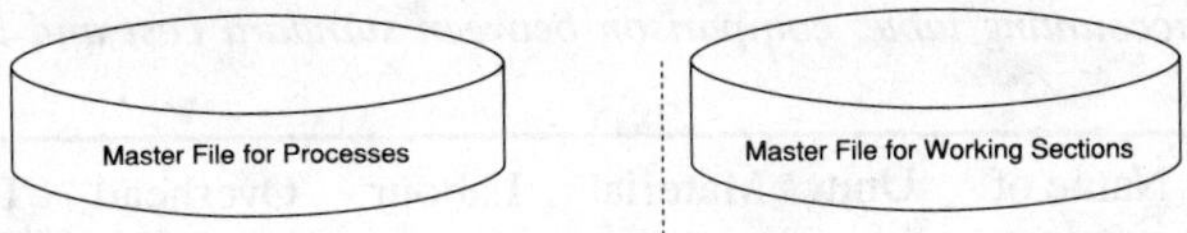

[Data Contained]

Component	Process No.	Description of Process	Standard Hours	Working Section No.	Working Section No.	Wage Rate	Ratio of Indirect Cost
Z	1,000	Component Y & X	2	1001	1001	¥ 50	¥ 70
X	2,000	Component Y & W	1	2001	2001	¥100	¥180
Y	3,000	Cutting	1	2002	2002	¥ 90	¥200
	4,000	Grinding	0.5	2002		¥ 45	¥100

Figure 4. Data on manufacture engineering.

Table 2. Cost accounting table for product Z.

Component No.	Item	Material cost	Labour cost	Indirect Manufacturing cost	Total
Component V	Total	¥ 1000	¥ —	¥ —	¥ 1000
Component Y	Component V	1000	—	—	1000
	Process 3000		90	200	290
	Process 4000		45	100	145
	Total	1000	135	300	1435
Component W	Total	1000	150	150	1300
Component X	Component Y	1000	135	300	1435
	Component W	1000	150	150	1300
	Process 2000		100	180	280
	Total	2000	385	630	3015
Product Z	Component X	2000	385	630	3015
	Component Y	2000	270	600	2870
	Process 1000		100	140	240
	Total	4000	755	1370	6125

Decision support system based on cost comparison

When the part number is input into the computer, the single-level parts list is used to look up the standard and latest values of the materials, labour and indirect manufacturing costs (see Table 3). It can readily be seen from the table that for the materials cost alone, there is a difference of 160 between the standard and latest costs. This indicates that the price of the component parts has risen 160 since the standard cost was determined. In the calculated cost table, the piled-up cost is shown as both 'This Level Costs' and 'Total Costs'. However, in this example, all the parts are assembled to produce a double door set, so that the process costs are 'This Level Costs' and the cost of all the parts is added to the 'This Level Costs' to give the 'Total Costs' of the double door set. Unlike the production instruction number system, it can be seen that the production instruction number of a part is not attached to the order forms for all the component parts when using the MRP production management control system.

Consequently, the cost calculation for a product manufactured under MRP is carried out by a 'piling-up' calculation through the levels of a multi-level parts list. To take an example, the cost of sub-subordinate parts are calculated and accumulated into the cost of the subordinate parts which they make up, and then

Table 3. Cost accounting table: comparison between standard cost and latest cost (from Fujimoto, 1980).

Component No.	Name of product	Unit	Material cost	Labour cost	Overhead	Total	
AA—01	Double Door Set	set	1240·65	751·50	1608·65	3600·65	Total cost
			·00	46·00	69·00	65·00	This level
	Subordinate Component Quanity			Latest Cost			
AA—02	Right door	1 set					
AA—03	Left door	1 set	610·02	342·00	748·00	1700·02	
AA—08	Centre post	1 set	569·22	342·00	748·00	1659·22	
AA—13	Hinges	6 each	43·31	21·50	43·50	108·31	
AA—14	Screws	31 each	15·00	0·00	0·00	15·00	
			3·10	0·00	0·00	3·10	
			0·00	46·00	69·00	115·00	This level
			1400·65	751·50	1608·50	3760·65	Total cost

the costs of these parts are compiled into the cost of the parent part which they constitute. This method of making piling-up cost calculations using multi-level parts tables has the merit that it can determine both the standard and latest costs of a product.

The advantages of using this new cost management system can be summarized as follows:

(1) Monthly balances are facilitated by the standard cost calculation.
(2) The system helps determine cost management objectives. Cost fluctuations are apparent from the differences between standard cost and latest cost, and these figures provide a valuable aid to cost management of each part.
(3) The system is an aid in meeting the target prices set at the cost planning stage, as it readily provides estimated costs (latest costs) due to design changes, equipment installation and changes between in-house and subcontracted manufacture at the product planning stage.
(4) The latest cost figures allow the business department to get a grasp of profit margins to determine management strategy, such as the ideal mix of product.
(5) The system aids the business department in determining the sale price of products by providing latest cost figures to carry out profit and loss calculations on ordered products.

Conclusion: my profit growth management method

After the management of a company has drawn up a profit plan, it is extremely important that they work out a management strategy based on management information supplied from performance evaluation criteria when implementing the plan. In many ways the clearest way of determining the relative value of management is the relation between market share and the ratio of profit to capital. In the case of a closed corporation other factors, including the administrative capabilities of the management, the power of the management to protect the corporation against the competition, credit-worthiness of the management, and the personal assets of the management are particularly important. In the case of the management of a non-closed corporation other factors also apply in addition to these, such as political, social and economic factors affecting the company and the general environment. It is also very important that the management has the right personality to permit it to act rationally to protect the company in the face of rapid depression and other economic uncertainties.

The following methods are available for the quantitative evaluation of the management:

(1) analysis of the flow of funds from fund statements,
(2) the static liquidity determined from the level of the liquidity ratio, and the dynamic liquidity determined from the asset turnover period, debt turnover period and the ratio of current income to current expenses,
(3) profitability level determined from the ratio of profit to capital.

These indices are very important for the individual, concrete appraisal of a company. In particular, the following four items before and after tax must be investigated when considering the profitability (item 3) above:

(a) the standard gross capital current profit rate,
(b) the standard operating capital current profit rate,
(c) the standard owed capital current profit rate,
(d) the standard capital current profit rate.

For example, if the capital interest and financial expenses are approximately 10% of the gross capital current profit and operating capital profit rate, and the risk is 3%, each profit rate should be 13% after tax. Before tax, assuming that the tax rate is 50% for capital interest and financial expenses, 23% profit should be gained in each case. Since 20% of gross capital of Japanese industries is owed capital, capital interest on owed capital is 10%. Thus, the standard gross capital current profit rate is required to be 5% after tax and 7% before tax, assuming that the tax rate is 50%.

The standard owed capital current profit rate is 25% after tax, 35% before tax. The standard profit rate to the capital is (if the capital proportion is 15%), 33% after tax and 46% before tax.

The figures from the theoretical calculations shown above may be extremely high. In determining companies' criteria for profit increase management, it is important to carry out management control in accordance with relative values and absolute financial values.

Bibliography

Fujimoto, K., 1980, Serving the big manufacturers: how to cope with short lead time and changing delivery schedules.

Hall, R.W., 1981, Driving the Productivity Machine: Production Planning and Control in Japan. APICS.

Hall, R.W. and Vollman, T.E., 1978, Planning your material requirement. *Harvard Business Review*, **56**(5), 105–112.

Monden, Y., 1983, *Toyota Production System* (Industrial Engineering and Management Press, Institute of Industrial Engineers).

Mori, M. and Harmon, R.L., 1980, Combining the Best of the West with the Best of the East—MRP and KANBAN working in Harmony. APICS.

44

Optimum surface feet in flexible manufacturing systems workstations

Layek Abdel-Malek

Department of Industrial Engineering, New Jersey Institute of Technology, Newark, NJ 07102, USA

Abstract

This chapter studics the effect of metal cutting economics on the cost of products in a flexible manufacturing environment. Although the literature in the area of flexible manufacturing is extensive, it only emphasizes the effect of in-process inventory, routing and scheduling of jobs on the performance of such systems. The results of this study show that metal cutting economics is a dominant factor in minimizing the cost per part. Open and closed networks of queues are used to model the different types of flexible manufacturing systems. Taylor's formula is utilized to evaluate the tool life.

Introduction

With the increasing acceptability of flexible manufacturing systems (FMS) in the workplace and the large amounts of capital invested in their equipment and operation, utilization of such systems becomes an important factor in determining their economic feasibility. Researchers aware of this fact spared no effort in the development of analytical models and simulation techniques to evaluate the performance of these systems and identify the parameters governing their performances. Stecke (1983) was among the first to define a set of production planning problems that must be addressed for efficient use of an FMS. These problems are machine grouping, machine loading, part selection, part routing, and resource allocation. Mathematical as well as heuristic solutions for these problems are presented in her work.

In this chapter we focus on an important measure of performance which has not yet been sufficiently addressed, that is the surface feet of metal cutting workstations in a flexible manufacturing environment. Groover (1987) states that 60% of the equipment in today's FMS are for metal cutting. Mainly such systems are composed of numerically controlled milling, turning and drilling machines. So, it is eminent that the operational conditions of such machine tools affect the economics of the whole system.

Central to the performance of machine tools is the metal removal rate, because this rate, in surface feet, directly affects the tool life. Taylor (Boucher, 1987) empirically showed that the tool life is inversely proportional to the cutting velocity (surface feet). With minor variations and refinements, the formula developed by Taylor in 1907 remains the base for metal cutting economics analysis.

The major objective of this work is to find the operational conditions of the different machine tools in a flexible manufacturing environment, mainly the surface feet, in order to minimize the production cost per unit.

Analysis

Several researchers have shown that queuing networks could be used to analyse the performance of FMS (Abdel-Malek in press; Chen *et al.*, 1988). In this chapter we use two models to describe the performance of different types of FMS. The first model is an open network of queues which is found to be appropriate for flexible transfer lines. A flexible transfer line (FTL) is defined as a series of numerically controlled machine tools in tandem joined together by an automated material handling system. Jobs in this type of system usually follow the same sequence of operations.

The second model is a closed network of queues. This model is suitable for FMS. Yao and Buzacott (1985) developed a closed queuing network model. They assumed in their model that the system model for an FMS could process Z part types, $z = 1, 2, \ldots Z$. The total number of parts from the different types, m, is always constant. That is, when a part is finished a similar part replaces it keeping the total number of parts in the system constant at m.

Ross (1985) argues that, if the processing time and the job arrival rate enjoy Markovian properties (see Abdel-Malek, preprints, for further details), the sojourn time (the flow-time), the time a job spends until it is fully processed could be expressed as:

$$W = \sum_{i=1}^{N} \frac{1}{\mu_i - \lambda} \tag{1}$$

for an FTL and

$$W = \sum_{j=1}^{K} W_m(j) = \sum_{j=1}^{K} \left[\frac{1}{\mu_j} + \frac{(m-1)\pi_j W_{m-1}(j)}{\mu_j \sum_{i=1}^{k} \pi_i W_{m-1}(i)} \right] \tag{2}$$

for an FMS

where λ is the jobs arrival rate for an FTL, μ_i is the effective mean processing time of a certain job on machine i, π_i is the steady-state probability (the percentage of time machine i is used in an FMS, note that the arrival rate to a certain node, machine, in a closed network is a function of π_i), N, K are the numbers of workstations in an FTL and FMS, respectively.

In the three succeeding sections we present the cost equation and its application to find the surface feet in FTLs and FMSs metal cutting workstations that is most economical.

The cost equation

As mentioned earlier, central to the cost equation is the formula developed by Taylor in 1907, relating the surface feet, V, to the tool life by

$$T = (\psi/V)^{1/\phi} \tag{3}$$

where T is the tool life in minutes; V is the surface feet; ψ, ϕ are constants depending on the cutting tool.

In this analysis the parameters of eqn (3), T, ψ, V, ϕ, will be given the index i, to refer to machine i in the system under consideration.

In general the processing time of a job on a certain machine tool, i, could be expressed as

$$t_{p_i} = t_{a_i} + f_i t_{c_i} + t_{l_i} \tag{4}$$

t_{a_i} is the average spindle time (machining time) of a product on machine i, t_{c_i} is the tool changing time of the worn-out tool on machine i (that is, the time required to replace a blunt tool by a new one), $f_i = t_{a_i}/T_i$ is the fraction of the tool lifetime that wears out during machining of a certain job on machine i, t_{l_i} is the time required to load/unload the job on machine i (in modern machine tools, fixtures are usually provided for the job to be processed next, thus reducing t_{l_i} significantly).

The dominant costs of a job at metal cutting station i are those of machining, tooling, handling, and tied-up capital (in-process inventory). In this work our objective is to minimize these costs which are affected by metal cutting conditions, that is, machining (spindle time), tool handling, and tied-up capital (in-processing inventory) costs. Hence, the cost per part equation of station i could be expressed as:

$$C_{p_i} = \left[C_{a_i} t_{p_i} + f_i C_{t_i} + h_i W_i\right] \tag{5}$$

where C_{a_i} is the cost of processing a job on machine i per unit time, C_{t_i} is the tool cost per edge used by machine i, h_i is the inventory carrying cost per part per unit time in station i, W_i is the expected time that a job spends in station i.

It should be noted that the cutting speed V_i, in a flexible manufacturing environment directly affects f_i, t_{p_i}, and w_i, that is the tool life, the machining cost, and the inventory carrying costs, respectively, at each machining station.

Minimum cost surface feet for an FTL

Consider an FTL with N workstations in tandem where jobs arrive at random from a Poisson stream with a rate λ. Consider further that out of this N workstations, there are K machine tools with processing times following negative exponential distributions with means expressed by eqn (4). Consequently, the metal removal cost equation could be given by:

$$C_p = \sum_{i=1}^{K}\left[C_{a_i} t_{p_i} + f_i C_{t_i} + \frac{h_i}{\mu_i - \lambda}\right]. \tag{6}$$

(Note that μ_i is the reciprocal of t_{p_i} given in eqn 4.)

To find the optimum controllable operational parameter of each metal cutting workstation, we consider the optimality condition of eqn (6) with respect to the velocity, i.e. $\partial C_p/\partial V_i = 0$. Keeping in mind that V_i (the surface feet of station i) is a function of the inverse of t_p, the differentiation yields the following expression for the optimum surface feet of station i.

$$V_i^* = \frac{\psi_i}{\left(\frac{1}{\delta_i} - 1\right)^{\phi_i} (C_{t_i} + t_{c_i} C_{a_i})^{\phi_i}} \left[C_{a_i} - \frac{h_i\left[t_{c_i}\left(\frac{1}{\phi_i} - 1\right)\left(\frac{V_i^*}{\psi_i}\right)^{1/\phi_i} - 1\right]}{[g\{V_i\}]^2}\right]^{\phi_i} \tag{7}$$

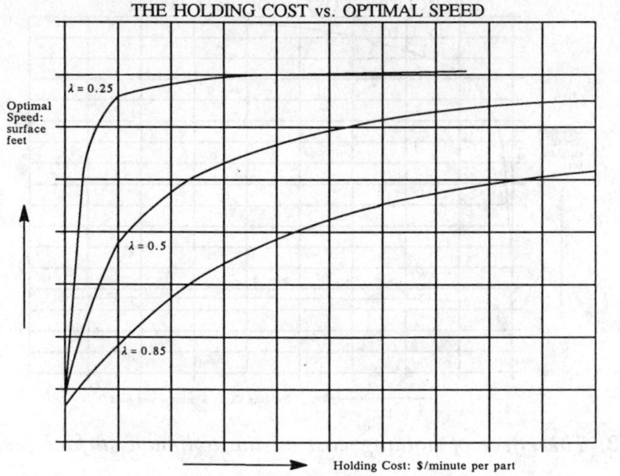

Figure 1. The effect of holding cost on the optimum surface feet in FTL.

where

$$g\{V_i^*\} = 1 - \lambda\left\{\beta/V_i^* + [\beta/V_2^* \cdot t_{c_i}(V_i^*/\phi_i)]^{1/\phi_i} + t_{l_i}\right\}$$

and β is a function of a particular job contour length (note that the machining time of a part, t_{a_i} can be expressed as a function of β/V_i).

Hence, the optimum velocity could be obtained by solving eqn (7) recursively.

Analysing the above equation it could be observed that as the inventory holding cost increases, the rate of change in the optimum velocity approaches zero. The upper limit for the optimum velocity is given by:

$$\lim_{h_i \to \infty} V_i^* = \psi_i \left[\left(\frac{1}{\phi_i} - 1\right) t_{c_i}\right]^{-\phi_i} . \tag{8}$$

Exhaustive numerical analysis was performed for a wide range of tool parameters, job types, inventory holding cost rates, machining costs and tool costs. The analysis showed that the optimum surface feet, velocity, approaches an upper limit as the work in process inventory hold cost increases. The analysed range of parameters showed the same pattern which is exhibited by Figure 1.

Minimum cost surface feet for FMS metal cutting workstations

An analysis similar to that of FTLs was performed for FMS metal cutting workstations. The results of the analysis, as expected, were basically like those obtained for FTLs. Equation (6) could be modified to express the metal cutting cost in a flexible manufacturing system as follows

$$C_{Pf} = \sum_{j=1}^{K} [C_{a_j} t_{p_j} + f_i C_{t_j} + h_i W_m(j)]. \tag{9}$$

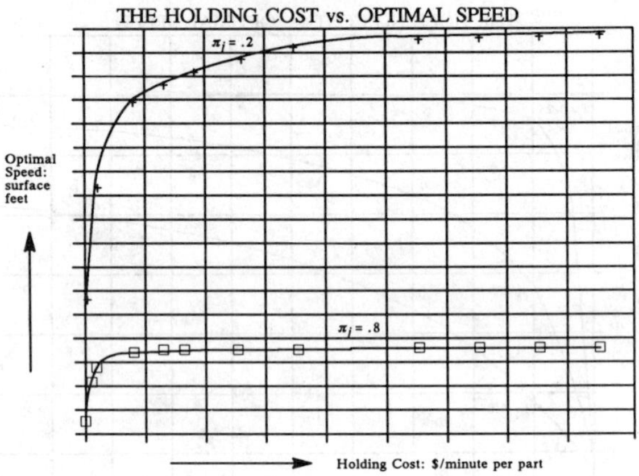

Figure 2. The effect of holding cost on the optimum surface feet in FMS.

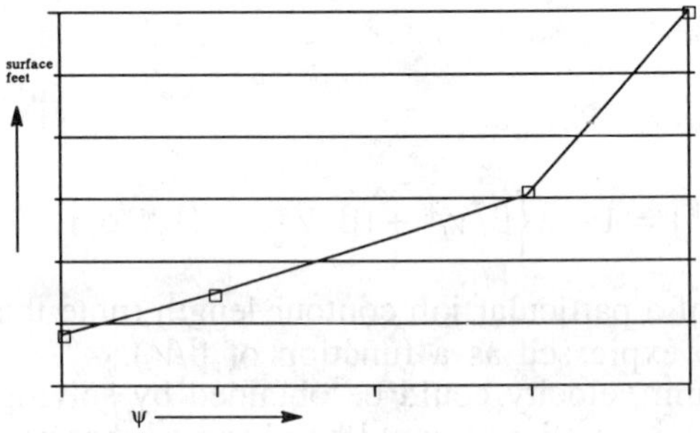

Figure 3. The effect of the tool constant ψ on the optimum surface feet.

As in an FTL, the optimum surface feet can be obtained by setting $\partial C_{p_f}/\partial V_j = 0$ for $j = 1, 2, \ldots, K$. A computer program was developed to obtain numerical values for the optimum velocity depending on the different operating parameters.

In this case, it was found also that when the inventory holding cost approaches infinity, the optimum surface feet has an upper limit which could be expressed by:

$$V^*_{\substack{j\\ h_l \to \infty}} = \left[\frac{g_j(V^*, \pi, \beta)\left(W_m(j) - \frac{1}{\mu_j}\right)}{\left(\frac{1}{\phi_j} - 1\right) t_{c_i}} \right]^{\phi_j} \psi_j \tag{10}$$

where $g_j(V, \pi, \beta)$ is a function of the velocity, V, steady-state probability, π_j, and job contour length, β_j, of that considered workstation.

The numerical computations for FMS systems are shown in Figure 2. The figure exhibits the asymptotic nature of the optimum surface feet as the inventory costs increase.

Further numerical results

We studied the effect of increasing the value of ψ, defined in the literature as the constant representing the velocity of cutting in surface feet per minute for one minute of tool life (see Doyle *et al.*, 1985). It is found, as shown in Figure 3, that using ceramic tools which have higher values of ψ, leads to relatively higher

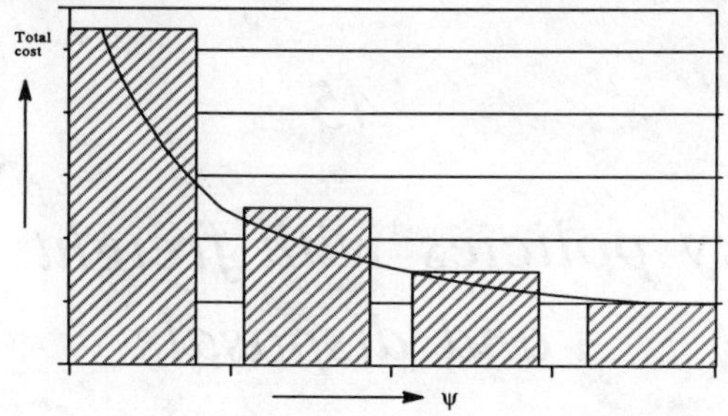

Figure 4. The effect of the tool constant ψ on the product cost.

operational surface feet. It is also found that the higher the optimum surface feet the lower the cost, provided that the tool cost does not increase sharply. Figure 4 shows the exponential decrease rate in the total cost as the value of ψ increases.

Conclusion

This work extends the preprints by Abdel-Malek. Its focus is determining the optimum surface feet in a flexible manufacturing environment. The objective is to minimize the unit cost per part which mainly consists of machining cost, inventory holding cost and tool cost.

We modelled flexible transfer lines and flexible manufacturing systems as open and closed queuing networks, respectively. Both analytical and numerical results showed that the central operational issue that governs the economic performance of this type of system is 'tool life'. Although the work in-process inventory cost affects the optimum surface feet, its effect diminishes as it increases.

In order to increase the efficiency of such systems, in addition to the ongoing research in the areas of machine loading, routing and reduction of work in-process inventory, researchers on the hardware side should continue to develop mechanisms that reduce the tool changing time t_c and tools with higher ψ. The investigation of these two parameters (t_c, ψ) showed their significant effect on reducing the unit production cost. It should be mentioned that ceramics, which modern tools are made of, have relatively high ψ values.

References

Abdel-Malek, L., in press, Minimum cost throughput of metal cutting work stations in a flexible manufacturing environment, *International Journal of Production Planning and Control*.

Boucher, T., 1987, The choice of cost parameters in machining cost models. *The Engineering Economist*, **32**, 3, 217–230.

Chen, H., Harrison, J., Mandelbaum, A., Van Ackere, A. and Wein, L., 1988, Empirical evaluation of a queuing network model for semiconductor wafer fabrication. *Operations Research V*, **36**, 2, 202–215.

Doyle, L.E., Keyser, C.A., Leach, J.L., Schrader, G.F. and Singer, M.B., 1985, *Manufacturing Processes and Materials for Engineers* (Englewood Cliffs, NJ: Prentice-Hall).

Groover, M., 1987, *Automation, Production Systems, and Computer Integrated Manufacturing* (Englewood Cliffs, NJ: Prentice-Hall).

Ross, S., 1985, *Introduction to Probability Models*, 3rd edn. (London: Academic Press).

Stecke, K.E., 1983, Formulation and solution of nonlinear integer production planning problems for flexible manufacturing systems. *Management Science*, **29**, 3, 273–288.

Yao, D. and Buzacott, J.A., 1985, Modeling a class of state dependent routing in flexible manufacturing systems. *Annals of Operations Research*, **3**, 153–167.

45

Inventory policies with freight discounts and disposals

F.J. Arcelus and J.E. Rowcroft

University of New Brunswick, PO Box 4400, Fredericton, New Brunswick, Canada E3B 5A3

Abstract

This chapter considers the effect of freight discounts and disposals on the inventory policies of a profit maximizing firm with price elastic demand and with price as a mark-up on the acquisition cost. In addition, the cost of placing an order is in part dependent upon a freight–cost structure, sensitive to the size of the order and upon the possibility of selling unwanted inventory at a lower price. A numerical example is used to illustrate the resulting inventory policies.

Introduction

A common problem facing potential buyers is the selection of the most suitable purchase price structure from among several quantity discount schedules. The problem is further compounded when discounts also occur in the transportation costs. Quite often, these opportunities are rejected as uneconomical, when in fact they may be beneficial if a secondary market is found for the excess inventory to be liquidated at less than full price or even at a loss. Even if the existence of such opportunities for disposing of unwanted stock is considered to form an integral part of the pricing decision in practice, its importance has not been duly recognized in standard inventory theory. This chapter represents a first attempt at closing the gap by modelling the inventory decision in the presence of purchase price and freight rate discounts and with the possibility of disposals.

The existing literature treats the various elements separately. The quantity discount problem is well known (e.g. Silver and Peterson, 1985; Dolan, 1987) and requires no further explanation, beyond observing that discounts may also affect the demand rate, a fact only recently incorporated in inventory models (e.g. Arcelus and Srinivasan, 1987; Arcelus and Rowcroft, 1989). Freight discounts have also been discussed extensively in the literature (e.g. Jucker and Rosenblatt, 1985; Lee, 1986; Dolan, 1987; Knowles and Pantumsinchai, 1988). The question of disposals has been studied by Sethi (1984) within a multiple price-break setting with no freight or demand rate changes.

For ease of exposition, we make the simplifying assumption of only one price break. This is not as restrictive as it appears, since, as Moorthy (1987) shows, it is sufficient for the case of homogeneous customers. Moreover, the extension of the model to the multiple price-break case is straightforward. Further, Dolan (1987) shows that, among these homogeneous buyers, those using some form of lot-sizing rule and downward sloping demand functions are more likely to be receptive to quantity discounts. Our model reflects these findings. Further details of the model appear in the next section, to be followed by a numerical example. A conclusion section completes the chapter.

The model

The model considers an all-units discount schedule which allows the buyer to purchase z units per year in lots of size q at a cost of c if q does not exceed the quantity break, q_0, and of c_0, otherwise, where

$$c_0 = r_0 c \qquad 0 < r_0 \leq 1. \tag{1}$$

Holding stock on hand is costed at a rate of h per unit value. Freight charges are modelled as part of the schedule in eqn (1). There is also a provision for a car-load type of fixed charge, which, when imposed, results in an increase in the ordering costs from the regular j per order to j_0. Disposals enter the model as a result of the assumption that the discount in eqn (1) does not yield enough of an incentive for the buyer to acquire the additional units. As a result, we allow for the possibility of purchasing q_0 units of which $q < q_0$ are sold at the regular unit price of p, where p is a mark-up, m, on unit cost,

$$p = mc \qquad m \geq 1. \tag{2}$$

The remaining $(q_0 - q)$ units are disposed of at a unit price of $\hat{c}$, with

$$\hat{c} = \hat{r}c \qquad \hat{r} \geq 0 \tag{3}$$

where $\hat{c} < c_0$ to avoid the case of increasing profit by buying specifically for immediate disposal.

On the basis of the study by Dolan (1988) alluded to earlier, the yearly demand rate, z, may be represented by a downward sloping function of p, with a price elasticity of G, i.e.

$$z = k_z p^G \qquad G \leq 0. \tag{4}$$

The effect on z of the discount depends upon the value of G and of the buyer's market strength. We start with the extreme case where demand is unaffected by the discount. This may be due either (i) to the existence of a totally inelastic demand ($G = 0$); or (ii) more likely, to the buyer's market strength, which allows for the adjustment of the mark-up so as to render p in eqn (2) and, hence z in eqn (4), unchanged in value. At the other extreme, a firm with no market power would be led to pass the entire savings on to its customers, by reducing the resale price accordingly. In general, the savings will be shared between the firm and its customers. Thus the effect on the mark-up may be represented by μ_0, where

$$m_0 = \mu_0 m \qquad 1 \leq \mu_0 \leq 1/r_0. \tag{5}$$

Using eqns (1) and (5), expressions (2) and (4) may be rewritten, respectively, as

$$p_0 = m_0 c_0 = \mu_0 r_0 p \tag{6}$$

and

$$z_0 = k_z p_0{}^G = z(\mu_0 r_0)^G. \tag{7}$$

Then combining (1)–(7), the yearly inventory cost function, $TC(q)$, consists of the sum of the purchasing, the ordering and the holding costs, as follows:

$$\begin{aligned} TC(q) &= cz + jz/q + hcq/2 \qquad \text{without discount} \\ &= c_0z_0q_0/q + j_0z_0/q + hq[c_0q_0 - \hat{c}(q_0 - q)]/2q \qquad \text{otherwise.} \end{aligned} \tag{8}$$

The corresponding revenue functions, $R(q)$, may be written as

$$\begin{aligned} R(q) &= pz \qquad \text{without discount} \\ &= p_0z_0 + \hat{c}(q_0 - q)z_0/q \qquad \text{otherwise.} \end{aligned} \tag{9}$$

To derive the second portions of eqns (8) and (9), recall that it is not beneficial to order more than q_0 units per order. Otherwise, there is no need for disposals. As a result, the only feasible discounted order size is q_0, of which only $q < q_0$ are needed. Hence, z_0/q orders of size q_0 must be placed, for a total yearly purchase of z_0q_0/q units, at a cost, from eqn (8), of $c_0z_0q_0/q$. From this purchase, only z_0 units are resold at the regular price of p_0. The rest, $(q_0 - q)z_0/q$, are sold off at the lower price, $\hat{c}$. The sum of these two types of sales yields the revenue function in eqn (9). In addition, it is assumed that disposals occur at the time of arrival. Thus, the average level of inventory at hand at any point in time is $q/2$, rather than $q_0/2$ and the value of those units equals the purchase cost of each order, c_0z_0, minus the receipts from disposals, $\hat{c}(q_0 - q)$. This justifies the expression for the holding costs in eqn (8).

Finally, subtracting eqn (8) from eqn (9) yields $PR(q)$, the profit function, where:

$$\begin{aligned} PR(q) &= c(m - 1)z - jz/q - hcq/2 \qquad \text{without discount} \\ &= z_0(p_0 - \hat{c}) - z_0[j + (c_0 - \hat{c})q_0]/q - h[c_0q_0 - \hat{c}(q_0 - q)]/2 \qquad \text{otherwise.} \end{aligned} \tag{10}$$

Using eqn (10), the buyer is confronted by two choices: either to buy $q_1 < q_0$ units at a unit cost of c or buy q_0 units at the discounted rate of c_0, of which only q_2 units per order are needed for regular sale. For the optimality conditions, the following results can be readily established.

Proposition 1. If feasible, the optimal values of q_1 and q_2, denoted by q_1^* and q_2^*, are given by

$$q_1^* = \sqrt{2jz/hc} < q_0$$

and

$$q_2^* = q_1^* \sqrt{(\mu_0 r_0)^G [j_0 + q_0 c(r_0 - \hat{r})]/j\hat{r}} < q_0.$$

Proposition 2. Select the discounted option if $PR(q_2^*) \geq PR(q_1^*)$, i.e. if

$$q_2^* \leq q_D = \{[(\mu_0 r_0)^{G+1} - 1]2mz - hq_0(r_0 - \hat{r}) - 2z[\hat{r}(r_0\mu_0)^G - 1] + 2hq_1^*\}/2h\hat{r}.$$

Propositions 1 and 2 depict a decision process which depends upon the various parameters related to the costs of disposals, $\hat{r}$, purchasing, r_0, and ordering, j_0, as well as to the markup, μ_0, and the quantity break, q_0. We now present a numerical example designed to illustrate the properties of the model and the sensitivity of the optimal solution to fluctuations in the parameters.

Table 1. Sensitivity analysis.

Case	VP [a]	Value of VP	q_1^*	$PR\ (q_1^*)$	$PR\ (q_2^*)$	q_2^*	q_D
1 [b]	Basic	Model	125·71	34·42	34·85	332·60	334·75
2	j_0	26			34·34	335·15	334·75
3		27			33·83	337·68	334·74
4	q_0	405			34·53	333·87	334·44
5		410			34·21	335·15	334·13
6	r_0	0·92			27·40	333·63	298·54
7		0·94			20·65	333·04	264·18
8	$\hat{r}$	0·82			35·90	307·64	314·86
9		0·84			37·38	281·82	295·92
10	μ_0	1·061			34·22	328·35	327·34
11		1·067			33·52	323·75	319·29

[a] VP, Parameter whose value differs from that of Case 1.
[b] Parameter values for Case 1 are: $j = 15$; $j_0 = 25$; $h = 0{\cdot}25$; $c = 1$; $G = -5$; $k = 1000$; $m = 1{\cdot}5$; $q_0 = 400$; $r_0 = 0{\cdot}9$; $\hat{r} = 0{\cdot}8$; and $\mu_0 = 1{\cdot}056$.

A numerical example

Table 1 summarizes the information needed for the analysis of this section. Case 1 is used as the standard of comparison. The other cases are characterized by the change in the value of the parameter specified in the VP column. For each case, Table 1 includes the optimal values for q_1 and q_2 given by Proposition 1 and their corresponding maximum profit. The value of q_D from Proposition 2 is also provided to illustrate its function as a yardstick for the selection of the optimal inventory policy. Interpretation of the results is straightforward. The attractiveness of the discount option decreases with increases in the cost parameters for that option, as shown by Cases 2 and 3 for j_0 and by Cases 6 and 7 for r_0. Increases (decreases) in the mark-up (Cases 10 and 11) will lead to higher (lower) resale prices, which will result in a detrimental (positive) effect on the demand rate and therefore on profits. Furthermore, since the return on disposals is independent of the demand rate, then the higher the value of $\hat{r}$ (Cases 8 and 9) the more profitable the secondary market is and hence the higher the likelihood of selecting the discount option. Finally, the size of the breakpoint q_0 (Cases 4 and 5) has a positive influence on the probability of rejecting the discount. This is due to the fact that, for a given demand rate and cost structure, higher and higher values of q_2^* are required to be sold at full price to recover the additional purchasing costs, with the corresponding increases in holding costs.

Conclusions

The integration of purchasing and stock control policies in the presence of secondary markets is an important practical problem, relatively neglected in the inventory literature. This chapter represents an attempt to fill this gap. An area of future research interest to the authors is the consideration of multiple price-break structures under various freight rate modes.

Acknowledgement

Financial support to the first author from the Natural Sciences and Engineering Research Council of Canada for the completion of this research is gratefully acknowledged.

References

Arcelus, F.J. and Rowcroft, J.E., 1989, Generalizing the price-break EOQ problem. *Engineering Costs and Production Economics*, **17**, 183–195.

Arcelus, F.J. and Srinivasan, G., 1987, Inventory policies under various optimizing criteria and variable markup rates. *Management Science*, **33**, 756–762.

Dolan, R.J., 1987, Quantity discounts: managerial issues and research opportunities. *Marketing Science*, **6**, 1–22.

Jucker, J.V. and Rosenblatt, M.J., 1985, Single period inventory models with demand uncertainty and quantity discounts: behavioral implications and a new solution procedure. *Naval Research Logistics Quarterly*, **32**, 537–550.

Knowles, T.W. and Pantumsinchai, P., 1988, All-units discounts for standard container sizes. *Decision Sciences*, **19**, 848–857.

Lee, C.Y., 1986, The economic order quantity for freight discount costs. *IIE Transactions*, 318–320.

Moorthy, K.S., 1987, Managing channel profits: comment. *Marketing Science*, **6**, 375–379.

Sethi, S.P., 1984, A quantity discount lot-size model with disposals. *International Journal of Production Research*, **22**, 31–39.

Silver, E.A. and Peterson, R., 1985, *Decision Systems for Inventory Management and Production Planning* (New York: Wiley).

46

Evaluation of flexible capital investments—with consequences for decisions and suppliers

Ingvar Persson, Carl-Henric Nilsson and Håkan Nordahl

Department of Industrial Management, Lund Institute of Technology, Box 118, S-221 00 Lund, Sweden

Abstract

Capital-back is a method for evaluating capital investments, especially taking into account the different investment proposals' abilities to meet radical changes, if for instance, the market of the product produced by the machinery suddenly drops. Capital-back implies that every investment is divided into one flexible part and one inflexible part. The requirement for the flexible part is the normal discount rate. The remaining surplus goes to the inflexible part of the investment. The Capital-back method also invites an analysis of who is most eligible for owning and administering the flexible part of the capital investment.

Introduction

The environment of companies today is much more dynamic than it used to be; this manifests itself in a higher degree of uncertainty on the market, shorter product life cycles and a higher demand for customization. Flexibility in manufacturing is therefore of substantial significance as a means of strategic advantage for companies in the manufacturing industry.

Certain types of machinery are better prerequisited than others to meet the demands for flexibility. Especially well equipped are, for example, Flexible Manufacturing Systems (FMS) and industrial robots. These advantages must be taken into account in the capital budgeting process. With the Capital-back method the ability to handle the most radical changes is evaluated.

By way of introduction, the significance of flexibility will be discussed. Flexibility is a wide concept and here its relationship to Capital-back will be illustrated. Capital-back itself and its characteristics will then be described through a comparison with the Pay-back method. Conclusions will also be drawn concerning Capital-back and its consequences for decision making in companies as well as the opportunities it creates for suppliers of flexible machinery, such as robots. Finally, a deeper study of the Capital-back method is carried out concerning sensitivity to the rates of interest, the lifespan of the machinery, and an evaluation of the results is done.

Flexibility of the machinery

Flexibility related to the design of the products and the flow of goods is well analysed. Several authors (Gerwin, 1983; Browne *et al.*, 1984) have discussed and made definitions of flexibility in the manufacturing process. These definitions have one major shortcoming: the analysis is only executed in two dimensions, the product dimension and the material flow dimension. This classification does not consider the possibilities of re-using the equipment in different manufacturing processes. A method which takes this prospect into account already at the acquisition stage implies that the flexibility is evaluated as well as analysed.

Flexibility concerning investment analysis of the manufacturing process is given three dimensions: (1) material flow, (2) products, and (3) machinery.

Capital-back (CB) is a method for evaluating the flexibility of the machinery. Flexibility of machinery means having the possibility of using the machinery for other purposes than originally intended, either inside the company or in some other manufacturing process outside the company.

CB implies that every investment proposal is split into two parts, one flexible and one inflexible. It is true that the different parts constitute one and the same investment, but from the point of uncertainty they are quite different. The inflexible part of the investment is denominated 'The Risky Investment'.

One part of the investment (G_f) is flexible enough to be used for some other purposes. This may happen if, for example, the market expectations are not fulfilled, or if the machinery becomes obsolete due to a radical change in the manufacturing process. Examples of this type of equipment are robots, CNC machines or parts of FMS. Due to the flexibility and hence lower degree of uncertainty a lower interest rate is acceptable for the capital invested in this type of equipment.

The rest of the investment constitutes those parts that are dedicated for a specific manufacturing process or product. These parts are inflexible and belong to the part of the investment called the risky investment (G_r). The parts belonging to the risky investment cannot be used to produce other products in other manufacturing processes. The residual value is zero irrespective of how long the set-up has been used. Examples of this type of component are fixtures and gripping appliances that are intended for a specific product. Included as part of the capital investment are also expenditures for projection and design as well as for the physical installation. These expenditures are usually not possible to re-utilize if the project fails and are hence a part of the risky investment.

The total investment (G) is the sum of the flexible part of the investment and the risky investment, $G = G_f + G_r$. This categorization of the parts that constitute the capital investment into one flexible part and one risky is in itself an important distinction. One possibility is to stop here and specify a measurement of profitability together with the size of the flexible investment or the risky investment.

Capital-back and Pay-back

With the Capital-back method (Nilsson and Nordahl, 1988; Persson, 1988) we advance further and the profitability and the flexibility are weighted to form a single measurement.

The need of a new method depends on, for example, the high Pay-back demands in the manufacturing industry. The object of Capital-back (CB) is to release flexible investments of the unreasonably high demands of a short Pay-back (PB) period, requested due to a high degree of uncertainty and short-term planning. The PB demands are especially high for rationalization investments where, for example, robots are of current interest. According to Hayes and Garvin (1982), 25% of the American companies had a demand for a PB period no longer than three years. Ten

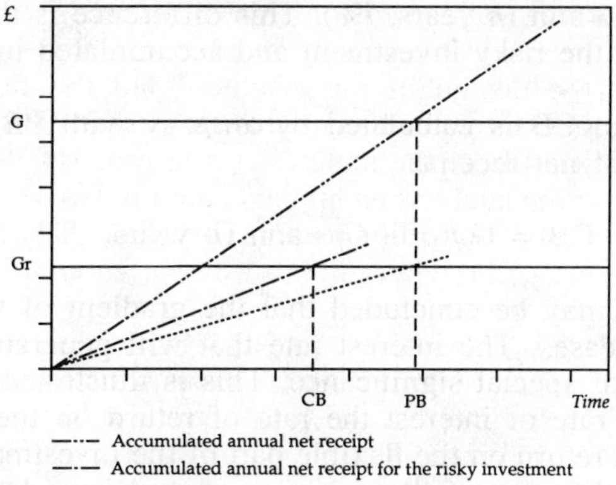

Figure 1. A comparison between Capital-back and Pay-back.

years earlier only 20% of the companies had this demand. PB demands in the Swedish industry range from 2·7–3·1 years (Yard, 1987). According to Yard its not quite clear if consideration has been given to the rate of interest or to the inflation. This means that taking these into consideration the possible interval in Swedish companies is a PB period somewhere between 2·4–4·3 years (Yard, 1987). Through our own research we have found PB demands as low as two years to be frequent.

PB emphasizes short-term planning; above all the method captures the liquidity aspect and the uncertainty. For flexible investments, like for instance, a robot unit excluding gripping appliances, etc. it can lead to the wrong investment decision if the uncertainty is focused by the high requirements for profitability. The advantage of CB is that it takes into consideration the uncertainty of the custom-made part and the requirement for profitability for the flexible part.

Calculation of the CB period thus implies that the flexibility and the profitability are weighted to form one single measurement. CB presents the PB period for the risky investment, while the flexible part of the investment generates a yield as high as the discount rate. The CB period can, alternatively, be described as the time it takes to consolidate the invested capital; this is the reasoning behind the CB concept. It is further understood that the flexible part of the investment is already consolidated through its flexibility.

CB is described in Figure 1. The horizontal lines show the capital investment, the lower line is the risky investment (G_r) and the upper line (G) is the total investment, $G = G_f + G_r$. The difference between the lines is thus the flexible part of the investment (G_f). The two lines proceeding from the origin, show how the accumulated annual net receipt for PB and CB grows as a function of time. The lines are straight due to the restriction that the annual net receipt is constant with time. The steeper curve is the accumulated annual net receipt for the total investment (a) in pounds (£). This amount accumulates with time and when it intercepts the upper line G, the PB period is reached.

When using CB, a cost for the flexible part of the investment is calculated. This cost reduces the accumulated annual net receipt thus levelling the line. When the lower line in the figure intercepts the risky investment, G_r, the CB period has been reached. The cost for the flexible part includes both depreciation and rate of interest. This can be calculated in different ways. The choice of calculation method affects the result. The most natural method to use in capital budgeting is the annuity method which provides a constant annual cost during the lifespan of the installation. If the interest rate is (i)% and the expected lifespan is (n) years the annual cost for the flexible part of the investment, according to the annuity method, is $G_f * \text{ann}$ (n years, i%). When this amount is reduced from the annual net receipt, the difference is: $a - G_f * \text{ann}$ (n years, $i\%$). This difference is then calculated as the annual net receipt of the risky investment and accumulated in the same way as in the PB method.

In analytical terms, CB is calculated by analogy with PB, i.e. the investment divided by the annual net receipt:

$$\text{CB} = G_r/(a - G_f * \text{ann}\ (n \text{ years},\ i\%)).$$

From the formula it may be concluded that the gradient of the line decreases as the interest rate increases. The interest rate that will generate a CB period equal to the PB period is of special significance. This is illustrated with the dotted line in Figure 1. At this rate of interest the rate of return on the risky investment is as high as the rate of return on the flexible part of the investment. (This calculation presupposes that the lifespan of the risky investment is as long as the lifespan of the flexible part of the investment).

The CB method is an especially valuable aid if the CB period is shorter than the PB period. If this is not the case the profitability for the risky investment is lower than for the flexible one.

The described calculation of the CB period is valid only if the annual net receipt is equal every year. If the annual net receipts vary, they have to be added together until the sum equals the amount invested in the risky part. Also this is analogical to the PB method.

The CB condition for the investment is then:

$$G_r = \sum_{x=1}^{\text{CB}} (a_x - G_f * \text{ann}\,(n \text{ years},\ i\%))$$

where a_x = annual net receipt year X (£), n = the lifespan of the flexible part (years), i = the discount rate of the company (%).

Consequences for decisions and suppliers

Consequences for the decision-making

Using the CB method will affect the investment process as well as the crucial decisions being made. CB presupposes that the components that constitute the capital investment are divided into one flexible part and one inflexible part. The process of dividing the components will force the decision makers to analyse as well as evaluate the flexibility of the different investment proposals. This process in itself is therefore important.

The rate of change in the manufacturing industry is constantly increasing. New models and new products are superseding each other at a rapid pace. This means that flexibility in the manufacturing process is, or ought to be, a strategic goal for companies in the manufacturing industry. CB is an excellent aid to incorporate the strategic importance of flexibility in the capital budgeting process. The CB method illuminates the importance of flexibility and promotes flexible investment-alternatives. Traditional capital budgeting techniques do not take flexibility into consideration but regard the flexible and the risky parts of the capital investment as equally uncertain.

Neither CB, nor any other capital budgeting technique gives any information about the degree of uncertainty that prevails *if* the inflexible, or risky, investment becomes obsolete. Neither will it tell anything about *when* the investment will be obsolete.

Financing and consequences for suppliers

Dividing the investment into one flexible part and one risky part will also have consequences for the financing of the capital investment. The risky investment will be regarded in the same way as the total investment is today. The company will buy the machinery and take all the risks. The flexible part on the other hand can be treated differently depending upon the nature of the company that is investing.

One possibility is that the company in a traditional manner invests in the flexible part as well as in the risky part of the investment. In this case the CB method will be a relevant capital budgeting technique to use when evaluating the proposal. Since the method assumes alternative uses for the flexible part, it is more consequent to consider the flexible part as a potential resource for all the different workshops in the company. Larger companies could buy the flexible part of investments on a company-wide basis. Proceedings such as this, with central administration, are already a reality in some companies today.

In reality different machinery has different degrees of flexibility. When CB is used the possibility exists to differentiate the rent according to the degree of flexibility. Doing this guides the different subdivisions toward more flexible machinery. Using CB will give the management an opportunity to release the right 'signals' early in the capital budgeting process. The subdivisions will accordingly receive an incentive to invest in flexible machinery.

An interesting alternative way to finance the investment is that the supplier of the flexible parts directly grants a lease of these parts. The supplier has a better knowledge of the market and is best fitted to find new users for the flexible machinery. The advantages of this procedure are greatest for smaller companies, lacking the possibility to find new users for the flexible machinery within their own company. The supplier's leasing firm will in this case take over the function of the central administration found in the larger company.

When working under uncertainty, an external leasing procedure such as the one described is favourable, especially for smaller companies. If the leasing procedure is used then CB is reduced to a PB calculation. The investment is the risky part and the flexible part is taken on lease thus charging the annual receipt with the rent.

Deeper analysis of the Capital-back method

The CB period is influenced by four factors:

(1) the flexible share of the investment,
(2) the level of the discount rate,
(3) the profitability of the total investment (measured with for instance the Internal Rate of Return (IRR) or the PB period),
(4) the lifespan of the flexible part of the investment.

There is no room to carry out a total analysis here, but we will illustrate how the different parameters influence the outcome. The flexible part of the investment has been chosen as main parameter. The effect on the CB period is then analysed for different levels of discount rate and profitability, and lifespan.

The discount rate level

In order to select discount rates we will proceed from earlier research. Those studies carried out in Swedish companies show that the average discount rate is 20% with no consideration given to taxation. The variation, however, is high. Yard (1987) found that the minimum discount rates varied between 10% and 30% and that the maximum rates amounted to 50%. Tell (1978) also found discount rates slightly

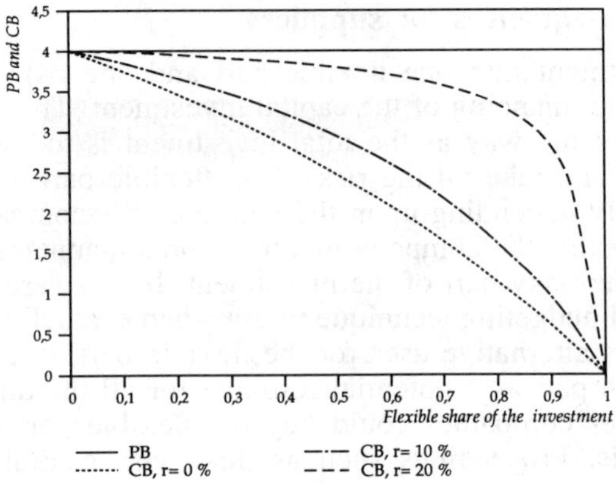

Figure 2. The CB period as a function of the flexible share of the investment. The discount rate is 0%, 10% and 20%. The PB period is four years and the lifespan of the flexible part is ten years.

below 20%. Neither could establish if the discount rates were real or nominal. According to Gitman and Forrester (1977) the discount rates in American companies were lower, in several cases between 10% and 15%. Hayes and Garvin (1982) mean that it is considerably higher.

With these results as a starting point the influence of the discount rate on the CB period is analysed for discount rates between 10% and 20%. In order to save space the total analysis is only carried out for a PB period of four years. This analysis is supplemented with a schematic description of what will happen if the PB period is shortened to two years. The extreme value of the zero per cent discount rate is also displayed for comparison. In the analysis it is presupposed that the annual net receipt is equal every year.

Figure 2 shows CB as a function of the share of flexibility in the total investment under the following restrictions:

— The lifespan of the flexible part of the investment is ten years.
— The profitability is expressed as the PB period, which is four years.
— The discount rate varies between zero and 20%.

The figure reveals that:

(1) The CB curves are decreasing which means that the CB period decreases when the share of flexibility increases. This is valid as long as the IRR is higher than the discount rate. When the share of flexibility approaches 100%, the CB period approaches zero.
(2) The CB period increases as the discount rate increases, which has been established earlier.

Now we will make a closer study of the curves starting with the discount rate of 20%. The CB period is here close to the PB period, which is four years. The significance of this is that the IRR is very close to the discount rate, in this case slightly above 20%. The share of flexibility has to be 70% or more in order to show significantly different periods for PB and CB.

If the discount rate is 10%, the CB period decreases more smoothly and the difference between PB and CB is more significant, even if the share of flexibility is low. This is representative of the CB method. A low discount rate generates a low annual cost for the flexible part of the investment. Thus the share of annual net receipt that goes to the risky investment increases. The CB period is then improved, i.e. becomes shorter.

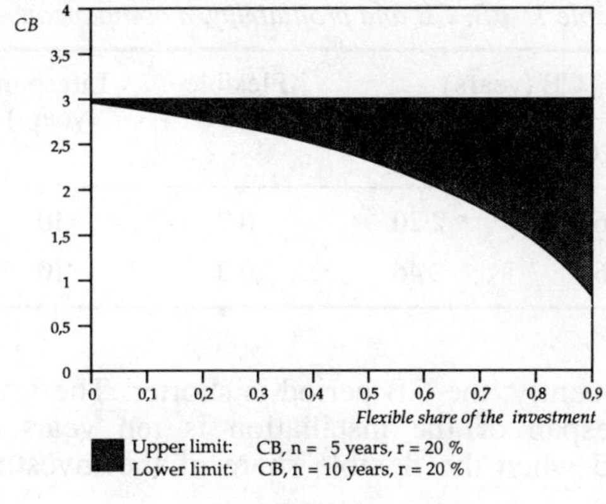

Figure 3. The CB period as a function of the flexible share of the investment. The lifespan of the investment is varied between five years (upper limit) and ten years (lower limit). The discount rate is 20% and the PB period is three years.

Profitability of the total investment

In Figure 2 the PB period is four years. If the PB period becomes any shorter, then the CB period becomes less dependent upon the level of the discount rate. This is due to a high IRR and the fact that the annuity method has been chosen to calculate the annual cost.

If the IRR is high, then the annual net receipt is high and, accordingly, the amount that goes to the risky investment. For instance, an investment with a PB period of two years has a very high profitability. If the lifespan of the total investment is ten years, IRR is close to 50%. If the demand for profitability of the flexible part of the investment is 20% then the remaining part of the IRR will be added on to the risky investment. This means that if the IRR is very high, then it does not matter much how high the discount rate is.

The result is also influenced by the annuity method. When using the annuity method the total cost of capital is distributed with an equal annual cost every year during the lifespan of the investment. This implies that the depreciation is lower during the first years. For instance if the lifespan of the investment is ten years, the first year's depreciation is 3·9% with a discount rate of 20%. CB is also compatible with other depreciation and rate of interest models. Those alternatives will, however, not be covered here.

The lifespan of the flexible part

The lifespan of the risky investment and the flexible part do not have to be equal. It is, for instance, possible that the lifespan of a robot is longer than that of manipulators and fixtures (Björkman and Ekdahl-Svensson, 1986). This will influence the IRR. If the flexible part of the investment consists of components with different expected lifespans, then the calculations will be affected too. In order not to complicate the analysis we will assume that all the parts of the investment have the same lifespan.

The dark field in Figure 3 shows the CB period for different lifespans of the flexible part of the investment. With a lifespan of five years, the CB period is very close to the PB period that is three years. This is dependent upon the fact that the discount rate and the IRR are very close to each other. In this example the IRR for the total investment is slightly below 20%. If the lifespan is longer, then the

Table 1. PB, CB and profitability, a comparison.

	PB (years)	CB (years)		Flexible share	Lifespan (years)	Internal rate of return (%)
		$r = 20\%$	$r = 10\%$			
A	4	3·61	2·20	0·7	10	21
B	3	2·67	2·46	0·3	10	30

IRR is higher. Consequently, the CB period is shorter. The lower limit reveals the CB period if the lifespan of the installation is ten years. The CB period is considerably shortened when the flexible share of the investment is increased.

Capital-back and priorities

An important factor is how the method orders the different investment proposals. We will look at the preferences when using CB compared to the preferences when using the IRR or PB. In Table 1 two different investment proposals, A and B, are compared. The profitability and the flexible share of the proposals are chosen in order to change the priorities when the discount rate is reduced from 20% to 10%. The flexible share of proposal A is high but it has a low IRR, 21%. This makes alternative A sensitive to the discount rate. Alternative B on the contrary has a low flexible share but a high IRR. This means that alternative B, is not so sensitive to changes in the discount rate.

For the competing investment proposals in Table 1 the CB recommendations are:

— A company with a high requirement for profitability, i.e. willing to take risks, will prefer proposal B. B has a high profitability, IRR 30%, and a high risk, 70% inflexible share.

— A company with a low requirement for profitability, i.e. risk-averse, will prefer proposal A. A has a low profitability, IRR 21%, and a low risk, 70% flexible share.

References

Björkman, M. and Ekdahl-Svensson, B., 1986, *Monteringsekonomi kostnadsmodeller för kalkylering och känslighetsanalys* (Linköping, Sweden: Linköping Studies in Science and Technology, no. 96).

Browne, J., Dubois, D., Rathmill, K., Sethy, S.P. and Stecke, K.E., 1984, Classification of flexible manufacturing systems. *The FMS Magazine*, April, 114–117.

Gerwin, D., 1983, *A Framework for Analyzing the Flexibility of Manufacturing Process* (School of Business Administration University of Wisconsin).

Gitman, L.J. and Forrester, J.R., 1977, Survey of capital budgeting techniques used by major US firms. *Financial Management*, Fall, 26–30.

Hayes, R. and Garvin, D., 1982, Managing as if tomorrow mattered. *Harvard Business Review*, May–June, 71–79.

Nilsson, C.-H. and Nordahl, H., 1988, *Investeringsbedömning av industrirobotar—värdering av Flexibilitet och Andra Svårkvantifierbara Aspekter* (Lund, Sweden: Institute f. Industriell Organisation, Lunds Tekniska Högskola).

Nordahl, H., Nilsson, C.-H. and Martins, G., 1988, Beurteilen der Investionen von Industrienrobotern. *Werkstatt und Betrieb*, **12**, 963–965.

Persson, I., 1988, Capital-Back—metod för riskanalys eller metod för Lönsamhetskalkylering. *Aktiv Industriautomation*, **1**, 6, 56–62.

Tell, B., 1978, *Investeringskalkylering i Praktiken* (EFI) (Lund, Sweden: StuderHitteratur).

Yard, S., 1987, *Kalkyllogik och Kalkylkrav—Samband Mellan Teori och Praktik vid Kravställande på Investeringar i Företag* (Lund, Sweden: Lunds University Press).

PART V
HUMAN FACTORS ISSUES IN PRODUCTION

To be successful in the future, companies will rely on a highly trained and motivated workforce. A number of approaches to work-organization have been tried in recent years to create socio-technical systems which make better use of the potential of employees, and give greater autonomy to work groups. In this section Lindér discusses the results of analyses of several such systems in the Swedish automobile industry and their contribution to total efficiency.

As the level of manufacturing technology increases so the nature of operators' jobs can change, placing greater emphasis on machine and system control. In this context Ogawa reports on a study to investigate the relationship between learning and perceptual motor load in a combined manual and decision task, and Gray *et al.* address some of the difficulties which may be encountered by programmers of industrial robots when using teach controls.

The safety and comfort of operators must always continue to be of major concern in the design of manufacturing systems, and three contributions are included in this area. Järvinen and Herranen describe an approach for reducing musculoskeletal risks in materials handling. Haslegrave reports on a study aimed at improving workplace design and working postures of maintenance workers, and Gregg and Corlett describe a multifactorial approach to the design of industrial seating.

SECTION V.1. ERGONOMICS AND HUMAN FACTORS

47

Sociotechnical evaluation of assembly systems in the Swedish automobile industry

Jan Lindér

Department of Industrial Management, Chalmers University of Technology, S-412 96 Göteborg, Sweden

Abstract

The competitive situation for Swedish car manufacturers has since the 1970s been characterized by increased demands on flexibility, high performance products and new work organizations. During the 1970s and 1980s a number of approaches were made to create new systems, often characterized by an attempt to use sociotechnical principles, in order to achieve a more attractive work content and higher efficiency. This chapter discusses the result of analyses made at some of these new plants. The analyses are made in the form of a comparative analysis between traditional systems (driven lines) and alternative production systems. The findings show how changes in the technical structure combined with work organization efforts significantly increase the total efficiency. The new systems are normally more flexible, and allow a work organization with more motivating work roles. As for the efficiency, increases in quality are significant. In total, most systems show a higher cost efficiency at the same time as increased flexibility.

Introduction

Efficient production system development is a function of both how well the internal factors are designed and how well the system is able to meet external demands. Suitable solutions will vary depending, for example, on the shape of the market and the product structure. This chapter discusses changes in production design aiming at both a better adaptation to external demands and increased efficiency. The point of departure is efforts made in the Swedish automobile industry during the 1970s and 1980s, where changed conditions generated by both the market and society have created a need for changes in production design.

From a theoretical point of view, these issues are of current concern within different types of theory postulation. This type of issue is treated in both the general management literature and research findings within sociotechnical theory and related fields. The overriding issue in this connection is how to create efficient systems from a production economy point of view while at the same time creating attractive work roles which, for the individual, offer something positive in return.

On the level of the individual, the issue of desirable work characteristics and their correlation with different result variables has been highly topical and theory in this area has gradually been refined (Turner and Lawrence, 1965; Hackman and Oldham, 1980). To date there are no clear guidelines for how to solve the problems of poor work content in complex industrial production systems with high demands on efficiency. The assembly of buses and trucks is one such area. These issues will be further treated below on the basis of empirical studies of Swedish plants.

Background

Since the early 1970s Swedish car manufacturers have made efforts to change the production design. The driving forces behind these efforts have been the need for increased flexibility, recruitment difficulties and quality problems (coupled with competence and motivation), among others. During the 1970s and 1980s various alternative concepts of car and truck assembly have been developed (Söderberg and Almgren, 1980; Engström and Karlsson, 1982), some of which have been put into practice. For some of the solutions extensive results of analyses have been published (Rationaliseringrådet, 1976; Agurén, 1984; Berggren and Holmgren, 1985), particularly in the case of Volvo Kalmarverken. However, no comprehensive comparison between the traditional systems and the newly developed Swedish systems has been made taking into account production economy efficiency and the quality of the work content.

The issues concerning an effective mix of social and technical/economic factors, given a certain level of competition and a certain product situation, are currently highly topical. Extensive development work is being carried out both within Volvo and Saab-Scania.

The aim of this chapter is to make a comparative analysis of traditional plants and the alternative systems built in the 1970s and 1980s. The analysis aims both at an evaluation of the potential of the technical design and at an assessment of how well the social systems are able to utilize this potential. An overall evaluation of the results of the analysis forms the basis of a discussion concerning the concepts now being put into operation.

Methodology

This chapter summarizes the results of various analyses made of Swedish car manufacturers. The analyses are based on comparative case studies of the traditional design (lines) and alternative systems. The study is based on an analysis of three plants for car assembly and two for truck assembly. In addition, two plants being started up have partly been studied.

The case studies were carried out from 1977 to 1988 comprising a period of analysis of approximately four months each. Some of the studies have been carried out by the author alone and some in co-operation with colleagues (Careborg and Lindér, 1977; Linder, 1984, 1985, 1988; Lindér and Johansson, 1986). In some cases the studies are based on analyses where the author has been project leader (Carlberg, 1985; Carlberg and Kullberg, 1985; Nilsson and Tunälv, 1985; Bergdahl and Johansson, 1987).

The data collection is primarily based on interviews and on an analysis of production technology data together with survey data. To the primary empirical results are also added data from other studies of the objects in question.

Evaluation of the systems is made in three dimensions.

Type of work organization—Description of organization and analysis of content of work roles.

System efficiency—The analyses of system efficiency based on development of the method for loss analysis described in Wild (1975) and Engström and Karlsson (1982). The method is based on a comparison between an ideal system (with no loss) and the actual results.

Flexibility—The point of departure is here an evaluation of flexibility described, for example, in Lindér (1984). In principle, this is done by separating the concept of flexibility in various dimensions and by making comparisons between different demands on flexibility and the characteristics of the system.

The results are given in relative numbers with the traditional system as a reference point.

Results

In this section the results of the analyses are briefly presented in the form of a comparative analysis between traditional systems and alternative systems.

Case A (Volvo Car Corporation)

Under this heading a comparison is made between Volvo Torslanda's assembly lines and the Kalmar plant. The results presented here are primarily based on Agurén (1984), Berggren and Holmgren (1985), Carlberg and Kullberg (1985), Nilsson and Tunälv (1985) and Lindér and Johansson (1986).

The traditional system

Flow and organization

In this case the technical base structure is a traditional mechanical assembly line with a cycle time of 2 min. The objects are high prestige passenger cars produced in a large number of model options. The work organization is based on a division of work into line operators, instructors, replacement personnel and foremen. Specially assigned personnel handle quality control and adjustments. Within each supervisor's area of responsibility a certain degree of job rotation has been implemented, which creates a certain degree of flexibility. The pay system is based on an agreement on performance in accordance with negotiated MTM rate.

The alternative system (The Kalmar Plant)

In this case the alternative is built on a base object flow where the objects are moved by auto-carrier. This system allows assembly carried out on a stationary object. Technically, the layout is based on a division into smaller units separated by buffers. The layout and the technical system forms the basis for different types of work organization. It is possible to carry out the operation from line to team work, where two men work together (doing the whole work content within the unit, so called dock assembly) as a basic unit. In the standard case (a pattern relatively quickly established after start-up) the possibility of dock assembly is not used but a somewhat simpler organizational principle based on work expansion and job rotation.

The greatest differences in Kalmar are assembly on a stationary object, cycle time 4–16 min, a more direct quality feedback and a larger work content.

At the time of the analysis of line versus Kalmar the same object was produced in both systems. The production mix, in part corresponding to complexity, is comparable, possibly with some advantage on the part of Kalmar. With respect to volume the line corresponds to about 1·5 Kalmar units.

Comparisons

Work organization

The measurements carried out both in-house and by various research groups at Volvo show a certain positive effect in Kalmar with regard to work content. The work characteristics differ somewhat in favour of Kalmar through somewhat larger variation (longer cycle time) and autonomy.

Flexibility

A comparison of flexibility shows that the systems to a great extent are comparable. Kalmar has an advantage in that the speed of production can be increased more quickly as stations can be added on within each unit and this can gradually be extended along the flow. A requirement is the carrier flow, enabling the speed to be varied for short sections. It is also technically possible to remove objects from the flow. From an administrative point of view, however, the disturbances are equally large in both systems.

System efficiency

An analysis of the degree of system efficiency in terms of balancing, handling and system losses (indirect personnel) viewed in relation to an entirely ideal system (no extra effort is needed over and above the time required according to an MTM analysis of the object), gives the following picture:

Handling loss	−1%
Balancing loss	−5%
System loss	−15%
In total	−21%

This analysis of the degree of efficiency for Kalmar in relation to line shows that Kalmar is more efficient, primarily with regard to costs of inspection and adjustment. The somewhat longer cycle time also allows a somewhat more efficient balancing.

Taken together, the differences between the systems are not large. However, the change in production technology and work organization has nevertheless proved efficient, both technically–economically and socially.

Case B (Volvo Truck Corporation)

The next comparison is based on an analysis of two systems for assembly of heavy trucks within the Volvo group. The data has primarily been collected from Carlberg (1985), Lindér (1985), Ergonomigruppen i Göteborg (1985) and Berggren (1986).

The traditional system

The traditional system is here a mechanical line. The cycle time is approx. 15 min. A certain degree of job rotation exists. Inspection, adjustment, instruction,

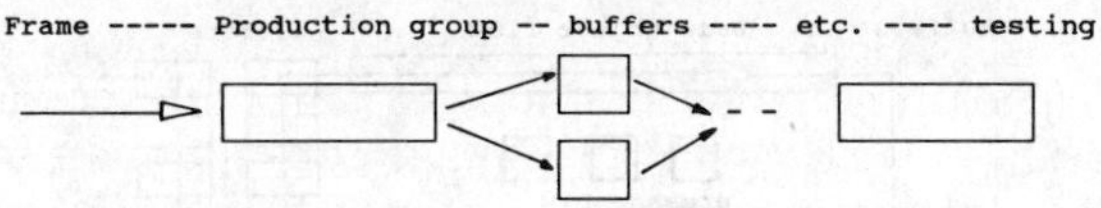

Figure 1. Flow structure for the alternative system.

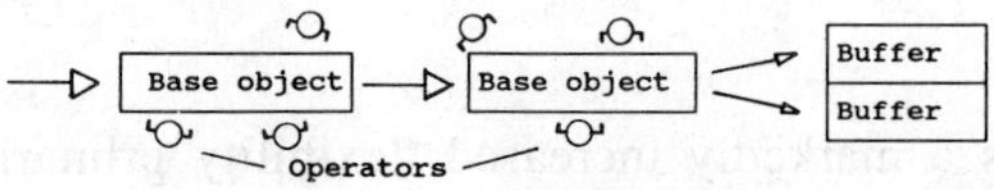

Figure 2. Production group.

etc. are carried out by specially assigned personnel. Pay is based on agreement on performance according to MTM standardized rate.

The alternative system

The alternative system in this case is a plant started in 1982. The plant is situated in the same region as the line plant and has an equivalent production capacity. Figure 1 shows the essential structure of the flow.

The production is organized into two parallel flows. Each flow is manned by serial-connected production groups separated by parallel buffer places. Parallel preassembly groups for motor and cabin are linked to the respective flow. The organizational base unit is the production group. This group is essentially described as shown in Figure 2.

Each production group has two base object stations at its disposal for work distribution. The group is led by a co-ordinator function called group representative (50% of the time). The group representative is responsible for local supervision as well as quality feed-back. The requirements for becoming a group representative are that the applicant should command assembly competence corresponding to the work of half of the group and that he undergoes special training. Such training is open to anyone subject to the approval of the management. Usually there are two to three group representatives for each group. The size of the group usually varies between five and nine members. The cycle time at the time of analysis was approx. 45 min, i.e. the work quantity in the groups was up to 45 × 9 min. The pay system puts a premium on both increased assembly competence (two steps) and responsibility for surrounding activities. The degree of responsibility the group as a whole is prepared to take (a requirement is that at least one member has the required competence) has an effect on the pay in three steps.

Comparison

Work organization

The data is extensive on how line work and work in the new plant is viewed by the workers (Berggren, 1986). Comparisons in relative terms, i.e. based on the views of individuals who have worked on the line, come out strongly in favour of the new system. There are no clear differences with regard to personnel turnover and rate of absenteeism, which was not expected, as these are more influenced by factors outside the actual work design. It should, however, be pointed out that the majority of those who left during the period 1982–5 applied for more qualified jobs (which they got). This was partly due to increased competence generated by working in groups.

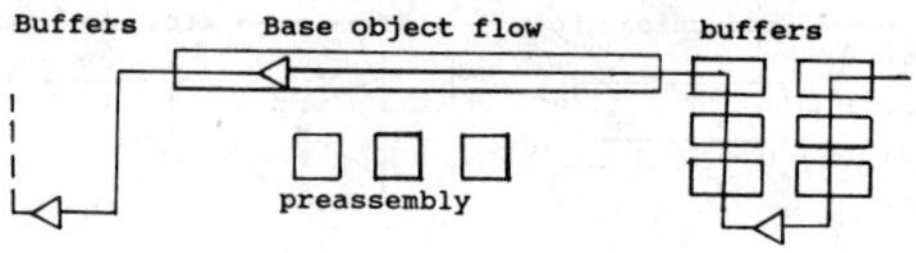

Figure 3. The new system (Miniline).

Flexibility

The new system has a markedly increased flexibility primarily in respect of the ability to produce different model options. The reasons for this are, for example, increased competence (which gives increased flexibility within groups as to different work distributions for different types of model options), speedy rebalancing within groups (group balancing is determined by the group representative and the production technicians) and variant levelling with the aid of buffers. The flexibility within the groups makes it also possible to redispose work tasks when disturbances occur. This has made the system less vulnerable to disturbances.

System efficiency

With the use of the same methodology as used previously the comparison turns out as follows:

Handling losses	–18%
Balancing losses	–7%
System losses	–105%
In total	–130%

This analysis of the degree of system efficiency in the new plant in relation to line shows an extremely large difference in efficiency. The new system is better from a production technology point of view and there is also a possibility of reduced costs to produce the approved quality. A complete assessment of the systems shows that the alternative design, including the changes in work content, responsibility and technology that have taken place, to a great degree has been successful. A decisive factor in this respect is that the increased degrees of freedom of the technical system have been utilized to create a more efficient social organization.

Case C (Saab cars)

This case refers to a comparative analysis carried out at ten-year intervals in a plant for car assembly. The first analysis was made in 1977 (Careborg and Lindér, 1977) and concerns a traditional line system with cycle times of approx. 3 min and the usual type of work distribution. Ten years later the same plant was analysed (Bergdahl and Johansson, 1987). It had then been rebuilt into a somewhat more flexible system called miniline (this was made step by step between 1982 and 1985). The name reflects the division of the line into short sections, each intended for 20–30 individuals. The new system is outlined in Figure 3.

An analysis of the potential of the system (on the basis of earlier research) shows that the technical structure allows work organization solutions of a markedly different kind from that of the line.

The key factors are: buffers allowing disconnection, geographical integration of preassembly and final assembly, the possibility of direct adjustment of defects and integration of inspection and adjustment work with assembly work (through team work). Each miniline can also vary its own speed within the limits of buffer capacity.

The outcome of the system far from matches the existing potential. The systems are compared below using the same model as before.

Comparison

Work organization
There is only a marginal difference with regard to work content. Job rotation is somewhat more extensive on the miniline but only for similar work tasks. The majority of the personnel are young people (< 24 years old). The status of the work is low, which results, among other things, in a 30% turnover of personnel.

Flexibility
The flexibility analysis shows that the miniline is more efficient only in respect of the ability to compensate for small disturbances occuring within an area. By using the buffers the other operators are able to continue work. The area where the disturbances occurred can then increase its speed in order to catch up in relation to the total flow. There are no other differences.

System efficiency
The outcome of the efficiency analysis is as follows:

Handling loss	+10%
Balancing loss	+11%
System loss	+11%
In total	+32%

Analysis of the degree of system efficiency for miniline in relation to line shows that the efficiency of the new system is lower, despite the fact that technically it has a higher potential. This is due partly to the fact that the work organization was not developed to meet the demands of the individuals (resulting in a low degree of work morale and job satisfaction) and partly to the fact that the system is overmanned (operator density is too high), which makes efficient balancing impossible. The latter is a result of one-sided emphasis on volume.

Conclusions

A comprehensive evaluation of the reported cases shows that it is possible to achieve a good overall economy in alternative systems, subject to a number of requirements being satisfied. The nature of the market and product must further be such as to cater to very exclusive customer demands. Mass production only is hardly profitable in the alternative systems. The level of training and demands from the workforce must also agree with the required level of knowledge and degree of responsibility.

A prerequisite for efficient utilization of the technical and administrative systems is far-reaching development of the work organization. Within the Swedish car industry, team work has been given pride of place. Different types of teams have proved to be one way of creating flexible forms of work which create opportunities for far-reaching job enrichment.

A well-developed team organization may result in both social and economic gains. Unless adaptation of the social organization takes place, the overall result may be negative, as the larger investments are not utilized. This is illustrated in Case C.

The next generation of systems

Continued development has taken place for the objects in question in respect of both market, i.e. performance and customer adaptation, and the demands from society. Therefore continued development of the systems is a highly topical issue. Flexibility, cost effectiveness and possibilities of attractive work roles are the desired qualities. Today, there are several ongoing projects aiming at solving these problems. Within the car industry two plants are being started up.

The basic elements of both these plants are continued division of responsibility and continued efforts towards parallelizing of flows. This increases the cycle times for operators and creates opportunities for both increased flexibility (model option) and efficiency (balancing). A parallelized flow is also conducive to more pronounced management by objectives (e.g. one group can be put in charge of a larger section of the processing), which forms the basis of job enrichment (Lindér, 1988). The difficulties inherent in a design with parallelized flow primarily concern learning and material supply. In the cases in question different solutions are being tested. In one of the plants a solution is being tested with prefabricated kits, i.e. all material is delivered to the assembly unit simultaneously with the body. Transports are handled by flexible carrier systems.

Learning has presented problems (Lindér, 1984) and therefore has been an obstacle. Modern educational research has shown that it is possible to expand the work content considerably, provided the work task is logically structured. Extensive practical tests show that this should be possible also in the case of car assembly (Engström *et al.*, 1988).

Even if it is too early to tell whether the new systems will be successful, they indicate a clear intention to continue the development of advanced sociotechnical systems.

References

Agurén, S., 1984, mfl Volvo Kalmarverken efter 10 år. Rådet för utvecklingsfrågor SAF-LO-PTK.M-gruppens förlag.

Bergdahl, A. and Johansson, K., 1987, Socioteknisk analys av personbilsmontering. *Industriell Organisation* (Göteborg: Chalmers Tekniska Högskola).

Berggren, C. and Holmgren, A., 1985, Volvo Kalmarverken och monteringsarbetets utveckling. Arbetsmiljölaboratoriet, Stockholm.

Berggren, C., 1986, Uppbrott från bandet? Arbetsmiljölaboratoriet, Stockholm.

Careborg, M. and Lindér, J., 1977, Socioteknisk analys av en driven line. *Industriell Organisation* (Göteborg: Chalmers Tekniska Högskola).

Carlberg, I., 1985, mfl. Socioteknisk analys av Volvo lastvagnars monteringsfabrik, Tuve. *Industriell Organization* (Göteborg: Chalmers Tekniska Högskola).

Carlberg, I. and Kullberg, P., 1985, Jämförande analys av arbetsorganisationen vid två produktionssystem för personvagnsmontering. *Industriell Organisation* (Göteborg: Chalmers Tekniska Högskola).

Engström, E. and Karlsson, U., 1981, Alternative production systems to line. A problem concerning material supply. *VI ICPR*, Novi Sad.

Engström, T. and Karlsson, U., 1982, Alternativ montering. *Industriell Organisation* (Göteborg: Chalmers Tekniska Högskola).

Engström T., Medbo, L. and Thunberg, K., 1988, Principer för och funktionering hos Volvo Uddevallaverken slutmonteringsfabrik. *Transportteknik* (Göteborg: Chalmers Tekniska Högskola).

Hackman, R. and Oldham, G., 1980, *Work Redesign* (Reading, MA: Addison Wesley).

Lindér, J., 1984, Flexible production organization. Paper presented at the 4th World Productivity Conference, Oslo.

Lindér, J., 1985, Analys och värdering av organisationsformen produktverkstäder. *Industriell Organisation* (Göteborg: Chalmers Tekniska Högskola).

Lindér, J., 1988, Oktoberrevisionen. Värdering av monteringskonceptet. *Industriell Organisation* (Göteborg: Chalmers Tekniska Högskola).

Lindér, J. and Johansson, M., 1986, Jämförande analys av line kontra parallelsystem. *Industriell Organization* (Göteborg: Chalmers Tekniska Högskola).

Nilsson, P. and Tunälv, C., 1985, Analys av förutsättningarna för flexibel personvagnsmontering. *Industriell Organisation* (Göteborg: Chalmers Tekniska Högskola).

Rationaliseringsrådet, 1976, Volvo Kalmarverken. Stockholm.

Ergonomigruppen i Göteborg, 1985, Samarbetsmodell vid ergonomiskt förändringsarbete i industrin. Göteborg.

Söderberg, S. Almgren, B., 1980, UARDA rapport no. 11, och 12. *Industriell Organization* (Göteborg: Chalmers Tekniska Högskola).

Turner, A. and Lawrence, P., 1965, *The Industrial Jobs and the Worker* (Boston: Harvard University Press).

Wild, R., 1975, On the selection of mass production systems. *International Journal of Production Research*, **13**, 5.

48

Effect of learning on perceptual motor load in a combined manual and decision task

Iiji Ogawa[1] and Abdul Raouf[2]

[1]*Department of Industrial Engineering, University of Windsor, Windsor, Ontario, Canada N9B 3P4, and* [2]*Department of Systems Engineering, King Fahd University of Petroleum and Minerals, Dhahran, Saudi Arabia 31261*

Abstract

The purpose of this study was to investigate the relationship between learning and Perceptual Motor Load (PML) in a two-stage Combined Manual and Decision Task (CMDT). The dual task method was applied to measure PML Index. Also, Heart Rate Variability (HRV) was measured as a physiological measure of PML. Twenty male subjects participated in the experiments under four different conditions, i.e. the combinations between random or sequential inputs and unimanual or bimanual activations. The results of PML Index showed that PML Index decreased as the number of trials increased; however, HRV decreased as the number of trials increased in three of the four major experiments.

Introduction

The significant increase in the involvement of both the senses and the central nervous system in most modern industrial jobs has subjected the worker to a great amount of mental stress and strain, thus causing the imposition of what is called Perceptual Motor Load (PML). This may induce failures in the proper functioning of the work mechanism and may even trigger a nervous breakdown in the human being. PML considered here can be defined as an index which measures the worker's spare mental capacity for a given task.

Using the secondary task performance as a measure of learning the primary task, early experiments in this area were administered by Bahrick *et al*. (1954). A method of measuring PML was established by Michon (1964, 1966) and Michon and Van Doorne (1967) using the dual task method. However, the scoring method did not sufficiently take into account statistical characteristics of PML as effected by the primary task.

On the other hand, many industrial control tasks can be described as Combined Manual and Decision Tasks (CMDT; Sadosky, 1968; Raouf and Mehra, 1974; Thomas *et al*., 1974; Raouf and Khare, 1975; Raouf and Sethi, 1976; Raouf and Arora, 1980; Raouf and Abada, 1982; Raouf and Ogawa, 1987; Raouf *et al*., 1987). As the trend towards the centralized control of industrial systems continues, such tasks are becoming increasingly more common and important. Raouf *et al*. (1984) applied Michon's scoring method to CMDT and found that information load and activation were highly significant factors affecting PML. In the absence of well-

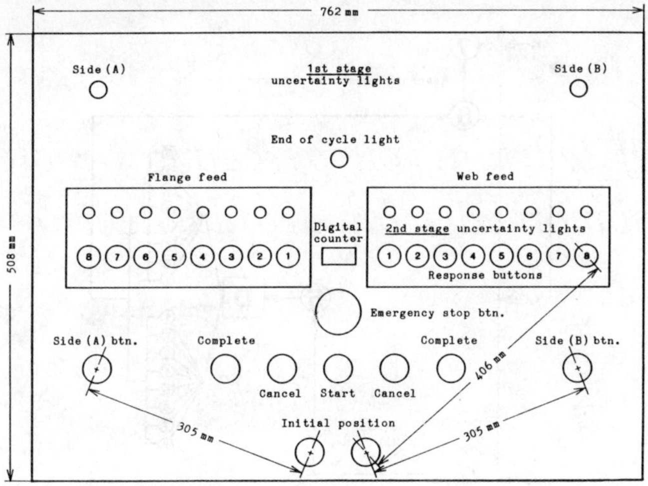

Figure 1. Layout of a two-stage CMDT simulator.

tested methods for calculating PML caused by various levels of CMDT Raouf *et al.* (1985), Raouf and Ogawa (1987) developed and tested a new scoring method for obtaining PML Index (PMLI) for CMDT. Ogawa (1988) concluded in his dissertation that PMLI is a better measure of PML in terms of correlation coefficients between performance time and PML.

This experimental study was conducted in order to investigate the relationships between learning and PMLI, and learning and Heart Rate Variability (HRV) changes over a certain period of time in a two-stage CMDT.

Method

Experimental set-up

The equipment used for this study consisted of the following:

(a) two-stage CMDT simulator,
(b) pedal,
(c) Grass Model 7 Polygraph,
(d) PDP 11/03 Minicomputer.

Primary task

The primary task, in which PML was measured, is basically a two-stage signal-response unit and a simulation of a semi-automatic punch press machine commonly used in the steel fabrication industry. A layout of the control panel of the simulator is shown in Figure 1. The panel was mounted on the top of a box which houses the electronic circuits, the programmed chips required to control the levels of information load, and a potentiometer to control the duration of the time lag. The box was fixed on the top of an adjustable stand. A more detailed description of the equipment is given by Ogawa (1988). A node diagram for the task can be expressed in Figure 2. Break points t_{3-4} and t_{5-6} determine the proportions of the decision time and the movement time within the decision component. The latter contained a combination of two-alternative choice elements and up to 16-alternative choice elements, respectively.

In case of an emergency, indicated by a continuous flashing of all lights of

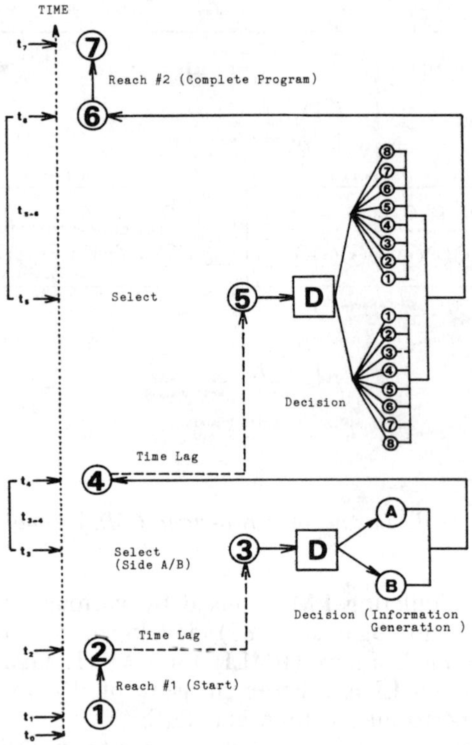

Figure 2. A node diagram for a two-stage CMDT simulator.

the second stage on the panel, the subject had to activate the emergency stop button as soon as possible and then the program cancel button(s). Activating the program cancel button(s) allowed him to start the cycle again. In all circumstances whereby the subject committed an error, he had to respond immediately to correct the error. The time which elapsed between the occurrence of the stimulus and the completion of the responses in the second stage was labelled Performance Time (PT) and was recorded in milliseconds on a time measuring and recording unit. The data were converted on a magnetic disk of a PDP 11/03 minicomputer for analysis.

Secondary task and heart rate variability measurement

The simple task of foot tapping was used as the secondary task to obtain the quantitative measurement of PML. Each subject was asked to tap a pedal by foot as regularly as possible at a constant rate of about one or two taps per second during the performance of the primary task. The pedal had a micro-switch and electric circuit to send a pulse to the PDP 11/03 minicomputer placed in an adjacent room through the Polygraph's pen driver in order to monitor the tapping intervals. Also, HRV was measured as a physiological measure of PML.

Graphical representation of data processing procedures is shown in Figure 3. For monitoring tapping intervals, recording each heart beat and the heart rate (beats/min), a Grass Polygraph was used. It essentially consisted of an EKG tachograph pre-amplifier with four DC pen driver amplifiers, of which two were used for heart rate and one for tapping intervals.

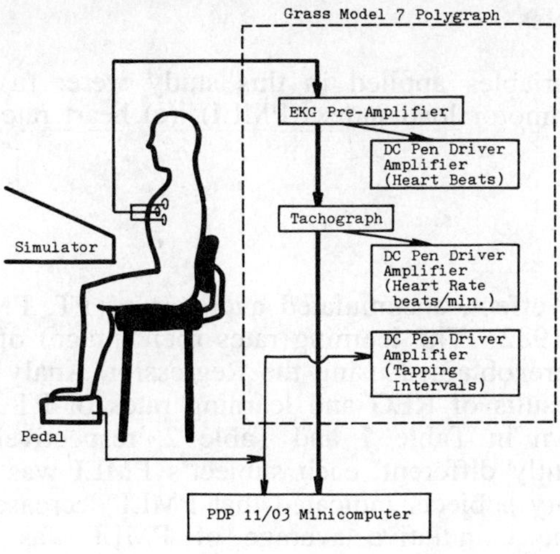

Figure 3. Graphical representation of data processing procedures for measuring PML.

A program was developed to store the tapping intervals and heart rate data on a magnetic disk of the minicomputer. The stored data were subsequently analysed to calculate PMLI developed by Ogawa (1988) and HRV using the scoring method suggested by Luczak and Laurig (1973).

Subjects

Twenty male students (all right-handed), between 20 and 30 years of age, were chosen to act as subjects for the experiments. All the subjects were in good physical condition and were interested in the study.

Experimental Design

Experimental conditions

The experiments consisted of the following four major studies:

(a) unimanual activation and random input,
(b) unimanual activation and sequential input,
(c) bimanual activations and random input,
(d) bimanual activations and sequential input.

Here, random input corresponds to the stimulative repetitions being varied randomly. On the other hand, sequential input occurs to the stimulative repetitions being varied in the form of sequential dependencies. Five subjects participated in one of the four experiments. A combined total of 20 subjects performed the experiments. Each subject was given a sufficient practice period for the secondary task before running the experiments. To assess the learning effect, each experiment consisted of nine runs of 30 cycles under the following condition:

(a) Information load: 3 bits in the case of random varied stimuli,
(b) Time lag: 1·5 s in the first stage and 1·0 s in the second stage,
(c) Probability of emergency signal occurrences: 0·03.

Therefore, each subject performed 270 cycles for this experiment.

Independent variables

The independent variables applied in this study were: (a) performance times (PT), (b) perceptual motor load index (PMLI), (c) heart rate variability (HRV).

Results

To see the learning effect, accumulated averages of PT, PMLI and HRV were calculated (Niebel, 1982). The learning rates (percentage) of PT at each experimental condition were obtained using the Regression Analysis (REG) procedure in SAS 824. The results of REG and learning rates of PT under four different conditions are shown in Table 1 and Table 2, respectively. Since individual PMLI was significantly different, each subject's PMLI was examined. Data for five out of the twenty subjects indicated that PMLI decreased as the number of trials increased. The cumulative average of PMLI was plotted against the number of trials for each subject as shown in Figure 4. In order to study the effect of learning on performance time, the cumulative average of PT was plotted against the number of trials for each subject as shown in Figure 5.

Figure 6 shows HRV versus the number of trials under four different conditions. It was observed that HRV decreased as the number of trials increased. The plot of PT versus the number of trials under four different conditions is shown in Figure 7.

Conclusion and Discussion

Within the constraints of the experiments, the following conclusions can be made:

(a) Differences between individual PML Indices are quite significant during the learning stage ($\alpha = 0{\cdot}01$).
(b) PMLI for five subjects decreases as the number of trials increases.
(c) HRV decreases as the number of trials increases in three of the four major experiments.

The results of PMLI show that PMLI decreases as the number of trials increases. This means that the more the subjects learn, the less they use their mental capacity. Considering performance time, this corresponds with the finding that suggests the more the subject learns, the less he uses his mental capacity (Bahrick *et al.*, 1954; Bahrick and Shelly, 1958). However, HRV decreases as the number of trials increases in three of the four major experiments. This can be interpreted to mean that, the more the subjects learn and understand, the more they concentrate on the task.

The above finding of HRV in comparison with PMLI is contradictory. Since the learning effect can be seen on performance time, it is clear that measuring PML did not affect performance time. This contradiction can be attributed to the difference between spare mental capacity and attention/concentration. When the operator learns, he tries many strategies to complete the task. At the same time, his mental capacity is full. Gradually, as the number of trials increases, he understands the task and starts to concentrate on the task.

It is hoped that the results of this investigation have more fully increased our understanding of learning effect on both human performance and mental capacity in a CMDT.

Table 1. Results of regression analysis of performance time under four different conditions.

Unimanual-Random

ANALYSIS OF VARIANCE

SOURCE	DF	SUM OF SQUARES	MEAN SQUARE	F VALUE	PROB>F
MODEL	1	0.03190150	0.03190150	860.778	0.0001
ERROR	7	0.000259429	0.000037061		
C TOTAL	8	0.03216092			

ROOT MSE	0.006087795	R-SQUARE	0.9919
DEP MEAN	6.770526	ADJ R-SQ	0.9908
C.V.	0.08991614		

PARAMETER ESTIMATES

VARIABLE	DF	PARAMETER ESTIMATE	STANDARD ERROR	T FOR H0: PARAMETER=0	PROB > !T!	VARIABLE LABEL
INTERCEP	1	6.89540847	0.004715517	1462.280	0.0001	INTERCEPT
R	1	-0.08779573	0.002992457	-29.339	0.0001	REPLICATION

Unimanual-Sequential

ANALYSIS OF VARIANCE

SOURCE	DF	SUM OF SQUARES	MEAN SQUARE	F VALUE	PROB>F
MODEL	1	0.08682841	0.08682841	12487.875	0.0001
ERROR	7	0.000048671	.00000695302		
C TOTAL	8	0.08687708			

ROOT MSE	0.002636857	R-SQUARE	0.9994
DEP MEAN	6.799755	ADJ R-SQ	0.9994
C.V.	0.03877872		

PARAMETER ESTIMATES

VARIABLE	DF	PARAMETER ESTIMATE	STANDARD ERROR	T FOR H0: PARAMETER=0	PROB > !T!	VARIABLE LABEL
INTERCEP	1	7.00578355	0.002042471	3430.052	0.0001	INTERCEPT
R	1	-0.14484343	0.001296148	-111.749	0.0001	REPLICATION

Bimanual-Random

ANALYSIS OF VARIANCE

SOURCE	DF	SUM OF SQUARES	MEAN SQUARE	F VALUE	PROB>F
MODEL	1	0.01706185	0.01706185	414.075	0.0001
ERROR	7	0.000288433	0.000041205		
C TOTAL	8	0.01735028			

ROOT MSE	0.006419089	R-SQUARE	0.9834
DEP MEAN	7.05984	ADJ R-SQ	0.9810
C.V.	0.090924		

PARAMETER ESTIMATES

VARIABLE	DF	PARAMETER ESTIMATE	STANDARD ERROR	T FOR H0: PARAMETER=0	PROB > !T!	VARIABLE LABEL
INTERCEP	1	7.15116935	0.004972133	1438.250	0.0001	INTERCEPT
R	1	-0.06420678	0.003155304	-20.349	0.0001	REPLICATION

Bimanual-Sequential

ANALYSIS OF VARIANCE

SOURCE	DF	SUM OF SQUARES	MEAN SQUARE	F VALUE	PROB>F
MODEL	1	0.02680390	0.02680390	644.170	0.0001
ERROR	7	0.000291270	0.000041610		
C TOTAL	8	0.02709517			

ROOT MSE	0.006450577	R-SQUARE	0.9893
DEP MEAN	6.921847	ADJ R-SQ	0.9877
C.V.	0.09319155		

PARAMETER ESTIMATES

VARIABLE	DF	PARAMETER ESTIMATE	STANDARD ERROR	T FOR H0: PARAMETER=0	PROB > !T!	VARIABLE LABEL
INTERCEP	1	7.03631872	0.004996523	1408.243	0.0001	INTERCEPT
R	1	-0.08047609	0.003170783	-25.381	0.0001	REPLICATION

Table 2. Learning rates of performance time under four different conditions.

Conditions	Learning rates (%)
Unimanual random	94·1
Unimanual sequential	90·5
Bimanual random	95·6
Bimanual sequential	94·6

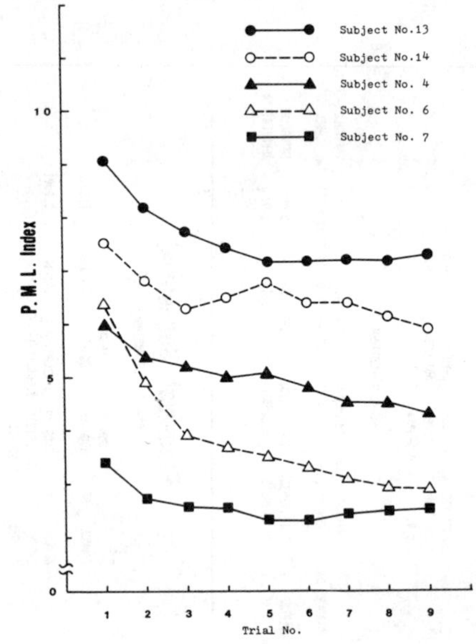

Figure 4. PMLI versus trial number for five subjects.

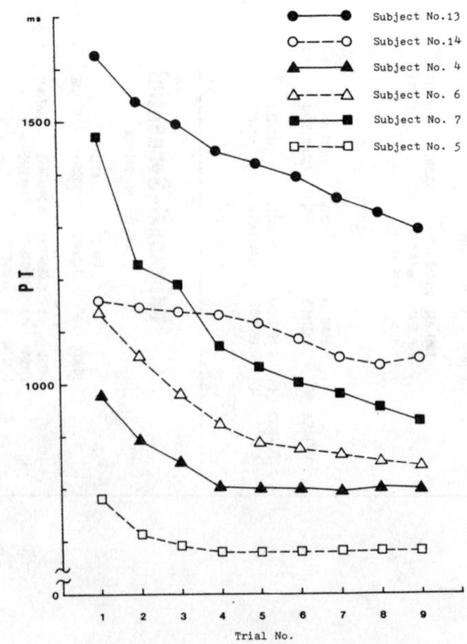

Figure 5. PT versus trial number for six subjects.

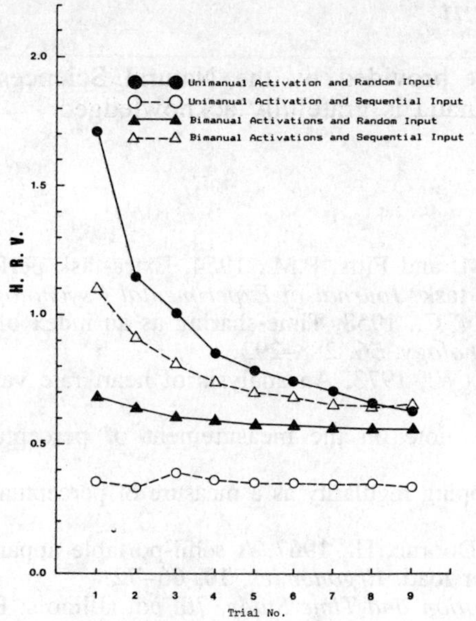

Figure 6. HRV versus trial number under four different conditions (averages plotted).

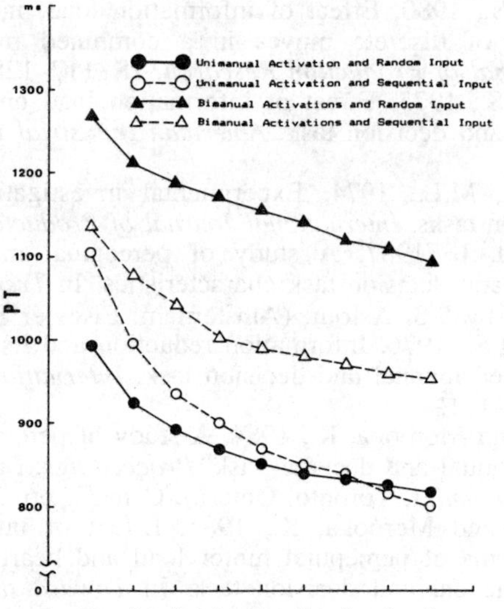

Figure 7. PT versus trial number under four different conditions (averages plotted).

Acknowledgement

The financial support provided by the Natural Sciences and Engineering Research Council of Canada is gratefully acknowledged.

References

Bahrick, H.P., Noble, M. and Fitts, P.M., 1954, Extra-task performance as a measure of learning a primary task. *Journal of Experimental Psychology*, **48**, 4, 298–302.

Bahrick, H.P. and Shelly, C., 1958, Time-sharing as an index of automization. *Journal of Experimental Psychology*, **56**, 288–293.

Luczak, H. and Laurig, W., 1973, An analysis of heart rate variability. *Ergonomics*, **16**, 85–97.

Michon, J.A., 1964, A note on the measurement of perceptual load. *Ergonomics*, **7**, 461–463.

Michon, J.A., 1966, Tapping regularity as a measure of perceptual motor load. *Ergonomics*, **9**, 401–412.

Michon, J.A. and Van Doorne, H., 1967, A semi-portable apparatus for the measurement of perceptual motor load. *Ergonomics*, **10**, 66–72.

Niebel, B.W., 1982, *Motion and Time Study*, 7th ed. (Illinois: Richard D. Irwin).

Ogawa, I., 1988, An investigation of human performance in a two-stage combined manual and decision task under different work-load conditions. Ph.D. Dissertation, University of Windsor, Windsor, Ontario, Canada.

Raouf, A. and Abada, S.A., 1982, Correctability function in combined manual and decision tasks. *International Journal of Production Research*, **20**, 65–72.

Raouf, A. and Arora, S., 1980, Effect of information load, index of difficulty direction and plane angles of discrete moves in a combined manual and decision task. *International Journal of Production Research*, **18**, 117–128.

Raouf, A. and Khare, S., 1975, Effect of information load on human performance in a combined manual and decision task. *American Industrial Hygiene Association Journal*, **36**, 76–81.

Raouf, A. and Mehra, M.L., 1974, Experimental investigations related to combined manual and decision tasks. *International Journal of Production Research*, **12**, 151–157.

Raouf, A. and Ogawa, I., 1987, A study of perceptual motor load as effected by combined manual and decision task characteristics. In *Trends in Ergonomics/Human Factors IV*, edited by S.S. Asfour, (Amsterdam: Elsevier Science), pp. 261–268.

Raouf, A. and Sethi, H.S., 1976, Information reduction and its effects upon performance times in a combined manual and decision task. *International Journal of Production Research*, **14**, 111–121.

Raouf, A., Ogawa, I. and Morooka, K., 1984, A study of perceptual motor load in a two-stage combined manual and decision task. *Proceedings, International Conference on Occupational Ergonomics*, Toronto, Ontario, Canada, pp. 34–38.

Raouf, A., Ogawa, I. and Morooka, K., 1985, Effect of information load on human performance in terms of perceptual motor load and heart rate variability in a two-stage combined manual and decision task. In *Toward the Factory of the Future*, edited by H.-J. Bullinger and H.J. Warnecke (Berlin: Springer), pp. 859–864.

Raouf, A., Ogawa, I. and Morooka, K., 1987, A study of human performance in combined manual and decision task under different work-load conditions. *Proceedings, International Conference on Production Research*, Cincinnati, Ohio, pp. 1354–1357.

Sadosky, T.E., 1968, Prediction of cycle times for combined manual and decision task. Ph.D. Dissertation, University of Michigan, Ann Arbor, Michigan.

Thomas, W.V., Hancock, W.M. and Chaffin, D.B., 1974, performance of a combined manual and decision task with discrete uncertainty. *International Journal of Production Research*, **12**, 409–425.

49

Workplace design and working postures for repair and maintenance tasks

Christine M. Haslegrave

Department of Production Engineering and Production Management, University of Nottingham, University Park, Nottingham NG7 2RD, UK

Abstract

In many industries, repair and maintenance tasks have to be carried out in restricted workspaces, or where access to components is difficult. There is an increased risk of injury in such awkward working postures. The hazards associated with these tasks and workplaces may be assessed using biomechanical models to evaluate the loads imposed on the body. These may also be used to set criteria for safe loads or force demands in the more complex workplaces. However, it is difficult to specify realistically the postures which will be adopted by workers in confined workspaces.

The present study was undertaken to measure the postures adopted for maximal force exertions in typical standing, kneeling and overhead working tasks. The posture records have been analysed to identify factors which determine the choice of posture, and to develop ways of predicting postures as input to biomechanical models used in workplace evaluation.

Introduction

In most industries, maintenance is a large and important area of work, even in highly automated systems. However, the various tasks are less structured than in many other jobs, are less predictable and in some cases may take place outside the normal working environment. Typically, repair and maintenance tasks involve the use of powered and non-powered tools or the movement and adjustment of equipment, and often have to be carried out in restricted workspaces or where access to components is difficult. The operators can be forced to adopt awkward working postures, frequently having to twist or stretch to reach fixings at arm's length, and have to apply relatively high forces in these postures.

The difficulties encountered in maintenance work were graphicaly illustrated by Seminara and Parsons (1982) in their studies of the ergonomic problems of power plant operation. They related how on one site the smallest man in the maintenance team often had to be called in to squeeze through gaps in machinery to access some components. Such access restrictions can severely reduce strength capability for lifting and exerting forces, and are known from epidemiological studies to have an increased risk of injury (particularly of back injury). Force applications can be particularly damaging if they are combined with lateral bending or twisting or if the work is carried out at arm's reach, and sudden or unexpected overloads also cause considerable risk (Chaffin *et al.*, 1977; Andersson, 1981; Westgaard and Aarås, 1984).

Workplace design

As part of the workplace design process, designers of machines and workplaces need ergonomic information which relates the layout factors to the strength capability of the operators and to the consequences of working in constrained postures in terms of the risk of injury. Traditionally this has been obtained from design handbooks, but will soon be incorporated within computer-aided design (CAD) systems by modelling the workplace and using biomechanical and strength prediction models to assess the demands on the operators while carrying out typical tasks. Biomechanical models evaluate the loads on the body in terms of the compressive load on the spine or torques at the major joints.

Since CAD systems now have full three-dimensional (3-D) capability, it is a natural extension to use this to assess the biomechanical consequences of work tasks, just as materials strengths are evaluated with finite element software. The full potential for incorporating biomechanical models in CAD systems has not yet been realized, although the components of the system already exist in workspace modellers which contain anthropometric models of the human operators, (e.g. SAMMIE; Bonney *et al.*, 1979), biomechanical analysis models (Schultz and Andersson, 1981) and force prediction models (reviewed in Chaffin and Andersson, 1984). In a maintenance application, the Crew Chief man model is being developed for the US Air Force (McDaniel and Askren, 1985).

The possibility of combining these in a CAD system was recently simulated and evaluated in a design for manufacture application (Örtengren *et al.*, 1990). This will eventually lead to an integrated analysis of many types of manual tasks (in materials handling and assembly, as well as in repair and maintenance work), and Örtengren *et al.* have shown how valuable this will be to the design teams. However, before such models become used as a routine part of the manufacturing design process, it is worth looking at some of the underlying assumptions in their new application, to ensure that their criteria are realistic for assessing both task capability and injury risk.

In their present applications, biomechanical models are generally used to evaluate and redesign existing workplaces, in which the operators can be observed and recorded while carrying out the tasks. When biomechanical models are used as predictive tools for workstation design, information is needed on the posture which would be adopted by operators using the workstation. This can be quite difficult to determine for asymetric tasks, and particularly for operators working in confined spaces, but it is important since both strength capability and injury potential depend to a large extent on the way that the task is carried out and on the layout of the workplace.

Strength variation with posture

Many physical factors determine the maximum force or torque which can be produced by a person, related to the workplace, environment and task situation as well as to the physical and cognitive characteristics of the operator. The model shown in Figure 1 has been developed to define the principal determinants of force capability in a whole body exertion (involving many muscle groups) which need to be incorporated in any biomechanical model.

In the workplace, the main factors include the geometrical layout (such as reach distance and location, and direction of exertion) and characteristics of

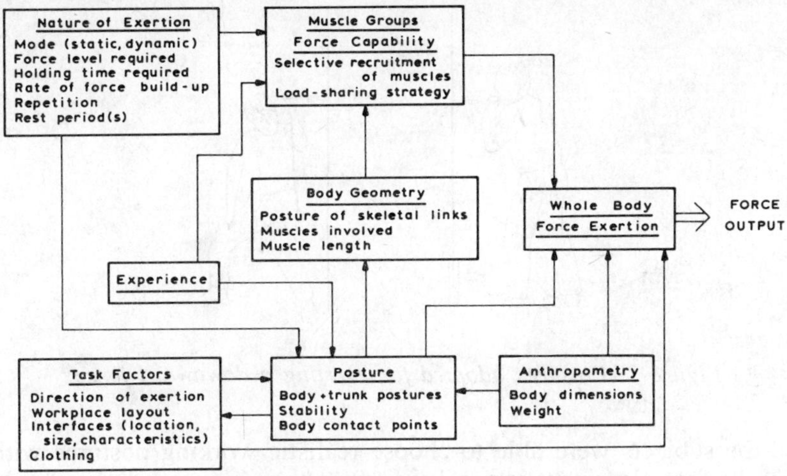

Figure 1. Model of whole body force exertion

interfaces with workpiece and floor. For the operator, they include the capacity of the individual muscle groups, use of body weight, joint strength and stability, and stability of the posture adopted for the task—as well, of course, as the individual variability in strength and anthropometric characteristics. The model indicates how the posture and task factors interact, without detailing the factors involved at the physiological level in the individual muscles.

Some of the factors have been well studied, so that biomechanical models can take account of the individual anthropometry and many task factors. Others, such as the longer term effects of holding time are beginning to be defined (Monod and Scherrer, 1965; Rohmert, 1973; Milner *et al.*, 1986), but it is not yet possible to model dynamic forces accurately.

The model shows that the influence of posture on force exertion is quite complex. It directly affects the force level through body contact points, stability of the posture, and the direction of exertion of the force, but also indirectly affects the force level through the use of body weight and the body geometry which determines the muscle groups which can be brought into play.

There are complex interactions between task factors, choice of posture and the strategies which may be developed by operators through experience and training. As yet, very little is known about the factors influencing the choice of posture by the individual for a particular task. The effects of variations in posture on the resultant strength capability are not trivial, and a recent study has shown that strength capability may have been considerably underestimated in certain circumstances (Tracy and Gray, 1989).

Analysis of posture

Much further study will be needed in understanding these relationships. In the present project, a study was undertaken to record the postures adopted for maximal force exertions in some typical maintenance situations while standing, kneeling and working overhead. Strength capability was found to be particularly affected by the location and direction of force application in these postures, and this variation should be incorporated in design guidelines for fixtures and fittings which are used in maintenance.

A CODA optical scanner was used to record the postures adopted during the

Figure 2. Postures adopted for exerting a downward force.

tests, and the subjects were able to choose realistic working postures without the constraints on foot placement which have often been imposed during experimental studies. At the same time, the maximal isometric force which they were able to exert (single-handed) was measured in six directions on a strain-gauged testpiece. Details of the equipment and procedures can be found in Haslegrave *et al.* (1987).

The data from the tests have been collected in a data base which is being used to develop design guidelines for different types of work, but it is also being used to investigate some of the underlying techniques or strategies for force exertion and to look at the consequences of working in these postures, particularly how the changes in posture affect the loads transmitted within the body—on the spine and the joints—and whether safe loads may be exceeded if the person has to exert his full muscular strength in some postures.

The postures adopted for the different tasks varied considerably, since operators change their posture to gain the best leverage for applying forces in different directions, and to use the strongest muscle groups. For instance, in using a handtool such as a spanner or a screwdriver, the operator moves close to the job and carefully positions his arm at the best angle to apply a high force. The posture therefore varies with the task that is being carried out, and the actual force the person is able to exert varies with both the direction of exertion of the force and with the reach distance.

The most apparent features of the observed working postures were that they were highly asymmetrical, quite distinct from the symmetrical and standardized postures which have been adopted for much of the strength testing reported in the literature and likely to be used by industrial designers. Further analysis of the tests is being carried out to identify factors which influence an operator's choice of posture, and to develop ways of predicting postures as input to biomechanical models used in workplace evaluation.

A large variability in posture was also observed between the different subjects (who were all experienced miners) carrying out the same task. Figure 2 shows an example of two alternative postures which were used for exerting a downwards force on a testpiece at shoulder height. Both postures produced the same high level of force, but had different biomechanical consequences. In this case, the difference was seen in a high elbow torque (80 Nm) for the second posture where the operator was turned nearly sideways with feet braced well apart. The elbow torque in the first posture (where the subject faced forwards and bent slightly forward with his shoulder close to the testpiece) was calculated to be approximately 30 Nm.

During the project some inexperienced operators were also tested, and differences were found between the experienced and inexperienced groups in their approaches to carrying out forceful tasks and in the postures they chose to adopt. This may be one source of the individual variability in injury susceptibility. Some inexperienced operators observed during the tests appeared to have little perception of the risks involved in handling loads. Manual labourers can learn by training and experience to handle heavy loads and (perhaps) safer and less stressful handling techniques, although the injury statistics indicate that they do not always do so. Reference to Figure 1 indicates some of the factors which might be used in learning improved methods: knowledge of the capacity of muscle groups which can be brought into play, proper use of body weight, awareness of the possibility of slipping, and the need to maintain balance and to make sure that they have a stable posture in case the object on which the force is being exerted moves. There are in addition other considerations (such as postural discomfort) which may affect the posture that a skilled worker actually chooses to adopt, given that he has a choice of sharing the load between different muscle groups in order to reduce the effort or to avoid fatigue.

Presumably, the experienced operator integrates information on all these factors and then chooses a compromise to produce the maximum force. This involves complex body movements and many muscle groups, so that there is likely to be considerable individual variability and probably no unique best way even for one individual.

Application to computer-aided design

The model in Figure 1 shows the logical relationships between the input data on strength and posture which can be incorporated in biomechanical and strength prediction models for workplace analysis and CAD systems.

It is necessary to build assumptions into CAD biomechanical models concerning, for instance, load sharing between muscle groups. However, at present the criteria used to define working posture for input to a model are not well defined. In strength prediction models, the postures are determined by optimization techniques giving the body posture which can theoretically generate the highest joint torques. In view of the degree of variability observed in the present tests, the models must take account of the postures which are actually adopted by operators in industry (at all levels of experience and training), and not just the theoretically optimum posture. Even experienced operators were found to adopt quite different postures for the same task, with considerable differences in the body loads resulting from the postures and consequently also in the potential level of injury risk.

This is particularly important in setting criteria for safe loads or force demands in the more complex workplaces where postures are constrained by awkward or restricted access. In existing workplaces, it is possible to carry out trials to observe and record the working postures which are being used, but this is difficult at the design stage and when strength prediction models are being used. In order to do this, it will be necessary to establish more comprehensive posture data bases to match the current knowledge on strength capability which is available for use in biomechanical models.

Acknowledgements

This project was funded by the ACME Directorate of the Science and Engineering Research Council.

References

Andersson, G.B.J., 1981, Epidemiological aspects of low-back pain in industry. *Spine*, **6**, 1, 53–60.

Bonney, M.C., Case, K., Porter, J.M. and Blunston, C.A., 1979, Man–machine interaction in work systems. *International Journal of Production Research*, **17**, 6, 619–629.

Chaffin, D.B. and Andersson, G., 1984, *Occupational Biomechanics* (New York: Wiley), pp. 221–227.

Chaffin, D.B., Herrin, G.D., Keyserling, W.M. and Foulke, J.A., 1977, Pre-employment strength testing in selecting workers for materials handling jobs. *DHEW (NIOSH) Publication*, No. 77–163 (Cincinnati: NIOSH).

Haslegrave, C.M., Tracy, M. and Corlett, E.N., 1987, Industrial maintenance tasks involving overhead working. In *Contemporary Ergonomics 1988*, edited by E.D. Megaw (London: Taylor & Francis), pp. 197–202.

McDaniel, J.W. and Askren, W.B., 1985, *Computer-aided Design Models to Support Ergonomics*. AAMRL Report TR-85-075, Harry G Armstrong Aerospace Medical Research Laboratory, Wright-Patterson Air Force Base, Ohio.

Milner, N.P., Corlett, E.N. and O'Brien, C., 1986, A model to predict recovery from maximal and sub-maximal isometric exercise. In *The Ergonomics of Working Postures*, edited by E.N. Corlett, J. Wilson and I. Manenica (London: Taylor & Francis), pp. 126–135.

Monod, H. and Scherrer, J., 1965, The work capacity of a synergic muscular group. *Ergonomics*, **8**, 329–338.

Örtengren, R., Liew, M., Torgnysson, M., Glimskär, B., Mynerts, G. and Dandenell, R., 1990, Ergonomic CAD systems in work design. An investigation of requirements and limitations in a practical application. In *Work Design In Practice*, edited by C.M. Haslegrave, J.R. Wilson, E.N. Corlett and I. Manenica (London: Taylor & Francis).

Rohmert, W., 1973, Problems in determining rest allowances. *Applied Ergonomics*, **4**, 91–95 and 158–162.

Seminara, J.L. and Parsons, S.O., 1982, Nuclear power plant maintainability. *Applied Ergonomics*, **13**, 3, 177–189.

Schultz, A.B. and Andersson, G.B.J., 1981, Analysis of loads on the lumbar spine. *Spine*, **6**, 1, 76–82.

Tracy, M.F. and Gray, M., 1989, Block-laying in the construction industry: a biomechanical evaluation. In *Contemporary Ergonomics 1989*, edited by E.D. Megaw (London: Taylor & Francis).

Westgaard, R.H. and Aarås, A., 1984, Postural muscle strain as a causal factor in the development of musculoskeletal illness. *Applied Ergonomics*, **15**, 3, 162–174.

50

A study of the occupational safety and ergonomic aspects of material flow

J. Järvinen and S. Herranen

Technical Research Centre of Finland, Safety Engineering Laboratory, PO Box 656, 33101 Tampere, Finland

Abstract

Minimizing in-plant materials handling saves production costs, improves safety, and reduces the risk of over-exerting the musculoskeletal system. The material flow at the assembly department of a tyre factory was studied as a part of a research programme, which was aimed at determining how the problems caused by musculoskeletal diseases could be reduced at work. The material flow was surveyed by the aid of a computer program. The transport quantities and values of components and operations were examined and the problems were identified. Material flow diagrams were drawn up on the basis of the program output. Alternatives to locate new machines were evaluated by simulating the material flow of three alternatives. The total transport trip distance was reduced by 28% when comparing the chosen alternative with the initial layout. Traffic on the busiest route was reduced. Proposals were made to reduce and reorganize the material flow at the assembly department.

Introduction

It is obvious that minimizing in-plant materials handling directly saves production costs (Buffa, 1983; Turner *et al.*, 1978). In addition, reducing the number of material transports, manual handling and lifting between and within workstations improves safety and reduces the risk of over-exerting the musculoskeletal system. A large number of material transports also cause hazards in in-plant pedestrian traffic (Planning Guidebook '88, 1988; Priest, 1985).

Material transports, handling and intermediate storage should already be minimized at the design stage of a production system. It is also possible to reorganize the material transports when planning changes to existing production systems (Priest, 1985).

The material flow at the assembly department of a tyre factory was studied. The study was a part of a research programme of the Finnish Work Environment Fund, the aim of which was to determine how the problems caused by musculoskeletal diseases could be reduced at the workplace. The material flow was surveyed by the aid of MASI computer program.

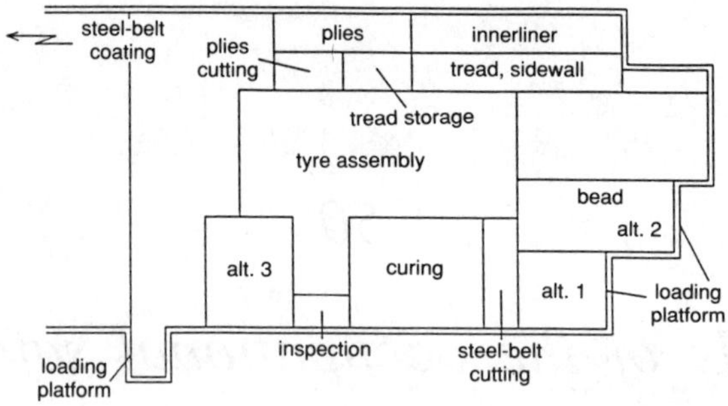

Figure 1. Layout of passenger car radial tyre production facilities.

Description of the manufacturing process

Passenger car radial tyres consist of the following components: inner liner, plies, beads, side walls, tread and belts. Each of these components is manufactured separately and then combined with the others in the final assembly process. The layout of the passenger car radial tyre production facilities is shown in Figure 1.

The components are brought together at the tyre building machines. The carcass of the tyre, including beads, plies, side walls and an inner liner, is constructed on a carcass assembly machine. Tread and belts are assembled on a belt assembly and joining machine. The carcass is transferred to the joining machine on a roller conveyor. The two subassemblies are then joined together and the process results in an uncured tyre. The new tyre building machines have been developed further and in them two subassemblies can be built and joined together.

The uncured tyres are transferred to an intermediate storage by a belt conveyor and loaded to specific carts. The carts are then transported from the storage to curing presses by fork-lift trucks.

Handling and transporting of the tyre components

All the components are loaded onto their specific carts after initial manufacturing on the machines. The carts are transported to the tyre-building machines or to the intermediate storage areas by fork-lift trucks. The inner liner, plies and side walls are driven to reels in the carts. Treads are cut into pieces of precise length and loaded automatically into the cart. The carts containing beads are pushed manually to the carcass assembly machines.

At the time of the study the steel-belts were manufactured in two phases, both of them in different locations. Steel wire and rubber were mated together and then transported to cutting machines. After cutting the belts were driven to reels in the transportation carts, which were transported to the belt assembly and joining machines by fork-lift trucks.

MASI computer program

MASI is a computer program that 'transports' the product components sequentially between workstations and simulates the workflow of each component, including storage locations along the way. The program was created at the Tampere

University of Technology as part of a degree thesis. The programming language used was FORTRAN IV+.

As an input the program requires data of components (amounts, weights), operations and locations of routes and materials handling points (operations, storages). The program calculates transport quantities as the total trip distance transported, ton kilometrage, number of transports, average transport distance, total weight of transports, different kinds of values and quantities for comparison of layout alternatives, etc.

So-called equal materials handling points, e.g. all carcass assembly machines in this case, are loaded randomly according to a uniform distribution.

Simulation of the material flow

The analysis of the material flow was based on the production data of one 'normal' 24-h period, i.e. three shifts. To simplify the simulation the component data (weight and dimensions) on one 'typical' tyre were used instead of the data of all 21 different models and sizes manufactured during the period.

The transport quantities and values of each component and each operation were examined and the problems were identified. Different layout alternatives for locating new steel-belt machines were evaluated. The new machines include also cutting. The input data for layout were modified and the transport quantities and values were calculated for each alternative using MASI program.

The computer used for the simulation was MicroVAX II. Material flow diagrams were drawn up on the basis of the program output using AutoCAD computer-aided drafting software. Bottle-necks and other problems of transports can be identified by aid of the diagrams.

Results of the simulation

The production of the 24-h period required about 37 km of transports. The total ton kilometrage (weight of load × transport trip distance) was 329 tkm. The total transport trip distances and ton kilometrages of each component are shown in Figures 2 and 3.

The total transport trip distance of the most problematic component, steel-belt, was 14·8 km. The high values of the transport trip distance and the ton kilometrage were caused by the long transport distances and by manufacturing the steel-belts in two phases, both of them in different locations.

Bead carts were pushed manually 3·3 km (80·5 m/transport) during the 24-h period. An alternative route for the bead cart transport was proposed that would shorten the transport trip distance into 2·2 km (53·6 m/transport). The high ton kilometrage value of the tread was caused mainly by the heaviness of the component.

The three possible alternatives to locate the new steel-belt machines were evaluated by simulating the material flow of each alternative. Differences between transport quantities and values of the layout alternatives were caused by changes in the steel-belt transports. The total transport distances and ton kilometrages of the initial layout and the alternative layouts are shown in Figures 4 and 5.

As an example of the output of MASI program a comparison of so-called transport coefficient A for different layout alternatives is shown in Figure 6. The coefficient depends on an average transport trip distance and on the area of the rectangle containing all material handling points. The transport coefficient *A*

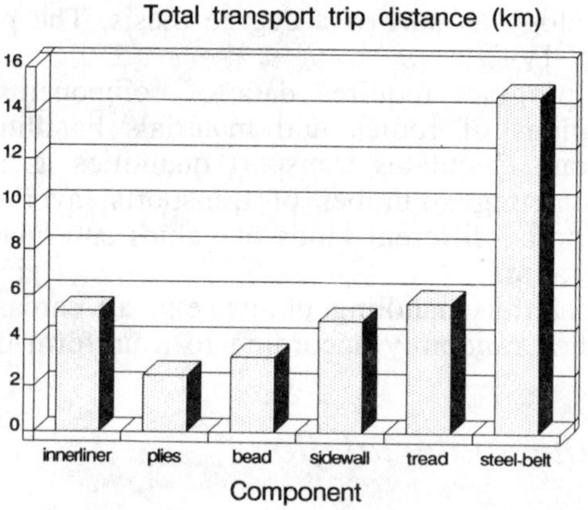

Figure 2. Total transport trip distances of each component.

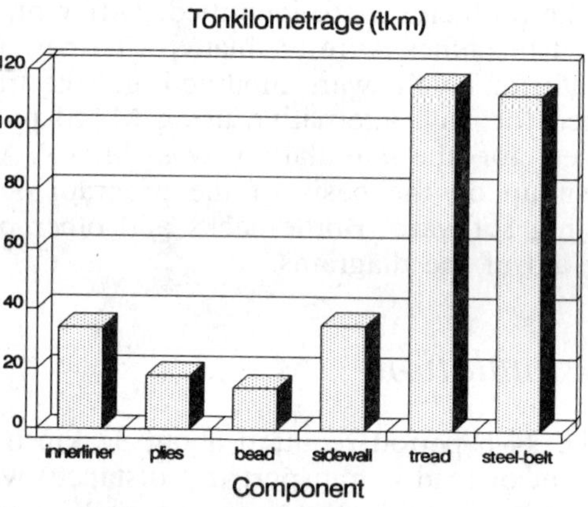

Figure 3. Ton kilometrages of each component (= total weight of transported components × trip distance).

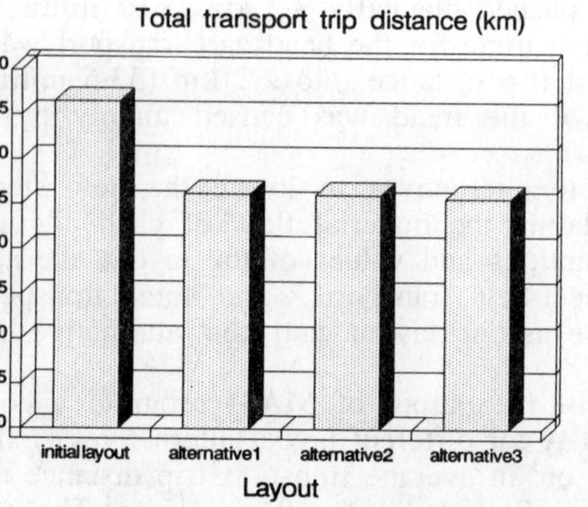

Figure 4. Total transport trip distances of different layout alternatives.

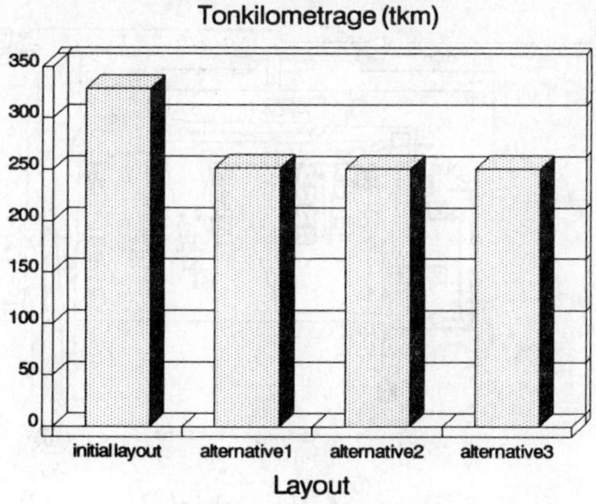

Figure 5. Ton kilometrages of different layout alternatives.

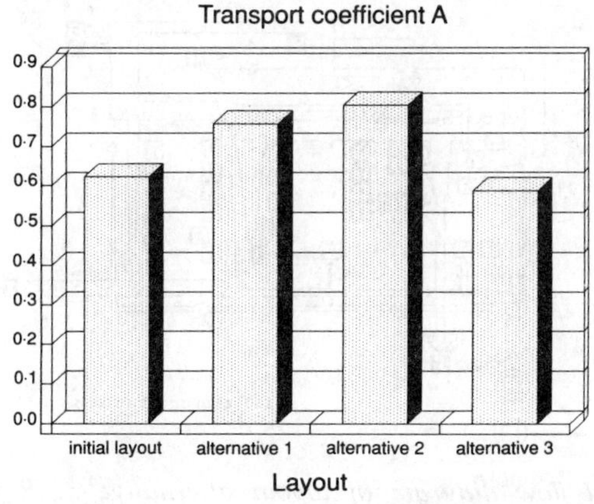

Figure 6. Transport coefficient A of different layout alternatives.

helps to compare the location of machines in different layout alternatives. A lower value of *A* refers to better location of the machines. The layout alternative 3 was the most advantageous from the material handling aspect at the initial volume of the production. However, alternative 3 requires an extra transport of the steel wire. When the need for the steel wire increases 60% or more, which is predicted to happen, alternatives 1 and 2 will be superior to alternative 3. Alternative 1 was chosen and implemented.

As a result, the transport trip distance reduction of the steel-belt was 69%, when comparing the chosen layout alternative with the initial one. The total transport trip distance, i.e. all components and the production of three shifts, was reduced 28% and the total ton kilometrage, 21%. In addition, traffic on the busiest route was reduced. Pedestrians were also guided to use a less busy route and suggestions were made to reduce and reorganize the material flow at the assembly department.

Examples of the material flow diagrams are shown in Figures 7 and 8.

Conclusions

This study and previous experiences prove that there are often unnecessary material transports in manufacturing processes which can be eliminated with

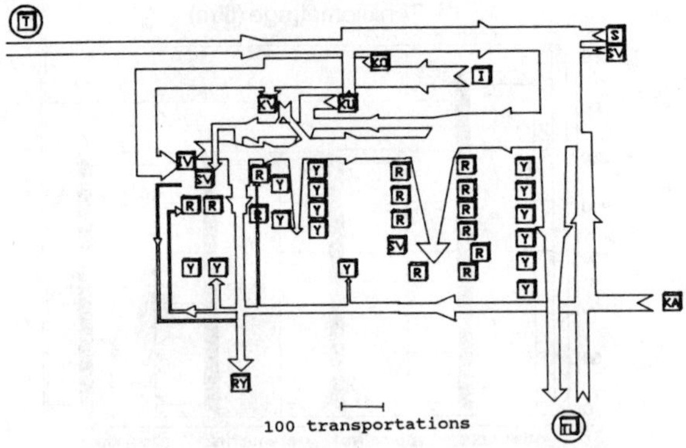

Figure 7. The material flow diagram of the initial layout: R = carcass assembly, Y = belt assembly and joining, T = steel-belt coating with rubber, TL = steel-belt cutting.

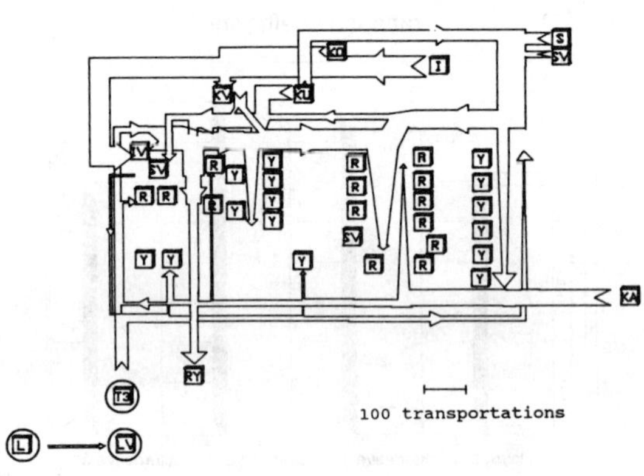

Figure 8. The material flow diagram of layout alternative 3: R = carcass assembly, Y = belt assembly and joining, T3 = steel-belt coating and cutting.

relatively minor measures. The reduction of material handling operations decreases the risk of musculoskeletal disorders and back diseases.

Material flow simulations are worthwhile in developing manufacturing technology, purchasing machines and planning new plants, when the layout still can be improved easily and with low costs.

References

Buffa, E.S., 1983, *Modern Production/Operations Management* (New York: Wiley).

Planning Guidebook '88, 1988, *Modern Materials Handling* **43**, 3 (Cahners Publications).

Priest, J.W., 1985, Ergonomic changes in work place can improve the productivity of production operations. *Industrial Engineering*, **17**, 7, 40–43.

Turner, W.C., Mize, J.H. and Case, K.E., 1978, *Introduction to Industrial and Systems Engineering* (Englewood Cliffs, NJ: Prentice-Hall).

51

Experimental evaluation of control selection using different robot teach pendant designs

S.V. Gray, C.S. Syan and J.R. Wilson

Department of Production Engineering and Production Management, University of Nottingham, Nottingham, UK

Abstract

This chapter addresses some of the difficulties that may be encountered by industrial robot programmers. These include a reversal of the control–display relationship when the robot is viewed from different positions. As part of a series of experiments, this chapter presents work carried out to evaluate the selection of controls in the JOINT programming mode. Two teach control designs were used and subjects were asked to select the appropriate controls for robot movement viewed from different orientations. The results showed that, although no differences in overall errors were found between the teach controls, all the subjects used strong stereotype associations to select the control for a perceived robot motion. Furthermore, the subjects perceived the movement of the robot in different ways and as a result selected different controls for the same move. This has implications for the design of teach controls which can provide safer and easier robot programming.

Introduction

There is currently great diversity in the design of teach pendants for different types of industrial robots. Recent studies show that teach pendants vary on a whole range of aspects; e.g. size, weight, layout, number of keys, display type, labelling and method of motion control (Parsons, 1986; Cousins, 1988). An issue of growing concern is that there should be some standardization in teach pendant design, in particular the motion controls used to drive the robot arm. In general there are two distinct designs of motion controls currently in use: a joystick control or push-buttons. Examples of each design are shown in Figure 1. These examples show differences in control type and there are many variations of each group currently used to control industrial robots.

Several problems relating to robot motion control have been highlighted by recent International Standards (JISHA, 1985; HS/G, 1988; ANSI, 1988). The most important are:

(1) The complexity of the control task caused by changes in human–robot orientation.
(2) The appropriate and unambiguous labelling of motion controls.

These problems are interdependent and are fundamental to the cause of errors in robot motion control. The main reason for concern is that if an error is made

(a) *(b)*

Figure 1. Two teach pendant designs currently in use: (a) using a joystick for motion control and (b) using toggle switches.

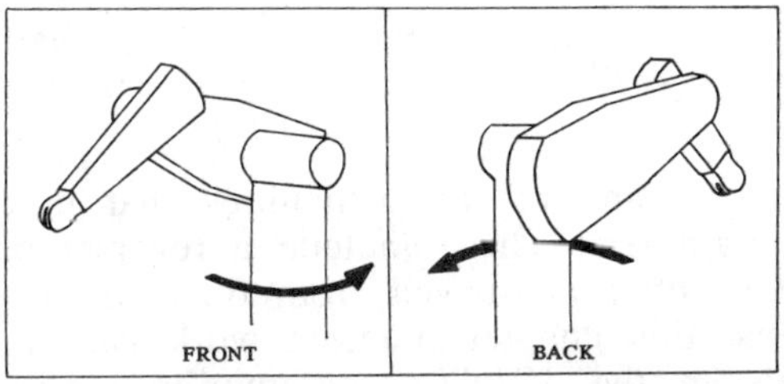

Figure 2. Rotation of the robot waist when viewed from the front and back orientations.

there is a risk of injury to the programmer. This is because he/she may be in close proximity to the robot arm when programming. The complexity that arises is in remembering whether moving the robot arm in a particular direction (e.g. right) requires the same control for various programmer orientations. Figure 2 shows how the same move of the robot arm appears different when viewed from different positions. This has direct implications for control labelling as 'right', written on the teach control, would clearly be confusing.

This research is aimed at addressing these problems, by experimentally assessing the control errors which are made in different positions, using different teach control designs.

Experimental details

Equipment

A PUMA 560 (Mk 1) anthropomorphic robot was used. Movement sequences were pre-taught using the VAL (I) programming language. Two mock-up teach boxes were constructed to represent the controls relevant to move the robot. These were a joystick design and a toggle switch design (see Figure 3). These were linked to a BBC microcomputer via analogue communication channels. The computer was used to recognize and record the control selections made with each teach box. An input hand held switch was linked to the robot controller via the input channels and was used to initiate each individual robot movement as and when required.

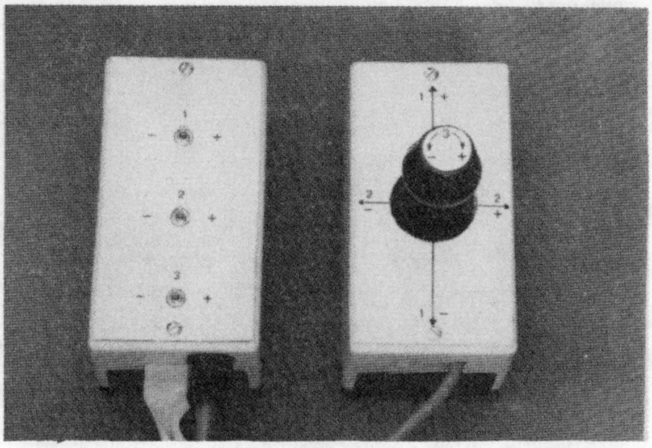

Figure 3. Control boxes used in the experiment.

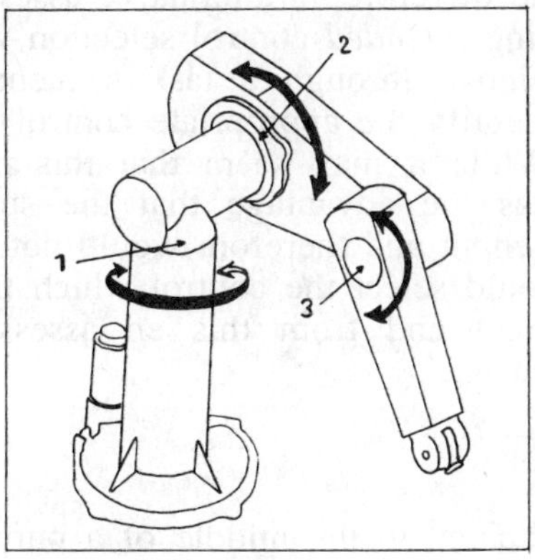

Figure 4. Movement of the robot in JOINT mode (major axes only).

Subjects

Twelve postgraduate engineering students volunteered to participate in the experiment. Their ages ranged between 22 and 30 years. All had similar academic background (sciences) and none had experience in robot programming.

Methodology

The experiment presented in this chapter uses the robot in the JOINT programming mode. This mode can provide more detailed information on the types of errors made in robot motion control than other programming modes (e.g. WORLD and TOOL).

In order to limit an otherwise complicated experimental design, only the three major axes of movement were examined. The joints used in the experiment are shown in Figure 4, each joint can be rotated in either direction about its fixture independently of other joints.

As stated in the equipment section, the mock-up control boxes could not be used actually to drive the robot. This was because the necessary software for interfacing was not fully developed. This prevented a direct evaluation of control

Figure 5. Experimental set-up.

selection performance, and therefore an alternative assessment measure had to be used; a method of assessing *intended* control selection was developed. The robot was preprogrammed to move through a task sequence and the experimental subjects were asked to identify the appropriate control on the teach pendant for each robot movement. While it may seem that this method of assessment is perhaps unrealistic, it has the advantage that the subjects receive no direct feedback of their control input and therefore would not be able to learn by 'trial and error'. Thus, they would select the control which they intuitively associated with each robot movement and from this an assessment of control–display compatibility could be made.

Procedure

The PUMA robot was situated in the middle of a purpose-built cage. This was constructed so that the subjects could gain access to all four sides of the robot (see Figure 5).

At the beginning of the experiment the subject stood at the left side of the robot and was shown how the three joints moved in each direction. The subjects were instructed that they would view the robot movements at each of four positions (front, right, left and back), and from each position they would be shown a sequence of joint moves. A sequence consisted of six separate moves; one move of each joint in each direction. A different sequence order was presented from each position. During each move the subject was required to select and activate the appropriate control on the teach box that they thought would have produced that motion. These control selections were recorded on the computer and later compared with the correct controls needed to produce each move. After each individual move the robot was stopped and the subject was asked to describe in their own words what movement had occurred in terms of (a) the joint that had moved, and (b) the direction it moved in. This experimental task was repeated at each position using both teach boxes for all subjects. The presentation orders of teach control, position, and movement sequence were counter balanced and randomly allocated to subjects.

Results and discussion

A two factor (2×4) ANOVA with repeated measures design was used to assess the effect of the experimental factors on errors made. The factors were defined

Table 1. Control selection errors made with each teach control box at each viewing position.

	VIEWING POSITION				
	FRONT	RIGHT	LEFT	BACK	TOTAL
Joystick pendant					
Total no. of robot moves	72	72	72	72	288
Direction errors	48	41	35	33	157
Control errors	7	7	6	7	27
Total errors	53	48	41	40	184
Toggle switch pendant					
Total no. of robot moves	72	72	72	72	288
Direction errors	45	46	40	40	171
Control errors	3	0	0	1	4
Total errors	48	46	40	41	175

by teach control (joystick and toggle), and subject–robot position (front, right, left and back).

Error scores

The error scores for each teach control are shown in Table 1. It can be seen that most of the errors made, irrespective of teach control, were 'direction errors'. These are errors caused by the correct joint control being selected but moved in the wrong direction. The other type of errors are 'control errors' whereby the wrong joint control was selected. However, because so few errors of this type were made, the analysis was carried out on total error scores.

The analysis of variance showed no significant differences in overall error scores for either teach control design. In addition, there were no overall differences as a result of the subject's viewing position. However, closer examination of individual subject's performances did reveal some trends. Figure 6 shows the proportion of responses that were: correct control selections, direction errors and control errors, for each teach control. As can be seen, the subjects had different performance patterns and it was found that they could be grouped according to their performance pattern. Figure 7 shows the same data after grouping.

The performance of the subjects was found to be directly related to their own verbal descriptions of robot movement. For example, subjects 1 and 5 made virtually 100% direction errors with both pendants and these subjects were found to have given consistent descriptions of the same moves from all orientations. However, they had selected a control–display relationship for the +/– controls that was the complete reversal of that actually used for the PUMA robot, but they were consistently incorrect. Subjects 4, 6, 8 and 10 had also been consistent in their descriptions but they had chosen the correct controls for joints 2 and 3 but were incorrect on joint 1. Subjects 11 and 12 were correct for joint 1 but incorrect for joints 2 and 3. All of these subjects, therefore, may not have selected the controls incorrectly if they had been trained or if they had received some feedback of the correct control–display relationship. The subjects of interest in this experiment were 2, 3, 7 and 9 who had been inconsistent in their description of moves when viewed from different positions, as they always described movement relative to themselves. These subjects made correct selections from the left side and back of the robot but made direction errors from the front and right side. Subject number 3 also accounted for almost all of the control errors made with the joystick.

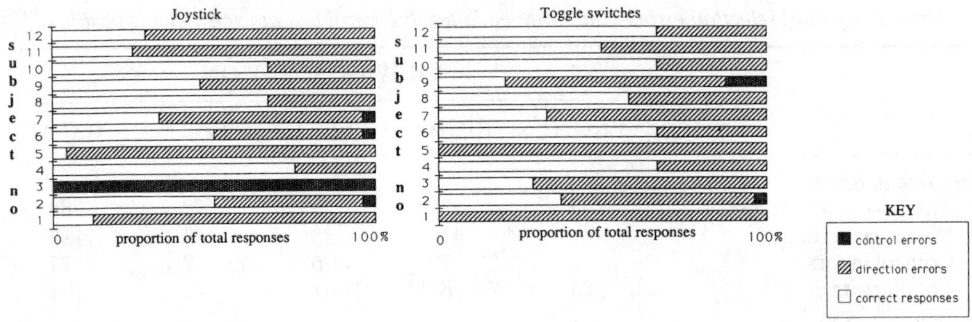

Figure 6. Individual subject's performance patterns with teach control showing the proportion of their responses that were: correct control selections, direction and control errors.

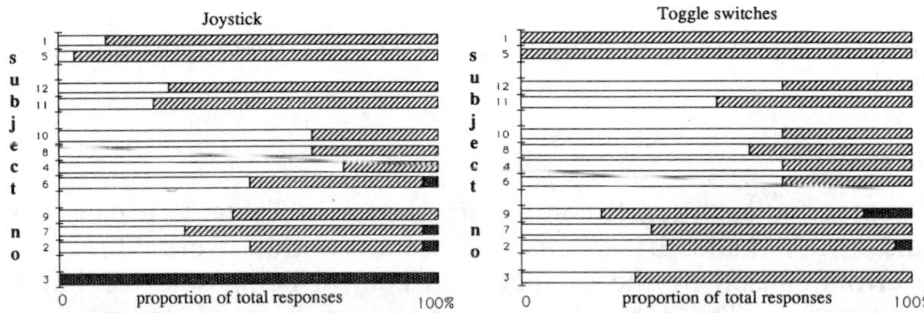

Figure 7. Individual subject's performance data after grouping. Key as in Figure 6.

Unfortunately, there were not enough subjects in each group to allow for re-analysis of the data according to the grouping. However, interpretation of the results could still be made.

Stereotype control–display associations

One of the reasons for the lack of uniformity in subject performance patterns was that the subjects used different terminologies to describe robot movement (e.g. right/left, clockwise/anticlockwise, up/down, forward/back, towards/away). In addition, they had chosen different reference points from which movement was related; relative to the left side of the robot, or the front, or relative to the subject's current viewing position. The effect of these descriptions on control selection could be measured according to expected stereotypes. Table 2 shows the number of times that the +ve and –ve controls were selected when certain descriptive terms were given.

It can be seen from Table 2 that there were strong stereotype associations between perceived motion and control selection. This result shows that the subjects' 'perception' of robot movement (i.e. how they chose to define directions and what reference point was used), directly influenced their selection of controls.

Subject preferences

There was some difference in subject preference for the teach controls. Seven of the subjects (58%) said that they would prefer to use the joystick for control of robot motion but that they had found the toggle switch easier to learn on. Two

Table 2. The description–control associations for all subjects using both teach controls.

Description of direction of movement	Control selected	
	+	–
Right, clockwise, up, forward, towards	270 (90%)	30 (10%)
Left, anticlockwise, down, back, away	22 (8%)	254 (92%)

of the subjects preferred the toggle switches and the remaining three subjects had no preference for either control.

Conclusions

Although no significant statistical differences were observed in error scores for the two teach pendants, there were some trends in subject performance.

(1) There were strong stereotype associations between perceived motion and control selection. For example, subjects selected the '+ve' control for movements described as right, forwards or clockwise. This resulted in different types of errors made as a result of the perceptual strategy used by each subject.
(2) The subjects who were consistent in their description of movement were either correct or incorrect (dependent on their initial reference point) but were consistently so in all orientations. The subjects who were inconsistent in their description were correct in only one orientation and made control selection errors in the other orientations.
(3) No particular differences were found between the teach pendant designs, because the subjects tended to describe the motion first and then look at the pendant for the appropriate control. However, there was a higher preference for the joystick control even though these subjects found the toggle controls easier to learn on.

In view of the strong stereotype associations used for the control–display relationship, the results showed that the majority of subjects did not perceive robot movement in the same way that the Unimate manufacturers have designed for the PUMA anthropomorphic robot. This would suggest that the '+ve'/'–ve' controls used for labelling should be reversed on some joints. However, it seems that the solution is more complex as different subjects made control selection errors on different joints.

References

ANSI (American National Standards Institute), 1986, Proposed American National Standard of human engineering design criteria for hand held control pendants. ANSI/RIA R15. 06–1986 (Broadway, NY).

Cousins, S.A., 1988, Development of a human engineering design standard for robot teach pendants. In *Ergonomics of Hybrid Automated Systems I*, edited by W. Karwowski, H.R. Parsaei and M.R. Wilhelm. (Amsterdam: Elsevier), pp. 429–436.

HS/G (Health and Safety Executive), 1989, Industrial robot safety HS(G) series, HS/G 43 (Broad Lane, Sheffield, UK).

JISHA (Japan Industrial Safety and Health Association), 1985, An interpretation of the technical guidance on safety standards in the use, etc. of industrial robots (Tokyo).

Parsons, H.M., 1986, Human interfaces in industrial robotics. *Proceedings of the 19th Annual Meeting of the Human Factors Association*, Canada, August, pp. 189–192.

52

A multifactorial approach to the design of an industrial seat

H. Gregg and E.N. Corlett,

Institute for Occupational Ergonomics, Seating Research Group, Department of Production Engineering and Production Management, University of Nottingham, University Park, Nottingham NG7 2RD, UK

Abstract

The problems of industrial seating are outlined, with an indication of the consequences of inadequacies in seat design for the users. A procedure is then given for the assessment of seating, with a brief description of the techniques. The form of seat required in industry and business to permit comfortable sitting at a range of working heights is presented, and the results of an evaluation of seats to this design in a textile plant are given.

Introduction

Industrial seat design is task dependent. The importance of properly designed seats may be appreciated by considering that an estimated three quarters of all work in industrial countries is sedentary (Treaster, 1987), and that poor seating potentiates (amongst other things) back pain, the cost of which to the National Health Service in Britain was 156 million pounds in 1982, according to a report published by the Office of Health Economics (Wells, 1985).

The factors of importance in seat design are both varied and interactive. Optimization in one area will often be at the expense of another. For example, it is possible to increase the comfort, or at least, decrease the discomfort associated with a seat for some users by countouring its shape during the upholstering. However, to do this is to render the seat less suitable for other users of a different somatotype. Such contouring also increases the postural constraint imposed by the seat on the sitter, a point discussed more fully later. Therefore it will be seen that no single factor can be used to determine the specifications of a work seat, and the importance of an approach which embraces many factors becomes clear.

Physiological and anatomical specifications, postural criteria, measures of the sitter's performance at work using the seat, and subjective evaluations are the main procedures traditionally employed in seat design and evaluation, and are described here in the context of the development protocol.

Table 1. Measurement parameters used to evaluate the response of the sitter to the experimental seat.

Sitter's response	Measurement
Discomfort	General discomfort questionnaires
	Chair feature checklists
	Magnitude estimation (discomfort)
	Cross modality matching (discomfort and pressure)
	Body part discomfort assessments ('discomfort lists')
	Fidget frequency
Posture adopted	Films, etc.
	Posture principles (checklists)
	Lordosis (presence/absence)
	Ease of ingress/egress
	Dimensional measurements
	Reach envelopes
	Stability measurements

Assessment criteria

It was required to create a sit-stand seat for business and industrial applications. Such a seat enables its user to sit or stand at work with the minimum change in posture, and therefore with the minimum interference to the task. The design specifications and assessment criteria were developed from theoretical considerations dictated by the workplace and the physiology of the sitter.

The seat's development was a three-fold process, comprising a theoretical model (Eklund and Corlett, 1986), laboratory studies and simulations, and field trials. Validation of the seat was continuous throughout the development programme, since the three processes were allowed to interact in an iterative fashion.

Assessment criteria, and their consequent measurements fall into two groups, subjective (what the seat feels like), and objective (whether it provides appropriate support for what its user wants to do). Table 1 is a summary of the response of the sitter, and consequent measurement parameters.

Subjective assessments

The difficulties in measuring comfort are well known, since 'comfort' means different things to different people. Indeed Lueder (1983) stated that as yet we do not have a universally accepted operational definition of comfort. However, the concept of 'industrial comfort' has been used by Corlett and Bishop (1976) in the development of their technique for assessing postural discomfort. This represents comfort as a threshold level, below which an individual would not be distracted in his or her work. The measurement of levels of discomfort therefore gives an indication of departures from this threshold, and can be used to evaluate improvements.

A similar view was expressed by Branton (1969) in that comfort is not sought for its own sake, but rather a state is sought which is optimal for the pursuit of other purposes. 'A seat may be measurably "inefficient" to the degree to which it interferes with the primary activity' (Branton, 1969). Branton argues that the seat characteristics are a source of 'noise' interfering with the 'signals' emanating from other sources competing for the subject's attention. Thus the measurement of discomfort is an indicator of the appropriateness of a given design. Furthermore, the location of such discomfort reveals what is wrong with a given posture (Van Wely, 1970).

Subjective assessments represent user acceptability which makes them of particular value in the evaluation of seat prototypes. Departures from an assumed ideal situation for each parameter of measurement were examined by conventional scaling techniques and feature checklists. However, because these methods only generate ordinal data (Stevens, 1975; Pepermans and Corlett, 1983; Gregg and Corlett, 1989) the methods of cross modality matching and magnitude estimation were also employed to gauge the magnitude of improvements.

Postural monitoring

The definition of good sitting postures must include the demands of the task (Corlett, 1983), a point which is often not fully appreciated. These depend upon the overall arrangement of the workstation, and the principles of good working postures (Corlett, 1983) were embraced by the design specifications (Corlett and Gregg, 1987).

Physiological and anatomical factors

Apart from the various techniques for analysing posture itself, its effects on the body also form the basis of assessment regimes. These are made up of factors such as spinal loading with respect to lumbar lordosis/kyphosis, minimization of postural constraint coupled with adequate support and stability, ease of ingress and egress, and reduced static muscle loading.

Lumbar lordosis

It is well known that sitting tends to lead to increased spinal loading especially in the lumbar region. A number of factors contribute to this, but basically the move from standing to sitting in the absence of a backrest extends the hamstring and gluteal muscles, causing the pelvis to tilt backwards (Andersson *et al.*, 1979). This produces kyphosis with a consequent rise in lumbar loading.

Posterior pelvic rotation can be reduced by increasing the angle between the thighs and trunk during sitting, so the contour of the seat surface is important. Clearly a forward sloping seat at an appropriate height can achieve this, but the horizontal component of body weight generated at the seat surface will cause the sitter to slide off unless equal and opposing forces can be applied by the feet, or by friction. The former increases the weight taken by the feet, which rapidly causes discomfort if it exceeds approximately 1/3rd of body weight (Eklund *et al.*, 1982), whereas a high friction surface on the seat is also an inadequate solution because if has no effect at the skin/clothing interface.

Shear force may also be counteracted by the use of shin pads as on the Balans chair, but this introduces a level of postural constraint inappropriate for the sit-stand application. Saddles too have been used as sit-stand seats. Whilst they introduce little or no shear force, they are only really suitable if the sitter is wearing trousers.

The answer to this problem is to ensure that body weight always acts in a direction perpendicular to the support surface, whatever its height. This was achieved in the Nottingham Seat by a combination of contouring and an adjustment mechanism providing appropriate tilt for any given height.

Minimization of postural constraint and static muscle load

It has long been argued that a good posture is one which can be changed. This requires minimal postural constraint commensurate with the requirements of the task. However, a balance has to be reached between this and the level of support required for stability. Branton (1969) noted that the seated individual is intrinsically unstable and that to maintain an upright posture requires muscle activity. Part of the importance of periodic changes in posture is that they allow different muscle groups to provide this function, so that no single group becomes fatigued.

The level of muscle activity involved depends upon the stability of the sitter which depends upon the degree of support offered by the seat. This itself tends to constrain the posture. Thus there is an optimum level of support above which posture change becomes impeded, resulting in fatigue in supporting musculature, and below which muscle activity must increase in order to maintain stability. A subject's perception of his or her stability is one measure by which to approach this optimum.

Anthropometry

It will now be apparent that no single factor may be used to generate industrial design specifications. The many British Standards dealing with the relationship between the physical dimensions of the sitter and the workstation should not be considered to identify fully design specifications, especially where such Standards refer only to static linear anthropometric data, since sitting is an essentially dynamic activity. Whilst it is self evident that if the seat dimensions do not match those of the sitter this is bound to lead to discomfort, a comfortable though inappropriate seat may be worse than no seat at all. Therefore linear anthropometric dimensions must be augmented by the measurement of reach envelopes and other dynamic parameters associated with each particular task.

Stability

The degree of postural stability which a seat must provide is related to the task. Thus, when reach envelopes are specified for a particular workstation, an indication of the required stability across each envelope is useful. Pilot studies were performed to evaluate the feasibility of an index for functional stability.

The location and design of a backrest also influences stability. Only a small percentage of body weight is carried by the backrest, Swearingen *et al.* (1962) estimated this to be around 4% for a backrest raked at 15 degrees to the horizontal. Despite this, the backrest provides the means by which the sitter locates him or herself in the seat during ingress, and helps to stabilize the lumbar motion segment, thereby decreasing muscle activity. In addition, on tasks which require the application of vertical forces, the backrest provides a brace against which to push.

A sit-stand seat for industry

This research will be described in detail in future papers, but in brief the seat shape developed from the theoretical model (Eklund and Corlett, 1986) was initially tested in a number of forms generated by the vacuum consolidation process. Each form was assessed by subjective measurements. A purpose built,

Table 2. Magnitude estimation results: group geometric means[a] (n = 10).

Seat feature	Conventional seat	Change	New seat
Comfort	1·09	+0·3	1·39
Stability	1·37	+0·15	1·52
Weight on feet	1·11	+0·4	1·51
Weight on seat	0·97	+0·24	1·21
Backrest position	0·80	+0·83	1·63
Backrest angle	0·76	+0·87	1·63
Overall seat shape	0·84	+0·6	1·44
Seat angle	0·88	+0·57	1·45
Seat length	1·06	+0·38	1·44
Seat width	1·19	+0·19	1·38
Seat padding	0·92	+0·28	1·26
Pressure distribution under thighs	0·79	+0·58	1·37
Ease of rising	1·07	+0·43	1·50
Suitability to job	0·89	+0·60	1·49
Ease of adjustment	0·53	+1·14	1·67

[a] The group geometric mean represents the mean of the logs (base 10) of individual results. Thus, for instance, in comfort measurement, the original difference between the two measures (1·09 conventional seat, and 1·39, new seat) may be found by taking the antilogs., i.e. 12·3 and 24·5. A higher number indicates an estimated improvement in the factor by the subject.

fully adjustable chair interfaced to a computer was used to generate and analyse force profiles during ingress and egress over the complete range of height and backrest positions, and the effect of these upon discomfort, subjective pressure assessments and stability ratings was examined during short term and prolonged periods of sitting. Subject selection was on the basis of the target user population where practical, and the effect of changes in the slope of the seat front on ease of ingress and egress, and on subjective pressure estimates was examined.

Prototypes were built and evaluated at appropriate workstations (Table 2) and following modification suggested by these field trials, the new seat, for which patent applications have been filed, was produced commercially under licence.

References

Andersson, G.B.J., Murphy, R.W., Ortengren, R. and Nachemson, A.L., 1979, The influence of backrest inclination and lumbar support on lumbar lordosis. *Spine*, **4**, 1, 52–58.

Branton, P., 1969, Behaviour, body mechanics and discomfort. *Ergonomics*, **12**, 2, 316–327.

Corlett, E.N., 1983, Analysis and evaluation of working posture. In *Ergonomics of Workstation Design*, edited by T.O. Kvalseth pp. 1–18 (London: Butterworth).

Corlett, E.N. and Bishop, R.P., 1976, A technique for assessing postural discomfort, *Ergonomics*, **19**, 2, 175–182.

Corlett, E.N. and Gregg, H., 1987, Developments in the design and evaluation of industrial seating. In *Production Research: The State of the Art*, edited by A. Mital pp. 1350–1353 (Amsterdam: Elsevier).

Eklund, J.A.E. and Corlett, E.N., 1986, Experimental and biomechanical analysis of seating. In *The Ergonomics of Working Postures*, edited by E.N. Corlett, J. Wilson and I. Manenica, pp. 319–330. (London: Taylor & Francis).

Eklund, J.A.E., Houghton, C.S. and Corlett, E.N., 1982, *Industrial Seating. Report on Some Pilot Studies*. Internal Report, Department of Production Engineering and Production Management, University of Nottingham.

Gregg, H. and Corlett, E.N., 1989, Subjective assessment techniques: where to draw the line? In *Contemporary Ergonomics 1989. Proceedings of the Ergonomics Society Annual Conference*, Reading, edited by E.D. Megaw, pp. 344–348 (London: Taylor & Francis).

Lueder, R.K., 1983, Seat comfort. A review of the construct in the office environment. *Human Factors*, **25**, 6, 701–711.

Pepermans, R.G. and Corlett, E.N., 1983, Cross modality matching as a subjective assessment technique. *Applied Ergonomics*, **14**, 3, 169–176.

Stevens, S.S., 1975, *Psychophysics*, edited by G. Stevens, pp. 134–136. (London: Wiley).

Swearingen, J.J., Wheelwright, C.D. and Garner, J.D., 1962, *An analysis of Sitting Areas and Pressures of Man.* Report No. 62.1, US Civil Aero-Medical Research Institute, Oklahoma City.

Treaster, D., 1987, Measurement of seat pressure distributions. *Human Factors*, **29**, 5, 563–575.

Van Wely, P., 1970, Design and disease. *Applied Ergonomics*, **1**, 5, 262–269.

Wells, N., 1985, *Back Pain.* ISSN 0473 8837 (Office of Health Economics).

PART VI
MANUFACTURING SYSTEMS AND AUTOMATION

Improvement programmes in the area of manufacturing systems and the growth and development of automated production have had far reaching effects throughout manufacturing industry. The efficient use of workforce and machines in production is critical to the success of any organization.

In his paper Mital compares the performance and costs of manual versus flexible assembly in four different countries and identifies the factors influencing the choice of assembly method. Koether examines the implications on logistics and materials supply, of using flexible manufacturing systems (FMS) in assembly.

In modern manufacturing environments industrial robots are being used increasingly in assembly operations. Mohammadion and Larsson describe the use of a sensory feedback system to overcome errors in the positioning of a robot arm. In the area of flexible transfer systems Fujii *et al.* present a routing control method for automated guided vehicles by the shortest path using the concept of time-windows. Kaspi and Tanchoco discuss the concept of optimal unidirectional flow paths of automated guided vehicle systems.

SECTION VI.1. ROBOTICS

53

Precise assembly by use of reference measurement and sensory feedback

A. Mohammadion and C.N. Larsson

Department of Design, Manufacture and Engineering Management, Strathclyde University, UK

Abstract

Today's assembly robots are provided with various types of end effectors and precision wrist devices for assisting robotic assembly. They also need more sophisticated functions as standard in order to cope with the uncertainties in the position and orientation of both robot end effector and parts presented to the robot for the assembly. Both sensory data and intelligence capabilities provided by the robot algorithms are important and are the main keys to the successful assembly.

This chapter describes an automatic assembly method using an error recovery process which repositions the robot arm using sensory information derived from inductive position sensors to create a new frame of reference which permits the assembly to be accomplished. In order to confirm the concept of the proposed method, a series of experimental tests was carried out during which the robot was programmed to insert a round peg into an accurately bored hole with very small clearance, i.e. ± 0·02 mm, in an assembly fixture carried on a pallet of an indexed conveyor track.

Introduction

Robots are no longer employed only for non-critical tasks such as paint spraying and spot-welding and their applications for more sophisticated work such as inspection and assembly have increased in the last few years. The greater the number of demands for the use of robots in the these areas becomes, the more in-built sophisticated functions are expected from them. In practice, however, assembly seems to be the most difficult job for a robot to perform. This is due to the great dexterity required by most assembly operations. So far, if precision placement is also required, most robotic assembly is now performed with the aid of sensory information derived from contact, tactile or vision sensors.

As early as 1977 efforts were made to overcome inherent poor positioning repeatability and accuracy by the use of compliant end effectors. An example of the use of compliance without sensory feed back was the introduction of Remote Centre Compliance (Drake *et al.*, 1977) and the concept was extended to include sensors in Instrumented Remote Centre Compliance (Seltzer, 1979). Since that time considerable improvement has been achieved in this field. Different approaches have been made using these various types of sensors for providing a certain amount of adaptability. An example has been reported (Merlet and Yoshikawa, 1986) of an advanced robotic assembly station which uses several types of sensor. A vision-sensing device was used to acquire and place disorientated work pieces and a laser range-finder for determining their positions and orientations, and when the pieces were grasped an active compliance device with force sensors was used for the assembly. A similar approach had been made at the University of Delft in The Netherlands, to develop a flexible assembly cell which was equipped with a number of different sensing devices for assistance in assembly tasks (Spaa *et al.*, 1984).

There are two limiting factors in the provision of multiple sensors, namely space and cost. The use of television cameras for error correction in electronics component placement machines is now commonplace. By searching for fiducial marks on the printed circuit board correction can be made for errors in the respective positions of the board and placement head, whether the error be due to the head movement mechanism or the board delivery system. A television camera can give information about position and orientation in a plane, i.e. with three degrees of freedom, and this is well suited to printed circuit board assembly where surface mount components are placed on a flat board or pin in hole components have their short legs inserted in holes in a thin board. In either case rotation about axes in the board plane is not critical. Board height is critical but this is either fixed or sensed indirectly by the insertion head using force sensors. For assembly of three-dimensional engineering components up to six degrees of freedom might need to be catered for, although design would normally reduce this number for any particular mating of parts. To measure and compensate for six degrees of freedom at least two television cameras in known relationship to each other are required, but generally two would not be sufficient because calculation of the position and orientation of objects in some parts of the field of view would lack the necessary precision. Although very small cameras are now available there must be practical difficulties in arranging for two cameras to be mounted on an end effector. Also when taken together with the frame stores and computing power necessary for analysis of the pictures, television cameras are expensive.

Probably most of the above-mentioned methods have met the requirement for one particular task, but unfortunately not much attention was given to providing more intelligence capability for the industrial robot by means of improving both its own sensory abilities and software. In recent years, research has been directed towards manufacturing robots with more built in sensory devices and more software, so that, for example, important information such as dimensional changes in the robot elements caused by various parameters such as thermal expansion (Mooring *et al.*, 1985; Mohammadion and Larsson, 1988), can be detected and corrected either by an external controller or by the robot software. During the same period the sophistication of algorithms within industrial robots' operating systems has improved enormously so that sensor information can be acted upon in real-time to give continuous correction of errors detected.

In this chapter most attention has been given to describing the implementation of a reference measurement method for precise insertion assembly. The chapter concentrates on describing the advantages of the assembly method that first takes a reference measurement by means of sensory data, then repositions the robot arm before the insertion phase.

Use of a local frame of reference to assist assembly

The effectiveness of industrial robots in assembly is mainly determined by their positioning repeatability and accuracy. Achieving these with a computer-controlled manipulator in which the system accepts user-demanded positions of the robot arm (points) in the form of the location and orientation of the tip of the current tool relative to a specific frame of reference, requires more sophisticated software than

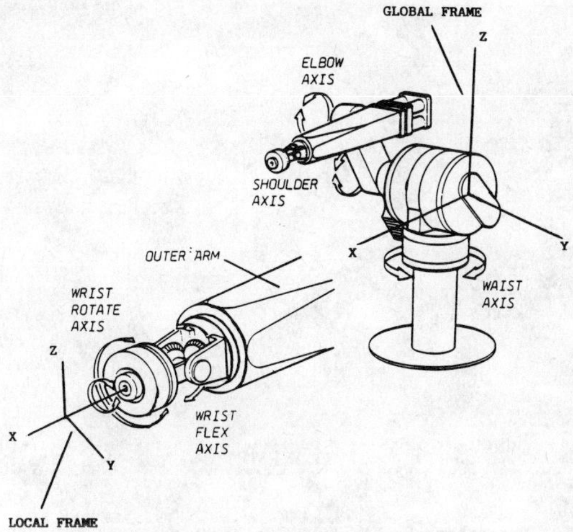

Figure 1. Frames and axes defined by the robot system.

if the arm position is specified only by the joint angles. With earlier robots, the positions of the tool or end effector tip required by a program were taught by manually operating the controls until the desired position had been achieved. The robot position was then defined by its joint angles which were stored and called up later during the program execution. Present generation robots have software which allows the tool tip position and orientation in some frame of reference to be specified by the user. The software algorithms then calculate the joint angles necessary to achieve the required position. There is the possibility of error as the internal calculations rely not only on the readings of internal position potentiometers or optical encoders, but also on the accuracy of the values of the dimensions of the robot components stored in a calibration file within the system. The major dimensions such as arm lengths and offsets are not easy to measure but will have been entered in the file by the robot manufacturer. The dimensions of the tool relative to some fixed reference such as a faceplate have to be supplied by the user and may require considerable metrological skill to obtain with sufficient accuracy. The problem of achieving the accuracy necessary for assembly of parts with small clearances is severe for those robots with a revolute geometry and six degrees of freedom, particularly if some or all of the joints have to turn through large angles to take the end effector from the robot calibrate position to the assembly fixture.

The software of the Rediffusion robot which was used for this work recognizes three different types of frame of reference. The one frame of reference which never changes is called the global frame. Its origin is located in the centre of the robot and is based on three X, Y and Z axes (Figure 1). The programmed point in the global frame is the centre of the faceplate at the end of the arm. A tool frame is recognized which has its origin at the centre of the faceplate, and once the position and orientation of the tool tip within this frame is specified in a calibration file, the tool tip becomes the programmed point in the global frame. Finally it is possible to create a local frame with its origin and orientation, with respect to the global frame, anywhere and at any angle within the envelope of the robot. In this work an assembly fixture was taken to exist in a local frame. The position and orientation of that frame were found by inductive sensors, fastened to the gripper tool, measuring the distance to target studs on the fixture, during a search subroutine in the assembly program. It was thus possible to calculate the movement of the tool needed to place a pin in a hole in the fixture, from the geometry of the fixture without knowing its position in the global frame beforehand. Also this meant

Figure 2. Robot in goal position for creating frame of reference.

that successful assembly relied on the accuracy of calculated robot movement over only a relatively small distance, i.e. from target point to hole in the assembly fixture. The existence of any geometric deviations such as dimensional error in the components of the robot, accumulative error in encoder readings and misalignments between the robot global frame of reference and the tool frame were compensated by the use of a local frame. In effect the robot taught itself the position and orientation of the local frame by means of sensors, and then the joints only had to move through small angles to put the end effector into the assemble position. Creating a local frame can be therefore a useful method for making short precise movements and making these movements with respect to different origins on separate occasions.

This feature provided by the robot algorithms in conjunction with the sensory information also provided compensation for uncertainty in position of the assembly fixture. In the present work the assembly fixture was mounted on a pallet on a stop–go conveyor track. The variability in the position of the stopped pallet was many times greater than the clearance of the pin to be assembled in hole in the fixture.

Application of the method

A test of the application of the method of creating a local frame with an origin determined by sensors on the end effector, took the form of a series of experimental tests during which the robot was programmed to pick up a round pin of 23·96 mm diameter from a compliant holder on the conveyor track, and insert it in a hole of 24 mm in an assembly fixture on a conveyor track track pallet (Figure 2). The location of the pin in the gripper was ensured by using V-block jaws of measured dimensions. The variability of the track was over a range of 0·18 mm.

With the robot arm extended almost to 1 m for assembly the clearance between pin and hole of 0·04 mm was equivalent to a subtended angle of approximately 9 seconds of arc at the waist axis. The Rediffusion robot uses optical encoders which give a resolution of 13·5 seconds with stepping motor interpolation of 1·35 seconds so the machine was sufficiently sensitive to achieve placement of the pin in the hole. However, an absolute movement of some 120 degrees from the power up calibrate position to the assembly position with a +/− 4·5 second accuracy was beyond sensible expectation. To find the origin of the local frame of the assembly fixture, three mutually perpendicular inductive sensors with a range of 1 mm were fastened to an aluminium shell on the end effector. These were used by the robot using a search sub-routine to first detect and then to measure the distance to the three target studs on the assembly fixture containing the hole. It was expected that the orientation of the pallet would not vary as the conveyor track was precision built for assembly work and the experiments were therefore originally designed with the intention of just using the three sensors to find the position of the pallet. It was soon found that the orientation did vary significantly from experiment to experiment and so as a matter of expediency dial gauges were incorporated in the tool holder to measure the skew. While the distance between the inductive sensors was fed into the assembly program automatically by means of an a/d interface, the readings from the dial gauges had to be entered manually.

The sequence of events for an assembly was that once the pallet was in position the conveyor controller flagged the robot to start the search routine. The robot moved the end effector to bring the sensors close to the target studs on the fixture and the search routine moved the arm until they were within measuring range. At this point the dial gauge readings were input and the robot calculated the origin of the local frame of reference. The arm then moved to place the gripper holding the pin to a point above the hole and finally lowered the pin into the hole. Full details of the experimental procedure have been recorded elsewhere (Mohammadion, 1988).

Conclusions

The assembly method involving delivery of components by the use of a conveyor track followed by recalibrating the position of the robot arm relative to the pallet, before insertion, operated satisfactorily. Creating a local frame of reference that permitted the assembly to be accomplished with the use of a reference measurement eliminated the need for the large movements with respect to the robot global frame. By this means precise and effective short movements were achieved. The use of sensory information provided the flexibility needed to overcome the uncertainties in the positions and orientations of the parts presented to the robot for the assembly.

References

Drake, S., Watson, P. and Simunovic, S., 1977, High speed robot assembly of precision parts using compliance instead of sensory feedback. *Proceedings of the 7th International Symposium on Industrial robots*, Japan.

Merlet, J. and Yoshikawa, T., 1986, An advanced robotic assembly station using multiple sensors. *Proceedings of the 7th International Conference on Assembly Automation*, Switzerland, pp. 351–358.

Mohammadion, A., 1988, *Robot repeatability and intelligent location of end effectors*, PhD Thesis, University of Strathclyde, Glasgow.

Mohammadion, A. and Larsson, C.N., 1988, The use of local temperature control to improve the positioning repeatability of an industrial robot. *27th Matador Conference*, Manchester.

Mooring, B., Bingham, D. and Hoffman, J., 1985, Enhancement of repeatability during warm-up for the IBM 7565 robot. *Computers in Engineering*, **1**.

Seltzer, D., 1979, *Use of sensory information for improved robot learning*. Society of Manufacturing Engineering Technical Paper, MS79-799, Michigan.

Spaa, J., Zimmerman, N. and Singe, U., 1984, Sensory controlled robot assembly and inspection station: hardware configuration. *14th International Symposium on Industrial Robots*, Sweden, pp. 231–240.

54

Image-based cell calibration in a graphical off-line programming environment

Heikki Aalto,[1] Juho Joensuu[2] and Ilpo Kauhaniemi[2]

[1]*Helsinki University of Technology, Laboratory of Automation Technology, and* [2]*Technical Research Centre of Finland (VTT), Laboratory of Engineering Production Technology, Finland*

Abstract

This paper describes an automatic cell calibration technique for graphical off-line programming based on the procedure for calibrating off-the-shelf cameras (Tsai, 1987). The compensation program compares defined points of a known structure, e.g. the object currently being worked on. These points are calculated in two different ways: from the CAD-data base and from the charge-coupled device (CCD)-camera image. Both sets of data are calculated in the corresponding tool centre points (tcp)-position relative to the work object. The camera is attached to the robot wrist. This allows posing errors up to six degrees of freedom (d.f.) to be compensated. The procedure was tested on a hydraulic block assembly. This kind of calibration method is suitable for the local compensation of articulated robots (6 D.F.) and could be used in remote, adaptive robot control, e.g. for assembly.

Introduction

Off-line programming was developed to meet the challenges of robot programming in integrated factories. The programmer can type in the robot program without using the actual production robot. Especially when using high-level three-dimensional (3-D) graphical off-line programming systems it is possible to use various common computer-aided design (CAD)-data bases and hence integrate various computer-integrated manufacture (CIM)-activities such as product design, production planning and control, and robot programming. In the Laboratory of Engineering Production Technology at VTT a 3-D graphical off-line programming system 'GRASP' (Graphical Robot Application Simulation Package) was used. The program uses a high object-level programming language which allows the user to optimize the robot cell layout, program the robot using a single universal language, make collision tests and simulate the program cycle.

Once the program has been simulated and runs perfectly on the computer it is translated into the robot-oriented language using a special post-processor. All the post-processed information, both logical data and movement and positioning instructions, is transmitted to the robot or cell controller over a local area network (LAN) or similar network.

For demanding applications, like assembly, the positioning accuracy of real robots is not sufficient. There are deviations between the model-frames and the real world frames, i.e. pose errors, which should be eliminated.

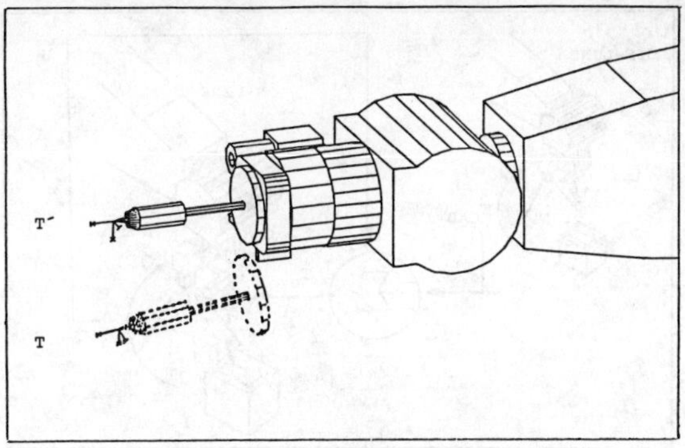

Figure 1. A pose-error in off-line programming.

Pose error in off-line programming

The 3-D graphical off-line programming system uses a kinematic modeller. The user can create the computer models of special manipulators or use the robot library. There are about 20 different robot models in the library. The models are made according to the manufacturers' specifications. The robot models are controlled as precisely as the real robots.

When the robot program made off-line is started in the robot controller the real robot will not go exactly to the off-line commanded positions. This is due to inaccuracies in the fixtures and the robots. These errors are mainly caused for the following reasons:

— the robots are individuals as a result of manufacturing tolerances,
— the robot's structure is not rigid,
— the control resolution for the robot's movements is not sufficient,
— the word length in the robot control processor is too short,
— there are always measuring inaccuracies in robot cell installation.

Figure 1 shows the situation. The posing error between poses T and T' can be realized as a transformation matrix T_e. The tcp was commanded to go to the pose T with respect to an object by the off-line program but the tcp was located to the undesired position and orientation T'. This was caused by one or even all of the reasons listed above. If the robot cell had just been installed, the main reason would probably be the measuring inaccuracies.

Either an attempt can be made to eliminate the sources of the errors or to measure the posing errors and compensate for their effects. This compensation means that only linear frame errors are corrected. This is called cell calibration. If the calibration is carried out as part of the robot program and without any manual interruption, the technique is called automatic cell calibration.

The method of cell calibration used at the Technical Research Centre of Finland (VTT)

The technique uses the CCD-camera image of a known structure, e.g. the work object. The procedure was originally invented at IBMs T.J. Watson Research Center by Roger Y. Tsai (1987) and was used for the 3-D camera calibration of off-the-shelf TV cameras and lenses.

Two methods are used to calibrate robot cells at VTT: calibration of the world

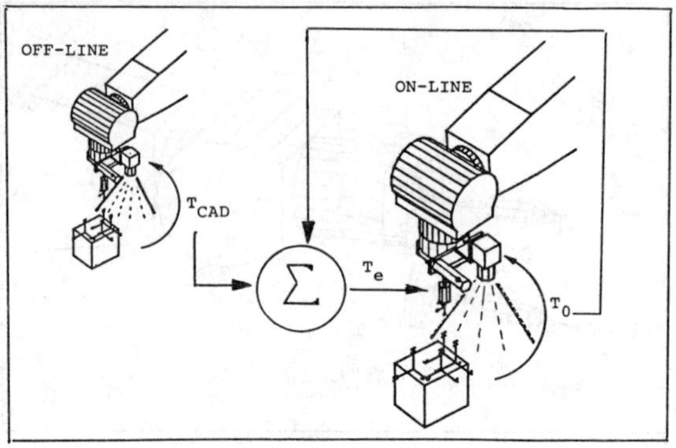

Figure 2. The automatic image-based pose-error compensation system.

frame and calibration of the object frame. The object frame calibration is shown in Figure 2.

World frame cell calibration is carried out semi-automatically. The idea is to update the layout model to fit the actual situation. A special tcp-tool is set to predefined positions in the robot cell, see Figure 1, and the layout model is updated according to this rough location data. This procedure fixes the robot cell in the x, y, z-space with respect to the world reference.

The second approach, robot cell calibration according to the work object frame, will now be discussed in detail. The calibration is done for all workpiece frames. For example, during assembly, when the assembled work object is changed, all the new picking and placing positions have to be calibrated for the new work object coordinate system.

The method in words

The characteristic points of the captured image of the work object are compared with the corresponding points in the 3-D CAD-system (GRASP) calculated from the work object data and robot geometry. The compensation matrix is calculated from the difference between the captured image points of the work object in the real robot cell and the corresponding points of the CAD-model of the work object.

The method in terms of matrices (Figure 2)

The transformation from the workpiece coordinates to the camera coordinates, T_1, is calculated from the camera image. The position and orientation of the camera coordinate system is known in robot coordinates and can be transformed to the tool frame of reference using transformation T_2. The transformation from the work object coordinates to the tool frame is:

$$T_0 = T_1 T_2.$$

The same transformation at the known position of the robot tcp calculated from the CAD-data is T_{CAD}.

Because of the posing errors the transformations T_0 and T_{CAD} are not exactly the same. The position and orientation, i.e. the pose in terms of the CAD-based tool frame of reference can be converted to the real tool frame by using transformation T_e. This transformation is the posing error and can be calculated from the equation:

$$T_{CAD} = T_e^{-1} T_0,$$

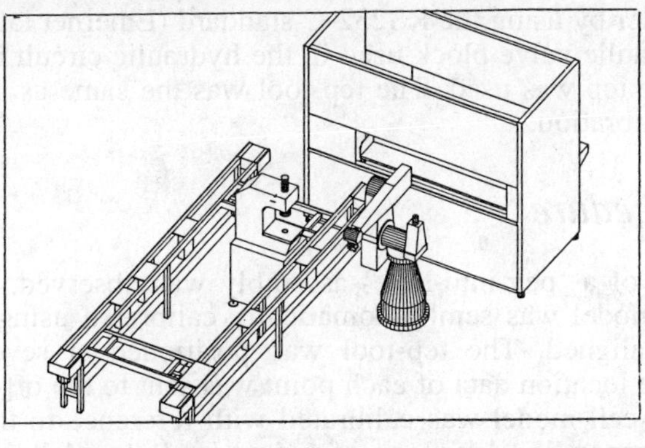

Figure 3. An automatic flexible assembly cell at Technical Research Centre of Finland.

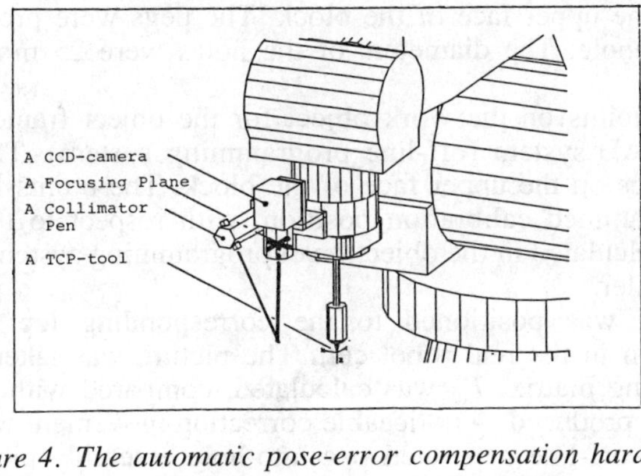

Figure 4. The automatic pose-error compensation hardware.

which gives

$$T_e = T_0 \, T_{CAD}^{-1}.$$

The posing error is the difference between the model-based and the real tool frame of reference, e.g. the work object coordinate system.

The test cell

This research work has been carried out on a flexible, automatic assembly cell which has been used in various assembly pilot-cases and was originally planned to meet the requirements of flexible assembly for small batches (see Figure 3). The cell is controlled by the cell controller which supervises all the machines in the cell: the robots, the warehouse, the conveyor and the hydraulic press. Communication is done via direct serial links. The cell communicates with the higher systems in the hierarchy via the Ethernet-network. The off-line programming system was connected to the cell via this LAN in late 1989.

The compensation program was installed in a PC-AT computer, the cell controller, using MS-C-language. The CCD-camera is a Panasonic WV-cd-50 with f 8 mm lens. Triangular measurement is used to avoid measurement discrepancies in the *z*-direction. For this purpose a laser collimator pen, a Philips 515CQL-A, was selected. The collimator pen was installed on the robot wrist with the CCD-camera at an angle (Figure 4). A Unimation Puma 762 with six degrees of freedom was used. The off-line programming system was connected to the robot controller and to the cell controller by using the RS232-C standard (Ethernet late 1989). The work object was a hydraulic valve block used in the hydraulic circuit of a tractor. In this research, only one tcp was used. The tcp-tool was the same as that used for semi-automatic cell calibration.

The test procedure

The functionality of a 'peg-into-hole' assembly was observed.

First, the cell model was semi-automatically calibrated using the tcp-tool. The robot wrist was aligned. The tcp-tool was positioned at several points in the assembly cell. The location data of each point was sent to the off-line programming computer and the cell model was calibrated with reference to the world frame by updating the corresponding locations of points in the model.

The off-line program was written for the peg-into-hole assembly of a hydraulic block. The program consisted of the assembly sequence for four peg-into-hole routines on the upper face of the block. The pegs were programmed to enter 50 mm into each hole. The diameters of the holes were 25 mm and of the pegs 24·8 mm.

The reference points on the work object for the object frame calibration were selected in the CAD-system (off-line programming system). These were the six centres of the holes on the upper face of the block. The points formed the matrix T_{CAD} in the programmed calibration position, with respect to the tcp-tool model frame. The data calculated in the object level programming system was downloaded to the cell controller.

The real robot was positioned to the corresponding tcp position, i.e. the calibration position in the real robot cell. The picture was taken in this position. The real positioning matrix, T_0, was calculated, compared with T_{CAD} and the correction matrix, T_e, produced. A noticeable correction movement was detected. After the correction the off-line generated peg-into-hole assembly program was run.

It was noted that the assembly worked according to the program with the holes nearest to the tcp-tool in the calibration position. With the holes beyond the nearest ones slight sticking was observed. However, the tcp-tool managed to enter all the five holes to the pre-programmed depth.

The accuracy of the technique depends on the camera used. In this project the positional accuracy was estimated to vary from $+0{\cdot}2$ mm to $-0{\cdot}2$ mm.

Conclusions

The program compensates for the errors of position and orientation in the robot with respect to the work object in three dimensions. Thus, articulated robots can be calibrated for demanding assembly off-line program tasks. Using this kind of technique allows the camera system also to be used for identification, inspection, etc. Another benefit is that the program runs on a PC and produces the correction in the form of a frame. The technique is independent of the robot controller, and hence can be adapted for different makes of robots. The next step in developing the compensation program is to include more tcps to improve flexibility. So, the different tools could be calibrated separately with respect to the work object. Also, a six d.f. force sensor is to be attached to the robot wrist.

Acknowledgements

This project was funded by The Technology Development Centre (TEKES), Valmet Tractor Works and Valmet Paper Machinery.

References

Halme, A., Heikkilä, T. and Torvikoski, T., 1986, An interactive robot control system. *Proceedings of the IASTED Conference on Robotics and Artificial Intelligence*, Toulouse.

Joensuu, J., 1988, The calibration and control of 3D pointing device (Master Work).

Paul, R.P., *Robot Manipulators: Mathematics, Programming and Control.*

Pugh, A., *Robot Vision.*

Storr, A. and McWaters, J.F., *Off-line Programming of Industrial Robots.*

Tsai, R.Y., 1987, A versatile camera calibration technique for high-accuracy 3D machine vision metrology using off-the-shelf TV cameras and lenses. *IEEE Journal of Robotics and Automation* **RA-3**, 4.

The Laboratory of Automation Technology, 1989, *The Development of Interactive Robotics.* Report, Technical University of Helsinki.

55

A distributed information–control structure for robotic systems

Gen'ichi Yasuda[1] and Keihachiro Tachibana[2]

[1]*Department of Mechanical Engineering, Faculty of Engineering, Nagasaki Institute of Applied Science, 536 Aba-machi, Nagasaki 851-01, Japan and*
[2]*Department of Industrial and Systems Engineering, Faculty of Engineering, Setsunan University, 17-8 Ikeda-nakamachi, Neyagawa, Osaka 572, Japan*

Abstract

This chapter discusses a distributed information–control structure for real-time robot control and its implementation on a network system of computing devices with point-to-point communication links. A control mechanism for co-operative motion control based on physical modularity and parallel processing capabilities is presented. Then an information–control structure for implementation of the control mechanism is designed. The structure consists of a number of modules which execute their own assigned task and communicate with each other. The functions and features of processes which operate concurrently and communicate through channels in each module are described, and are implemented using microprocessor-based computing devices, Inmos transputers, and occam which is an efficient programming language for describing concurrent processes.

Introduction

The present day advanced manufacturing systems are distinguished by multi-degree of freedom mechanisms in which two or more controlled elements interact kinematically and dynamically. Robotic systems, such as manipulators or robotic vehicles, which are representative of such manufacturing systems, must be designed to perform a set of tasks which need co-ordinated motion control of a number of mechanical elements not only with high accuracy but at high speed. Control systems for such robotic mechanisms necessitate not only sophisticated control of a single robotic element but also motion co-ordination among these elements. However, dynamical equations describing a robotic mechanism to handle an object or assemble complex parts are largely complex and only approximate. For the co-ordinated motion control of robotic mechanisms under uncertain or unknown environmental conditions, a conventional supervisory control scheme using a tree-structured computer architecture seems to be inefficient. Each local control unit needs to compensate various types of disturbances intelligently by communicating with other units quickly and frequently. Therefore an optimal information–control structure to perform parallel processing involved in direct control tasks must be investigated so that the real-time performance is guaranteed.

Furthermore, from the standpoint of production systems engineering, especially with respect to application software development, the following features are also important:

(1) The control system must include all interrelated programmable functions ranging from task specification to on-line programming for real-time motion control as a unified entity.
(2) The system must be easy to understand hardware and software architecture of the total system and to construct user's application programs in detail.
(3) The system must be cost-effective and adaptable to different robotic mechanisms, tasks and environments.

These features implicate the need for a modularized computer architecture to develop complex software for real-time robot control. In this chapter we first discuss a distributed processing control mechanism involved in the co-ordination of a robotic system. The control mechanism possesses a functionally hierarchical structure of robotic task operation. However, in addition to the hierarchical nature, physical modularity and parallel processing capabilities are considered as the two important factors in the control system design. To exploit the parallelism involved in the horizontally related computational tasks of the described control algorithm fully, a distributed information–control structure for real-time co-operative motion control is presented. The structure consists of a number of modules that communicate with each other by sending messages. The structure can be efficiently implemented by making use of microprocessor-based computing devices, transputers, and occam which is an efficient programming language for describing concurrent processes in transputers. Because the emphasis of this study is on the design of the computer system, the detailed structure of the system is described.

Distributed processing for co-operative motion control

Functionally hierarchical nature of robotic motion control

The robotic motion control in general consists of two distinct steps which are hierarchically related: motion planning and motion execution. Motion planning specifies key set points on the robotic motion task. The actual path between the set points remains undetermined. Motion trajectories may be generated between the key points and are described by a large number of set points along the path. These cartesian set points must be transformed into corresponding joint positions, and then sent to the motion execution part for driving the joint motions. Motion execution is to implement a feedback control algorithm, which calculates the required torque for each robotic joint to move it to the next set point position.

From the functionally hierarchical nature of robotic motion control we investigated a distributed processing architecture with parallel processing (Yasuda and Tachibana, 1987). The system is organized in the following three functional modules: (1) user interface, (2) motion planning, and (3) motion execution. The first module performs robotic task specifications using a computer-aided design (CAD) system for geometrical representation of controlled objects. The second module plans motion trajectories and performs coordinate transformations. The third module performs direct control of robotic actuators and sensors.

In the following section we focus on the design of a parallel processing control algorithm for the motion execution part. The control mechanism detects the error between the desired states and the actual states of each joint, and produces adequate torques in accordance with the errors. The time interval between two consecutive joint torque outputs, is constrained by the robot dynamics and required to be in the order of milliseconds. In addition to the feedback control, the motion execution part must monitor the conditions of various sensors in the working environment in real-time. Thus some distributed control algorithm with parallel processing is desirable to exploit the parallelism involved in the motion execution part.

Control algorithm for parallel processing

The dynamic behaviour of robotic mechanisms is described by means of a set of non-linear differential equations, but only approximately since it is based on several simplifying assumptions, such as lumped masses, frictionless joints, etc. Furthermore, some parameters depending on factors such as the weight of the object being manipulated are unknown. In this chapter, from the viewpoint of distributed parallel processing, we revise the performance-adaptive self-organizing control algorithm with the non-linear control structure by Saridis and Stephanou (1977). The control algorithm implements the following two principles to perform kinematical and dynamical co-ordination considering variable, uncertain or unknown parameters of controlled elements:

(1) anthropomorphic trajectory control based on optimal control theory,
(2) on-line compensation for dynamical couplings among controlled elements, and other external disturbances.

The control problem is decomposed into several subsystems, each corresponding to one mechanical degree of freedom. The equation of motion of the ith subsystem is described as:

$$\ddot{\theta}_i = f_i + u_i \tag{1}$$

where θ_i is the actual actuator or joint angle, u_i is the input quantity which is obtained from the vector of actual input torques by a non-linear transformation (Saridis and Stephanou, 1977), and a nonlinear term f_i represents the influence of gravity on the subsystem dynamics and coupling effects from other subsystems. Then a suboptimal structure of the subsystem controller is represented as:

$$u_i = L_i(z_i - x_i^d) + N_i f_i^a(z_{i-1}, z_i, z_{i+1}) \tag{2}$$

where x_i^d is the vector of target states (the actuator angle and its velocity), z_i is the vector of current observed or estimated states, f_i^a is an approximate function of the nonlinear term f_i. L_i and N_i are adjustable gains to minimize a performance criterion, so that the first term of eqn (2) is the optimal control for the hypothetical isolated subsystem,

$$\ddot{\theta}_i = u_i \tag{3}$$

while the second term is a compensating input to reduce the effect of the nonlinearities. The control algorithm for each subsystem is summarized as follows:

(1) The controller obtains the current values of states from the neighbouring subsystems.
(2) It calculates the linear term of eqn (2) using the current error and the non-linear term using the current values of states.
(3) The algorithm obtains input quantities from the neighbouring subsystems and calculates the required torque from input quantities by an approximate non-linear transformation using the current values of states.

The above three steps, and besides sending current values of its own states and input quantity to the neighbouring subsystems, must be executed at each servo interval. Then, after the target states are attained, the following on-line search is executed:

(4) It changes values of adjustable gains L_i and N_i to improve the performance criterion.

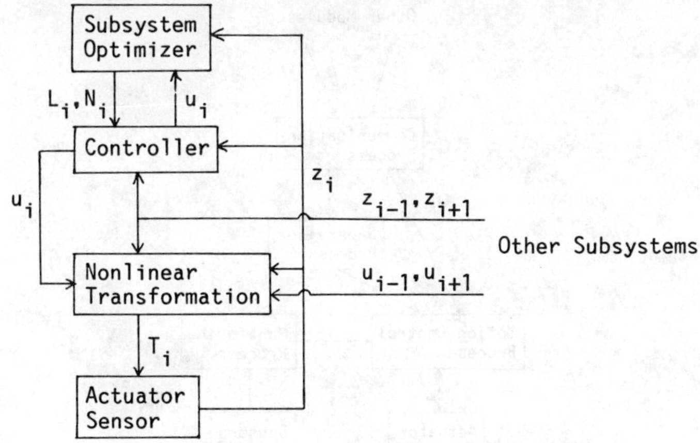

Figure 1. Structure of subsystem controller.

Some values of weighting matrices of a quadratic form performance criterion represent relative importance of the states and the input, and are specified as control parameters. The resultant structure of each subsystem control algorithm is shown in Figure 1.

Information–control structure using communicating modules

Overview of the structure

We examined information flow within the above-mentioned distributed processing architecture. Consequently a distributed information–control structure for a system consisting of a number of modules was designed. Each module performs its own local information processing with equal hardware and software capabilities. In the total system no central supervisor or master–slave relationships exist. The co-operative activities of these modules are conducted as a group through mutual information exchange.

To realize the functional hierarchy of robotic task operation, each module is assigned one of the following three types of tasks: (1) user interface, (2) motion planning, (3) motion execution. A number of modules which are assigned motion execution perform the above-mentioned control algorithm cooperatively in parallel.

Function of the modules

Each module must be equipped with specific capabilities for execution of its own assigned task. For example, motion execution modules must be equipped with input/output interfaces for direct control of robotic actuators. Furthermore, it must be equipped with communication capabilities to neighbouring modules. Therefore each module must have intelligence to perform these two main subtasks successfully. This integration is performed by the local supervisor in each module. So it is justified that, for example, each motion execution module has the following four processes: the supervisor process, the motion control process, the monitoring process and the communication process. The information–control structure among processes in each motion execution module is shown schematically in Figure 2. The main functions of each process are described below.

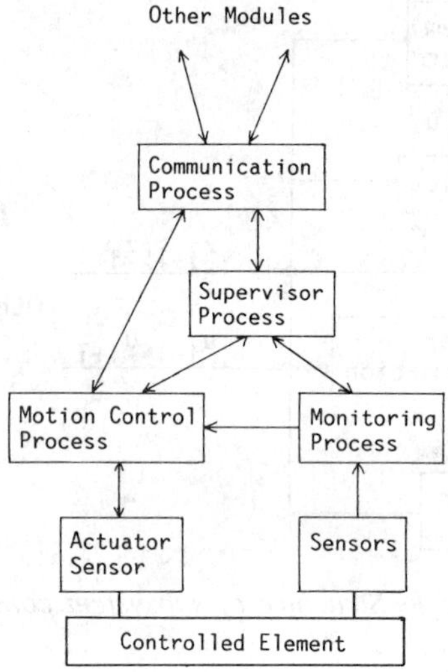

Figure 2. Information–control structure of motion execution module.

Supervisor process

This process receives task specification data (set point and control parameters) from the user interface module through the communication process. Then it activates the motion control process for motion execution and the monitoring process, and sends robotic task data to the motion control process. This process supervises all other processes in the module during motion execution, while it receives status data of the neighbouring modules through the communication process. It also receives information with respect to conditions of the actuator and sensors from the motion control process and the monitoring process.

Motion control process

This process is activated by the supervisor process and performs motion control. It obtains current values of actuator's states of neighbouring modules through the communication process, then calculates the required torque which is sent to the actuator through the output port at each servo interval.

Monitoring process

This process monitors the status of the controlled element and its environmental conditions through specific sensing devices. When the status or conditions are summed up or abnormal, this process sends information to the supervisor process or the motion control process.

Communication process

This process communicates with communication processes in the neighbouring modules, sending or receiving task specification data, status data and current values of actuator's states for use in control algorithm execution by the motion control process.

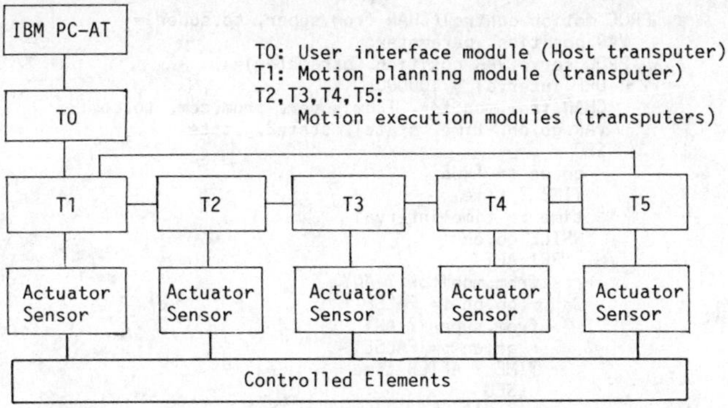

Figure 3. System configuration of experimental system.

Implementation on a system of transputers

Features of the transputer as a real-time control module

The information–control structure of the distributed processing control software for robotic systems can be constructed from a network system of transputers which operate concurrently and communicate with each other through point-to-point communication links. The control software is implemented in a programming language for transputers, occam, as a collection of the four processes operating concurrently and communicating through named channels (May and Shepherd, 1985).

In the transputer, a concurrent process takes transitorily one of the following two states, which are switched by the microcoded scheduler: (1) active, (2) inactive. Active processes are being executed or waiting to be executed, and are held on a linked list of process workspaces. A process is in an inactive state when it is waiting to input or output, or waiting for the timer. Message-passing between two processes is achieved by means of a channel which is implemented by a single word on the same transputer or by a point-to-point link on different transputers. The process which first becomes ready must wait and be in the inactive state until the second one is also ready. Thus the activities of concurrent processes on the same module or on different modules are mutually controlled using message-passing based on synchronous communication.

The system configuration is shown in Figure 3. Each motion execution module with a transputer (Inmos T414 Microprocessor) is connected in a ring. The user interface module for robotic task specification is also used as the host computer for software development, and is connected with a common personal computer IBM PC-AT. A system control program ultimately intended for a network of modules is compiled and transferred to each module, in advance of execution of a robotic task, by the software development system.

Implementation of the control system

Control software including concurrent processes can be easily written in occam. Occam programs are built from three primitive processes: (1) assignment, (2) output, (3) input. The primitive processes are combined to form a 'construct'. Concurrent processes are defined as a PAR construct which means processes executed in parallel. A construct which is built from components executed sequentially is defined as a SEQ construct. A construct is itself a process and may be used as a component of another construct.

```
PROC motion.control(CHAN from.super, to.super)=
  VAR position, parameter:
  PROC servo(VAR position, parameter)=
    DEF interval = 100000:
    CHAN from.monitor, from.super, from.com, to.com:
    VAR go.on, time, state1, state2, state:
    SEQ
      go.on := TRUE
      TIME ? time
      time := time+interval
      WHILE go.on
        PRI ALT
          from.monitor ? ANY
            go.on := FALSE
          from.super ? ANY
            go.on := FALSE
          TIME ? AFTER time
            SEQ
              PAR
                from.com ? state1; state2
                to.com ! state
              SEQ
                ...subsystem controller program
                time := time+interval:

  WHILE TRUE
    SEQ
      from.super ? position; parameter
      servo(position, parameter):
```

Figure 4. Example program of motion control process.

The four processes in each motion execution module can be concurrently executed and are defined as a PRI PAR construct, since the supervisor process must be given a high priority. The basic control sequence of each process is described as follows. Initially the motion control process and the monitoring process execute 'input message' instructions and then are switched in the inactive state, while the supervisor process and the communication process can be active. Each supervisor process receives robotic task specification data from the user interface module and the motion planning module through the communication process in the same module, then activates other processes in the module by sending messages.

During motion execution the supervisor process receives messages from the communication process, the motion control process or the monitoring process. According to the content of the received message, the supervisor process calls a subprocess. This procedure is summarized as a collection of if–then rules and defined as an IF construct in occam.

The motion control process is activated or inactivated by messages from the supervisor process. This process must execute the control algorithm, for example, at a cycle of 10 ms during motion execution. If the monitoring process detects the information which indicates that the robot must be stopped, then it directly sends a stop command to the motion control process. So 'input message' instructions from the supervisor, the timer and the monitoring process are alternatively executed in the motion control process, which is defined as a PRI ALT construct.

The communication process acts as an agency for message exchange among processes on different modules, because a transputer is equipped with only four communication links. It receives an 'input message' from the supervisor process, the motion control process or neighbouring modules. For example, it receives current values of actuator's states from the motion control process, then sends the same values to the neighbouring modules. Messages from communication links are interpreted, then are sent to their destinations. These procedures are defined as a PRI PAR construct of SEQ constructs which execute the 'input message' in the beginning. Each main process in the above four processes is iteratively executed for real-time motion control, which is defined as a WHILE construct.

Using the above programming facilities, an example program of the motion control process in each motion execution module is generally described in occam as Figure 4, including the servo subprocess which is called from the main process at the beginning of feedback control.

Conclusions

We have presented a distributed information–control structure for co-operative motion control of robotic systems, and described its implementation on a computer network using communicating modules, transputers.

Each module, which constructs our distributed computer system, has its own local operating system. However, there is no runtime supervisory operating system which sits on top of these local operating systems. So system co-ordination is performed through local message exchange which is controlled by the information–control structure in each module. From this viewpoint we have discussed the functions and features of the four processes which are concurrently executed in each module, and message exchange between these processes.

The information–control structure with the parallel processing computer architecture may be an optimal system structure for multiple co-ordinating robots as well as for a single robotic system in terms of the real-time system performance, costs and other constraints.

References

May, D. and Shepherd, R., 1985, Occam and the transputer, In *Concurrent Languages in Distributed Systems*, edited by G. Reijns and E. Dagless (Amsterdam: Elsevier), pp. 19–33.

Saridis, G. and Stephanou, H., 1977, A hierarchical approach to the control of prosthetic arm. *IEEE Transactions on Systems, Man, and Cybernetics*, **SMC-7**, pp. 407–420.

Yasuda, G. and Tachibana, K., 1987, A multimicroprocessor-based distributed processing system for advanced robot control. *Proceedings of the 9th ICPR*, pp. 1926–1933.

SECTION VI.2. AUTOMATION

56

Development of a programmable parts feeder for rectangular parts

Patrick S.K. Chua[1] and Eric Sung[2]

[1]*School of Mechanical and Production Engineering and* [2]*School of Electrical and Electronics Engineering, Nanyang Technological Institute, Nanyang Avenue, Singapore 2263*

Abstract

This work is concerned with the development of a prototype programmable parts feeder, named PROFEED, which employs a vibratory bowl for parts singularization and two conveyor belts and trip devices for orientation of rectangular parts. Diffusive fibre optic sensors were used for feature recognition and their output characteristics for various features and surface conditions of parts were studied. Consistent conveyor belt speed and surface conditions of parts were the important parameters affecting the feeder's performance.

Introduction

Automatic parts feeding and orientation has been realized as a critical aspect of production systems (Eade, 1989). The concepts of flexibility, universality and programmability are now being explored in the domain of parts feeding and orientation (Goodrich and Maul, 1986). Such systems would be extremely beneficial to production systems with short runs or frequent change-overs due to the increasingly diversified and dynamic market demand.

In tackling the problem of feeding and orientation of the myriads of parts encountered in manufacturing, one approach is to break up the problem into smaller areas and concentrate on one of these areas. A survey (Boothroyd and Dewhurst, 1982) revealed that about 60% of parts are rotational and about 30% are basically rectangular (Figure 1). In this work, a flexible and programmable feeder, named PROFEED, has been developed for the feeding and orientating of rectangular parts.

The operation of PROFEED

Microprocessor-controlled, PROFEED (Figure 2) operates as follows:

(a) The vibratory bowl singularizes the parts from bulk and feeds them to the main belt conveyor singly via a singularizing mechanism at the bowl outlet.

(b) When the part reaches the main conveyor it is pushed against the side of the inclined deflector blade. Due to this action, the part travels along the side of the blade in any of its six rusting aspects.

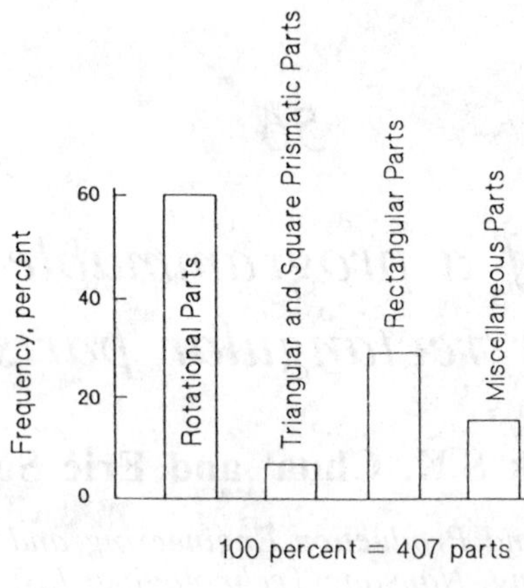

Figure 1. Relative frequency of component basic shape classes.

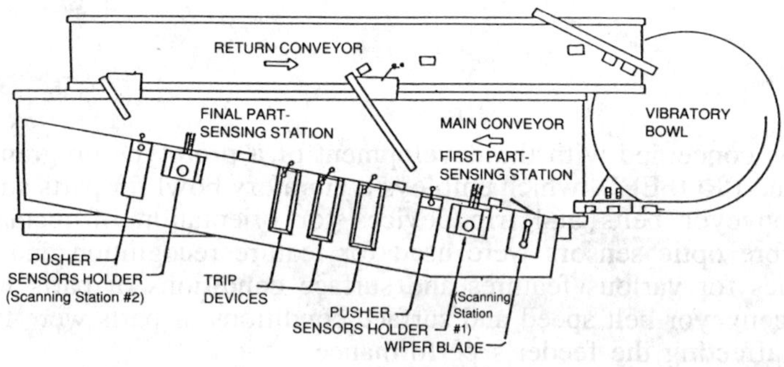

Figure 2. Plan view of programmable feeder, PROFEED.

(c) The part will pass under a wiper blade which will topple any part that is resting on any of its four narrow sides (unstable orientation) and parts which are stacked on top of one another. The wiper blade height is adjustable by the microprocessor via a stepper motor to suit different part sizes.

(d) The part will pass under any array of four optical fibre sensors at scanning station no. 1 for identification of part orientation. The microprocessor will sample the output of the sensors and attempt to match the resultant signatures with the stored signatures of the known orientations in its memory. If the matching fails, the microprocessor will activate a pusher which will return the part to the return conveyor belt and back to the vibratory bowl for re-circulation.

The pusher designed uses two double-acting pneumatic cylinders (Figure 3). The smaller cylinder is used to retract the pusher blade when the pusher is returning to its starting position so that the blade will not hit any part which is arriving at the ejecting station. The larger cylinder is used to move the whole housing of the blade.

(e) The part whose orientation has been identified will move down the main conveyor. If the part needs to be orientated to the final orientation, trip devices will be activated to rotate the part by 90° anti-clockwise each time. The number of trip devices activated depends on the initial input orientation of the part. The three trip devices are each driven by a stepper motor and triggered by the presence of parts at proximity sensors.

(f) The second scanning station which is the final inspection station checks for the correct orientation only. If the orientation of the part is still incorrect, the microprocessor will activate the final pusher to reject the part onto the return conveyor and back to the vibratory bowl for re-circulation.

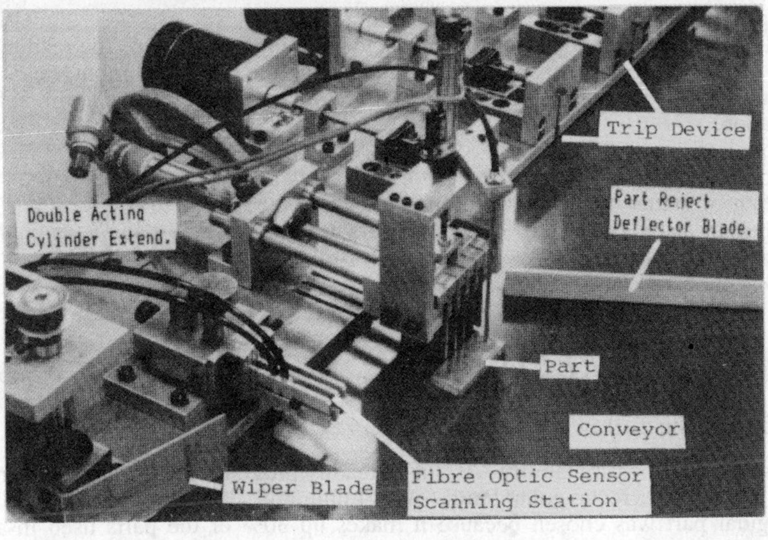

Figure 3. The pusher is activated, pushing the rejected part to the return-part belt conveyor.

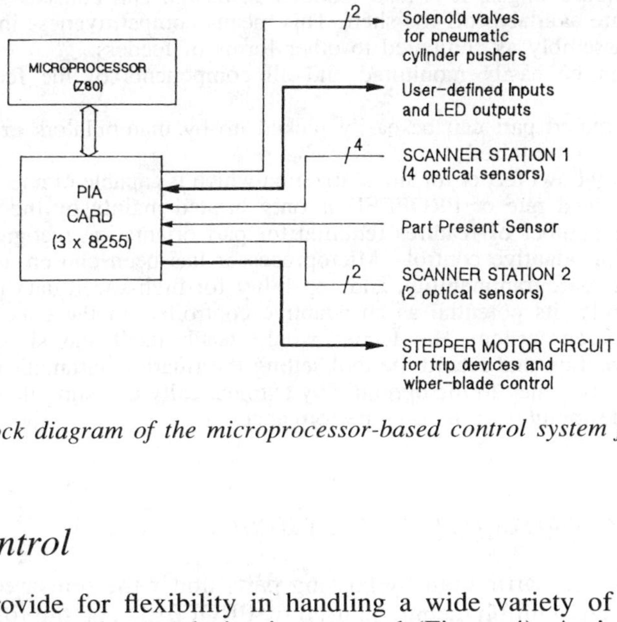

Figure 4. Block diagram of the microprocessor-based control system for PROFEED.

Feeder control

In order to provide for flexibility in handling a wide variety of different parts, a microprocessor-based control circuit was used (Figure 4). A single board computer, the Microprofessor Z-80 kit, was chosen for ease of program development (Leventhal, 1987) without requiring any expensive development system. Via a peripheral interface board (PIA) which contains three 8255, the Microprofessor can communicate, monitor and control the PROFEED system (Liu and Gibson, 1986). PLC is not suitable for PROFEED because of the need for data processing which could be very involved when dealing with two-dimensional depth map obtained from optical line scans needed for the recognition of orientation of more complex parts.

The entire control program is resident in the Microprofessor RAM. The program execution multi-tasks through the various functions namely:

(a) Sequential control of the pneumatic actuators.
(b) Sampling of the four optical sensors at scanner station no. 1. Since the reading of each of the optical sensors is separated into a 2-bit code, the combined input yields a 1 byte signature. This is then compared with a set of template signatures to determine the part's orientation.
(c) Communication with the user.
(d) Control of the stepper motors for the trip devices and wiper blade.
(e) Sampling of status of the part-presence sensor.
(f) Sampling of status of the two optical sensors at scanner station no. 2 for confirmation of correct orientation.

Features of PROFEED

PROFEED has the following features:

(a) Flexibility and programmability—It can be readily programmed to handle different parts of the same basic rectangular family with minimum set-up time and re-tooling. The rectangular part was chosen because it makes up 30% of the parts used in assembly and the reorientation tooling is simpler compared to other part-families such as rotational parts.
(b) Besides being flexible and programmable which enables it to respond rapidly to frequent product or batch changes, it is also modular in design and consists mainly of components that are available off-the-shelf. This means competitiveness in terms of costs and ease of assembly as compared to other forms of feeders.
(c) Parts flow can be easily monitored and all components of the feeder are easily accessible.
(d) The final orientated part can be easily picked up by manipulators or robots and no special tracking is required.
(e) With a vibratory bowl feeder for singularization which is capable of one part per second, the maximum feed rate of PROFEED is only limited mainly by the microprocessor speed and the number of features required for part orientation recognition.
(f) Its potential for adaptive control—Microprocessor has been chosen as the controller because of its reprogrammability, and capability for high-speed data processing and, most importantly, its potential as an adaptive controller. In the concept of adaptive controlled feeding system, the feeder would teach itself the signatures of each orientation, program and record the tool setting information automatically. The feeder will adjust its efficiency to the optimum by automatically adjusting its tool setting and feed rate while monitoring its own performance.

Performance of recognition system

A series of tests was carried out by passing parts under the sensor to establish the reliability of the recognition technique used by PROFEED and the following results were obtained.

(a) There is a systematic sensor output pattern for different materials and surface conditions in terms of reflectance when the target distance is varied, as follows (Figure 5):

Aluminium ('mirror'-like > Aluminium (sandpapered) > Steel (sandpapered) > Steel (rough).

The sensor output reaches a maximum at a target distance of about 2 mm regardless of materials and surface conditions. This can be explained by the cone of emission and acceptance of the sensors (Figure 6). As the target distance is increased from zero, the two cones start to overlap which increases the intensity of reflected light. With further

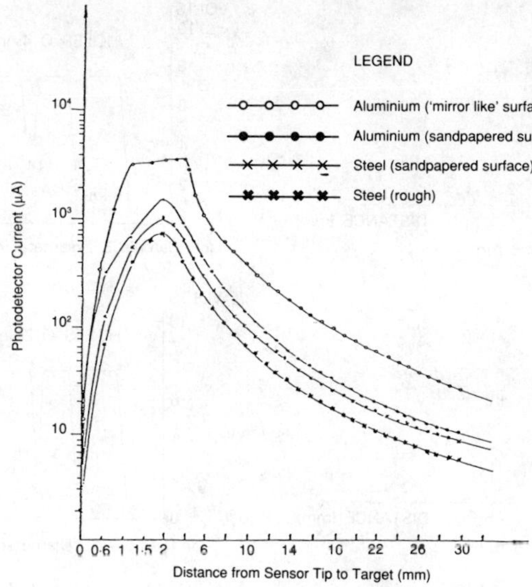

Figure 5. Diffusive fibre-optic output characteristics for different materials and surface finishes.

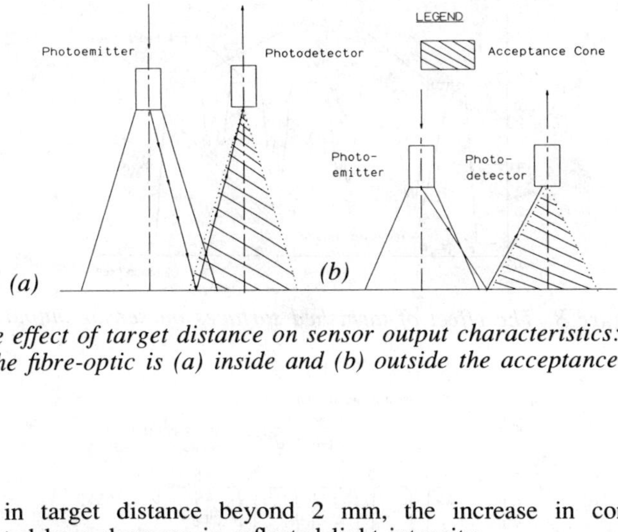

Figure 6. The effect of target distance on sensor output characteristics: the reflected light received by the fibre-optic is (a) inside and (b) outside the acceptance cone of the fibre-optic.

increase in target distance beyond 2 mm, the increase in cone overlapping is counteracted by a decrease in reflected light intensity.

(b) With regard to feature recognition [Figure 7(a)–(d)], the most noticeable characteristic is the 'V'-shaped output with a central pulse and the drop in output with increase in the depth. Smaller diameter holes have a higher output than larger holes of similar depth except for the 2 mm diameter holes which also do not have the pulse in their output. This phenomenon can be explained by the fact that the intensity of reflected light is dependent upon the target distance and the domain in which the emitted light can enter and reflected light be received. The smaller the diameter, the lower the emitted light intensity received in the hole and the lower the reflected light intensity which results in the absence of the pulse in the sensor output characteristics.

(c) Figure 8 provides a measure on how a tarnish surface can give a deception that a hole exists. The recognition system may label a wrong orientation to such a part or reject it as unrecognizable. Figure 9 shows that the sensor is capable of detecting even a blind hole of diameter 1·5 mm.

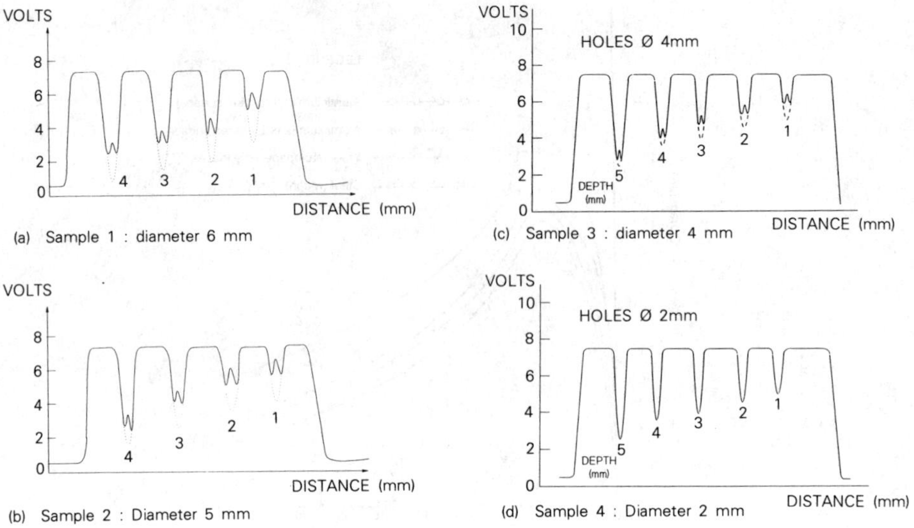

Figure 7. Sensor output characteristics for various diameters and depths of blind holes on parts.

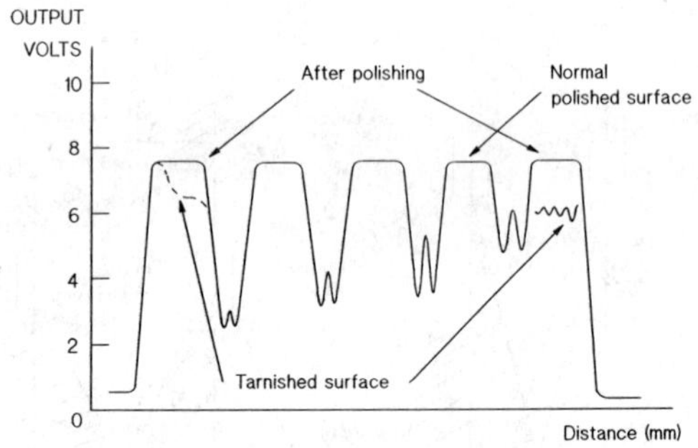

Figure 8. The effect of tarnished surfaces on sensor output.

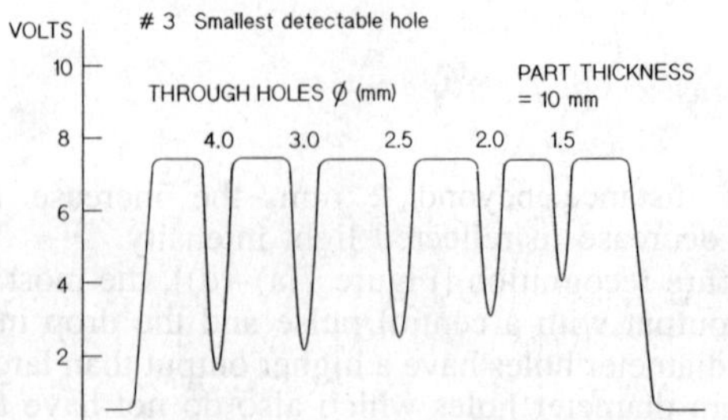

Figure 9. Sensor output characteristics for a series of through holes.

Conclusion

A microprocessor-controlled prototype programmable parts feeder, which combines the high speed and efficiency of vibratory bowl in parts singularization with the stability of belt feeders for feeding and orientating rectangular parts, has been developed. Initial tests on the feature recognition technique used have shown that it is capable of even identifying small features and that with consistent surface quality of incoming parts, there is great potential for such a programmable feeder.

Further work will consist of more recognition tests on different and more complex features as well as enhancing the capability of the feeder in terms of efficiency, speed and adaptive control.

Acknowledgements

The authors thank C.H. Phoa for his invaluable contributions to this work. The award of an applied research grant by Nanyang Technological Institute to this work is gratefully acknowledged.

References

Boothroyd, G. and Dewhurst, P., 1982, Parts presentation and feeding for robot assembly. *Annals of the CIRP*, **31**, 377–381.

Eade, R. (Ed.), 1989, Automatic assembly: the key is orientation. *Manufacturing Engineering, SME Publication*, **102**, 3, 64–69.

Goodrich, J.L. and Maul, G.P., 1986, *Programmable Parts Feeders, Automated Assembly*, edited by Jack D. Lane (SME Publication).

Leventhal, L.A., 1987, *Z80 Assembly Language Programming*, International edn (New York: McGraw-Hill).

Liu, Y.C. and Gibson, G.A., 1986, *Microcomputer Systems: The 8086/8088 Family—Architecture, Programming and Design* (Englewood Cliffs, NJ: Prentice-Hall).

57

Flexible porous materials handling in a CIM environment

P.R. Nicholson and M. Sarhadi

Department of Manufacturing and Engineering Systems, Brunel University, Uxbridge, Middlesex UB8 3PU, UK

Abstract

A vacuum-based gripper for de-stacked fabric part handling has been developed. The gripper provides adaptive features, under program control, to match the shape of the parts being processed. Rapid grip and release actuation from a high-speed switched vacuum source developed for the application has been employed. Positive clamping is achieved with porous materials and flattening is found to occur with creased workpieces. Results obtained with the gripper integrated within a computer-integrated manufacturing (CIM) unit, implemented for apparel manufacture, are reported.

Introduction

The automated handling of rigid materials has received considerable attention (Chen, 1982). Robotic handling of flexible materials presents a greater challenge to automation and this chapter reports recent work into the adaptive handling of pre-cut fabric components.

Progress in limp fabric handling has resulted in a number of handling techniques based upon forms of needle actuated surface engagement, such as the Polytex device (AAMA, 1987; Schell, 1987; Hall, 1988), Gepec Hand (Littlewood, 1978) and the Crampton Soft Gripper (AAMA, 1987; Hall, 1988). These enable surface engagement at discrete points on fabric surfaces and are most suitable for small components. Disadvantages with the above are surface intrusion, surface marking and unreliability. The mechanisms are associated with numerous patents and prior art. Other types of gripper have been investigated: electrostatic (Monkman and Taylor, 1988) and freeze/thaw types (Guse, 1988) are examples.

Most of the above types of gripper are point gripping types and are thus not suited to picking up differing part types from operation to operation, unless means are provided to reconfigure the arrangement of the grippers.

Alternative methods for positioning using pin engagement in (TC)2, described by Disher (1987a, b), have been applied. Techniques for positioning using dropping such as plate positioning in the Jet Sew Clupicker (Brotherton and Tyler, 1986), and air flotation (Kurt Salmon Associates, 1979) have been developed.

To address large component handling, an adaptive vacuum-based gripper has been developed at the University of Belfast by Wright *et al.* (1986). Here, tubes connect a flat grip pad to fixed valves, requiring long runs of tubing with associated pressure drop at high flowrate and possible articulation restrictions due to the connecting tubes.

In the approach adopted at Brunel, adaptive vacuum clamping is used for fabric part transfer, but the grip pad employs an integrated matrix of vacuum valves. The system was integrated into an assembly cell application.

The assembly cell application

The assembly cell automated an activity normally carried out by a single operator in the garment assembly industry. This performed the attachment of edge binding tape to batches of fabric parts and was implemented with four modules. The first comprised a Ply Destacking Module, capable of destacking fabric parts from a batch of pre-cut components and transferring each to an alignment module. The destacking module is described in detail by Sarhadi *et al.* (1986). An Alignment Module formed the second module, facilitating edge alignment. The module accepted components from the destacking module, and using an alignment table articulated in X, Y and θ axes, moved fabric parts into a known mechanical alignment for subsequent open loop operations. The third module comprised an adaptive vacuum based ply transfer module, described here, for pick and place between the alignment and edge binding modules. Finally an edge binding attachment module was used to attach binding tape along a specified edge of a fabric part. This module accepted a component transferred from the adaptive vacuum-based ply transfer module and effected edge binding operation using an adapted industrial sewing machine and a synchronized feed conveyor belt. The assembly cell is described in further detail by Nicholson (1986).

Design of the vacuum source

Requirements for the gripper vacuum source included high flowrate to accommodate leakage through porous workpieces and rapid creation and removal of vacuum (less than 0·5 s) to facilitate high-speed clamp and release. Available sources were fan-based air exhausters, rotary and piston type pumps and air ejectors. For fans, air exhausters and pumps, vacuum build up is dependent on the mechanical inertia of the moving parts (several seconds) and hence bulky and expensive valves would be needed to provide switching. However, in the air ejector, high flowrate combines with high-speed development and removal of vacuum is possible. Construction of the air ejector body in lightweight materials and integration of the body as a gripper support member is feasible. It was thus regarded as a good solution for the source and a survey of available ejectors was carried out. Available ejector vacuum pumps included those tailored to standard high vacuum/low flowrate suction cup type applications and proprietary multi-stage units. The suction cup pumps provided inadequate flowrates and the multi-stage units require inbuilt one-way valves which would increase release time. It was thus necessary to design an ejector with the required specification in-house.

A diagram of the air ejector used is given in Figure 1. The ejector comprises a forcing nozzle (1), combining tube inlet cone (2), a combining tube (3) and diffuser cone (4). Compressed air fed from source (5) delivered through the forcing nozzle mixes with static air in the ejector, transferring momentum and creating an induced airflow in the inlet cone, resulting in a vacuum, which is routed to the vacuum outlet (6). Vacuum development is limited only by the speed of development of the forcing nozzle airflow, which can be designed to meet the 0·5 s clamping criteria. Vacuum removal is effected rapidly upon switching off the forcing jet as the ejector is open to atmosphere. Kastener and Spooner (1949)

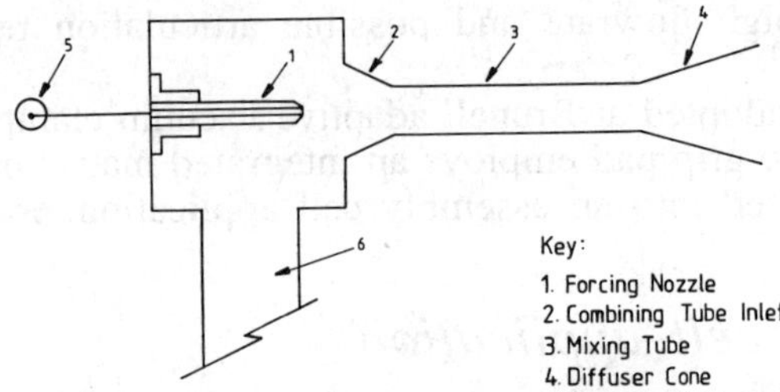

Figure 1. Air ejector components.

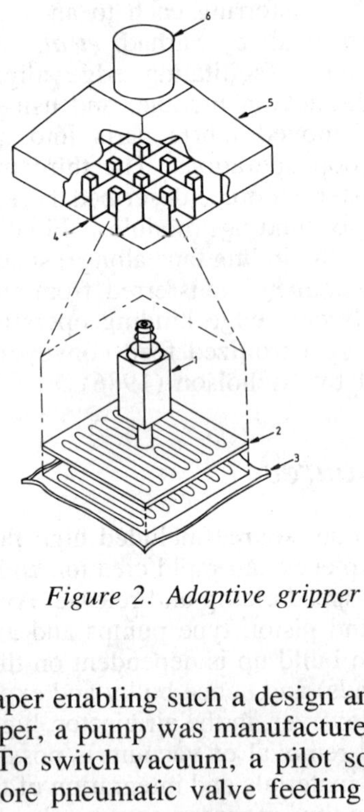

Figure 2. Adaptive gripper matrix valve.

provide a design paper enabling such a design and based upon calculations derived from the design paper, a pump was manufactured to supply a high flowrate at low terminal vacuum. To switch vacuum, a pilot solenoid valve was used to operate a half-inch wide bore pneumatic valve feeding the air ejector nozzle.

Implementation of the prototype matrix gripper pad

Adaptive gripping is required where part size or shape alters between handling operations such as in the assembly of two dissimilar parts, or where grip or release must occur in the vicinity of other free components such as the removal of a fabric part from a cutting table. Here the active grip area must match the shape of the component to be transferred. If this is not done, air leakage around the component will shunt the attained vacuum, reducing the clamp pressure and the associated airflow may cause turbulence, possibly disturbing the gripped component or the surrounding components already laid.

In the prototype system, a matrix vacuum pad architecture was adopted, but to remove the connecting tubes used by Wright *et al.* (1986), valves to operate the matrix were sited directly behind the pad grille. A single valve unit is shown in Figure 2, with its position in the matrix shown. The valves were designed and fabricated in house, their construction based upon a 12 V solenoid (1) and spring return valve plate lifting mechanism. A seal was made by mating the valve plate (2) grille with the pad grille (3). The grilles comprised slots of equal pitch but offset such that the pad grille blanked the plate grille when actuated. A matrix of 20 cells in a 4 × 5 configuration was chosen to make up the gripper (4).

The complete gripper pad was connected to the vacuum source via a duct (5) and a 4 inch diameter flexible pipe (6). Each valve was mounted in a grid arrangement upon a support frame attached to the duct, and electrical connection made through its side. Solenoid spring rates were adjusted to withdraw the valve plate when the solenoid was not actuated.

The transporter system

The transfer operation in the assembly cell application required pick and place between two fixed points in the same plane and so a specific solution was used. A two-axis hard automation transfer mechanism satisfied the articulation requirement: articulation by robot would provide more flexible positioning in X, Y and θ axes but would be more expensive. The resulting mechanism comprised a support frame possessing two degrees of freedom in cartesian X and Z axes, allowing the pad to raise, lower and move laterally between endstops. Pneumatic rodless cylinders were used to power the mechanism. Valves and adjustable flow restrictors controlled the cylinder ports providing speed control and reed switch limit sensors, actuated by the magnetic cylinder piston, indicated the limits of travel of the axes. Positioning accuracy of 0·002 inch was claimed for the cylinders within the stroke of an axis but here the unit was driven in an endstop operated mode. The cylinders provided variable rate pneumatic dampers at their endstops to halt the mechanism in a regulated manner. The cylinders were driven such that line supply pressure was applied to both ports when stopped. To cause the pad to move, one port would be depressurized through an associated speed control flow restrictor. The articulation air supply was decoupled from the vacuum source pneumatic supply and conditioned with a filter–dryer–lubricator unit.

Implementation of the controller and software

A modular controller was designed to drive the gripper and articulation so that evaluation could be carried out within an assembly cell or in a stand-alone mode. A block diagram of the system is shown in Figure 3. The controller functioned as a peripheral to the supervisory cell controller and was implemented with a PC XT computer (1) and a proprietary Input/Output subsystem (2), providing high-speed digital control via opto isolated Input/Output modules (3). This subsystem provided an interface to the grip pad (4), articulation valves (5), articulation sensors (6), vacuum on/off control (7), and interface to the cell controller (8). An emergency stop facility (9) was included in the controller interface. The Input/Output subsystem was connected by a high-speed control bus (10) (500 ns read–write cycles), enabling it to be located remote from the computer. The overall cost of such a unit was considerably less than an industrial programmable logic controller (PLC) unit, and software development could be carried out with more support and flexibility than in a PLC version.

The interface between the assembly cell controller and the adaptive vacuum transfer controller was implemented with a simple digital control interface, using a group of control lines. Signals to start one of several pre-defined transfer cycles,

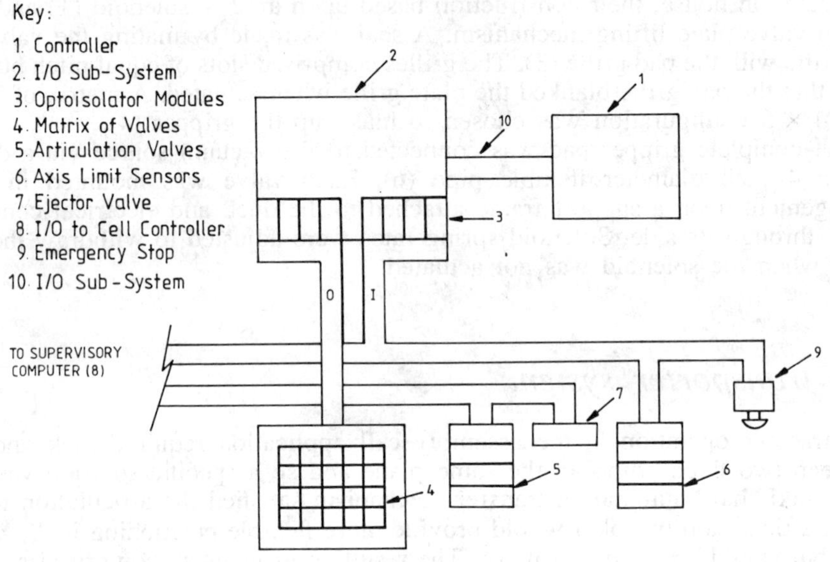

Figure 3. The controller arrangement.

indicate active processes and fault conditions and to reset/initialize the controller were provided.

The Input/Output subsystem, modular power supplies, flyback diodes for the pad solenoids and connectors were housed in a 19-inch industrial casing.

Application programming comprised a single task coded in Microsoft C. A system design was prepared for the application program and coding implemented directly linking with Input/Output subsystem driver software. The application program was debugged with Microsoft CodeView and alpha-tested. A graphically-based operator interface was provided to display the pad form and allow experiment with the programmed shape. The control task served to load the gripper with one of several pre-defined pad forms and service the articulation control during transfer.

In operation the program polled the digital control interface until a start cycle command was asserted, starting the transfer cycle. Axis movements were checked against limit switch positions to identify end of travel. The articulation moved to the pick up position in the *X* axis initially then dropped to its lowered position above the component. The valve matrix form was loaded and vacuum turned on. The pad was then raised and moved to the release position, lowered and vacuum turned off. During operation, the digital control interface would indicate the status of the operation to the supervisor computer including a faulted status, upon which the supervisor could act. An emergency stop function was also implemented for safety as high forces were developed by the pneumatic cylinders during transfer. Should the emergency stop be issued, the articulation ports would close, causing the transfer mechanism to stop directly.

Results

The transfer mechanism was operated in a test run with 500 parts within the assembly cell described above and the following results were obtained.

Speed of operation

The following times were obtained: (i) time to set the pad was measured as 100 ms while time to clear the pad was 150 ms, (ii) time to grip was 200 ms while time to release was 300 ms, (iii) clamp and release times were considerably less than the transfer time at 4·20 s. In the edge binding application, a schedule was constructed from the overall timing and sequencing requirements of the workstation. The transport stage was found to lie off the critical path in the sequence, indicating its operating speed was satisfactory.

The controller

The control program operated correctly and the emergency stop features were validated. The interface to the cell supervisor operated correctly. It is intended to remove the hardware flag interface and replace it with a flag system implemented in network software or with a direct serial link to the supervisor.

Workpiece handling and reliability

At this stage of development there is insufficient data to provide a statistically meaningful reliability figure. Within the 500 part run, no failed operations were observed. Thus a larger (possibly 5000 component run) would be required to determine reliability more accurately.

The gripper was found to operate wholly reliably within the test. No mechanical disturbance of the part was found other than flattening of components occasionally laid with creases by the upstream destacking unit. Those components which were creased tended to be flattened. Normal components were consistently gripped properly.

A 4 × 5 matrix of gripper cells proved sufficiently flexible to process a limited range of workpieces used in the manufacture of men's briefs. This was partly due the simple nature of the fabric parts used. For parts with more irregularity, a finer resolution matrix would be required to reduce air loss and associated turbulence.

Transportation accuracy

Positioning accuracy of 1/10 inch for normal garment assembly pick and place tasks is required (AAMA, 1987) and this was attained. The pneumatic damping arrangement at the transporter limits of travel allowed the mechanism to halt in a damped manner without disturbing the workpiece.

Conclusions

The adaptive gripper was found to operate as required in conjunction with the vacuum source and met the clamp and release timing criteria. The articulation operated satisfactorily after decoupling the air supply from the air ejector supply, and delivered a 4·2 s transfer time, regarded as satisfactory.

Acknowledgements

The authors express their acknowledgements to ACME of SERC and to the industrial partner, Jockey PLC, for their support of this work.

References

AAMA (American Apparel Manufacturers Association), 1987, *Automated Limp Fabric Handling Devices*. Final Report, April 1987, The Limp Fabric Task Group of the Apparel Research Committee of the American Apparel Manufacturers Association.

Brotherton, A.A. and Tyler, D.J., 1986, Clupicker performance and flexible apparel automation. *Hollings Apparel Industry Review*, **3**, 2, 15–34.

Chen, F.Y., 1982, Gripping mechanisms for industrial robots: an overview. *Mechanism and Machine Theory*, **17**, 5, 299–311.

Disher, M., 1987a, Sewing techniques. *Manufacturing Clothier*, August, 8–11.

Disher, M., 1987b, Unit Production systems: investing in automation: Japans (TC)2 system. *Manufacturing Clothier*, January, 11–15.

Gilles, M., 1990, Programmable logic controllers—architecture and application (Chichester: John Wiley).

Guse, R., 1988, Freeze/thaw technique part of automated handling study. *Knitting International*, July, 90.

Hall, M., 1988, Picking up the pieces. *Knitting International*, August, 58.

Kastner, L.J. and Spooner, J.R., 1949, An Investigation Of the Performance And Design Of The Air Ejector Employing Low Pressure Air as the Driving Fluid.

Kurt Salmon Associates, 1979, The KSA report on technology as applicable to the textile industry. December.

Littlewood, K.J., 1978, Method and Apparatus for Handling Fabric Workpieces. UK Patent Number 1, 533, 536, 29 November 1978.

Monkman, G.E. and Taylor, P.M., 1988, Electrostatic grippers, principles and practice. *Proceedings of the International Symposium on Industrial Robots*, April, pp. 193–200.

Nicholson, P.R., 1986, Automation of garment assembly processes. Ph D Thesis, University of Durham.

Sarhadi, M., Nicholson, P.R. and Simmons, J.E., 1986, Advances in gripper technology for apparel manufacturing. *Proceedings of the Institute of Mechanical Engineering*, paper C372/86.

Schell, F. (Polytex), 1987, Gripper Head, USP 4, 687, 241: 18 August. Priority Application: Switzerland, 6114/84, 21 December 1984.

Wright, E.J., Curran, J.P.J. and Armstrong, P.J., 1986, Robotics handling of composite materials in a lay up cell. *Proceedings of the Institute of Mechanical Engineering*, Paper C365/86.

58

Optimal flow path design of unidirectional AGV systems

Moshe Kaspi and J.M.A. Tanchoco

School of Industrial Engineering, Purdue University, West Lafayette, IN 47907, USA

Abstract

This chapter describes an alternative formulation of the Automated Guided Vehicles (AGV) Flow Path Layout (FPL) problem which was first formulated by Gaskins and Tanchoco (1987) as a zero-one integer programming problem. A computationally efficient procedure is proposed which is based on the branch-and-bound technique. An algorithm for satisfying the reachability condition for nodes in the AGV flow path network is also presented. A simple illustrative example is discussed to demonstrate the procedure and a more complex problem is also given.

Introduction

Many of today's modern factories use Automated Guided Vehicles (AGVs) to move materials throughout the factory. These AGVs automatically navigate along a fixed guide path which is usually defined by a wire embedded in the floor, by strips of flourescent tape, or by other guidance methods. They are dispatched and routed by a transport system controller via signals transmitted over the wire in the floor or over a radio channel.

One of the important design variables for efficient operation of an AGV system is the guide path layout design. The guide path layout has a direct effect on the operational performance of the system due to the degree of congestion induced by imbalances in the material flows at traffic intersections and other points of potential congestion. Because of blocking delays, the required number of vehicles is also impacted.

The AGV guide path layout problem was first formulated in Gaskins and Tanchoco (1987) as a zero-one integer linear programming problem. The problem is formulated as a node-arc network where the nodes represent pick-up/delivery stations and aisle intersections and the arcs are the guide paths which connect the nodes. The objective function for this formulation is the minimization of total transportation distance, i.e. flow volume times distance. Two types of constraints are given—a connectivity constraint to guarantee that an arc entering a node has an arc leaving the same node, and a reachability constraint to assure that each node is reachable from any other node. The formulated problem was solved using a multi-purpose optimization package (Cohen and Stein, 1978). For practical problems, however, the number of zero-one variables needed tends to be very large and computational efficiency becomes an important issue. This chapter describes an alternative formulation of the problem and proposes a computationally efficient procedure which exploits the characteristics of the material flow from–to chart. The procedure proposed is based on the branch-and-bound technique. The selection of a starting point is critical for efficient implementation of the procedure developed. A simple illustrative example is discussed to demonstrate the procedure and a more complex guide path layout problem is given.

The primary motivation for the work reported in this chapter is based on ongoing research on *Intelligent Transport Systems* at Purdue's Engineering Research Center for Intelligent Manufacturing Systems. The goal of this project is the development of a very flexible transport system that will support the transportation needs of tomorrow's factories. It is envisioned that such a transport system would be serviced by free-ranging vehicles, i.e. vehicles which follow 'virtual' flow paths (Gaskins *et al.*, 1989). Thus, the flow path configuration would depend on the structure of the material flow system operating at any given time. In terms of the specifics of the flow path network, we must be able to change the direction of the flow path as the need arises. Equally important is that the computation required to obtain a new flow path layout must be in the reasonable range. It is noted that the term 'flow path' is used, instead of 'guide path', from this point on to reflect this flexibility in the direction of flow of vehicles. The term also reflects the fact that the methodology described in this paper is equally applicable to other materials handling systems such as electrified monorails, tow-line conveyors, and other fixed-path materials handling systems.

Model description and definitions

The objective of the Flow Path Layout (FPL) problem is to find the direction of flow for each link in the guide path network. The guide path network represents the pick-up/delivery stations, aisle intersections, and the aisles within a manufacturing facility. Unidirectional flow of vehicles along each aisle is assumed in this study.

Let the nodes represent pick-up points, delivery points and aisle intersection points, and let the arcs represent possible direction of flow between two adjacent nodes. Each arc is assigned a length equal to the distance (or travel time) between the nodes it connects. The flowrate between pick-up and delivery stations is described by a from–to chart. Once the node–arc network and the from–to chart have been obtained, the following variables and parameters can be defined:

n = number of entries in the from–to chart,
f_{lm} = flow intensity from pickup node l to delivery node m,
d_{ij} = length of arc i–j (the distance from node i to an adjacent node j),
Y_{lm} = path length from pickup node l to delivery node m,

$$X_{ijlm} = \begin{cases} 1 & \text{if arc } i\text{–}j \text{ is included in the path from pickup node } l \text{ to delivery node } m \\ 0 & \text{otherwise,} \end{cases}$$

$$Z_{ij} = \begin{cases} 1 & \text{if arc } i\text{–}j \text{ is directed node } i \text{ node } j \\ 0 & \text{otherwise.} \end{cases}$$

Analytical model

The objective function for the unidirectional FPL problem is to minimize the total distance that vehicles have to traverse to satisfy the total transportation requirements, i.e.

$$\text{Minimize} \sum_{l,m} f_{lm} Y_{lm}.$$

The constraint set includes the following:

C1. Path length from pickup node l to delivery node m

$$\sum_{i,j} X_{ijlm} d_{ij} = Y_{lm} \qquad - \text{¥} l, m;$$

C2. Ensure the path from pickup node l to delivery node m is feasible

$$X_{ijlm} \leq Z_{ij} \qquad - \text{¥} l, m \qquad \text{¥} i, j;$$

C3. Unidirectionality constraints

$$Z_{ij} + Z_{ji} \leq 1 \qquad - \text{¥} i, j;$$

C4. At least one input arc

$$\sum_{i} Z_{ij} \geq 1 \qquad - \text{¥} j;$$

C5. At least one output arc

$$\sum_{k} Z_{jk} \geq 1 \qquad - \text{¥} j;$$

C6. One output arc from pickup node l using the path from node l to delivery node m

$$\sum_{k} X_{lklm} = 1 \qquad - \text{¥} l, m;$$

C7. One input arc to delivery station m using the path from pickup node l to node m

$$\sum_{k} X_{kmlm} = 1 \qquad - \text{¥} l, m;$$

C8. Number of input arcs equals the number of output arcs

$$\sum_{i} X_{ijlm} = \sum_{k} X_{jklm} \qquad - \text{¥} l, m \qquad - \text{¥} j.$$

The main difficulty in finding the solution to the above problem is the large number of variables required for realistic size problems. For example, more than 10 000 variables are required for a 10 pick-up/delivery station problem with a total of 30 nodes representing the pick-up/delivery stations and aisle intersections. A natural approach to use is the branch-and-bound procedure.

Branch-and-bound approach

The specific technique used is the branch-and-bound with depth search first and backtracking rather than the jumptracking type of approach. Using the backtracking method a feasible complete solution (not necessarily optimal) is obtained very quickly and the required memory is much less as compared to the jumptracking method. The proposed approach involves eight steps. Each of these steps is described later, but first, some additional definitions are introduced.

Additional definitions

UB = upper bound, i.e. the current (known) best value of the objective function. The initial value of UB is set at infinity. Any time a feasible complete solution is obtained with a value less than UB, then the value of the upper bound UB is updated.

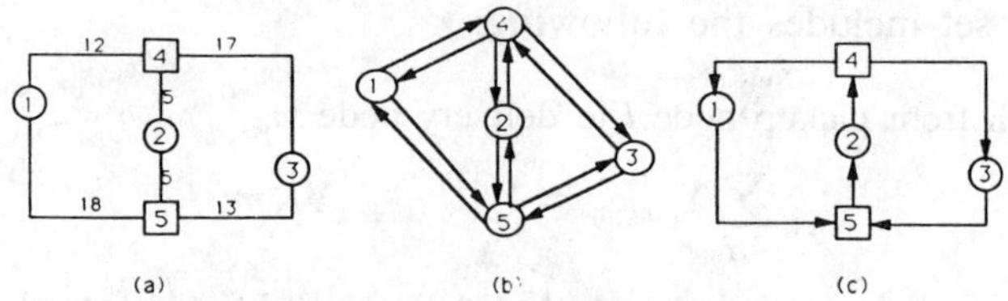

Figure 1. (a) Example layout, (b) node–arc network, (c) optimal flow path.

Table 1. From–to flow chart for Example 1.

FROM	TO		
	1	2	3
1	—	10	15
2	20	—	ε
3	5	10	—

LB_k = lower bound of branch k is the best value of the objective function with constraint C2 ignored for all arcs in {U}. The lower bound LB_k is used to label the branches in the search process. Any time a lower bound of a certain branch is greater than (or equal to) the upper bound UB, then this branch is bounded.
$\{D\}$ = the set of directed arcs.
$\{U\}$ = the set of undirected arcs.
$\{A\}$ = the set of all the arcs, i.e. $\{U\} \cup \{D\} = \{A\}$, $\{U\} \cap \{D\} = \phi$.
$\{A'\}$ = the set of all the arcs connected to a pick-up/delivery node.

Search procedure

For clarity's sake, the proposed branch-and-bound method is explained through a simplified numerical example, denoted as Example 1. The departmental layout considered is shown in Figure 1(a), and the material flow from–to chart is given in Table 1. Note that an ε value equal to a very small number is entered as the flow value from node 2 to node 3 to ensure that all pick-up/delivery stations are reachable from any other pick-up/delivery station.

Step 1. Initialization

The corresponding node–arc network is shown in Figure 1(b). The procedure is initiated by determining set $\{A\}$. Initially, $\{A\}$ = {1–4, 4–1, 1–5, 5–1, 2–4, 4–2, 2–5, 5–2, 3–4, 4–3, 3–5, 5–3}. Since all the arcs are currently undirected $\{U\}$ = {A} and $\{D\} = \phi$. The upper bound UB = ∞.

Step 2. Branching

The branching process is initialized using the arcs that are directly connected to pick-up/delivery nodes, i.e. arc i–j, arc j–i ε $\{A'\}$. It is only after all arcs in $\{A'\}$ are exhausted that the other arcs in $\{A\}$ are considered.
Additionally, for the two branches connecting nodes i and j

(i) $Z_{ij} = 1$, $Z_{ji} = 0$ or $Z_{ij} = 0$, $Z_{ji} = 1$ where Z_{ij}, Z_{ji} ε $\{D\}$
(ii) $\{U\} = \{U\} - \{i - j, j - i\}$.

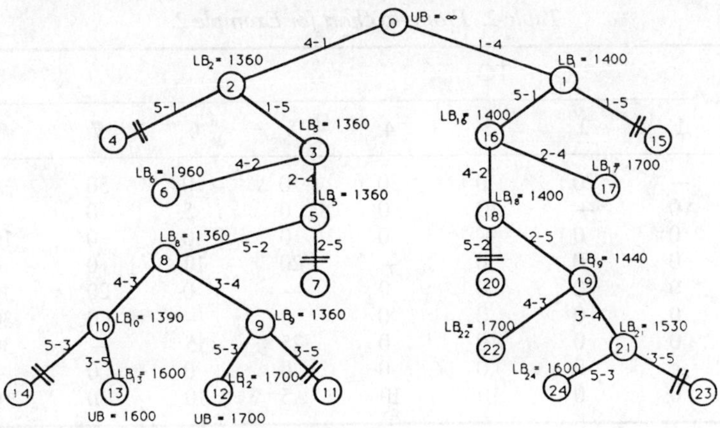

Figure 2. Search tree for Example 1.

For the example, the branching is initialized with arcs 1–4 and 4–1. As shown in Figure 2, this initial branching corresponds to branch k = 1 (Z_{14} = 1, Z_{41} = 0) and branch k = 2 (Z_{14} = 0, Z_{41} = 1).

Step 3. Labelling

For each of the two new branches, the FPL problem is solved with constraint set C2 ignored for all arcs in $\{U\}$. Actually, the procedure looks for the shortest path Y_{lm} from each pickup node l to the delivery node m considering both the arcs in $\{D\}$ and $\{U\}$. Then the lower bound of each branch, $\text{LB}_k = \Sigma f_{lm} Y_{lm}$, is calculated.

For branch k = 1 (Z_{14} = 1, Z_{41} = 0), the shortest paths are: Y_{12} = 17, Y_{13} = 29, Y_{21} = 23, Y_{31} = 31, Y_{32} = 18, and the lower bound LB_1 = 1400. For branch k = 2 (Z_{14} = 0, Z_{41} = 1), the lower bound is LB_2 = 1360.

Step 4. Setting bounds

Branch k is bounded if:

(i) lower bound LB_k > upper bound UB;
(ii) no feasible path from a pick-up node to a delivery node is obtained; or
(iii) for any node, all the arcs emanating from the node and all the arcs entering the node are in $\{D\}$, and constraint sets C4 and C5 are violated.

Step 5. Branch selection

Considering the two new branches (see the branching procedure), the one with the lowest LB is selected. The information on the other branch (LB, $\{U\}$, $\{D\}$) is recorded. If both branches are bounded, the backtracking procedure (Step 7) is invoked.

The lower bound of branch k = 1 is LB_1 = 1400 and for branch k = 2, LB_2 = 1360. Thus, branch k = 2 is selected, see Figure 2. The branching process is continued with LB_3 = 1360 while branch k = 4 is bounded since $\{D\}$ = $\{Z_{41}$ = 1, Z_{14} = 0, Z_{51} = 1, Z_{15} = 0) violates constraint set C5.

Step 6. Updating of upper bound

When constraint set C2 is not violated, then the solution obtained in Step 5 (branch selection) is a feasible complete solution to the FPL problem. If the value of the

Table 2. From–to chart for Example 2.

FROM	TO								
	1	2	3	4	5	6	7	8	9
1	—	20	0	30	0	0	50	0	20
2	0	—	10	0	0	5	0	0	5
3	0	0	—	0	0	0	0	10	10
4	0	0	0	—	20	10	0	0	0
5	0	0	0	0	—	0	20	30	0
6	0	0	0	0	0	—	0	30	0
7	0	0	0	0	25	15	—	30	0
8	ε	0	0	0	0	0	0	—	0
9	0	0	10	10	5	10	0	0	—

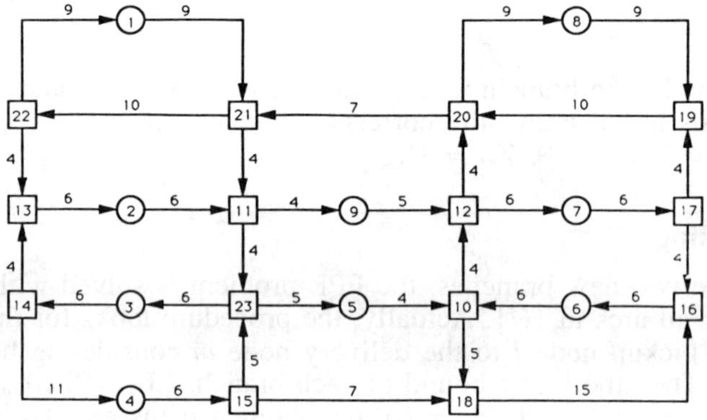

Figure 3. Optimal flow path for Example 2.

objective function for this solution, LB_k, is less than UB (the current best value of the objective function), then UB is updated, i.e. UB = LB_k.

For the example, UB_{12} = 1700 and none of the arcs' direction violates any constraint, so it is a feasible complete solution. Until this point, UB has been equal to ∞, so UB is updated to be equal to 1700.

Step 7. Backtracking

The backtracking procedure is invoked any time a feasible complete solution is obtained. The backtracking procedure returns to the source branch. If a previously not selected branch of the source, i.e. a sibling branch, is available (i.e. it is not bounded and it has not been selected before), then the procedure continues through this branch. If the sibling branch is not available, then backtracking is performed again.

Referring to Figure 2, branch k = 12 represents a feasible complete solution, so the procedure returns to its source, branch k = 9. The sibling branch, branch k = 11, is bounded so the backtracking procedure continues to branch k = 8. Since branch k = 10 is available, it is selected; and the branching selection procedure (Step 5) is executed.

Step 8. Termination

When backtracking reaches the root source ($k = 0$), and both branches $k = 1$ and $k = 2$ are no longer available, then the search is terminated. If UB is still infinite, then there is no feasible solution to FLP problem; otherwise, UB is the optimal solution.

The optimal flow path layout for the example problem is shown in Figure 1(c).

Example 2

A real size problem of nine pick-up/delivery stations with a total of 23 nodes and 66 arcs was also solved. The from–to chart is presented in Table 2. The optimal flow path is shown in Figure 3. The arc lengths given in Figure 3 are based on vehicle travel time units. The total value of the objective function for Example 2 is 10 170. The first feasible solution is obtained after 33 branching iterations in 57 s with an objective function value of less than 1% of the optimal solution. The optimal solution is obtained after 17·2 min while it took 27·5 min to ensure optimality.

Acknowledgement

The work reported in this chapter was partially supported by the NSF/Purdue University Engineering Research Center for Intelligent Manufacturing Systems under Grant No. CDR 8803017.

References

Cohen, C. and Stein, J., 1978, *Multi-purpose Optimization System User's Guide, Manual No. 320*. (Northwestern University, Evanston, IL: Vogelback Computing Center).

Floyd, R.W., 1962, Algorithm 97—shortest path. *Communications of the ACM*, **5**, 245.

Gaskins, R.J. and Tanchoco, J.M.A., 1987, Flow path design for automated guided vehicle systems. *International Journal of Production Research*, **25**, 5, 667–676.

Gaskins, R.J., Tanchoco, J.M.A. and Taghaboni, F., 1989, Virtual flow paths for free-ranging automated guided vehicle systems. *International Journal of Production Research*, **27**, 1, 91–100.

59

Task planning for the flexible manufacture of sheet metal enclosures

N.W. Hird, C.F. Earl, P. Holland and R.J. Stamp
Department of Engineering, Bristol Polytechnic, Bristol, UK

Abstract

A considerable proportion of sheet metal components are produced in low volumes where the planning from design to manufacture may account for a large percentage of the production time. With the aid of an automatic manufacturing cell, incorporating a press brake and an industrial robot the requirements for task planning are investigated. The planning is separated into several key stages each of which is discussed along with the importance of an effective strategy to link them. The planner is capable of the synthesis of a realizable method of manufacture from the design of the component and the subsequent formulation of a robot handling routine.

Introduction

The planning and setting up of the majority of low volume manufacturing processes may account for a considerable percentage of the production time. The ability to automate this planning task can benefit low volume manufacture by reducing the time from design to manufacture and may facilitate the introduction of additional flexible automation. One such area which may benefit from task planning is in the fabrication of sheet metal components, where it has been identified that a high proportion of the production is in small batches, typically less than 100 (Gossard *et al.*, 1980). The manufactured components range in complexity from simple shapes formed from parallel bends, including building lintels and industrial light fittings to three-dimensional, box-like enclosures, such as computer cabinets and vending machines.

To facilitate the development of the task planner an experimental manufacturing cell consisting of a Bronx 50-25 UH CNC/VDU upstroking press brake, a Cincinnati T^3-776 industrial robot and an IBM RT PC computer has been established. The necessary communication protocols and hardware enhancement required to integrate the cell have been implemented. A reconfigurable vacuum gripper has also been constructed.

Background

The methods employed in the manufacture of sheet metal components are largely dependent upon batch size. In high volume production, as encountered in the automobile industry, sheet metal components are generally produced by a series of forming and shearing operations using power presses. In medium volume production of sheet metal components, for example consumer 'white' goods, a method of piercing and blanking pre-cut flat sheets before wiping the bends is employed. Both of these processes, because of their long production runs, are able to utilize a high level of automation but lack flexibility on component size, material thickness and bend angles.

The low volume manufacture of sheet metal components is usually performed by pressing a series of straight line bends on a pre-cut flat sheet between 'V'-shaped tools and relocating the partially formed component between bends. This is generally achieved on a press brake where the operator is responsible for the machine setting and the manipulation of the component.

Press brakes are capable of producing a large range of components, from a variety of materials, and with the use of computer numerical control (CNC) offer considerable flexibility in the manufacture of sheet metal components. This flexibility, in relation to other forming methods, demands an equally flexible handling and planning ability and has hitherto hindered the success of attempts at press brake automation (e.g. Danel, 1982).

Description of the task

The design of a component will contain details of bend angles and flange lengths, together with information on critical dimensions and tolerances, and based on this information it is possible to synthesize an appropriate method of manufacture. Crucial to the success of the operation is the selection of appropriate press brake tools and the order in which the bends are pressed. Certain bend orders will produce a configuration of the partially formed component where a collision between the press brake and component will occur and further work on the component is prevented. The choice of a bend sequence and the selection of suitable press brake tools are tasks usually performed by a skilled operator. Once a bend sequence has been established the press brake can be programmed with instructions on component bend angles and positions. Details of the inter-bend manipulations are also recorded. The manipulation instructions for the human operator require no greater detail than phrases in the form 'turn round front to back' or 'turn over top to bottom'. A change in orientation of the part developed component may require it to be temporarily withdrawn from the press brake, and since a typical separation between the top and bottom tool is only 50 mm this may necessitate a complex set of moves.

Task planning

The aim of the task planner is to produce the optimum method of manufacture based on the design of the sheet metal component. The generation and optimization of the method will be influenced by the specific resources and layout of the manufacturing cell. These resources include the range of press brake tools available, the type of gripper equipment and the working envelope of the robot. Information pertinent to production, such as limits on cycle time and batch sizes, will also influence the planner. With a knowledge of the component design, the production criteria and the resources available, it should be possible to synthesize the optimum method of manufacture.

Restrictions imposed on the task planner will be dependent upon a wide range of factors, which vary from the size and shape of the individual components to the production batch size. The restrictions may affect specific areas of planning or alter the approach taken. For instance, when planning for components which will buckle due to their own weight, the position of grip must account for this added constraint and be situated at a point where the robot is able continuously to support the workpiece, throughout the bend forming operation. For components which are sufficiently rigid to support their own weight, and where there is little risk of distortion, this restriction does not apply. The batch size of components may also influence the planner output. For batches of several hundred units, economical production may only be possible if the cycle time is short and the planner is then called upon to produce a method to achieve this. In those cases where batch size is small, possibly numbering only one component, a long computation time to find the method with the shortest cycle time would realize little gain.

The total constraints of the system may prohibit the successful synthesis of a task plan. An important element of the planner is to identify and overcome any failures. The error recovery may reinitiate a particular aspect of the planning, to try and overcome the point of error, or in more difficult cases solicit information from an external source. Certain planning areas will necessitate the specification of particular resources to avoid failure. One such area is the generation of a bend sequence for components which are not formed solely from parallel bends. In this case it is necessary to specify the positions of gaps in the press brake tools which will permit a collision free bend sequence. Other areas of failure may only be apparent when the component manipulation is evaluated, and may be corrected by the use of additional cell fixtures, such as a device to allow the robot to release the component or, regrip it another location. The ability to modify the resources available will increase the probability of the optimum sequence being produced by the task planner.

For the planned task to succeed it is imperative that the simulated sequence of operations is performed accurately. This may require the planner to initiate certain sensor-guided robot motions to ensure that the component is positioned correctly. A technique for the in-process inspection and adaptive control of the formed bend angle is under development, which will allow the off-line task plan to be executed.

It is not possible to give a complete assurance that a generated task plan will succeed, as a predicted path may produce a singularity in the robot workspace, due to joint alignment, which will cause unexpected behaviour of the robot and the subsequent failure of the plan. However, by monitoring the success of plans it may be possible to identify the cause of such failures and update the planner accordingly.

The problems associated with the planning for the flexible manufacture of sheet metal components can be split into three main areas;

(1) input of the component design, resources available and production restrictions,
(2) generation and optimization of the bend-folding sequence, and
(3) generation and evaluation of the inter-bend manipulation.

The output of the planning system is dependent upon the mutual satisfaction of each of the constraints associated with these areas. The initial research has focused on the specific requirements of the above areas for rigid parallel bend components.

Task input

In a fully integrated sheet metal manufacturing system the design of the sheet metal component will originate within a computer-aided design (CAD) data source, from where information on angles, dimensions and tolerances may be retrieved. For 'box' type components it is important to identify the individual fold lines and determine how they affect the component configuration. For parallel bend components the design is described by a linked series of bend vectors. The magnitude of the vector represents the distance between adjacent bends and the direction is determined from the bend angle. Information on the material used is required to enable calculations on the degree of component distortion. The material will also affect the available methods of grip, for example a magnetic device would be unsuitable for an aluminium component and a vacuum gripper could not hold a perforated sheet.

In conjunction with the component design the task planner will have geometric models of the resources available, and details of production requirements.

Bend sequence generation

To calculate all the permutations of the bend-folding order exhaustively, in order to find all possible sequences, would be computationally expensive when the component has more than a few bends. The bend-folding orders are restricted to those which produce intermediate component configurations that do not collide with the press brake. This restriction prohibits the folding of certain bends last, and this factor is utilized by initiating the search from the final component configuration and selecting one of the bends it would be possible to form last. A bend is only possible if the component is free from interference before and after the bend is formed. Interference detection is achieved with a line intersection algorithm to compare the positions of the partially formed component and the press brake. The reverse order bend sequence is thus determined by progressively unfolding the component until the flat sheet is reached. If a configuration is produced where there are no possible last bends, the most recent unfolded bend is reset and a new one selected. This recursive process is repeated until all geometrically feasible sequences have been found.

Even though the strategy of unfolding limits the search area, certain components can be formed by thousands of different bend sequences, which results in a long computation time.

Bend sequence evaluation

Each of the sequences found will determine the order in which the bends are formed and consequently how the developing component is manipulated between pressing operations. These sequences are dynamically sorted with respect to the complexity of inter-bend manipulation. Those sequences which require the least amount of manipulation are favoured for further analysis as this will minimize additional planning problems associated with robot motion.

There are two types of inter-bend manipulation: one is a translation between two points and the second requires a change in component orientation. As the translation motion is required to thread the component between the press brake tools the complexity of this manipulation is dependent upon the number of formed bends to pass through the tool gap. A change in orientation usually requires a translation move to remove the component from the press brake temporarily. There are two types of component manipulation required to change the orientation, one switches the top and bottom surfaces and the other reverses the position of the front and back edges. These manipulations can be performed singularly or in combination to achieve a required component orientation. The ease with which these are performed is dependent upon how the component is gripped and the manipulator configuration. For a particular gripper and manipulator combination it is possible to attain a measure of the complexity of manipulation from the number and type of orientation moves and the amount of translation required.

The bend sequence generation and evaluation will produce an optimum sequence for a particular press brake and tooling set. The sequence selected will be used in the subsequent planning and optimization for gripping position and robot trajectory,

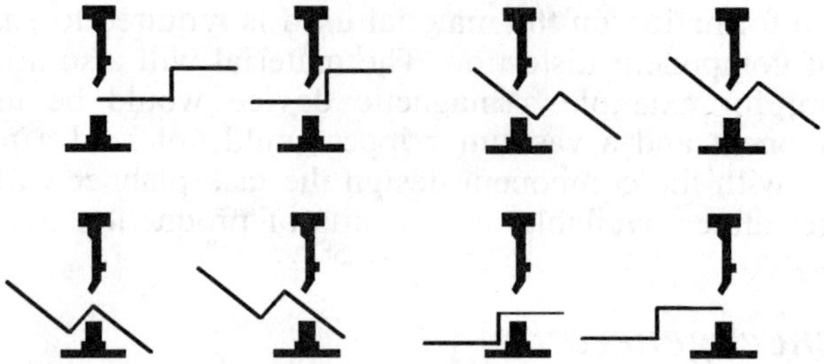

Figure 1. Component insertion.

to produce a suboptimal solution. As only rigid components are considered the remaining planning deals with the individual inter-bend manipulations, and does not consider the problems associated with continuously supporting the component.

Component motion

Once the optimum bend sequence has been established it is possible to analyse the component motion. The problem is to determine if the motion is possible and if so to generate the path. This type of motion planning is pertinent to robot task level programming and has been extensively studied (Yap, 1987).

If it is possible to thread the component through the press tooling it can be achieved by a combination of straight line moves, along the line of the enclosed flange, and rotations about the the midpoint of the tooling (see Figure 1). The motion is decomposed into the series of sub-moves required to reach intermediate bend lines. The method adopted generates the maximum straight line motion permissible towards the sub-target position. If this position is not reached the component is rotated by the minimum degree which will allow the sub-target to be reached. This process is repeated until all intermediate bends have been crossed and the goal position attained.

The component locations required to to perform this manipulation are recorded with reference to the press brake and are therefore robot independent. For a sequence to be feasible each of the required inter-bend manipulations must be possible. If any manipulation causes the sequence to fail, by virtue of a collision with the press brake, a new sequence may be selected, or a different method of positioning the component attempted, such as removal from the sides of the press brake.

Gripping position

Choice of gripping position is critical to the performance of the operation as it establishes the link between component motion and that of the robot. The positions available are restricted to areas which can be accessed at the start and finish of an inter-bend manipulation and will allow adequate support of the component. It is preferable to select a gripping position which is mutual for the start and finish of a manipulation. This is not always feasible and in such cases positions which permit the component to be regripped are chosen.

The gripper used in the present implementation is a vacuum device and the feasible positions are only restricted to the areas between folds which are of sufficient dimensions to accommodate the gripper. These areas are then assessed for possible interference of the gripper with the component or with the press brake.

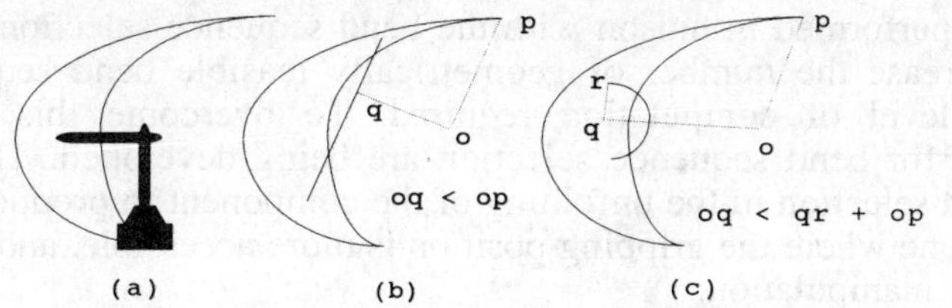

Figure 2. Robot workspace.

Robot trajectory

A knowledge of the manner in which the component should be moved and the position on its surface where it is gripped will enable the required robot trajectory to be generated and assessed. Validation is achieved by an analysis of the robot workspace for coordinate positions and orientations. The trajectory is defined from a transformation of the component coordinates and therefore follows the same series of straight line moves and rotations, as this motion is confined to a two-axis frame, in the current experimental implementation, the analysis is simplified.

The robot used, a Cincinnati T^3-776, is a common design of a six revolute axis manipulator, where there are three joints coincident which form a wrist. This wrist point can be used as a reference for the position of the tool endpoint and is independent of orientation, which simplifies the calculations as position and orientation can be dealt with separately. For a motion to be possible the start and endpoints of the planned move must be in the workspace boundary and the required orientation at these points must also be possible. If these conditions are satisfied then further calculations are performed to ensure the whole move between the points does not impinge on the workspace boundary.

For parallel bend components the inter-bend manipulations are confined to a two-dimensional plane. The experimental cell is aligned so the plane of the component motion is along the '*X*' axis of the robot coordinate frame, which allows the workspace to be defined by four intersecting arcs [see Figure 2(a)]. The expressions defining these arcs are used to assess the wrist positions. Evaluation of the motion between the two points is determined by different techniques dependent upon the type of motion, either a straight line move or a change of component orientation. Straight line moves are feasible if the points on the line which pass closest to the centres of the arcs which describe the interior boundaries are within the workspace [see Figure 2(b)]. It is possible for the robot to undertake a rotation move of the component, so the component may be threaded through the press brake tools, if one of the following conditions is met: the sum of the radius of rotation with one of the interior boundary radii is less than the distance between the centres, or if the line joining the two arc centres is not crossed.

For the configuration of the robot used all orientations are possible on a spherical envelope apart from a cone of 120 degrees around the axis of the fourth joint. Orientations where the angle between the axis of the gripper and the fourth joint axis is greater than 60 degrees will therefore be possible. A method to ensure each discrete point of a move does not violate the orientation condition is under investigation.

Implementation of the task planner

To enable the planner to synthesize an optimum or near optimum solution it is required that the planning of component motion, gripping position and robot trajectory are performed in unison with the bend sequence selection. The effect of this is to decrease the number of geometrically feasible bend sequences and to increase the level of computation required. To overcome this computational problem rules for bend sequence selection are being developed. These rules will guide the bend selection in the unfolding of the component to produce a more open shape, that is one where the gripping position is more accessible, and will minimize the inter-bend manipulation.

Planning routines to generate a bend sequence, the inter-bend component manipulation and the corresponding robot programme have been developed. The routines are written entirely in the 'C' programming language. A means of generating a graphical simulation of the planned task has also been developed as an aid to programming. The current work is concentrated on the linking of the individual planning modules and their enhancement.

Conclusions

The research has established that it is feasible to generate a task from the design of a sheet metal component automatically. The approach of guiding the planning through successive levels of geometric reasoning is suitable for other areas of manufacture where either the environment is changing or, as in this case, the workpiece grossly alters form during the execution of the task.

Acknowledgements

The authors acknowledge the support and advice from the ACME Directorate of SERC (Grant GR/E 06619), and industrial collaborators Verson HME Ltd, Broxton Engineers Ltd, British Aerospace PLC and TI Jackson Ltd (subsequently GEC Creda).

References

Danel, F., 1982, A specially designed robot for press brake operation. *6th International Conference on Industrial Robot Technology*. (Paris: IFS).

Gossard, D., Hardt, D.E., McClintock, F.A., Alison, B.T., Stelson, K.A. and Gu, I., 1980, *Sequential Forming of Sheet Metal Parts*. (USAF Wright Aeronautical Laboratories).

Yap, C.K., 1987, How to move a chair through a door. *IEEE Journal of Robotics and Automation*, **3**.

60

Manual versus flexible assembly: a cross comparison of performance and costs in four different countries

Anil Mital

Ergonomics and Engineering Controls Research Laboratories, Department of Mechanical, Industrial and Nuclear Engineering, University of Cincinnati, Cincinnati, OH 45221, USA

Abstract

The use of robots to automate subassemblies and assemblies (flexible assembly) is becoming increasingly popular. Reductions in manufacturing and assembly costs and improvements in material flow are cited as the major benefits of designing products suitable for automated assembly. Recent works, however, have demonstrated that when products are designed for assembly (DFA), the manual assembly option may be economically more desirable. This chapter further compares the costs of manual and flexible assembly methods. Labour costs and interest rates in four different countries (Japan, UK, USA, and Germany) are additional factors considered in this comparison. The cross comparison of costs in the four countries suggests that while robotic assembly methods may be economically more desirable in one country, the manual mode may be economically more desirable in other countries. The interest rate, production volume, labour costs, and human performance are identified as the major factors that influence the decision to choose a particular assembly method.

Introduction

Automation is revolutionizing manufacturing world-wide. The drive to automate, however, is not limited to manufacturing operations only. Increasingly, we come across industries that have automated their materials handling and product assembly or subassembly operations. While considerable progress has been made in automating basic manufacturing and materials handling operations, progress on assembly automation has been slow. One of the major problems confronting engineers is the basic design of the product; products are either not designed for automated assembly or are poorly designed.

The present trend in product design favours assembly; DFA products are increasingly becoming popular. Over the past few years certain design guidelines have also been adopted. For instance, snap-in retainers are a common feature of many products designed primarily for automated assembly. Besides the design simplicity and assembly convenience, DFA products tend to lower materials and inventory costs, by virtue of fewer components, and materials handling costs. It is also believed that DFA products have lower assembly costs. This belief, however, has no proven basis.

On the contrary, one recent report (Boothroyd, 1988) suggests that when DFA products are manually assembled, the cost of assembly is no more than the cost when such products are assembled with the help of robots (flexible assembly method) or automatically by hard-core dedicated equipment (automatic assembly method). Another study (Mital *et al.*, 1988) demonstrated that for DFA products manual mode of assembly is cheaper than the flexible assembly mode. As a matter of fact, Mital *et al.* point out that choice of assembly method for DFA products is based on such factors as labour rate, production volume, learning effect and equipment reliability as these are the factors that determine the economic desirability of a particular assembly method.

This work further explores this issue by comparing performance (assembly) times and unit assembly costs of robotic and manual methods. Two DFA products (compressor head valve assembly and floppy disk drive assembly) are considered and the costs of assembling them manually and with the help of robots in four different countries (Japan, UK, USA, and Germany) are compared. The rationale for this comparison being: the economic desirability of a particular assembly method is a function of local costs and may differ from country to country. Thus an assembly method, while economically desirable in one country may not be economically desirable in other countries.

Since interest rates, labour costs, equipment costs and performance times are known to differ from country to country, these were expected to influence assembly costs significantly. Therefore, in all performance time and unit cost determinations local data were used. These data were obtained from a variety of sources including newspapers, International Labour Office (ILO) publications, personal communications with colleagues in the countries considered, and various industry sources. The major sources are identified.

The DFA products and their assembly sequence

The two DFA products used in the comparison were: cylinder head valve assembly (Figure 1) and floppy disk drive assembly (Figure 2). The cylinder head valve assembly has 12 elements (exhaust valve leaf, leaf lock, two rivets, valve plate, valve body, gasket, suction valve plate, and four screws) and 300 assembled units are needed each hour (Romeo and Camera, 1980). The assembly sequence is as follows: pick valve plate and place on rivets already fed and placed in the rivetting machine, pick bent exhaust valve leaf and place it on valve plate, pick leaf lock and place it on valve plate and leaf, perform rivetting, pick up all rivetted components, rotate the subassembly 90°, and place it upon valve body and gasket located at another station—all pick up and placement operations are done by one robot arm; pick and place gasket on valve body, pick suction valve plate and place it upon the rivetted subassembly, pick up the entire group of parts and place it upon four preplaced screws in a pneumatic screwing machine, complete screwing operation, pick up and place the assembled group on a belt—all pick up and placement operations are done by the second robot arm. Special feeders and chutes are required for feeding parts.

The floppy disk drive subassembly has eight component parts (Albo, 1983) (Figure 2): spindle (bottom centre), flanged bearing (above spindle), preload spring (above bearing), plain bearing (above spring), drive pulley (above plain bearing), the base casting (upper left), the right diskette guide (below casting), and the left diskette guide (below right guide). The assembled drive is shown in the upper right corner of Figure 2. The sequence of assembly is as follows: insert the flanged

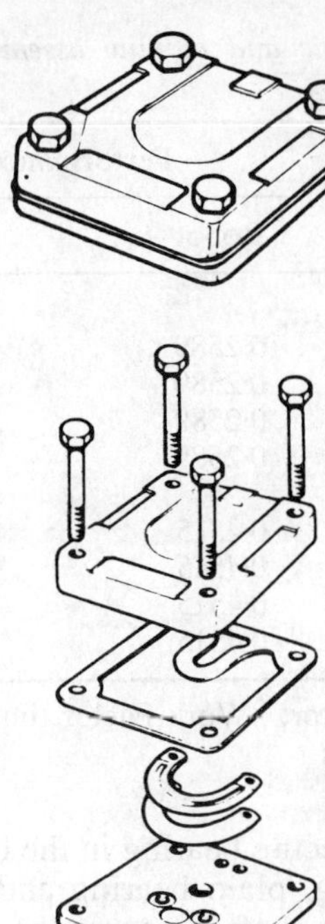

Figure 1. Compressor head valve assembly and the components (from Romeo and Camera, 1980).

Figure 2. Floppy disk drive component parts (from Albo, 1983).

Table 1. Estimated cycle times for robotic and manual assembly of compressor head valve subassembly and floppy disk drive assembly.

DFA product	Performance Time (Minute)	
	Robot	Human
1. Compressor head valve subassembly		
Germany	0·2589	0·1960[b]
Japan	0·2589	0·1715[c]
UK	0·2589	0·2450[a]
USA	0·2589	0·2450[a]
2. Floppy disk drive assembly		
Germany	0·4515	0·3290[b]
Japan	0·4515	0·2891[c]
UK	0·4515	0·4113[a]
USA	0·4515	0·4113[a]

[a] Based on MTM; [b] Based on Work Factor; [c] Work-Factor time estimate reduced by 10%.

bearing in the casting spindle bore, secure bearing in the bore, insert spindle through the flanged bearing, place the spring, plain bearing and pulley on the shaft from the other side, force the pulley on the shaft to secure the spindle assembly by using a press, and finally attach the two plastic diskette guides by forcing cast-on studs into mating holes in the base casting. Two hundred units are needed per hour.

Performance (assembly) time estimation

The assembly cycle times for robotic assembly method were estimated from RTM (Paul and Nof, 1979). For the manual assembly method, MTM (Maynard *et al.*, 1948) was used for estimating cycle times in the UK and USA (in both these countries MTM is very popular; K. Knott, personal communication). Work Factor (Konz, 1983) was used to estimate performance times of German workers (D. Overly, personal communication; W. Rameirez, personal communication). Performance times obtained by Work Factor were reduced by 10% to obtain performance times of Japanese workers (Rameirez, 1985). Table 1 shows the manual and robotic assembly cycle times for both DFA products in all four countries.

PUMA assembly robot was used for assembling both products. Given the size of the various components and the equipment, workstation layout for assembly was configured. From this, distances moved by the robot and human arms were determined. These distances and the type of movements were used to estimate cycle times. As an example, Table 2 shows the details of assembly sequence and RTM and MTM elements for the compressor head valve assembly.

Assembly cost estimation

From the assembly cycle times and production volumes, annual costs for the two assembly methods were estimated. The hourly earnings of workers were estimated from ILOs Year Book of Labour Statistics (1987). Table 3 gives the summary of worker hourly earnings and interest rates in each of the four countries (interest rates were taken from the Wall Street Journal). The number of work hours per week was also determined from ILOs Year Book of Labour Statistics. It should be noted that even though the actual number of hours worked per week differed from country

Table 2. Assembly sequence and RTM and MTM elements for assembling compressor head valve assembly.

Assembly sequence	RTM element	Time[a]	MTM element	Time[a]
1. Pick valve plate and place on rivets (robot arm 1)	R2 100	26	R30C	26·7
	ST 0·5	3	G1A	2
	GR 0·5	4	M30C	30·7
	M2 75	27	P1 S	5·6
	ST 0·5	3	RL1	2
	RE 0·5	4		
2. Pick bent exhaust valve leaf and place it on valve plate (robot arm 1)	R2 75	26	R30C	26·7
	ST 0·5	3	G1A	2
	GR 0·5	4	M30C	30·7
	M2 75	27	P1 S	5·6
	ST 0·5	3	RL1	2
	RE 0·5	4		
3. Perform rivetting		30		30
4. Pick up rivetted component, rotate 90 degrees and place it upon valve body and gasket located elsewhere (robot arm 1)	R2 30	23	R10A	8·7
	ST 0·5	3	G1A	2
	GR 0·5	4	T 90	5·4
	Rotate assy.	20	M30C	30·7
	M2 100	29	P1 S	5·6
	ST 0·5	3	RLI	2
	RE 0·5	4		
5. Place gasket on valve body (robot arm 2)	R2 100	26	R30C	26·7
	ST 0·5	3	G1A	2
	GR 0·5	4	M30C	30·7
	M2 75	26	P1 S	5·6
	ST 0·5	3	RL1	2
	RE 0·5	4		
6. Pick up suction valve plate and place it upon rivetted subassembly (robot arm 2)	R2 100	26	R30C	26·7
	ST 0·5	3	G1A	2
	GR 0·5	4	M30C	30·7
	M2 75	27	P1S	5·6
	ST 0·5	3		
	RE 0·5	4		
7. Pick up all grouped parts and place them upon 4 screws (robot arm 2)	R1 1	10.5	G1A	2
	ST 0·5	3	M30C	30·7
	GR 0·5	4	P1 S	5·6
	M2 100	29	RL1	2
	ST 0·5	3		
	RE 0·5	4		
8. Complete screwing operation		40		40
9. Pick up and place subassembly on the belt (robot arm 2)	R2 100	26	R10C	12·9
	ST 0·5	3	G1A	2
	GR 0·5	4	M30C	30·7
	M2 100	29	RL1	2
	ST 0·5	3		
	RE 0·5	4		

[a] Time is in TMU (1 TMU = 0·0006 min).

Note: Elements 1 and 4 are done concurrently by the robot arms. Manually, element 5 is performed while rivetting (element 3) is being performed. Distances are in centimetres for RTM and inches for MTM.

Table 3. Summary of wage and interest rates in Japan, UK, USA and Germany.

Country	Interest rate (%)	Hourly wage rate (US dollars)
Germany	6·500	9·13
Japan	3·375	14·27[a]
UK	13·000	6·78
USA	12·000	10·00

[a] Computed from monthly wages ($2379·14) and assuming 2000 hours per year.
[b] Since the number of work hours per week and the number of work days per year varied from country to country, but not very widely, to simplify comparison it was assumed that total number of work hours per year was 2000.
Note: 1 US Dollar = 128·6 Japanese Yen; = 1·86 West German Mark; = 0·5726 British Pound.

Table 4. Summary of annual assembly costs (US dollars) for compressor head valve subassembly and floppy disk drive assembly.

DFA product	Robotic assembly	Manual assembly
1. Compressor head valve subassembly		
Germany	24 507	19 671
Japan	21 891	29 892
UK	31 802	18 494
USA	29 895	26 522
2. Floppy disk drive assembly		
Germany	33 047	21 821
Japan	29 506	29 779
UK	43 201	20 014
USA	40 568	28 854

to country (on an average 40·4 in Germany to 41·6 in UK), for convenience sake work-weeks in all four countries were considered to have 40 hours. For the same reason, the number of work-days per year was also considered to be the same in all countries (in reality the number of actual work-days in each country differs slightly due to different national holidays, etc.).

From these data, the annual assembly costs for the two DFA products were determined. Table 4 summarizes the assembly costs for both robotic and manual assembly methods in each of the four countries.

As an example, the cost calculations for assembling compressor head valve subassembly in the United States are shown below.

Robotic Assembly Method

Assembly cycle time = 0·2589 min

Production capacity/workstation = 1/cycle time = 3·8625 units/min
= 231·75 units/h

Production volume = 300 units/h

This means that if only one work shift is operated per day, two workstations will be needed. The least investment alternative, however, would be to have two shifts per day (approximately 11 h/day).

Fixed cost for the workstation:

Robot (PUMA assembly robot) = \$37 000 (Unimation, Pittsburgh, USA)
Feeders (5) = \$85 000 (@ \$17 000/feeder)
Rivetting and screwing fixtures = \$2000
Bins and work bench = \$1000
Miscellaneous (foundation, etc.) = \$10 000
Total = \$135 000 = Total investment.

Variable cost for the workstation:

Maintenance = \$2000
Energy = \$4000
Total = \$6000

Annual cost = 6000 + 135 000 [A/P .12,10] (assuming investment life to be 10 years)
= \$29 895

Manual Assembly Method

Assembly cycle time = 0·245 min
Production capacity/workstation = 1/cycle time = 4·0816 units/min
= 244·896 units/h
Production volume = 300 units/h

Therefore, two workstations will be needed if only one shift is run per day. With one workstation, the operator would need to work 10 hours per day or 1·25 shift. The least investment alternative would be to have one workstation and to run the operation 10 hours each day.

Fixed cost for the workstation cost would be:

Bins (9) = \$450
Work bench = \$500
Rivetting and screwing fixtures = \$2000
Total = \$2950

Variable cost would be:

Labour cost = \$25 000/year
Utilities = \$1000/year

Annual cost = 26 000 + 2950 [A/P .12,10] = \$26 522

Discussion and conclusion

The costs of assembling two DFA products, manually and with the help of a robot, in four different countries, were determined in this chapter. As shown in Table 4, the manual assembly option was found to be more economical in Germany, the UK and the USA while the flexible assembly option was cheaper in Japan. The difference in costs between the manual and automated options was relatively small in Germany as compared to such cost difference in either the UK or the USA. When the initial investment increased, for example in the case of floppy disk drive assembly, the flexible assembly option was still cheaper in Japan. The increased investment, however, resulted in greater economic advantage for manual assembly in Germany, the UK, and the USA. This was in spite of the fact that Japanese workers took less time to complete the assembly task.

As shown in Table 3, interest rates and hourly wage rates in all four countries are substantially different. Low interest rates in Japan lead to lower investment costs. Higher hourly rates also favour automation. Relatively smaller hourly wages and higher interest rates in western countries tend to favour manual assembly. Germany, which has the lowest interest rate among the three western countries considered here and also has relatively lower hourly wages compared to Japan and the USA, could opt for either the manual or the automated option especially if the initial investment is not very high.

The cost analysis clearly shows that if one country chooses a particular method of work and finds it economically beneficial it does not mean the same method would also be economical in other countries. The impact of local factors such as interest rates and hourly wages could be very profound and must be taken into consideration prior to choosing a particular option (manual or flexible assembly in this case).

Changes in production volume would also affect the final decision. Certainly if the production volume increases to a point that a complete second or a third shift is needed, the flexible assembly method would become more economical. In the UK and the USA also, if the production volume increases to a point that additional worker hours would be required, the advantage of the manual assembly method would diminish. However, if the demand increases to a point that new investment in an additional robot and workstation becomes necessary, the economic advantage would again favour the manual assembly method.

In conclusion, the decision to choose a particular method of assembling a product depends upon a number of factors. The most important factors that appear to influence the choice of method appear to be interest rate, hourly wages, human performance and production volume.

References

Albo, R.T., 1983, Robotic assembly of floppy disk drives. *Proceedings of the 13th International Symposium on Industrial Robots and Robots 7*, **1**, April, Chicago, Illinois, pp. 5·1–5·11.

Boothroyd, G., 1988, Making it simple—design for assembly. *Mechanical Engineering*, February, 28–31.

Konz, S.A., 1983, *Work Design: Industrial Ergonomics* (Columbus, OH: Grid Publishing).

Maynard, H.B., Stegemerten, G.J. and Schwab, J.L., 1948, *Methods-Time Measurement* (New York, NY: McGraw-Hill).

Mital, A., Mahajan, A. and Brown, M.L., 1988, A comparison of manual and automated assembly methods. *Proceedings of the Institute of Industrial Engineers Integrated Systems Conference*, pp. 206–211.

Paul, R.P. and Nof, S.Y., 1979, Work methods measurement—a comparison between robot and human task performance. *International Journal of Production Research*, **17**, 277–303.

Romeo, G. and Camera, A., 1980, The DEA pragma assembling system—a couple of applications in compressor subassemblies. *Proceedings of the 10th International Symposium on Industrial Robots and 5th International Conference on Industrial Robot Technology*, March, Milan, Italy, pp. 491–500.

61

A routing control method of automated guided vehicles by the shortest path with time-windows

Susumu Fujii,[1] Hiroaki Sandoh[2] and Ryusuke Hohzaki[3]

[1]Department of Systems Engineering, Kobe University, Nada, Kobe 657, [2]Department of Business Administration, University of Marketing and Distribution Sciences, Nishi-ku, Kobe 673, and [3]Graduate School of Science and Technology, Kobe University, Japan

Abstract

The present study proposes a routing control method of automated guided vehicles (AGVs) in a manufacturing system which can cope with a larger scale transfer system with many AGVs. When an AGV receives a travelling request from one point to another in the manufacturing system, it is necessary to determine a path for this AGV which does not cause interference with any other AGV. This study regards the above problem as the shortest path search problem on a network consisting of arcs and nodes with time-windows, and an algorithm for the path search problem is proposed. Numerical examples are also presented.

Introduction

Automated guided vehicles (AGVs) are recognized as an essential component to realize a flexible transfer system in a manufacturing system. As the scale of a manufacturing system becomes larger, the scale of the transfer system will also become larger in the number of AGVs and in the scale and complexity of the runways of AGVs. In such a situation, the control of many AGVs will become very difficult since the possibility of interference among AGVs increases causing many deadlocks and blockings of AGVs.

Fujii and Sandoh (1987) proposed a routing algorithm for the above situation where the problem is formulated into a kth shortest path finding problem, and then into a set of several linear programming problems and a combinatorial one. Fujii *et al.* (1988) also proposed three routing strategies based on the previous study, and compared the efficiency of the strategies. The present study introduces a concept of time-windows which regulates AGVs' arrival and starting times at some locations, and regards the routing problem as a shortest path finding problem on a network with time-windows. Based on these ideas, an alternative routing algorithm is proposed, which has an ability to indicate the infeasibility of problems that the algorithm previously proposed lacks.

Setting time-windows

In this study, runways of AGVs in a manufacturing system are represented by a network consisting of arcs and nodes in the same way as in Fujii and Sandoh (1987). In the following, a set of nodes and that of arcs in a network are denoted by V and A, respectively, and $G = (V, A)$.

A conventional time-window (Solomon and Desrosiers, 1988) is defined by a time interval (t_1, t_2) $(t_1 \leqq t_2)$ placed at nodes, indicating that an AGV must leave the nodes within the specified time intervals. On the other hand, time-windows in this study are placed on arcs as well as on nodes, and constrain an AGV both to start and to arrive at the node or the arc within the specified time-windows. Moreover, each node and arc can have several time-windows in this study. It is to be noted that these additional restrictions change the characteristics of the problem, and that the algorithm developed in the studies on shortest path finding problems with conventional time-windows cannot be applied in this study.

Let us consider a manufacturing system which has N AGVs, $C_1, \ldots, C_N$. When an AGV, say C_1, receives a travelling request from its current node to another in the network and the other AGVs are travelling to fulfil their own travelling requests, a path for C_1 must be determined in a manner such that C_1 does not interfere with any other AGV and the travelling time of C_1 becomes the minimum. If the path of C_i and its time schedule on the path are specified for $i = 2, \ldots, N$, the time-windows for C_1 can be obtained by the following rules:

(a) Time-windows at a node which any AGV except C_1 passes by—If an AGV C_i $(i \neq 1)$ passes a node with a time-window (t_s, t_e), the other AGVs including C_1 are prohibited to travel passing the node during the time interval $(t_s - \varepsilon, t_e + \varepsilon)$, where ε is pre-determined considering the size of an AGV to avoid collisions between any two AGVs. This indicates that the time-windows for the other AGVs at this node are $(-\infty, t_s - \varepsilon]$ and $[t_e + \varepsilon, +\infty)$.

(b) Time-windows at a node where an AGV except C_1 arrives and stays—If C_i $(i \neq 1)$ reaches a node at time t_a and stays there after that, the time-window to be placed at the node for the other AGV is $(-\infty, t_a - \varepsilon]$.

(c) Time-windows on an arc—Let us consider the case C_i $(i \neq 1)$ travels along arc (i, j) on the condition that the AGV starts from node i at time a_i and arrives at j at time $a_i + d_{ij}$, where d_{ij} denotes the travelling time of the AGV along the arc (i, j). In this case, the time-windows to be imposed on arc (i, j) for the other AGVs are not necessary since the time-window at node i is active, while the time-windows to be placed on arc (j, i) become $(-\infty, a_i - d_{ji} - \varepsilon]$ and $[a_i + d_{ij} + \varepsilon, +\infty)$ for the other AGVs.

Time-windows of the network for C_1 can be acquired by setting time-windows at nodes and arcs of C_i $(i \neq 1)$ following the above rules and by obtaining the intersection of the time-windows of all the AGVs except C_1.

Modified and shifted time-windows

Let us express a set of time-windows placed at node i as

$$T(i) = \{T_{ik} = (t_1^{ik}, t_2^{ik}); k = 1, \ldots, n_i\} \tag{1}$$

where n_i denotes the number of time-windows T_{ik} in $T(i)$ and

$$t_1^{ik} \leqq t_2^{ik}, k = 1, \ldots, n_i \tag{2}$$

$$t_2^{ik} < t_1^{i(k+1)}, k = 1, \ldots, n_i - 1. \tag{3}$$

Likewise, a set of time-windows of arc (i, j) is expressed as $M(i, j)$.

For a family K of sets which consists of any sets of time-windows, we introduce four operators, MIN, $\oplus$, $\ominus$ and $\otimes$ which are defined as

$$\text{MIN}\, T(i) = t_1^{i1} \tag{4}$$

$$T(i) \oplus r = \{(t_1^{ik} + r,\ t_2^{ik} + r); k = 1, \ldots, n_i\} \tag{5}$$

$$T(i) \ominus r = \{(t_1^{ik} - r,\ t_2^{ik} - r); k = 1, \ldots, n_i\} \tag{6}$$

$$T(i) \otimes U(j) = \{(t_1^{ik},\ t_2^{ik}) \cap (u_1^{jq},\ u_2^{jq}); k = 1, \ldots, n_i, q = 1, \ldots, n_j\} \tag{7}$$

where $T(i) = \{(t_1^{ik},\ t_2^{ik}); k = 1, \ldots, n_i\} \in K$ and $U(j) = \{(u_1^{j1},\ u_2^{jq}); q = 1, \ldots, n_j\} \in K$, and r is a real number. Furthermore, one relational operator is introduced which is defined as

$$T(i) \gg U(j) \Leftrightarrow t_1^{i1} > u_2^{jn_j}. \tag{8}$$

We here introduce concepts of the *modified time-windows* and the *shifted time-windows* for arcs and nodes, respectively, which are useful in solving the routing problems.

A set of *modified time-windows* of (i, j) which is expressed as $M'(i, j)$ is defined as:

$$M'(i, j) = M(i, j) \otimes (T(j) \ominus d_{ij}). \tag{9}$$

If $M'(i, j) = \phi$, it is signified that any travelling along (i, j) is prohibited all the time. In case C_1 travels along (i, j), it is unnecessary to consider the constraint $T(j)$ by considering $M'(i, j)$.

When C_1 is allowed to be idle on a time-window W at node i and travels along (i, j), the time at which C_1 can arrive at j is on a time-window in a set of time-windows expressed by

$$T_i(j) = \{[W \otimes M'(i, j)] \oplus d_{ij}\}. \tag{10}$$

Equation (10) means that $T_i(j)$ is obtained by finding the intersection of W and $M'(i, j)$ and by shifting the intersection to node j. Hence, the time during which C_1 can be idle within a time-window $T_{jk} = (t_1^{jk},\ t_2^{jk})$ at node j becomes $[\text{MIN}(T_i(j) \otimes T_{jk}),\ t_2^{jk}]$ if $T_i(j) \otimes T_{jk} \neq \phi$ for $k = 1, 2, \ldots, n_j$. The set of time-windows obtained in this manner is called a set of *shifted time-windows*.

The routing algorithm described in the next section makes use of these concepts of the modified and the shifted time-windows.

Routing algorithm

Notation

d_{ij}	time necessary for an AGV to travel along (i, j),
s	start node of C_1,
e	destination node of C_1,
t_0	time allowed for C_1 to start from s,
$T(i)$	set of time-windows of node i,
n_i	number of time-windows in $T(i)$,
$M(i, j)$	set of time-windows on arc (i, j),

m_{ij}	number of time-windows in $M(i, j)$,
$E(i)$	$\{E_{ik} = (e_1^{ik}, e_2^{ik}); k = 1, \ldots, n_i\}$, set of shifted time-windows of node i,
$\nu_k(i)$	node from which the kth shifted time-window at node i is obtained
$M'(i, j)$	set of modified time-windows on (i, j),
μ_{ij}	number of modified time-windows in $M'(i, j)$,
$L(i)$	shortest travelling time from i to e in case time-windows are ignored.

Algorithm

The algorithm developed in this study using a branch-and-bound method aims to find a path for C_1 which does not interfere with the other AGVs and which reveals a minimum travelling time for C_1. For this purpose, the algorithm generates shifted time-windows at nodes in sequence starting from s, so that some of them construct paths from the starting node s to the destination node e.

The following theorem is easily proven in the problem considered here.

Theorem 1

If there exists a feasible path for C_1 which includes a loop such that C_1 starts from node i at time u_i and returns to the same node at time v_i and that u_i and v_i are in a same time-window, there also exists a feasible path which does not include such a loop and reveals the same travelling time. □

The algorithm described below is based on the above theorem, i.e. all paths including loops as mentioned in the above theorem are eliminated in the process of path search.

Step 1 If $(k, i) \in A$ for any $i, k \in V$, generate the sets of modified time-windows;

$$M'(k, i) = M(k, i) \otimes \{T(i) \ominus d_{ki}\}.$$

Step 2 Define the shifted time-windows of node s as:

$$E(s) = T(s) \otimes [t_0, +\infty) \text{ and } \nu_k(s) = 0,\ k = 1, \ldots, n_s.$$

Initialize the shifted time-windows of all the other nodes as:

$$E(i) = \phi \text{ and } \nu_k(i) = 0 \text{ for all } i \neq s,\ k = 1, \ldots, n_i.$$

Step 3 Let U denote the set of nodes which are connected by directed paths from s to e. If $U = \phi$, then go to *Step 7*. Otherwise, go to *Step 4*.

Step 4 Let L denote a set of nodes having shifted time-windows in U, i.e.

$$L = \{i \in U \mid E(i) \neq \phi\}.$$

If $L = \phi$, then go to *Step 7*. Otherwise, find i_0 and k_0 which are given by

$$e_1^{i_0k_0} + L(i_0) = \min_{\{i \in L, E_{ik} \neq \phi\}} [e_1^{ik} + L(i)]$$

which plays a role of a lower bound for the travelling time in the branch-and-bound method.

In case $i_0 = e, e_1^{i_0k_0}$ is the minimum travelling time from s to e and thus go to *Step 8*. When $i_0 \neq e$, if

$$[e_1^{i_0k_0}, e_2^{i_0k_0}] \oplus L(i_0) \gg T(e) \text{ or } E_{i_0k_0} \oplus L(i_0) \gg T(e)$$

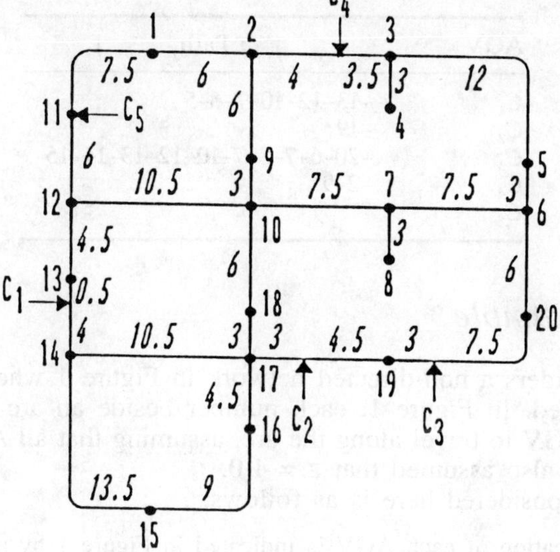

Figure 1. Network model.

then go to *Step 7*, otherwise go to *Step 5*.

Step 5 Generate the shifted time-windows of nodes in U connected to i_0 by the following procedure:

for any node p in U where $(i_0, p) \in A$, if

$$[e_1^{i_0k_0}, e_2^{i_0k_0}] \gg M'(i_0, p) \text{ or } E_{i_0k_0} \gg M'(i_0, p)$$

then eliminate the arc (i_0, p) from the original network since it is forbidden to travel along the arc by the time-window limitation. When the above relationship does not hold, generate a shifted time-window or shifted time-windows to node p as follows:

obtain $$T_j(p) = [E_{i_0k_0} \otimes M'(i_0, p)] \oplus d_{i_0p}$$

and if $T_j(p) \cap T_{pk} = \phi$, then revise the shifted time-window E_{pk} as

$$E_{pk} = \{\text{MIN}[T_j(p) \otimes T_{pk}], t_2^{pk}\}, k = 1, \ldots, n_p.$$

When node p is located in a path from the starting node s to node i_0, the following processing is necessary at node i_0.

If both the shifted time-windows, one is previously generated at node p and the other is newly generated at p in the above, are included in a same time-window T_{pk} in $T(p)$, the smaller one is removed with the according revision of $\nu_k(p)$.

Step 6 Remove $E_{i_0k_0}$ from $E(i_0)$ and go to *Step 3*.

Step 7 Stop. There exists no feasible solution.

Step 8 By backward tracing from node e to s, the shortest travelling time path without any interference can be obtained. Stop.

In the above algorithm, it is to be noticed that *Step 7* indicates the infeasibility of problems.

Table 1. Paths of AGVs.

AGV	Path
C_1	–13–12–10–7–6–5
C_2	–19
C_3	–20–6–7–8–7–10–12–13–14–15
C_4	–2–9
C_5	?

Numerical example

This section considers a non-directed network in Figure 1 where five AGVs, C_1, …, C_5 are installed. In Figure 1, each number beside an arc indicates the time required for an AGV to travel along the arc, assuming that all AGVs run with the same speed. It is also assumed that $\varepsilon = 1{\cdot}0$.

The situation considered here is as follows:

(1) The current location of each AGV is indicated in Figure 1 by an arrow, which was randomly determined along with its destination node.
(2) The path for each AGV is shown in Table 1.
(3) AGV C_5 is newly requested to travel from the current node to node 7, and the problem here is the determination of a path for C_5.

In the above situation, the numbers of time-windows placed at nodes and arcs were 95 and 68, respectively, which generated 94 modified time-windows on arcs. The algorithm described in the last section provided C_5 with a path of node sequence, 11–12–10–18(7·0)–10–7, where the number in parentheses signifies the idle time of C_5 at node 18. The computation was conducted on IBM 3083 VM/CMS and the central processing unit (CPU) time spent was 29 ms.

The proposed algorithm was also applied to 100 different problems similar to the above problem on the condition that 10 AGVs were installed into the network in Figure 1, which resulted in 50 ms of the mean CPU time.

References

Fujii, S. and Sandoh, H., 1987, A routing algorithm for automated guided vehicles in FMS. *Proceedings of 9th ICPR*, **2**, 2261–2267.

Fujii, S., Sandoh, H. and Hohzaki, R., 1988, Routing control of automated guided vehicles in FMS. *Proceedings of USA–Japan Symposium on Flexible Automation*, **2**, 629–636.

Solomon, M.M. and Desrosiers, J., 1988, Survey paper: time-window constrained routing and scheduling problems. *Transportation Science*, **22**, 1–13.

62

Assembly sequence planning using operation networks

Hans-Jörg Bullinger and Uwe A. Seidel

Fraunhofer Institute for Industrial Engineering (IAO), Nobelstrasse 12c, D-7000 Stuttgart 80, Germany

Abstract

Computer-supported planning methods operate on computer internal models of objects. For Computer-Aided Process Planning, a model of tasks and processes is one of the prerequisites. Like each model it must be designed in view of the requirements (and level of sophistication) of the methods. Due to our development of more sophisticated planning methods we felt the need for a more elaborated model. To stress the fact that we have emphasized the different structural aspects of processes we call our model Operation Network (OPNET). This chapter covers aspects of modelling, generating, assessment and use of Operation Networks.

Introduction

Between a product idea and the successful marketing of a product many different planning activities have to be carried out in a systematic way by a number of people in an enterprise (see Ammer *et al.*, 1986). Manufacturing engineers are responsible for how (method) and by what technological means a product is manufactured and assembled. After many years of concentrating on optimizing manufacturing (part making), currently assembly problems are coming increasingly into focus. Typical research topics are, for example, designed for assembly (DFA), robotics, flexible assembly systems (FAS).

Today's products may contain hundreds and thousands of parts and many different technologies. Product structure and, in connection with it, the product's assembly process (sequence) will significantly influence cost and quality, and therefore success, of a product.

Assembly sequence planning

One important task of an assembly planner is to define all operations (tasks, actions, processes) necessary for the assembly of a specified product. He has to define assembly operations like gripping, insertion and joining as well as non-assembly operations like reorientation, feeding, quality control and transport. Then, all operations have to be mapped into an assembly system consisting of manual, (flexible) automated or hybrid assembly stations. This is a typical demand/supply problem, which has to be solved under different planning constraints and goal functions (see Figure 1).

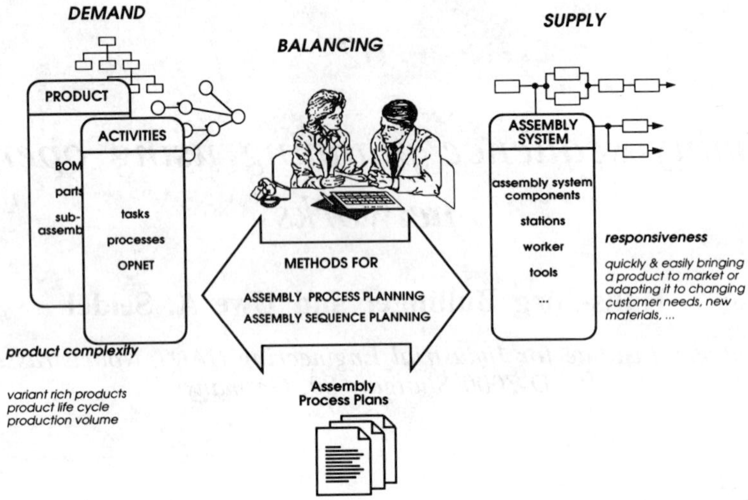

Figure 1. Assembly sequence planning—a demand/supply problem.

It is important to know in which sequence these operations have to be carried out. Very often a product allows different sequences of operations for its assembly. The responsibility of the planner is to find the best sequence under a given set of planning constraints, e.g. production volume, quality, available technology in the shop floor. In manufacturing (part making) applications an optimal sequence of operations is relatively easy to find, as many automated planning systems bear witness. Very often the operations form a chainlike sequential structure. In comparison to this, it is typical for assembly processes to be highly complex and non-sequential.

Before assembly, all parts can be considered more or less separately, but in assembly all parts and subassemblies are joined and must fit and function together. As Whitney *et al.* (1988) states: 'assembly is inherently integrative'.

Some known models of operations

In this chapter we consider some well-known models of operations.

While Petri nets (e.g. see Peterson, 1981; Glüer and Schmidt, 1988) have been proven useful for the description of assembly systems we neglect them in this paper. We omit also, for the sake of briefness, a discussion of artificial intelligence (AI) based models which have been developed in the research field of Autonomous Robots and Automatic Planners (see Figure 2).

OR models

Operations Research (OR) methodologies have made use of (and have further refined) Graph Theory to describe problems of project planning and optimization. Examples are travelling salesman, flow-shop, job-shop, PERT, GERT, CPM, MPM. Common to all these methodologies is the use of graphs (or nets) to describe activities and their interrelation (Noltemeier, 1975). Basically a graph is a set of nodes and edges, each edge connecting two nodes. The various methods differ in how they model activities and their interdependencies.

In Critical Path Method (CPM) graphs, nodes describe events (e.g. start, stop), while directed edges (forming arrows) describe actions and the interrelation (pre-

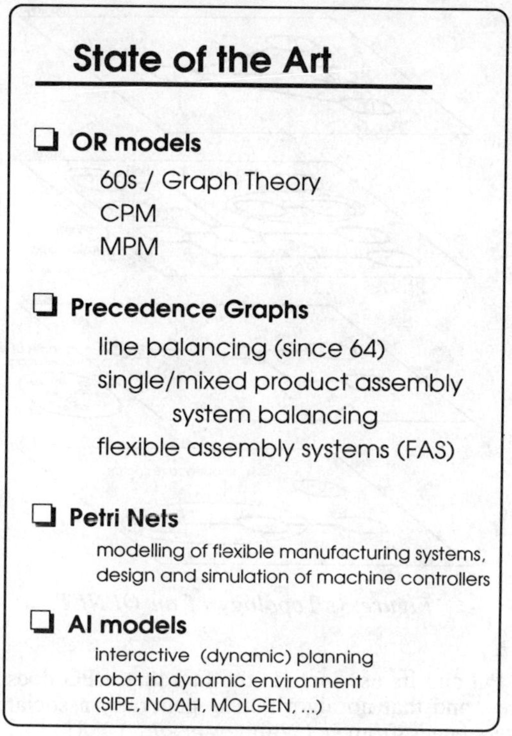

Figure 2. Some well-known task models.

cedence) between events. Edges are weighted to represent time or any other cost factor an action will consume. CPM is historically the oldest method; it is widely used but rather simple.

In comparison MPM (Metra Potential Method) is much more powerful in allowing expression of the coupling of concurrent actions, as well as the overlapping of successive actions. MPM models are focused on actions represented by nodes, while edges represent precedence relations.

Precedence graphs

Already in 1964 the problem of mapping assembly tasks into an assembly line was analysed and solved by using precedence graphs (PG) (or: precedence diagrams, as in Prenting and Battaglin, 1964). In a PG nodes represent tasks and edges represent precedence relations. In this context, PG basically are used to describe operations and operation precedence interrelations of a product assembly. PG and MPM graphs are very similar. Ammer (1985) has a discussion of similarities and differences.

Using PG to describe assembly tasks, the problem of line balancing can be stated in terms of homomorphical transformations of PG. Many authors have developed higher sophisticated balancing techniques, considering also the problem of mixed product assembly lines (see for example Görke, 1978; Schad, 1986; Ránky, 1986; also F. Fremerey, in this volume).

Critique of precedence graphs

PG have many shortcomings. In this chapter we will point out four problems, which stimulated our development of OPNETs:

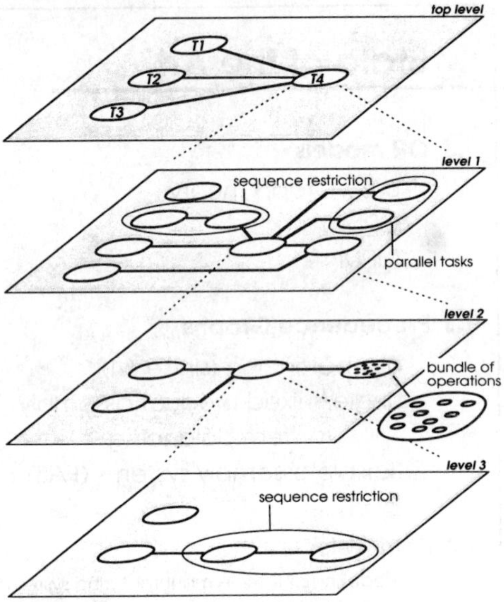

Figure 3. Topology of an OPNET.

(1) Shortcomings of PG and its use include the fact that a PG does not always contain all possible sequences, and that no algorithm is generally associated with either creating it or extracting sequences from it (Whitney *et al.*, 1988).
(2) A major problem is complexity. A great share of today's consumer products usually contain more than 100 components, which means that a PG may contain hundreds of activities. It is almost impossible for an engineer to master this complexity, if presented in an unstructured ('flat') traditional PG. There are two basic mechanisms to break down and master complexity: modularization (locality principle) and abstraction.
(3) Another problem of all well-known process and sequence planning methods using PG is that they cannot adopt to the level of detail of a planning process. It should be possible to use models at different planning stages, running from rough to fine. This requires PG with different, but compatible, levels of granularity.
(4) The model of PG is only capable of representing precedence relations between activities. In the real world there are more and different relations between activities and to other objects (e.g. equipment, worker, parts). Higher sophisticated planning methods should make use of such relations, using semantic networks to model them.

Operation networks

As we pointed out earlier, the variety of different sequences in which a product can be assembled is so tremendous that it is impossible for a planner to define them all explicitly. This is a result of the combinatorial nature of the problem (see e.g. Harary and Palmer, 1973). Therefore we recommend the definition of a more generic structure, out of which all valid sequences can be generated automatically. This structure is basically defined as a modular network incorporating different levels of abstraction. It should reflect only constraints set by the product itself. From an assembly planner's point of view such a structure is easier to describe.

Topology and Syntax of OPNETs

The topology of an OPNET graph is defined as follows (see Figure 3):

(1) An OPNET is a graph.
(2) Each OPNET graph consists of a set of labelled nodes and a set of labelled edges.

(3) Each node can be a node or an OPNET subgraph; if a node is another OPNET subgraph, a hierarchical structure will result, providing levels of abstraction.
(4) Each node, edge and (sub)graph is an object and has a set of attributes (e.g. text, values) attached that can be used as general repositories for information. Further, each object can have a logical string attached to it. With respect to a set of external controlled parameters, the string will define if an object is valid or not.
(5) Each edge connects two objects (nodes); it can be directed or non-directed. The graph build by all nodes and all directed edges will conform to PG syntax: it is finite, directed, not self-referring, has no parallel edges and is cycle-free (i.e. it is a digraph).
(6) Each object (node) also can have directed and non-directed relation(s) to objects of other models within a given planning system.

Semantics of OPNETs

The basic semantics of an OPNET are as follows:

(1) The OPNET graph of a product/subassembly describes all operations which have to be carried out to assemble the product/subassembly.
(2) Nodes and edges represent operations and relations between operations, respectively.
(3) An operation can be subdivided (decomposed) into sub-operations, forming a hierarchy of operations and OPNET subgraphs; the relation between an operation and its sub-operations is 'consists of'.
(4) Attribute sets attached to nodes and (sub)graphs can be used to store information about an operation, like standard time (value), task description (text), assembly relevant attributes (classification), etc. Logical strings define to which model in a mixed-model OPNET an operation belongs to. These strings can be derived from product configuration data.

    Edges can have (semantically) different meaning and importance. For example, a precedence relation can be the result of the physical (geometrical) situation of an assembly (due to the product design), but it can also be enforced by the assembly process (due to the production design). The reason for defining a precedence relation is stored and used by some of the planning methods. The non-directed edges are attributed with the type of relation they represent (IS-A relation). Logical strings define to which model in a mixed model OPNET a relation belongs to.
(5) A directed edge represents a sequence (precedence) relation between two operations. Non-directed edges represent other (non-precedence) relations between operations (internal relations; e.g. defining bundles of operations, concurrent tasks or sequence enforcement) or relations to objects of other models in the planning system (external relations; see below).
(6) Each object (node) can also have relation(s) to objects of other models within a given planning system, e.g. a relation between an operation and a tool (which is used), or, an operation and a subassembly (where the operation is carried out) (see Figure 4).

Generation of OPNETs

The assembly sequence (or more general, assembly structure) is determined by (1) the relational structure of a product/subassembly, (2) by geometry and (3) by the physical constraints of an assembly. Different attempts were made to generate possible assembly sequences from product description data.

Known generation methods

In the literature different assembly sequence generation methods are described. Here, we briefly sketch and describe three methods. Whitney *et al.* (1988) offer a modified Bourjault's method which generates all possible sequences. They give an example of an assembly of 11 parts, for which more than 41 000 sequences were generated. This took about 40 s computing time. They judge that their method is feasible for very simple products of 10–15 parts.

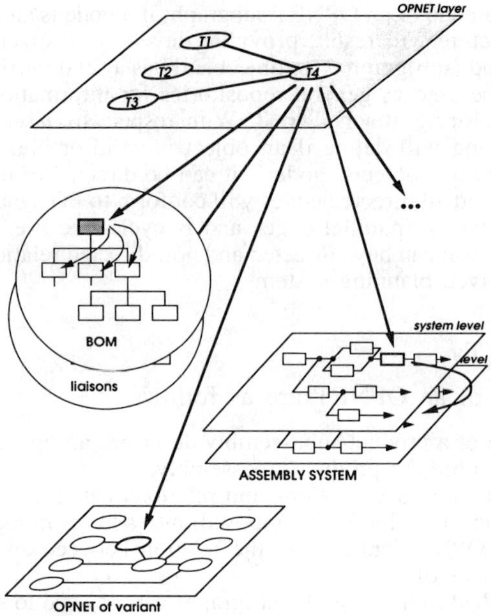

Figure 4. External relations of an OPNET.

Frommherz and Hornberger (1988) report research work about the use of PG for robot action planning. They argue that in each planning phase new precedence relations (PR) emerge. They give some rules (based on penetration, stability, side-effects) by which possible PR can be generated out of a CAD model. Their software system needs more than 48 min CPU on a VAX 8700 for the Cranfield Assembly Benchmark (17 parts).

Starting from a three-dimensional (3-D) model (EUCLID) of a product, Weule and Friedmann (1987) follow the approach of CAD data reduction, liaison analysis, interpretation of technological data and simulated disassembly to generate possible PG. There is no information given about the performance of their system.

Product design determines which assembly sequences are physically possible. In our project, the typical product size ranges between hundreds (HF amplifier) and ten-thousands of components (car final assembly) components. Even if products are broken down into subassemblies of significantly smaller size, none of the mentioned methodologies could handle the assembly sequence planning problem in acceptable time. Besides that, none of these methods provides mechanisms for breaking a product into subassemblies. Furthermore, it is still a fact, that only the minority of all products are stored in a CAD system in an adequate 3-D representation. In general, minimal support is given by the methods to master the complexity of a product and of its planning problem.

OPNET generation method

A method was developed to generate a first draft of an OPNET out of the product structure, which is presented to the system in form of a structured bill-of-material (BOM). The method translates the relational structure of the components directly into an OPNET structure. (Methods for direct reasoning from CAD data, simulation of disassembly in CAD, etc. are outside the scope of this project, see discussion below.) A more detailed OPNET structure can be generated using information about the set of liaisons assigned to each subassembly. The engineer can elaborate on

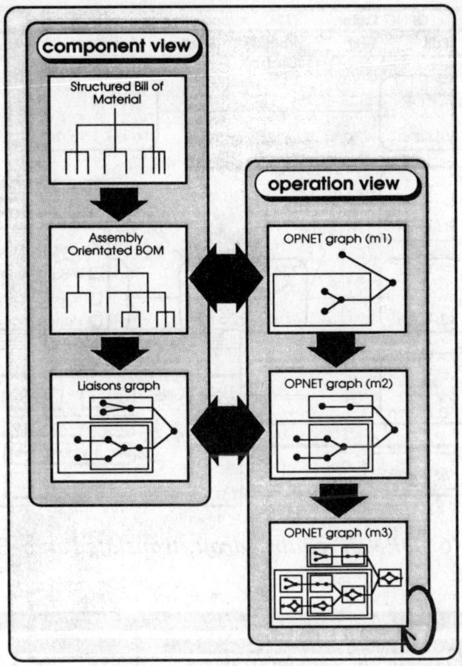

Figure 5. Product model.

this initial structure taking the production design process into consideration (see Figure 5).

In our methodology, a structured BOM and the corresponding OPNET are just two different views of our product model. Therefore relations between parts/subassemblies correspond to relations between operations/OPNET subgraphs. This conveys much power to methods for replanning and similarity planning.

Assessment of OPNETs

It is obvious that the choice of a good assembly sequence is vital for designing and running an economic assembly system. We have developed some heuristic formulas to supply the planner with indices, giving a rating how 'good' an operation structure under investigation is.

The assessment of an OPNET can be based both on structure and data of the elements of the graph. Depending on the level of description (elaboration) of a given OPNET, the formulas consider structure, standard times and assembly relevant attributes of each task. We have defined the following three indices: (1) nesting, (2) granularity and (3) weighted nesting. All three indices were defined with a mixed product assembly system balancing problem in mind. Figure 6 shows the draft of a prototype window layout (translated into English).

The rationale of the three heuristic indices is as follows:

(1) The assignment of tasks to assembly stations, robots or workers is heavily restricted by PRs between tasks. Each PR is a constraint and has a negative effect on the demand/supply mapping process. Parallel operation structures are generally preferred to more sequential ones. Accordingly, the nesting index gives an idea how 'sequential' an OPNET is. This index can be calculated even if no standard times are supplied.

(2) With a view to optimization heuristics used in balancing methodologies (e.g. largest candidate rule), an OPNET is more favourable, if the mean value of the standard times

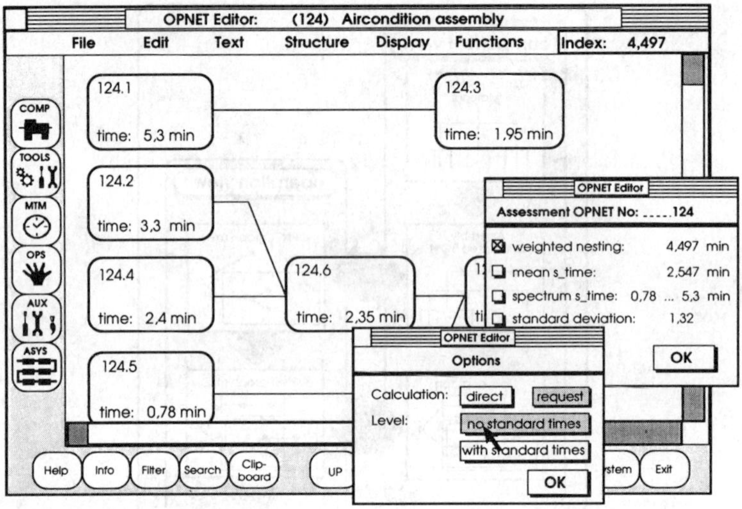

Figure 6. OPNET-editor (draft, translated into English).

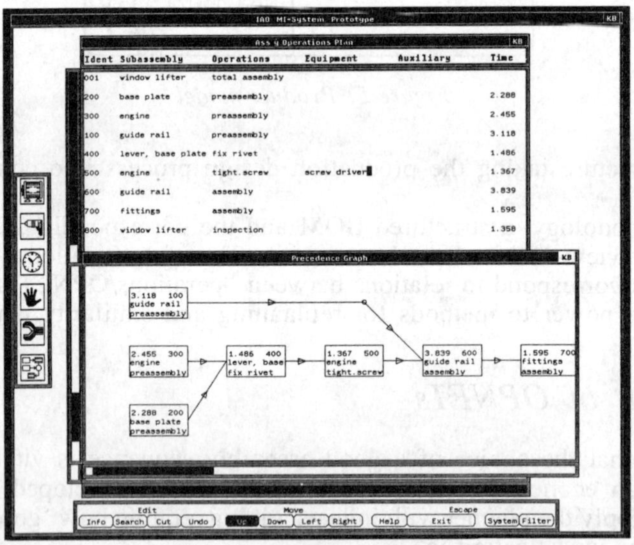

Figure 7. Interactive operation network editor (prototype).

is small and the deviation is high. The granularity indices give information about mean value, standard deviation and spectrum of standard times.

(3) The weighted nesting index is calculated similarly to index (1), but it takes all standard times into consideration. Thereby 'bottle-necks' in an OPNET structure will add an adequate penalty to the index.

OPNET editor and interfaces

For handling operation structures an Interactive Operation Network Editor was developed. It integrates methods and procedures for network generation, modification, queries and evaluation under a uniform graphical man machine interface.

The editor follows the object oriented approach applying direct manipulation techniques (see Figure 7). The OPNET editor is part of a large CAPP system which has also model-mix balancing and sequencing methods for flexible assembly systems. Well-defined interfaces are provided to product planning systems (CAD, product design) and assembly process planning systems (CAPP; assembly equipment databases; Methods-Time Measurement (MTM) modules, etc.).

Status of work and acknowledgements

Background to this research work was a project carried out with and supported by four German companies. In 1990, a software company in Germany will start marketing an Integrated Assembly Planning and Optimizing System (IMPOS), which incorporates a derivative of the OPNET editor described in this paper. A substantial part of our work was funded by the German Ministry for Research and Technology (BMFT) which is gratefully acknowledged.

References

Ammer, E.-D., 1985, *Rechnerunterstützte Planung von Montageablaufstrukturen für Erzeugnisse der Serienfertigung* [CAPP for series production goods], Doctoral Thesis, University of Stuttgart (Berlin: Springer).

Ammer, D., Dungs, K., Seidel, U.A. and Weller, B., 1986, *Systematische Montageplanung—Handbuch für die Praxis [A Systematic and Pragmatic Approach to Assembly Planning]*, edited by H.-J. Bullinger (München: Carl Hanser).

Frommherz, B. and Hornberger, J., 1988, Automatische Erzeugung von Vorranggraphen [Automatic Generation of Precedence Graphs]. *Technische Rundschau*, No. 24, pp. 54–61.

Glüer, D. and Schmidt, G., 1988, The application of Petri-net techniques in modelling, simulation and control design for flexible manufacturing systems. *Automatisierungstechnik* **36**, 12, 463–471.

Görke, M., 1978, Rechnerunterstütztes Verfahren zur Leistungsabstimmung von Mehrmodell-Montagesystemen [Computer-aided mixed product assembly line balancing], Doctoral Thesis, University of Stuttgart.

Harary, F. and Palmer, E.M., 1973, *Graphical Enumeration* (New York: Academic Press).

Noltemeier, H., 1976, *Graphentheorie mit Algorithmen und Anwendungen [Graph Theory with Algorithms and Applications]* (Berlin: Springer).

Peterson, J.L., 1981, *Petri Net Theory and the Modelling of Systems* (Englewood Cliffs, NJ: Prentice-Hall).

Prenting, T.O. and Battaglin, R.M., 1964, The precedence diagram: a tool for analysis in assembly line balancing. *Journal of Industrial Engineering*, **15**, 208–213.

Ránky, P.G., 1986, *Computer Integrated Manufacturing: An Introduction with Case Studies* (Englewood Cliffs, NJ: Prentice-Hall).

Schad, G., 1986, *Entwicklung und Einsatz eines interaktiven Verfahrens zur Leistungsabstimmung von Montagesystemen [Development and application of an interactive system for assembly line balancing]*, Doctoral Thesis, University of Stuttgart (Berlin: Springer).

Weule, H. and Friedmann, T., 1987, Rechnerunterstützte Produktanalyse in der Montageplanung [CA product analysis in assembly planning]. *VDI-Z*, **129**, 12, 59–63.

Whitney, D.E., De Fazio, T.L., Gustavson, R.E., Graves, S.C., Cooprider, K., Holmes, C.A., Klein, C.J., Lui, M. and Pappu, S., 1988, *Computer-aided Design of Flexible Assembly Systems*. Final Report, CSDL R-2033, The Charles Stark Draper Laboratory Inc., NSF Grant 84-17949.

63

Cause-based failure recovery approach for automated manufacturing cell components

Young B. Moon[1] and Colin L. Moodie[2]

[1]*Department of Mechanical and Aerospace Engineering, Syracuse University, Syracuse, NY 13244, and* [2]*Purdue University, West Lafayette, IN, USA*

Abstract

Failure recovery in automated manufacturing cells has been attempted through a simple repetition of the previous operation, a finite state table look-up, and rule-based expert systems; however, failure recovery requires various sources of knowledge as well as flexible and powerful control strategy. Furthermore, the cause of failure should be identified lest the same failure should occur repeatedly. This chapter describes a failure recovery approach adopting a multiple blackboard architecture. A cause-based recovery strategy is proposed to demonstrate the opportunistic control aspects of the approach. The system has been implemented in an object-oriented language, Smalltalk-80.

Introduction

Since simple mechanized production machines were devised about two centuries ago, continuous efforts have been made to control manufacturing components automatically. However, for each progressive step the components of the controlled system become more numerous and complex. Accordingly, the control system must have become more sophisticated to cope with this increased complexity. The control system now needs to be furnished with functions such as interpreting current circumstances, diagnosing problems, and recovering from failures as well as its original functions of monitoring and co-ordinating the components.

One of such advanced functions of the control system is the ability to deal with errors, failures and unexpected events in automated manufacturing environments. Approaches to deal with such failures can be categorized into two groups: preventive approach and recovery approach. In the preventive approach, attempts are made to preclude all the possible failures as well as to increase the reliability of the controlled system. On the other hand, necessary recovery actions are sought and taken after a failure is identified in the recovery approach.

Even though both types of approach should be adopted supplementally, complete elimination of errors or unexpected events is extremely difficult because of the dynamic and noisy nature of manufacturing environments. The environments are sometimes so sensitive that even small deviations can cause failures. Typical dust and metal chips on the shop floor may also create problems. Co-ordination between components can be another problem source. Moreover, a failure in a more reliable system can have more disastrous consequences than in a system with frequent breakdowns. Not only are preventive approaches necessary, but recovery approaches are essential.

Concepts of using blackboard architecture and various knowledge sources in failure recovery have been proposed before (Moon and Moodie, 1989). In the proposed framework, recovery operations in the proposed framework are accomplished via interactions of four blackboards: (1) for diagnosis, (2) for remedy, (3) for supporting actions, and (4) for control. Recovery actions are determined incrementally and opportunistically based on information residing in each knowledge source.

In this chapter, a recovery strategy adopted in the framework is described and discussed.

Cause-based failure recovery approach

Knowing the exact cause of a failure is sometimes essential for recovery activities (Srinivas, 1977, 1978). A typical example is a tool breakage problem. The apparent phenomenon is a tool breakage. A simple-minded remedy for the situation is to replace the broken tool by an undamaged one; however, the correction only does not solve the problem because the ultimate source of the failure may reside elsewhere. For example, the real source of the problem may be the wrong material selection. Unless the source is identified and corrected, the same failure may occur repeatedly. Therefore, the question is how to find out the ultimate cause and necessary remedial actions.

A typical approaches to find out the cause is to gather all the possible state information available and to analyse it. However, sometimes available information is not enough to determine any cause. Also, some of the gathered information is inaccurate. Nevertheless, more and better information can be acquired as we activate supporting equipment to gather further the necessary data. The problem is then to determine what kind of additional information we need.

The idea of cause-based recovery is to use some plausible causes as active knowledge sources within the framework. Instead of interpreting given evidence to find a cause, searching is carried out for the facts which endorse the likelihood of particular causes. Thus the searching area for new facts is limited to evidence satisfying certain plausible causes. During the recovery period, each cause knowledge source either suggests a set of facts supporting its consequences or becomes evaluated with additional evidence.

Causes and cause knowledge sources

Two important elements in the cause-based failure recovery are cause objects and cause knowledge sources. The cause object includes various information describing a particular cause while the cause knowledge source contains knowledge invoked by new events on blackboards.

Cause objects of a particular failure are classified according to the nature of the corresponding errors. That is, the causes are categorized into logic error, precondition error, constraint error, operation error and function error. Each cause is represented as a particular instance of one of the five cause classes (Goldberg and Robson, 1983). Three major pieces of information are attached to every cause: evaluation criteria for candidate causes, detection method of a cause, and knowledge sources for remedial actions.

Evaluation criteria

The evaluation criteria are used when there is more than one candidate plausible cause available at any stage during the reasoning process. Three criteria adopted are probability, cost and priority.

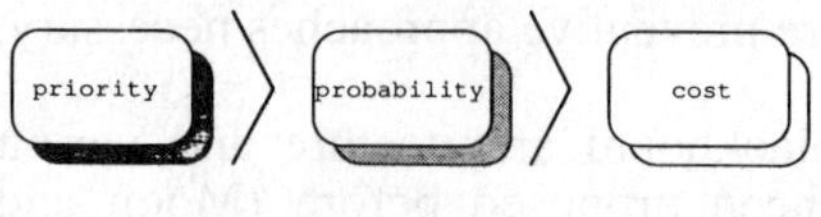

Figure 1. Criteria precedence.

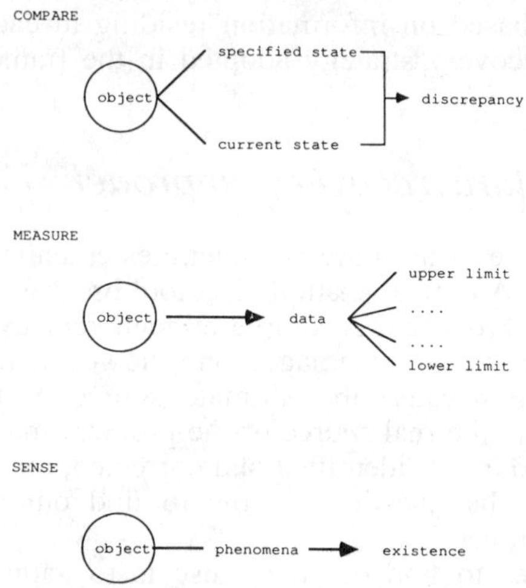

Figure 2. Ways of determining a failure cause.

A manufacturing cell is assumed to have been in operation for m time units, and there are n plausible causes for a failure. The number of failure occurrences due to a cause i during m time units is denoted as x_i. $x_1, x_2, \ldots, x_n$ are assumed to be independently distributed according to a Poisson distribution with parameters, $a_1 = \mu_1 m$, $a_2 = \mu_2 m$, ..., $a_n = \mu_n m$. Then, the maximum likelihood estimators of μ_is are x_i/m. The estimator for the probability of each cause was derived where the last failure of manufacturing cell operations occurred at time t_l, the current failure occurred at time t_c,

$$\exp\left(-(t_l - t_i^*)\frac{X_i}{m}\right) - \exp\left(-(t_c - t_i^*)\frac{X_i}{m}\right).$$

Using this estimator, it is reasonable to examine the plausible causes in order of the value of the estimated probability. This assertion was proved for cases of one or more causes involved (Moon, 1988).

The second evaluation criterion, cost, is used only when two or more plausible causes have the same probability. That is, the cost criterion plays the role of breaking a tie among the first selected causes.

Priority overrides both probability and cost. That is, causes with a non-zero priority are examined first before causes with only cost or probability are examined. Among causes with non-zero priority, ones with higher values are examined first.

The precedence relationships among the three criteria are expressed in Figure 1.

Detection method

Methods for determining the exact cause of a failure are grouped into three categories: compare, measure, and sense (Figure 2). The compare category

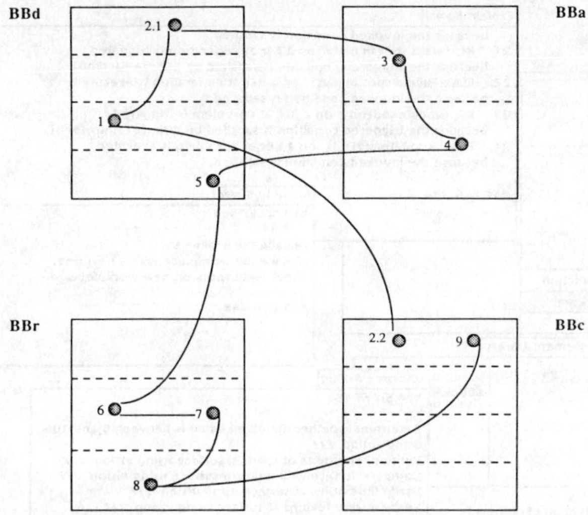

Figure 3. Basic operation for cause-based recovery: (1) beginning Bob, (2·1) hypothesis Bob, (2·2) 'PROBLEM', (3) search evidence, (4) search actions, (5) evidence, (6) remedy type, (7) search remedy, (8) remedy actions, (9) 'END'.

represents methods which compare the current state with the desirable state. For example, the cause 'selection of a wrong part' is determined by comparing the current specification of the part with the desired specification. The measure category represents such methods which measure aspects to determine that it is normal. For example, a dull tool is identified by measuring its wear. The criterion for this type is whether the measured data are within the desirable range. The sense category represents methods which sense objects in order to collect further information about the situation. For example, a vision system is used to find an oil spill on a part. The criterion for this type is whether a certain phenomenon is present or not.

Procedure of cause-based recovery

The recovery procedure consists of two important steps: finding causes and generating remedial actions. Once a failure phenomenon is posted on a blackboard, some cause knowledge sources are invoked. As a result of this, the same number of hypothesis objects as invoked knowledge sources is posted on a blackboard for selection. Each hypothesis is examined according to a suggested order until one or more causes are identified. Then, for the second step, desirable recovery actions are pursued depending on the content of remedial scheme of cause knowledge sources. A more detailed procedure is presented in Figure 3.

In Figure 3, small dots represent blackboard objects. The number which represents the order of an action appears near the resulting blackboard objects in the figure. The procedures can be explained as follows:

(1) IF there is a new object, and no other objects on any blackboard THEN post an object of 'Problem' on the blackboard for control, post a hypothetical object on the blackboard for remedy, from an invoked knowledge source.
(2) For the hypothetical object, search evidence.
(3) IF there is evidence satisfying the hypothetical object, THEN post an object requesting a set of remedial actions.
(4) For the selected object, search remedy.
(5) Execute the remedial actions, post an object of 'END' on the blackboard for supporting actions.

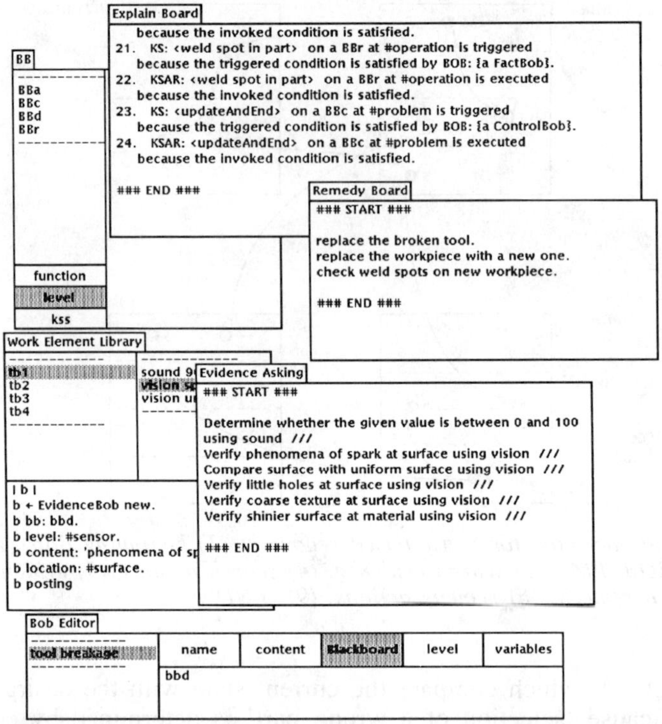

Figure 4. Various boards and editors.

Results

The proposed framework and the cause-based recovery approach were implemented in a prototype system developed in Smalltalk-80. The current implementation includes various editors as well as windows displaying intermediate and final results of failure recovery (Figure 4).

Test data were obtained from the results of doctoral work by Chaya Garg who proposed a knowledge elicitation framework, and obtained knowledge about tool breakage/change from the field workers (Chaya, 1987). Particularly, knowledge concerning tool breakage is adopted for demonstrating the cause-based failure recovery approach. The test results produced a reasonable diagnosis and remedy for 26 different causes of tool breakage (Moon, 1988).

Conclusion

The need for knowing the exact cause is realized by employing a cause-based recovery strategy. Several important issues are addressed in this strategy. First, it provides a way to check the results of events which have previously occurred. Causes are usually past events which occurred before the current failure has been detected; therefore, to trace and determine what previously happened is very important in failure recovery. The cause-based recovery addresses this problem by partially using evidence remaining from past events. Secondly, multiple causes can be identified. The proposed approach provides a better chance of identifying multiple causes than the approach of interpreting signals to decide causes. Theoretically, if an infinite, or a very large number of causes are used for the evaluation criterion, most involved causes will be disclosed. Thirdly, the strategy utilizes the relative frequency of each cause by means of statistical analysis. In this way, historical information of failures is also considered in the recovery procedure.

References

Garg, C.K., 1987, Development of a methodology for knowledge elicitation for building expert systems, Ph.D. Thesis, Purdue University.

Goldberg and Robson, 1983, *Smalltalk-80, the Language and its Implementation.* (Reading, MA: Addison-Wesley).

Moon, Y.B., 1988, A framework for failure recovery of automated manufacturing cell components. Ph.D. Thesis, Purdue University.

Moon, Y.B. and Moodie, C.L., 1989, A framework for failure recovery in a manufacturing cell. *International Journal of Advanced Manufacturing Technology*, May.

Srinivas, S., 1977, Error recovery in robot systems. Ph.D. Thesis, California Institute of Technology.

Srinivas, S., 1978, Error recovery in robots through failure reason analysis. *Proceedings of National Computer Conference*, pp. 275–282.

SECTION VI.3.
FLEXIBLE MANUFACTURING SYSTEMS

64

Determining and controlling the quality of prismatic parts machined in a flexible manufacturing cell environment

Graham T. Smith

School of Manufacturing and Mechanical Engineering, Birmingham University, Edgbaston, Birmingham, B15 2TT, UK

Abstract

In order to maintain the quality of prismatic parts machined in a flexible manufacturing cell (FMC), due attention must be given to the effects of cutting tool wear, thermal drift and machine tool calibration. If these factors have been overcome, then methods of part assessment by either in- or post-inspection needs consideration and depends upon a number of interrelated factors: part mix, cycle time, part geometry and so on. An area often overlooked, or at best superficially considered, is that of workpiece fixturing and a conflict of interest arises between accuracy of part location and fixturing flexibility—these are mutually exclusive conditions which afflict part set-up on turning and machining centres alike. Yet another consideration is that of condition monitoring during part production, through the judicious use of tool force monitoring and breakage detection, via adaptive control techniques.

If all of these monitoring methods, or at the very least some, are present then confident untended machining can be successfully exploited. Typical of this automatic part manufacturing technique can be gauged by the authors' experiences developing a prismatic rotational FMC at the Southampton Institute, where a fully industrial amalgamation of machine tools, sensing devices and other peripheral items are interfaced.

Introduction

Conflicting opinion about the failure or success of Flexible Manufacturing Systems (FMS) has recently been mentioned. On the one hand, there are reports that praise the merits of FMSs listing up to 85 distinct benefits to be gained from their implementation (Challis, 1988) with others equally unanimous in their support (Sugden and Antriss, 1988), whereas some observers tend to take a more pessimistic view of why companies are turning away from these large-scale FMSs (Dwyer, 1988; Childs, 1989). Taking a positive stance with regard to the successful operation of FMSs will be the view taken here. Bearing in mind the comments made recently, about; 'how European manufacturers are being forced to adopt FMS concepts to survive, as they provide a means of productive capability which is inherently flexible and in this manner will offset the greater penetration of our markets in the future by Far Eastern countries … as a manufacturer's future will shortly depend upon how successfully it can demonstrate product differentiation in the World Market' (Dwyer, 1988). Prior to discussion of methods of controlling and maintaining the quality of products machined in an FMS it might be worth establishing a working definition of these systems and cells. The report commissioned by the US. Department of Defense by the Manufacturing Technology Information Analysis Centre, defines a Flexible Machining System (which equates to an FMS), as simply; 'a manufacturing line that includes at least two machine tools that can change the shape or features of a machined part. These machine tools are linked together by a material handling device and are controlled by a computer' (Kochan, 1989). Most recently in Japan, there has been a positive move away from the large scale FMSs in favour of the cell concept (Childs, 1989) and the theme of quality and its control of prismatic parts using a group technology approach will set the scene for this chapter.

The problems of control of the quality in parts manufactured during untended machining in an FMC

It was observed (Smith, 1989 and others) that machine tools in the stand-alone or cell configuration when run in the minimally manned or untended condition with the absence of an operator, create a considerable number of problems that must be overcome if the machining system is to function satisfactorily. These problems arise in performing the monitoring and service functions which would normally be administered by the operator, who would normally: monitor the cutting tools' performance and condition, replacing any worn or defective tools by interrupting the cutting cycle; assessing the quality of the workpiece during machining; changing speeds and feeds if required; responding to any unusual conditions that are seen or heard during the cutting operation; measuring the component after machining and adjusting the machine tool if necessary. Whereas, in the unmanned situation, suitable monitoring systems must be installed to provide a degree of artificial intelligence necessary to mirror the experience of fully-skilled operators and their instinctive reactions, whilst providing the type of expertise associated with human involvement.

There is a considerable number of monitoring systems currently available that can be used in untended machining conditions and some of these will be discussed later in the chapter.

Whenever a machine tool is used in an FMC, or for that matter in a stand-alone mode under variable conditions of usage at, or near, its process capability (Cp < 1·1), which is frequently the case of late in precision engineering, then an uncompensated machine tool is prone to produce some scrap parts. If one considers the case of a vertical machining centre in the FMC which is subjected to infrequent usage, due to the part mix and in the main is machining bosses and holes from wrought material in the uncompensated condition, then the likely outcome is the effect shown in Figure 1.

A machine tool producing bosses by circular interpolation methods at or near the process capability of the machine tool [Figure 1(a)] when uncompensated for tool wear and thermal drift, will eventually produce scrap. This is caused by a combination of the lack of tool wear compensation and the machine tool's structure heating and cooling during and after the machining cycle. Expansion of the 'C-type' frame typically found in vertical machining centres is a recognized problem and companies such as KTM use thermal sensors in the bed, throat and head of the machine and after plotting the hot and cold temperature ranges typically found in these machines, the software in the control will compensate the machine tool by adjusting the *X*, *Y* and *Z* axes continously. In the case of Figure 1(a), the outside

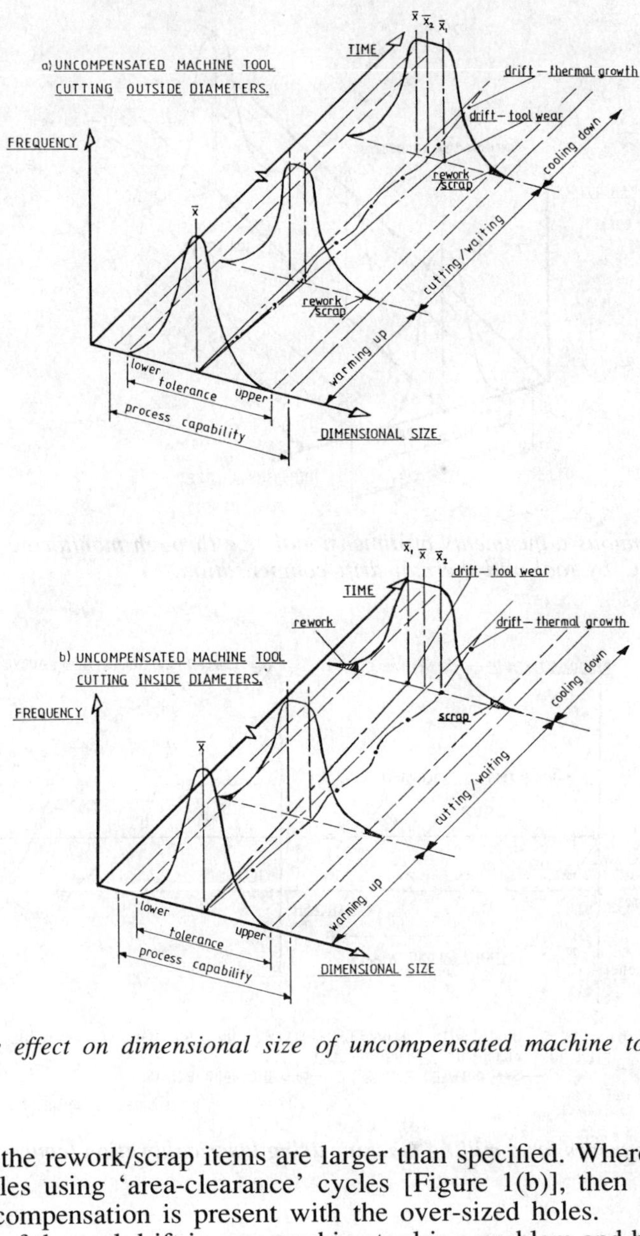

Figure 1. The effect on dimensional size of uncompensated machine tools in an FMC environment.

diameters of the rework/scrap items are larger than specified. Whereas, in the case of milled holes using 'area-clearance' cycles [Figure 1(b)], then true scrap will occur if no compensation is present with the over-sized holes.

The effect of thermal drift in any machine tool is a problem and by the judicious use of sensors placed at strategic positions on the machine it is possible to minimize thermal growth effects. The use of strain gauging techniques and tool wear compensation devices reduce scrap to virtually zero.

It is usual for the tool sensor to up-date the tool offset table in the machine's computer numerical control (CNC), whereas the thermal growth sensors can be used regularly to up-date the machine's fixture offsets periodically, in a similar fashion. Figure 2 shows just such a monitoring system in action; the effects of thermal drift are overcome and only tool wear becomes a problem, which is further reduced if 'sister tooling' is automatically selected once flank wear has reached the prescribed limit (normally VB 0·7 mm).

N.B. This thermal drift problem is based on a uniform thermal growth and is a rather simplistic view. In reality, thermal growth will affect one axis more than

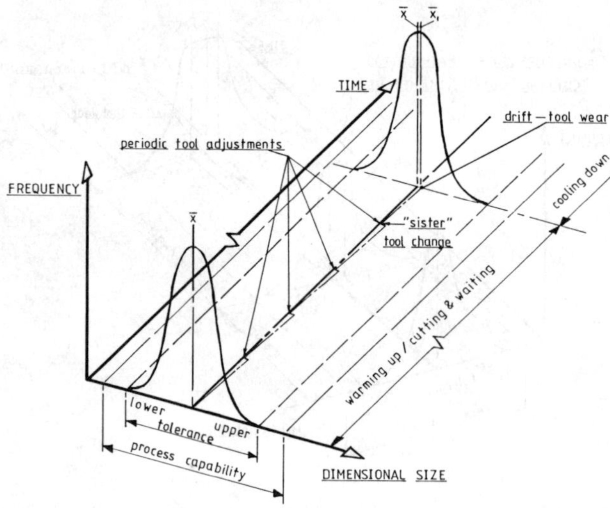

Figure 2. Continuous adjustments of dimensional size through monitoring of the machine tools in the FMC by tool and thermal drift compensation.

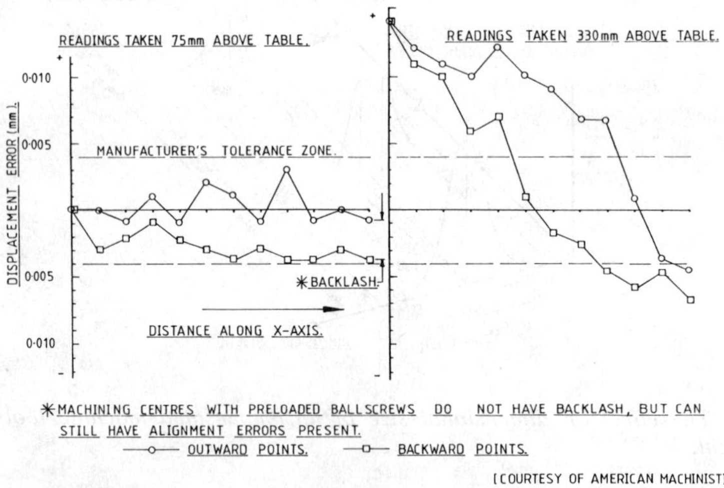

Figure 3. X-axis traces for a milling machine using laser calibration (courtesy of American Machinist*).*

another and cause the positional accuracy to be affected, rather than hole geometry and this is the reason why the fixture offsets are up-dated for the machine tool's thermal drift.

A problem which is always likely to be present is the effect of the machine drifting-out of its calibrated alignment. In a recent survey by the Hewlett Packard Co. (Christy, 1989), it was stated that most machine tool alignments are taken at a height of 75 mm above the table and under these conditions the machine tool is within the manufacturer's recommendations. However a different picture emerges (Figure 3) when readings are taken at 330 mm above the table and in most cases the machine tool is often well outside the tolerance zone recommended by the manufacturers. This is unfortunately the position where workpiece fixturing is usually situated and means that even on a new machine tool scrap can be present.

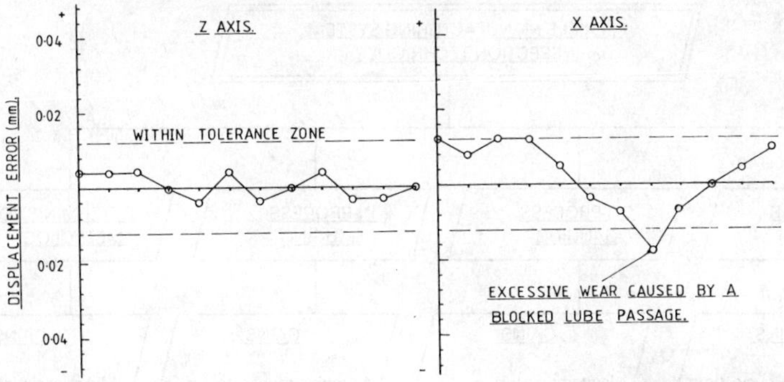

Figure 4. A machine tool after one year's use (courtesy of American Machinist*).*

In the Hewlett Packard survey mentioned earlier, their engineers found that 88% of the machine tools tested were out-of-specification upon arrival at the company's premises (Christy, 1989). Even after re-calibration on-site, a year after alignment testing, many of the machine tools were producing results that were out-of-tolerance on some axes when assessed again by laser calibration methods (Figure 4).

This problem of calibrating machine tools periodically is not the norm and one wonders how many machines in an FMC or otherwise, produce poor part quality or scrap, simply because they need re-calibration? In fact after alignment testing, certain parts of the table and axes can be well within the manufacturer's tolerance zone and are euphemistically known as its 'sweet spot', and high accuracy workpieces are often machined in these regions (Christy, 1989). This technique is practically understood by metrologists when using making-off tables, that are known to have high accuracy regions on their surface, for the measurement of high quality parts, once these tables have been calibrated in laboratory.

Workpiece monitoring through gauging and inspection techniques in an FMC

The quality of parts produced in an FMC environment can be enhanced by a variety of workpieces gauging and inspection methods and it is usual for any system to have one, or more of these methods present (Figure 5).

One of the most popular methods is to use touch-trigger probing to interrogate the workpiece and if discrepancies arise then the fixture offsets are up-dated after a 'canned' probing cycle has finished. In a similar fashion the tools can also be assessed for tool wear using the tool-setting probe which is also based on this omni-directional switch technique and the tool offsets are 'refreshed' as frequently as the programmer feels that up-dating is necessary.

The problem with any on-process gauging technique is it uses up valuable cycle-time and unless a turning operation is the requirement, then the need for in-process gauging methods is somewhat limiting. This is particularly relevant in machining centre's case where the use of vision-based systems cannot be readily accommodated (Smith, 1989).

The correction of workpiece errors is usually attained by the post- or near-inspection methods, as they overcome the problems of utilizing cycle-time, yet allow through a closed-loop measurement function, the up-dating of tool/fixture offsets. This technique is historical in its processing of part data and obviously occurs after a component or batch is completed, where it usually samples parts on

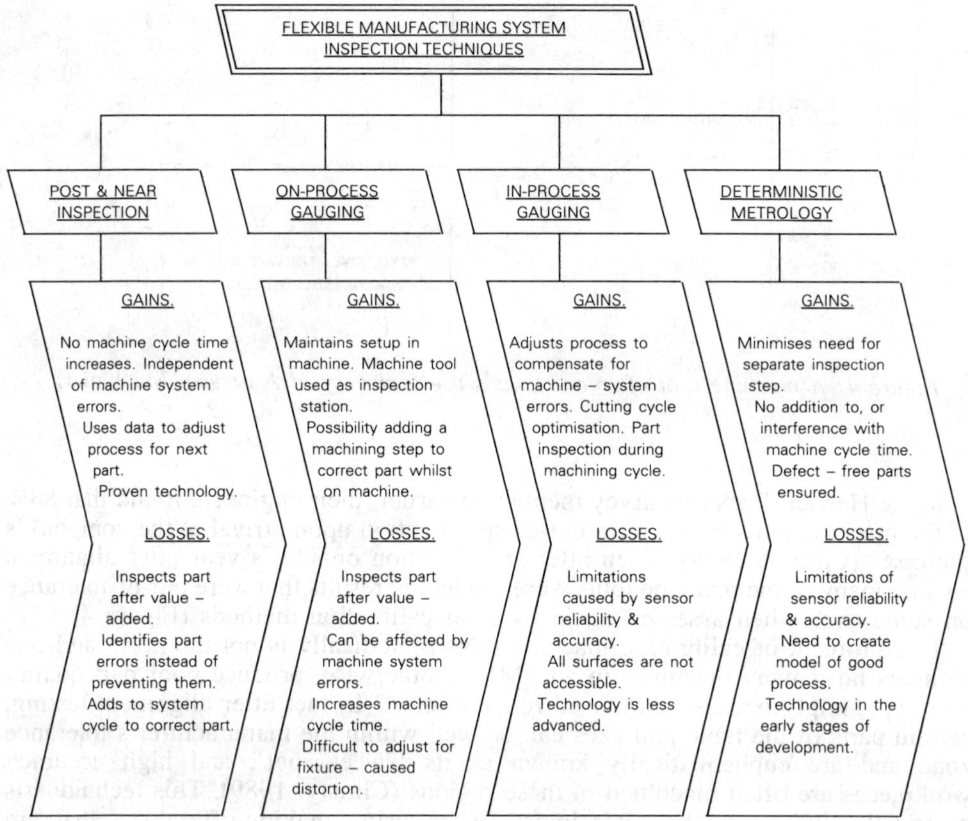

Figure 5. Techniques applied to FMSs to maintain the quality of parts by different metrological methods and their advantages and limitations (courtesy of the FMS Magazine*).*

a statistical basis. Typical equipment used includes receiver gauges with some form of closed-loop facility where up-dates based on Statistical Process Control (SPC) techniques occur, or by using CNC Coordinate Measuring Machines (CMM) with suitable interfaces to the machine tools CNCs.

Recently, much talk has been centred around the Deterministic Metrology techniques (Smith, 1989, and others), whereby error prediction and corrective action are developed and the idea is to try and anticipate machining errors in the FMC in 'real-time', then correct them. The thoughts behind this newly developing method, are to cut a good part and so eliminate the need for further inspection. Thus a detailed and accurate mathematical model is required which can be developed to show the error-producing parameters and their effects and interactions during machining of the part. The method is still in the experimental stage and requires further refinement before it can be used effectively in a production environment.

The problem with most of these metrology applications for assessment of parts produced in FMCs, is that the component is usually measured at conditions of the ambient temperature, humidity and to a lesser extent atmospheric pressure causing unacceptable variations. This means that part quality will suffer according to the degree of variation of the measurement parameter. One method which is becoming popular lately is to use the 'footprinting' technique (Smith, 1989, and others) as a means of overcoming these measuring inconsistencies. In simple terms, the part measured by the inspection device is noted for any errors present and then taken to the metrology laboratory and assessed for any part dimensional variations. The

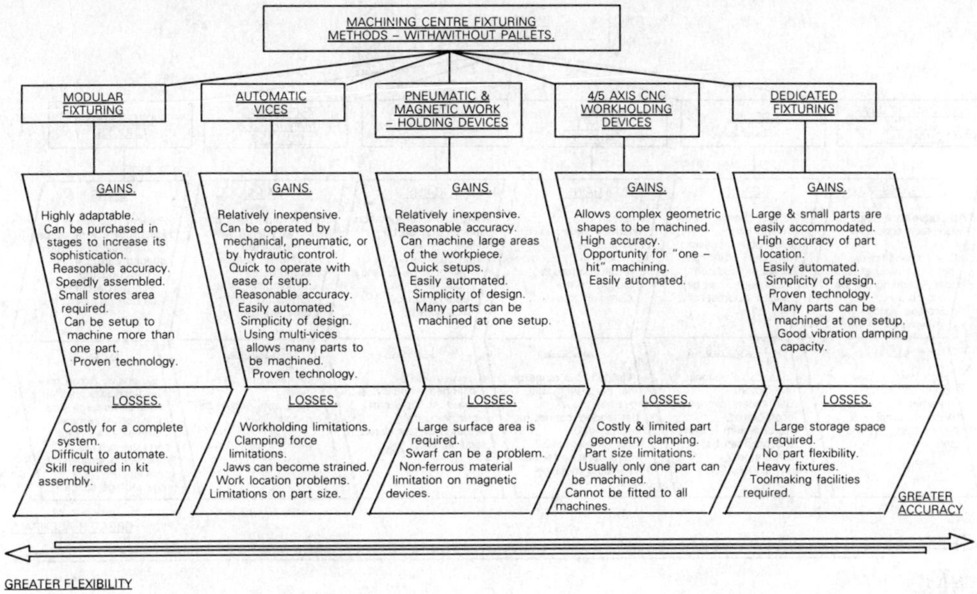

Figure 6. Workpiece fixturing for machining centres and their advantages and limitations.

variation between the cell inspection device and the metrologicial measurement in the laboratory is determined, then the inspection equipment in the FMC is up-dated with the data from the metrology tests and in this manner the fluctuations in metrological environment are overcome. The 'footprinting' is assessed periodically, based on a statistical model and in this manner frequent up-dating occurs.

Its main limitations lie in the fact that extra time and manpower are required to maintain this technique and data gathered will always be historical—assuming of course, that the metrology equipment within the FMC cannot have its own environmental stabilization present, in the form of an automatically enclosed temperature, humidity and pressure room.

Workpiece fixturing methods on machining centres

Many companies of late have a vast array of workholding equipment available to aid them in increasing productive capabilities within an FMC (Figure 6). The choice of equipment used in most cases is dependent upon the likely batch sizes, their physical dimensions, adaptability, but, strangely, rarely on accuracy. It is often the case, that part inconsistencies are caused by inadequate clamping and workholding methods, rather than machining problems or material variations. If a company requires high accuracy parts regardless of the batch size, then the only way to ensure the tolerances are held on the part is by using dedicated fixtures. The problems of restraint and location will always be there with the adaptable fixturing methods, whereas the more permanent workholding methods offer greater part security and dampen vibrations caused by the cutting action and hence quality of the part is assured.

Workholding methods on turning centres

What was said about the machining centres, is equally true in the case of part location and restraint in turning centres (Figure 7). There is a distinct trade-off

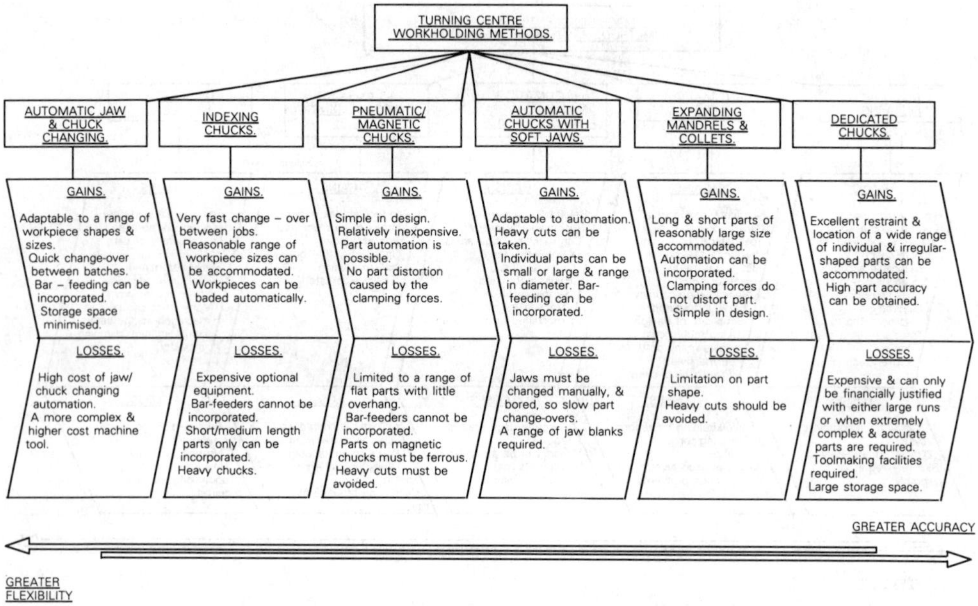

Figure 7. Workholding methods for turning centres and their advantages and limitations.

between workholding flexibility and the accuracy of the part produced as Figure 7 shows. Once again, this only becomes a problem when workpiece tolerances approach the process capability of the machine tool and under such conditions the only alternative is to use dedicated chucks, and similar methods regardless of the batch size.

Tool condition monitoring in an FMC

Given that the workpiece is constrained and located precisely in its workholding device, then the other major variable in the satisfactory production of a part, is the efficiency of the cutting process. Smith (1989) states that part quality is affected by many fixed and variable cutting tool parameters, when machining components, assuming that the most efficient cutting tool types and grades have been selected for the manufacturing processes within the FMC—which is often not the case!

The tool's edge condition must be monitored in some manner in order to ascertain that machining efficiency is not sacrificed. There are a number of ways in which the edge condition can be assessed and these range from simple inexpensive tool breakage proximity sensors, to highly sophisticated Acoustic Emission (AE) equipment of considerable expense which is available. In between these two extremes, lies a whole host of tool health monitoring equipment such as touch-trigger tool probes, feed-force sensors (usually of the strain gauge type situated on the turret), or ballscrew nut assembly, etc; others include adaptive control techniques, typified by the torque-controlled machining (TCM) method. These vastly differing tool monitoring methods are present simply to improve and in many cases extend the efficient cutting life of the tools, while optimizing the cutting parameters of feed or speed, in the case of adaptive control techniques. If the edge condition in the untended machining state can be maintained and extended over predictable tool life, then high quality workpieces will result.

In most European companies, the general trend is to optimize the cutting conditions, so that feeds and speeds are of optimistic rates—approaching 100% tool loading, which gives a shorter tool life, but a greater productive throughput of parts. They rely on large tool libraries with 'sister' tooling being present, or in certain cases where much used cutters occur across a range of parts, triplicating these tools. The alternative approach to this tool problem is that typified by a Japanese company, Mazak Machine Tools, Worcester. The tooling philosophy adopted here is to run the cutters at between 70 and 80% of the optimum rate. The company argues that although productive capacity is affected the edge condition is maintained considerably longer and part quality is improved in their large-scale FMS. This quality improvement is brought about by a number of factors: cutting edge quality is predictable through lighter cuts and part quality is improved by less abusive cutting regimes which would affect the surface integrity (Smith, 1989); dimensional characteristics such as the geometric tolerances are held more consistently as less stress is present in the dedicated fixtures; less down-time of each machine tool occurs caused by tool crashes/breakage. These factors and efficient work-in-progress methods actually increase the throughput of parts while simultaneously improving part quality, making subsequent re-work and scrap almost non-existent. There must surely be a moral here for a more prudent technique of the optimization of cutting parameters, based not on productive throughput of parts, but on increased quality through consistent cutting conditions coupled to improvements in process planning methods.

A typical FMC using some workpiece and tool monitoring systems for prismatic part production

The FMC shown in Figure 8, which has just been completed and is in its first phase of development, has been designed by Cincinnati Milacron Company and the staff at Southampton Institute. It comprises well-proven FMC technology and can machine up to 64 prismatic parts in a variety of batch sizes at any time, using a group technology approach.

The limitation in the number of parts is caused by the CNCs on each machine tool having only 32 possible part-program identifiers at any one time; others however can be selected once the identifier for the part is free to be swapped to another part-program.

The robot can be scheduled to carry parts to any machine in the cell and has a self-centering double gripper with offset jaws and proximity work sensors. The turning centre has Krupp Widia Widiaflex modular quick-change tooling with a Reinshaw touch-trigger probing for workpiece sensing. The machining centre also has present the Widiaflex quick-change modular tooling and a centreline multi-spindle drilling head with tooling protected by adaptive control through torque controlled machining sensors. The chuck and fixture for turning and machining centres, respectively, are pneumatic and hydraulic in operation—catering for a range of prismatic parts.

The cell controller can schedule parts onto the CNC CMM which has software enabling it to 'soft-scan' the parts using a motorized touch-trigger probe, which can up-date the tool's offsets through suitable software and feedback loop. At any time, the robot can be called to check the part quality on the CMM which allows for a degree of manual intervention into the untended machining cell.

Tools are pre-set using an optical pre-setting device which allows inserts to be correctly seated and tool coordinates to be up-dated manually, or automatically in each machine tool. Thus part quality can be controlled from the tool pre-setting stage, then through the on-machine gauging facilities and lastly the post-process gauging assesses the final part quality, prior to removal from the cell.

As only two machine tools—the turning and machining centres—play an actual role in producing parts, all the peripheral equipment simply allows the system to

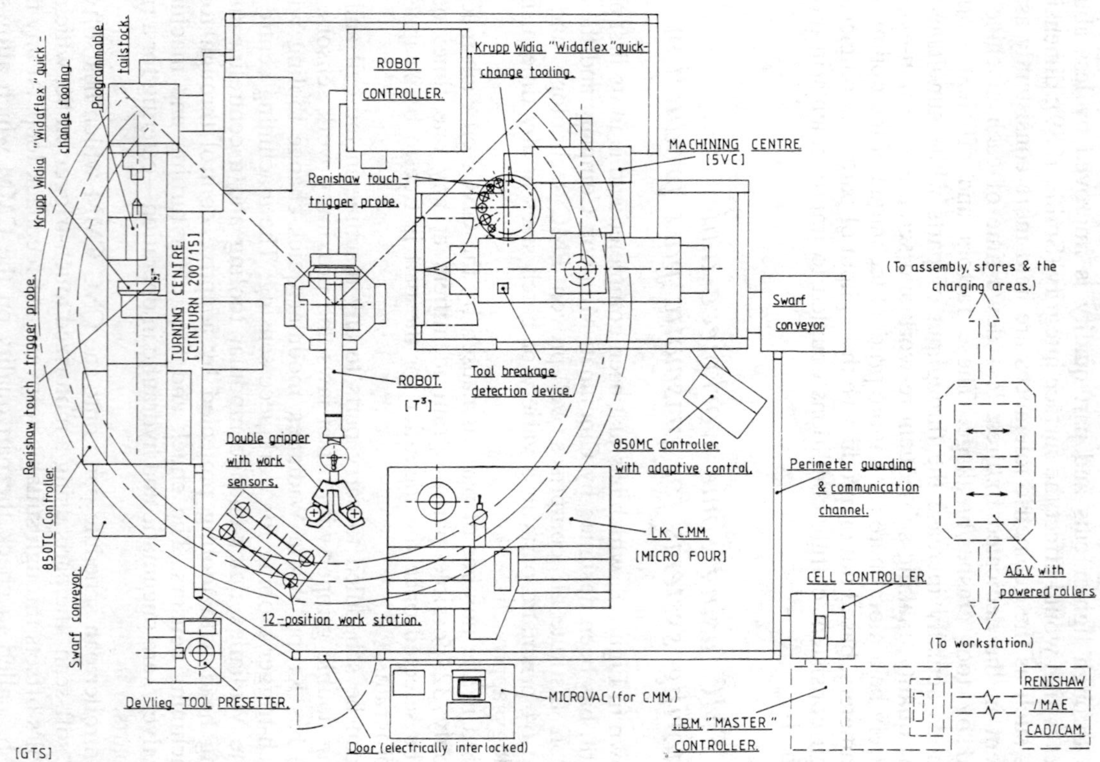

Figure 8. Southampton Institute's FMC (Cicinnati Milcron Equipment) Phase One. Phase Two shown by dotted lines (now under construction); Phase Three not shown.

manufacture parts of consistent quality by maintaining feedback loops to these machine tools. The robot is present simply to transport the components between each process. These are the vast indirect costs, but are necessary extra items required for unmanned machining.

Concluding remarks concerning Southampton's FMC and some experiences worth consideration

When designing and commissioning complex equipment such as an FMC of industrial proportions, it is often the case that success or failure of such equipment is not caused by the technological problems of 'interface hand-shaking' exercises, but by more mundane and easy-to-remedy problems which should have been considered at the outset! As such, these everyday overlooked considerations will at best only impede the FMCs satisfactory operation and at worst, cause real problems during its subsequent use. Typical of such minor 'considerations' are:

(i) Will the machinery go through external workshop doors satisfactorily?
(ii) Is the air supply capable of coping with the high demands of air volume? *Note*: this is not the same as pressure and may require considerable further expenditure to overcome!
(iii) Is the electrical power supply of the shop able to withstand the demands of this equipment, in addition to the resident machine tools?
(iv) Will the floor-loading cope with the high loads imposed over relatively small surface areas? If not, then at best frequent machine recalibration will be necessary and at worst, new foundations are needed.
(v) Coolant supply is a major problem and careful monitoring of its condition and replenishment is necessary, if the cutting tools are to operate satisfactorily during intended use.
(vi) Can access be easily attained to each element within the machine tool, for preventative maintenance and error recovery?
(vii) Are the technical staff sufficiently trained in all aspects of machine tool operation and programming, but far more important, error recovery in 'automatic mode' of operation? This is very significant as errors in-cycle can take considerable time and experience to overcome successfully—this training is a high indirect cost!
(viii) Can the organization support the use of such an FMC and financially justify the day-to-day consumable cost of running the system?
(ix) If the answer to (viii) is 'No', then can this equipment be utilized in a 'stand-alone' mode, or will taking one machine tool out of commission reduce/limit its use?
(x) Is it possible to expand the FMC to a full system, by incorporating peripheral devices, such as tool pre-setters and CAD/CAM stations, or will the workshop environment limit its use through lack of temperature control (for measuring equipment), excessive vibrations influencing other equipment in the vicinity and other such considerations.

If these points are considered early on in the development of such highly sophisticated equipment, then it is more than likely that a successful FMC implementation will result. Failure to consider them means that the system will not perform to its intended specification and 'loss of face' results for those in intimate association with the project!

References

Challis, H., 1988, Applying flexible manufacturing systems. *Flexible Manufacturing*, September, 25–29.

Childs, J.J., 1989, Japan moves away from large-scale FMS. *The FMS Magazine*, January, 41–43.

Christy, D., 1989, What a laser can say about your machine tool. *American Machinist*, January, 53–54.
Dwyer, J., 1988, Failure to implement FMS. *Automation*, December/January, 21–22.
Kochan, A., 1989, Flow FMS. *Automation*, September, 43–44.
Smith, G.T., 1989, *Advanced Machining—The Handbook of Cutting Technology* (IFS).
Sugden, P., 1988, The rise, fall and re-birth of FMS. *Automation*, September, 43–44.

65

A flexible system performance analysis in probabilistic and degraded machining conditions

L. Zavanella, G.C. Maccarini and A. Bugini
Universita' degli Studi di Brescia, Facolta' di Ingegneria Meccanica, Via Valotti 9, 25060 Brescia, Italy

Abstract

In the area of manufacturing plants, the improvement of flexible systems and the availability of computerized control instruments have determined an increasing interest in the optimal utilization of the resources such as the optimal number of items of each tool family population i involved in the workpiece (wp) production (Pereira and Carrie, 1987; Zavanella and Gentili, 1988). The authors have for several years taken an interest in this subject achieving some interesting results (Maccarini *et al.*, 1987a, b, 1988; Zavanella and Gentili, 1988; Zavanella *et al.*, 1988).

Now a further improvement can be developed by considering the tool life as a stochastic expression (Ramalingam and Watson, 1977); for this purpose the model of a flexible manufacturing system (FMS) (four identical workstations machining four different wps) drawing the necessary tools (six per wp) from the same tool populations ($1 \leq i \leq 10$) will be considered (Figure 2 and Appendix).

Tool life and machining times

Notation

- T_i useful life of tool type i,
- $t_{i,j}$ cutting time of tool type i when machining a type j workpiece,
- $f_i(t)$ probability density function of tool type i,
- $R_i(\alpha)$ desired reliability, i.e. the reliability value set at the planning stage, when considering $f_i(t)$, and associated with a $t_{\alpha,i}$ value for the examined tool population by means of eqn (2),
- $R_{an,i}(\alpha)$ analytical reliability, i.e. the reliability value obtained by means of an analytical procedure,
- $R_{a,i}(\alpha)$ actual reliability, i.e. the reliability value obtained by means of a simulative procedure,
- $T_{\text{rep},i}$ established replacement time for tools of family i,
- $T_{\text{mach},i}$ recorded machining time of each tool belonging to family i.

In a stochastic model each tool can be described by some time values:

(1) T_i which is the true tool type i life depending on its stochastic distribution, if no previous tool replacement takes place;

(2) $T_{rep,i}$ which is the imposed (calculated) tool change time as the reliability $R_i(\alpha)$ has been reached; due to the flexible conditions, the tool faces the machining of a random wp arrival and may reach its $T_{rep,i}$ while cutting.

On the other hand the tool life check is executed after the chip removal cycle (it is useless to fulfil this request earlier because of the stochastic nature of the tool life): this fact suggested the testing of different tool replacement policies to the authors (see next section).

The final machining time $T_{mach,i}$ is calculated as

$$T_{mach,i} = \sum_{k=1}^{N} t_{i,j_k} \text{ where } N = \text{number of jobs performed on the } j_k\text{th workpiece} \quad (1)$$

and is associated with a reliability level $R_i(T_{mach,i})$; in our flexible environment the $T_{mach,i}$ value is a function of the possible combinations with repetition of the machining times and, furthermore, should be as close as possible to the established $t_{\alpha,i}$ value, in order to guarantee by means of the following equation:

$$\int_0^{t_{\alpha,i}} f_i(t)\mathrm{d}t = 1 - R_i(\alpha) \quad (2)$$

the desired $R_i(\alpha)$ reliability for each tool population i.

Tool replacement policies

The outlined procedure leads to the analytical evaluation of the reliability of each tool family i as a function of the adopted tool replacement policy and a suitable security margin could be the appropriate choice of one of the following three strategies:

(a) replacement of tool type i when $T_{mach,i} \geq (t_{\alpha,i} - T_{max,i})$, where $T_{max,i}$ is the estimated maximum machining time for tool type i: the remaining tool life guarantees the longest machining job to be performed at a reliability level equal or superior to the one desired;
(b) replacement of tool type i when $T_{mach,i} \geq (t_{\alpha,i} - T_{min,i})$, where $T_{min,i}$ is the estimated minimum machining time for tool type i;
(c) tool replacement when $T_{mach,i} \geq t_{\alpha,i}$ (case c).

The analytical reliability

The flow diagram of Figure 1 illustrates the adopted procedure for the calculation of the 'analytical' reliability, where the need to associate a weight to each wp succession probability (a function of the production mix composition) is evident.

The estimated reliability is based on the data input of the $f_i(t)$ parameters (normal distribution) and on the $T_{rep,i}$ values. Furthermore, the following data input is requested:

(a) $t_{i,j}$ values (Table 1): the jobs are performed on different surfaces of the same wp, so that an unforeseen breakage does not affect the subsequent operations;
(b) productivity mix: the percentage of each wp type j over the total equals its probability entry to the line (random arrival).

Once the initial value of the variables JJ (counter) and TOTAL-COMBINATION (total number of combinations examined at the previous stage) have been started, the true calculus begins.

For each combination examined the total machining times are known and, consequently, the relative $t_{i,j}$ figure is added to each of these values, thus obtaining

```
                         LEGEND
WRKPC_FAM      ⇒  Workpiece family (index j)
TOT_WRKPC_FAM  ⇒  Global number of workpiece families
AVG_R          ⇒  Analytical reliability (average reliability)
M.ED_TIME      ⇒  Utilized tool life

For each tool family:
1.  Read the data sets of:
  a) wp machining times t_i,j (M.IN_TIME)   c) tool pdfs
  b) wp entry probability (PROBIN)          d) T_rep,i values (REPL_TIME)
2.  Set the initial value of:
  a) JJ=0                      c) AVG_R=0                 e) OLD_PROB(1)=1
  b) TOTAL_COMBINATIONs=1      d) OLD_M.ED_TIME(1)=0
3. | For I=1,2,...,TOTAL_COMBINATIONs
   |  1. | For WRKPC_FAM=1,2,...,TOT_WRKPC_FAM
   |     |   JJ=JJ+1
   |     |   M.ED_TIME(JJ) = OLD_M.ED_TIME(I) + M.ING_TIME(WRKPC_FAM)
   |     |   PROBABILITY(JJ) = OLD_PROBABILITY(I) · PROB_IN(WRKPC_FAM)
   |     |   IF M.ED_TIME(JJ) > REPL_TIME
   |     |       AVG_R = AVG_R + PROBABILITY(JJ)·[ 1 - ∫_0^M.ED_TIME f_i(t)·dt ]
   |     |       JJ=JJ-1
   |     |   END IF
   |     |______
   |__________

    IF JJ > 0
        TOTAL_COMBINATIONs=JJ
      | For I=1,2,...,JJ
      |     OLD_M.ED_TIME(I) = M.ED_TIME(I)
      |     OLD_PROBABILITY(I) = PROBABILITY(I)
      |______
        GOTO  3
    END IF
4.  WRITE  AVG_R
5.  STOP
```

Figure 1. Flow diagram.

Table 1. Tool family parameters and relative machining times.

Tool type	Machine times (min) Workpiece types				Probability density function character (min)		Handling time	Pre-set time	Check time	Total no.
	1	2	3	4	AVG	RMS				
1	7	0	13	9	60	20	12	8	1·5	39
2	0	0	27	0	120	40	16	2	0·5	21
3	0	17	0	10	80	26	17	3	0·5	21
4	15	17	0	0	80	26	10	7	1·3	24
5	0	6	0	17	80	26	10	5	1·0	39
6	9	13	0	17	80	26	3	5	1·0	21
7	27	17	15	0	120	40	4	3	0·5	24
8	7	0	5	4	36	12	7	3	0·5	63
9	0	4	3	0	36	12	4	3	0·5	27
10	14	0	5	14	60	20	13	6	0·9	33

a new combination set equal to the previous one multiplied by the number of wp types workable by the tool. The probability of a certain resulting combination is finally calculated by multiplying the likelihood relative to the old combination by the probability of arrival of each new wp type.

At this point the computed $T_{\mathrm{mach},i}$ values are checked: if these rise above the imposed limits, the reliability factor is calculated relative to the total time obtained: the program terminates when all possible combinations have exceeded the imposed time limit $T_{\mathrm{rep},i}$. The analytical reliability is the sum of all the reliability factors previously found, weighted against the probability factor associated with them: Table 2 shows the results obtained by the adopted procedure and compares them with the data obtained by the simulation technique.

Table 2. Comparison between the desired reliability and the values obtained by means of the analytical (A) and the simulative (S) procedures (average and actual reliability).

Desired reliability	Policy (a)		Policy (b)		Policy (c)	
	A	S	A	S	A	S
99·0	99·4	99·4	99·2	99·3	98·1	98·0
95·0	97·6	97·0	95·7	94·5	91·7	90·3
90·0	94·9	94·4	91·4	90·9	84·9	84·0
70·0	80·1	80·3	72·7	72·4	59·9	59·2
50·0	64·6	63·2	54·5	53·1	41·0	40·6
30·0	43·6	44·1	34·3	34·8	22·5	21·5

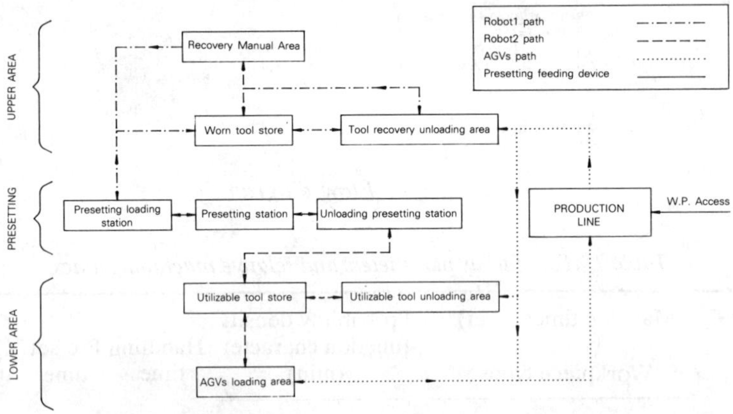

Figure 2. Plant layout and connection network. Transporters: two robots (1 and 2) and two AGVs, both six-tool size. Manual tool recovery area: two operators resharpening or replacing tools. Pre-setting area: two stations each fed by a 48-capacity two-way wheel. Production managed by means of tool-kit strategy (Zavanella et al., 1988).

The simulations

A model of the production line and of the tool management area was implemented on a PC (Figure 2), in order to allow the calculation of the actual reliability, referring to the machining of 1000 wps (250 per type randomly created): the resulting value is calculated as:

$$R_{a,i}(\alpha) = 1 - \frac{N_{\text{broken tools } i}}{N_{\text{replaced tools } i}}. \tag{3}$$

The results in Table 2 underline the good matching of the two procedures. The true effectiveness of strategies (a), (b), and (c) may be evaluated by comparing the total machining time for the four wp types and N entities successfully machined of the 1000 entered: any evident difference arises for the final machining time when $R(\alpha) \leq 0{\cdot}95$ (Figure 3). The figures quoted refer to the behaviour of the entire tool pool (312 entities), being the optimal weight of each tool family previously determined (Zavanella and Gentili, 1988): the comparison between the three policies' data shows a higher number of broken tools for policy (c) (Figure 4), with

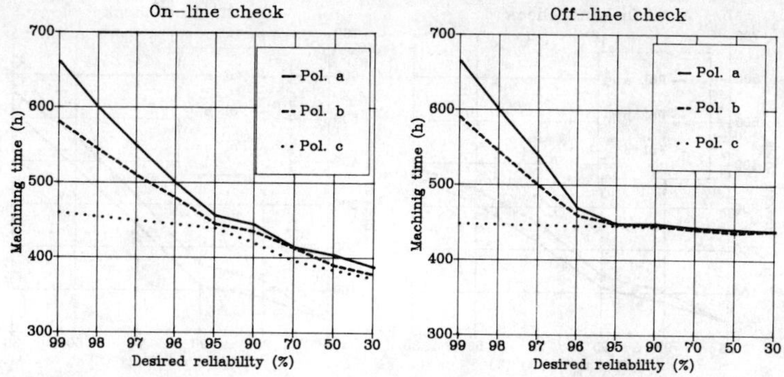

Figure 3. Final machining time (1000 workpieces) vs. the desired tool reliability for different checking and tool replacement strategies.

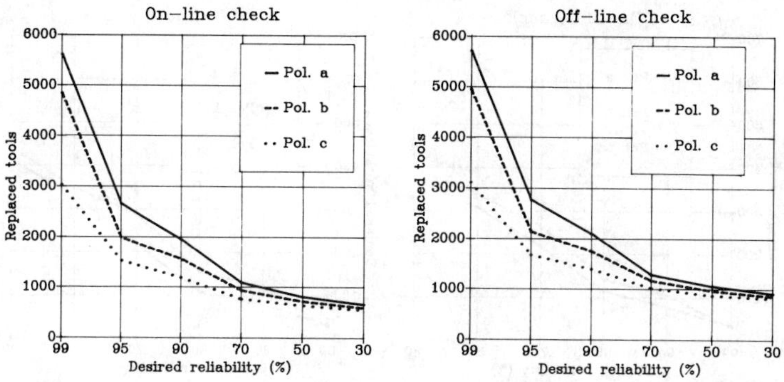

Figure 4. Number of tools to be replaced for each checking and tool replacement policy vs. the desired reliability.

the inevitable increase in workload for the tool management area and the consequent delay in the necessary tool-set departures.

Checking strategies and conclusions

In the case study under examination, the opportunity of an 'on line' tool condition control was examined, instead of an 'off-line' check, utilizing the machining centre as a measurement station (the tool check times are reported in Table 1).

We can note that the on-line check allows a significant reduction in the final machining time (Figure 3): this fact suggests the non-convenience of on-line checks for machining systems operating at high $R(\alpha)$ values. The same drawing highlights that strategy (c) determines an approximately horizontal line for both the checking policies, showing the suitable performance of the tool management area (a lower number of entities transported to the tool restoring area) but, on the other hand, only a change in the number of circulating tools and/or in the tool room layout can positively modify the curves of policies (a) and (b). The final number of replaced tools is lower in the case of on-line check (Figure 5) and the number of scrapped wps (Figure 6) is shown to be higher for the off-line strategy.

In conclusion, the present chapter gives rise once more to the question of the tool criticality evaluation in a manufacturing system, showing how the simulative approach permits an accurate analysis of the industrial process, while the general indications of a possible trend can be simply deduced by an analytical procedure.

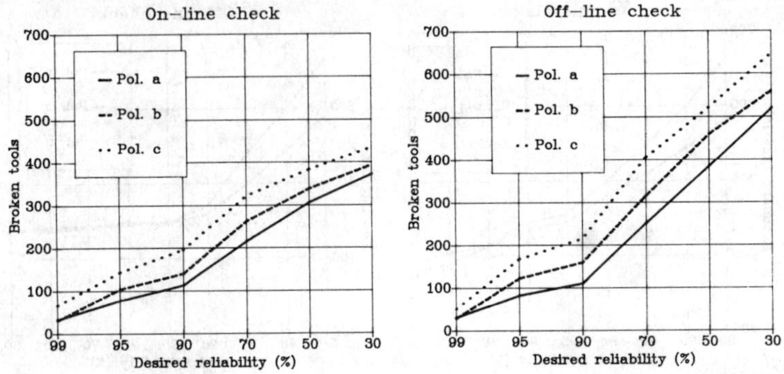

Figure 5. Number of broken tools vs. the desired reliability for different checking and tool replacement strategies.

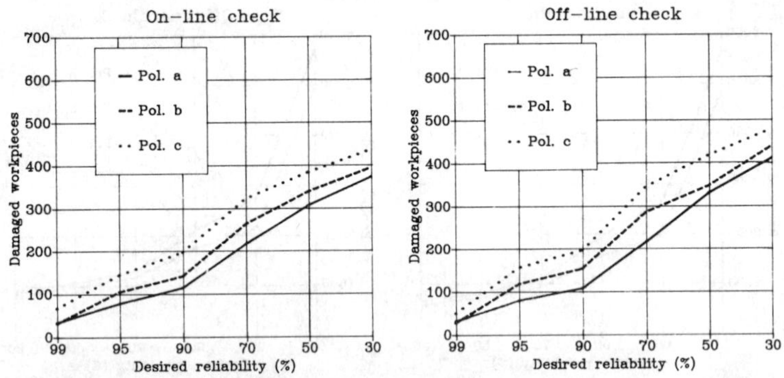

Figure 6. Comparison of damaged workpiece number while varying checking and tool replacement policies vs. the desired reliabilities.

Appendix

The tool management area model includes tool pre-setting and tool restoring stations with the necessary magazines: these resources are connected with the four machining centres cell of the production line by means of two six-tool size automatic guided vehicles (AGVs); the plant resources are (Maccarini *et al.*, 1987a, b; Zavanella and Gentili, 1988; Zavanella *et al.*, 1988):

— two operators, restoring the worn or broken tools by resharpening or replacing them, and two pre-setting stations, where the restored tools are checked and measured and sent to the live tool store;
— the live tool store, furnishing the AGV loading area with the proper tools: the tools leave the above-mentioned area assembled in predetermined kits, the composition of which is a function of the necessary tool equipment for the wp machining cycle completion;
— the worn tool store, where the tools coming back from the production line are placed, awaiting the restoring job by the operator;
— two ROBOTs, connecting the previously defined stations and the AGV loading and unloading area, and two AGVs, connecting the tool management area and the production line: the input and output buffers of the stations are managed by means of simple First In First Out (FIFO) criteria.

Each pre-setting station is fed by a 48-capacity bi-directional wheel. Selection rules of the queues at the loading and unloading stations of the wheel itself and of the pre-setting have been modified in order to identify the best management.

References

Maccarini, G., Giardini, C., Zavanella, L. and Bugini, A., 1987a, Different kind of tool room models for an FMS: a simulation approach and analysis. *International AMSE Conference*, Karlsruhe, Vol. 4, pp. 87–99.

Maccarini, G., Tameni, S., Zavanella, L. and Bugini, A., 1987b, Planning and choosing the tool room type for an FMS: the simulations and analyses. *6th International Conference on Flexible Manufacturing Systems*, Turin, pp. 223–234.

Maccarini, G.C., Zavanella, L., Rovetta, A., Wang, Q. and Gentili, E., 1988, Tool store management of FMS: analysis and simulation of an inventory system. *3rd International Conference on Computer Aided Production Engineering*, University of Michigan, pp. 609–618.

Pereira, D.T.S. and Carrie, A.S., 1987, Simulation of tool flow within a flexible manufacturing system. *6th International Conference on Flexible Manufacturing Systems*, Turin, pp. 211–222.

Ramalingam, S. and Watson, J.D., 1977, Tool life distributions. *Journal of Engineering Industry, Trans. ASME*, series B, **99**(3), 519–531.

Zavanella, L. and Gentili, E., 1988, Influence of dispatching rules on the tool flow of an FMS tool room. *3rd International Conference on Computer Aided Production Engineering*, University of Michigan, pp. 137–146.

Zavanella, L., Maccarini, G.C., Giardini, C. and Gentili, E., 1988, Evaluation of tool flow between management area and workstations: the simulation analysis. *International AMSE Conference*, Istanbul, Vol. 4B, pp. 57–68.

66

Effective work organization in flexible automation

P. Seppälä and E. Tuominen

Institute of Occupational Health, Topeliuksenkatu 41 a A, SF-00250 Helsinki, Finland

Abstract

Current trends in organizing production in manufacturing firms are discussed on the basis of the findings obtained from nine Finnish engineering companies, that had introduced advanced flexible technologies in recent years. It is concluded that in order fully to utilize new technology it must be accompanied by a flexible organization with multi-skilled, motivated and responsible personnel.

Introduction

The market environment has become increasingly more competitive for enterprises in the 1980s. Customers are demanding more individual and flexible services and customized products. Engineering companies have responded to this market pressure by introducing flexible automation. This means utilization of computer numerical control (CNC) technology and flexible manufacturing systems (FMS).

A crucial question arises: What is the most effective way to organize work in order to attain the potential benefits offered by advanced manufacturing technology? A factory without workers is the 'dream of an engineer'. Behind this thinking is the so-called technocentric view that human beings are unreliable and disturbing components in a technically advanced and complex system. One so-called human-centred view, which has received empirical backing recently, implies that flexibility and effective utilization of advanced manufacturing technology can be achieved only by capable, flexible and motivated operators (Brödner, 1986).

This theme was treated in a study conducted in nine Finnish engineering companies which had implemented flexible technology in recent years in order to enhance their competitiveness. Advanced technology was used for the part manufacture of different types of machines such as diesel engines, tractors, rock drills, lifts and heavy duty trucks. The firms had implemented systems of different sizes applying CNC-technology: (1) flexible manufacturing systems or cells (FMS/FMC), (2) flexible modules (CNC machining centres with pallet changer) and (3) robotized machining cells.

The aim of the study was to find out the motives and strategies in the introduction of advanced manufacturing technology and its implications for work organization and job contents. We were especially interested in pinpointing the kind of organization structure and division of labour which was most effective, and in determining how these aims coincided with the goals of meaningful, motivating and challenging jobs.

The data were collected by means of job analyses, interviews and questionnaires. A total of 35 job analyses were made, and 68 operators of 21 systems as well as 35 supervisors, production technicians or production managers were interviewed.

Trends in production organizations

Developments prior to modern technology

Already in the 1970s many Finnish manufacturing firms started to reconsider and reorganize their manufacturing systems. At that time most of the machine shops were organized according to the functional principle, which resulted in lead times of several months or weeks and a lot of capital was invested in the work-in-process and in large material storages. Along with the introduction of NC machines and cellular manufacturing philosophy, a number of firms launched projects where production control systems were simplified by building manufacturing cells, which were loading points (in terms of production planning and control) instead of separate machine tools. A machining cell made a batch of parts ready for assembly for example in a week or a few days. In this way the total lead time was cut from several months or weeks into weeks or days. Typically, a machining cell consisted of one NC machine (a so-called key-machine), whose running time was maximized, and other machine tools needed for machining a part ready for assembling. The machining cell was manned by a small group of skilled metal workers who performed the tasks independently.

Modern technology and organization philosophy

Apart from this trend from functional layout to production cells on the shop-floor level, the manufacturing firms still had traditional hierarchical and rigid organizations with specialized staff located in their own organizational units. It was not until the beginning of the 1980s when, hand in hand with the information technology and advanced manufacturing systems, changes began to happen in the entire production organization.

The changes in market situation, increased competition, demand for customer orientation and flexibility coincided with the rapid development of new technologies. These developments prompted the management to renew their organizations. The product shop is the dominating principle according to which manufacturing firms are currently organized. This is a continuation of the cell philosophy, i.e. the manufacture is organized on the basis of group technology and part-families into organizational units, which have all the machines and other resources needed for making a certain product or a part ready for assembling.

The investments in advanced technology have usually been part of the plant reorganization. A robotized machining cell or FMS/FMC has been a way to achieve the flexibility demanded by an increased number of product variants and variation in the capacity needed.

Division of labour and job contents

An important aspect in the new work organization is the way tasks and responsibilities are allocated. The traditional organization was based on an elaborate vertical and horizontal division of labour. In manufacturing firms functions such as work planning, production scheduling, method planning, quality control and inspection, tooling, material orders and maintenance were carried out by specialized organizational units. Before the age of distributed interactive computer systems it could take several weeks for papers to move from the sales department to the manufacturing department, and for the production of a batch to get started. Correspondingly, a lot of time was consumed during the manufacture by machine set-ups and separate inspection phases after each machining stage. Furthermore, often the tasks of machine set-up and operating a machine were allocated to a skilled setter and an unskilled operator, thus decreasing the effective running time of the machines.

Presently, when flexibility is demanded from both the organization and the technology, it is evident that the old way to allocate tasks is no longer feasible or effective. Consequently, changes have been made at all organizational levels. After the transition to product shop organization, most of the functions carried out by separate organization units have been transferred to the product shop as close as possible to the production itself. This means that the head and the staff of a product shop are now in charge of the functions of work planning, method planning, material orders, etc. related to the parts or products made in his/her department. Especially when the Just-in-Time (JIT) philosophy is adopted, the product shop must have direct contacts to subcontractors, material suppliers and customers.

Supervisor's role

The supervisor's role in a product shop using advanced technology is more to co-ordinate and organize than to supervise and control the details of a task performance. He/she ensures that material and other resources are available and co-ordinates the work of the production cells. The multi-skilled operators take care of the details of work in the machining stations.

Job contents at workstations

After the decision of going to the product shops and cellular manufacturing systems utilizing advanced and flexible technology, it has been a logical solution to man the cells with skilled workers who are motivated to learn new things and willing to assume new responsibilities. In all of the cases studied the management was of the opinion that the cell crew collectively should be able to do basically all the tasks needed for the effective operation of the systems. In all the systems the following tasks were therefore carried out by the cell crews: (1) correction and editing part programs, (2) entering and manipulation of machining schedules, (3) machine set-ups and pre-setting and changing of tools, (4) inspection and quality control of the processed parts, (5) loading and unloading workpieces, (6) maintenance and minor repair of machines and (7) monitoring and disturbance correction.

Although the tasks listed above were usually allocated to a cell crew as a whole, in many cases all of the group members were not equally capable of performing all of the tasks needed. This was mainly due to the original differences in the skills and background of the operators recruited to the new systems. One reason was also that when the systems were enlarged and new workers were recruited the newcomers did not get adequate training. The importance of adequate training and a homogeneous skill level became evident in one of the FMSs studied, when long down-times occurred due to the polarized task structure.

Human aspects of new organizations

Parallel to the technological developments and adoption of group technology, cellular manufacturing systems, and group work at the beginning of the 1970s, much interest and research were focused on the quality of working life. Emphasis was placed especially on job content, variety of work, qualifications, autonomy and

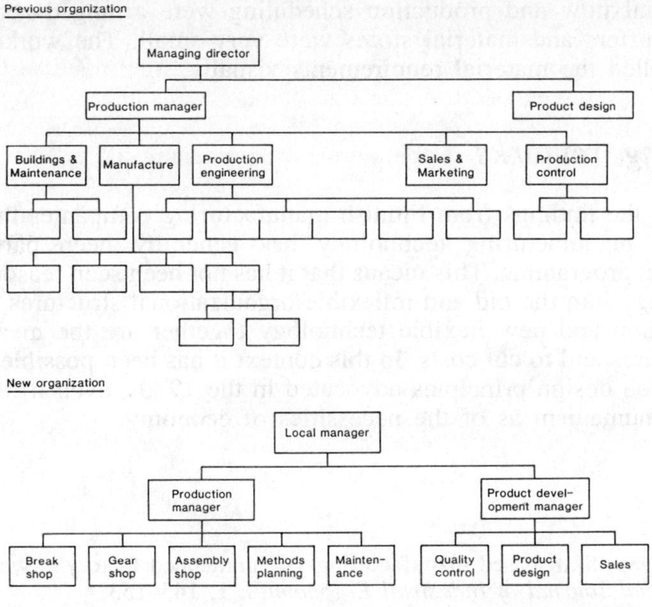

Figure 1. Previous and new organization of an axle works.

social contacts at work. In order to solve the problems of repetitiveness, monotony, pacing and isolation, different job redesign methods such as job rotation, job enlargement, job enrichment and autonomous groups were introduced.

If we look at the principles of organizing production in modern manufacturing plants we can see that they fit very well within the scope and ideas of job redesign. From the point of view of production efficiency and flexibility, it is reasonable to enlarge job contents, increase autonomy and favour multi-skilling which makes work more variable, interesting and challenging. In fact, in order to achieve high performance work teams it is necessary to enlarge jobs and increase qualification demands and autonomy (Buchanan, 1987).

A case: new axle works

One of the nine cases studied was a recently built plant manufacturing axles for heavy duty trucks. The axle factory was a fairly independent unit which manufactured different axle systems for the trucks made by the firm itself as well as for trucks made by other companies. The new factory had about 100 employees and an annual production of 3500 axles.

The old factory was run by a hierarchical and segmented functional organization (Figure 1) and mainly old conventional machinery. After the decision to develop and make axle production more profitable, it was decided to build an entirely new plant. The leading principles in the planning were the application of new organizational ideas and the utilization of advanced technology. The result was a factory which consisted of three supervisory areas and in which the manufacture was organized into four product shops on the basis of the main parts of an axle: housing, brakes, drive gear and hubs. The product shops consisted of machining cells utilizing CNC technology such as FMS/FMC, machining centres with automatic pallet magazines and robotized cells. The machining cells were located around the assembly area, where an automated guided vehicle (AGV) was used to transport axles between different work stations.

The material flow and production scheduling were arranged according to JIT principles. Buffers and material stores were very small. The workers and supervisors controlled the material requirements visually.

Concluding remarks

According to the findings from Finnish manufacturing companies the introduction of advanced manufacturing technology has generally been part of a larger modernization programme. This means that it has not been seen reasonable to install new technology into the old and inflexible organizational structures. Both flexible, flat organization and new flexible technology together are the means to react to market pressures and to cut costs. In this context it has been possible to materialize the humane job design principles advocated in the 1970s, even if not so much on the basis of humanism as of the necessities of economy.

References

Brödner, P., 1986, Skill based manufacturing vs. 'unmanned factory'—which is superior? *International Journal of Industrial Ergonomics*, **1**, 145–153.

Buchanan, D., 1987, Job enrichment is dead: long live high-performance work design. *Personnel Management*, **19**, 41–43.

67

FMS in assembly and its specific demands on logistics and material supply

Reinhard Koether

Fachhochschule München, Postfach 20 09 27, D-8000 München 2, Germany

Abstract

The logistic demand for reduction of throughput time to minimize bound capital is optimally fulfilled by the assembly line. However, assembly line rarely meets flexibility demands caused by variants due to customers' demands and national restrictions. Especially in the automotive industry, where buffers are expensive because of the product's size, the lack of the assembly line's flexibility causes high extra costs for model-mix loss, remake and balance delay. Consequently, assembly systems with flexible assembly sections such as (i) parallel assembly lines, (ii) automated assembly with parallel manual bypass and (iii) parallel, single work centres, were developed and realized in industry.

To meet logistic demands in flexible assembly systems one of the following measures has to be taken: (1) reduce the lead time of the preassembled module, (2) change the assembly structure, or (3) synchronize pre-assembly and final assembly by sequencing buffers. A method is shown to find a compromise between high flexibility and economic material flow.

Demand on flexibility

Nearly all products to be sold on international markets are offered in a broad number of variants due to national legal restrictions and to customers' demands. Modular design and flexibility in production technology and control make it possible to produce 'tailored' products under serial production conditions. The assembly, the last production step, has to work up all the variants and is forced to produce Just-in-Time to customer demands. The high value of parts and products demands reduction of throughput time to minimize bound capital.

This logistic demand, to minimize the assets bound in material is optimally fulfilled in a 'flow system' like an assembly line, as the throughput is combined with the transport and the value adding operations. However, assembly line rarely fulfills flexibility demands, especially in the automotive industry, where buffers are expensive because of the product's size. High extra costs for model-mix loss, remake and balance delay are caused by lack of flexibility of the assembly line.

Consequently, modules of assembly systems with higher flexibility and with automated assembly sections were developed; such modules can be (i) automated sections with manual bypass, (ii) parallel lines, and (iii) parallel single work centres without line pace. These modules are realized in the final assembly and in pre-assembly.

The list shows these modules with increasing flexibility. Unfortunately the logistic advantages are simultaneously decreasing. Thus a closed loop for rework disturbs the sequence; consequently the pre-assembled modules will not meet the right final product. Therefore a compromise has to be found between high flexibility and economic material flow. To plan a flexible assembly system and its logistics, a method was developed which has the following steps:

Step I Find the assembly structure which fits to

— flexibility,
— automation,
— modules of the product.

Step II Analyse this structure from the logistic view:

— classify the modules of the assembly systems on behalf of the sequence variability,
— find the variance in throughput time of the lead product,
— find the lead time consisting of times for throughput through pre-assembly, transport, information and safety lead time.

Step III Design the material flow and the information flow to synchronize the assembly sections of the lead product with the assembly section of its modules.

— Find the information point (I-point) in the assembly flow of the lead product that informs the pre-assembly systems about the sequence of modules to produce or to deliver.
— Find the buffer capacity and access mode decoupling pre-assembly from final assembly.

Step IV Plan the capacity and the capacity control for each assembly section.

Each step of the methodology can end with a loop back to the step before. The desired assembly structure of a section, for example, can be of such high flexibility that enormous buffer capacities are needed to make sure that each pre-assembled module will meet its specific lead product in final assembly. An example from the automotive industry shows the application of this planning method.

Logistics in assembly

The objective of logistics in production is to manage the material flow throughout the factory. In assembly material can be of different types: (1) material to be assembled which can be supplied in containers or in single modules and (2) work-in-progress.

Material in containers is easy to supply with a two-container system: one container can be exchanged while the other container is used as storage for the material to be assembled. The empty container is also an order for new material.

For the logistic objective, to reduce work-in-progress the assembly line is the optimal solution:

— Every assembly step is combined with transport to the next assembly station, batches are split to minimum transport batch size 1.
— Products are conveyed through the materials.
— A determined throughput time and sequence of products makes it easy to control and synchronize the production flows of the main product and its pre-assembled modules.

Furthermore the flow through the assembly line is used to control the material supply of pre-assembled modules. These modules contain a large part of the final product's variants. In other words, the type of engine depends on the destination of the car. So each item of one type of module like engine, seat, dashboard and doors, has to be married to a specific body. In the automotive industry this material supply of modules is controlled by the sequence of the bodies which are the lead products in final assembly. The sequence of the lead products informs the store or the pre-assembly which sequence of variants of a module has to be delivered Just-in-Time to the final assembly line.

The lead time is the time period between the I-point (identification point) and the marriage point.

The easiest way to control the sequence bound supply of the modules is the assembly line but a line is the most inflexible assembly system, which rarely meets the flexibility demands of a broad number of variants to assemble in model-mix.

Analysis of the assembly structure

For flexibility reasons it is desirable to separate the line into modules such as:

- assembly lines with section buffers and station buffers,
- parallel assembly lines,
- automated assembly with parallel manual bypass,
- assembly sections activated on demand (i.e. for rework or special options),
- closed loop for rework,
- parallel, single work centres.

Figure 1 lists the basic types of assembly structures and the variation of their throughput times.

The line's throughput time is determined and the same for each workpiece. The branches of any type of parallel assembly line have their determined throughput times too, but they can differ from branch to branch thus changing the sequence of the workpieces assembled. Sequence changes can be easily controlled and forecasted if the throughput times are known. Throughput through a multi-parallel assembly section with independent single work centres is probabilistic. Organizational restrictions can limit the variations, improving the control in variation of the sequence after such a section.

A change of sequence does not really harm the supply of the modules if a time period can be found for each module where the sequences of neither the lead product nor the module are altered during the lead time of the module. Then the beginning of the final assembly line may not be the I-point for all modules.

Problems occur if the sequence of either the lead product or the pre-assembled module are changed between the I-point and the marriage point. Then one of three measures has to be taken:

(1) reduce the lead time,
(2) change the assembly structure,
(3) resequence the assembly flow to the sequence of the I-point with a sequencing buffer.

The lead time consists of time for:

- throughput through pre-assembly or picking the product's module out of the store,
- information processing,
- transport of the product's module to the final assembly including time to form transport batches,
- safety lead time.

Measures to reduce lead time can be found for each element of the lead time. Reducing the size of the transportation batches reduces the lead time in consequence. An automated transport with an overhead conveyor can be a technical realization for this logistic measure. A change in assembly structure leads back to *Step II* of the methodology. *Step III* leads to the adequate design of a buffer to

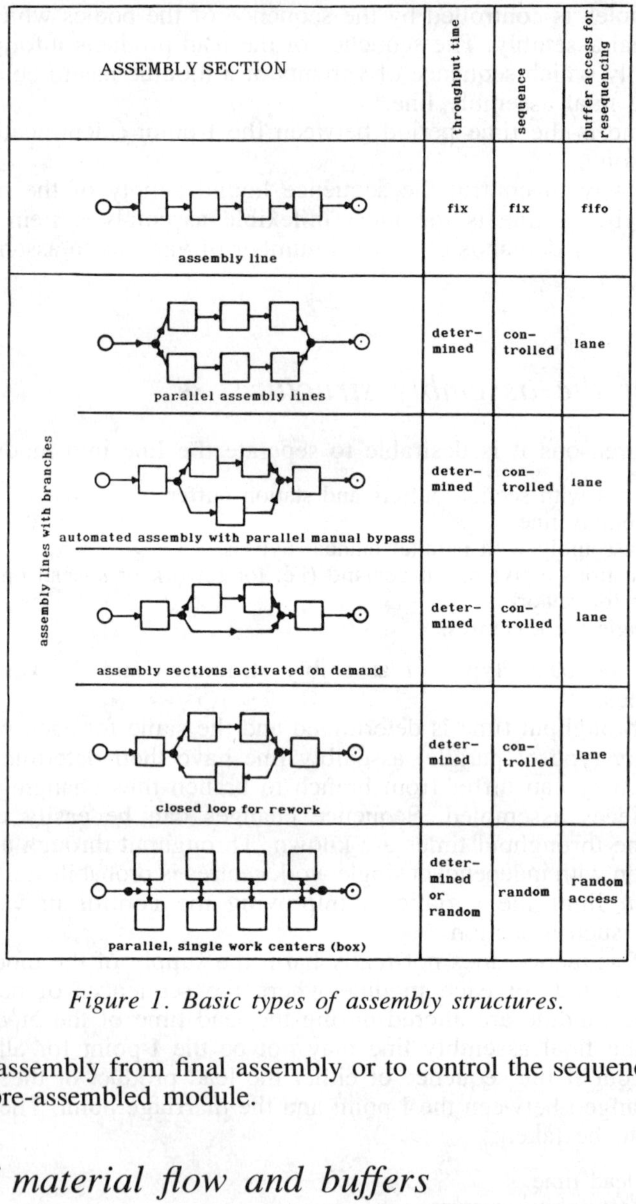

Figure 1. Basic types of assembly structures.

decouple pre-assembly from final assembly or to control the sequences of the lead product and pre-assembled module.

Design of material flow and buffers

The buffer access depends on the allowed variation in sequences of assembled products. This variation can appear in the lead product's flow (i.e. the final assembly of the car body) or in the pre-assembled module's flow (the pre-assembly of the engine) or in both.

The buffer access can be:

— sequence bound [First in First out (FIFO)] if there is no variability in either the sequence in final assembly or the sequence in pre-assembly (both assembly lines);

— parallel lanes with fixed sequence per lane but random access on the first items of each lane. This buffer is necessary to smooth the variance in throughput times of parallel assembly lines in final assembly or in pre-assembly;

— only a buffer with random access can resequence the products while there are independent, decoupled single work centres in final assembly or pre-assembly between the information point and the marriage point.

An adequate buffer to resequence depending on the assembly structure is also listed in Figure 1. Laying out the buffer's connecting assembly and pre-assembly, the second step is to define the capacity of the buffer, the maximum number of items to store. The buffer's capacity depends on:

(1) the desired reliability of the complete assembly system by calculating with reliability control methods,
(2) the wanted flexibility to smooth the maximum permissible lag in throughput time through final assembly or pre-assembly.

The third step of the buffer planning activities is to find the place in the product flows for such a buffer. It can be situated either in the flow of the lead product, in the flow of the pre-assembly, or in both flows. A buffer in the flow of the lead product, the final assembly, is favourable if several flows of pre-assembled modules can be synchronized with the flow of the final product just by resequencing the final assembly sequence. A buffer in the flow of the pre-assembly, where modules smaller than the final product are to be handled, causes less investment for transportation engineering; this is favourable if resequencing of the final product could only synchronize one or a few flows of modules to the lead product. Buffers in the flows of pre-assembly and final assembly make sense if resequencing of the lead product allows the Just-in-Time supply of some different modules. To meet the correct sequence of the lead product it may be necessary also to resequence pre-assembled modules if they are assembled in a more flexible structure than a line.

Capacity planning

The fourth step of the methodology is the capacity planning and control of the assembly sections. The methods used in this step are:

— line balancing to define each worker's and each assembly station's workload, material to use, tools for assembly and the physical access to product (i.e. from underneath or from the top),
— defining the models to be assembled on one line of an assembly with parallel lines,
— sequencing of the models to be assembled to minimize model-mix losses,
— control of an assembly section with multi-parallel stations, run as single work centres (boxes).

Many methods on these topics have been published, e.g. Bullinger *et al.* (1985), Koether (1985, 1987), Letters (1985). The results can be summarized. Line balancing must be controlled by the planner but with support of a computer. The information about the assembly and the product, above all the precedence diagram, is hard to draw for complex products like cars and even harder to keep up to date. Therefore the planner has broader knowledge about the assembly task than the information coded for computers.

For model-mix reasons the variability in each line of parallel assembly lines should be reduced. The production program in each line must be found by classifying the products and finding similar products to be assembled in one line. The similarity can be expressed in terms of similarity of material and similarity of assembly time. As assembly time increases with increasing number of parts to build in, normally the assembly time in the selected section is a good function for expressing similarity of products and can be used in a cluster analysis to group the models.

For assemblies with a high investment in space and machinery, such as automotive assembly, the capacity planning problems of a section with assembly boxes can only be solved by simulation. Questions to be answered are:

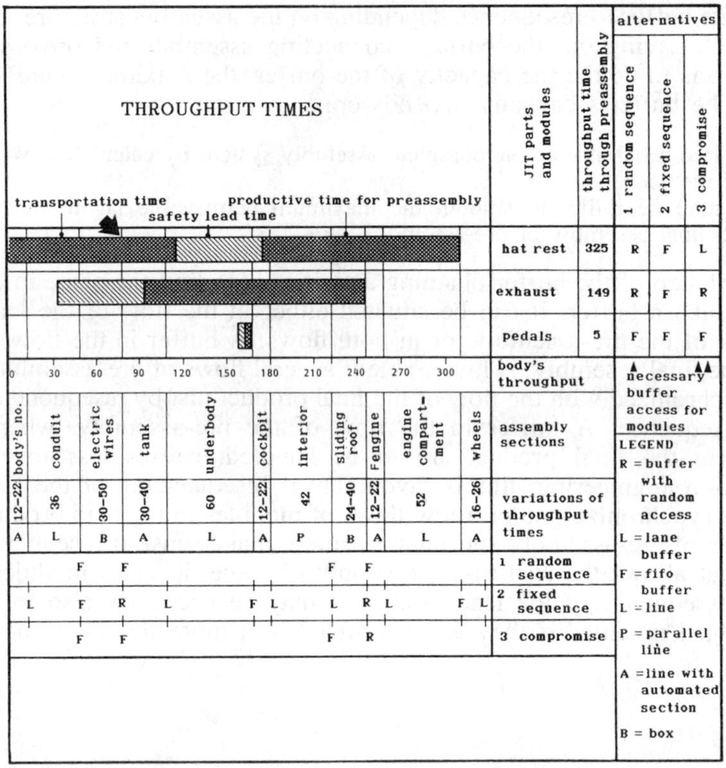

Figure 2. Example of throughput times through an automotive assembly (extract).

— order administration: How is the administration of the bodies to assembly organized (i.e. FIFO, priorities)?
— box control: Are there dedicated boxes for special options or universal boxes for all possible options?
— carrier control: How are carriers attributed to the transport orders (bodies to bring to one box, ready assembled bodies to fetch)?

Example

Figure 2 shows a selection of the throughput time analysis of an automotive assembly. In the lower part of the figure the structure of the final assembly and the resulting variation of the throughput times through these sections are shown. The upper part of the figure shows the throughput of three pre-assembled modules.

The assembly sections of the final assembly, their type of structure and their throughput times are given in Table 1.

The first strategy is the random sequence of the lead product with no resequencing. Then the supply of the pedals is no problem as their lead time is within a line section. The exhaust must be stored in a buffer with random access and the capacity to smooth the maximal possible variance of throughput times of the lead product of at least 56 min. A similar calculation could be made for the hat rest, but here it is recommended that the lead time is altered. Transportation time can be reduced by reducing the size of transportation batches.

Only FIFO-buffers for the modules are needed if the lead product is resequenced after each assembly section, but a high investment in transportation has to be made

Table 1. Final assembly sections, throughput times and structure.

Section name	Throughput time (min)	Structure
Stamp of body's number	12–22	Automated with bypass
Conduit	36	Line
Electric wires	30–50	Box
Tank	30–40	Automated with bypass
Underbody	60	Line
Cockpit	12–22	Automated with bypass
Interior	42	Parallel lines
Sliding roof	24–40	Box
Engine	12–22	Automated with bypass
Engine compartment	52	Line
Wheels	16–26	Automated with bypass

for the final assembly. Buffers with random access like high rack stores for bodies are expensive solutions for realizing the Just-in-Time delivery of pre-assembled modules.

The optimal solution (strategy three, compromise) is to resequence after the section where the sliding roof is built in and adjusted. If the security time for the supply of the exhaust can also be reduced, no random access buffer will be needed for modules.

Conclusions

The methodology shows that flexible assembly systems must be designed by a team of production planners, material flow specialists, transportation engineers and production controllers. The layout of the assembly structure with its lines, automated sections, pre-assemblies, and the transportation and control systems must be planned simultaneously to find an optimal compromise between logistics and flexibility.

References

Bullinger, H.-J., Schad, G. and Lentes, H.-P., 1985, INTAKT—interactive assembly line balancing. In *Toward the Factory of the Future, Proceedings of the 8th International Conference on Production Research and 5th Working Conference of the Fraunhofer-Institute for Industrial Engineering (FhG-IAO)*, University of Stuttgart, 20–22 August (Berlin: Springer), pp. 773–777.

Koether, R., 1985, Improving productivity in model-mix-assembly. In *Toward the Factory of the Future, Proceedings of the 8th International Conference on Production Research and 5th Working Conference of the Fraunhofer-Institute for Industrial Engineering (FhG-IAO)*, University of Stuttgart, 20–22 August (Berlin: Springer), pp. 761–766.

Koether, R., 1987, Design of a sequencing buffer for model-mix-assembly in the automotive industry. In *Proceedings of the IXth International Conference on Production Research*, 17–20 August, Cincinnati, Ohio, USA (Cincinnati: University of Cincinnati).

Letters, F., 1985, Simulation as an aid to assembly planning. In *Toward the Factory of the Future, Proceedings of the 8th International Conference on Production Research and 5th Working Conference of the Fraunhofer-Institute for Industrial Engineering (FhG-IAO)*, University of Stuttgart, 20–22 August (Berlin: Springer), pp. 751–756.

Optimal part routing in flexible manufacturing systems

Claudio Arbib and Luca Perugini

Dipartimento di Ingegneria Elettronica, Università di Roma 'Tor Vergata', Via Orazio Raimondo, I-00173 Roma, Italy

Abstract

Parts in a Flexible Manufacturing System (FMS) should be routed in order to match suitable requirements. In particular, useful features are a balanced distribution of the workload among the manufacturing centres and the minimization of part transfers from one centre to another.

For medium/large productions, if operations are to be executed on all parts of the same type in a fixed order, the problem of balancing the workload via a suitable routing can be solved as a linear program (Arbib *et al.*, 1988). Furthermore, if the number of such operations amounts to u, then the number of different routes needed to balance the workload is $O(u)$, and it is possible to choose in polynomial time, among the feasible routes, the ones which guarantee movement minimization. However, the number u of operations can be too large to make the number of required routes easy to manage.

In this chapter we give almost linear algorithms to solve the routing problem above and show that the number of routes needed is always bounded by the number of machines in the system.

The routing problem

The production model here considered can be described as follows:

(i) Production is organized in lots (batching). Each lot contains parts of possibly different types. Given a lot, we denote as $n(t) >> 1$ the number of parts of type t it contains. A fixed sequence of operations, consisting of $u(t)$ operations, is to be executed on all parts of type t.

(ii) The system consists of m machines. Each machine is able to perform a set of operations. The time τ_i required by the ith operation of a sequence is known a priori, and does not depend on the machine where the operation is performed.

(iii) The transportation system links the machines to each other and to the input/output devices, so that any routing is allowed. Transportation times are known a priori.

Batching approaches to global flow management and part type selection are considered in many applications (see e.g. Whitney and Gaul, 1985; Hwang, 1986; Rajagopalan, 1986). The assumption $n(t) >> 1$ for each part type t is generally reasonable in medium/large productions. Though alternative sequences of operations may in some cases be used to process a part, in most practical cases a single (e.g. the most likely) sequence is considered in order to reduce the problem complexity.

In the considered model, a machine can be as well regarded as a group of workstations with the same, or different, capabilities. The basic assumption is, however, that different centres have the same speed in performing the same operations.

Apart from some paradoxical situations where throughput maximization may be achieved by workload unbalancing (see e.g. Stecke and Solberg, 1985), balancing is generally referred to as a desirable property in many contexts (Berrada and Stecke, 1986; Arbib *et al.*, 1988; Schriber and Stecke, 1988; Agnetis *et al.*, 1989): in particular, the optimality of balancing machine workloads in order to maximize the system expected production has been proved for several types of FMS (Stecke and Morin, 1985). Actually, for medium/large productions, to maximize the system throughput is in practice equivalent to minimizing the completion time χ of all jobs on all machines. In these contexts, χ typically decreases depending on several factors, among which:

— χ decreases as the workload of the busiest machine decreases;
— χ decreases as the transportation system utilization decreases.

On these grounds, the following problem can be defined:

Routing problem For each lot, find a sequence of machines (route) to be visited by each single part, such that (i) the workload (total active time plus unavailability) is equally divided among the machines; and (ii) the total amount of transfers of parts from one machine to another is minimized.

The solution approach here considered is that of finding a solution achieving objective (ii) in a set of solutions for which condition (i) is satisfied. In other words, we look for balanced routings which are minimal with respect to part movement. The problem of finding a balanced routing (not necessarily minimal) is here referred to as the *Balancing problem*. This approach has been first developed by Arbib *et al.* (1988) for the case of chain precedence relation among operation, and successively extended by Agnetis *et al.* (1989) to the case of tree structured precedence relation.

A balanced routing can be defined in several ways, depending on the particular definition of workload balance. One possibility is the following:

Definition 1 A workload distribution is said to be balanced if it minimizes the workload w of the busiest machine.

Notice that in general it is possible to find unbalanced solutions corresponding to a globally lower amount of part transfers. A routing is obtained by identifying for each part type a family of routes, or manufacturing paths, that is a family of sequences of machines to be visited by disjoint subsets of parts of the same type, under the constraint that each all operations must be executed on each part in the given sequence. Observe that, after parts have been routed in the system, their scheduling is a standard Job Shop problem (Coffman, 1976).

Arbib *et al.* (1988) proved that if a perfectly balanced solution exists, then the number of manufacturing paths needed to implement it is less than or equal to the number of machines in the system. In the next section we extend this result to the case in which a perfect workload balance cannot be achieved. In Arbib *et al.* (1988) it is also indicated how to choose the manufacturing paths off-line in order to minimize the total amount of part transfers. In the last section we prove that such an optimal set of manufacturing paths can be found with no further computational cost.

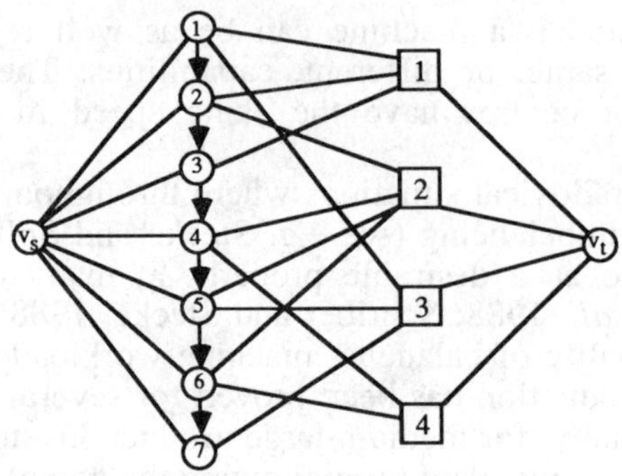

Figure 1. The problem of minimizing the maximum machine workload is formulated in terms of the minimum network cost.

The balancing problem

For each lot, we solve the balancing problem separately for each different part type. By Definition 1, this approach may in general introduce an error (for instance, when the bottle-neck machine is the same for every part type). However, the workload surplus of the bottle-neck machine(s) can be taken into account by means of a suitable down-time vector, as will be clear in the following.

We let M denote the set of machines in the system (with $|M| = m$), N a set of parts of the same type to be processed (with $|N| = n$), and U the set of operations to be executed on all such parts (with $|U| = u$). Operations in U are a priori sequenced according to an operation chain, i.e. a digraph with vertex set U and arcs from operation i to $(i + 1)$, $i = 1, \ldots, (u - 1)$. To indicate which operations can be executed by which machines we make use of a bipartite graph $B = (U, M, E)$, where

$$E \equiv \{(i, j) | i \in U, j \in M, \text{ operation } i \text{ can be executed by machine } j\}.$$

Let τ_i denote the time needed for performing the ith operation on a single part, and set $T = \Sigma_{i=1\ldots u}\tau_i$. An m-vector r, expressed in time per single part to be processed, will account for machine down-time, with $R = \Sigma_{j=1\ldots m} r_j$.

According to Definition 1, the *Balancing problem* can be formulated as follows: let us transform graph B into an acyclic digraph H by adding a source node v_s, connected to all the nodes in U, and a sink node v_t, connected to all the nodes in M and by directing all the edges in B from U to M (see Figure 1).

A capacity is given for each arc of H : τ_i for arcs leaving v_s; $+\infty$ for arcs in E; $w - r_j$ for arcs entering v_t.

If we associate the flow variables ϕ_i, x_{ij}, ψ_j with the above capacities, we can formulate the problem as a particular Capacity Assignment problem on digraph H:

$$\begin{aligned} & \min \quad w \\ \text{s.t.} \quad & \phi_i = \Sigma_j a_{ij} x_{ij} = \tau_i \qquad i = 1, \ldots, u \\ & \psi_j = \Sigma_i a_{ij} x_{ij} \le w - r_j \qquad j = 1, \ldots, m \\ & x_{ij} \ge 0 \qquad \forall (i, j) \in E \end{aligned} \tag{1}$$

where $a_{ij} = 1(0)$ if $(i, j) \in E$ (if $(i, j) \notin E$), x_{ij} is the duration of operation i on machine j (as computed per single part) and $(w - r_j)$ is greater than or equal to the busiest machine workload.

A way to obtain a feasible flow distribution on H is for example that of solving the following Hitchcock problem

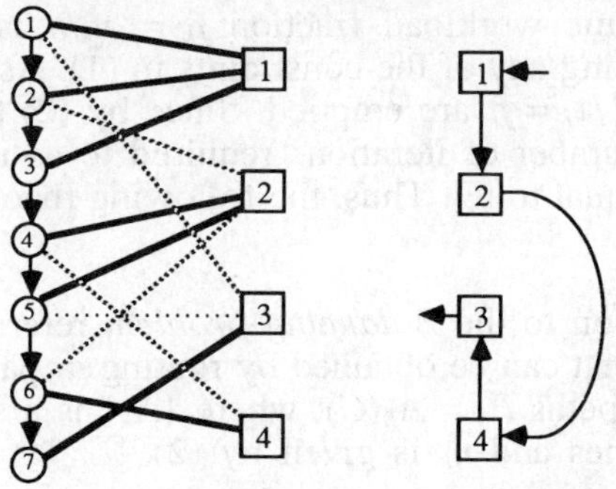

Figure 2. A manufacturing path corresponds to a semi-assignment of operations to machines.

$$\begin{aligned} &\min \quad cx \\ \text{s.t.} \quad &\Sigma_j a_{ij} x_{ij} = \tau_i \qquad i = 1, \ldots, u \\ &\Sigma_i a_{ij} x_{ij} = (T + R)/m - r_j \qquad j = 1, \ldots, m \\ &x_{ij} \geq 0 \qquad \forall i \in U, \forall j \in M \end{aligned}$$

where $c_{ij} = 0$ if $(i, j) \in E$, $c_{ij} = 1$ otherwise. With this position, if $cx = 0$, then x is optimal for problem (1); if on the other hand $cx > 0$, then x is infeasible, since we have $x_{ij} > 0$ for some $(i, j) \notin E$. In this case, a feasible solution to problem (1) is immediately obtained by moving the infeasible flow to arcs in E.

An optimal solution can now be efficiently computed as follows: let j be a node in M such that $\psi_j \geq \psi_k$, $\forall k \in M$. If U_j is the neighbour set of vertex j on B, let

$$\psi_h = \min \{ \psi_k | (i, k) \in E, i \in U_j, x_{ij} > 0 \}.$$

A new feasible solution to problem (1) with a lower cost is obtained by setting $x_{ij} := x_{ij} - \Delta$ and $x_{ih} := x_{ih} + \Delta$, with $\Delta = \min \{x_{ij}, (\psi_j - \psi_h)/2\}$. The optimum is reached when no flow can be moved from the busiest machine to any other. A trivial time bound to the above procedure is $O(um|E|)$.

Let now x^* denote an optimal solution to (1). In the model proposed by Arbib *et al.* (1988), a way for distributing the workload among machines according to x^* via a set of independent semi-assignments of operations to machines is described: a semi-assignment of operations to machines is defined as a subset A of E such that any vertex in U is the endpoint of exactly one edge in A. A semi-assignment A transforms the operation chain C into a manufacturing path $P = A(C)$ on M (see Figure 2).

Let us consider the set E' of all the edges (i, j) such that $x^*_{ij} > 0$. E' may always be regarded as the union of some semi-assignments $\{A_k\}$. By the equality constraints in (1), x^*_{ij}/τ_i can be interpreted as the fraction of the ith operation to be executed by the jth machine if we want to achieve workload balance. If $(i, j) \in A_k$, then we route n_k parts out of N on manufacturing path $A_k(C)$, with

$$n_k / n = \min_{(i,j) \in A_k} x^*_{ij} / \tau_i \tag{2}$$

and

$$\Sigma_k n_k = n. \tag{3}$$

By doing so, the same workload fraction $f_k = n_k/n$ is distributed among all machines without violating any of the constraints in (1). As a consequence, all arcs $(i, j) \in A_k$ such that $x^*_{ij}/\tau_i = f_k$ are emptied. Since by (2) there is at least one arc in this condition, the number of iterations required to assign the whole workload is clearly less than or equal to $|E|$. Thus, the following theorem holds (Arbib *et al.*, 1988):

Theorem 1 A solution to the *Balancing problem* referred to a single set N of parts of the same type that can be obtained by routing n_k parts out of N on no more than $|E|$ manufacturing paths $P_k = A_k(C)$, where $\{A_k\}$ is a set of semi-assignments of operations to machines and n_k is given by (2). □

Actually, we may in practice assume $|E| = O(u)$, since the set of machines which can perform the same operation is typically small if compared with M. Nevertheless, this is not yet a good bound to the number of manufacturing paths needed. A better bound is provided by the following theorem:

Theorem 2 No more than m different manufacturing paths are necessary to distribute the workload among machines so that the workload of the busiest machine is minimized. □

PROOF: Let $w^* = \{w^*_j\}$ be the workload corresponding to an optimal solution to (1). Problem (1) can be reformulated as an Hitchcock problem with supplies $\{\tau_i\}$ and demands $\{w^*_j\}$. A basis solution $\{x^*_{ij}\}$ to this problem contains exactly $(u + m - 1)$ variables with non-zero flow. When we route n_1 parts on P_1, we subtract the same workload fraction f_1 to each operation, and the workload fraction still to be assigned is $(1 - f_1)$ for each operation. By iterating this procedure, after no more than $(m - 1)$ iterations we have exactly u arcs having the same non-zero flow. Hence, at the mth iteration, we make all these arcs empty with a single semi-assignment. The number of manufacturing paths cannot therefore be greater than m,

QED.

Balanced routing with movement minimization

For any given optimal solution to problem (1), the set of manufacturing paths necessary for a balanced routing can be arbitrarily chosen. Arbib *et al.* (1988) proved, however, that in order to achieve movement minimization an effective policy is to choose the routes in non-decreasing order of length. A complete algorithm to achieve workload balance with movement minimization can be therefore coded as follows:

```
program ROUTING (INPUT: a flow distribution {x*_ij};
                 OUTPUT: a routing {Π, {n_k}})
begin
    given every arc (i, j) in E a weight w[i, j] := x*_ij/τ_i;
    repeat
        choose a min length manufacturing path A_k(C);
        add it to Π;
        compute n_k as n min_{(i,j)∈A_k} {w[i, j]};
        set w[i, j] := w[i, j] − (n_k/n) for all (i, j) ∈ A_k
    until w[i, j] = 0, ∀(i, j) ∈ E
end ROUTING.
```

By Theorem 2, with an arbitrary choice of manufacturing paths Program ROUTING would end in time O(um). In the following, we show that a balanced set of minimum length manufacturing paths can as well be computed in O(um).

To compute the length of a manufacturing path, it is necessary to know which operations can be done on the same machine without interruption. This information is given by a family $\mathfrak{I}$ of discrete intervals on the operation set $U = [1, \ldots, u]$, where we assume that operation i is the ith in the operation chain. Let $f : \mathfrak{I} \to M$ be a function associating with each interval $I = [i, j]$ in $\mathfrak{I}$ a machine which can perform operations from i to j without interruptions. If $\mathfrak{I}' \equiv \{I_1, \ldots, I_p\}$ is a covering of $\mathfrak{I}$ such that operation 1 belongs to I_1 and operation u to I_p, then a corresponding manufacturing path is given by $\Pi \equiv \{f(I_1), \ldots, f(I_p)\}$.

We say that an interval $I \in \mathfrak{I}$ is maximal if it is not contained in any interval in $\mathfrak{I}$. Clearly, a minimum length manufacturing path corresponds to a family of maximal intervals. We have the following result:

Lemma 1 If $\mathfrak{I} \equiv \{I_1, \ldots, I_p\}$ is a family of maximal intervals on $U = [1, \ldots, u]$, then $|\mathfrak{I}| = p \leq u$.

PROOF: If the size of every interval in $\mathfrak{I}$ is k, then $|\mathfrak{I}| = (u - k + 1)$, hence $|\mathfrak{I}| \leq u$. Assume that intervals in $\mathfrak{I}$ are differently sized, with the greatest of size k; let be given a total ordering on $\mathfrak{I}$, so that $I = [i, i'] < J = [j, j']$ if $i < j$ and let I be a size k interval such that the interval J immediately on its right (or its left) is of size $h < k$. Since $|I \setminus J| \geq 2$, we can substitute I with a maximal interval I' such that $|I'| = h$. By iterating such a procedure, we get a family $\mathfrak{I}'$ with the same cardinality as $\mathfrak{I}$, and such that all sizes of its intervals equal the size of the smallest maximal interval in $\mathfrak{I}$; the thesis follows

QED.

On the basis of Lemma 1, we can prove the following theorem:

Theorem 3 Algorithm ROUTING ends in time $O(um)$. □

PROOF: By Theorem 2, no more than m iterations of the 'repeat' cycle of Program ROUTING will occur. Thus, we have only to show that each of these iterations requires $O(u)$ time.

At each iteration of the 'repeat' cycle, the arcs of B with non-zero flow define a family $\mathfrak{I}$ of maximal intervals on U. To represent this family, we construct a data structure consisting of a list L, each element $[a, b, c]$ of which contains the following three pieces of information: (a) machine identificator, (b) first operation and (c) last operation of the interval. [List L can clearly be constructed in time $O(um)$.] The following operations are defined on L:

FIND: find a covering of U with a minimum number of intervals;
UPDATE: update L after an interval has been split into two sub-intervals.

A FIND operation can be performed as follows: let $e_0 = [a_0, b_0, c_0]$ be the first element of L and set a current depth d to c_0; while $b_i \leq d + 1$, scan L forward. Then, set d to c_i and iterate the procedure starting from current element e_i, until the end of L is reached. By Lemma 1, this operation requires $O(u)$ time. If we store the information (a) for each current element (including the first), we clearly get a manufacturing path of minimum length. At the end of the kth iteration of the 'repeat' cycle in Program ROUTING, as many UPDATE operations occur as arcs in A_k are emptied: in fact, if the flow of arc (i, j) is set to zero, then the interval $I = [p, q]$ of $\mathfrak{I}$ containing operation i and such that $f(I) = j$ is split into two sub-intervals $[p, i - 1]$ and $[i + 1, q]$, since operation i can no longer be performed on machine j.

An UPDATE operation can be performed as follows: let $e_i = [a_i, b_i, c_i]$ be the element of L corresponding to the interval to be split, and $e_i' = \{a_i, b_i, c_i']$, $e_k'' = [a_i, b_i'', c_i]$ the elements replacing e_i. To insert e_i', scan L and check whether e_i' is not contained in an interval preceding e_i : if so, no insertion is needed and we must simply drop e_i; if not, insert e_i' immediately before e_i and drop e_i. No further scanning is necessary: in fact e_i' will not contain any element in L, since $e_i' \subset e_i$ and all elements in L are maximal. The same procedure applies to insert e_i''. The whole operation clearly requires $O(u)$ time,

QED

References

Agnetis, A., Arbib, C., Lucertini, M. and Nicolò, F., 1990, Part routing in flexible assembly systems (Univ. Roma La Sapienza, RAP. 05·89), and *IEEE Transactions on Robotics and Automation*, **6**, 6.

Arbib, C., Lucertini, M. and Nicolò, F., 1991, Workload balance and part transfers minimization in flexible manufacturing systems (University Roma La Sapienza, RAP. 06·88) and *International Journal of Flexible Manufacturing Systems*, **3**(1), 5–25.

Berrada, M. and Stecke, K.E., 1986, A branch and bound approach for machine load balancing in flexible manufacturing systems. *Management Science*, **32**, 10, 1316–1334.

Coffman, E.G., 1976, *Computer and Job-shop Scheduling Theory* (New York: Wiley).

Hwang, S., 1986, Part selection problems in flexible manufacturing systems planning stage. In *Proceedings of the 2nd ORSA/TIMS Conference on Flexible Manufacturing Systems: Operation Research Models and Applications*, edited by K.E. Stecke and R. Suri (Amsterdam: Elsevier), pp. 297–309.

Rajagopalan, S., 1986, Formulations and heuristic solutions for parts grouping and tool loading in flexible manufacturing systems. In *Proceedings of the 2nd ORSA/TIMS Conference on Flexible Manufacturing Systems: Operation Research Models and Applications*, edited by K.E. Stecke and R. Suri (Amsterdam: Elsevier), pp. 311–320.

Schriber, T.J. and Stecke, K.E., 1988, Machine utilizations achieved using balanced FMS production ratios in a simulated setting. *Annals of Operations Research*, **15**, 229–267.

Stecke, K.E. and Morin, T.L., 1985, The optimality of balancing workloads in certain types of flexible manufacturing systems. *European Journal of Operational Research*, **20**, 68–82.

Stecke, K.E. and Solberg, J.J., 1985, The optimality of unbalancing both workloads and machine group sizes in closed queuing networks of multiserver queues. *Operations Research*, **33**, 882–910.

Stecke, K.E. and Suri, R. (Eds), 1985, Flexible manufacturing systems: operation research models and applications. *Annals of Operations Research*, **3**.

Whitney, C.K. and Gaul, T.S., 1985, Sequential decision procedures for batching and balancing in FMSs. *Annals of Operations Research*, **3**, 301–316.

69

A computer-based system for automatic cutting tool selection

C.S. Syan

Department of Production Engineering and Production Management, University of Nottingham, Nottingham NG7 2RD, UK

Abstract

This chapter reports progress so far in the development of an expert system for automatic cutting tool selection (ACTS) that assists the design engineer. The system also advises on the optimal cutting parameters for machining purposes. Prolog was selected as the programming language because of its suitability and the comparative ease with which the required rules can be constructed and manipulated. The development of the system concentrated on machining a hole, the commonest of all machining operations. On the basis of this, the cutting tools incorporated in the system currently include centre drills, drills, reamers and boring tools.

Introduction

Through the creative work of the design engineer, the market need for a product is translated into drawings and a bill-of-materials. In the design activity, the engineer has to consider many conflicting and complex issues. The product must function in the most efficient and economical manner within the constraints that apply. The major constraint is cost, although other factors such as safety, pollution and legal requirements often have to be considered. What sets apart the successful designer is the ability to design an efficient product which can be economically produced.

All engineers know that design governs product performance, but rarely appreciate the impact it has on manufacture. Consideration of manufacturing problems at the design stage is the major means available for reducing manufacturing costs and increasing productivity. Too often the benefits of such design revisions are realized too late. If the product is already in production, the issue is complicated by factors such as the investment committed to tooling and equipment and the remaining product life expectation; whereas an early identification of improved design for manufacture can maximize the benefits.

The design process involves two distinct types of activity: (1) the conceptual design and (2) refinement and development phases. The conceptual design is the creative part, which is not well understood, and is likely to be the domain of design engineers for the foreseeable future. However, the refinement and development phase involves evaluation, analysis and optimization of the design options. Hence, this second stage is better understood and lends itself to automation by the application of computers.

Various attempts have been made in the past to make manufacturing process information more readily available to designers, but these have not been taken up widely. The advent of the Knowledge-Based Expert Systems and the increasing popularity of computer-aided design (CAD) systems in industry suggest that a solution to the problem may be within reach by integration of these systems (Swift *et al.*, 1986).

One of the most challenging problems is to design products for economic manufacture. Besides the conventional process planning and process selection problems, this includes economical and efficient use of cutting tool resources. We have addressed this problem in order to provide the designer with early advice on the availability of the required cutting tools and the implications for the subsequent manufacture.

Nature of the problem

In order to specify the optimal cutting tools, the designer needs to have the following knowledge:

(a) understanding of the capabilities of the range of cutting tools,
(b) ability to match the design requirements to the capabilities of the cutting tools,
(c) availability of the required tools for the product manufacture,
(d) knowledge of the national/international and/or company cutting tool standards and policies.

Most computer-aided process planning (CAPP) systems being developed nowadays are mainly expert systems, e.g. the CUTTECH (Barkocy and Zdeblick, 1985), Excap (Darbyshire and Davies, 1984) and Gari (Descotte and Latombe, 1981) systems. A typical function often present in structure of CAPP systems is tool selection for easy scheduled cutting operation. These embedded cutting tool selection programs are accessed through the hierarchical structure of the system, for internal use only. These cutting tool selection modules within the CAPP systems are very limited, as they only consider the geometrical compatibility of the workpiece and differences between roughing and finishing tools. This selection of cutting tools is extremely important as the decision will affect the surface quality, operation cost and the availability of the tool in-house and the associated tooling time and cost penalties for erroneous decisions. This problem could be solved if the design engineer had direct access to the automatic tool selection module early in the design process.

Because of the deficiencies in the above CAPP systems, Giusti *et al.* (1986) proposed a modular approach. They developed the PICAP system, which entails the cutting tool selection module, for turning tools, named COATS. This module could also be used as a stand-alone tool selection system. The survey carried out by Eversheim and Schulz (1985) has shown existence of 55 CAPP systems, but no generalized system exists which is able to solve all problems connected with all types of machining.

As discussed above the cutting tool selection problem is not trivial, as:

(a) tool selection depends upon practical experience of the process planner or tool engineer, and often choice criteria cannot be obtained by an exacting and deterministic analysis;
(b) there are great numbers and diversity of tools available. Tool manufacturers supply large variations of tools and holders as well as variants of the combinations of the two;
(c) there is a constant stream of new tool topologies, materials and geometries coming onto the market. These add to the problem of keeping up to date with the developments in the field.

Table 1. Parameters important in cutting tool selection.

1. Diameter of hole
2. Length of hole
3. Type of hole, e.g. blind, through, counterbored, tapered, grooved, etc.
4. Number of holes
5. Surface finish
6. Concentricity
7. Tolerance on diameter
8. Material to be machined
9. Hardness of material
10. Grade of material
11. Material condition, e.g. normalized, hardened, tempered, precipitation hardened, etc.
12. Tool material

Expert systems

The basic concepts and structure of expert systems, and their ability to use high-level knowledge, have been discussed for many years (Michie, 1974; Feigenbaum and McCorduck, 1983). These systems are computer programs that differ from the conventional procedural programs in two distinct ways. The first is a structural difference, that is, the knowledge used in problem solving is separated from the inference procedures used for its application. The second difference, as the name suggests, is that these systems also use heuristics or expert rules to solve problems rather than just data and calculation procedures. Their application to design engineering problems has also been proved in design for assembly (Swift *et al.*, 1984) and surface treatment/coating selection (Syan *et al.*, 1987). A larger expert system R1, for example, configures DEC Vax computers directly from sales orders (McDermott, 1982). For more information regarding such systems, the reader is referred to a useful survey by Bramer (1981).

Methodology—system description

The many machining operations which normally occur on the shop-floor make the problem of tool selection a fairly complex one. For that reason it was decided that the initial development of the system would concentrate on machining a hole, the commonest of all machining operations. The prime aims for the initial system development were (1) to incorporate traditional machining expertise into an expert system which could be used by both the novice an by the expert in machining principles for cutting tool selection, and (2) to produce an interactive, stand-alone system, which could ultimately be incorporated into a CAD/computer-aided manufacture (CAM) environment.

The complexity of tool selection for the simplest of operations, such as machining a hole, can be seen by the number of machining parameters that have to be satisfied before a selection can take place (see Table 1).

The structure of the ACTS expert system is illustrated in Figure 1. The modular approach allows further data bases, specific to the manufacturing organization, to be added to the system. Obviously, the corresponding procedures and knowledge for their selection will also have to be incorporated into the system. Currently there are four data bases of cutting tools, comprising centre drills, twist drills, reamers and boring bars. These have been compiled using tool suppliers' catalogues and the relevant British standards (see Figure 1).

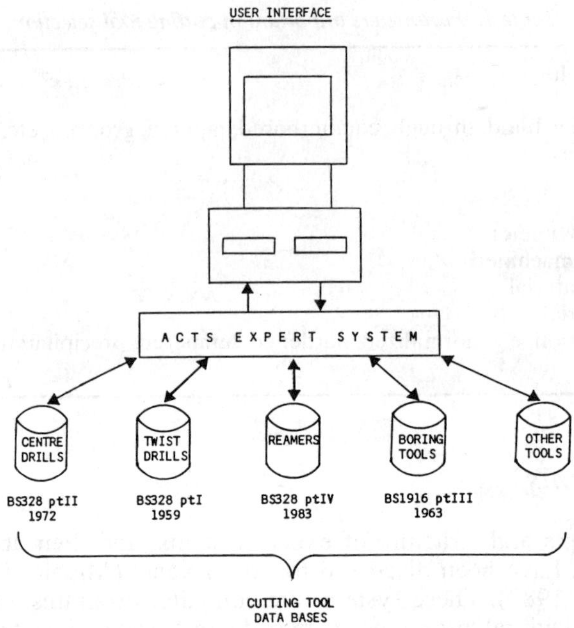

Figure 1. Structure of the ACTS program.

The system can be used directly as a tool store management system. This allows the user to access the tool store data bases directly and check availability of the cutting tools required for an operation. This mode of usage entails the user selecting the appropriate option from the top-level menu (see Figure 2). The relevant data base is initialized, the user can then query the data base by following the subsequent dialogue.

In this 'query' mode, the user is allowed to give particulars of the tool and the system checks for its availability. If the tool is in store, the user is supplied with the tool number and the quantity in stock, otherwise a message is displayed stating that the tool is not in stock.

In automatic selection mode for an individual type of tool, for example twist drills, the user is guided in supplying the required information in order to select the optimal cutting tool by the use of the system's tool selection knowledge. Both the 'Query' and 'Automatic' modes are available for all the types of cutting tools contained in the data bases via the top-level menu.

If the user is not familiar with the cutting tool capabilities and their associated machinable part features, the system can be used fully automatically. Again, an option is selected (option 5 in Figure 2) from the main menu, the system interacts with the user and selects the cutting tool or a combination of cutting tools, suitable for producing the part feature. This enables the non-expert to select cutting tools that are immediately available and are capable of producing the accuracy and surface finish requirements of the intended feature.

The rules that are used for selecting various cutting tools are expressed as situation → action pairs, e.g.

IF ⟨antecedent 1⟩, THEN ⟨consequent 1⟩,

– " – , – " – ,

⟨antecedent n⟩ ⟨consequent n⟩.

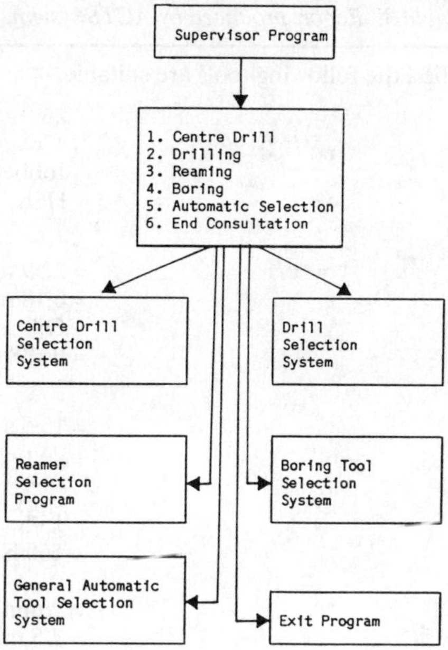

Figure 2. Top-level options of the ACTS system.

Table 2. Example of a Reamer selection rule.

IF	depth of hole is = < hole depth in data base, and feature is a blind hole and there is a straight flute reamer in the data base
THEN	straight flute reamer (X) is suitable

For example one of the rules used by the system for selecting a reamer is shown in Table 2. (X) in this rule denotes a variable that can be instantiated to a particular reamer in the data base during the consultation.

The data bases contain facts and data relating to the cutting tools. The data, for example, regarding drills are stored as follows:

drill (diameter, flute length, diametrical tolerance, feed rate (related to the hardness of the workpiece material), drill material, type of drill (e.g. jobber), overall length of drill).

The information regarding the tool usage is also stored in the data base. This information relates to the capabilities of the tool as well as the related cutting parameters, e.g. for the drill usage data the facts are stored as:

usage (material to be machined, grade of material, range/value of material hardness, material condition (e.g. hardened), cutting speed, coolant requirements).

All the above data are available from the tool suppliers or can be calculated using standard procedures. Some of the items in the data base, therefore, are themselves calls for invoking calculation procedures in order to obtain such values.

Table 3. Report produced by ACTS system.

From the information supplied the following tools are suitable:	
Tool	
Drill number	426
Drill type	Jobber
Drill material	HSS
Cutting parameters	
Cutting speed	= 22·9 m/min
Feed rate	= 0·18 mm/rev
Coolant required is	Oil
Concentricity attainable (+/−)	= 0·138 mm
Machining time and cost	
Total machining time	= 15·5 s
Total machining cost	= 24 pence
based upon:	
Direct labour cost of	£5/h
Overheads cost of	£50/h
One operator per machine	
Time taken per regrind	= 2 min
Time between regrinds	= 7·5 min

The ACTS system advises the user of the suitable cutting tools in the form of a tool report. A typical final report is shown in Table 3. This shows the chosen tool, material, cutting parameters, time for processing and the cost for the operation. The ground data used for the calculation of time and cost are also displayed for clarity. Currently the ACTS system contains about 80 rules regarding the selection of the four types of cutting tools. The data bases contain 4 centre drills, 300 twist drills, 150 reamers and 125 boring tools.

The system is constructed using the declarative type programming language Prolog (Clocksin and Mellish, 1981). This is an Artificial Intelligence (AI) language, based on the horn clausal subset of first order predicate logic. The built-in pattern-matching and backward-chaining mechanisms are used directly by the ACTS system. These facilities considerably reduce the time and effort required in developing the necessary functions for building such systems. The particular dialect of Prolog used is the Salford University LISP/PROLOG system version 67, and this is implemented on the PRIME 9650 mini-computer.

Currently, the ACTS system only offers a limited explanation facility, i.e. the user is afforded the response 'why' for any request for data. This response displays the current rule being used, as an explanation of its line of reasoning.

Further developments

The system is in its early development stage. The progress to date has confirmed the usefulness of a smart tool selector for use as a stand-alone system by design engineer, tooling engineer or shop floor personnel. It also offers a direct link to the tool stores system, enabling up-to-date information to be utilized by the process planner or the design engineer.

Eventually, it is envisaged that the system will be linked to a CAD system for automated analysis of designs in the early stages. This could provide the designer with early feedback on the practicalities and difficulties of subsequent manufacturing with regard to tooling decisions. The system also lends itself to integration into existing CAPP systems to improve their capability and usefulness. A manufacturing process selection system also needs to be developed; this could further enhance the usefulness of the ACTS system. A process selection system, still in the research stage, is being developed by Hon and Ismail (1987). This approach could offer complete tooling and processing consideration for the design in its early stages. The data bases for the complete range of cutting tools need to be built up, e.g. grinding tools, turning tools, milling tools.

In addition there are areas of development regarding the system facilities that need to be enhanced. For example, the explanation facility needs to be more comprehensive and hence more useful. Because the data bases of tools are fairly rigid and fixed they provide possibilities of automated updating and consistency checking. The next phase of development will address most of the above points.

Conclusions

Many CAPP systems have been developed, but none are comprehensive enough for wide industrial applicability. This work, whilst recognizing the great difficulty in producing complex, comprehensive and useful systems, has attacked a manageable part of the total problem. Even this section of tool engineering, which is relatively well defined, has shown itself to be non-trivial in its solution. Such comprehensive expert system modules together with the conventional analysis packages (e.g. finite element analysis), can be embedded within the CAD systems and offer possibilities of real 'design' systems in the future. These modules can act as expert consultants, analysing the design as it is being created at the CAD terminal. The ACTS system is still in its experimental stage, and needs more effort to turn it into a viable and useful tool. The time taken to produce prototype expert systems has been reduced to a matter of few man-months by the availability of good and relatively inexpensive system-building tools. However, the time and resources required for developing these prototypes into useful systems to solve real problems should not be underestimated.

References

Barkocy, B.E. and Zdeblick, W.J., 1985, A knowledge-based system for machining operation planning. *Carbide and Tool Journal*, **17**, 4, 25–30.

Bramer, M.A., 1981, A survey and critical review of expert systems research. *Proceedings of the British Computer Society Conference*, pp. 486–519.

Clocksin, W.F. and Mellish, C.S., 1981, *Programming in Prolog* (Berlin: Springer).

Darbyshire, I.L. and Davies, B.J., 1984, EXCAP: an expert systems' approach to recursive process-planning. *Proceedings of the 16th CIRP International Seminar on Manufacturing Systems*, Tokyo, pp. 127–128.

Descotte, Y. and Latombe, J.-C., 1981, GARI: a problem solver that plans how to machine mechanical parts. *Proceedings of the 7th Joint Conference on Artificial Intelligence*, Vancouver, Canada, pp. 766–772.

Eversheim, W. and Schulz, J., 1985, Survey of CAPP systems. *1st CIRP Seminar on CAPP*, Paris, January, **34**, pp. 607–613.

Feigenbaum, E.A. and McCorduck, P., 1983, *The Fifth Generation: Artificial Intelligence and Japan's Challenge to the World* (Reading, MA: Addison-Wesley).

Giusti, F., Santochi, M. and Dimi, G., 1986, COATS: an expert module for optimal tool selection. *Annals of the CIRP*, **35**, 1, 337–340.

Hon, K.K. and Ismail, H., 1988, A knowledge-based system for the selection of hole-making process. In *Expert Systems in Engineering*, edited by D.T. Pham, pp. 444–451, (London: IFS).

McDermott, J., 1982, Domain knowledge and the design process. *Design Studies*, **3**, 2, 31–36.

Michie, D., 1974, *On Machine Intelligence* (Edinburgh: University Press).

Swift, K.G., Matthews, A. and Syan, C.S., 1986, The potential for intelligent knowledge-engineering design. *Proceedings of the 13th CAM-I Annual Conference, Man or Machine—A Choice of Intelligence*, USA, pp. 5-21–5-30.

Syan, C.S., Matthews, A. and Swift, K.G., 1987, Knowledge-based expert systems in surface coating and treatment selection for wear reduction. *Surface and Coatings Technology*, **33**, 105–115.

Swift, K.G., Matthews, A. and Syan, C.S., 1986, The potential for intelligent knowledge-based CAD in manufacturing engineering. *Proceedings of the International Conference on Computer-aided Production Engineering*, Edinburgh, pp. 47–53.

70

A comprehensive group technology system for cellular manufacture

G. Harhalakis,[1] I. Minis,[2] R. Nagi[1] and J.M. Proth[3]

[1]Department of Mechanical Engineering and Systems Research Center, [2]Department of Mechanical Engineering, University of Maryland, College Park, MD 20742, USA and [3]Institut National de Recherche en Informatique et en Automatique, INRIA-Lorraine, 4 rue Marconi, Technopole Metz 2000, 57070 Metz, France

Abstract

This chapter presents a comprehensive Group Technology (GT) system that arranges the production equipment of a machine shop into cells, and partitions the entire set of parts into families. Each part-family is assigned to a single cell, in order to optimize the part flow within the shop. The system consists of a two fold heuristic procedure. During the first step, a bottom-up aggregation algorithm is used to form the manufacturing cells by minimizing the inter-cell part traffic. The solution is further refined during the second step of the procedure, whereby the significance of each work centre to its corresponding cell is validated. The system considers both the utilization of parts and the sequence of operations specified in part production routings. Capacity constraints which are encountered in every industrial application can also be addressed by a pre-processing algorithm currently under development. This will be used to prepare the input data of the proposed heuristic. The system has been applied to a large-scale industrial project that involved 134 work centres and 1186 make-to-stock parts. The results obtained yielded 58·7% savings in inter-cell part transfers.

Introduction

Cellular manufacturing has led to increased throughput in batch manufacturing systems, resulting from considerable reductions in transportation, set-up and queue times. Furthermore, the decomposition of the shop to smaller, decoupled subsystems facilitates both scheduling and shop floor control. These functions no longer involve the entire shop but a number of independent, less complex cells and their corresponding part-families. Additional advantages of the cellular layout concept include savings in materials handling costs.

Several methods have been developed to group work centres of a given shop in manufacturing cells, that are capable of processing parts with given production routings. These methods are classified into two major types. The first includes algorithms that are based on cluster analysis, such as the bond energy method developed by McCormick *et al.* (1972), the rank order clustering method proposed by King (1980), the direct clustering algorithm of Chan and Milner (1982), and the

Garcia Proth method (GPM) developed by Garcia and Proth (1985, 1986). The common objective of this class of algorithms is to maximize the number of operations performed to each part-family, within its corresponding cell.

In the second class of grouping methods, work centres are progressively aggregated into cells depending on the degree of 'similarity' among them. This class includes the single linkage algorithm developed by McAuley (1972), the method of Leskowsky *et al.* (1987), who used the Average Common Part Weighing Metric, and the Knowledge-Based Group Technology (KBGT) system proposed by Kusiak (Kusiak, 1988; Kusiak and Wadood, 1988). It is noted that the objective of the last three algorithms is not expressed by mathematical functions.

Most of the above methods, however, do not address the following important issues encountered in any common industrial practice.

The sequence of operations—The objective of a typical matrix clustering approach is to confine the majority of operations performed on a given part within a single cell, regardless of operation sequence. However, the sequence of operations in a part production routing has an important effect on materials handling costs and times. If, for example, an intermediate operation of a part is performed in an external cell it will necessitate two inter-cell transfers, as opposed to an operation that is either the first or the last one in the routing of the part. Furthermore, a sequence of consecutive operations in an external cell involves the same materials handling effort as a single outside operation.

Non-consecutive operations on the same machine—It is not uncommon that a given machine is used more than once for processing a part. The methods that are based on the part-machine incidence matrix cannot address this case, since each non-zero element of the matrix represents a single operation.

Distribution of functionally identical work centres—This problem is particularly important in the case of a functional machine shop, where there exist entire departments of functionally identical machines. The distribution of this equipment to cells must be performed in such a way that: (i) the cells are decoupled, and (ii) the workload of each machine is balanced.

A comprehensive Group Technology system that directly addresses the first two issues is presented in this chapter. The objective of this system is to arrange the equipment of a machine shop into cells, in order to minimize the total number of inter-cell transfers necessary for the manufacture of the entire set of parts. Furthermore, each part is assigned to the cell where most of its operations are performed. The system consists of a two-step heuristic procedure. The first step uses a bottom-up aggregation procedure to form the manufacturing cells. The second step validates the assignment of each work centre to its corresponding cell. In addition, a pre-processing step is also proposed to address the case of functionally identical machines. It determines the parts to be processed by each of these machines in a way that machine capacities are not exceeded. Furthermore, this step determines the most 'favourable' set of production routings for the decomposition of the system.

The next section presents the proposed system, and outlines the algorithms used for both the bottom-up aggregation procedure and the local refinement procedures. The application of the system to a large industrial project is then presented and the results obtained are compared with those of two other representative GT methods. Then an enhancement of the system to cater for the distribution of functionally identical work-centres and capacity constraints is discussed. Finally, the conclusions of this study are summarized.

The proposed group technology system

The production equipment and the manufactured parts of the machine shop under consideration are represented by the sets $M = \{m_1, m_2, \ldots, m_m\}$ of m machines, and

$P = \{p_1, p_2, ..., p_n\}$ of n part types, respectively. It is noted that each element $m_i \in M$ represents a unique work centre of the shop. Furthermore, a unique routing r_i corresponds to each part $p_i \in P$. It identifies both the work centres that are used for the manufacture of p_i and the sequence of operations performed.

The problem consists of finding a partition, $C^* = \{c_1, c_2, ..., c_w\}$, of the set M into w subsets, or classes, such that the material handling effort or the traffic of parts between these classes is minimized. To account for the material handling effort, we associate with each part p_i a weight u_i, which may represent the volume of production, or the number of batches, or a cost factor that corresponds to p_i.

For the routing of r_k of part p_k, let x_{ij}^k, represent the number of times any machine belonging to c_j is the immediate successor of any machine belonging to c_i in any given partition C^0. Then the traffic t_{ij} between two distinct classes c_i and c_j can be defined as follows:

$$t_{ij} = \sum_{k=1}^{n} u_k (x_{ij}^k + x_{ji}^k).$$

The Normalized Inter-Class Traffic, T_{ij}, between c_i and c_j is given by:

$$T_{ij} = t_{ij}/(n_i + n_j)$$

where n_i, n_j represent the cardinals, or number of work centres, of subsets c_i and c_j, respectively. Note that the traffic between classes is symmetric, i.e. $t_{ij} = t_{ji}$.

Using the above notation, the cellular layout problem is formulated as follows:

$$\text{MIN} : \sum_{i=2}^{w} \sum_{j=1}^{i-1} T_{ij}$$

$$\text{S.T.} \qquad n_k \leq N;\ k = 1, ..., w$$

where N represents the maximal number of work centres in a cell.

A two fold heuristic is proposed to reach a good, if not optimal solution. During the first part, a bottom-up aggregation procedure is used to group the work centres into cells in order to minimize the inter-cell part traffic. The solution is then refined by the second part of the procedure where the assignment of each work centre to its corresponding cell is validated. Both parts of the method are described below.

The bottom-up aggregation procedure

At the beginning of the algorithm each machine is placed in a separate class. Then, at each step of the minimization procedure, the Normalized Traffic is calculated for all the pairs of classes, that, if aggregated, will result in a class of size less than or equal to the specified limit N. The two classes c_i and c_j, between which the Normalized Traffic, T_{ij}, is maximum are aggregated into a single class. Thus the total inter-class traffic in the system is reduced by t_{ij}.

The traffic between classes is now revised according the following rules:

(1) The traffic between two unaffected classes remains the same.
(2) The traffic between an unaffected class and the new aggregate is the summation of the traffic between the former and the components of the aggregate.

The procedure is continued until it is either not possible to form any feasible aggregation due to the cell size limit constraint, or the traffic between each of the existing classes is zero (perfect decomposition).

The algorithm operates on a lower triangular 'traffic' matrix with elements t_{ij}. Once the best feasible aggregation is identified, the matrix is revised according to the operators (1) and (2). The outcome of the procedure depends on two parameters:

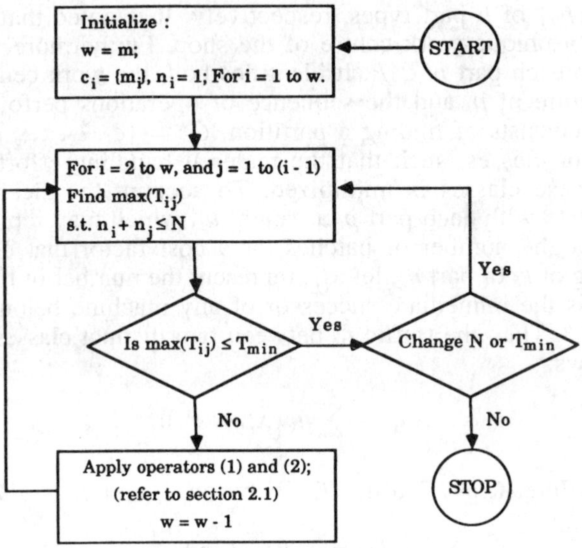

Figure 1. Block diagram for the aggregation algorithm.

(i) *N*, the cell size, and (ii) T_{min}, the minimum value of Normalized Traffic that permits aggregation. The parameter T_{min} ensures that only significant aggregations are performed, and cells with weaker links remain unclassified. The flow chart of the algorithm is presented in Figure 1.

The local refinement procedure

The output of the bottom-up aggregation procedure must be re-examined to validate the assignment of each work centre to its corresponding class. This is necessary since once a work centre is assigned to a given class, it cannot be reassigned to a newly formed one, even though it might be more appropriate to the latter in terms of reduction in inter-cell part transfers. At each step of the refinement procedure a given work centre is considered as a separate external entity, and its traffic with each of the existing classes is evaluated. This work centre is then assigned to the class with which its interaction is the most significant. Although work centres usually remain in their original class, some reassignments are possible. This process is repeated for each work centre. It is noted that reassignments may alter the cell size and consequently may violate the size constraint. A detailed account of the local refinement algorithm is given by Harhalakis *et al.* (1990).

Only after the formulation of manufacturing cells is finalized, the entire set of parts is partitioned into families, each corresponding to a given cell. Each part is assigned to the cell where the maximum number of operations required for its manufacture are performed. The merit of such an assignment is realized if modifications to part routings are allowed in order to revise operations currently performed outside their respective cells. This task is more feasible when a reasonably small number of operations is considered for modification.

Industrial application

Company background

Kop-Flex, Inc., manufactures flexible shaft couplings, clutches and related rotational parts. Product lines consist of mechanically-flexible gear couplings, and a wide range of material-flexing couplings, which incorporate either elastomer materials or metal membrane elements.

The company manufactures almost 1200 make-to-stock parts. In addition, an average of 1500 make-to-order parts are in process at any given time. All manufactured parts are cylindrical and span a wide range of sizes and shapes. The machine shop is functionally arranged and includes over 200 work centres, such as lathes, gear-shapers, drills and milling machines.

Case analysis

The make-to-stock part base was used as the foundation of the analysis, since; (i) it is almost static over time, (ii) strong similarities exist between make-to-stock and make-to-order parts with respect to both design features and processing requirements, and (iii) common work centres are used for the manufacture of both these segments of the part population.

The data required for the case analysis include; (i) the list of work centres, (ii) the production routings of all make-to-stock parts (1186), (iii) the production volume of each part over a certain time period (16 months).

After the completion of the data collection and validation phase the following steps were followed:

(i) Central facilities, such as packing, storage areas and work-ups, which are visited by almost any part in the system, were removed from the routings under consideration. All inspection workstations were also ignored, since all inspection operations that are not performed in-process require special metrology equipment, which cannot be assigned and physically moved to a particular cell.

(ii) Work centres that process most of the manufactured parts, such as saws and rough-out drills, were assigned to a special area of the shop and were not included in the analysis, since they are visited by almost every part-family. The incorporation of this equipment into cells would cause unnecessary high inter-cell traffic.

(iii) The problem of distributing identical machine tools (initially identified as one work centre) to different cells was treated by artificially assigning an equal number of parts to each of these identical machines, which can now be considered as distinct work centres. It is noted that such an assignment is completely artificial and directly influences the formation of cells. This 'artificiality' will be removed by the proposed enhancement outlined in the next section.

The output of the above three-step procedure is a set of 1185 production routings that utilizes 134 work centres. This set, together with the production volume data of all 1186 parts were used as the input to the proposed GT system. For purposes of comparison, two other representative methods, one from each class of grouping algorithms discussed in the introduction, were also used to form the manufacturing cells for the present industrial project: (1) The Enhanced Garcia Proth Method (EGPM) (Nagi, 1988), that forms cells and part-families in order to maximize the number of operations performed to each part-family within its corresponding cell; and (ii) the Knowledge Based Group Technology system (KBGT) of Kusiak (Kusiak, 1988; Kusiak and Wadood, 1988) that forms cells and part-families in such a way that each part-family is exclusively manufactured by work centres in its corresponding cell, or by bottle-neck machines, which do not belong to any cell.

The solutions obtained were evaluated with respect to the reduction of inter-class transfer of parts. The measure used to quantify the merit of each solution is given by:

$$e = \frac{F - I}{F}$$

where F and I represent the total number of transfers required for the manufacture of all parts in the functional and the cellular layouts, respectively. All part transfers

Table 1. Summary of results.

	EGPM		KBGT		ICTMM	
	No. of cells	Machines per cell	No. of cells	Machines per cell	No. of cells	Machines per cell
	16	8	1	7	9	8
	1	6	2	6	2	7
			2	5	1	6
			5	4	3	3
			1	3	4	2
			7	2		
Total no. of cells	17		18		19	
Machines removed		None		68		25
Parts removed		None		711		None
Efficiency (%)		32		21		58·7

between work centres of the same cell were considered local, and were not accounted for the calculation of I.

The cell size limit was set to eight work centres per cell for all three methods. The results are summarized in Table 1, where ICTMM identifies the proposed system as an Inter-Class Traffic Minimization Method. The KBGT system excluded from the cellular solution 51% of the available work centres and 60% of the parts. This is consistent with the objective of the KBGT system, since the parts excluded cannot be manufactured entirely in any single proposed cell. Hence, the machines excluded are used exclusively for the manufacture of these excluded parts.

In contrast, no parts were excluded from the solutions of either EGPM or the proposed method. Twenty five work centres were not placed in any cell by ICTMM, since they shared a low number of common parts with the cells formed. The proposed system offered a substantial (58·7%) reduction in the total number of inter-class transfers. It must be emphasized however, that the efficiency measure, e, represents the objective of ICTMM and is not directly related to the objectives of the other two methods. None the less, the comparative results obtained still provide a useful insight to alternate cellular layout approaches.

Enhancement of the proposed system

The problem of distributing functionally identical work centres to different manufacturing cells is not directly addressed by the proposed system. As mentioned earlier this problem is currently handled by artificially assigning an equal number of parts to each of these work centres. This assignment, however, directly influences the solution obtained, and may cause a high number of unnecessary part transfers. In addition, the capacity available of a given work centre may be exceeded, resulting in a non-feasible solution.

A pre-processing algorithm, which operates on the production routings prior to the application of the GT system, is proposed here, in order to: (i) distribute the parts processed by the functionally identical work centres in a way that favours the decomposition of the system; (ii) satisfy the capacity constraint. This algorithm is briefly outlined below:

For each part $p_i \in P$, a set $R_i = \{r_i^1, r_i^2, \ldots, r_i^{\lambda(i)}\}$ of $\lambda(i)$ alternate routings is defined as follows: if p_i is processed by a work centre type that corresponds to k functionally identical machines, then k alternative routings are included in R_i, each corresponding to a distinct machine member of M. The pre-processing algorithm selects the preferred routing(s) for each part p_i among the elements of the corresponding set R_i. The underlying idea is to maximize the similarity between the routings of parts that already display similar processing requirements.

The similarity between routings r_i^k and r_j^l of parts p_i and p_j is quantified by the following measure:

$$K[i, j, k, l] = \frac{1}{2} + \frac{a_{11} - (a_{12} + a_{21})}{2m}$$

where a_{11} represents the number of common machines used for the manufacture of both parts, according to the above routings; a_{12} represents the number of machines used exclusively by p_i; and, a_{21} the number of machines used exclusively by p_j.

The objective of the pre-processing algorithm is given by:

$$\text{MAX}: \sum_{i=1}^{n} \sum_{k=1}^{\lambda[i]} \sum_{\substack{j=1 \\ j \neq i}}^{n} \sum_{l=1}^{\lambda[j]} K[i, j, k, l] n_i^k n_j^l$$

where n_i^k, n_j^l represent the number of units of parts p_i, p_j, to be made according to routings r_i^k, r_j^l, respectively.

The production volume required for each part and the capacity available on each machine impose the following constraints on the algorithm:

$$\text{C1}: \sum_{k=1}^{\lambda[i]} n_i^k = U_i$$

where U_i represents the production volume of part p_i, and n_i^k represents the number of units of p_i manufactured according to routing r_i^k.

$$\text{C2}: \sum_{i=1}^{n} \sum_{k=1}^{\lambda[i]} \Theta_{ik}^q n_i^k \leq \Theta_q; \qquad \text{for } q = 1, 2, \ldots, m,$$

where Θ_q is the capacity available on machine m_q, and Θ_{ik}^q is the processing time of part p_i on machine m_q, when manufactured according to routing r_i^k.

The output of the pre-processing algorithm constitutes the input to the system discussed earlier.

Conclusions

A comprehensive Group Technology system that can address most aspects of the cellular layout problem is presented in this chapter. The system arranges the equipment of a machine shop into cells, in order to minimize the total number of inter-cell transfers necessary for the manufacture of the entire set of parts. In addition, each part is assigned to the cell where most of its operations are performed. Thus, the machine shop is decomposed into a number of smaller, less complex, and almost independent systems. The proposed method caters for the sequence of operations specified in production routings, and considers repeated, non-consecutive operations performed on a part by the same work centre.

The system was applied to a large-scale industrial project that involved 134 work centres and 1186 parts. Two other representative GT methods were also employed for the same application. Although all three methods were applied under identical conditions, the present GT system yielded a considerably higher reduction in inter-cell part traffic.

Currently, the system is being enhanced to facilitate the distribution of functionally identical work centres into appropriate cells. A pre-processing algorithm is proposed to select the most appropriate set of routings for a good decomposition of the machine shop, while satisfying the capacity constraint.

Acknowledgements

This research has been partially supported by the Engineering Research Center program NSFD CPR88003012, University of Maryland, and the Institut National de Recherche en Informatique et en Automatique (INRIA-Lorraine, France). Acknowledgements are also due to the Maryland Industrial Partnerships (MIPS) program and the Kop-Flex company for funding and other support.

References

Chan, H.M. and Milner, D.A., 1982, Direct clustering algorithm for group formation in cellular manufacturing. *Journal for Manufacturing Systems*, **1**, 1.

Garcia, H. and Proth, J.M., 1985, Group technology in production management: the short horizon planning level. *Applied Stochastic Models and Data Analysis*, **1**.

Garcia, H. and Proth, J.M., 1986, A new cross-decomposition algorithm: the GPM. Comparison with the bond energy method. *Control and Cybernetics*, **15**, 2.

Harhalakis, G., Nagi, R. and Proth, J.M., 1990, An efficient heuristic in manufacturing cell formation for group technology applications. *International Journal of Production Research*, **28**, 1.

Harhalakis, G., Hilger, J., Nagi, R. and Proth, J.M., 1989, The density approach: a tool for the design of manufacturing cells and part families. *Applied Stochastic Models and Data Analysis* (in press).

King, J.R., 1980, Machine-component grouping using ROC algorithm. *International Journal of Production Research*, **18**, 2.

Kusiak, A., 1988, EXGT-S: a knowledge-based system for group technology. *International Journal of Production Research*, **26**, 1.

Kusiak, A. and Wadood, I., 1988, Knowledge-based system for group technology. *1st International Conference on CIM, RPI*, Troy, NY, 22–25 May, 1988.

Leskowsky, Z., Logan, L. and Vannelli, A., 1987, Group technology decision aids in an expert system for plant layout. *In Modern Production Management Systems* (Amsterdam: Elsevier).

McAuley, J., 1972, Machine grouping for efficient production. *Production Engineering*, **51**, 53.

McCormick, W.T., Schweitzer, P.J. and White, T.E., 1972, Problem decomposition and data re-organization by a clustering technique. *Operations Research*, **20**, 993.

Nagi, R., 1988, Selection and layout of facilities for cellular manufacturing systems. M.S. Thesis, University of Maryland.

PART VII
EXPERT SYSTEMS AND PRODUCTION SIMULATION

Expert systems and production simulation

The development of 'expert' or 'knowledge based' systems has brought the achievements of artificial intelligence research to the point of commercial application and exploitation. Fahnrich and Thines review and describe the state of the art of knowledge based production applications, based on documented literature in the United States, Great Britain and Germany. Kwok and Carrie present an expert systems approach to the design of flexible manufacturing systems (FMS). They attempt to bring together the software tools, analytical models and simulation techniques, and integrate them in an expert system.

Simulation modelling is a powerful tool in the field of production systems, assisting in areas such as planning and decision making. The applications considered in this section include such elements as simulation of production disturbances involving human intervention at danger zones by Häkkinen *et al.* Gallwey and O'Muirthile report on simulation of work-time distribution for a paced assembly line whilst MacCarthy considers the use of simulation to aid short term decision making.

SECTION VII.1. ARTIFICIAL INTELLIGENCE

71

Intelligent control in automated manufacturing environments

Nobuto Nakamura, Katsushi Noma, and Morio Isohashi

Department of Industrial and Systems Engineering, Faculty of Engineering, Hiroshima University, Shitami, Saijo-cho, Higashi-Hiroshima 724, Japan

Abstract

This chapter describes the application of advanced artificial intelligent approaches, especially blackboard architecture, to manufacturing control. An emulator which moves as if it were a real system is first generated on a computer. Then the Blackboard Cell Control System (BCCS), based on a blackboard composed of knowledge sources, and a blackboard controller are built to control this emulator in the real-time mode. The emphasis is on response to disruptions, i.e. machine failures or other abnormal occurrences, and the corrective action of BCCS is demonstrated by an example.

Introduction

The control of an automated manufacturing system is very complex with a huge number of interactive relationships between NC machines, parts, tools, and carts. One approach that can cope with the complexity is that of breaking the control into a hierarchy such as factory, shop, cell, and equipment levels (O'Grady, 1986).

In this research we deal with cell-level control. The major functions of the cell-level control are to schedule jobs, machines and other resources, and to order operation commands at the equipment level to achieve the goals from the shop level. But in real life, the control systems must allocate machines or carts to the corresponding parts in an ever-changing situation. Furthermore, they have to cope with events that cannot be considered at the shop level, such as machine failures, new job arrivals, and so forth. In such circumstances it is expected that there will be unanticipated deviations from the original schedule. The control function therefore has the responsibility of executing real-time corrective actions to adhere to the original schedule as much as possible. To achieve such real-time control many researchers have recently been interested in using expert systems, but many of these investigations are still at the prototype stage. However, the Intelligent Cell Control System (ICCS) (O'Grady and Lee, 1988) uses blackboard architecture to indicate a new control system.

In this research, we are concerned specifically with modelling of the cell control system, based on the blackboard control method concept (Hayes-Roth, 1985).

First, we generate an emulator which moves as if it were a real system. Secondly, the Blackboard Cell Control System (BCCS) is generated, which can control this

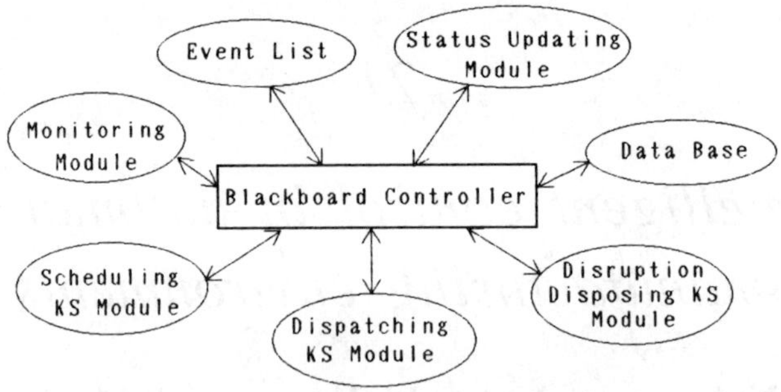

Figure 1. Blackboard Cell Control System (BCCS).

emulator in real-time on Apollo 3010 in DOMAIN commonLISP. Finally, the feasibility of the BCCS truly controlling the emulator is clarified.

Blackboard Cell Control System

The blackboard system is a knowledge representation/inference method which has been described by Hayes-Roth (1985). This system is usually divided into working memories, several related production systems, and a control mechanism which controls a system starter, so as to co-operate with various hierarchies or different kinds of knowledge by having a common data area called the blackboard. We try to apply the concept of the blackboard to the cell real-time control, and call the acquired system the Blackboard Cell Control System (BCCS; see Figure 1 for its concept). The BCCS is constructed from an architecture which is composed of some modules and where overall control can be achieved by passing appropriate messages among the blackboard controller and the modules around it. The major functions of the BCCS are:

(1) Scheduling: the BCCS schedules resources within the cell so as to perform the goals or orders set by the shop level.
(2) Operation dispatching: the BCCS generates detailed operations to order to the equipment level.
(3) Monitoring: the BCCS receives and classifies the feedback information from the equipment level.
(4) Disruption disposing: the BCCS recognizes unanticipated phenomena in the cell and copes with those problems.

The procedures or knowledge sources (KSs) which achieve these functions are included in the modules around the blackboard controller, that is, a scheduling KS module which has several scheduling or rescheduling rules, a dispatching KS module which generates detailed operations, a disruption disposing KS module which has some KSs for machine failures, late jobs, etc., a data base, a status updating module, an event list, and a monitoring module.

The blackboard controller repeats the following steps:

(1) Updating the event list by using the status updating module and information from the monitoring module.
(2) Choosing an event from the updated event list.
(3) Executing the chosen event.

Each KS module is represented by several production rules, and the BCCS currently contains about 260 rules.

Functions of each module

Monitoring module

The role of the monitoring module is basically to interface with the higher level (in this case shop level) and the lower level (in this case equipment level). In this research, we emphasize the interface with the equipment level. The initial information from the shop level is: an original schedule, operation sequences of parts, due-dates, etc., and after operations have started, the module receives any information about new job arrivals.

The feedback information from the equipment level is: the arrival cart at its destination, the finished operation at each work station, NC machine failures, the finished repairs, the finished tool change, etc. The orders to the equipment level are: moving a cart, executing an operation, etc.

Status updating module

The status updating module checks the following items using information from the monitoring module and the event list, and then updates the event list.

- Checking the cart status and the cart queue, selecting a cart, and updating the cart status; when a cart becomes idle, the module allocates it a new job and updates the cart status.
- Deciding whether to call a cart to the main buffer; when there are some parts in the main buffer, the module decides whether a cart should be called to the main buffer considering operation status of parts, NC machines, inspecting station, etc.
- Deciding whether to call a cart to the washing station; when there is a finished part in the washing station, the module decides whether a cart should be called to the washing station, and checks if the part has already finished all its operations.
- Deciding whether calling a cart to the tool buffer; when there is an NC machine needing its tools to be replaced, the module judges if a cart should be called to the tool buffer.
- Judging whether the tools of an NC machine could be replaced; when there is a cart carrying the tools of an NC machine, the module judges if the tools of the NC machine could be replaced.
- Checking whether all scheduled operations have finished; the module checks whether all scheduled operations have already finished.

Event list

The event list has much information from the monitoring module and some events have been generated in the status updating module. The events contain cart queues (at NC machines, the main buffer, the washing station, etc.), some parts at the main buffer, and executed operations, etc.

Scheduling KS module

The scheduling KS module has some KSs for original scheduling and rescheduling. The original schedule essentially should be made at the shop level, but in this research, it is made in this system as information from the shop level. This original scheduling is made by using SPT or S/OPN. The rescheduling KSs are used when a disruption occurs.

Dispatching KS module

The dispatching KS module has several KSs to generate detailed actions, such as passing a part between a workstation and a cart according to a previously determined schedule. For example, the part passing rules are basically as follows:

Rule 1: When a part is put to the workstation from a cart,
IF • the cart destination is the workstation
• the cart has a part
• there are no parts or free spaces in the workstation
THEN • move the part to the workstation from the cart
• when the workstation is an NC machine, a washing or inspecting station, the workstation immediately starts operation
• cart is idle
Rule 2: When a part is moved to the cart,
IF • the cart destination is the workstation
• the cart has no parts
• there is a part in the workstation
• when the workstation is an NC machine, a washing or inspecting station, the workstation has finished the part operation
• the next destination is idle or has a free space
THEN • move the part to the cart
• the workstation is free or idle
• change the cart destination to the part next destination.

However in the NC machine a part is put on the NC buffer, which can have one part. And since there are 10 buffers in the main buffer, the passing rules have still more searched free spaces for putting a part. Also since there are four load–unload stations, it is necessary for rules to select the station.

The operation rules of the NC machine are as follows:

Rule 3: After putting a part on the NC buffer,
IF • the NC machine is not down
• there are no operating parts
• the NC machine is not waiting tool-change
THEN • start the part operation
• the NC machine is operating
• the NC buffer is idle
• record operation start
Rule 4: When the operation is finished,
IF • the NC buffer has a pre-operation part
• the NC machine is not waiting tool-change
THEN • exchange the part on the NC machine and one on the NC buffer, and start the operation
• call a cart
• record operation finish and start
Rule 5: When the operation is finished,
IF • the NC buffer has no parts
THEN • move the part to the NC buffer
• the NC machine is idle
• record operation finish
• call a cart

Moreover there are some rules such as for NC machine failures and tool-change related to the NC machine. This dispatching KS module contains about 70 rules similar to the ones above.

Data base

The data base in the BCCS contains the status of jobs and workstations, which are composed of property lists in commonLISP. For example, there are machine

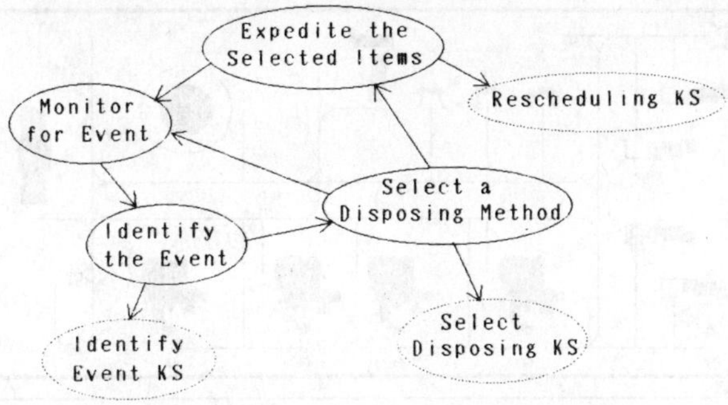

Figure 2. Control flow.

number, operation status, operation parts, expected operation finish time, remaining operations and tool number in the data related to an NC machine.

Disruption disposing KS module

The BCCS copes with NC machine failures, new job arrivals, late jobs, and crowded jobs at a certain NC machine when the disruption occurs. The BCCS decides disposal methods according to the control flow (see Figure 2). These disposal methods are in the disruption disposing KS module, and motivated by the information from the monitoring module. In case of an NC machine failure, at first, it receives the machine number and the degree of the failure, and confirms operation status and tools of all NC machines. Next, it decides the method of relocating tools and rescheduling jobs. In the case of a late job, whenever one operation of a part finishes, the module judges whether it is late or not. If the job is late, the job priority is changed into 'important' and the BCCS reschedules all the remaining jobs. In the case of a new job arrival, the module judges whether it is urgent or not; if the job is urgent, the job priority is changed into 'important' and the BCCS reschedules all the remaining jobs to include the new job.

Emulator

In this research, we referred to a flexible manufacturing system (FMS) at Mt Orab in Cincinatti Milacron Ltd. (Meredith, 1988) as the emulator model and generated a graphic emulator in DOMAIN C and 2D GMR on Apollo 3010 (see Figure 3).

The emulator is composed of four identical NC machines, four load–unload stations, three carts, one tool-load station, one inspecting station, one washing station, a main buffer (capacity of ten parts), three tool buffers, and one pallet pool.

The operation status of the NC machines are represented by changing their colour in this emulator. The carts can move in one side direction at the central oval, and can move in both directions at others. NC machine failures can be broken out by a Poisson distribution or file input. The interface between the BCCS and the emulator is made by using input and output files.

An example

To show the feasibility and availability of the BCCS, we practised with the following example. The original schedule is shown in Figure 4 in the case where there are ten jobs, each job having three operations. The mean operation time is

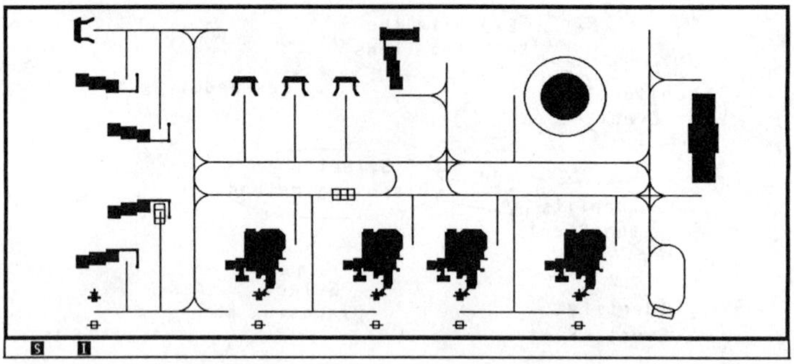

Figure 3. FMS emulator.

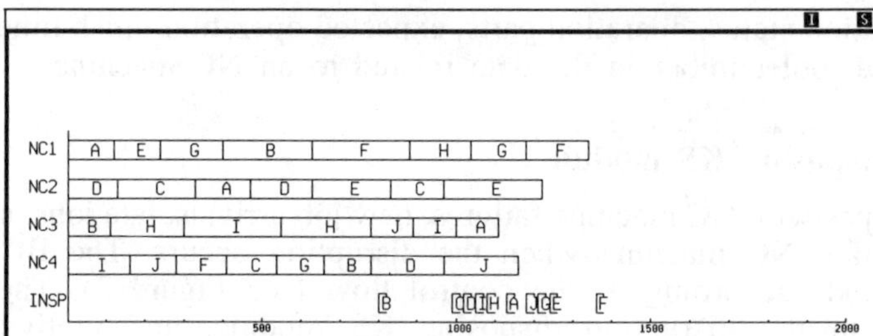

Figure 4. Original schedule.

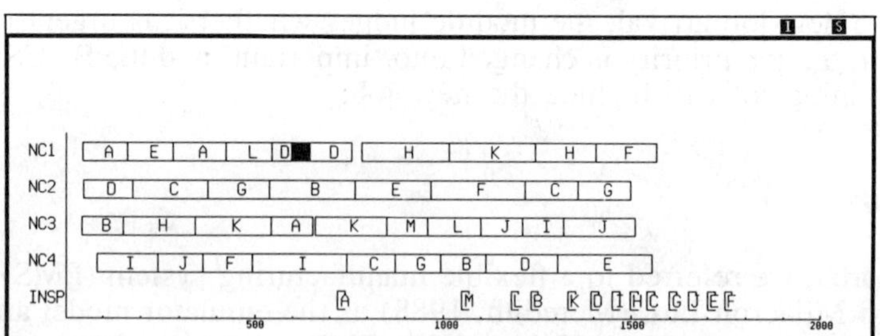

Figure 5. Operation result.

Table 1. Operation result.

	Late KS	SPT	S/OPN
Mean tardiness	74·4	90·6	93·2
Total production time	1801	1827	1831

169 and the mean due-date is 1550. In this case, three new jobs arrived at 150, 300, 500 respectively and tool damage occurred in NC1 at 644.

The BCCS rescheduled on the arrival of the new jobs, and judged that job A became late at 166 and job K at 544. The result is shown in Figure 5. Table 1 shows the results in case of using disruption disposing KS for late jobs, and that of only using SPT or S/OPN. (However when a new job was introduced, both of them were rescheduled by the S/OPN rule and the part priority.) The blocked parts in Figure 5 show tool replacement. From this result, it can be seen that the BCCS can completely perform all the operations and get a better performance than that obtained without using the KS for judging the late job.

Conclusion

In this chapter we generated the Blackboard Cell Control System (BCCS) on Apollo 3010 in DOMAIN commonLISP to carry out intelligent real-time control at the cell level in an hierarchical control system. The BCCS was constructed using blackboard architecture, which has several advantages including modularity, simplicity and parallelism. We showed its feasibility by controlling the FMS emulator; this system can cope with disruptions such as NC machine failures, late jobs and so forth. Also, it is relatively easy to add new knowledge because the knowledge in the BCCS is modularized, enabling knowledge stock enlargement.

References

O' Grady, P.J., 1986, *Controlling Automated Manufacturing Systems* (London: Kogan Page).

O'Grady, P.J. and Lee, K.H., 1988, An intelligent cell control system for automated manufacturing. *International Journal of Production Research*, **26**, 5, 845–861.

Hayes-Roth, B., 1985, A blackboard architecture for control. *Artificial Intelligence*, **26**, 251–321.

Meredith, J., 1988, Installation of flexible manufacturing system teaches management lessons in integration, labor, cost, benefits. *Industrial Engineering*, **20**, 4, 18–27.

72

Linking shop floor and management systems in real-time

U.S. Bititci,[†] A.S. Carrie and C.R. McMillan

Department of Design Manufacture and Engineering Management, Manufacture and Engineering Management Division, University of Strathclyde, Glasgow, UK

Abstract

Cell management systems provide a bridge linking plant level production management systems to shop floor systems. They provide an interface between the planning systems which are responsible for plant-wide planning functions and shop floor systems which operate in real-time. This chapter presents a study of currently available cell management systems in order to identify the variety of functions offered. The majority of those reviewed provide both planning functions and real-time functions. Operational decisions are carried out at the planning level and executed at the real-time control level. The limitations of the available cell management systems have been identified. The principal ones are the lack of real-time decision making ability to react to disturbances and the lack of diagnostic systems of cell level. A brief review of current research is included and further research needs are highlighted.

Introduction

Cell management systems provide a bridge between shop floor and engineering and management functions such as production planning and control, maintenance, design, process planning and industrial engineering. Cell management is therefore a critical transition point within the management hierarchy where plans are converted into real-time actions (Figure 1). The cell management system receives manufacturing plans and instructions from systems residing at plant and centre level, processes this data and translates it into instructions to control the shop floor systems (i.e. equipment and operators) in real-time. It also captures production information from the shop floor and presents it as management information to centre and plant level systems.

Previous studies on various industrial installations on both fully automated and human-centred systems have identified the essential functions of cell management systems as:

[†] Present address: The Strathclyde Institute for Computer Integrated Manufacture, Exchange House, 229 George Street, Glasgow G1 1RX, UK.

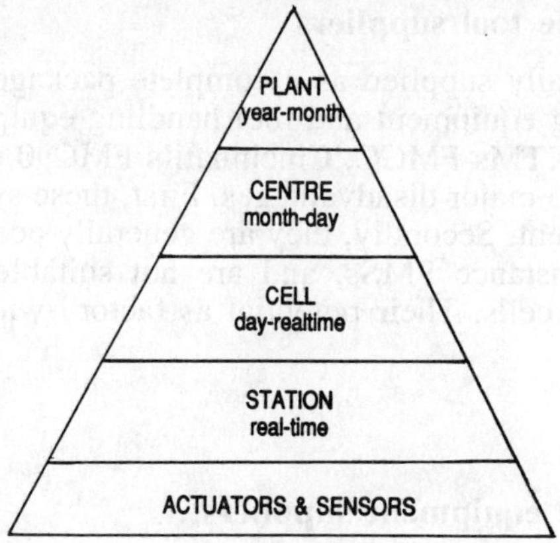

Figure 1. Levels of management and control.

— material management,
— work centre management,
— tool management,
— maintenance management,
— cost management.

A variety of off-the-shelf cell management packages, which posses many features of the functions mentioned above, are now available from equipment suppliers and software houses. All of them view the cell management system at two levels; the planning level and the control level. The planning level is responsible for the translation of instructions from plant and centre levels into work-to-lists for each work centre created, using pre-defined process plans and scheduling algorithms. The control functions simply execute the work-to-lists by controlling the activity of each work centre and the flow of tooling and material. This separation of planning and control functions within the cell management system leads to various operational difficulties and makes the system inherently inflexible.

Cell management: the state of the art

A survey of various proprietary cell management systems currently available has been carried out. Some of these systems are well established in the market place while the others have recently emerged. In general it was found that those systems offered by Flexible Manufacturing Systems (FMS) equipment suppliers are established systems. These systems have been developed as a natural progression from stand-alone machine controllers and are in wide use throughout the industry as FMS control systems. More generic systems, not specific to one manufacturer's hardware have recently begun to emerge. These generic systems are in most cases products of systems and software houses.

The analysis of the survey suggests that the cell management systems currently available may be categorized under the following headings:

— machine tool supplier specific systems,
— control equipment supplier specific systems,
— generic systems,
— generic shells.

The nature of each type is discussed with comments on their functional capabilities and limitations.

Systems from machine tool suppliers

These systems are usually supplied as a complete package including the machine tools, material handling equipment and tool handling equipment where applicable. Typical examples are KTMs FMCC, Cincinnati's FMC90 and Scharmann's ESCO systems. They have two major disadvantages. First, these systems are limited to the supplier's own equipment. Secondly, they are generally configured to manage fully automated cells, for instance FMSs, and are not suitable for managing human-centred manufacturing cells. Their potential as factory-wide management systems is therefore limited.

Systems from control equipment suppliers

These systems are also commonly used in industry, but again they are typically used as FMS controllers and are not necessarily suitable for managing human-centred cells. As with the previous category they are limited to work with the supplier's own range of products, such as PLCs and computer numeric control (CNC) controller units. This limitation may not be as severe as in the previous case since most manufacturing equipment, such as CNC machines, is usually available with a range of controllers. However, recent experience (Bititci, 1988) shows that even within a single control equipment manufacturer's product range, compatibility problems may arise due to controllers not being identical. One of the most commonly used systems in this category is the Siemens FMC 20·32 system, which is widely used in European FMS installations.

Generic systems

Over recent years several generic systems, such as Simacs, Factomate, Intercim and Factorial, have emerged from systems and software houses. These are systems which are configured to carry out most of the tasks outlined, such as material management, work centre management and tool management. They do not have a specific interface to shop floor systems but rely on control devices, such as CNC controllers, robot controllers and PLCs, as well as operators, to carry out shop floor tasks. All the control devices and operators are treated as systems to which instructions or part programs can be downloaded. For instance, just as a part program can be downloaded to a CNC machine, an instruction may be similarly sent to an operator interface. Although this permits management and control of virtually any equipment, as well as human-centred systems, the amount of control available is limited by the capability of the attached devices. The majority of the systems in this category make use of data networks to achieve the necessary communications. A major disadvantage is the cost associated with the specific protocols which need to be developed to link devices on to the system's data network.

All of these systems are structured systems with specific functional characteristics. The end user is expected to supply the data in order to make the system complete. If the system's functionality is not what the user requires, it is usually not possible to modify those rules and procedures which affect the functionality of the system. It is not uncommon to find cases where the end users opted for a tailor-made system after examining the off-the-shelf options (Carrie and Bititci, 1988a); this is due to two reasons. The first is the incompatibility of the shop floor equipment with the off-the-shelf system, any modification being too expensive. The second reason is the inadequate functionality of the system.

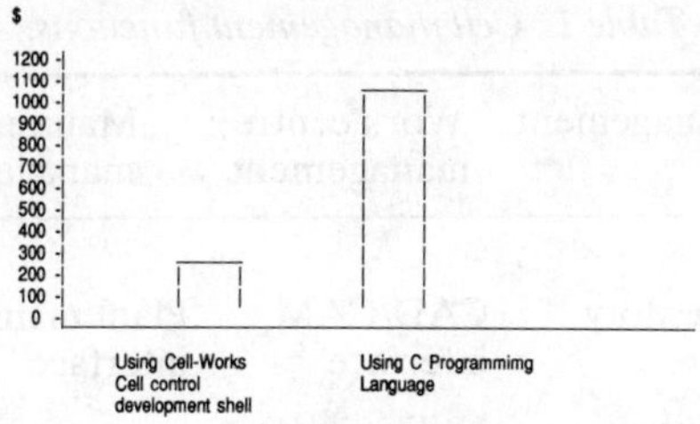

Figure 2. Cost comparison between tailor-made and generic shells.

Generic shells

A new generation of cell management systems is now emerging. These are cell management shells. They provide all the interfaces that would be expected from a cell management system, in a modular fashion. If a cell consists of two machine tools with Siemens controllers, a Puma robot, a Texas Instruments programmable logic controller and two Allen-Bradley barcode readers, the user simply purchases the appropriate driver modules. These provide a fully functional off-the-shelf interface between the attached devices and the shell. The shell itself is a cell management and control development environment which enables the user to develop the desired functionality, basically by stating operational rules. The drivers, transparent to the user, make use of the instructions generated by the application specific rules and data supplied to control shop floor activities. An example of such a system is Cell Works—a cell management development environment offered by Fastech Integration Inc.

Generic shells provide a single environment for complete cell integration tailored to meet the requirements of a specific application. It is claimed that they dramatically reduce software development and integration costs. Savings of between 4 : 1 and 10 : 1 have been claimed when compared with tailor-made applications using conventional programming languages. Figure 2 shows the cost comparison for a management system to manage and control a complete printed circuit board assembly cell consisting of insertion machines, robots, a materials handling system, a wave soldering machine, cleaners and testers.

Specific features of cell management systems

The survey included a range of cell management systems from each category. Table 1 summarizes the functionality of the systems surveyed. The generic features associated with these systems have been identified and subdivided into planning and control functions. Although all of the systems surveyed address a majority of the functions in one way or the other, each one employs a different approach. The following sections discuss the various ways in which each function is tackled.

Material management

The functionality of the systems surveyed with regard to material management is summarized in Table 2. One very important function which resides at the planning level is scheduling. In the simplest form of a cell control system, the operator provides the system with a work-to-list, consisting of component identification, quantity required and priority. At the control level the sequencing function uses this work-to-list, together with process plans previously supplied, to sequence work to the appropriate workstations. The process plans supplied to the cell provide a

Table 1. Cell management functions.

Material management	Tool management	Work centre management	Maintenance management	Cost management
Planning functions				
MRP interface	Tool inventory control	CAD/CAM interface	Plant maintenance interface	Tool costing
Process plan importing	Tool requirements planning	Utilization reporting		Machine costing
Scheduling	Tool preparation			Operator costing
Control functions				
Control of materials handling system	Tool supply	Machine interface	Cell diagnostics	
Work tracking and identification	Tool identification and tracking	Operator interface	Machine diagnostics	
	Tool data transfer	Status monitoring		
	Tool condition monitoring			

Table 2. Summary of material management functions.

	MRP interface	Process planning	Scheduling	Control of of MHS	Work identification and tracking
Scharmann ESCO 1 2 & 3	Data import	Pre-defined manual	Max spindle utilization	Dedicated PLC	Status mapping
KTM FMCC	N/A	Pre-defined manual	Manual operations sequence	Dedicated PLC-RGV	Status mapping
Cincinnati FMC 90	N/A	Pre-defined manual	3 Level capacity/ resource/ sequence	Dedicated PLC-RGV	Status mapping
Siemens FMS 20·32	Manual entry	Pre-defined manual	Max spindle utilization	Dedicated PLC	Status mapping
SD LTD FACTOMATE	Data import	Pre-defined manual	Due-date/ order priority	As any workstation	Limited to equipment
SCI-CON SCIMACS	Data import	Pre-defined import	Due-date/ resource cost	As any workstation	Limited to equipment
INTERCIM Corp. INTERCIM	Data import	Pre-defined import	External	As any workstation	Limited to equipment

detailed description of the route a workpiece should follow including alternative resources.

In more sophisticated systems a works order is imported to the cell management system, through an interface to the materials resource planning (MRP) system. The scheduling algorithms provided will then generate an operations schedule, accounting for due-dates, order priorities, machine loads and tooling availability. Typically in such a system, scheduling within the cell management system exists in three levels: (1) capacity scheduling, (2) resource scheduling, and (3) sequencing. When a works order is received from the plant level system, the first task is to schedule this works order within the limits of the cell's capacity, for a given planning horizon. Scheduling at capacity level involves analysis of the total cell capacity including tools, fixtures and pallets to allocate and start and finish times for each component. This information is then passed to Resource Scheduling. Resource Scheduling carries out resource allocation to satisfy the works order. The typical planning horizon here is one shift or one day. As a result of resource scheduling a detailed production schedule for the given planning horizon is generated together with an operator requirements plan. This information is then used by the sequencing function for real-time control of resources by deciding upon the best part to sequence next on to a workstation. This is usually achieved using algorithms to maximize resource utilization.

At the control level the systems provided by the machine tool and control equipment suppliers are usually associated with dedicated materials handling systems, such as Rail Guided Vehicles (RGV), Automated Guided Vehicles (AGV) and conveyor systems. This permits a dedicated controller to be configured to meet the system's requirements. Therefore, the systems by the equipment manufacturers tend to demonstrate comprehensive capability for controlling material handling systems. Through this feature they are also capable of tracking work by mapping the status of each workpiece. Generic systems, however, are not associated with a dedicated materials handling system. The materials handling system controller is treated as any other workstation. Therefore the system's capability is dependent on the capabilities and limitations of selected controllers.

Tool management

Table 3 summarizes the tool management functions for the systems surveyed. In early manufacturing systems the issue of tool management was largely ignored. During recent years the subject of tool management has received considerable attention from both academics and industrialists. In the light of the recent developments the majority of the modern cell management systems provide some tool management capability.

At the planning level, tool management functions are tool inventory control, tool requirements planning and tool preparation. Tool inventory control tends to be largely associated with plant level systems such as TOMAS, MICROSS 8000 tooling module and AMAPS/Q tooling module. In most of the cell management systems which were reviewed, the tool inventory control function exists almost as an element of the tool requirements function. The majority of the systems assume that a particular cell is serviced from a central tooling store, i.e. they adopt a centralized tooling strategy. Scharmann's ESCO systems support both centralized and decentralized tool storage strategies, where a tool store is self-contained within the cell.

Tool requirements planning is a central function within tool management. Its main task is to convert the production schedule into net tool requirements for both durable and consumable tools. In doing this it requires information on the tool sets required by each operation, existing prepared tools and partially-used tools already

Table 3. Summary of tool management functions.

	Tool inventory control	Tool requirements	Tool preparation	Tool supply	Tool identification and tracking	Tool data transfer
Scharmann ESCO 1 2 & 3	Centralized and decentralized	Individual machines	Assembly instructions	AGV or robot, manual	Mapping	DNC
KLM FMCC	N/A	Individual machines	Preparation List	Manual	Mapping	Manual
Cincinnati FMC90	N/A	Individual machines	Preparation List	Manual	Mapping	Manual
Siemens FMS 20·32	N/A	Planned tool changes	Preparation List	Robot Manual	Mapping	Manual/ Auto
SD LTD FACTOMATE	N/A	Per workstation Limited to M/C[1]	Graphics	Manual Auto	Mapping	LAN
SCI-CON SCIMACS	N/A	Per workstation Limited to M/C	Preparation List	Manual Auto	Mapping	LAN
INTERCIM Corp. INTERCIM	N/A	Per workstation Limited to M/C	Preparation List	Manual Auto	Mapping	LAN or DNC

[1] M/C: machine controller

on the workstation. Based on this information, it then generates a net tool requirements list for individual machines. At least this is the case when the system is supplied by the machine tool or control equipment manufacturers. In the majority of cases the machine tool controller stores the information on the status of the tools on its tool magazine. The cell control system merely takes a snapshot of the current tool configuration for tool requirements planning purposes. Therefore in the case of generic systems the capability of the tool requirements planning functions is limited by the capability of individual controllers or the system design parameters.

One set of outputs from tool requirements planning is the tool preparation instructions to the tool room for tool pre-setting and fixture configuration. The reviewed systems, in their simplest form, produced a tool preparation list which consists of a list of the tools required, expressed in tool identification numbers. More sophisticated systems are capable of providing the operator with graphical or textual assembly instructions for each tool.

Tool management functions at the control level are tool supply, tool identification and tracking, and tool data transfer. Most of the generic systems are capable of controlling both automated and manual tool supply mechanisms. They are capable of downloading tool removal and replacement instructions to an operator terminal for manual tool supply or to the controller of an automated device for automated tool supply. Equipment suppliers' systems do not enjoy this flexibility. However, some suppliers offer a variety of options for tool handling such as AGVs, robots or manual methods.

Most systems employ a mapping technique to identify the location and status of each tool. The mapping technique used is greatly influenced by the capability of the individual workstation and the communication facility between the workstation and the cell management system. Where the workstations demonstrate the appropriate communications capability through direct numerical control (DNC)

Table 4. Summary of work centre management functions.

	CAD/CAM interface	Utilization reporting	Machine interface	Operator interface	Status monitoring
Scharmann ESCO 1 2 & 3	N/A	Historical reports	DNC	Limited dialogue	All status in real-time
KTM FMCC	N/A	Historical reports	DNC	LEDs Push-Buttons	Comprehensive
Cincinnati FMC90	N/A	Production summary	DNC	Comprehensive	Comprehensive
Siemens FMC 20·32	Yes Sinec-Hi	Historical reports	SINEC-HI LAN	Terminal/ keypad	Comprehensive
SD LTD FACTOMATE	YES LAN	Historical reports	CAM-WAY LAN	Comprehensive User defined	Comprehensive Limited to M/C
SCI-CON SCIMACS	Yes LAN	Historical reports	LAN	Comprehensive User defined	Comprehensive Limited to M/C
INTERCIM Corp. INTERCIM	Yes LAN	Historical reports	Factory-net LAN	Graphics Comprehensive	Comprehensive Limited to M/C

links or Local Area Networks (LAN), the mapping is done by taking a snapshot of the tooling configuration at each workstation. Otherwise mapping is done at the cell management system level by recording which tool was allocated to which workstation.

The tool data transfer function serves two purposes. One is to support the tool identification and tracking function and the other is to enable the transfer of tooling data (tool lives, maximum force, torque, etc.) to individual work centres. This function is again limited to the capabilities of individual workstations. Some systems provide an automatic data transfer facility, through LANs and DNC links. Others rely on the operators to input the tool data manually at each workstation.

Work centre management

Table 4 summarizes the work centre management functions associated with the systems studied. With current technology, any cell management system should be able to import graphical data and post-processed part programs created by the plant level computer-aided design/computer-aided manufacture (CAD/CAM) systems. The CAD/CAM interface facility seems to be only available on those systems which make use of LANs for communication purposes. The majority of the machine tool supplier specific systems make use of DNC links to establish communication between the shop floor and the cell management system primarily for downloading part programs and communicating tool data where applicable.

The operator interface is particularly critical where a cell consists of a number of human-centred workstations. Appropriate instructions and alarm conditions need to be communicated to the operators. In addition, the operators should be able to provide information on the status of their workstation to the management system. In basic systems, the operator interface is achieved through an array of light-emitting diodes (LEDs) and push-buttons. This is more suitable for automated systems where only limited operator dialogue is required. More sophisticated operator interfaces provide comprehensive facilities, where the dialogue is user defined and communication with the operator is by means of colour graphics terminals.

Most cell management systems, of whatever type, provide a comprehensive status monitoring facility. However, in the case of the generic systems the capability of the system may be limited by the capability of each work centre and the design parameters of the manufacturing system. All systems demonstrate the capability to summarize this status information and present it as historical reports.

Limitations of existing cell management systems

All the systems discussed are limited, in that they make use of pre-defined process plans, which tend to associate an operation with a visit to a particular machine tool. This operation is usually identified with a part program, which in turn makes use of a standard set of tools. This leads to the development of static schedules, such as the production schedule and the tool replenishment schedule. Although the system is capable of reacting to minor disturbances through its sequencing algorithms, due to the static nature of these schedules, they lack the flexibility required to deal with major disturbances. For instance, if a critical resource unexpectedly goes down, those parts which are scheduled to visit that particular resource will have to be delayed until the resource is available again. Although it is possible to reschedule the part to an alternative resource, this will be done with the aid of an operator.

In modern systems, disturbances which may affect the performance of a cell may be identified under two headings: internal disturbances and external disturbances. An internal disturbance may be defined as a disturbance to the current schedule due to an unexpected event within the cell, as in the example above. An external disturbance, however, is inflicted upon a cell due to a disturbance in a related cell. The current trend is towards cellular manufacturing, where a number of cells supply each other based on Just-in-Time (JIT) principles, and synchronized manufacturing. In this situation there is a high degree of inter-dependency between the cells; the requirement for real-time communication between individual cells is an issue which needs to be addressed.

In principle, it could be argued that both the planning and control features associated with the existing cell management systems provide acceptable levels of support providing that the system functions according to plan. However in general, disturbances other than minor ones, result in a situation where the system cannot react in real-time. This is mainly due to static process plans which in turn lead to standard operations, part programs and tooling.

Another limitation is the lack of maintenance management functions. Modern manufacturing cells require a large amount of investment. Therefore once operational they need to be kept at top form in order to minimize down-time. Despite this, only a limited number of cell management systems address maintenance management. Some systems allow the utilization reports to be communicated to plant level systems and some provide planned maintenance schedules to be generated at cell level. None of the systems surveyed address the real-time diagnostics issue, as discussed by Bititci and Ross (1988). In order to maximize availability of manufacturing cells, real-time systems capable of identifying fault conditions and prompting the corrective action are required.

The research issues

Over recent years a considerable amount of research and development effort has been spent on artificial intelligence (AI) applications to solve manufacturing control problems. This work has largely been targeted towards process planning (Khorami *et al.*, 1988). Other work has concentrated on scheduling (Bensena *et al.*, 1988) with a view towards integrating theoretical scheduling knowledge, empirical knowledge (such as priority rules) and practical knowledge available from the schedules. Work has also been carried out to define a cell controller architecture such as the blackboard based system put forward by O'Grady and Lee (1988). However, these applications have not been discussed in a cell management context, which is necessary in order to minimize the limitations discussed in the previous section.

This discussion identifies the limitations of existing systems due to lack of integration between the planning functions and control functions. This problem is inflated by limitations imposed on day-to-day operations arising from static process plans. Process plans are created on a once-per-design basis, usually well in advance of the actual production, without regard to the conditions existing when the part is actually manufactured. To overcome this limitation further research is needed to bring the process planning activity to cell control level and support it by an expert system. To gain the full benefits of such a system the process planning and scheduling functions at cell level should be able to operate without the restriction imposed by the standard tools, part programs and fixtures. Therefore, research into AI-based support functions such as part program configuration and fixture configuration from adaptable standard modules is required.

The lack of functionality with regard to real-time diagnostic systems is recognized. For some time researchers have been interested in the diagnostics problem and have applied knowledge-based systems to address this issue. Such systems usually relied on interaction with a user to obtain information. The more recent work in this area of diagnostics indicates growing research interests in automatic monitoring and recognition of fault conditions. Bititci and Gohari (1989) discuss the maintenance and diagnostic requirements of a manufacturing system. The requirements are categorized at plant, cell and station levels. At plant level the requirements are for maintenance planning tools for which software packages are readily available. At the station level, most of the modern intelligent equipment, such as CNC machine tools and robots, is equipped with comprehensive, proven diagnostic facilities. Current research also examines automatic recovery of individual workstations from fault condition (Barnes *et al.*, 1983; Hardy *et al.*, 1983).

On the whole, very little development work addressed the diagnostic issue at the cell level. Further research is required to develop diagnostics methodologies at cell level in the context of cell management systems.

References

Barnes, D.P., Lee, M.H. and Hardy, N.W., 1983, A control and monitoring system for multi-sensor industrial robots. *3rd International Conference on Robot Vision and Sensory Controls*, Cambridge, Mass, USA, November, pp. 471–479.

Bensena, E., Bel, G. and Dubouis, D., 1988, OPAL-A multi knowledge based system for industrial job-shop scheduling. *International Journal of Production Research*, **26**, 5, 845–861.

Bititci, U.S., 1988, A strategy for designing integrated manufacturing systems. *4th International Conference on Computer Aided Production Engineering*, Edinburgh, UK, November, pp. 327–334.

Bititci, U.S. and Gohari, F., 1989, PLC based diagnostic systems. *Proceedings of Sunderland Advanced Manufacturing Technology Conference*, Sunderland, UK, March.

Bititci, U.S. and Ross, A., 1988, Diagnostic systems for FMS. *Proceedings of FMS-7*, Stuttgart, Germany, September, pp. 201–211.

Carrie, A.S. and Bititci, U.S., 1988a, Tool management; a major challenge to integration. *23rd European Conference on Production and Inventory Control*, Birmingham, UK, November, pp. 58–74.

Carrie, A.S. and Bititci, U.S., 1988b, Tool management in FMS for integrated information and material flow. *International Conference on Integration Information and Material Flow*, Cambridge, UK, September, pp. 301–308.

Hardy, N.W., Lee, M.H. and Barnes, D.P., 1983, Knowledge engineering in robot control. *Proceedings of Expert Systems 83*, Cambridge, UK, December, pp. 70–77.

Khorami, M.T., McMillan, C.R. and Mowforth, P., 1988, A generative integrated process planning system (GIPPS v 1·1): unexpected material benefits. *4th National Conference on Production Research*, Sheffield, UK, pp. 162–168.

O'Grady, P., and Lee, K.H., 1988, An intelligent cell control system for automated manufacturing. *International Journal of Production Research*, **26**, 5, 845–861.

73

The application of an AI-based vision system to robotic assembly

D.J.M. Elliott and B.J. Griffiths

Manufacturing and Engineering Systems Department, Brunel University, Uxbridge, Middlesex, UK

Abstract

This chapter describes a vision system which is used in conjunction with a horizontal belt feeder to supply medium-sized piece parts to a robotic assembly cell. The vision system is a hybrid one which combines traditional feature extraction with an artificial intelligence technique based on N-tuple analysis. The system is able to track, determine the orientation and inspect a variety of piece parts which, for example, have complex internal features. Two-dimensional plate-type parts as well as three-dimensional parts have been investigated. The advantage of the vision system is that the recognition speed is independent of part or image complexity.

Introduction

In manufacturing and assembly systems the aim has always been for flexibility to cope with a range of possibilities and hence robotic assembly systems are being developed. One such system is the 'Windsor' robotic assembly cell (Williams and Lill, 1987). This cell is designed to provide 90–95% of the hardware and software requirements for the assembly of many different products within certain size and mass limitations. Such systems rely on feeder devices which must be equally flexible and provide components in a reliable and predictable manner. A variety of feeder types are needed to deal with a variety of parts. This paper concentrates on the design of a low-cost vision system used with a medium-sized parts feeder based on a horizontal belt. Flexibility in parts handling is provided by bars, trips, slides and wiper blades suspended just above the feeder belt which provide primary ordering of parts, a recirculation system for reject and purging and also a vision system for tracking, orientation and inspection.

The feeder system

The horizontal belt feeder is shown in Figure 1. It is positioned with its exit end just within reach of the assembly robot. At the other end of the feeder there is an elevating hopper to lift parts from a lower hopper to the belt as required. The parts are transported along the belt until they reach the angled blade where they are ordered and pass by the camera viewing station. The viewing system determines the orientation and using the pivoting blade (flipper) can either reject the part or

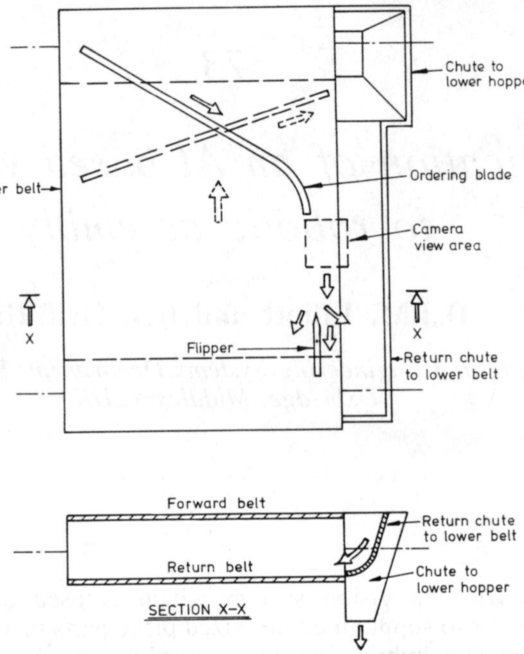

Figure 1. The horizontal belt feeder.

file it into a lane position. Rejected parts fall off the belt edge and are directed to the return belt below by a simple angled continuous chute. At the other end the lower clearing blade wipes these recirculated parts back into the chute to the lower hopper. The recirculation is completed as the elevating hopper lifts them to the belt level again. The blades are pre-set and parts can be rejected at any point. The recirculation system permits purging and rapid parts changeover.

Vision systems

Vision systems have been successfully employed for many years to recognize and identify artefacts. In engineering terms the most popular have been the feature extraction-based systems which are essentially feature calculation. Less well used, or indeed understood, are the artificial intelligence (AI) or neural mimic approaches using N-tuple classifiers (Braggins, 1987). The system used in this case is a hybrid one which combines in a modular manner subsets of both systems which can be matched together to provide the best solution to a vision problem.

Feature extraction vision

These techniques are most convenient for geometrically definable parts. The principle of operation is that an image is located and windowed. The image is traced and a variety of analysis techniques can be applied from simple matching to complex calculations. Feature extraction techniques give a more than satisfactory performance in many circumstances and provided parts are of a low to medium complexity can be reasonably economical in terms of memory. The technique is essentially feature measurement and largely sequential in nature, consequently the processing time is directly related to image complexity. For complex images or those with many internal features the technique is slow. Back or through lighting may also be required.

AI vision

Conceptually the algorithms used mimic the operation of a group of neural brain cells hence the titles ‘Artificial Intelligence’ or ‘N-Tuple Classification’. The technique has been successfully applied in fields like facial recognition, but not in traditional manufacturing engineering. The N-tuple technique (or in engineering terms pixel batch sampling) is particularly suited to complex external and internal shapes since the computation time is independent of complexity. The only commercial available N-tuple classifier is the Wizard system as demonstrated by Aleksander (1983).

With regard to the hybrid vision system, much work has been done on the tuple size and the number of tuples and discriminators to provide the highest accuracy and efficiency whilst avoiding saturation. The system can also be used to analyse any form of captured data from whatever source.

The hybrid system

The hybrid system has been given the name Windowed Image Shape Extraction or WISE. The equipment consisted of a vidicon camera, a semi-intelligent frame store and a conventional PC and monitors. The camera is positioned above the belt to monitor the appropriate area. The captured image is edge traced, the centre determined and a window of predetermined size, shape, form and configuration placed around the part. The N-tuple algorithm is then employed on a binary or grey scale as appropriate. Discriminators are trained corresponding to various orientations and positions of a part and at several levels as is necessary such that in the recognition phase the orientation is given by the greatest discriminator response.

Simple arrow shape example

As an example of the hybrid system capability, consider the simple arrow shape shown in Figure 1. Although this is a plate type part with only two natural resting aspects the system has been proven with 3-D shapes. The arrow shape discriminator responses are shown in Figure 2. The full 360° orientation can be covered by eight off discriminators trained at 45° spacing giving an interrogation time of about 0·3 s. Should a free orientation vision system be required for any reason, a multi-layer classifier with cascaded results and/or interpolation would provide the orientation sensitivity required. This represents the ‘free’ orientation situation but when used in conjunction with the horizontal belt feeder the blades will provide primary orientation, alignment and ordering so in this case only two discriminators would be required.

The system also allows checking or inspection since an alteration to the image will be shown as a discriminator response attenuation. Although the individual discriminator responses will be reduced, the response pattern or histogram is unchanged and the orientation capability is unaltered. The level of the responses is an indication of quality as Figure 3 shows where there is a relationship between response level and alteration to image. This illustrates that the system can be used for searching, tracking, orientation and quality testing. A variety of engineering parts, assemblies, identifiers and texts have been successfully used with the system. One such part is a remote control unit case and the remainder of this chapter shows how it is fed by the system.

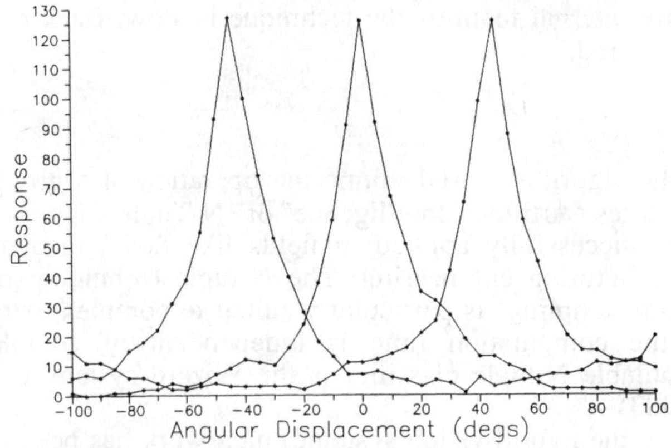

Figure 2. Discriminator responses for the hollow arrow shape.

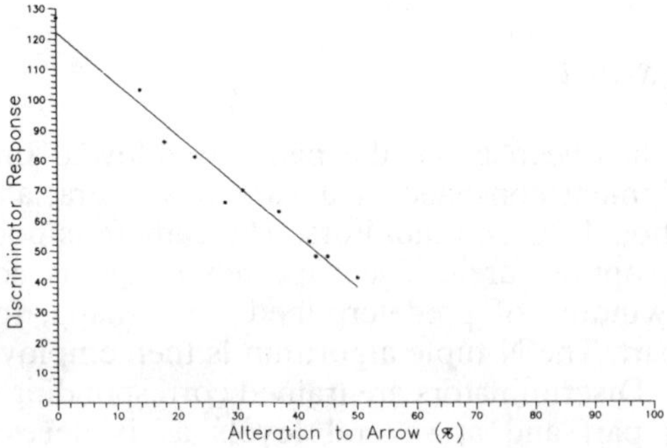

Figure 3. Discriminator response against percentage alteration to hollow arrow shape image.

Remote control case

The remote control handset case consists of a shallow rectangular cup with dimensions of 10 × 60 × 120 mm. It has various rectangular holes in it for buttons and lights, etc., with printing above the holes. With regard to the feeder the blades are pre-set actively to orientate the parts into one of four orientations, these being: face up or down and forwards or backwards. The vision system tracks the cases, determines their orientation, rejects those fed face down, inspects those face up (and rejects if required), and finally informs the robot of the status and position of the acceptable cases.

Figure 4 shows the tracking response from the four discriminators and the maximum response is indicative of the correctly orientated case. The system could equally well use some form of trigger to capture the operator frame. The image is inspected and each feature is sequentially checked by special windowing algorithms. This determines the correct number and size of the button holes. Further windowing can check for edge defects and printing. Figure 5 shows the response from a number check used as a basis for inspecting the printing. The accompanying histogram shows the discriminator response pattern for the number six. The response from discriminator six is the highest although the responses from zero, three, five and eight show a small sensitivity because of the shape similarity six has with them. In all cases the response from the trained discriminator was at least

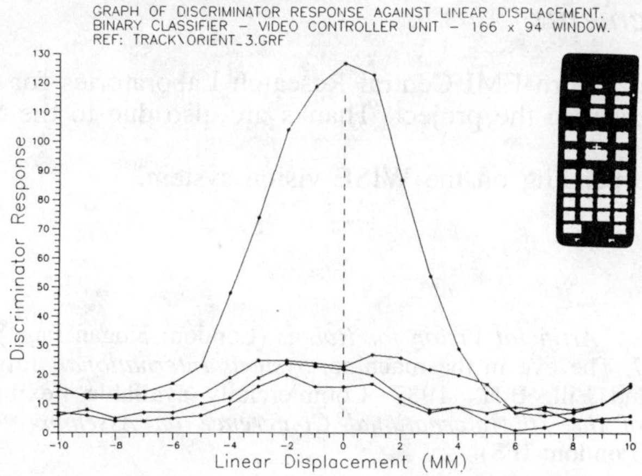

Figure 4. Tracking response for remote control case.

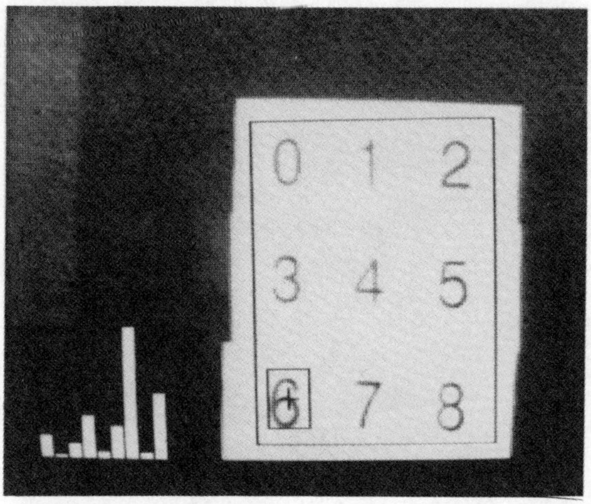

Figure 5. Histogram response to numbers.

twice the nearest response. This illustrates the ability of the system to recognize not only letters but also words, diagrams and codes.

The advantage of this hybrid method is that it depends upon a reaction to shapes and patterns rather than feature calculations. The image complexity is irrelevant, the response time is invariant and discriminator cascading allows interrogation at different levels and complexity. The system is limited by optics rather than mathematics. It can just as easily deal with rectangular piece parts as complex printing.

Conclusions

A medium parts feeder has been developed which uses a low-cost, PC-based vision system to track, orientate, check and feed medium-sized piece parts to a robotic assembly cell. The vision system is a hybrid one which combines traditional feature calculation and neural inspection techniques to provide the optimal solution to a recognition problem.

Acknowledgements

The authors thank Thorn EMI Central Research Laboratories for the support and encouragement given to the project. Thanks are also due to the SERC.

Note: Patents are pending on the WISE vision system.

References

Aleksander, I., 1983, *Artificial Vision for Robots* (London: Kogan Page).

Braggins, D., 1987, The eye in the machine. *Systems International*, July.

Williams, A.M. and Lill, B.H., 1987, Commercially available flexible assembly cell. *Proceedings of the 7th International Conference on Assembly Automation*, May, Birmingham (London: IFS).

SECTION VII.2. EXPERT SYSTEMS

74

Knowledge-based systems in computer-assisted production

K.-P. Fähnrich, G. Groh and M. Thines

Fraunhofer-Institut für Arbeitwirtschaft und Organisation, Postfach 80 04 69, Nobelstrasse 12, D-7000 Stuttgart 80, Germany

Abstract

With the development of 'expert systems' or 'knowledge-based systems (KBS)', the efforts of Artificial Intelligence (AI) research attained commercial application. While there are presently only few investigations concerning the extent of this transfer, assessment of the extent of manufacturing applications has been mostly the result of speculation.

This chapter describes the state of the art of KBS production applications in a total survey of documented literature of systems in the USA, UK and Germany.

The reasons leading to computer assistance and eventually to integrated solutions are discussed in detail. For the individual task areas in production the computer-assisted technological systems are assigned under the terms CAE/CAD (construction and design), CAP (production planning), PPS (production control), CAM (production of parts) and CAQ (quality assurance). No corresponding terms have emerged for the sectors of assembly, maintenance, distribution and customer service. Solutions which are to connect automation islands and provide as well as realize a permanent flow of information without media breaks are summed up under the term 'CIM' (Computer-Integrated Manufacturing).

Methodology of investigation

In this investigation the variables developer organization and developer country, industries in which systems are used, state of development, field of employment in the company, type of task and time of publication are explored.

Concerning developer countries, the main emphasis is laid on the USA, Germany and the UK. Here a total survey of the systems documented in the literature can be assumed. Concerning the industries where systems are used, a classification according to the categories of the Statistisches Bundesamt[1] (Federal Office of Statistics) has been applied. For the state of development, a life cycle model in seven steps based on Hayes-Roth *et al.*[2] and Harmon and King[3] has been defined. The field of employment in the company has been defined according to the structure in chapter 2 of this publication, and the type of the task according to the type definition given in Chapter 1. A more precise discussion of methods may be found in an earlier publication[4] by the authors.

A computer-based collection of entries on 829 expert systems amply documented in literature served as raw material. From among the total number, 345 systems

MERTENS	CLANCEY	STEFIK (HAYES-ROTH)	IAO-CLASSIFICATION
	Identify:	Interpretation	Interpretation
Selection			Selection
Diagnosis	Diagnose	Diagnosis	Diagnosis
	Monitor	Monitoring	Monitoring
	Control	(Control)	
Planning	Design:	Planning	Planning
Configuration	Plan	Design	Configuration
Expertise	Configure		
	Prediction	Prediction	
Decision			Decision
Consulting			

Figure 1. Frequently documented classification schemes of knowledge-based systems.

were not in the production trade, 52 systems were not developed for the processing trade and 179 systems could be classified as production external through the variable field of employment in the company. Through this strict definition of the term production area a total number of 253 systems could be identified as a spot check for the investigation. The models for evaluation and survey have been based on the procedure model from Seelig.[5]

Knowledge-based systems can be classified according to various aspects

An investigation cannot be better than its categories. Therefore in the following section, the classification on which the investigation is based is explained first of all. The type definitions of knowledge-based systems found in the literature can be divided into cognition psychological, system theoretical as well as engineering-oriented functional ones. In Figure 1 the classification schemes frequently documented in the literature are compared with the categories employed in the investigation; in Table 1 the definitions of the categories employed can be found.

As early as the 1950s type definitions of problem classes in human information processing were analysed. A résumé and critical retrospective of these investigations can be found in Newell and Simon.[6] Clancey[7] has put these problem-oriented approaches in more concrete terms under a more system theoretical aspect in the type definition of knowledge-based systems and thus achieved a systematization of the Hayes-Roth's type definition, which is widely spread in the literature. An application-oriented extension of the 'classic' categories was achieved, for the first time in the German speaking area, by Mertens in his paper on the state of the art of knowledge-based systems.[8]

Table 1. *Classification of knowledge-based systems — categories used in the study.*

Classification of knowledge-based systems	Example
Interpretation System makes statements according to a model representation of the environment	CRIB at ICL interprets failures in hardware and software on the basis of an internal model
Selection Selection of elements from a great number of alternatives considering given marginal conditions	Expert system for key selection at SODECO-SAIA
Diagnosis System recognizes faulty system states	Id. EST by FhG/IAO diagnoses failures in CNC machine tools
Monitoring System, that recognizes faulty system states and reacts in real-time	PDS at the Westinghouse Electric Corp. diagnoses and monitors online production processes
Planning System generates action sequences in order to achieve a given target state	ISIS supports production planning at the Westinghouse Turbine Comp.
Configuration System develops arrangement of elements which fulfills the conditions given	XCON configurates computers at Digital Equipment
Decision System undertakes decision finding considering given marginal conditions	IMS of the Carnegie-Mellon University provides support in strategic market decisions

Differentiation of computer-assisted production into areas of function and corresponding tasks

After the possibilities of the classification of knowledge-based systems presented in the preceding section, there now follows an outline of the production areas and the corresponding tasks in which the area of discrete production in particular will be considered. The structure presented in Figure 2 is especially marked in the case of a customer-oriented manufacturer of individual items or small series. The production here is roughly divided into construction, work preparation, production and assembly. Prevailing in the development and construction are, according to Wiendahl,[9] the tasks: technical processing of offers, product development, operation tool development and planning, process planning, internal company standardization, as well as administration of the resulting information. The final products are drawings and parts lists.

Production planning and production control, instead of which the terms 'work planning and work control' are often employed, can be summarized in the terms work preparation or production preparation, respectively. Here planning of materials, process and time as well as costs is effected. Among the long-term tasks in this area are factory planning, method planning, investment planning and work environment design. The operations list together with the corresponding assembly plans constitute the most important result of this function area.

Production control consists of rough planning, in the sense that orders are

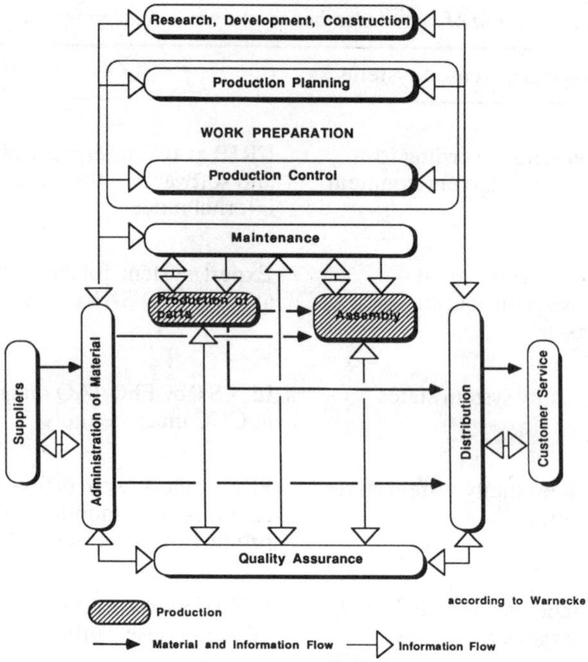

Figure 2. Differentiation of production areas.

included in the planning, the medium-term planning tasks of determining the lead times, determination of an order completion date, accounting of capacity use and capacity levelling as well as short-term planning, which effects machine manning and sequence planning. Also assigned to this area, for the purposes of this text, is administration of materials with its planning and controlling functions in the fields of storage, transport, requirement, stock and procuring.

The production itself, here called manufacture, can be divided into the sectors manufacture of parts and assembly. Within the manufacture of parts, tasks of handling, controlling, transport, storage and manufacture can be distinguished.[11] Excerpts of assembly functions can be divided into tasks of assembling, adjusting, monitoring, handling as well as certain special functions.[12] Due to their importance concerning the analysis of the state of the art of knowledge-based systems, the sectors of 'maintenance' and 'quality assurance' have been treated separately. 'Quality assurance' can be divided into quality planning, quality controlling and quality check.[13] Maintenance and manufacture are also closely related via means of operating. Here we must differentiate between the three sectors of service, inspection and repair.[14]

Later the reasons leading to computer-assistance of these functions and, last of all, to integrated support solutions[15] are discussed in detail. For the individual task areas in production discussed above, the computer-assisted technological systems are assigned under the following terms:

— CAE/CAD (construction and design),
— CAP (production planning),
— PPS (production control),
— CAM (production of parts),
— CAQ (safeguarding of quality).

No corresponding terms have emerged for the sectors of assembly, maintenance or distribution and customer service. Solutions, which connect automation islands and

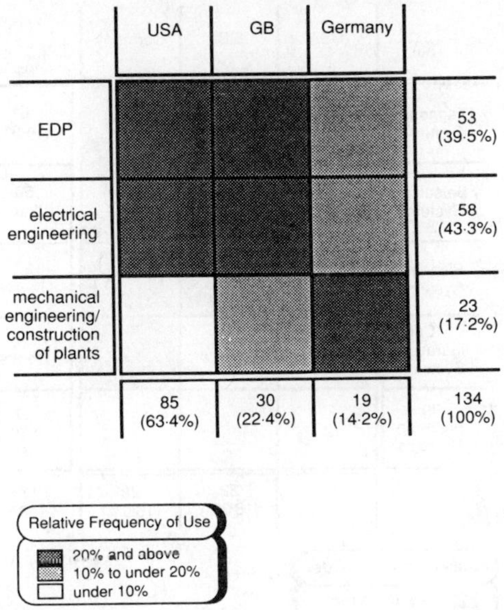

Figure 3. Developers of knowledge-based systems in various industries.

provide as well as realize a permanent flow of information without media breaks, are summed up under the term CIM (Computer Integrated Manufacturing).

Bullinger has introduced into the literature a company-wide way of looking at the problems.[16] In this chapter, reference is also made to the information technological dimensions of the problem: computer-assisted manufacture is increasingly understood as a distributed, information-processing system.

Results of the empirical exploration with regard to countries and industries

Almost two thirds of the systems contained in the spot check were developed in the USA, 32 of the systems documented in the literature were recorded in Germany, which means that this country, in view of its economic strength, is slightly underrepresented.

Development of expert systems is considerably influenced by commercial developers. Projects where universities and research co-operate with commercial developers are under-represented. This applies especially to Germany. In the case of the USA and the UK two waves of development of expert systems are manifest, in accordance with the thesis[17] frequently discussed in the literature that development projects of knowledge-based systems are subject to a 'wave cycle'. This phenomenon cannot be found in Germany. Here a normal product life cycle can be stated. The main emphasis lies at the moment with system implementation. A certain discrepancy in time between Germany on the one hand and the USA and the UK on the other hand can be determined.

Figure 3 shows marked differences between the various countries concerning the industries in which the systems are employed. In the USA, development projects are mainly concentrated in the EDP industry and electrical engineering. The percentage of systems in mechanical engineering and construction of plants is low. Great Britain develops mainly in the area of electrical engineering, but has also a high share of systems in the EDP industry. The majority of the systems developed in Germany are found in mechanical engineering and construction of plants. The

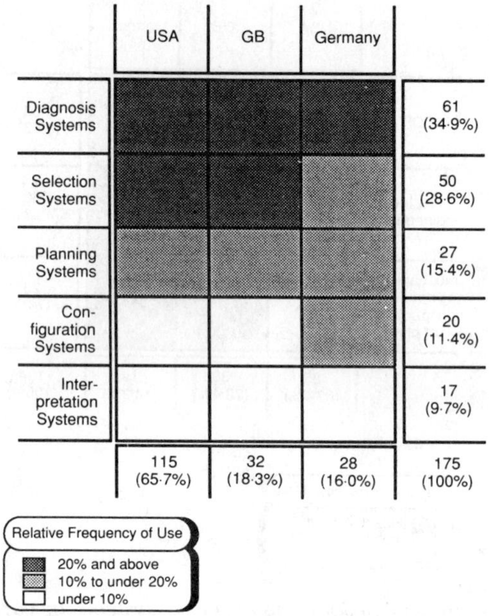

Figure 4. Types of knowledge-based systems realized by countries.

often used argument that German development projects are reduced to 'copies' of successful American systems has to be considerably revised: in view of the consolidated positions of German mechanical engineering and construction of plants in the world markets, system developers have proved themselves able to meet the market requirements and even act as forerunners in some of these areas.

Figure 4 shows that diagnosis and selection systems as well as configuration systems have been realized most frequently. All countries investigated have recently been working intensively on planning systems.

A direct comparison between the state of development of single system types shows the vast differences represented in Figure 5. Diagnosis systems were already developed at the beginning of the 1980s under commercial aspects and are exceptionally far developed today. Planning systems, however, were not projected before the second half of the eighties and are therefore under-represented in more advanced states of development.

In Figure 6, it can be seen that diagnosis systems are employed comprehensively in various industries to find causes for errors. As can be expected, configuration systems are found in the EDP industry and predominantly selection systems in electrical engineering. The percentage of planning systems in mechanical engineering and construction of plants is, however, surprisingly high.

Knowledge-based systems diffuse with varying velocity into various company sectors

Figure 7 shows the state of the art of knowledge-based systems in the individual company sectors defined earlier. In this section this very irregular distribution is investigated in detail from the engineering point of view, so that, in the next section an outlook of how knowledge-based systems could be employed in the future can be achieved.

In the sector 'construction and development' CAD selection systems are mainly

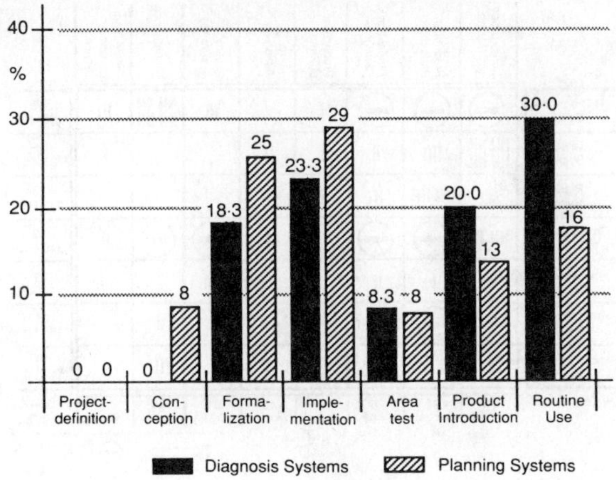

Figure 5. Comparison of the state of development: planning systems versus diagnostic systems.

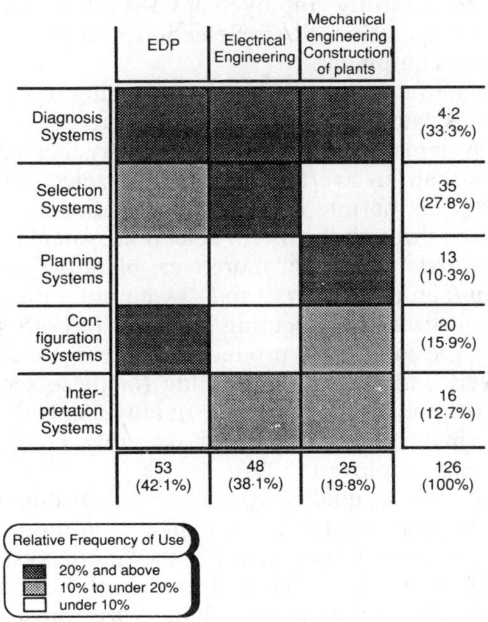

Figure 6. Types of knowledge-based systems in various industries.

employed for the activities 'Designing' and 'Detailing'. Moreover, in this company sector, there are interpretation systems that simulate designed products on a model basis. These have, however, not been developed as far as systems with configurative tasks, due to the higher demands on the model representation of the 'environment'. Such configuration systems are employed in the processing of offers, where individual solutions are projected according to customer specification. Systems supporting the variant construction in the sectors of mechanical engineering and construction of plants show clearly that such tasks do not occur exclusively in the EDP industry.

Typical tasks such as planning of requirements and materials are algorithmically structurable and therefore sufficiently manageable. Knowledge-based planning

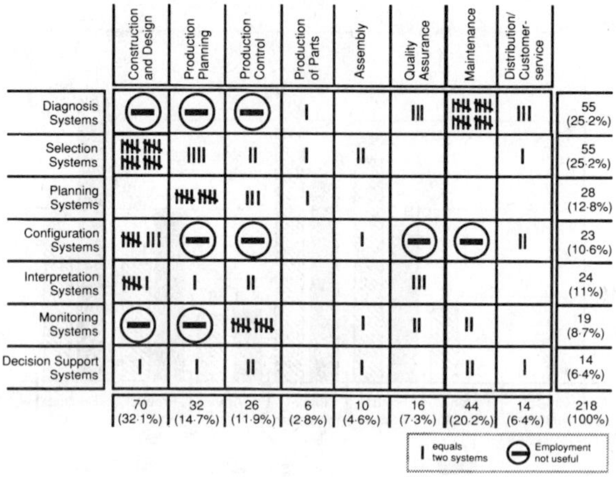

Figure 7. State of the art of knowledge-based systems in company sectors.

systems are, however, successfully employed at CIM interfaces such as CAD/CAM or in the drawing up of operations lists. Selection systems are mainly employed in the planning of work equipment.

The central task of manufacture control is the recording of operating and machine dates (recording of actual dates) and their control through the target data given. The necessity of real-time systems for this task explains why these control systems have been developed to less than an average degree. The selection of the route of the production order—partly taking into account maintenance intervals with long-term orders—can be optimized through the use of selection systems. In the determination of an order completion date there are, moreover, planning systems employed.

In the sectors 'manufacture of parts' and 'assembly', the information flow is operationalized and transferred to executing systems. In both sectors conventional systems largely satisfy the needs. AI products from robotics, sensoric and image processing are, however, undoubtedly interesting for these sectors. The sectors of production machine maintenance or product quality control that are directly or indirectly related to the manufacture of parts or assembly, constitute, however, areas where expert systems are employed.

Knowledge-based systems in quality assurance (CAQ) diagnose possible causes for errors in defective products and develop explanation models for the errors that have occured through interpretation of data. For the future, systems are conceivable here that are linked with manufacture control in a control circuit. Beside diagnosis systems, control systems are also projected in quality assurance that are to recognize defective products in real-time.

Ninety per cent of all systems in the sectors of 'maintenance' are diagnosis systems with the aim of minimizing times of standstill and optimizing maintenance intervals. Maintenance planning with knowledge-based planning systems is a new approach, for which at the time of data collection, no publications could yet be found. Here, however, a future application area for knowledge-based systems can certainly be located. In addition, at the interface to the CAQ sector, control systems are planned.

The company sector 'distribution/customer service' has been openly defined. Thus configuration systems have not been attributed to the CAD sector in the case of firms that effect processing of offers in the customer-related sector. In this category, there are also diagnosis systems employed in the work outside the office or via remote diagnosis.

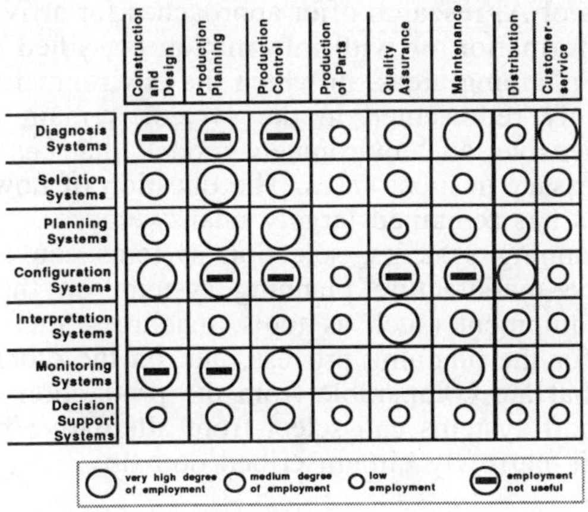

Figure 8. Prospect: use of knowledge-based systems within the next five to ten years.

Prospect: theses concerning the future state of the art of knowledge-based systems in computer-assisted production

After discussion of the real state, prognoses are now given with respect to the possibilities of the use of knowledge-based systems in the next five to ten years. The prognoses are the result of the evaluation of semi-standardized interviews carried out with experts in this field; the results are shown in Figure 8.

Thesis 1 In the sector of computer-assisted construction and development (CAD) selection and configuration systems will also dominate in the future.

In the CAD sector, configuration systems for the support of the conception and design work will be employed using components developed earlier. Such systems promise to be especially successful in the construction of variants. Selection systems can be used for finding 'similar' constructions, for new constructions and variant constructions. The further development of selection systems in recent years will in future facilitate the solution of far-reaching planning tasks with knowledge-based technology, for selection from a large number of alternatives precedes the planning process, so that a gradual establishment of systems with complex planning tasks can be effected via selections systems. Thus planning systems can support the whole project planning, for example, in the CAD sector with the construction of complex and large plants where a large number of people have to co-operate in a co-ordinated way.

Thesis 2 In the sector of production planning development of efficient planning systems with a complex structure of tasks remains the predominant task.

Knowledge-based systems offer the possibility of arriving at a satisfactory solution quickly and without complete information in the sector of process planning using previously selected work equipment. If a complete model of the systems to be planned exists it is possible to carry out planning tasks with conventional means straight away. Calculation in the model and development of various alternatives is then, at last, a question of computer performance and time available. In sectors, however, where it is necessary to react to short-term changes or irregularities there is seldom sufficient time available for calculating a planning procedure all over again. Here, methods of AI research offer approaches for arriving at solutions even without complete information or with information classified as 'uncertain'.

But medium-term planning areas, in which the environment of employment has to be comprehensively represented in the system, remain problematic: if task planning is comprehensive, an 'environment model' that has to be represented in the computer is necessary in most cases. The question of how such decisions can be built into systems has remained largely unanswered.

In addition to planning systems, selection systems can also be employed at various levels in the manufacture planning system: on the one hand, for the selection of work equipment (such as tools, machines, etc.), in order to do the 'preliminary work' for the planning process, but, on the other hand, also in order to recognize parts that are comparable from the production engineering point of view. Decision support systems can select from alternative planning variants the 'optimal' plan in this narrowly circumscribed domain.

Thesis 3 Control systems and planning systems are joined together in production control to form control circuits which react in real-time to discrepancies from initial values.

In the sector of production control, planning systems for the detailed planning of data from production planning are employed. Here, especially time-related and capacity-related marginal conditions must be taken into account. Together with the planning systems, control circuits can be established that are able to react quickly to faults.

Decision support components can help here, even in the case of discrepancies from the planned procedure, with the search for corrected values. Due to their complexity, decision support systems are, however, only suited for small, narrowly defined tasks, even for medium-range periods. In the cases of long-term and extensive planning tasks, the problem lies in the high degree of complexity of the task.[18] As in the modelling of the environment, it is very difficult to transform the criteria that are crucial here into formal models. Moreover, the problem fields are characterized by many marginal conditions and a considerable interconnection of the various criteria and marginal conditions.

Selection systems can be employed in the detailed task of assigning certain persons, certain machines or certain materials to given orders.

Thesis 4 Intelligent robots, sensor-equipped manufacturing facilities as well as environment-sensitive transport systems are increasingly employed in the manufacture of parts and in assembly.

Robots increasingly contain intelligent sensory equipment which makes it possible for them to form a general idea of their surroundings. This is important for programming optimization, for increased flexibility as well as for safety reasons. The same applies, to a lesser degree, to machine tools.

Problems with configuration and/or selection are present in the planning of robot use or in robot configuration and the sensors needed.

Automated transport vehicles (ATV) or conveyor vehicles become more environment-sensitive through sensoric equipment. Quality assurance and maintenance constitute additional areas of employment for knowledge-based systems.

Thesis 5 Knowledge-based systems will gain increasing importance in computer-assisted quality assurance.

Selection systems can be employed when selecting examination plans, examination tools and appropriately qualified examiners. Control systems in the process assure a quick reaction in the case of critical discrepancies from the initial quality. The quality of the parts produced can be checked via installed diagnosis systems. The changes of quality characteristics ascertained can be processed via an interpretation system. The aim here is to draw conclusions from the changes in quality concerning the change in procedure required.

Thesis 6 Maintenance remains the most important employment area of diagnosis systems.

In the maintenance sector diagnosis systems, in co-operation with efficient control systems, can prevent damage in resources and/or rapidly determine the causes of faults as well as suggest possible repair measures. A co-operation of control systems with interpretation systems and planning systems installed after them offers the possibility of optimizing the whole maintenance planning depending on the current state of the resources.

Robots, machine tools and transport systems, already discussed above, will be the main areas of employment here.

Thesis 7 In distribution the drawing up on an offer is made quicker and more effective by configuration systems.
The company sector 'customer service' will undergo important changes through new diagnosis concepts.

In distribution, interpretation systems can transfer the customers' requirements to system solutions and thus be employed as the means of support for distribution employees. Based on this, configuration systems can draw up offers more effectively and quickly, taking into account economic considerations.

In the sector of 'customer service', diagnosis systems will be increasingly employed. The concept of remote diagnosis has already been touched on above, another diagnosis concept favours the portability criterion in the form of portable diagnosis PCs. Planning systems, in combination with selection systems previously installed can select the appropriate service employees and spare parts probably needed.

Thesis 8 In all company sectors of computer-assisted production front ends will be used much more intensively.

Knowledge-based help systems—more generally, intelligent front ends—will be increasingly developed in the near future as will knowledge-based tutorial systems. The need for active as well as passive assistance will grow continually with increasing qualification requirements at the place of work.

On the other hand a whole new field of application, namely that of training, is possible because of the explanatory abilities of knowledge-based systems. Due to their self-explanatory abilities, knowledge-based systems can be used for training and further education.

In various areas of application, different system types dominate: trend statements concerning the state of the art of knowledge-based systems in the administrative office

Whereas statements concerning the state of the art have up to now been concerned with the sector of computer-assisted production, trend statements concerning the employment of systems in the office sector are made in this conclusion which, however, unlike the application study in production, is not based on a complete

	Office	Staff/ Administration	Finances Accounting	Management	Distribution Customer Service	Other (e.g. data bases)	
Selection	+++	++	++ ++	+++	+++ +++	+++	21 (24·4%)
Decision Support	+	+	+++ +++ +++ ++	++ ++	++	++ ++	23 (26·7%)
Planning	+++		+	+++ ++	+		10 (11·6%)
Interpretation	++		+++	++		+	8 (9·3%)
Diagnosis					+++ ++		5 (5·8%)
Monitoring					+++		3 (3·5%)
Help	+++ +++		+		+	+++ ++	13 (15·1%)
Intelligent Assistant	++			+			3 (3·5%)
	17 (19·7%)	3 (3·5%)	20 (23·3%)	15 (17·4%)	18 (20·9%)	13 (15·1%)	86 (100%)

+ equals one system

Figure 9. Types of knowledge-based systems within various office sectors.

investigation, but on a spot check of 90 systems. In Figure 9, the frequency of the system types already defined in the production analysis is assigned to various office sectors.

Nearly 20% of all the systems in the spot check were planned for typing pools and offices. In this sector, however, only few executable systems could be located in circumscribable domains, as, for example, the planning of appointments and business trips. Less well developed were front ends and systems whose tasks are difficult to describe, such as work involving electronic or physical letters or translation of documents.

In the personnel and social sector, the task profile can either be recorded algorithmically and can therefore be solved with conventional data processing, or it is a matter of processes so personal and sensitive that it will hardly be possible to handle them with knowledge-based systems in the foreseeable future. Solely in further education within companies can initial stages be discerned at the moment.

The financial and accountancy system, on the other hand, is characterized by a heuristic problem-solving behaviour, especially in investment decisions. In these sectors a higher than average number of executable systems could be found which is most cases can be subsumed under decision support.

On the management level, classified systems have been less developed than on average. Decisions here are frequently made involving 'common sense' which stands in the way of circumscribing the problem area (domain). Systems designed for the modelling or long-term planning of global company decisions will not be realized in the foreseeable future, as has been discussed in the last section.

In 'distribution/customer service', the systems projected are mostly characterized by a narrowly confined domain. The tasks of 'diagnosis' and 'control' can be found exclusively in distribution involving a scope extending from vehicle diagnosis to dispatch control.

The 'Other' category contains for the most part systems providing active or passive assistance for the user of large data bases.

The development projects of knowledge-based systems, which focus on the sector of the 'administrative office', in contrast to the investigation in production, have been considerably influenced by developers from universities in the investigated countries USA, UK and Germany. Moreover, a very important part is played by decision support and assistance systems in the office sector, i.e. systems to which only little importance for this sector of employment could be assigned today in the investigation on the state of the art of knowledge-based systems in production.

75

Handling heuristic control logic in the domain of a manufacturing simulation

G. Fozzard,[1] J. Spragg,[2] D. Tyler,[2] D. Sinclair[2] and D. Parker[2]

[1]*The Cimcentre, Leicester Polytechnic, Leicester, UK,* [2]*Department of Clothing Design and Technology, Hollings Faculty, Manchester Polytechnic, Old Hall Lane, Manchester, UK*

Abstract

Constructs provided within conventional simulation tools do not facilitate the encoding of human decision making, a major component of many manufacturing systems. Knowledge-based systems (KBS), however, are appropriate for the modelling of supervisory behaviour using rule-based problem solving techniques. The linking of KBS to a manufacturing simulation places constraints on the design of the computerized model, in that the rules controlling the manufacturing process must be externally accessed. In addition, the model must report data to the external KBS, so that relevant rules may be selected and implemented. The control software differs from classic expert system shells in that it is data intensive, non-consultative and capable of autonomous default reasoning. Furthermore, fast inference speeds are required, in order to avoid major delays in the execution of the simulation. The principles of software design are presented and illustrated by reference to the simulation of garment assembly lines.

Introduction

Within the UK Clothing Industry, the dominant system of manufacture for medium to high volume garment production is essentially flow manufacture, with high levels of inventory buffering to offset commonly occurring perturbations.

In building a model of this dynamic production environment, it became apparent that the role of the production line supervisor could not be adequately represented within the bounds of a conventional simulation tool. An overall view of the responsibilities of a supervisor is provided by Chuter (1988) and includes line balancing, organizing working methods, meeting production targets, work scheduling, quality control and many other personnel duties.

To include supervision in a model of clothing manufacture, a software architecture has been developed that allows the 'physical' production line and its supervisory 'control logic' to be modelled using two distinct approaches. Physical aspects, including work flow, process times, machine breakdowns and other stochastic elements, are modelled within a conventional simulation language having an appropriate worldview. Supervision is modelled using a Knowledge-Based Systems (KBS) software architecture that has been designed to interface with the physical model.

The combined model, essentially a multi-paradigm, computer-embedded system (Oren, 1986), is highly flexible and may be used to simulate a variety of flow line configurations.

Modelling human decision making

Modelling supervisory control of a production system is dependent on mimicking human inductive reasoning. This process is extremely complex and often the rules as to how decisions are made can only be partially distilled. Such rules, termed heuristics, are obtained by reasoning from past experience since no algorithm exists or is relevant. To handle heuristic knowledge, the control system must be capable of reasoning about action. This requirement gives rise to the 'hard' problems of Artificial Intelligence (AI). These have been identified as:

The frame problem—The difficulty of inferring all those things that do not change when actions are performed and time passes (McCarthy and Hayes, 1969).

The ramification problem—It is unreasonable explicitly to record all the consequences of actions. It would take extraordinary computational power to do this (Finger, 1987).

The qualification problem—The number of preconditions–conditions for each action are immense (McCarthy, 1977).

Implementation of a model of inductive reasoning is further complicated because human perception is essentially a parallel activity.

To address these problems, quasi-homomorphism (q-morphism) models have been identified as a possible solution (Holland *et al.*, 1987). Q-morphisms represent the supervisor's knowledge as a layered set of transition functions which map her conceptual model states (her mental model of the production line), coupled with categories of actions, to subsequent conceptual model states. Within a q-morphism, a hierarchy of levels is provided with broader categories of actions at the top to provide default expectations. The signalling of an exception at any level will invoke a lower level within the q-morphic model.

Q-morphisms allow several aspects of clothing production supervision to be modelled. Their hierarchical structure allows the system to make rough predictions on the basis of incomplete knowledge. They allow decisions to be made within a noisy environment where there is much obsolete or irrelevant information. As the model evolves, useful categories of actions may be retained as default rules.

The cognitive control system

It is possible to implement default hierarchies by encoding a population of rules as binary strings of a fixed length. Such systems are called Classifier systems (Holland, 1986), and have the following characteristics.

Parallelism

Large numbers of rules arranged in parallel, called classifiers, must be activated simultaneously before an action is fired.

Message passing

Classifiers are similar to rule-based systems in that they consist of a condition part and an action part. However, classifier systems depart from orthodox rule-based systems in that the condition part of the classifier is encoded as a number of parallel rule conditions, represented by strings of *k*-length, which must be satisfied simultaneously before the action part of the classifier, also encoded as a string of *k*-length, is activated. Messages are generated from detector bits embedded in the domain environment (simulation model in this case) and are internally generated by the activation of classifiers. Once a classifier becomes active, its action part is posted onto a message list as a new, internally generated, message. A message list is a data structure whose purpose is to give 'strengths' to posted messages. The strengths assigned to messages are determined by repeated cycles of the system: messages already on the message list activate other classifier conditions which post new action parts onto the message list. Strengths determine which posted message best represents the parallel rules pertaining to the current situation in the world. Once an appropriate classifier has been recognized it will be used to bring about some transformation of the environment.

The messages consist of strings of *k*-length over a two-letter alphabet (1, 0), where '1' indicates a category is active, and '0' indicates a category is inactive. Classifier conditions are specified by strings of length *k* over a three-letter alphabet (1, 0, #). The letter '#', incorporated into parallel rules, solves the ramification problem by playing the role of a 'don't care' symbol. The occurrence of this symbol within a classifier string means that its corresponding position within a message string can be a '1' or a '0'. For example, the message '00110011' matches the classifiers '##110011' and '0##10###'. The action part of a classifier is also a string of length *k* using the '#' symbol. In this context however, the symbol plays a 'pass through' role. Whenever the '#' symbol occurs in an action part of a classifier, the corresponding bit in the message satisfying the condition is passed through to the outgoing message. So, if the action part of a classifier is '00##1111' and the message which satisfies that action's condition is '00101111', then the new message that gets posted on the message list is '00101111'.

Message passing, parallel rule-based systems, offer the possibility of simulating human induction. Each activated classifier can be looked upon as a tentative hypothesis about some aspect of the simulation, which must compete with other plausible hypotheses currently activated. At each cycle of the system, the bit positions which the system finds useful are passed through to new messages for the classifiers to attempt to match against their encoded mental models. Classifiers can represent both the temporal relations necessary to express relationships between current and future states of the simulation, and associations between objects/actors of the simulation. Within our system both general and domain knowledge are encoded in a network of classifiers. These classifiers vary enormously in the complexity of their conditions and actions, representing categories that range from perception to highly sophisticated supervisory decisions based on 'stage setting' actions of earlier system cycles.

Linking the control system to the simulation model

The role of the computer in a simulation project is to maintain a computer representation of a conceptual model. This representation is commonly called a computerized model. Programming this model requires an extensive library of supporting routines that is usually provided by a simulation tool having a particular 'worldview' (Hooper, 1986).

Within a specific worldview, each simulation tool provides a set of constructs that should facilitate the rapid programming of the computerized model. However, it is common to see a trade-off between the ease of use of high-level constructs and the flexibility of low-level code, making the selection of a simulation tool a complex task (Chrystall and Kaye, 1987). If the simulation tool is to link up with

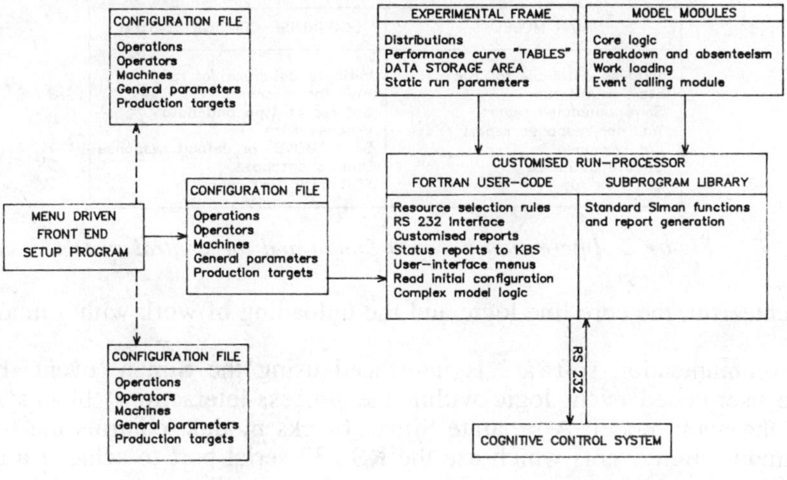

Figure 1. Software design of the Siman-based model.

external control software, the selection process must also address the requirement for both systems to share some form of common data base.

Implementations of data sharing may involve accessing a common data file or communicating data to replicate a common data subset within each system. Ironically, the flexibility required to implement this data base is provided by most general programming languages but not actually within simulation tools. A tool must be selected that will allow user-access to lower-level features of the machine operating system, possibly through linking in code developed in standard programming languages. Some general purpose simulation languages provide this facility.

The Clothing Production Model has been developed using the Siman/Cinema EGA Professor's package version 3·51 running under MS-DOS. Siman is a hybrid discrete-continuous simulation language offering both process-interaction and event-scheduling strategies (Pegden, 1985). It allows access to low-level machine routines through the use of 'user-code' in the form of Fortran or C. This code is separately compiled and then linked to the run processor to provide customized simulation software. Because of memory limitations when using Siman/Cinema under MS-DOS, the cognitive control system has been designed to be physically external to the simulation and a communications protocol has been developed to share common data between two separate computers. An overview of the design of the Siman-based model is shown in Figure 1.

The model has been designed to offer maximum flexibility without the need to recompile, or even link, any kind of source code. Central to this flexibility is the user-code developed in Microsoft Fortran version 4·01 which has been used to 'redefine' the role of the Siman experimental frame within the production model.

Usually the experimental frame defines a specific set of conditions under which output data are to be generated. In the production model developed it is used only to allocate data storage areas and the actual production configuration is loaded at run-time using a small Fortran subroutine. A user-friendly, front-end program may be used to set up a number of configurations and save them to file so that any particular one may be loaded into the general model 'shell'.

All Siman model code has been developed using a modular programming strategy to isolate functionally separate components. This has brought the usual benefits to model verification activities and has also been found convenient for validating the complex, multi-faceted model. Separate modules control line loading, breakdown

SIMAN MODEL	COGNITIVE CONTROL SYSTEM
Initialise database for run Start model run Send scheduled report Wait for response report Get response Update database Continue run	Initialise database for run Wait for report Get report type and data Process data Send "MOVE" or default response Update database Wait until next report

Figure 2. Interaction between Siman and the control system.

and absenteeism, the core line logic and the unloading of work with a final quality control.

The communication software is interfaced using the Siman 'event' block to integrate user-coded event logic within the process-interaction ('blocks') world-view of the main model. A separate Siman blocks module controls the timing of the communication events which use the RS 232 serial port to achieve transfer of data at 9600 Baud. An option at any stage of a run allows the model to be run in interactive mode where the user may manually change the operating conditions.

During a production run, the cognitive control system must be provided with data about the state of the simulation run. By declaring Siman variables as common and by using calls to the Siman subprogram library, any facet of the current state of production may be streamed to the control system. However, it is inefficient to transfer the whole Siman data base every time a single state change occurs. Reports have been designed to be specific and are scheduled to occur at appropriate times. This is a subject of current experimentation.

At present the control software is provided with information about the size of inventory buffers, the occurrence of machine breakdowns, and any operator absenteeism. Inventory checks are scheduled by default to occur every hour although the user may change this frequency. Reports on breakdowns and absenteeism are called whenever a change occurs in machine status or operator attendance, respectively. Reports are distinguished by preceding them with a specific header code and they are also terminated with an 'end' code to indicate to the control software that it should act on any information just received.

Once a report has been sent, the Siman model suspends execution until the control software returns a coded instruction. This may be a simple 'do nothing' or it may indicate that an operator is to be moved from one operation to another, in which case the Siman data base is changed to effect this. The whole cycle typically takes less than one second, a large slice of which may be attributed to serial communications. The interaction between the two systems is illustrated in Figure 2.

Example of operation

To explain the operation of the combined system, an example is used in which an operator at a particular operation goes absent for a number of days. Figure 3 is used to illustrate the execution cycle of the system under these conditions.

The stages in the execution cycle would be as follows:

(1) The absence is generated by a stochastic process within the Siman-based model.
(2) The send-absentee event is called from Siman resulting in a stream of characters being sent to the cognitive control system. On receipt of the 'end of message' code the stream forms a single status report within an input buffer.
(3) The status report is encoded into a bit stream to form an externally generated message. Bit positions encode the occurrence of the absence and mark other relevant categories such as other operators having appropriate skills to cover the operator.

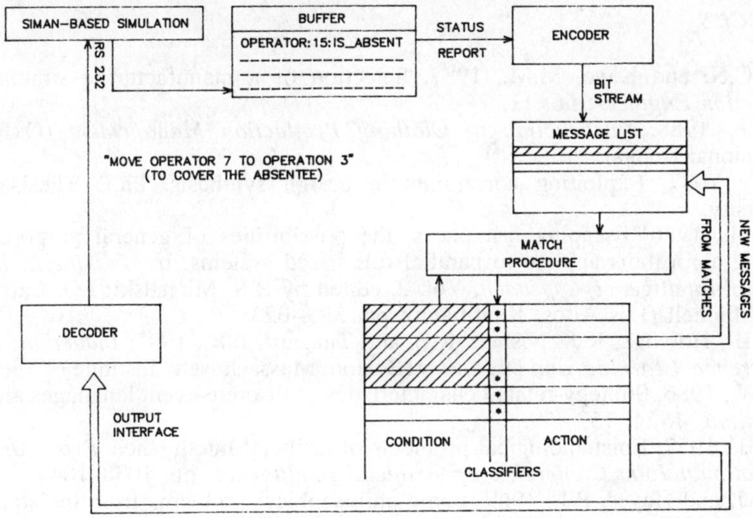

Figure 3. The cognitive control system showing absence message.

(4) The message is then compared to the rules embedded in the condition part of all other classifiers. All matches are recorded and weighted according to the suitability of their actions.
(5) The action part of each matching classifier is posted onto the message list as a new, internally generated, message. These messages contain predictions relating to the possible effector actions associated with operator 15 being absent. The goal of the system is to discover the action that best fits the unique state of the simulation at this particular point in time. In this example the classifiers relating to moving other operators are given priority.
(6) The message with the highest weight is used to execute an effector action. The effector action is first decoded into an appropriate form and then streamed back to the Siman-based model. In this case a code is sent that will update the Siman data base in the manner 'MOVE OPERATOR 7 TO OPERATION 3'.

Note: This example illustrates only one possible outcome of the decision process. If the system could not find another operator with suitable skills to replace the absentee then a 'stage setting' action at a lower level of the default hierarchy of classifiers would be triggered. Possibilities include summoning the mechanic to set up the absentee's machine for an alternative operation.

Conclusions

(1) Classifier systems have been found to be a suitable software architecture in which to model the complex activity of supervising clothing production.
(2) In order to link the control system to a model of the physical production line it is necessary to allow sharing of some form of common data base. This requirement places constraints on the selection of a simulation tool in that access to low level aspects of the host operating system is required.
(3) Modular programming methods have been found to be highly convenient for the verification and validation of the multi-paradigm model and for later experimentation. The investigation of different control strategies is straightforward because the external control software is effectively another module within the system that can be modified in isolation from other components.

References

Chrystall, C.N. and Kaye, M.M., 1987, Selection of a manufacturing simulation tool. *Production Engineer*, **66**, 11, 24–28.

Chuter, A.J., 1988, *Introduction to Clothing Production Management* (Oxford: BSP Professional Books).

Finger, J.J., 1987, Exploiting constraints in design synthesis. Ph.D Thesis, Stanford University.

Holland, J.H., 1986, Escaping brittleness: the possibilities of general purpose machine learning algorithms applied to parallel rule-based systems. In *Machine Learning: An Artificial Intelligence Approach*, Vol. 2, edited by R.S. Michalski, J.G. Carbonell and T.M. Mitchell (Los Altos: Kaufmann), pp. 593–623.

Holland, J.H., Holyoak, K.J., Nisbett, R.E. and Thagard, P.R., 1987, *Induction: Processes of Inference, Learning, and Discovery* (Boston: Massachusetts Institute of Technology).

Hooper, J.W., 1986, Strategy-related characteristics of discrete-event languages and models. *Simulation*, **46**, 4, 153–159.

McCarthy, J., 1977, Epistemological problems of artificial intelligence. *Proceedings of the International Joint Conference on Artificial Intelligence*, pp. 1038–1044.

McCarthy, J. and Hayes, P.J., 1969, Some philosophical problems from the standpoint of artificial intelligence. *Machine Intelligence*, **4**, 463–502.

Oren, T.I., 1986, Artificial intelligence and simulation. In *AI Applied to Simulation SCS Simulation series*, **l8,** 1, 3–8 , edited by E.J.H. Kerckhoffs, C.C. Vansteenkiste and B.P. Zeigler, (San Diego: Society for Computer Simulation).

Pegden, C.D., 1985, *Introduction to Siman* (State College: Systems Modeling Corporation).

76

An expert system: tool replacement in flexible manufacturing

C. Giardini, G. Maccarini and A. Bugini

Universita di Brescia, Meccanica, Dipartimento di Ingegneria, Via Valotti 9, 25060 Brescia, Italy

Abstract

The evaluation of the optimal production life of a cutting tool as well as the consequent economic cutting speed, is a relatively easily solved problem when the tool is considered as having a stochastic behaviour (damaged pieces and broken or accelerated worn tools may result from the established conditions) and it is employed in a series of jobs requiring the same cutting parameters (Ravignani *et al.*, 1979; Sekulič, 1982; Giardini *et al.*, 1988b, 1989a). Based on this assumption it is possible to find out how tool reliability influences the cost per piece. Today it is also very important to test the reliability of cutting tools by machining different pieces under the various conditions that occur in a flexible manufacturing system (FMS). In this case the forecasting of the tool production life is a relevant, but unknown, figure which must be updated by means of the data received from the production line, considering both tool breakage and ordinary tool replacement (L. Zavanella *et al.*, this volume).

The authors tried to develop a system capable of finding the productive tool life (i.e. the ordinary tool replacement time that minimizes the cost per piece) machining several pieces in a flexible system, by using every new datum of tool life and of tool replacement obtained from the workstations. This system is therefore constantly able to update its knowledge, confirming itself as an expert system (ES).

Introduction

Notation

C	machining cost per component,	d	depth of cut (mm),
C_0	unit labour and overhead cost,	m	feed exponent in Taylor's law,
Ca	tool cost,	r	depth exponent in Taylor's law,
Rw	penalty cost,	n	time exponent in Taylor's law,
Tc	cutting time (min),	$R\alpha$	reliability,
tc	tool changing time (min),	Tp	productive tool life (min),
ta	auxiliary time (min),	$T\alpha$	tool life at level of conf. α (min),
v	cutting speed (m/min),	Tr	average of tool lives $< T\alpha$ (min),
f	feed (mm/rev),	tf	penalty time (min).

It has been demonstrated (Ravignani, 1978; Sekulič,1982; Gardini *et al.*, 1988a,b, 1989a,b; Pacagnella *et al.*, 1988; L. Zavanella, this volume) that the reliability of a cutting system greatly influences the production time (Tp) and the cost of manufactured pieces (C/C_0); these figures can be expressed as follows (Ravignani, 1978; Sekulič, 1982; Gardini *et al.*, 1988b):

$$Tp = R\alpha \cdot T\alpha + (1 - R\alpha) \cdot Tr \tag{1}$$

$$C/C_0 = Tc \cdot \left\{1 + \left[tc + \frac{Ca}{C_0} + \left(tf + \frac{Rw}{C_0}\right) \cdot (1 - R_\alpha)\right] \Big/ Tp\right\} \tag{2}$$

where $R\alpha$ is the reliability of the tool operating in a stochastic environment while tf and Rw are assigned as the penalty time and the penalty cost, respectively. For Tf it is supposed that:

$$tf = \sum_{1}^{n-1} {}_jTc_j + Tc_n/2.$$

Equation (2) refers to a tool that constantly machines the same kind of pieces with the same cutting parameters. The Tp value that minimizes the C/C_0 term is called the optimal production life [Tp_{opt} that is the time after which the tool must be replaced, i.e. the ordinary maintenance interval ΔT_{man} (Giardini *et al.*, 1988a, 1989b; Pacagnella *et al.*, 1988)]. However as the unexpected failures are not all avoidable the aim of a programmed production consists in controlling the situation, trying to counterbalance the negative factors.

Tool behaviour

To describe the correct tool behaviour, the probability density function (pdf) of tool life (Sekulič, 1982; Gardini *et al.*, 1988b) must be a priori assigned, giving the new experimental data of the tool life, straight from the production line, the task of updating the knowledge of the system.

In the present chapter the authors, from preference, refer to the probability cumulative curve of tool life as more suitable for taking into account the interrupted trials (tools replaced before breakage) (Kapur and Lamberson, 1977). The expected tool life ($T\alpha$) and the corresponding reliability ($R\alpha$) may be linked by using the experimental data as Figure 6 clearly shows.

Flexible environment

As already shown (Pacagnella *et al.*, 1988; Giardini *et al.*, 1989a), if the machines belong to a flexible line, each tool may machine different pieces; in this case the cutting time for a tool should be recalculated for the new working conditions using Taylor's law as follows:

$$Tc_i^* = Tc_{i-1} \cdot \frac{v_{i-1}^{1/n} \cdot d_{i-1}^{r/n} \cdot f_{i-1}^{m/n}}{v_i^{1/n} \cdot d_i^{r/n} \cdot f_i^{m/n}} \tag{3}$$

while the total working time becomes:

$$Tt^* = \sum_{1}^{n-1} {}_j T_{c_j}^* + Tc_n \tag{4}$$

where n corresponds to the total operation performed by the tool under consideration. This tool will therefore be changed when the relationship: $Tp_{\text{opt}} \leq Tt^*$ is verified.

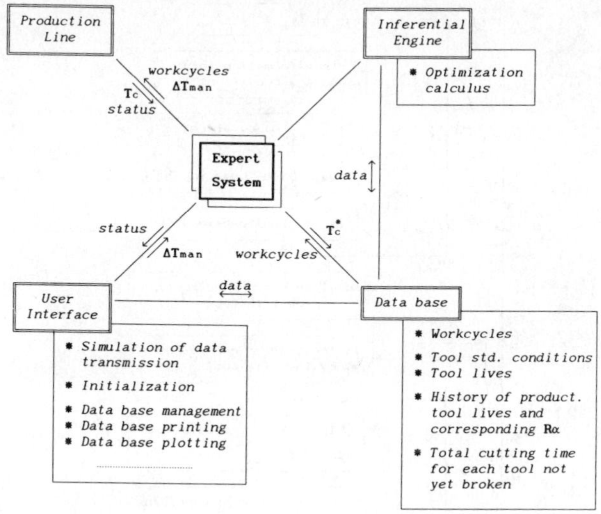

Figure 1. Block diagram of how the ES is linked with the line and its data bases; data exchanged between various subsystems.

In order to simplify the data management, times are transformed into standard conditions which correspond to those found in some current industrial catalogues (Sandvik Coromant, 1987).

The expert system

A system capable of updating tool life data bases was developed recalculating the optimal cutting conditions each time a tool substitution (broken or worn or ordinary replaced tool) occurs. For this purpose the program implemented on a PC was linked to the production line actually simulated on another PC. The corresponding block diagram is represented in Figure 1 where the principal elements of the developed system and the exchanged data can be found. In this figure it is also possible to see how the ES knowledge is built up by means of three data bases, each containing:

— the total machining time of each tool until breakdown,
— the standard conditions and the tool replacement times or tool lives for the whole set of tools,
— the workcycles of the workpieces.

BASIC is the language used in this first part of ES development.

The ES flow diagram is represented in Figure 2 where we can see how every new datum arriving from the line is recorded by the ES itself after its transformation into standard conditions. If the tool has to be replaced, the ES updates the cumulative function and calculates the new optimal conditions (cutting speed and productive time) which are found minimizing the cost per piece for each piece. After having calculated the weighted average programmed maintenance interval (ΔT_{man}) the ES sends this value to the line with the new whole set of workcycles. The significance of maintenance intervals is represented by the piece presence percentage in the production mix. The corresponding effective reliability $R\alpha$ is then associated to the calculated ΔT_{man}.

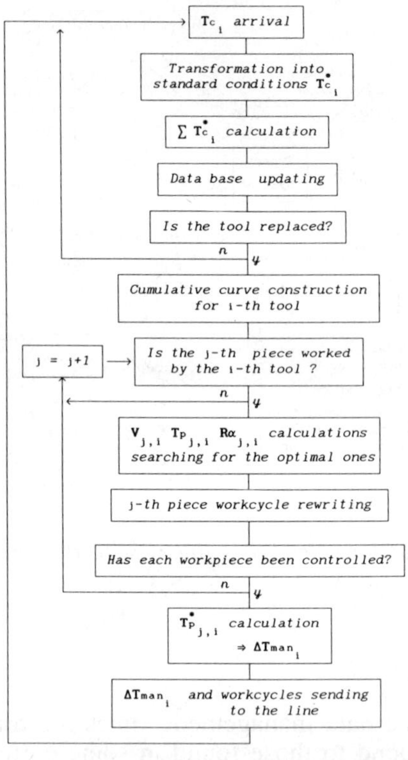

Figure 2. Flow chart of the program.

Simulated system and working hypotheses

In order to test the capability of the ES to understand and forecast tool behaviour better, the authors have tried to join the developed program to a simulative one in which three equivalent machines (turning) work five different pieces using eleven different tool types (Figure 3, Tables 1 and 2).

At the beginning the ES has no experience and therefore an approximated tool life datum must be introduced into the data base; in their turn the initial workcycle cutting parameters are based on indications found in the literature (Di Caprio, 1980; Sandvik Coromant, 1987). To improve the knowledge of the ES from the outset some tool breakages are intentional.

Other important hypotheses are:

— the pieces are supposed to enter the line randomly,
— in each workcycle a tool machines once only (more than once—the times must be reduced to a single pass with opportune working conditions),
— the piece on which the tool breaks down is always scrapped,
— tool status is checked at the end of *each* cutting operation,
— the tool is changed when its total working time, under standard conditions, is greater than the programmed tool life.

The block diagram of the simulated line is reported in Figure 4. The model was developed by using SIMAN language with an external Fortran routine able to manage the communication procedure and to bring the workcycle of the pieces up to date.

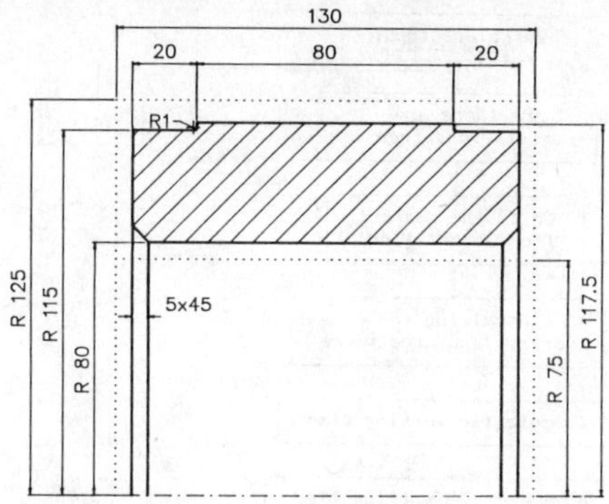

Figure 3. Workpiece number 1 drawing.

Table 1. Workcycle for workpiece number 1 (penalty cost (Rw/C_0), 30; stainless steel) (see Figure 3).

Tool number	Cutting speed (m/min)	Depth of cut (mm)	Feed (mm/rev)	Length of the cut	Diameter of the piece
1	150	3·50	0·40	240	243
2	160	2·25	0·30	240	155
3	230	0·50	0·10	280	235
6	230	0·50	0·10	120	160
7	170	2·25	0·20	140	195
8	230	0·50	0·10	70	195

Table 2. Tool types employed to carry out the whole set of operations.

No.	Tool type[a]	ISO code	Tip code	Material
1	Ext. roughing	PCLNR2525M19	CNMG190616	CG425
2	Int. roughing	S25TCTFPR16	TPUN160308	CG425
3	Ext. finishing (r = 1)	PCLNR2525M12	CNMG120408	CT515
4	Ext. finishing (r = 3)	PCLNR2525M19	CNMG190624	CT515
5	Int. finishing (r = 1)	S16RSDUCR07	DCMM070208	CT515
6	Int. finishing (r = 3)	S50WPCLNR19	CNMM190624	CT515
7	Head. roughing	PSKNR2525M12	SNMA120408	CG425
8	Head. finishing	PCLNR1616H12	CNMG120408	CT515
9	Ext. groove	R151.21.2525–50	[–]	[–]
10	Ext. groove	R151.21.3232–60	[–]	[–]
11	Int. groove	R156.3–202022	R156.3–22–400	GC435

[a] Ext., external; Int., internal.

Results

The ES was tested during a production of 1000 pieces. The simulation took about five hours.

As the simulative procedure continued, the ES improved its knowledge by refining the cumulative curve with new data of tool life and tool replacement times.

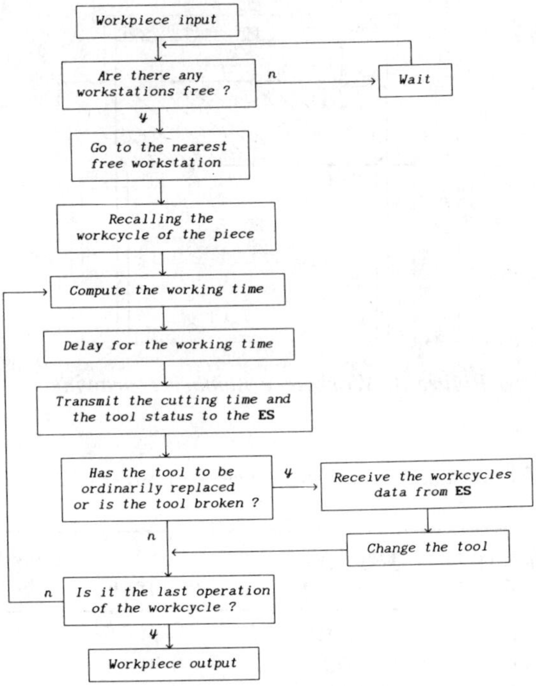

Figure 4. Block diagram of the simulated line.

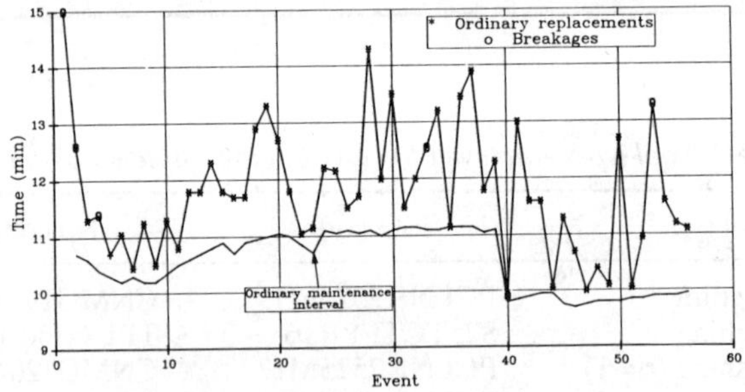

Figure 5. Behaviour of data obtained from the line (substituted and broken tools) and corresponding ordinary maintenance interval for tool number 1. Note the discontinuity where a premature failure occurs.

Only a few of the data were breakage data as Figure 5 shows. Once new data were available, the experimental cumulative curve was closer to the theoretical one. In Figure 6 a comparison between the cumulative curve after 10 and after 56 data is reported. The second is obviously smoother than the first: increased knowledge of the system improves the reliability of the operations (Figure 7).

Figure 5 also shows how the ES reacts to premature failure by reducing the ordinary maintenance interval.

A particular mention is required for Figure 8 where the longer cutting time (Tc) generates a more discontinuous curve with a large number of broken tools and consequently damaged and rejected pieces. In this case it seems most important to work with a higher reliability than the one calculated by ES, changing the policy of on-line control of the tool life (L. Zavanella *et al.*, this volume).

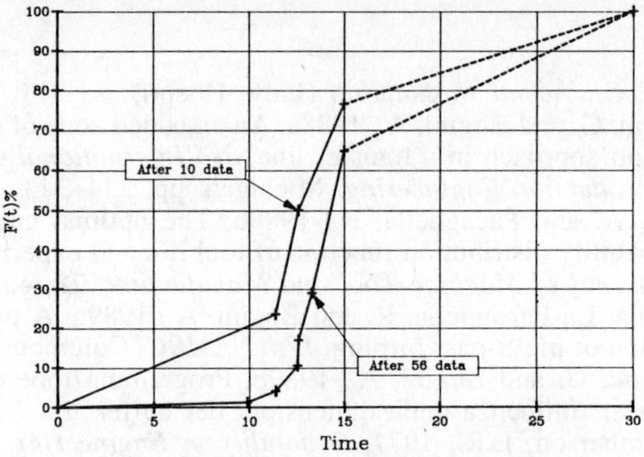

Figure 6. Improvement of ES knowledge as production continues for tool number 1 in terms of cumulative curve. F(t) % is the cumulative probability of tool breakage.

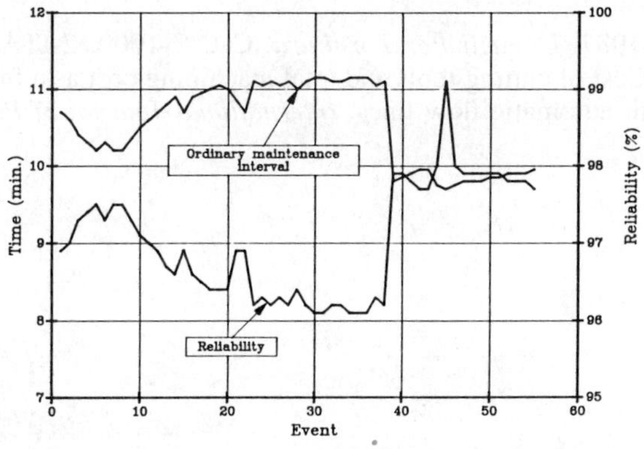

Figure 7. Reliability behaviour for tool number 1 as production continues.

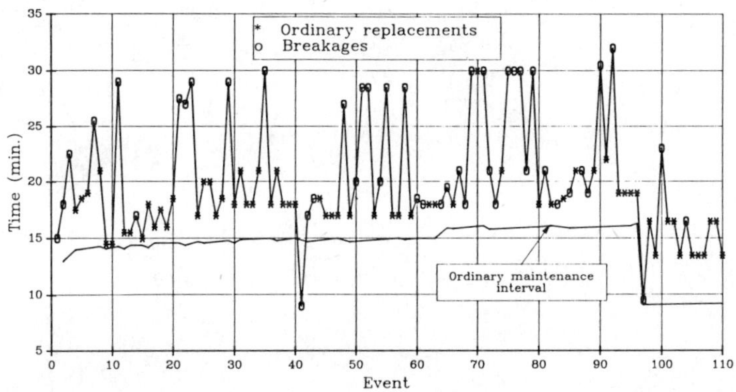

Figure 8. Behaviour of data obtained from the line (substituted and broken tools) and corresponding ordinary maintenance interval for tool number 3. Note the discontinuity of the curve corresponding to the tool life values.

References

Di Caprio, G., 1980, *Gli Acciai Inossidabili* (Italy: Hoepli).

Giardini, C., Maccarini, G. and Bugini, A., 1988a, An extended zone of minimum production cost by simulation approach in a transfer line. *3rd International Conference on Computer-Assisted Production Engineering*, Michigan, pp. 334–343.

Giardini, C., Bugini, A. and Pacagnella, R., 1988b, The optimal cutting conditions as a function of probability distribution function of tool life and experimental test numbers. *International Journal of Machine Tool and Manufacture*, **28**, 4, 453–459.

Giardini, C., Zavanella, L., Pacagnella, R. and Bugini, A., 1989a, A probabilistic approach to the optimization of multi-pass turning. *17th NAMRC*, Columbus, Ohio, pp. 254–259.

Giardini, C., Maccarini, G. and Bugini, A., 1989b, Programmazione della vita utensile in un sistema transfer: influenza delle dimensioni dei buffer. *1st UTENSIL*, Italy.

Kapur, K.C. and Lamberson, L.R., 1977, *Reliability in Engineering Design* (New York: Wiley).

Pacagnella, R., Giardini, C. and Restelli, G., 1988, Optimization of tool changes in working of workpiece families by simulation. 3rd *International Conference on Computer-Assisted Production Engineering*, Michigan, pp. 37–46.

Ravignani, G.L., Zompi, A. and Levi, R., 1979, Multi tool machining analysis. Part 2 economic evaluation in view of tool life scatter. Trans. *ASME*, J. of Eng. for Ind., **101**, 237–240.

Sandvik Coromant, 1987, *Utensili Per Tornitura,* Cat. C-1000: 2-ITA, 243–26l.

Sekulič, S.S., 1982, Cost of cutting tools and total machining cost as a function of the cutting tool reliability in automatic flow lines. *International Journal of Production Research*, **20**, 2, 187.

77

An expert system approach to the design of flexible manufacturing systems

M.C.K. Kwok[1] **and A.S. Carrie**[2]

[1]Department of Industrial Engineering, University of Hong Kong, Pokfulam Road, Hong Kong and [2]Department of Design, Manufacture and Engineering Management, University of Strathclyde, Glasgow G1 1XJ, UK

Abstract

This chapter presents an expert system approach to the design of flexible manufacturing systems (FMS). When designing an FMS, the designer is faced with the formidable task of considering a vast number of possible configurations and numerous interacting design parameters which will affect the operation of the FMS. An attempt is made here to bring together various software tools, analytical models and simulation techniques and integrate them in an expert system to support the FMS designer to evaluate a number of different alternatives. The approach adopted for such an integration is also described.

Introduction

Flexible manufacturing systems (FMS) are constructed from a number of engineering modules and some of the variables and options available are listed as follows:

- Operations—machining, inspection, washing, etc.
- Machines—turning centres or machining centres, all alike or dissimilar, stand alone or grouped, etc.
- Materials handling system (MHS)—conveyor, robot, rail guided cart or automated guided vehicle (AGV), etc.
- Load/unload—pallet shuttle, robot, manual, etc.
- Buffers—local buffers, generalized buffers, etc.
- Pallets—quantity, number of parts per pallet, etc.
- Fixturing—variety, type, etc.
- Tooling—variety, quantity, loading method, changing method, tool capacity of machine tool, etc.

The intention here is not to address all the above aspects of FMS design. Rather, the FMS design problem in this context is restricted to configuration problems such as determining the appropriate number of machine tools of each type, the capacity of the materials handling system, the type and size of buffers, and so on. Even in this restrictive sense, the FMS designer is often faced with a formidable task of evaluating a great number of possible configurations. It will be difficult for him to consider or vary the different interacting design parameters in order to arrive at the configuration which will give the desired system performance figures.

It is believed that a knowledge-based approach is most appropriate in this type of situation. An attempt is therefore made here to bring together various software tools, analytical models and FMS simulators and to integrate them into an expert system FMS-XD (Flexible Manufacturing System eXpert Designer). This will facilitate the FMS designer in the process of configuring FMS and evaluating various possible alternatives.

Hardware and software environment

FMS-XD is developed using ENVISAGE, an expert system shell running under the VAX/VMS operating environment. ENVISAGE is a rule-based shell and contains the inference engine for rule-based control processing as well as some user interfaces for better interaction between the user and the expert system. It supports both backward and forward chaining.

The knowledge base constructed for the FMS design problem can be roughly divided into the following areas:

(1) *Configuration knowledge*—Knowledge about the design procedure, the various engineering modules and how an FMS is constituted.
(2) *Analysis algorithms and simulators*—External routines for capacity calculations, analytical modelling and simulation of FMS.
(3) *Experience from existing FMS installations*—A library of some existing FMS configurations held in a data base which the designer can consult during the design process.
(4) *Interpretation knowledge*—Knowledge of how to interpret the results from an analytical model or a simulator.

Configuration knowledge and interpretation knowledge are coded in the knowledge representation language provided with the shell, while the VAX Rdb/VMS relational data base management system is used to set up the library of existing FMS configurations. As for the analysis algorithms and simulators, these consist of a mixture of VAX FORTRAN source codes and other proprietary software tools.

Since external routines are involved in FMS-XD, interfaces must be present, and VAX FORTRAN and PASCAL are used to program these interfaces. The ENVISAGE shell consists of a compiler and an executive. All the knowledge after compilation by the ENVISAGE compiler will be stored as a model and called up by the ENVISAGE executive when a consultation is made. The model can be linked to and communicate with external data bases and application programs coded in other programming languages. FMS-XD is developed such that the ENVISAGE executive will sit on top of the other programs and will be the main program being called up when consultation starts. The various elements of the expert system will be discussed in detail in the next section.

FMS-XD is developed on a VAX 11/750 computer running under the VMS operating system, but the proprietary simulator used is running on an IBM-PC, hence facilities for transferring ASCII files between the VAX and the IBM-PC must be present. A VT100 or compatible terminal is also required for consultation with the expert system.

Elements of the expert system

The major elements of the expert system, FMS-XD, are shown in Figure 1 and explanation of the individual elements given in the separate subsections to follow.

Input area

The knowledge about the design process is coded in rules in a rule base, and in this area the designer has to decide on the two main components of an FMS, that is:

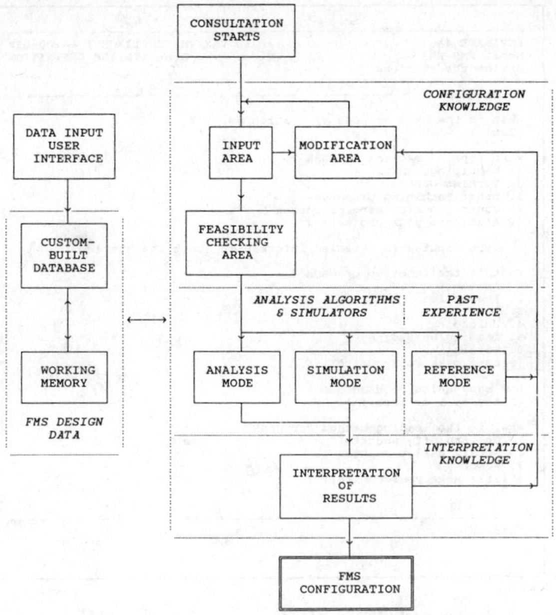

Figure 1. Major elements of the expert system, FMS-XD.

(1) the required stations, be it machining stations, inspection stations, load/unload stations, etc.
(2) the materials handling system required to connect the different stations.

This is achieved by posing a series of appropriate questions to the user of the expert system. The questions are designed in such a way that depending on the answers provided, details regarding the station or the MHS will also be asked for, such as the type of buffer used, the number of buffer positions, whether pallets are used or not, etc. A sample of the questions asked during a consultative session is presented in Figure 2.

Such a process ensures that the design conforms to a set of pre-defined constraints and uses a forward chaining strategy through the set of rules contained in the rule base to arrive at a specific FMS configuration.

Modification area

This area is present to make provisions for editions if the user is not satisfied with the original input, in which case he can re-answer some of the questions or restart a new session again. Only when he is satisfied with the input will these data be saved in an external file, which contains all the contents regarding the current state of the model. This file can be treated as the working memory or data base of the expert system, and it can be modified, saved or restored for later use whenever necessary.

The working data base of the expert system only contains a small portion of the data, the configuration data, for the FMS design problem. Another custom-built data base which houses the other data pertaining to a specific FMS design mission is also present. This type of data is referred to as the mission data. As shown in Figure 1, communications between the data bases are in both directions, and the various modules in the expert system draw the data from these two complementary data bases during execution. Further details about them are given in the section on data handling.

```
ENVISAGE 4A                              SD VAX/VMS ENVISAGE : 4A.00-038
Model FMS-XD                                   Area STATION_DEFINITION
Define the stations

What is the name or code of station No. 3 ?
(Text):  WASH

What type of machine is WASH ?
1. Machining centre
2. Turning centre
3. Other machining processes
4. Other non-machining processes
(Please make your choice): 4

     Now input more detailed information about the station

What is the function of WASH ?
1. Assembly
2. Inspection
3. Loading
4. Unloading
5. Loading/Unloading
6. Washing
(Please make your choice): 6

How many units of WASH used?
(Range is 1 or more): 1

What is the loading method for WASH?
1. On- and off- shuttle
2. Rotary pallet changer
3. Robot
(Please make your choice): 3

Command?
```

Figure 2. Sample consultative session for the input area.

Feasibility checking area

The components of the FMS are identified in the input stage but such a configuration may not be feasible, so simple rules for checking the feasibility of configurations are followed here. Some of the IF–THEN rules found in this area can be verbalized as follows:

(1) IF an FMS with linear tracks is used,
THEN on- and off-shuttles OR generalized buffers are needed to avoid blocking.
(2) IF an FMS with linear tracks is used,
THEN the AGV or rail cart must travel bi-directionally.

The output from this area will be in the form of advice to the designer, who can modify the model accordingly by returning to the modification area if he so wishes. After the user is satisfied with the present configuration, he can then proceed to any one of the modes, namely: analysis, simulation or reference modes, depending on the expertise of the designer and which stage of the design he is at.

Analysis mode

This mode is concerned with the capacity calculations and the evaluation of the FMS configuration using analytical models. Routines are built in here to calculate the capacity of the FMS under consideration, and the CAN-Q model (Solberg, 1980) is also included to give indications of the system performance measures, such as production rate, utilization figures and bottle-neck stations, at steady-state operation of the FMS. CAN-Q was chosen, because despite the simplicity of the algorithm, it can be used to model a variety of FMS configurations and has been found to be reasonably accurate in predicting the steady-state behaviour of such systems (Co and Wysk, 1986). The appropriate data are extracted from the two data bases mentioned earlier and the input file for execution of this module automatically generated. The whole process is transparent to the user, with the system performance values of the configuration under question being passed on to the interpretation knowledge area where they will be interpreted. The analysis mode is especially useful in the initial stages of the design where preliminary performance values can be obtained very quickly.

Simulation mode

A proprietary simulator, MAST (Lenz, 1986), that runs on an IBM-PC is used in this module. As in the analysis mode, data are extracted from the data bases of design data and by means of a simulation code generator, the input data file for MAST is provided. This file has to be downloaded to an IBM-PC for the simulation run, and after the simulation, the appropriate files have to be uploaded again back to the VAX for the interpretation of simulation results.

Simulation models more accurately represent the FMS set-up and in the case of MAST, a reasonable amount of detail can be incorporated, such as variation in number of pallets, different pallet shuttle arrangements, inclusion of scheduling rules, batching and the like. Consequently, this stage will be useful when dealing with the detailed design of the FMS (Browne *et al.*, 1985) after the configuration has been decided. The system behaviour can be carefully observed by varying different design parameters. This combination of expert system and simulation models can be classified as co-operative (O'Keefe, 1986), where both elements will co-operate in the task of arriving at a good FMS design.

Reference mode

A library of selected existing FMS configurations is collected here. Under each FMS installation, typical information included in the library are items such as the type of FMS, the number of stations involved, the MHS used, the machining operations undertaken and the type and number of pallets used.

The VAX Rdb/VMS relational data base management system is used to house all the information contained in the library of FMS configurations. Depending on the current state of the system proposed in the configuration knowledge area, the appropriate data can be extracted from the data bases containing the FMS design data and similar FMS configurations can be retrieved and displayed. The user of FMS-XD could treat this as some past experience that he can consult when configuring the FMS.

Interpretation knowledge

The results from the CAN-Q model are analysed here. The system performance measures of an FMS design are compared with the target objectives, and advice regarding the actions to be taken can be provided to the user, who then either modifies the design or accepts it and proceeds to other modules. It should be noted that the output results from analytical models are average values and are only rough indications of the performance of the FMS.

Simulation modelling is a more faithful way of evaluating FMS configurations, but usually produces a vast amount of statistical output data at the end of each simulation run. The MAST simulator is no exception, and mechanisms must be available for the interpretation of such results. In FMS-XD, the OUTPUT processor of the SIMAN simulation language (Pegden, 1987) is used to process the MAST simulation output, with various key performance measures determined and statistical inferences made. These can then be passed on to the diagnosis area for FMS design for further interpretation.

The principal function of this diagnostic area is to identify the potential problems of an FMS design from the simulation output, and then estimate the cause of such anomalies with the intention of proposing possible solutions to eradicate them. The potential inefficiencies of an FMS design include over-utilized machines, bottle-neck workstations and inadequate loading/unloading stations (Mellichamp and Wahab, 1987). The knowledge of the diagnostic system is represented as rules in a rule base, of which two examples are:

(1) IF the utilization of the machine is low,
AND the utilization of the buffer is high,
AND the utilization of the MHS is high,
THEN the MHS is the bottle-neck.
(2) IF the utilization of a machine is high,
AND it is uneconomical to duplicate the machine,
THEN try machine pooling.

Backward chaining is used in this diagnostic area which is in fact an intelligent back-end (Wichmann, 1987) for the MAST simulator. The weak points of a design may be highlighted, with appropriate advice on actions to be taken. The user can then proceed to rectify the situation by modifying design parameters. However, if the designer is satisfied with the current configuration, he can terminate the session and consultation ends here.

Data handling

To handle the large amount of input data for the FMS design problem, some form of data base management system must be present. Furthermore, two types of input data can be distinguished. The configuration data, such as number of stations and their types, MHS used, layout of the FMS and buffer arrangements, keeps on changing during the consultative process. On the other hand, the mission data, which includes mainly the production requirements for each family of components, and the routing information, will remain unchanged throughout.

As mentioned earlier, the configuration data is stored in the working memory or data base of the expert system and reflects the current state of the model. The data is derived from the consultative session, and no extra effort is required for data input. A custom-built data base has been developed on the VAX computer to house the mission data using the composite data declaration statements RECORD and STRUCTURE supported by VAX FORTRAN. Declaration of multi-field records and access to record components through field references are then possible in FORTRAN programs and these are explained in Figure 3. The question and answer session in ENVISAGE for data input is very inefficient when dealing with a large volume of input data such as routing information. An extra user interface has been developed for the input of mission data. This is a form filling-in type of input method and is programmed using the Screen Management Routines offered by the VAX Run Time Library. A sample screen is presented in Figure 4.

The two data bases are complementary to each other and communications are in both directions. In addition, specific interfacing programs are also included to generate the input data files for running CAN-Q or MAST. For instance, if the MAST simulator is required by the expert system, the specific interfacing program for MAST will be executed, and the appropriate data extracted from the custom-built data base added to the data in the working memory. All this information will in turn be written on to another file that conforms to the MAST input format.

Discussion

In the development of FMS-XD, emphasis has been placed on the modular approach. The analytical model, simulator, data bases, configuration knowledge area and diagnosis area are all independent modules. These islands of knowledge are

SOURCE CODE

```
STRUCTURE /OPER/
      INTEGER*4      REC_NO
      CHARACTER*1    CODE
      CHARACTER*10   ID
      CHARACTER*10   STATION
      CHARACTER*5    PROC_TIME
END STRUCTURE

RECORD /OPER/ COMPT_OP10
```

ARRANGEMENT OF DATA ITEMS

Record	COMPT_OP10				
Fields	REC_NO	CODE	ID	STATION	PROC_TIME

RECORD AND FIELD REFERENCING

```
Examples of referencing at different levels:
  1.  COMPT_OP10            Complete record
  2.  COMPT_OP10.STATION    The STATION field within
                            record COMPT_OP10
```

Figure 3. Use of structure and record statements in VAX FORTRAN.

```
                              FMS MISSION DATA

Edit Terminators:

FUNCTION   ADD   QUIT   INTERCHANGE   DELETE   SAVE   SUB-LEVEL   HELP
KEY        PF1    -         PF3         PF4      ,        0        PF2
----------------------------------------------------------------------

       OPERATION DATA

   ID        STATION              PROCESSING
                                  TIME
   --------------------------------------
   OP10      STATION 2            30
   OP20      STATION 3            20
```

Figure 4. User interface for data input.

integrated together to form the expert system. Data bases which house all the FMS design data have been created and communications between different modules are, as far as possible, through the use of interfacing programs which transfer ASCII files amongst the modules. The interfacing programs serve also as templates whereby new interfaces can be constructed, and consequently, the modules can be replaced with better algorithms or knowledge bases with relative ease.

The architecture of FMS-XD is a form of hybrid system with two types of elements: symbolic and numeric. Symbolic elements such as configuration knowledge and interpretation knowledge are coded in the knowledge representation language of ENVISAGE, whereas numeric elements are analytical models and simulators written in FORTRAN. This hybrid arrangement has proved to be very effective when dealing with the FMS design problem, where the knowledge comes from various sources. FMS-XD has been applied to the design of FMS with linear tracks, but this could easily be extended in future to cope with other FMS configurations.

References

Browne, J., Chan, W.W. and Rathmill, K., 1985, An integrated FMS design procedure. *Annals of Operations Research*, **3**, 207–237.

Co, H.C. and Wysk, R.A., 1986, The robustness of CAN-Q in modelling automated manufacturing systems. *International Journal of Production Research*, **24**, 1485–1503.

Lenz, J.E., 1986, *MAST User Manual* (Wisconsin, USA: CMS Research).

Mellichamp, J.M. and Wahab, A.F.A., 1987, An expert system for FMS design. *Simulation*, **48**, 201–208.

O'Keefe, R., 1986, Simulation and expert systems—a taxonomy and some examples. *Simulation*, **46**, 10–16.

Pegden, C.D., 1987, *Introduction to SIMAN* (State College, Pennsylvania, USA: Systems Modelling Corporation).

Solberg, J.J., 1980, *CAN-Q User's Guide* (School of Industrial Engineering, Purdue University, USA).

Wichmann, K.E., 1987, Trends in the development of simulation software tools for analyzing manufacturing systems. *Proceedings of the 3rd International Conference on Simulation in Manufacturing* (London: IFS Publications), pp. 40–48.

78

Development of a knowledge-based system for synchronizing production in a two-stage, geographically decentralized manufacturing process

W. Robert Terry and Thomas W. Matz

Ohio University Department of Industrial and Systems Engineering, Athens, OH 45701, USA

Abstract

A knowledge-based decision support system (DSS) for production planning for a highway maintenance organization producing traffic control devices is discussed. The organization is two-stage, with a central facility for initial manufacture and a number of systems to perform final assembly. The system design is described; an algorithm generates plans that minimize the deviation from a desired goal while minimizing labour cost. A subsystem of the DSS, the shop production planning system (PPS), and its five constituent functional primitives are described in some detail.

The approach discussed is thought to be applicable to other ill-structured problems associated with large-scale systems.

Introduction

This chapter describes the development of a knowledge-based decision support system (DSS) for production planning. The production system under investigation manufactures a vast array of different product models and is structured in two stages. The first stage is a central facility for initial manufacture of components. Production within this facility is organized in a group technology format. In the second stage there are *n* identical systems which perform a final assembly operation. Each of these final assembly systems is responsible for servicing customer demands in one of *n* geographically decentralized marketing areas.

Production systems with the characteristics described in the above paragraph are quite common. One example would be an international automobile manufacturer making electronic fuel injection systems in a central facility and shipping them to assembly plants in several different countries. A second example occurs in military logistics when replacement parts are manufactured in a central facility and then sent to a number of geographically decentralized maintenance facilities. A third example occurs in highway maintenance organizations. In this case traffic control devices such as signs and signals will typically be manufactured in a central facility and shipped to a number of geographically decentralized units which are responsible for installing the devices at specific locations on roadways.

The task of the DSS is to synchronize production between the two stages in the process. Synchronizing production in this way is a complex problem that is rarely analysed adequately. The problem, when recognized, is usually viewed as an inventory problem. It is assumed that buffers of inventory are needed between the stages. Optimal ordering and stocking policies are determined in an effort to reduce inventory carrying costs.

The growth of artificial intelligence (AI) has had great impact on problem solving and decision support capability (Simon, 1987). Problems that have been difficult to solve with traditional methods can now be formulated as AI representations. One significant problem solving mechanism in AI is search (Pearl and Korf, 1987). Search is a universal technique used in most problem solving activity. When direct algorithms are not available, a search over possible solutions is used to find the best solution. Search has widespread applications for those problems in decision making and allocation of resources that require incremental solution construction (Nardi and Paulson, 1986), hypothetical reasoning (Gevarter, 1987) and/or simultaneous exploration of alternative assumptions (Luger and Stubblefield, 1989).

The rest of this chapter is aimed at describing a system whose function is to synchronize the production of traffic control devices in a highway maintenance organization. The maintenance organization is structured into two stages. The first stage is the manufacture of traffic signs in a central shop. The second stage is the installation of all traffic control devices by several districts.

System design

Figure 1 shows a portion of the distributed DSS for one district and the shop. A forecasting model is used to generate demand for the planning modules of the system. Given this demand as input, the next module of the DSS (RCAPS) plans the activities of the districts. An algorithm is used to generate plans that minimize the deviation from a desired goal of the number of sign replacements while minimizing labour costs. The sign installation plans for the districts collectively determine the demand on the sign shop. Given this demand, the shop production planning system (PPS) totals demand for signs from each district. It then assigns production to satisfy the demand. Any demand that cannot be met by the shop is communicated back to the districts. The districts adjust their plans accordingly.

Since demand is seasonal, both planning modules must use detailed rather than aggregate production planning. In order to synchronize production between the two stages, the shop must know exactly what products are needed in the next planning period. Thus, all facilities need to solve the production smoothing problem in a multi-product fashion.

The following paragraphs focus on one subsystem in the DSS, the shop production planning system. It is described down to the functional primitive level. Logic similar to that in the shop production planning system is used to generate a feasible roadway plan in the roadway crew activity planning system as well. However, the alloted space does not permit discussion of the entire decision support system. The interested reader may refer to Matz (1989) for a complete description of the roadway planning system and the synchronization algorithm.

Production planning system

The production planning system is decomposed into five functional primitives (Figure 2). The algorithm begins by totalling demand for each sign into production orders. This grouping of sign orders over all markets enables larger lot sizes to be run. Production orders are then further grouped (during the search) with other orders

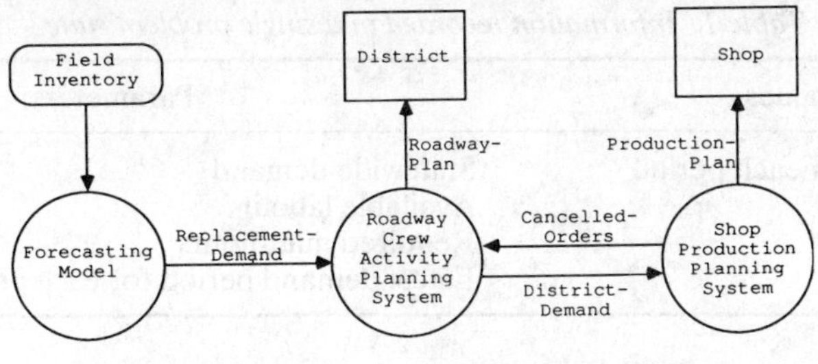

Figure 1. DSS for synchronizing production.

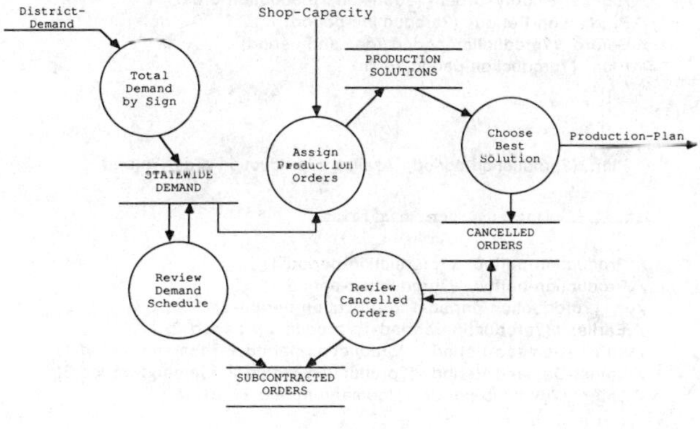

Figure 2. Shop production planning system.

of the same material, size, shape, and colour to form production jobs. This grouping allows for still larger lot sizes at the upstream operations in the sign shop. The feasible solutions generated in the state space are evaluated and the best one is chosen as the production plan.

Total demand by sign by period

The first function totals the demand over the entire system of district units. Each district has demand for a number of different signs. As demand arrives from each district it is grouped into production orders. These production orders are placed in the file called statewide demand. When a production order is created, it is assigned a priority. This priority is the average of all component sign orders in the production order.

Review demand schedule

The second function is one that is done by the sign shop superintendent. The entire demand on the shop over the planning horizon is printed out at this time. The sign shop superintendent has the option to remove any production orders for subcontracting. Some large orders consisting of a unique sign design may be better produced by a subcontractor.

Assign production orders

The third process is the search. The search begins with an empty production plan as an initial state. States are generated as the problem solver assigns production

Table 1. Information recorded in a single problem state.

Decision variable values	Parameters
Production plan for each period	Statewide demand
Cancelled orders	Available labour
	Required man-hours
	Latest demand period for each production period

```
Production-Order Assignment Operator

 Current-Demand-Period (?demand-period)
∧Demand (?demand-period,?demand)
∧Highest-Priority-Order (?demand,?production-order)
∧Production-Period (?production-period)
∧Before (?production-period,?demand-period)
∧Plan (?production-period,?plan)
   ⇓
 Plan (?production-period, Assign (?production-order,?plan))

Satisfy-Demand-in-Order Constraint

 Production-period (?production-period-1)
∧Production-period (?production-period-2)
∧≠ (?production-period-1,?production-period-2)
∧Earlier (?production-period-1,?production-period-2)
∧Latest-Demand-Period (?production-period-1,?demand-period-1)
∧Latest-Demand-Period (?production-period-2,?demand-period-2)
∧Later (?demand-period-1,?demand-period-2)
   ⇓
 False

Labor Hours Constraint
 Department (?department)
∧Period (?period)
∧Required-Manhours (?department,?period,?r-m)   ⟹ False
∧Available-Manhours (?department,?period,?a-m)
∧> (?r-m,?a-m)
```

Figure 3. Search operators.

orders to feasible production periods. The problem solver branches on the assignment of each production order to a production job in a feasible period in the production plan. A feasible period would be any period that comes before the period in which the production order is due. Assignment involves merging into an existing production job (if possible) or creating a new production job. The information recorded in a single problem state is shown in Table 1. To update the required amount of labour, a demon is attached to the production plan variables. When production is assigned to a particular production period, the demon fires. This firing causes a method to compute required labour hours given this new level of production.

The logic that generates the state space and constrains the search is shown in Figure 3. Each operator contains a lower-level code embedded in high-level predicate logic language. The search used here is a brute-force, rather than heuristic, search. It is important to expand each node, because it is never known which one will lead to the best solution. Producing a production order a few periods earlier may incur more inventory holding cost, but it may enable grouping the order in with another production job; thus allowing for more production.

To limit the search space, assignment takes place in chronological order, beginning with the first demand period. Within a period, production orders are

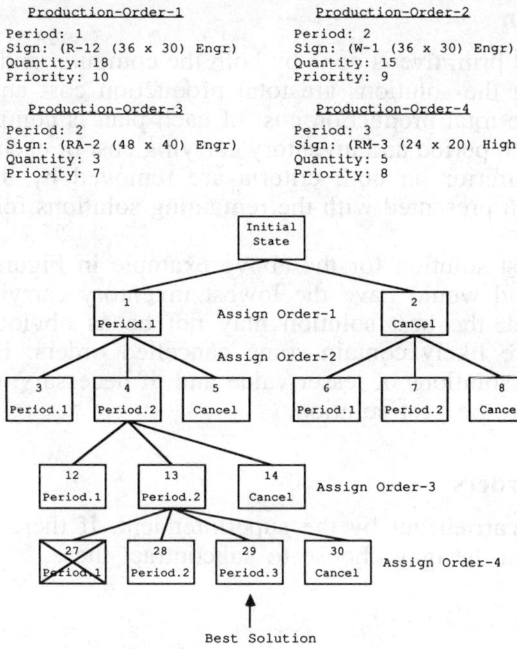

Figure 4. Example state space representation.

assigned in order of priority. The highest priority order is scheduled and then removed from statewide demand. Two constraints are used to prune the search space. First, if the required man-hours exceed the available man-hours in a period, the state is made infeasible.

Secondly, a state is made infeasible if production goes out of order. That is, if production in one period, *P*, is satisfying demand in another period, *D*, while at the same time production in a period later than *P* is satisfying demand in a period which is prior to *D*, the state is made infeasible. This prevents solutions that have, say, production in Period. 2 of an order not due until Period. 6 and production in Period. 5 of an order due in Period. 5. The main reason for doing this is to contain the state space to a feasible size.

To keep the search going in the event that all branched states are infeasible, the production order is cancelled and another state is created. This is accomplished by having an extra production period called Cancelled-Orders. At each branching, one of the nodes created has the assigned production order in Cancelled-Orders. This node is always feasible, since cancelled orders do not require any man-hours. Branching can continue from this node; thus allowing production orders of lower priority to be tried. The state space is completed when all production orders are assigned.

The sign shop search is illustrated with the following small example (Figure 4). The demand from the first week is assigned first. There are two possibilities for Production-Order-1. It is either produced in Period. 1 or cancelled, as depicted by the creation of States 1 and 2. There are two production orders due in Period. 2. The higher priority of these orders is Production-Order-2. It is assigned first, then Production-Order-3 is assigned. A lack of space prevents showing all states created by assigning Orders 3 and 4. The labels for each state are numbered in order of their creation. Production-Order-4 is assigned last. Assigning Production-Order-4 to Period. l from State l3 causes a contradiction. State 27 has Order-4 being produced before Order-2 and Order-3. The Satisfy-Demand-in-Order constraint matches and prunes the search space at State 27.

Choose best solution

The fourth functional primitive is done by both the computer and the user. The two criteria in evaluating the solutions are total production cost and total amount of cancelled orders. The total production cost of each plan is computed by summing the labour cost of each period and inventory carrying cost of each production order. Solutions that are inferior on both criteria are removed by the algorithm. The superintendent is then presented with the remaining solutions form which to make a selection.

Intuitively, the best solution for the above example in Figure 4 is State 29. It cancels no orders and would have the lowest inventory carrying cost. As more demand is scheduled, the best solution may not be as obvious. Then the only feasible solutions are likely contain some cancelled orders. Here is where the algorithm to remove solutions of lesser value and, if necessary, the aid of the user are needed to choose the best solution.

Review cancelled orders

The last process is carried out by the superintendent. If there are any cancelled orders in the solution, he may choose to subcontract these.

Conclusion

A decision support system is an effective tool for ensuring profitability in multiple facility systems. The DSS described here makes great use of state space representation, search, rules and demons. Demons and procedures enable significant amounts of information processing to take place when nodes of the space are created. Constraints are quite effective for limiting the network to only feasible solutions. The overall approach is amenable to many ill-structured problems associated with large-scale systems.

References

Gevarter, W.B., 1987, The nature and evaluation of commercial expert system building tools. *Computer*, **20**, 24–41.

Luger, G.F. and Stubblefield, W.A., 1989, *Artificial Intelligence and the Design of Expert Systems* (Redwood City, CA: Benjamin/Cummings).

Matz, T.W., 1989, *A decision support system for synchronizing manufacturing in a multifacility production system.* Master's Thesis, Ohio University.

Nardi, B.A. and Paulson, E.A., 1986, Multiple worlds with truth maintenance in AI applications. Intellicorp Technical Article Reprint from *Proceedings of the Seventh European Conference on Artificial Intelligence*, pp. 437–444.

Pearl, J. and Korf, R.E., 1987, Search techniques. *Annual Review of Computer Science*, **2**, 451–467.

Simon, H.A. *et al.*, 1987, Decision making and problem solving. *Interfaces*, **17**, 5, 11–31.

SECTION VII.3. SIMULATION

79

Obtaining resolution IV designs using Taguchi's orthogonal arrays

Kimball E. Bullington,[1] James N. Hool[2] and Saeed Maghsoodloo[2]

[1]*Micro Motion, Inc., 7070 Winchester Circle Boulder, CO 80301 and*
[2]*Industrial Engineering Department, Auburn University, AL 36849, USA*

Abstract

One concern expressed in the literature about Genichi Taguchi's methods for experimental design is that many designs given by Taguchi are not optimal in the sense that they are not the maximum resolution possible for a given number of main effects and array size. This chapter addresses that concern by demonstrating that Taguchi two-level orthogonal arrays that are fractional factorial designs consist of fold over designs (contained in the odd columns) and the two-factor products of fold overs.

The result of recognizing this characteristic is a simple method for guaranteeing resolution IV designs without adding to or changing Taguchi's arrays. This allows the experimenter to use Taguchi's orthogonal arrays without aliasing main effects with two-factor interactions. Sequential design is also permitted by use of the fold over columns.

Introduction

Some of Taguchi's designs are not of the maximum possible resolution for a given number of main effects and orthogonal array (OA) size. Taguchi (1988) does not address the concept of design resolution in his writings and yet it seems to have the type of simplicity and intuitive appeal that would fit well within Taguchi's techniques.

Fractional factorials

A full 2^k factorial design contains every possible treatment combination of k factors at two levels. The two levels may represent high or low levels, presence or absence, etc. of a particular factor. There are 2^k possible treatment combinations of k factors each at two levels. Two-level designs are especially useful for screening factors worthy of more detailed investigation.

A full 2^k factorial yields estimates for $2^k - 1$ effects. A 2^{k-p} fractional factorial will not permit independent estimates for $2^k - 1$ effects, but the usefulness and economy of fractional factorials make them desirable even though each effect estimated is actually 'aliased' with $2^p - 1$ other effects. It is desirable to ensure that main effects are aliased with effects that should be unimportant, namely higher-order interactions. Maximizing resolution accomplishes this, though compromises in resolution may be necessary to reduce the number of runs required or to permit sequential analysis. Resolution IV designs may in many cases be the best compromise between accuracy of estimates and economy of experimentation.

Design resolution

Box and Hunter (1961) coined the term 'resolution' to classify designs. This chapter deals primarily with resolution IV designs. Main effects in resolution IV designs will be free of other main effects and two-factor interactions.

Taguchi's (Taguchi *et al.*, 1987; Taguchi, 1988) recommended OAs are often not the maximum resolution for a given array size and number of main effects. Until now, no simple method for creating resolution IV designs from Taguchi's OAs has been suggested. Any method should retain the ease of use and intuitive appeal that have made Taguchi's methods acceptable to industry. The fold over concept is the key to the new method for assuring that designs are resolution IV.

Fold over designs

Box and Wilson (1951) state that designs can be created where estimates of main effects are free of two-factor interactions by 'folding over' the original design. The fold over is generated by conducting a second set of N experimental runs at the opposite level of the first set of N runs; that is, the '1s' are replaced by '2s' and vice versa.

Box and Wilson (1951) prove that folding over a design yields a design with main effects free of two-factor interactions as aliases and this proof is given in their appendix 2. In terms of Box and Hunter's (1961) commonly used terminology of design 'resolution', the fold over of a resolution III design guarantees a resolution IV design.

Webb (1968) conjectured and Margolin (1969) proved that all minimal resolution IV designs must be fold overs. A minimal resolution IV design is a resolution IV 2^{k-p} design with only $N = 2k$ experimental runs, hence all resolution IV 2^{k-p} designs must have at least $2k$ runs.

Discussion

The fold over nature of OAs can be quickly demonstrated. The generation of the simplest Taguchi OA from the most fundamental factorial design, the 2^1, follows. The 2^1 of Table 1(a) allows the experimenter to estimate the effect of the single factor as well as the mean of the experiment. The column of 1s is added in Table 1(b) to represent the mean. It could also represent the identity vector. The fold over of Table 1(b) is combined with itself to obtain Table 1(c).

The next step may be simplified by substituting '+' and '–' for the '1' and '2' of Taguchi's nomenclature. Table 2(a) is obtained from Table 1(c) by simply using this substitution. Table 2(b) is generated from 2(a) by the following transformation:

— column 1 of 2(a) becomes column 1 of 2(b),
— column 2 of 2(a) becomes column 3 of 2(b),
— column 2 of 2(b) is obtained by multiplying columns 1 and 3 of 2(b).

Table 2(b) is the L_4 Taguchi OA, as shown in standard form in 2(c), except for the fact that '+' is used to indicate level 1 of a factor, and '–' is used to show level 2 of a factor.

Table 1. Fold over of a 2^1 factorial.

	(a) 1	(b) 0 1	(c) 1 2
1	1	1 1	1 1
2	2	1 2	1 2
3			2 2
4			2 1

Table 2. Creation of an L_4 OA from the fold over of a 2^1 factorial.

	(a) 1	2	(b) 1	2	3	(c) 1	2	3
1	+	+	+	+	+	1	1	1
2	+	−	+	−	−	1	2	2
3	−	−	−	+	−	2	1	2
4	−	+	−	−	+	2	2	1

The Taguchi L_8, L_{16}, L_{32}, L_{64} and L_{128} OAs can be obtained by combining the mean column and the fold over of the next smaller OA to get the odd columns and then multiplying the odd columns to obtain the even columns. In every case the next smaller OA and its fold over are contained in the odd columns and the two-factor products are in the even columns. The fold over of the mean will be in column one.

Practical implications

Recognizing the fold over nature of Taguchi OAs yields a simple method for assuring resolution IV if the OA contains at least $2k$ treatment combinations. The experimenter guarantees that all main effects are free of two-factor aliases by merely assigning all main effects to odd columns of an L_N OA of size $N \geq 2k$.

The procedure for assuring resolution IV designs imposes no additional restriction on the size of the design. Webb (1968) proved that the smallest resolution IV designs for k factors must contain at least $2k$ experimental runs. Since Taguchi's OAs that are also regular fractional factorials all have twice as many experimental runs as odd columns, the procedure of assigning main effects to odd columns only is not overly restrictive.

Procedure for assigning main effects to columns of an OA to assure a resolution IV design

(1) Select k main effects.
(2) Select the smallest OA (L_N) with $N \geq 2k$ experimental runs.
(3) Assign main effects to odd columns using the TOIBTC to assure that the desired interactions may be studied.

An example follows.

An experimenter plans to study seven factors designated A, B, C, D, E, F and G. It is difficult to change the levels of factor A. The number of experimental runs

Table 3. L_{16} OA showing a column assignment for a resolution IV 2^{7-1} OA.

							L_{16} (2^{15})								
			A		B		C		D		E		F		G
	1	2	3	4	5	6	7	8	9	10	11	12	13	14	15
1	1	1	1	1	1	1	1	1	1	1	1	1	1	1	1
2	1	1	1	1	1	1	1	2	2	2	2	2	2	2	2
3	1	1	1	2	2	2	2	1	1	1	1	2	2	2	2
4	1	1	1	2	2	2	2	2	2	2	2	1	1	1	1
5	1	2	2	1	1	2	2	1	1	2	2	1	1	2	2
6	1	2	2	1	1	2	2	2	2	1	1	2	2	1	1
7	1	2	2	2	2	1	1	1	1	2	2	2	2	1	1
8	1	2	2	2	2	1	1	2	2	1	1	1	1	2	2
9	2	1	2	1	2	1	2	1	2	1	2	1	2	1	2
10	2	1	2	1	2	1	2	2	1	2	1	2	1	2	1
11	2	1	2	2	1	2	1	1	2	1	2	2	1	2	1
12	2	1	2	2	1	2	1	2	1	2	1	1	2	1	2
13	2	2	1	1	2	2	1	1	2	2	1	1	2	2	1
14	2	2	1	1	2	2	1	2	1	1	2	2	1	1	2
15	2	2	1	2	1	1	2	1	2	2	1	2	1	1	2
16	2	2	1	2	1	1	2	2	1	1	2	1	2	2	1

required is $N \geq 2 \times 7$. The L_{16} is the smallest OA with $N \geq 14$ so it is chosen for the design. One possible assignment of factors to columns is shown in Table 3.

This assignment results in a resolution IV design. No assignment was made to the first column in order to permit the screening study of all seven factors in an $L_8(2^7)$ prior to completing the second half of the design.

Comparison of the treatment level combinations of the first eight runs of the columns with main effect assignments in Table 9 to the treatment level combinations in the L_8 of Table 4 indicates they are identical. Obviously, the seven main effects can be studied in the resolution III first half of the L_{16} before proceeding with the second half of the design.

Sequential analysis using OAs, particularly when $N > 2k$, is an area that could be developed further with emphasis on the fold over nature of the designs. When $N = 2k$, the main effect assigned to column 1 cannot be studied in the $L_{N/2}$ OA.

Assigning main effects to odd columns will result in $N/2 - 1$ even columns that are available for certain two-factor interactions. A maximum of $N/2 - 1$ two-way interactions can be studied, which are determined by the TOIBTC. Another result of confining the two factor interactions to even columns is that this method will not generate resolution V designs. Taguchi's published designs for L_{16} (2^{5-1}) OAs should be used if a resolution V design is desired. The use of Taguchi's resolution V designs for five main effects does prohibit the type of sequential analysis described in the previous example. [The tendency to use grand designs rather than a sequential approach was cited by Montgomery (1988) as a shortcoming of Taguchi's methods for experimental design.]

The method outlined in this article to assure resolution IV designs by merely limiting main effect assignments to odd columns is easy to apply and should result in real benefits to industrial experimenters. This procedure will always yield a resolution IV design, but resolution IV designs are also available that have both odd and even column assignments.

References

Box, G.E.P. and Hunter, J.S., 1961, The 2^{k-p} fractional factorial designs—Parts I and II. *Technometrics*, **3**, 311–351, 449–458.

Box, G.E.P. and Wilson, K.B., 1951, On the experimental attainment of optimum conditions. *Journal of the Royal Statistical Society, B*, **13**, 1–45.

Margolin, B.H., 1969, Results on factorial designs of resolution IV for the 2^n and 2^n3^m series. *Technometrics*, **11**, 431–444.

Montgomery, D.C., 1988, Experiment design and product and process development. *Manufacturing Engineering*, **101**, 3, 57–63.

Taguchi, G., 1988, *System of Experimental Design* (White Plains, NY: UNIPUB/Kraus International Publications; and Dearborn, MI: American Supplier Institute).

Taguchi, G. and Konishi, S., 1987, *Taguchi Methods Orthogonal Arrays and Linear Graphs* (Dearborn, MI: American Supplier Institute).

Webb, S.R., 1968, Non-orthogonal designs of even resolution. *Technometrics*, **10**, 291–299.

80

System performance in parallel assembly systems with sequence restrictions due to synchronized feeding of materials

Mats I. Johansson

Department of Transportation and Logistics, Chalmers University of Technology, S-41296 Gothenburg, Sweden

Abstract

This chapter discusses system performance in parallel model-mix assembly systems with sequence restrictions due to synchronized feeding of materials. If these systems have several steps each with parallel channels, the assembly objects must be matched with the material, i.e. presented to each step in a predetermined sequence. This requires sequencing buffers between steps.

Synchronized feeding of materials is becoming increasingly common as Just-in-Time (JIT) delivery and internal subassembly of components are introduced. Such components are called synchronous components. This chapter presents a classification of components with respect to their frequency in the production. Two simulation studies are also presented. The first shows that a large part of the idle time is caused by sequence restrictions. The second study compares two types of production planning for systems with sequence restrictions. The use of separate sequences for each step is compared to the straightforward solution of using a fixed sequence throughout the system.

Introduction

The products manufactured by many companies today are becoming increasingly customer-specific, which, with the demands this places on the products means that more part numbers are needed to produce a complete product range. The number of variants increases both at product and component level.

A component is defined as one item of a part with unique specifications, for a specific purpose. This differs from a part (with part number), which can be identified from its own specifications. A variant component here refers to a component which, having passed a defined production step in respect to a later production step, has an alternative component or can be excluded in this later step.

The existence of variants will have an effect on many functions and costs within a company. This chapter, however, deals only with certain characteristic aspects of various classes of variant components, after a presentation of a classification of components with regard to how unique they are in the production.

There is a trend in assembly industries towards unpaced flows of assembly objects with a high degree of parallelism. An unpaced flow here means that the operator controls when an object is taken into or despatched from the workstation (see e.g. Dar-El and Mazer, 1987). There can be a number of purposes behind this, including the possibility of more varied and meaningful work (Karlsson, 1979) and the reduction of losses in the assembly system (Wild, 1975). One of these losses, balancing loss, has its origin partly in the existence of variant components and is caused by differences in product structure causing in their turn differences in assembly time. This creates utilization problems.

A further trend of importance for the problem is the development towards JIT systems. This term here refers more precisely to sequential deliveries from external suppliers. These supply systems often assume a predetermined and only slightly variable sequence of assembly objects in those positions in the assembly system where material and assembly object meet, or that the material is re-sorted. This demand conflicts with the way in which, in parallelized systems, one would like to run the systems to reduce problems in connection with the assembly time distribution expanding effect of the variant components, and also with the wish to enable the operator to vary his working pace. The demand that the sequence be known is valid also for many of the larger units subassembled outside the assembly object flow.

Design principles for sequencing buffers have been studied by Koether (1987), in connection with sequence balancing of lines. Udomkesmalee and Daganzo (1989) discuss the buffer sizes necessary to be able to recreate an object sequence.

This chapter presents the results from two simulation studies. The first study shows that a large part of the idle time is caused by sequence restrictions. The second compares two types of production planning for systems with sequence restrictions. The use of separate sequences for each step is compared to the straightforward solution of using a fixed sequence throughout the system.

Classification of components

Engström (1983) divides a product's components into the classes 'general' and 'specific'. Given a certain selection of objects general components are those common to all objects in the population. Specific components are also known as variant components, of which there are three types; 'replace' (replace another component), 'extra' (extra function) and 'synchronous' components (intended for a specific assembly object). This classification cannot unambiguously define a component.

All components in a defined product structure can be grouped into component groups. Such a component group comprises components which for a certain purpose are interchangeable in different products.

With regard to how unique the components are in the production, a classification can be made, which in this respect for a given population of products describes a component in a given position in the production system.

Each component is classified on the basis of three variables:

(1) the frequency with which this component group occurs,
(2) the number of individuals in the component's component group,
(3) the component's synchronicity.

The frequency of occurrence can be one of the two values 'occurs on each object in the population' and 'does not occur on each object in the population'.

The number of individuals in the component group can be divided into the classes 'one component' and 'more than one component'.

As regards synchronicity a division is made into three classes, 'not synchronous', 'practically synchronous' and 'really synchronous'.

Synchronous components are characterized by their being intended for a specific

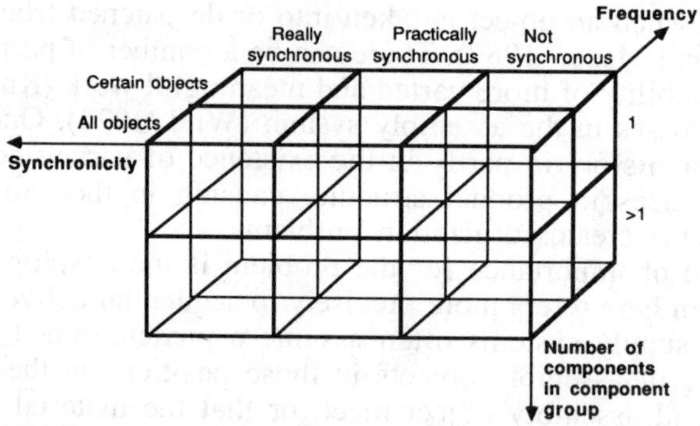

Figure 1. A classification of components with regard to how unique the components are in the production.

object and in the practical context not permissible on more than one object. Practical synchronicity is often a consequence of flow-related restrictions while real synchronicity is a consequence of production-related restrictions. A really synchronous component is unique and there is no guarantee that any other component would be able to be fitted and the demands on the final product still be met. A part number in this latter case has lost the function of identifying on which objects the component is to be fitted. A practically synchronous component is not unique in the same strict sense, but in the practical context it cannot be exchanged for another component with the same part number without considerable disruption.

The above classification covers all components, including those which are not variant components. These are characterized by their not being synchronous, occurring on each object and being alone in their component group.

It is important to note that a component's classification is valid in a certain position in a defined production system with a defined population of final products.

Characteristics of the component classes in manual assembly systems

A characteristic feature of components included in component groups which do not occur on all objects is the time expanding effect they cause in assembly. This time expansion is best explained by the marked positive intervariance between the number of parts and assembly time (Johansson, 198l; Engström, 1983). These papers show that the number of parts is the product-specific variable that best predicts assembly time.

In Table 1 data are presented from a product having 41 product variants grouped in four product families and each product consisting of a large number of components. Only those cases where there existed a time difference between the final products have been counted as observations. All components not represented on every object led to time expansion, while the majority of components which are members of component groups with more than one individual gave none.

For component groups containing more than one individual there are a number of characteristic features at different stages in the production process. In assembly these can take the form of space and learning problems. An observation (e.g. from the automobile industry) is that these components account for a large proportion of the increase in part numbers. As Johansson (1985) states, there is rarely a freedom of choice in the number of components in such a component group. This has been studied under the headings of 'the assortment problem' and 'commonality problem' by many authors, e.g. Dogramaci (1979).

Table 1. The number of components of each type that led to time expansion, and its average size (in seconds).

Type of component	No. of observations	ΔT (S)
Not on every object	71	56
> 1 component in the group	8	20
Common components	0	0

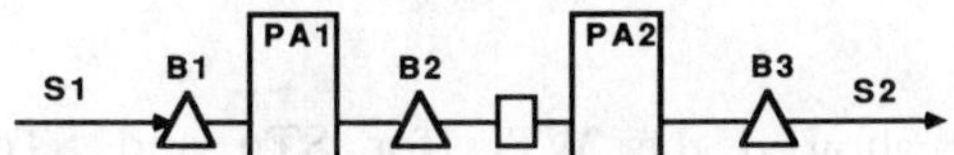

Figure 2. Schematic description of the studied system.

On a traditional assembly line problems related to synchronous components are often small, since the objects normally maintain their sequence throughout the flow. If the flow, on the other hand, is parallelized and such parallel steps are linked in series through buffers, problems can arise by the objects being allowed to 'overtake' each other. Variants and variations in working pace will add to this. Matching material and basic object can be achieved by sorting material and/or basic objects.

One possibility is to apply the principle known as planned sequence breaking. The fact that each object's time demand in the different production steps has an expected value is exploited, in order to plan separate sequences of objects in the different production steps.

Impact of sequence restrictions

The system model

The system studied is part of a larger system in which a relatively complex product with many variants is assembled. Figure 2 gives a schematic description of the flow in the system.

(S) denotes unpaced series flows and (B) buffers. All the buffers have a capacity of six objects. The object proceeds from B1 to one of ten parallel workstations in parallel area 1 (PA1). On the way to PA2, a workstation is passed where principally other work than assembly is performed. PA2 consists of six parallel work teams, each with a system capacity of four objects. Assembly within the team is performed in three steps, which means that the fourth place is used as a change-over area. The station times, depending on variant and workstation, vary between 30 and 60 min.

The objects must leave each buffer in the same sequence. The objects can, however, enter the buffers in different sequences, which means that the buffers have the possibility of sorting. All buffer places are accessible. There are, furthermore, rules governing when an object is allowed to leave a workstation to enter a buffer. If too many objects with too high 'sequence number' are allowed to enter the buffer, it can become impossible to re-sort to the prescribed sequence. The simulation was made with the 'CAD-motion' simulation program on an IBM AT machine.

Several factors were varied and a number of results obtained. Interesting result data in this context includes the degree of utilization of the buffers and idle time for the operators, expressed as a per cent of the workstation assembly time, here referred to as working time loss (WTL). As regards PA2 part of the WTL was treated separately, i.e. that which could be attributed to workstations being blocked by objects not being allowed to enter a buffer, despite free buffer capacity, through having too high a sequence number. Two separate distributions were used for the station times; both had the same average time but differed through the times in the one case having been expanded. ('STn' and 'STe', respectively.)

In the model it was assumed there was no variation in the operators' working pace, but a constant arrival intensity to B1 and a constant exit intensity from B3.

Each run corresponded to 80 h of production, of which the first 10 h constituted a warm-up period. The remaining 70 h were divided into blocks of 5 h. Random sampling of the autocorrelation factor indicated independent observations.

Results

The results for PA2 showed the WTL for STn and STe to be 12·1% (28 observations) and 13·7% (42 observations), respectively. The proportion of these losses attributable to the sequence breaking risk was 17·4% and 20·7% (mean 19·3%), respectively.

The degree of utilization in the buffers was low. B2 was on average utilized to 9·1% and 25·6%, respectively, for the two levels of station time. The corresponding values for B3 were 27·0% and 34·3%.

It was also seen that if PA2 was under-utilized with work WTL remained at approximately the same level, while the proportion caused by the risk of breaking the sequence increased to 85%. The reason for this was that the utilization of B3 increased to 89%.

Comparison between fixed sequence and planned sequence breaking

The system studied

In this section a simulation is used to compare the system performance of the alternative solutions of allowing a possibly broken sequence to be re-established as it was originally (FS model) and of assuring a predetermined calculated sequence between each production step (PSB model), i.e. planned sequence breaking.

The studied mix-model assembly system produces eight different variants and consists of two production areas (PA1 and PA2) in series, with an intermediate buffer with all places accessible for entry and exit. The buffer has a capacity of four objects. Within each production area there are six parallel workstations. The objects are taken into PA1 in a predetermined given sequence and PA1 has continuous access to objects, i.e. never needs to wait for an object. The objects can always leave PA2.

In the FS model the assembly objects are taken into PA2 in the same sequence as to PA1. To maintain this order with a finite buffer volume rules are needed to define the conditions under which an object is allowed to leave PA1. When an object has spent its allotted station time in PA1, the number of objects in the buffer is checked. If there are at most two objects (B-2, where B is the buffer capacity) the object may exit PA1 and enter the buffer. If there are three objects in the buffer, a check is made to ascertain if the next object in line for PA2 is in the buffer. If this is the case, the object is allowed to fill the vacant buffer place. If not, the object has to wait in PA1 until this condition is fulfilled.

The PSB model differs from the other only insofar as the sequence of objects into PA2 is calculated and is not the same as the sequence into PA1. The sequence is calculated on the following premises:

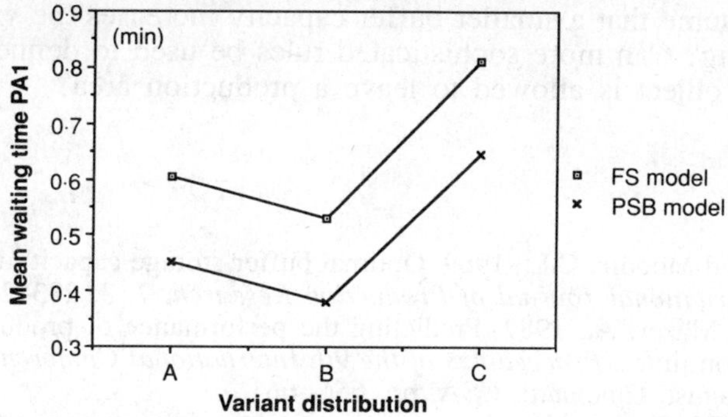

Figure 3. Mean waiting time in PA1 before entering the buffer.

(1) A new object is taken into PA1 when a workstation is vacant.
(2) Each object is assigned a station time corresponding to its norm time, i.e. variation in the operators' working pace is not considered.
(3) The exit sequence from PA1 as calculated above constitutes the planned entry sequence to PA2.

The system is completely balanced, which is achieved through the average assembly times for each variant being of equal length in both production areas. These average times are spread in accordance with a normal distribution (conditioned between ±3·5 standard deviations) with a coefficient of variation of 0·20. Other authors have used higher values (e.g. Anderson and Moodie, 1969) and other distributions (Slack and Wild, 1975). The distribution used can be motivated if a functioning collective system is assumed, in which the operators' variation in working pace to a certain extent is compensated within the work group.

Three different distributions of variants were simulated. In all three cases the variants' assembly times varied between 18 and 30 min, with a mean value of 24 min. Distribution A was closest to a normal distribution, while distribution C was uniform. Distribution B lay between these.

The simulation was made with the help of GPSS V, on an IBM 3081K. Each run corresponded to the production over two shifts preceded by a warm-up period corresponding to half a shift. At the beginning of the warm-up period a number of objects were generated directly into the buffer, so as to initiate it and obtain a more effective warm-up.

Result and discussion

As an indication of the system performance, measurement was made of the average time after completion in PA1 an object had to wait because all conditions for entry to the buffer had not been fulfilled. This is shown in Figure 3.

The result of the simulation indicates that separate sequences for each step in a parallelized assembly system can be preferable to the solution of using a fixed sequence throughout the system. Further research is necessary, however, to prove the reliability of these figures. The method used here was not suitable for testing this reliability, because of the amount of work involved in calculating the PA2 sequence in the PSB model.

A considerable number of questions and hypotheses arise, which can be subjects for further research. What would the result be if more sophisticated sequence calculations were used? How do different kinds of operators' work time distributions affect the result and is it true that, relatively speaking, a greater distribution between the variants' norm times is not to the advantage of the PSB model? Is it reasonable to assume that a smaller buffer capacity increases the value of planned sequence breaking? Can more sophisticated rules be used to define the conditions under which an object is allowed to leave a production area?

References

Anderson, D.R. and Moodie, C.L., 1969, Optimal buffer storage capacity in production line systems. *International Journal of Production Research*, **7**, 3, 233–240.

Dar-El, E.M. and Mazer, A., 1987, Predicting the performance of production line having uniform station times. *Proceedings of the 9th International Conference on Production Research,* August, Cincinatti, USA, pp. 656–661.

Dogramaci, A., 1979, Design of common components considering implications of inventory costs and forecasting. *AIIE Transactions*, **11**, 2, 129–135.

Engström, T., 1983, *Materialflödessystem och serieproduktion*. Ph.D. Thesis, Institütionen för Transportteknik, Chalmers Tekniska Högskola, Göteborg.

Johansson, M.I., 1981, *Konstruktionsanalys av några objekt som bedömningsbas för förändrade produktionssystem*. Institutionen för Transportteknik, Chalmers Tekniska Högskola, Göteborg.

Johansson, M.I., 1985, Effectiveness of production systems—a problem concerning information to the design department. In *Toward the Factory of the Future*, edited by H.-J. Bullinger and H.J. Warnecke (Berlin: Springer), pp. 616–621.

Karlsson, U., 1979, *Alternativa produktionssystem till lineproduktion*. Ph.D. Thesis, Sociologiska Institutionen, Göteborgs Universitet.

Koether, R., 1987, Design of a sequencing buffer for model-mix-assembly in the automotive industry. *Proceedings of the 9th International Conference on Production Research*, August, Cincinnati, USA.

Slack, N.D. and Wild, R., 1975, Research notes—production flow line and collective working: a comparison. *International Journal of Production Research*, **13**, 4, 411–418.

Udomkesmalee, N. and Daganzo, C.F., 1989, Impact of parallel processing on job sequences in flexible assembly systems. *International Journal of Production Research*, **27**, 1, 73–89.

Wild, R., 1975, On the selection of mass production systems. *International Journal of Production Research*, **13**, 5, 443–461.

81

Simulation of work-time distributions for a paced assembly line

Tim J. Gallwey and M. O'Muirthile
University of Limerick, Ireland

Abstract

A ten-station paced assembly line for a household product was studied by making video recordings of the operators. Times obtained from these were used in a simulation model using MicroSAINT. Empirical, Normal and Exponential distributions were tried. Measures used were the percentage of simulated products exceeding the cycle time at each workstation and the idle times of the operators. The Normal distribution gave results similar to the Empirical but the Exponential gave very different results. The observed times were mostly skewed to the left, gave a poor fit to the Normal and Exponential distributions, and fitted well to the Weibull. It is suggested that the Weibull parameters might be related to scores on psychomotor tests and that different times for different operators on the same task need to be considered.

Introduction

Despite the criticisms of research workers and industrial investigators paced assembly lines remain a popular method of organizing production. Predictable outputs can be achieved, after a fashion, and management is relieved from such tasks as motivating and disciplining the operators. But there are well-known problems of alienation and boredom so that Volvo (for example) devised something else.

Other problems which may result are inadequate quality levels and an output level which is less than the maximum rate theoretically achievable. Both of these latter problems result from the variability in work-times of individual operators on the one hand and variations between operators on the other. These can be alleviated either by increasing the cycle time or by providing buffer stocks between operators. The first causes a loss of output, the second ties up working capital and is no longer acceptable in the age of Just-in-Time (JIT). Therefore other strategies are required, but in order to evaluate other strategies it is necessary to be able to use work-times which bear a reasonable relationship to those which will actually occur. Predetermined Motion Time Systems could be used but make no allowance for variability which must therefore be estimated. Alternatively, observed times must be used. The values obtained can then be used in a simulation model to compare strategies.

For convenience it is desirable to use standard statistical distributions but then the question is which to use. The studies presented by Dudley (1968) gave times for unpaced work that were usually skewed to the left with something like a Log-Normal distribution, whereas in paced work the distributions came close to being

Normal. Sury (1964) found that his times were not Log-Normal but that by using the reciprocal transformation on them a Normal distribution was obtained for unpaced work. Hillier and Boling (1966) examined the Exponential distribution and Knott and Sury (1987) compared times on unpaced tasks to the Log-Normal, Gamma, Beta, and Normal distributions without obtaining statistical significance on any of them. It is possible that both the Log-Normal and Exponential require some occurrence of times of zero duration. This is obviously impossible, but its equivalent can be achieved by subtracting the shortest time from all the others.

To examine this question in some depth it was decided to analyse data collected from a real assembly line and to compare a variety of distributions by means of simulation. It was felt that a more adaptable distribution than those mentioned above might offer a promise of better results. The actual station work-time distributions were therefore to be compared with several standard distributions.

Method

Data collection

Video recordings were made of the operators on a ten-station line which assembled a household product. Each recording lasted an hour or more so that on the cycle time of 27 s a large sample of times was obtained for each station. Subsequently the recordings were viewed so that individual element times and station cycle times could be measured by stopwatch. From these times a FORTRAN program compiled frequency distributions to use as Empirical distributions for a computer simulation model.

Simulation

The simulation language used was MicroSAINT (Laughery, 1989) which is intended to be easy to use and offers routines for the Normal, Exponential, Gamma and Uniform distributions. Some manipulation was needed to enable it to sample from the empirical distributions. Simulation runs were made for the equivalent of 100 products, and five runs were made for each configuration with different seed numbers. Shortage of time precluded more runs since each took about 3 h on the AMSTRAD 1512 microcomputer used. The main measure of interest here was the percentage of products which exceeded the allowed time at individual workstations. These would indicate probable quality problems since, in theory at least, the operator cannot catch up subsequently.

Conditions examined

The percentage of excess times was obtained for the Empirical distributions, and for Normal and Exponential distributions with the same parameters as the actual data. Cycle times used were the official figure of 27 s as well as 26, 25 and 24 s with a buffer of one unit between stations. Due to the nature of the conveyer such a buffer was an inherent part of the operation.

Statistical comparisons

Work-time distributions were tested using SPSS (Hull and Nie, 1981) to see if they conformed to Normal or Exponential distributions, or whether the Logarithm, Reciprocal or Square Root transformations would change them to conformance with the Normal. Furthermore, conformance to the Weibull distribution was examined by means of a FORTRAN program using NAG subroutines (Numerical Algorithms

Table 1. Time values for the first nine workstations.

Work-station	Mean (s)	Standard deviation	Coefficient of variation	Skewness	MTM time
1	23·11	4·75	0·205	1·214	23·75
2	21·03	4·07	0·194	0·434	22·81
3	16·48	3·26	0·198	0·503	17·36
4[a]	77·78	11·37	0·146	−0·344	70·88
5[b]	31·55	4·93	0·156	−0·250	41·81
6	22·56	3·35	0·148	1·090	23·81
7	24·75	5·20	0·210	0·767	25·60
8	24·98	5·75	0·230	1·299	24·98
9	16·75	2·06	0·123	0·539	21·05

[a] Four operators worked in parallel.
[b] Two operators worked in parallel.

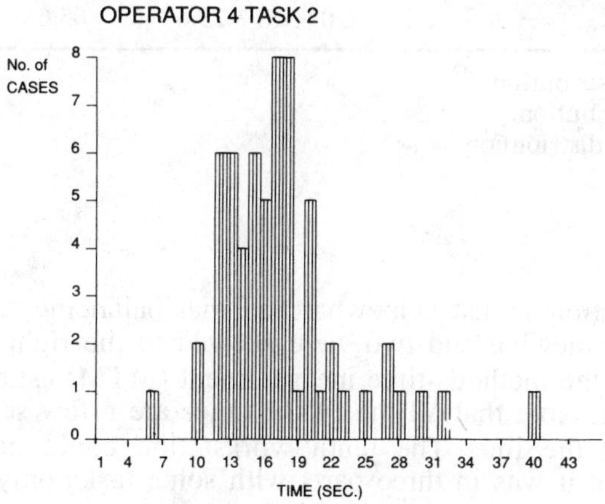

Figure 1. Example of work-time distribution.

Group, 1987). Since the Weibull can reproduce the Exponential, Rayleigh and Normal distributions, and gives Log-Normal-type distributions in between, it was seen to offer considerable potential. For this purpose the cumulative function was transformed by means of double logarithms to achieve a straight line form and was plotted using Median Ranks for the proportions of the cumulative distribution (O'Connor, 1985). For this purpose the shortest time for each element was taken as the location parameter (GAMMA) for its Weibull distribution and subtracted from all the other times.

Results

Workstation times

Despite Dudley's finding that pacing causes the work-time distributions to be Normally distributed, in this study they were generally skewed to the left as shown by Table 1 and the example in Figure 1. A possible explanation for this is that generally the operators completed their tasks in less than the cycle time (see Table 1) so that their work was to some extent self-paced. It can also be seen that the

Table 2. Percentage of products exceeding the allowed time at the end of each workstation for three work-time distributions.

Workstation number	Line Configuration (cycle time, buffer)					
	27,1 Distribution			24,1 Distribution		
	Emp	Norm	Exp	Emp	Norm	Exp
1	35·4	35·0	36·6	57·6	58·4	16·8
2	6·2	0·4	32·4	9·4	1·4	20·0
3	0·4	0·0	19·0	6·0	0·8	24·4
4	0·2	0·0	15·6	5·2	2·4	35·6
5	0·0	0·0	28·0	20·8	23·6	23·6
6	1·0	0·4	32·8	40·8	44·8	14·4
7	3·4	0·6	42·2	53·8	64·4	9·6
8	5·4	0·4	30·2	67·0	79·2	9·4
9	4·0	0·0	29·8	63·2	61·6	10·8
10	4·8	0·0	25·6	65·6	52·8	15·2

Emp = Empirical distribution.
Norm = Normal distribution.
Exp = Exponential distribution.

coefficient of variation varied somewhat and that balancing was less than exact. Skewness varied somewhat and two were skewed to the right. Most of the times were very close to the methods-time measurement (MTM) estimates but two were somewhat different. Note that all the MTM times are a few seconds less than the cycle time set for the line. The tenth workstation could not be used in this comparison because it was in three parts with some tasks only performed on one unit in four.

Percentage exceeding cycle time

The percentage of products exceeding the allowed time varied by distribution and cycle time as shown in Table 2. (Results for cycle times of 25 and 24 s were intermediate between these so are not shown). In all the cases excess is very high for the first workstation but it is thought that this may be due to a modelling artefact.

A workstation 0 was included which notionally worked at exactly the cycle time selected without any variability. It is notable that for a 27-s cycle time the Normal distribution gave virtually no components over the time whereas the Empirical gave up to 6% excess. The Exponential on the other hand gave values from 15·6 to 42·2% so is wholly deceptive. When the cycle time was shortened to 24 s the Empirical and Normal gave large percentages exceeding the allowed time but the Exponential gave appreciably smaller percentages, once again being deceptive.

Idle time

Operator idle time was examined for the three distributions and two cycle times used above. Once again the Normal gave results quite similar to the Empirical but results from the Exponential differed considerably (Table 3).

Table 3. Mean idle time (seconds) per operator per product produced, for three work-time distributions.

Workstation number	Line Configuration (cycle time, buffer)					
	27,1 Distribution			24,1 Distribution		
	Emp	Norm	Exp	Emp	Norm	Exp
1	2·68	2·48	5·54	0·69	0·50	4·33
2	4·01	4·34	5·99	2·88	2·22	5·89
3	9·08	9·79	8·85	6·27	7·00	8·75
4	28·98	29·25	30·25	20·83	23·54	27·12
5	21·36	21·48	25·23	17·32	19·23	22·34
6	1·02	1·53	14·59	0·26	3·55	11·45
7	1·20	2·83	9·50	0·26	3·00	5·22
8	4·54	4·23	6·45	1·79	2·59	5·52
9	10·58	8·91	14·13	9·25	11·19	13·72
10	2·71	17·89	11·79	2·25	14·26	17·55

Abbreviations as Table 2.

Work-time distributions

The distributions of work-times were found to differ markedly from the Exponential and in only two cases did the probability exceed 0·9 for the fit to the Normal. The transformed values did not fit the Normal distribution either although in each case one value exceeded a probability of 0·9 of fitting it (Table 4). These differences can be explained by the results for the Weibull which gave very high correlations to the data.

Note that the Weibull becomes the Exponential when Beta has a value of 1·0, the Rayleigh when Beta is 2·0 and approximates the Normal when Beta is 3·4 or more. Two Beta values were just greater than 1·50, two values were close to 2·0 and two approximated the Normal, but these last two are for different workstations from those transformed values which approached the Normal. It should also be noted that the Beta values for individual elements within workstations varied quite markedly (e.g. values from 1·90 to 3·89 were recorded for station 2).

Discussion

The difference between actual station mean cycle times and the cycle time for the line could be that the operators worked faster than normal due to being studied. This seems unlikely since the actual times achieved were generally close to the MTM values. Rather it must be assumed that the line was planned this way to ameliorate one of the shortcomings of paced assembly line work, despite giving a lower output.

Normal distributions of work-time are seen to give simulation results similar to those from the Empirical data. However the Exponential distribution gave markedly different results, both for products exceeding their allowed time and for operator

Table 4. Comparisons of actual times with standard statistical distributions.

Workstation number	Probability of fit (2-tailed)					Weibull	
	Distribution		Transformed vs. Normal			Values	
	Exp	Norm	Log	Reciprocal	SQRT	Beta	Correlation
1	0·000	0·143	0·719	0·892	0·372	1·528	0·965
2	0·000	0·900	0·545	0·200	0·816	3·301	0·982
3	0·007	0·771	0·992	0·613	0·913	1·775	0·973
4	0·000	0·959	0·651	0·260	0·852	1·551	0·866
5	0·000	0·248	0·036	0·002	0·109	2·867	0·970
6	0·000	0·244	0·536	0·717	0·388	1·880	0·991
7	0·000	0·019	0·749	0·121	0·644	4·059	0·956
8	0·000	0·050	0·361	0·930	0·176	2·034	0·987
9	0·014	0·761	0·234	0·032	0·472	2·376	0·991

Abbreviations as Table 2.

idle time. This may have arisen due to allowing times of zero duration rather than subtracting the shortest from all the others before taking the mean. The Weibull distribution appears as though it might give better results but it was not possible to try it in the simulation model. It also shows that the shapes of the distributions differed widely between different workstations and the largest value of the scale parameter (ETA) was found to be more than eight times the smallest value.

These latter figures suggest that the differences result from individual differences between operators. For too long have operators been lumped together as typical or 'average' human beings, despite overwhelming evidence to the contrary. Tests on a battery of psychomotor skills need to be conducted to account for these distribution differences. If successful they will enable us to establish much more precisely the work-time variances for assembly line operators. Fahy and Gallwey (1985) showed that a good relationship can be established between psychomotor test scores and MTM times so it should work here too.

Such an approach would, however, mean setting different times for different operators doing the same workstation task. That cuts right across the traditional approach and would be difficult to incorporate in practice, but it represents reality and it is only by trying to be realistic that advances can be made. To implement such an approach would require the abandonment of traditional assembly lines, an eventuality which ergonomists would welcome.

Conclusions

(1) In this study work-time distributions for paced work were skewed mostly to the left.

(2) The percentage of products exceeding the cycle time allowed at individual workstations was low, probably due to the cycle time of the line being some seconds greater than the mean work-times.

(3) The Normal distribution gave simulation results similar to the Empirical data but the Exponential gave results that differed considerably, probably due to an error in the data.

(4) Statistical tests of goodness-of-fit to the Empirical data were very poor for the Exponential distribution and mostly poor for the Normal.

(5) Very high correlations were obtained between the Empirical data and a straight line transformation of the Weibull distribution.

(6) Big variations in the Weibull shape parameter (Beta) might be accounted for by scores on a battery of psychomotor tests on the operators.

(7) The question of different times for different operators on the same task needs to be considered.

Acknowledgements

The hard work of Mr. M.G. Sweetman in painstakingly making the video recordings and the co-operation of a local manufacturer are gratefully acknowledged.

References

Dudley, N.A., 1968, *Work Measurement: Some Research Studies* (London: Macmillan).

Fahy, J. and Gallwey, T.J., 1985, An application of psychomotor tests to estimate worker performance. In *Toward the Factory of the Future*, edited by H.-J. Bullinger and H.J. Warnecke, Proceedings of the 8th International Conference on Production Research (Berlin: Springer), pp. 910–914.

Hillier, F.S. and Boling, R.W., 1966, The effects of some design factors on the efficiency of production lines with variable times. *Journal of Industrial Engineering*, **17**, 651.

Hull, C.D. and Nie, N.H., 1981, *SPSS Update 7–9: New Procedures and Facilities for Releases 7–9* (New York: McGraw-Hill).

Knott, K. and Sury, R.J., 1987, A study of work-time distributions on unpaced tasks. *IIE Transactions*, March, 50–55.

Laughery, K.R., 1989, *Manual for MicroSAINT* (Boulder, CO: Micro Analysis and Design).

Numerical Algorithms Group, 1987, *NAG FORTRAN Library* (Oxford, UK).

O'Connor, P.D.T., 1985, *Practical Reliability Engineering* (Chichester: Wiley).

Sury, R.J., 1964, An industrial study of paced and unpaced operator performance on a single staged work task. *International Journal of Production Research*, **3**, 21–102.

82

A microcomputer simulation software for scheduling production in a dynamic and flexible jobbing environment

Ricardo Sarmento Costa[1] **and Eduardo G.M. Jardim**[2]

[1]*Instituto Nacional de Tecnologia, Unidade de Produção por Computador and* [2]*Universidade Federal do Rio de Janeiro, Programa de Engenharia de Produção, Rio de Janeiro, Brazil*

Abstract

In the jobbing industry rescheduling of manufacturing operations is often necessary due to the desirable but unpredictable arrival of new orders. This prompts questions such as: What can be the earliest due dates for new orders? How can these be fitted in the current or planned shop floor programme? Will there be lateness or future capacity problems for orders already in production? If so, which are these?

In response, this chapter presents a microcomputer simulation program which has been developed for a jobbing manufacturer of eletro-mechanical equipment in Rio de Janeiro, Brazil. This system allows the creation and assessment of alternative scheduling solutions on an interactive basis.

Based on the finite capacity principle, the software includes the following modules: (1) Shop floor Modelling, (2) Bills-of-Materials, (3) Process Routing, (4) Order Release, (5) Order Progress, and (6) Scheduling, which comprises an algorithm that simulates the behaviour of future work-queues using different priority rules. The last two modules use the system data base to depict the current status of production orders and work centres.

Problem formulation

This chapter focuses on the solution of real scheduling and control problems in the jobbing industry, producing a wide variety of equipment and capital goods. The manufacturing process in question comprises universal machines and flexible production resources capable of performing a wide range of operations (e.g. machining, bending, welding, heat and surface treatment, painting, and assembly) some of which may be subcontracted.

Production depends directly on the arrival of customer orders and the factory often has to generate design or manufacturing data for its products. That is, in order to start production against a given order, it is necessary first to define the bills-of-materials as well as the manufacturing operations and their respective sequence.

Different products are made simultaneously and compete for the same plant facilities. They are produced in small quantities (usually on a one-off basis), with relatively large lead times, and in a non-repetitive way. The broad and changing diversity of parts, components, raw materials, manufacturing operations and routings makes the creation, monitoring and updating of work schedules intrinsically complex.

To carry out this function, the production controller has constantly to find answers to questions such as:

(1) Which is the work schedule which efficiently allows all the order delivery dates to be met. Or more precisely, to which work centres or machines, and in which sequences (times and priorities) should the purchasing, manufacturing and assembly operations of the various products be programmed?
(2) Will there be capacity problems related to the implementation of a determined work schedule? If so, which are these and how are they to be solved?
(3) Given that there is already a work schedule in execution at the shop floor, which delivery date should be set for a new customer order?
(4) If this refers to a priority product, how can its manufacturing operations be included in the current schedule? Will this cause delay of orders already in production?
(5) And finally, due to the randomness of demand and the inevitable occurrence of unexpected events (e.g. machine breakdowns, absenteeism, materials shortages), how should one deal with the frequent need to reschedule production?

Definition of objective

All these practical issues were examined in a dynamic and flexible jobbing environment. As a result a microcomputer software was developed to assist the generation, assessment, implementation, follow-up, and constant revision of work schedules. To cope with circumstances typically encountered in small and medium companies, the software was designed with two aspects in mind: (1) implementation free of rigid or sophisticated information system and organizational requirements and (2) ease of operation.

The central feature of the software is a discrete simulation algorithm. Based on different priority rules the algorithm allows the creation of alternative scheduling solutions and projects them into the future in order to evaluate their consequences in terms of meeting delivery dates, improving the use of facilities and reducing the work-in-progress.

By analysing the results of different simulation runs it is possible to choose the work schedule which best attains a set of pre-established objectives. On an interactive basis the software can also be used to explore in more detail a selected solution either to improve it further or to adapt it to new circumstances or priorities.

For the continuous rescheduling of production the simulation algorithm is supported by an information system. Through simple administrative procedures, the system updates the data base which depicts, at any given point in time, the current status of the plant (say, in terms of machine availability, work centre service queues, and position of items being processed). In this way the algorithm simulates succeeding events on the shop floor, starting always from a real and present situation (Costa, 1989; Jardim, 1982).

The approach

The scheduling method adopted is based on the finite capacity principle using the concept of forward planning which programmes tasks to their earliest possible starting date. Exceptions are made to items with high inventory costs since their acquisition or processing should be as near as possible to their utilization date.

Among the reasons which justify such procedure, three stand out: (1) being by its own nature a 'vendor of machine and craft time', the company often wishes

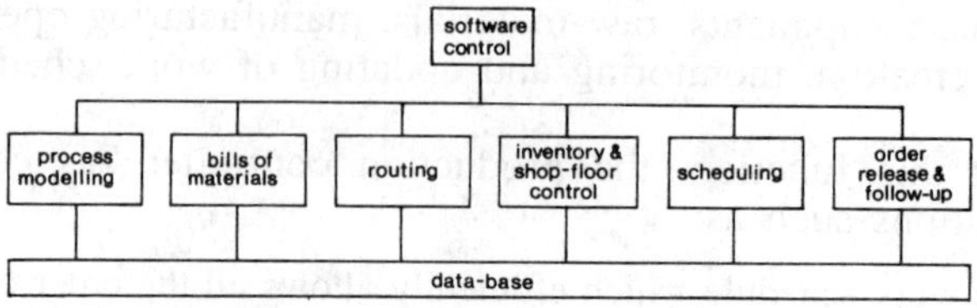

Figure 1. Main modules of the scheduling simulation software (SSS).

to have its critical facilities available as soon as possible to accommodate the unpredictable, but normally desirable, arrival of new orders; (2) within limits, overload capacity problems should be solved by means of the factory's own resources; i.e. work in excess or arrears tends to be carried forward, avoiding subcontracting or even overtime shifts; (3) by not using the conventional 'Materials Requirements Planning (MRP)' infinite capacity approach, the software provides feasible loading and sequencing solutions compatible with the various work centre capacities and technological constraints, a task which can be quite difficult to carry out on a routine and manual basis in a dynamic production environment.

Although the software developed is not specifically circumscribed into any of the modern production management techniques, it has some aspects of the 'Optimizing Production Technology (OPT)' strategy. Equally, it incorporates the MRP-II (Manufacturing Resources Planning) perspective in so far as the integration of various manufacturing programming and control functions are concerned.

In this respect, the software includes five other modules apart from work schedule generation, assessment and selection. As can be seen in Figure 1, they are: (l) process modelling, (2) bills-of-materials, (3) process routing, (4) inventory and shop-floor control and (5) purchase, manufacturing and assembly order release and follow-up.

Using the software

Figure 2 outlines the information framework within which the software is used. In this scheme, the production activities start with a Design Sector converting customer orders into drawings and product specifications. At this stage, the bills-of-materials are defined and fed into the computer system, determining the items which should be bought, manufactured or assembled and their respective component relationships.

These data, combined with those of the plant facilities, make it possible for a Routing Sector to enter the operation sequences and respective standard operating times into the system. In parallel, information from the Shop Floor continuously revises the software data base which describes the current manufacturing situation, particularly in terms of the service (purchasing, manufacturing and assembly) orders issued.

Up to this point all the information fed into the system forms an updated data base, which is used for the simulation of future production. This basically means determining for all customer orders on hand where and when their respective purchasing, manufacturing and assembly operations will take place for a given planning horizon and set of established priority rules (described in the next section).

Next, as mentioned before, the production programmer, by using different sets of priority procedures, can then generate and assess alternative schedule solutions and choose the most convenient one. The improvements or adjustments in the selected solution can be pursued through the identification and variation of critical resources.

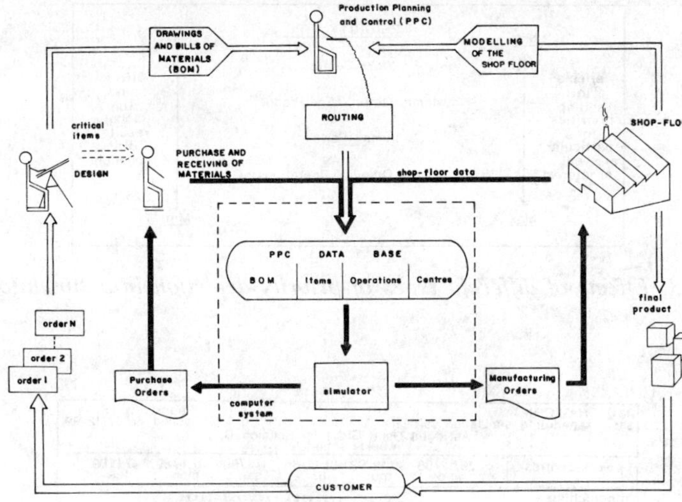

Figure 2. Simulation approach for jobbing production management.

Once there is a decision as to which schedule solution is to be used, the software then translates it into planned service orders which can be sent to the shop floor on a daily, weekly or monthly basis. These orders, together with other control reports, have been designed to take information to and from the shop floor and purchasing department in order to simplify and centralize the updating of the software data base. This procedure allows all data processing to be carried out on a single microcomputer within a Production Control Department.

Scheduling by simulation

Before each simulation run, in order to generate and assess a work schedule using an updated data base, it is possible to (re)define or confirm the system priorities. These can be four different types according to whether the focus of attention is (1) a service order or an operation, (2) an item or a component, (3) a customer order or a final product and (4) a machine or work centre.

Priority designation in the first three categories is optional while for the fourth it is mandatory for it establishes the work flow in the plant during the simulation run. In other words, this last set of priorities comprises the dispatching or loading rules (queue disciplines) which are the main system operating procedures. The most common of these (such as FIFO—First In First Out; SOT—Shortest Operation Time; and EDD—Earliest Delivery Date) are already incorporated in the software for user selection and work centre allocations.

The assignment of a priority to an operation, component or customer order can be done by simply referring these to pre-defined set of values.

When the four different priorities compete with each other, that referring to a service order will have precedence over the others and therefore this is the most effective way to favour the processing of a certain purchasing, manufacturing or assembly operation. None the less, its impact on the overall scheduling situation tends to be less significant since it only affects the execution of a specific activity at a certain point in the simulation time.

The next priority rule in order of precedence is that of an item which supersedes the customer order type which in turn overrides a work centre processing rule. But, in terms of global effects, the results of a simulation run are prone to be affected

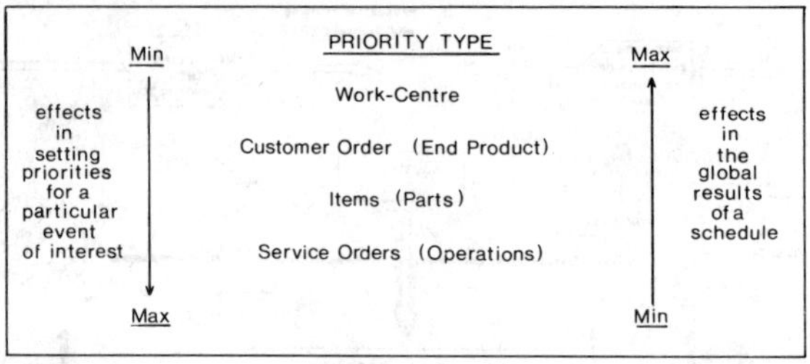

Figure 3. Effects of different types of priority depiction in a simulator run.

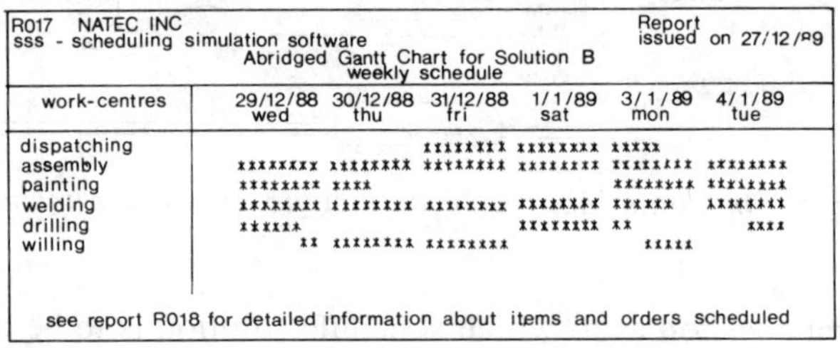

Figure 4. Example of software output.

in an inverse order, with the queue disciplines being the most relevant. Figure 3 sums up this priority sequence of decisions.

It is worth noting that as far as the production controller is concerned, to generate a work schedule from a current data base means simply to establish such priorities (all of whose respective default values have been previously defined). In this way the user avoids the laborious and complex task of identifying and analysing a large amount of data (e.g. lead times, work centre capacities, process routes, delivery dates, etc.) in order to determine where and when each operation should start.

This task is automatically carried out by the software, which registers during each simulation run the resultant starting (and ending) simulation time of each operation for the pre-established set of priorities. These (simulated) starting times for the operations can then be materialized in terms of planned service orders when, after trial runs, one of the priority sets is approved. This approval is based on reports and Gantt charts provided by the software which indicate the results for order lateness, machine utilization, machine overloading, and work-in-progress. See examples in Figure 4.

It should be noted that the software was not conceived to define, by means of statistical techniques, a pattern of work centre dispatching rules to be used by the shop floor when some specific scheduling objectives are to be attained according to typical manufacturing situations. This is commonly the focus of attention of the simulation studies of the dynamic job shop. The literature on this subject is vast, as reviewed by Conway *et al*. (1967) and Baker (1974).

The software differs from this classical approach in that it aims to analyse particular manufacturing situations, each stemming from a certain set of customer orders on hand and changing production conditions. For this, there is no reason why randomness and uncertainty should be added to the estimates of processing times.

Furthermore, the software also attempts to alleviate the burden of ordinary sequencing decisions (and consequent data manipulation) on the work centre foreman or operatives. The rationale is that this sequencing task can be transferred to the production controller who can apply as many despatching rules as necessary to achieve scheduling goals.

So, when scheduled manufacturing orders are issued to the shop floor, the work centre operatives do not necessarily have to worry about routine sequencing decisions, or bother about the (perhaps intricate) combination of priorities and rules behind them.

Other software features

Modelling the shop floor

The need to focus on the manufacturing process (machines and personnel) varies with the production planning and control activity under consideration.

As far as Process Routing is concerned, it is normal to consider the greatest possible set of (alternative) work centres and machines capable of performing the manufacturing operations relating to a certain order. However, this search for manufacturing flexibility may not be the main objective of Design staff who often have to consider specific machines with suitable process capabilities.

Shop floor management, in turn, with the ultimate responsibility for manufacturing operations, has to consider each machine in terms of current availability, performance and (use or loading) conditions. On the other hand, considering the process in detail may not be necessary or even useful for a plant manager who has to monitor the work flow and establish at any time the position of a given customer order.

The need to accommodate and integrate these different modelling perspectives is considered in the software which allows the representation of both the shop floor and the process routes at four distinct but unified levels. These are, in decreasing order of detail:

(1) *Work post*—The basic manufacturing entity of the simulation model, representing each machine or operator of a given Production Unit (e.g. one of the '*X, Y, Z,*' lathes);
(2) *Production unit*—A set of identical or interchangeable work posts belonging to the same work centre (e.g. all the '*X, Y, Z,*' lathes).
(3) *Work centre*—A set of similar Production Units which carries out the same type of manufacturing operation to which may be associated a common service queue (e.g. all the '*X, Y, Z,*' lathes plus other lathes of different types and sizes).
(4) *Area of responsibility*—A set of work centres under the supervision of the same foreman through which information flows from and to the shop floor (e.g. Machining Department with lathes, millers and drilling machines).

The manufacturing route of an item may be given in terms of each Production Unit through which it should be processed. This, however, reduces the degree of freedom for scheduling operations to alternative machines. Therefore, this procedure should only be adopted when design or technological characteristics so demand. In this sense, the most convenient way to define and enter a process route is by means of the work centres.

Bills-of-materials and routing

Figure 5 illustrates how the bills-of-materials and routings are represented in the software data base. A group of such graphs forms the set of customer orders on hand. Note that the operations needed to manufacture a given item are incorporated within the overall material structure.

Based on a given set of priority assignments, simulating the production process means basically to establish starting and ending dates for each of the operations of the customer orders on hand and to comply with three main conditions:

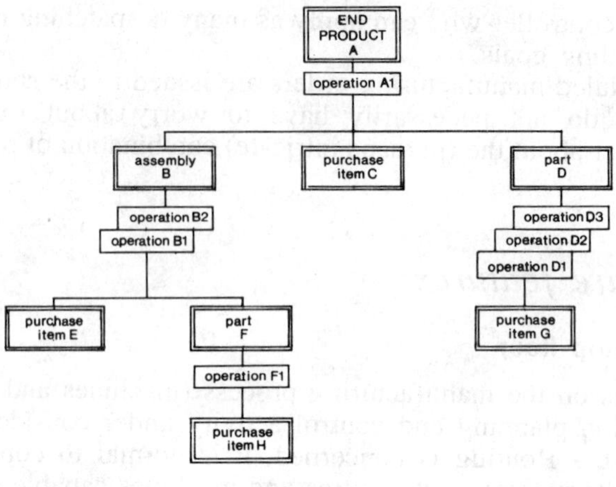

Figure 5. Combined representation of routing and bills-of-materials.

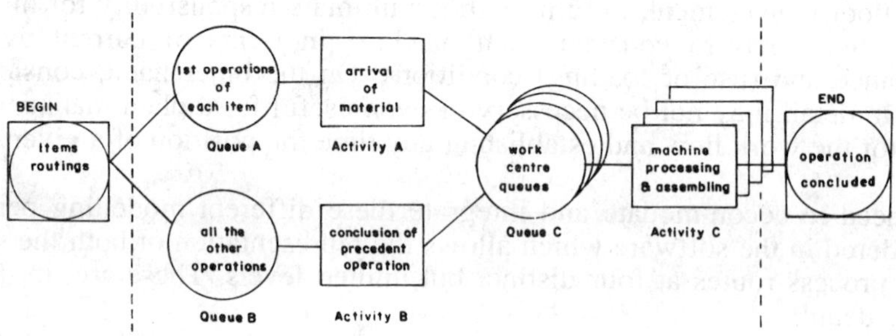

Figure 6. The life cycle of an operation.

(1) *Component restriction*, i.e. 'parent-item' (e.g. an assembly) cannot be made before its parts (e.g. subassemblies);
(2) *Technological precedence*, i.e. a manufacturing item has to be produced according to a predetermined sequence of operations;
(3) *Capacity limitations*, i.e. time is limited and machines have finite output rates (consequently, operations do compete for scarce plant facilities; queues therefore tend to be formed and priority decisions must be taken to process them).

The simulation algorithm

The algorithm developed uses the 'operation' (as exemplified in Figure 5) as the basic entity to be simulated. By definition, operations are the links between two modelling processes: (1) the shop floor (see 'Modelling the shop floor' earlier in this section) and (2) the various materials structures of the items to be made (see 'Bills-of-materials and routing' earlier in this section).

The underlying idea is that the dynamics of the whole manufacturing system can be fully and effectively represented by simply identifying (as simulation time goes by) the changing states of the production operations necessary to complete the orders on hand. The life cycle of an operation considering all the possible states through which it may pass, can be seen in Figure 6.

Briefly, this flow chart means that as the production controller defines manufacturing routes for a given product, two sets of manufacturing operations are updated, identified in Figure 6 as (abstracts) queues *A* and *B*. Queue *A* receives the first operation of all manufacturing routes specified in the bills-of-materials of every customer order. For example, referring to the items of Figure 5, this queue would receive the operations F1, D1, B1 and A1.

Each of these operations remains in the queue *A* until the arrival of the (purchased or manufactured) item necessary for its execution (activity *A*). When this happens the operation is then transferred to the queue of the work centre where it is going to be processed.

By contrast, the second queue (queue *B*) receives all the other manufacturing operations. In the example of Figure 5, this would mean operations D2, D3 and B2.

While queue *A* contains (first) operations, due to component restriction, queue *B* holds the operations which could not be carried out because of technological precedence. So, in this last case, an operation only changes its status through activity *B* (going to the queue of the work centre where it is going to be carried out) when the operation which precedes it in the manufacturing route is concluded.

Whether the operation started in queue *A* or in *B*, it finishes its life cycle through an activity *C*. This last activity, in turn, represents the manufacturing operation itself which can only take place when the respective machine is available and the operation in question has the greatest priority in the queue.

The algorithm developed, which accounts for the changing states of all the operations, includes conventional procedures of discrete event simulation, notably a simulation clock and routines which integrate activities *A*, *B* and *C*, controlling their conclusion and projecting ahead their starting dates.

The question which remains to be answered is how to simulate production using an updated data base which depicts the real manufacturing situation of the shop floor. The answer is to maintain an independent shop floor computer information system with the same logic and structure represented in Figure 6. In other words, when the computer system receives information that, for example, an item has arrived (activity *A*) or an operation has been concluded (activity *C*), it updates the current (real) status of all affected operations.

Thus, before each simulation run, the software first transfers the real and present operation states from system permanent files to real computer memory, therefore providing starting points for every simulation run. For instance, if the shop floor informs the system that a given manufacturing operation has started (activity *C*) the algorithm only contemplates its life cycle from this point onwards.

To sum up: to generate a work schedule means simply to register the resulting starting dates of all operations in activity *C* for a set of established priorities. It is also worth stressing a particular feature of the simulation solution as implemented: the algorithm does not focus on the physical flow of materials through machines and work centres. In this way it bypasses the difficulty of representing in a simple and structured form the wide variety of manufacturing route restrictions typical of job shop with assembly.

Software implementation

Being easy to operate, the software is suitable for IBM-PC type microcomputers. It was written in PASCAL and has been implemented in the factory which was used as the main reference for this chapter.

In this context, a typical run of the system involves around 2000 items, 4000 machine or assembly operations, and 6000 material or component relationships (of the type which is component of which). The software can deal with twice this volume of data.

The first system runs using real factory data took less than 3 min. This result demonstrates the practicability of using the trial-and-error process for creating and assessing different schedules in order to select one that suits permanent or temporary objectives.

Although the system is not yet used to its full potential in the plant, it has been operated by the factory managers with a data base related to past periods of production. The results obtained were considered by the firm as most satisfactory.

Final remarks

McKay *et al*. (1988) when surveying real jobbing production environments concluded that the conventional modelling methods for scheduling production have brought very little practical benefit for industry. They point out a number of causes which explain this gap between theory and practice.

Among such reasons one stands out and is perhaps the genesis of all the others: the vicissitudes and dynamism of real life which impinge on every objective and priority established. Often, unexpected circumstances are the rule rather than the exception. Thus, there is no way that a systematic planning method can survive without the active and decisive participation of the human element (end-user).

Bearing this in mind, the software developed aims basically to provide practical and updated 'snapshots' of what is happening on the shop floor. Based on such 'quantitative pictures', the production controller, using a simulation algorithm, can then rapidly create and assess new scheduling solutions in order to select, implement, follow-up and update the most convenient one.

References

Baker, K.R., 1974, *Introduction to Sequencing and Scheduling*, (New York: Wiley), pp. 213–233.

Conway, R.W., Maxwell, W.L. and Miller, L.W., 1967, *Theory of Scheduling*, (Reading, MA: Addison-Wesley), pp. 219–243.

Costa, R.S., 1989, Planejamento e Controle de Produção na Indústria Mecânica com Fabricação por Encomenda: Um Sistema Computacional de Simulação. M.Sc. Thesis, COPPE/UFRJ Universidade Federal do Rio de Janeiro, Brasil.

Jardim, E.G.M., 1982, Some Aspects of Modelling of Batch Production in a Gray Iron Foundry. Ph.D. Thesis, The University of Birmingham.

McKay, K.N., Safayeni, F.R. and Buzacott, J.A., 1988, Job-shop scheduling theory: What is relevant? *Interfaces*, **18**, 4, 84–90.

83

The simulation of a flexible delivery system for large-scale make-to-order manufacture

B. Wu and P.H. Lowe

Department of Manufacturing and Engineering Systems, Brunel University, Uxbridge, UK

Abstract

This chapter reports an investigation into the operation of an Order-Handling-Manufacturing-System (OHMS), which contains features of both batch and flow production. Its specific characteristics are the design and manufacture of major capital equipment to customer requirements, a long lead time of production and a heavy commitment of working capital against each customer contract. The investigation involved the analysis of the manufacturing system of a major UK company. A computer simulation model was developed which incorporated its key features and concentrated on the information and material flow patterns within the total system.

This chapter in particular presents the responses of the OHMS operation to customer contract uncertainty under different production control policies. The rules involved describe the uncertainty in an OHMS due-date environment in terms of the contract flow-time allowance, the deviation of customer delivery requirements, and the degree of due-date controllability. It was established that these variables were critical to the performance this particular type of manufacturing system.

Introduction

This chapter reports an investigation into the operation of an Order-Handling-Manufacturing-System (OHMS), which contains features of both batch and flow production. Its specific characteristics are the design and manufacture of major capital equipment to customer requirements, a long lead time of production and a heavy commitment of working capital against each customer contract.

The investigation involved the analysis of the manufacturing system of a major UK company producing underground mining equipment. A computer simulation model was developed which incorporated its key features and concentrated on the information and material flow patterns within the total system. The behaviour of the simulated system was then explored by utilizing real company data on contracts, procurement and operation lead times, and production costs. This chapter considers the OHMS responses to customer contract uncertainty under different production control policies (other aspects of the system behaviour, such as work-in-progress cost performance, are reported elsewhere). Uncertainty here is a function of the contract flow-time allowance, the standard deviation of customer delivery requirements and the degree of due-date controllability.

The system includes design and development as well as manufacturing operations. Figure 1 describes OHMS operations in function blocks using the $IDEF_0$ modelling technique (Machulak, 1985). Many of the unique characteristics and problems of OHMS are reflected in Figure 1. Although production planning and control may be more difficult when jobs are produced to order, rather than for stock, and the operations of such systems are affected by greater uncertainty, there has been little research in this field (Wu, 1988). With greater customer choice such modes of operation are becoming predominant in certain industries. As shown in Figure 1, contract negotiation is the first function in the activity chain. It must ensure that enough orders keep the plant loaded. The capability to accept and meet exacting delivery requirements may therefore become critical.

The due-date environment of OHMS operations

The ability to meet contract due-dates will not only be affected by production control rules, but also by the due-dates negotiated. A variety of due-dates rules can be used, such as (see Baker, 1984, for fuller treatment):

(1) CON (CONstant) : $d_i = r_i + K$
(2) TWK (Total Work) : $d_i = r_i + K \Sigma p_{i,j}$
(3) NOP (No. Operation) : $d_i = r_i + K \Sigma Q_{i,j}$
(4) SLK (SLacK) : $d_i = r_i + K + \Sigma p_{i,j}$
(5) PAW (Processing and Waiting) : $d_i = r_i + K \Sigma W_{i,j} + \Sigma p_{i,j}$
(6) FTB (Flow-Time Based) : $d_i = r_i + K \Sigma Q_{i,j} + Y \Sigma Q_{i,j} + \Sigma p_{i,j}$

where d_i : due-date of job i,
r_i : start date of job i,
$p_{i,j}$: standard processing time of job i on processor j,
$W_{i,j}$: waiting time of job i at operation j,
$Q_{i,j}$: operation flag = { 1/0 if job i is/is not processed by operation j,
X, K, Y : constants.

The properties of such rules have been well established, but mainly on the assumption that the manufacturer had total control of the due-date, i.e. that due-dates were internally assigned and deterministic. However, for an OHMS firm the completion date of a contract can be a major item of negotiation. To accommodate the resulting due-date uncertainty in negotiation, there is a case for corresponding assignment rules, with a limited degree of due-date controllability:

(7) CUB (CUstomer Based) : $d_i = r_i + R_i$
(8) CPB (ComPany Based) : $d_i = r_i + \Sigma Pe_{i,j} + \Sigma We_{i,j}$

where $Pe_{i,j}$: calculated processing time of contract i at operation j,
$We_{i,j}$: calculated waiting time of contract i at operation j,
$R_i = R(\mu_d, \sigma_d, i)$: a random variable (as far as the company is concerned) with mean μ_d and standard deviation σ_d.

With CUB a major customer has near control of contract due-dates, which will then be externally and, for the company, randomly selected. However, even here it is seldom that due-dates would be set without some participation by the company. This suggests that the distribution of flow-time allowances could be approximated by a normal distribution with a mean value of μ_d and standard deviation σ_d, where μ_d reflects the company's influence on the setting of due-dates; whilst the value of σ_d represents the uncertain outcome of negotiations. Alternatively, with a full order book, the manufacturer is better able to set the due-date, as per CPB rule, which is similar to PAW, i.e. the due-date is based on the actual flow time needed for a job to go through the system. However, in the CPB due-date rule defined here, the values of the processing and waiting times $Pe_{i,j}$, $We_{i,j}$ would be based on the

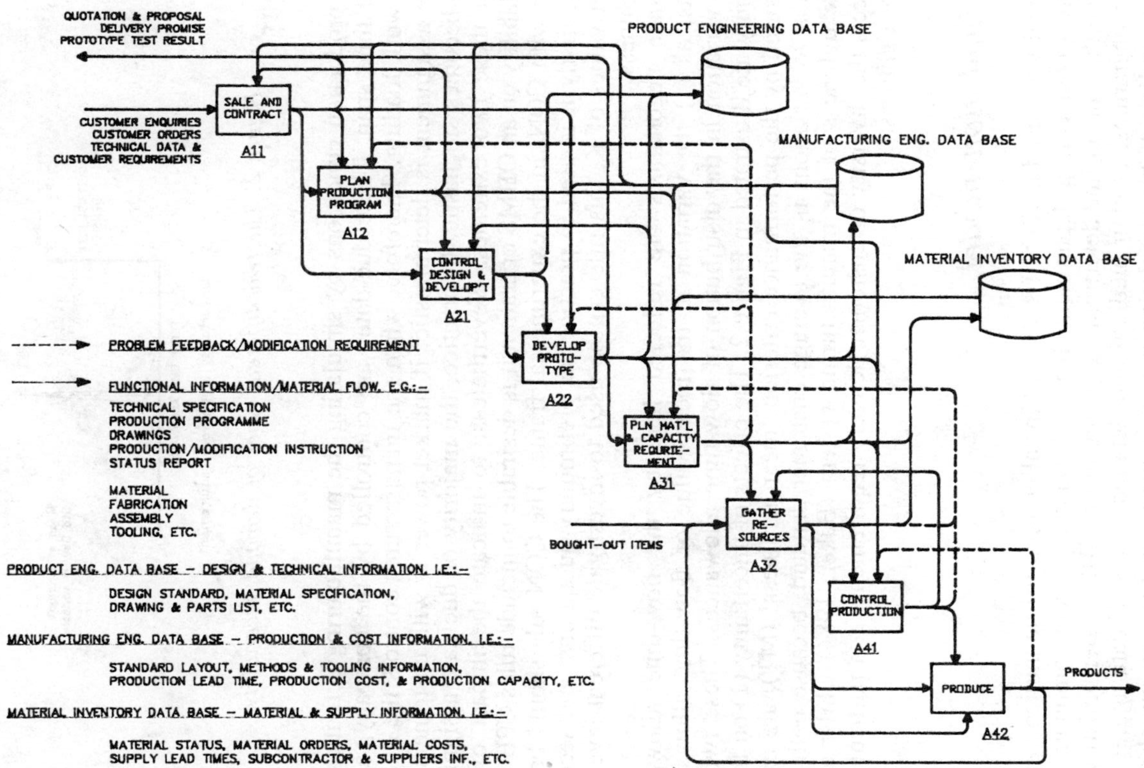

Figure 1. Simplified information and material flow model of OHMS operation.

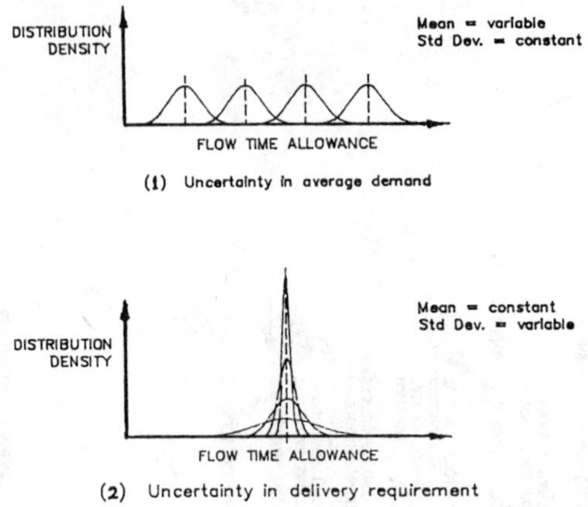

Figure 2. The two different types of demand uncertainty.

current production conditions by simulating the manufacturing activities over the required time span. These due-dates are controlled by the company, and reflect current work loading. Therefore, when the manufacturer controls the due-date the due-date assignment is deterministic; if market power is with customers, then the process becomes stochastic. In practice, the majority of due-date assignments fall between these two extremes. Nevertheless, to indicate the impact of due-date controllability on an OHMS operation it is desirable to include rules CPB and CUB, together with CON, in the experimental frame. The CON rule should be included because, firstly it is used in practice for preliminary quotations; and secondly, for different values of K, this rule can be used to investigate different levels of due-date tightness.

Due-date negotiations are reflected here by the flow-time allowances for contracts, as (1) the variation in the mean value of flow-time allowances, and (2) the variation in the distribution of flow-time allowances. These two types of uncertainty are illustrated in Figure 2. The effects of uncertainty (1) on the system, as measured by the performance criterion 'mean tardiness' (MT), are expected to be similar to variations in the average customer demand, because both affect the due-date tightness of the manufacturing system (Baker, 1984) whilst uncertainty (2) reflects the diversity of customer delivery requirements at a particular level of due-date tightness.

The simulation model

The model was based on FORTRAN and installed on a Prime minicomputer in the company's engineering department. It is a dynamic, discrete computer simulation suitable either for on-line deterministic planning or off-line deterministic/stochastic research operations, at both medium and detailed levels of data aggregation. The system boundary of the simulation model is based on the conceptual model, i.e. it simulates the manufacturing activities of the company within the four major OHMS functional areas.

The modelling frame, shown in Figure 3, consists of MODEL, INPUT AND CONTROL, and OUTPUT AND ANALYSIS subframes. For off-line simulation, the actual series of previous contracts was used. This 'real-systems' approach has been shown to be effective for the evaluation of total systems performance (Adshead and Price, 1986). The contract data at the required level of aggregation were extracted from past production programs, and incorporated actual production

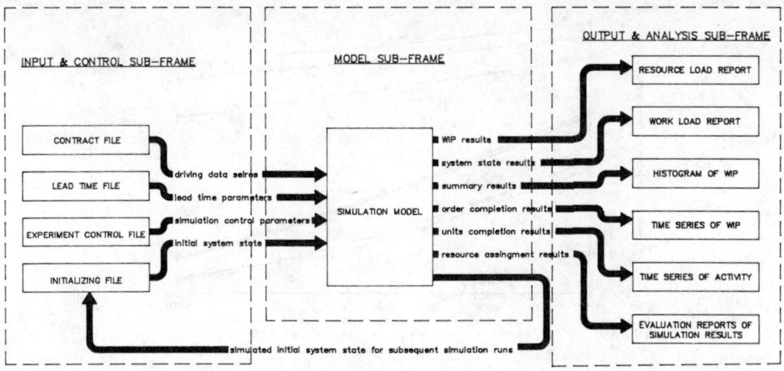

Figure 3. Overview of the modelling frame.

loads over six and a half years. Time-variant factors, such as structural system changes, changes in capacity, working practices, demand and products were analysed to justify the use of this data for the simulation model.

A series of simulations evaluated the system's responses to due-date uncertainties under various operating conditions and were designed for a wide range of contract due-date environments and production control policies. The following nine scheduling rules were built into the simulation model to plan for the contracts flow:

Rule 1 : FSFS—First come to System First Served,
Rule 2 : FCFS—First Come to workstation First Served,
Rule 3 : SI—Shortest Imminent operation time,
Rule 4 : LI—Longest Imminent operation time,
Rule 5 : EDD—Earliest Due-Date,
Rule 6 : LS—Least Slack
Rule 7 : SI1—Truncated SI with FSFS, waiting limits = 100 (h),
Rule 8 : SI2—Truncated SI with FSFS, waiting limits = 300 (h), and
Rule 9 : SI3—Truncated SI with FSFS, waiting limits = 500 (h).

Results and discussion

Variation in flow-time allowance

To experiment with different mean flow-time allowances, it is first necessary to obtain the mean flow time of the system. This was established at 2250 working hours by a simulation run, using the FSFS rule, and was confirmed as a realistic average contract flow time by the company. Five values: 1750, 2000, 2250, 2500, and 2750 working hours, were then chosen to obtain different levels of due-date tightness. It was noted that for the CON and CUB rules, if the parameters K and μ_d had the same value, then these rules would produce similar long run shop loading statistics and the same due-date tightness. Therefore, only the CON rule was used. The five levels of due-date tightness were then assessed against the nine production control rules.

The mean tardiness (MT) values from 45 simulation runs are listed in Table 1 and shown in Figure 4. As expected, for all nine rules, the MT value decreases as the tightness of due-date decreases. However, the shapes of the MT curves vary and fall into three groups:

(1) The SI, LI, and LS rules give relatively flat MT curves, suggesting that the effects of due-date tightness are less significant on their performance. When applied in practice these rules will seem more 'robust' than the others.

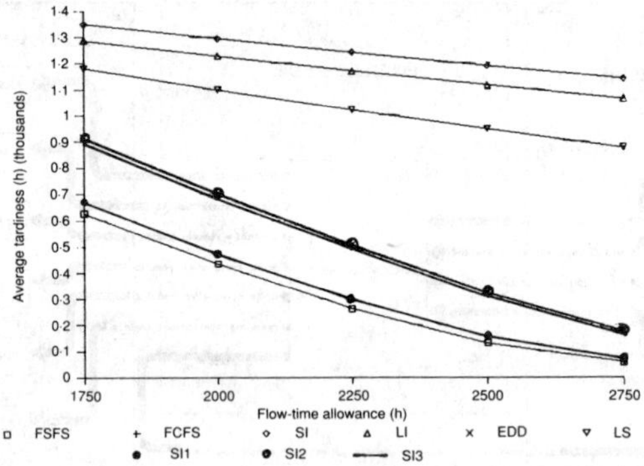

Figure 4. Effects of flow-time allowance.

Table 1. Effects of different flow-time allowance (due-date environment, CON).

Control rule	Tardiness (h) at Flow-time allowance				
	1750	2000	2250	2500	2750
FSFS	626·52	435·18	262·58	131·82	59·13
FCFS	911·78	698·05	501·93	321·68	175·10
SI	1349·74	1294·09	1240·91	1190·37	1140·86
SI-FSFS					
SI1	672·18	477·04	298·51	156·24	71·23
SI2	914·13	700·41	504·26	324·95	180·58
SI3	894·40	680·39	483·93	307·98	166·01
EDD	916·87	703·15	507·02	326·44	175·93
Least S	1181·34	1098·83	1021·83	947·58	875·80
LI	1287·77	1227·78	1169·18	1115·84	1064·13

(2) The second group include EDD, FCFS, SI2 and SI3, which are more sensitive to due-date tightness. They are less robust, but their performances improve rapidly as the due-dates are loosened.
(3) Finally, FSFS and SI1 are quite sensitive at the tight end, but become less sensitive at the loose end.

The different shapes of these curves suggest that crossover points might exist if the experimental range is sufficiently large. Different rules might thus become desirable at extreme conditions. However, it should be noted from Table 1 that within the experimental range of this study, only one crossover point is observed, i.e. between EDD and SI2 near the very loose end of the range. Furthermore, the differences in the MT outcomes between these two rules on both sides of the crossover point are small. It can be concluded therefore that for this particular system the relative performances of the production control rules will not be sensitive to the due-date tightness within the experimental range. Since in practice the company is unlikely to operate beyond the chosen range, this observation can also apply to it.

This finding seems to match previous work on different manufacturing systems. During their simulation study based on a single-machine system, Baker and Bertrand (1981) first observed that crossover points existed only when the due-dates were quite loose. Later, Baker (1984) reported a similar observation for a

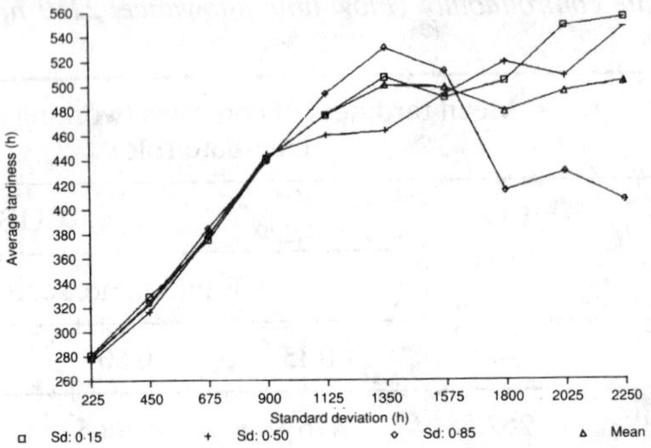

Figure 5. Effects of demand uncertainty.

Table 2. Effects of demand uncertainty (Control rule, FSFS; Due-date environment, CUB; Flow-time allowance, 2250 h).

Standard deviation (h)	Average tardiness (h)			
	Random no. seed			Mean
	0·15	0·50	0·85	
225	281·31	276·10	279·39	278·93
450	328·96	316·22	323·94	323·04
675	374·58	377·75	384·18	378·84
900	440·57	444·41	439·03	441·34
1125	476·18	459·65	493·63	476·49
1350	506·75	463·26	530·43	500·15
1575	490·34	493·23	511·60	498·39
1800	503·57	518·83	414·70	479·03
2025	548·61	507·24	429·88	495·24
2250	555·37	547·40	406·94	503·24

hypothetical job shop model. Since the method, data and simulation model used here are more complex, it suggests that this phenomenon is not confined to simple systems.

Variation in delivery requirements

This was investigated with the due-date rule CUB. An uncertainty coefficient V_d expressed the ratio of the standard deviation of the delivery demand, σ_d, to the mean flow-time allowance μ_d of 2250 hours. Ten values from 225 to 2250 were assigned to σ_d. Since the simulation runs were stochastic, three different streams of random numbers with seeds of 0·15, 0·50 and 0·85 were used for three independent replications.

Table 2 and Figure 5 give the system response to the uncertainty levels of delivery demand. The three different streams of random numbers produced similar MT values for low standard deviation. The MT increases rapidly with greater uncertainty revealing very sensitive initial system performance. However, with $V_d > 0{\cdot}5$ the system becomes much less sensitive to greater uncertainty of delivery requirements.

Table 3. Effects of due-date controllability (Flow-time allowance, 2250 h; standard deviation of CUB due-dates, 1125 h).

Control rule	Mean tardiness of contracts (working hours) Due-date rule					
	CPB	CON	CUB			
			Random no. seed			Mean
			0·15	0·50	0·85	
FSFS	0·00	262·58	476·18	459·65	493·63	476·49
FCFS	352·35	501·93	662·23	653·75	666·94	660·97
SI	1294·50	1240·91	1248·29	1272·66	1304·50	1275·15
SI-FCSFS						
SI1	99·29	298·51	500·23	487·13	520·83	502·73
SI2	355·64	504·26	666·53	657·70	679·44	667·89
SI3	331·98	483·93	648·43	640·32	652·63	647·13
EDD	340·43	507·02	434·49	442·43	432·60	436·51
LEAST S	1069·74	1021·83	1026·80	1079·63	1084·07	1063·50
LI	1223·59	1169·18	1212·61	1250·51	1236·70	1233·27

Variation in due-date controllability

The three selected due-date rules were compared and the effect of due-date control on the production system assessed. Here, the constant K of the CON rule and the mean flow-time allowance f_d of the CUB rule were set at 2250 h and V_d at 0·5. Since the last set of experiments indicated that the system performance deteriorated as the degree of the level of demand uncertainty increased until V_d reached the value of 0·5, the CUB due-date environment chosen here represented the extreme case when the customers' control of due-date would be most significant. Therefore, at the same level of due-date tightness, the due-date environment here shifts from one extreme (CPB), to the other (CUB); 45 simulations were carried out and the generated MT response is given in Table 3 and Figure 6.

Figure 6 shows that the lowest levels of MT were obtained under the CPB due-date environment and the highest with CUB for all but one of the control rules with better system performances (i.e. FSFS, FCFS, SI1, SI2, SI3). It confirmed the ranking:

$$\text{CPB} \rightarrow \text{CON} \rightarrow \text{CUB}$$

and strongly suggests that the system outcome of the order-handling company is a function of the due-date controllability.

In addition, Table 4 and Figure 7 show the MT outcomes for the three due-date environments against due-date tightness under the FCFS rule. The MT curve of the CPB environment is a horizontal line at the value given by the FCFS rule, because the due-dates assigned to contracts under the CPB environment are independent of due-date tightness. It can be seen from the MT curve of the CUB due-date environment that when the mean flow-time allowance in the system is small, the system outcome is much more sensitive to customer influence.

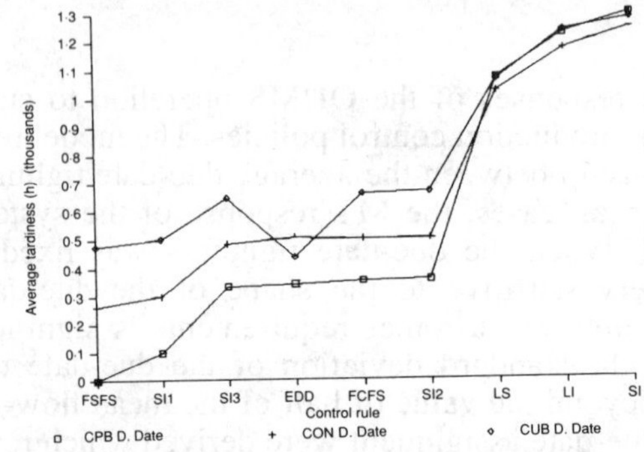

Figure 6. Due-date controllability.

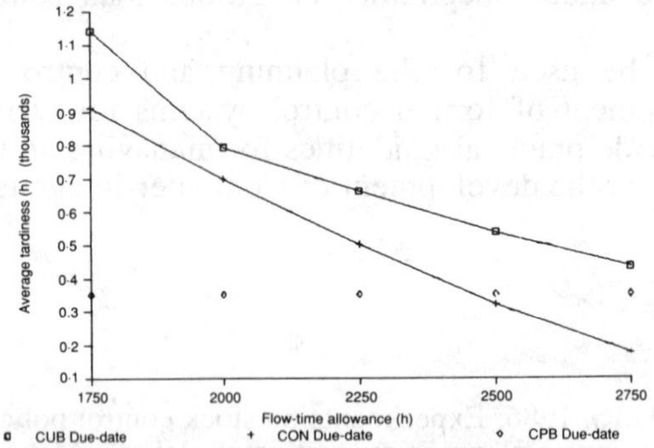

Figure 7. Due-date controllability (different flow-time allowance).

Table 4. Effects of due-date controllability with different Flow-time allowance (Control rule, FCFS; standard deviation of CUB due-dates, 1125 h).

Flow-time allowance (h)	Mean tardiness of contracts (h)					
	CUB				CON	CPB
	Random no. seed			Mean		
	0·15	0·50	0·85			
1750	930·63	1312·17	1183·33	1142·04	911·78	
2000	810·42	762·59	804·81	792·61	698·05	
2250	662·23	653·75	666·94	660·97	501·93	352·35
2500	524·77	524·80	562·59	537·39	321·68	
2750	418·73	445·74	441·14	435·20	175·10	

Conclusions

This chapter presents the responses of the OHMS operation to customer contract uncertainty under different production control policies. The model revealed a simple downward-sloping relationship between the average due-date tightness and the MT outcome of the system. In all cases, the MT response of the system increased as due-dates were tightened. When the due-date tightness was fixed at the average level, the system was very sensitive to the shape of the due-date distribution, indicating that the uncertainty of customer requirements is significant. The worst results were given when the standard deviation of the due-date distribution was allowed to reach and go beyond the value of half of the mean flow-time allowance.

Finally, three rules of due-date assignment were derived which represent the due-date environments with different degrees of company control over the contract due-date. As was expected, performance improved when the company had better control of due-date assignments. It was also established that the simple FSFS and EDD rules were more effective under uncertainty conditions than some of the more complex rules.

Such information can be used for the planning and control of production programs and the development of formal control systems for OHMS firms. The experiments not only provide practical guidelines for managing an OHMS system, but also have implications of the development of Computer-Integrated Manufacturing (CIM) systems.

References

Adshead, N.S. and Price, D.H.R., 1986, Experiments on stock control policies and leadtime setting rules, using an aggregate planning evaluation model of a Make-For-Stock shop. *International Journal of Production Research*, **24**, 5, 1139–1157.

Baker, K.R., 1984, Sequencing rules and due-date assignments in a job shop. *Management Science*, **30**, 9, 1093–1104.

Baker, K.R. and Bertrand, J.W.M., 1981, An investigation of due-date assignment rules with constrained tightness. *Journal of Operation Management*, **1**, 3, 109–120.

Machulak, G.T., 1985, An examination of the $IDEF_0$ approach used as a potential industry standard for production control system design. In *Automated Manufacturing, ASTM STP 862*, edited by L.B. Gardner, pp. 136–149.

Wu, B., 1988, An investigation into the operation of an order-handling-manufacturing system. Ph.D. Thesis, Brunel University, UK.

84

A simulation for scheduling in FMS

M.P. Rajith, C. Eswara Reddy and O.V. Krishnaiah Chetty

Manufacturing Engineering Section, Department of Mechanical Engineering, Indian Institute of Technology, Madras, 600 036, India

Abstract

Available literature indicates that simulation for scheduling in FMS was carried out considering either the breakdown of machines or palletization and its movements individually, but they were not combined to study the machine utilization and system's performance. Backtracking of jobs was also not considered. This chapter deals with a more realistic model relaxing many assumptions. A simulation program has been developed to assess the real-time scheduling and control algorithm for FMS, with an aim to investigate the effect of different priority rules, such as first in first out (FIFO), shortest processing time (SPT), total processing time (TPT) and total work remaining (TWR) on the queue length and machine utilization.

Introduction

A Flexible Manufacturing System (FMS) consists of a group of numerical control (NC) machines connected together by an automated workpiece handling system. It operates as an integrated system under computer control. FMS is capable of processing a variety of different part types simultaneously under NC program control at various workstations. In this case individual part features, routing through machines may be different. The operations and tooling required at each workstation may also differ.

The installation of FMS represents a large capital investment. In order to make the most effective use of this investment efficient scheduling of the FMS is required. FMS scheduling not only presents all the difficulties associated with job shop scheduling, but also has a higher degree of complexity due to the new variables involved. These variables may be the tools, fixtures and materials handling system that need to be scheduled together with the parts.

Taylor *et al.* (1980) developed global analysis of a multi-product, multi-line production system using Q-GERT modelling and simulation. This modelling and simulation include probabilistic branching network looping, multiple node realizations and queue nodes. They discussed its applications to line balancing, production planning, job design and system design. Mills (1982) and Cheng (1985) emphasized the importance of simulation in FMS and described the techniques and its industrial developments with applications. Manuelli and Guiducci (1985) built a simulator 'FASTSIM' which was able to evaluate the performance in steady and transient conditions of a manufacturing system as 'closed system'. It was possible to optimize the system performance by changing pallet-mix, transport system's capacity and the technological sequence of every piece, but with the assumption that there were no breakdown of machines.

Crookall (1985) examined the practical features of FMS and stressed the technical and economic advantages with the cost and financial performance modelling in the simulation of flexible manufacturing. They illustrated the planning for FMS with simulation and discussed the use of simulation at four different levels covering the design, planning and control of manufacturing systems. Barrett and Barman (1986) applied simulation to study the simplified flow-shop using SLAM-II. A set of scheduling rules was tried and the results indicated that the shop congestion and processing time variations affect the performance of the flow. Acre and Smith (1987) developed a model using SLAM-II. This was tested with a few traditional job shop scheduling rules and a few rules selected from computer science literature, assuming that machines do not breakdown.

Antonelli *et al.* (1986) applied hierarchical decomposition to a manufacturing cell and proved that a suitable choice of language like Ada makes possible the simulation, which can ultimately lead to development of a control software. Schriber and Stecke (1987) used mathematical programming to find out the input stream composition which maximizes the machine utilization in an FMS from a set of part type orders. They suggested simulation to estimate the degradation in the utilization depending on six different characteristics. Schriber and Stecke (1988) tried to achieve optimum machine utilization using balanced FMS production ratios in a simulated setting. They compared the theoretical machine utilization of the mathematical programming with the results achieved through simulation model. The study was concentrated on machine utilizations with FIFO and SPT. The experimental results were based on the assumptions that the palletizing, fixturing, defixturing, depalletizing and also machine breakdown times are negligible.

Miamon and Gershwin (1988) developed a 'dynamic scheduling and routing for FMS that have unreliable machines' in which they considered the short-term scheduling and routing on a global view of the system. The simulation results were obtained to verify the model for multiple routes for reliability parameters mean time to fail (MTTF) and mean time to repair (MTTR) for each machine. Choi and Malstrom (1988) developed a physical simulator which evaluates the work scheduling rules in FMS. The performance of decisión rule was dependent on the breakdowns of the machines in the system and the FMS operations were interrupted by the failures.

It is generally observed that the simulation was carried out considering either the breakdowns of machines or the palletization and its movement individually, but they were not combined together to study the machine utilization and the performance of the system.

Further, the simulation programs studied were developed with certain constraints while many of the models make FIFO sequencing rule (Taylor *et al.*, 1980). In reality, a host of other rules can be applied. In most of the simulation studies surveyed, there are several common but unrealistic assumptions which are made in order to simplify the study (Conway *et al.*, 1967; Doulgeri and Hibberd, 1985, Manuelli and Guiducci, 1985).

It is often assumed that

- a job visits a machine only once,
- transportation time between machines is not considered (instantaneous transfer to next machine after an operation has been completed),
- it is often assumed, for easier programming, that there is only one machine of each type in the shop,
- all operations can be performed by any of the system's identical machining centres,
- random disturbances in the system such as machine failures are not considered.

In this chapter a simulation model is developed relaxing all the above assumptions so as to provide an effective tool for scheduling and evaluating different types of control strategies.

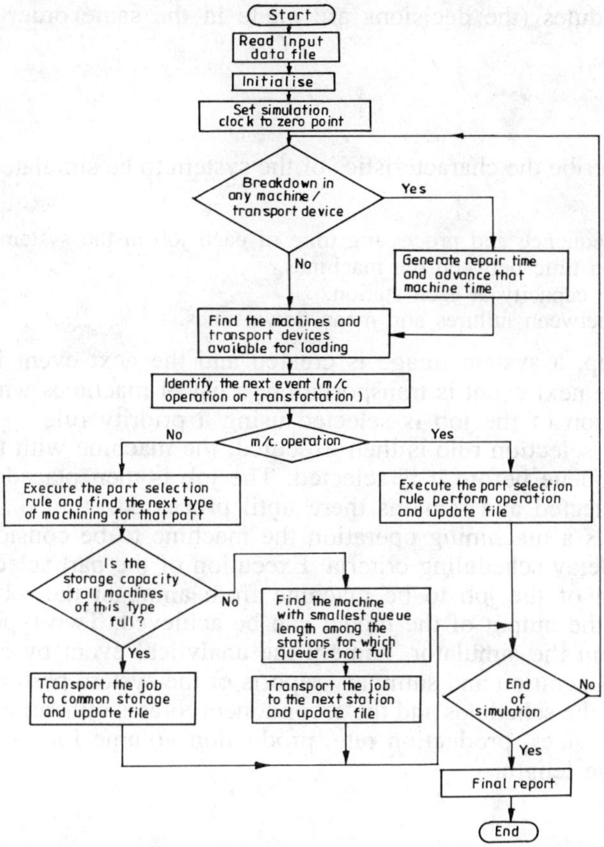

Figure 1. Decision flow chart.

FMS simulation model

An FMS is modelled in the present case as a system consisting of

(1) *M* machines (there can be more than one machine of the same type),
(2) a common storage of given capacity which also serves as a loading/unloading station,
(3) buffer storages for each machine with a given number of transport devices,
(4) a material transport system with a given number of transport devices.

There are *n* types of job $J_1 \ldots J_n$ processed in the system. Each job may have a different routing. The number of operations required to complete each job can vary from 1 to *N*, where *N* is a finite integer.

Backtracking of the job is allowed, i.e. a station may be visited more than once. Inter-arrival time of breakdowns and repair time of machines and transport devices are assumed to follow exponential distribution.

Program design

A generalized FMS model has been developed so that a system with given details can be modelled to allow a thorough evaluation of possible control strategies, pallet-mix and queue length.

A flow chart of this model is given in Figure 1. This program generates non-delay schedules (the decisions are made in the same order as they will be implemented).

Input

Data which describe the characteristics of the system to be simulated from the input include:

(1) machining sequence and processing time of each job in the system,
(2) transportation time between the machines,
(3) local storage capacity at each station,
(4) mean time between failures and mean down-times.

As a first step, a system image is created and the next event in the system is identified. If the next event is transportation, a list of machines which can perform the next operation of the job is selected using a priority rule.

The machine selection rule is then executed; the machine with the least number of jobs in the queue before it is selected. The job is then placed in the queue of the machine selected and remains there until processed.

If the event is a machining operation the machine to be considered is selected based on non-delay scheduling criteria. Execution of the part selection rule results in the selection of the job to be machine from amongst the jobs waiting in its queue. Finally the output of the model can be achieved. Two types of output can be obtained from the simulator, namely the analytical event by event description of the model's operation and summary reports of the system performance. The first report indicates the schedules and the second incorporates utilization of workstations and transport devices, production rate, production volume for each part type, and maximum queue length.

Case study

An example of the operation of the simulator is given below. The data are synthetic, however, the transport and operation times are realistic, as they are based on actual FMS. Earlier, Doulgeri and Hibberd (1987) suggested a simulation-based heuristic algorithm for FMS consisting of six stations with 1 and 3 and 2 and 4, but breakdown/failures of either machines or transport devices were not considered. In the present case, these breakdowns are considered and also the palletization. Backtracking of jobs is allowed.

The system consists of six stations of four different types and two transport devices. Stations 2 and 4 are identical to stations 1 and 3. There are four different parts. The input data are shown in Tables 1, 2 and 3. Table 1 gives the operation sequence and duration of the four parts existing in the system. Table 2 gives the matrix for transportation time, while Table 3 gives the decision rules which govern the control strategy of the system.

The simulation was carried out on SIEMENS-9748 system using FORTRAN-77. The simulator was run for a total of 500 min for each control strategy. Simulation was attempted considering one pallet for each part type in one case and two pallets per part type in the other case.

Results and discussion

In order to determine the effect of priority rules on the production volume, four different rules namely: first in first out (FIFO), shortest processing time (SPT), total processing time (TPT) and total work remaining (TWR), are used. SPT, TPT and

Table 1. Machining sequence of parts, machine tools to be used and operations times.

Part type	Priority[a]							
	First		Second		Third		Fourth	
	MT	OD	MT	OD	MT	OD	MT	OD
1	M1 M2	3·0	M3 M4	4·0	M5	2·5	M6	1·6
2	M1 M2	3·3	M3 M4	5·0	M6	0·7	M5	2·0
3	M3 M4	6·0	M1 M2	2·0	M5	3·3	M6	0·9
4	M6	1·2	M1 M2	3·6	M5	3·0	M3 M4	2·5

[a] MT, machine tool; OD, operation duration.

Table 2. Transportation time.

From	To						
	I/O	M1	M2	M3	M4	M5	M6
I/O	0	0·3	0·7	0·3	0·7	0·9	0·9
M1	0·3	0	0·3	0·3	0·7	0·7	0·9
M2	0·7	0·3	0	0·7	0·3	0·3	0·3
M3	0·3	0·3	0·7	0	0·3	0·7	0·3
M4	0·7	0·7	0·3	0·3	0	0·7	0·3
M5	0·9	0·7	0·3	0·9	0·7	0	0·3
M6	0·9	0·9	0·7	0·7	0·3	0·3	0

Table 3. Decision strategy.

Strategy	Remarks
1. First in, first out (FIFO)	Highest priority to the job that enters first
2. Shortest processing time (SPT)	Highest priority to the shortest processing time
3. Total processing time (TPT)	Highest priority to maximum total processing time
4. Total work remaining (TWR)	Highest priority to minimum work remaining

TWR rules depend on the processing times of the jobs at each machine whereas FIFO is independent of the processing time. It is found that when the number of pallets is less than or equal to the total number of machines in the system, priority rules do not have much effect on the production volume and hence FIFO can be used in such a case.

As the number of pallets increases, the jobs in the queue system also increase. So the production volume of each part type is affected by the priority rule. It is clear from Table 4, that using SPT, 24 parts of type 3 are produced, while priority

Table 4. Simulation results.

	1 pallet/fixture for each part type				2 pallets/fixtures for each part type				2 pallets/fixtures for each part type (restriction on length of the queue system)			
	1 FIFO	2 SPT	3 TPT	4 TWR	1 FIFO	2 SPT	3 TPT	4 TWR	1 FIFO	2 SPT	3 TPT	4 TWR
Production rate (Parts/h)	16·08	16·08	15·72	16·08	21·72	22·6	21·86	22·5	21·7	22·3	21·84	21·86
Total production	134	134	131	134	181	188	183	187	181	186	182	183
Part												
1	34	34	32	34	45	56	52	59	48	53	51	55
2	33	33	33	33	45	58	52	57	44	53	50	49
3	33	33	33	33	46	24	53	47	45	31	50	47
4	34	34	33	34	45	50	26	24	44	49	31	32
Utilization												
M1	42·8	42·8	40·0	42·9	54·5	61·7	54·9	56·0	55·6	56·0	57·0	56·9
M2	38·0	38·0	39·0	38·0	56·3	60·4	54·2	56·0	56·2	57·8	53·0	54·2
M3	45·0	59·0	67·0	59·0	81·0	78·5	87·5	88·6	80·3	78·0	83·5	84·6
M4	73·0	59·0	49·5	59·0	80·9	81·0	87·5	88·4	82·1	82·0	86·0	86·0
M5	73·0	72·0	71·0	72·0	98·0	98·0	98·0	98·3	98·3	98·0	98·3	98·0
M6	25·7	25·0	25·0	25·0	35·0	36·0	34·2	34·6	34·9	35·6	34·0	34·8
Transport R1	63·3	55·0	58·0	50·0	77·0	75·4	68·2	74·5	75·0	77·0	73·4	76·0
R2	63·4	50·0	58·0	55·0	78·0	77·0	69·0	74·5	76·9	80·0	76·0	76·0
Maximum length of queue system M1	1	2	2	2	2	3	3	3	3	3	3	3
M2	1	2	2	2	3	3	3	3	3	3	3	3
M3	2	2	2	2	3	3	2	2	3	3	3	3
M4	2	2	2	2	3	3	2	3	3	3	3	3
M5	3	2	3	2	6	5	7	5	3	3	3	3
M6	2	2	2	2	3	3	2	3	3	3	3	3

rule TWR produces 47 parts. Comparison of the strategies indicates that in the present case SPT gives the maximum rate of production followed by TWR.

Restriction on the maximum queue length at each machine also affects the production volume as can be seen in Table 4. In this case by restricting the queue length to 3, changes in product mix are observed. It is therefore, possible to adjust the production rates by controlling maximum queue length.

From the results it can also be seen that for a given priority rule, increase in number of pallets per part—from one to two—yields changes in production volume. Therefore, the choice of a scheduling strategy depends on many variables of the system. Each system deserves an exhaustive independent study of various control strategies using simulation and the maximum queue length permits one to select optimum control strategy and number of pallets required for each part type to achieve desired production volume.

Conclusion

A realistic simulation program has been developed to deal with scheduling in FMS. A number of priority rules can be checked and proper control strategies can be arrived at to match the production volume with the actual requirements. This simulation program can also be used to investigate the consequences of machine failures, changing the part mix, adding machines or transport devices to the configuration, etc.

References

Acre, E.S. and Smith, M.L., 1987, Simulation of a flexible manufacturing system—applications of computer operating system techniques. *Proceedings of the Annual Simulation Symposium*, pp. 205–215.

Antonelli, J.C., Volz, R.A. and Mudge, T.N., 1986, Hierarchical decomposition and simulation of manufacturing cells using ADA. *Simulation*, April, 141–152.

Barrett, T.R. and Barman, S., 1986, A SLAM-II simulation study of a simplified flow shop. *Simulation*, November, 181–188.

Cheng, T.C.E., 1985, Simulation of flexible manufacturing systems. *Simulation*, December, 299–302.

Choi, R.H. and Malstrom, E.M., 1988, Evaluation of traditional work scheduling rules in an FMS with physical simulator. *Journal of Manufacturing Systems*.

Conway, R.W., Maxwell, W.L. and Miller, L.W., 1967, *Theory of Scheduling*.

Crookall, J.R., 1985, Planning and simulation of FMS. *Annals of CIRP*, **34**, 2, 577–584.

Doulgeri, Z. and Hibberd, R.D., 1985, A simulation to develop and assess control algorithms for FMS. *1st International Conference on Simulation for Manufacturing*, pp. 185–193.

Doulgeri, Z. and Hibberd, R.D., 1987, The scheduling of FMS. *Annals of the CIRP*, **36**, 1.

Maimon, O. and Gershwin, S.B., 1988, Dynamic scheduling and routing for flexible manufacturing systems that have unreliable machines. *Operations Research*, **36**, 2, 279–-291.

Manuelli, R. and Guiducci, G., 1985, A simulation tool for FMS. *Proceedings of the 1st Conference on Simulation in Manufacturing*, Stratford, IFS, pp. 185–189.

Mills, R.I., 1982, Computer simulation—a feasibility and planning tool for FMS. *Proceedings of the International Conference on FMS*, pp. 185–186.

Schriber, J.T. and Stecke, K.E., 1987, Using mathematical programming and simulation to study FMS machine utilisations. *Proceedings of the Winter Simulation Conference*, San Diego, California, pp. 1–6.

Schriber, J.T. and Stecke, K.E., 1988, Machine utilisations achieved using balanced FMS production ratios in a simulated setting. *Annals of Operations Research*, **15**, 34.

Taylor III, B.W., Moore, L.J. and Hammesfahr, R.D., 1980, Global analysis of a multi-product, multi-line production system using Q-GERT modelling simulation. *AIIE Transactions*, **12**, 145–155.

85

Planning the implementation of JIT manufacturing with the aid of simulation modelling

M.D. Byrne

Department of Production Engineering and Production Management, University of Nottingham, UK

Abstract

The Just-in-Time (JIT) philosophy of manufacturing is increasingly being considered by manufacturing organizations, as a response to the increased pressure to supply high quality products with short delivery times and at low cost. These competitive pressures make it essential for the company's manufacturing resources, whether a highly automated or traditional manual system, to be operated in the most efficient and flexible way possible. One useful measure of the flexibility of a manufacturing system is the manufacturing lead times of its products: shorter lead times mean greater flexibility.

JIT is normally seen as an essentially experimental process of continuing small improvements. The main focus of JIT is to identify all forms of waste in the system and to eliminate them one by one. This is generally achieved by gradually reducing stock and time buffers in the system until a constraint of the system is identified. The cause of this constraint is then identified and eliminated, and the process continues to find the next constraint. However, one aspect of this process which discourages the introduction of JIT is the very real risk of disruptions to production as problems come to light.

This chapter proposes an approach to achieving lead time reductions without the need for the risks of disruption involved in the normal JIT approach. If a simulation model of the manufacturing system is used for the experimental aspects of implementation, rather than the real system, then many of the problems might be anticipated, and hence addressed, before production is affected. The approach is based on the use of a simulation model to evaluate the overall performance of the system under given conditions. Proposed enhancements to the system can then be evaluated to assess their impact before needing to implement the changes on the real system.

The approach is illustrated by its application to the problem of batch size determination and its effects on lead time performance. Results indicate that the approach may be suitable for application in a wide range of manufacturing companies.

Introduction

Competitive pressures in manufacturing companies are typically far greater than they were a few years ago. Today's more sophisticated market demands higher quality products with shorter delivery times and lower costs. These pressures, together with reducing product life cycles require the company's manufacturing operations to be organized and operated in the most effective and flexible way possible. In some industries increased use of automation and advanced technologies have helped to meet these pressures, but in many industrial sectors the capital investment required for such a solution is not available.

Whether highly automated or not, a major factor in increasing the responsiveness and flexibility of a manufacturing system is reduction of manufacturing lead times. The achievement of lower lead times can help to reduce costs and improve quality levels.

The philosophy of Just-in-Time (JIT) manufacturing, together with the complementary ideas of Total Quality Management (TQM) have been recognized as major factors contributing to the rapid rise of Japanese manufacturing industry during the past decades. There has been a surge of interest in these philosophies in Western countries hoping to emulate the Japanese success in manufacturing.

The JIT approach is characterized by systematic reductions of set-up time, batch size, lead time and work-in-progress (WIP) inventory. When a new problem is encountered which inhibits further such improvement, all effort is concentrated on solving that problem, thus enabling further progress. This same philosophy of continual improvement is applied to all areas of the organization (including suppliers), leading to greatly simplified control of the whole manufacturing process. This concept of simplicity is extensively discussed by Schonberger (1982).

Many of the ideas of JIT have been well received by Western industrialists and researchers. Over the past few years many reports of JIT implementations have emerged (see for example Voss, 1987).

Rather than the essentially experimental approach normally involved in JIT, with the associated risk of disruption, it might be advantageous if the effect of proposed changes could be predicted in advance. To predict the behaviour of a system under a given set of conditions a model of the system is required. For the purpose under discussion here some sort of conceptual model of the system rather than a physical model is appropriate.

Conceptual models may be categorized into analytical and numerical models. An analytical model consists of a series of equations which may be solved by some mathematical technique to obtain an optimal solution, or to find a general solution into which specific values may be substituted. However, it is normally possible to obtain an analytical solution only for a relatively simple system, or for a simplified representation of a more complex system. If the system of interest is too complex for an analytical model then a numerical model might be used, whereby the behaviour of the system under a specific set of conditions is calculated. Simulation is a commonly used technique of numerical modelling, and is particularly appropriate for manufacturing systems.

A common use of simulation for applications in manufacturing is the comparison of alternative approaches to the same problem (see Lee and Khumawala, 1974). Other examples involve using simulation to investigate the effects of uncertainty on manufacturing systems (Grasso and Taylor, 1984). The availability of easy to use simulation software, commonly with graphical animation capabilities, now makes simulation a tool which is very widely available to engineers and managers.

The analytical modelling approach contrasts with the simulation approach described above, in that it sets out to find a mathematical optimum operating point for the system under consideration. This is achieved by modelling the system by a set of mathematical expressions which can be manipulated to find an optimum in terms of a selected objective function (such as minimum cost or maximum profit). Operational research methods such as mathematical programming and queuing theory are examples of this type of approach. Despite the existence of a large number of analytical modelling approaches in the literature, there is little evidence of extensive industrial implementation of any of these methods. It may be concluded therefore that the basic restrictions of these analytical methods make them either too complex or insufficiently accurate or flexible to be usefully implemented in practical manufacturing systems.

The approach proposed in this chapter uses simulation to enable the effects of proposed changes to a manufacturing system to be investigated before they are physically implemented. This enables potential problems to be foreseen and avoided without causing major disruption to production. The approach can also be used to determine the most appropriate values of certain system parameters. It thus has great potential in aiding a continual improvement programme.

The search simulation approach

Central to the method proposed here is the use of a simulation model of the relevant manufacturing system to assess the effects of varying parameters of the system. The use of simulation has the great advantage that there is no inherent restriction on the complexity of the system to be studied. The level of detail to be included in the model may be chosen to meet such considerations as the degree of accuracy required from the model's outputs, the availability of time and other resources, and the aspects of particular interest in the system. The simulation approach therefore avoids the need to mould the real system into the rigid format which might be required if an analytical model was to be used.

Once a suitable simulation model of the system has been developed and validated, it can be used to evaluate the performance of the system with various possible values of the relevant parameters. Multiple simulations with varying parameter values can be carried out to seek the set of values leading to the best system performance.

For example, if the approach was to be used for the determination of the most appropriate batch sizes for a system, a convenient method of achieving this would be to choose some starting ('base') batch size for each product (possibly that currently implemented) and then select a new batch size using:

$$\text{new lot size} = k \times \text{base lot size}$$

where k is the batch size factor. The best value of k can then be sought experimentally, using the simulation model of the system.

The principle of this process is simple, but its practical implementation requires some procedure for systematic adjustment of the parameter values so as to discover the best system performance.

One way in which this adjustment process might be carried out is by use of a computer search algorithm designed to find the minimum value of a complicated function of several independent variables. In this case the system parameters of interest would be the independent variables, and the objective function would be the system performance as measured by the simulation output. An example of such a search algorithm which has been successfully applied to other aspects of manufacturing planning and control is the pattern search method of Hooke and Jeeves (1961). This method was used by Herschauer and Ebert (1975) in their methodology for selection of a job shop sequencing rule, and was also used in a modified form (the so-called 'adaptive pattern search') by Taubert (1978) in his search decision rule methodology. In view of this previous successful application,

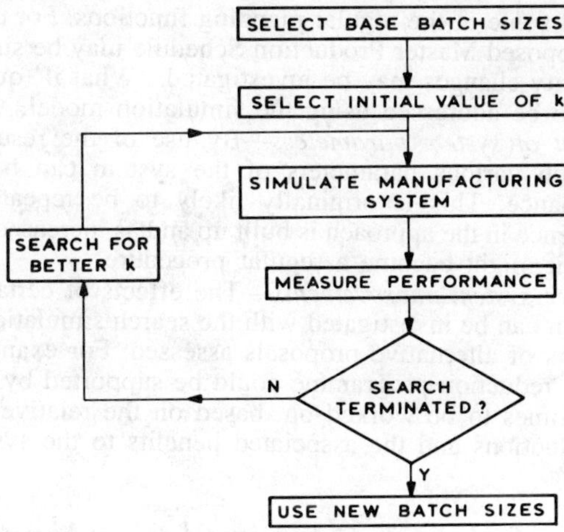

Figure 1. Search simulation approach for batch sizing.

Taubert's adaptive pattern search algorithm was selected for use in this 'search simulation' approach.

The objective function by which the search algorithm can assess the performance of the system may be based upon any characteristic or characteristics of the system. It is commonly convenient to use some measure of monetary cost (or profit) which combines a number of system variables (e.g. inventory levels and set-up times in a manufacturing system) into a common unit. An alternative function, appropriate for seeking improved flexibility, might be a measure of lead time.

The overall process, incorporating the search algorithm, as applied to batch sizing may be illustrated by a flow diagram (Figure 1). The search simulation approach may, in principle, be applied to any manufacturing system. An examination of its application is described below.

Use for JIT implementation

The idea of using a simulation model of a manufacturing system to evaluate the system's performance under various conditions can be extended to the JIT concept of continuous improvement. JIT normally involves making changes to the actual system and solving problems as they emerge. The alternative of using simulation to evaluate the effects of proposed changes in advance of their implementation could help to predict the problems which will occur, and hence enable the problems to be addressed before they cause disruption in the real system.

The use of simulation in this way requires a high degree of confidence in the realism of the simulation model being used. In view of this, a phased implementation scheme for the search simulation approach is proposed, which allows a gradual refinement and building of confidence in the model, until it can eventually be confidently used to aid a JIT-like continual improvement programme. The proposed stages of this phased implementation scheme are as follows:

(i) *Build and validate the simulation model*—The model being developed can initially be used independently of the existing planning and control system. Its outputs can be compared with the actual performance of the system and hence it can be gradually refined until a sufficient level of confidence in its performance has been built up. Regular updating and refinement of the model will still be required.

(ii) *Use for routine planning and control functions*—The validated simulation model can now be used to aid some of the regular planning functions. For example the effects on the system of a proposed Master Production Schedule may be simulated, or the effects of short-term priority changes may be investigated. 'What if' questions about specific situations may also be addressed using the simulation model.

(iii) *Use for refinement of system parameters*—By use of the results of the full search simulation approach various parameters of the system can be adjusted to achieve improved performance. This is originally likely to be repeated only occasionally, although as confidence in the approach is built up and as increased computational speeds become available it might become a regular procedure.

(iv) *Use for progressive system improvements*—The effects of certain proposed improvements to the system can be investigated with the search simulation approach and hence the relative benefits of alternative proposals assessed. For example a progressive set-up time/batch size reduction programme could be supported by using the approach to prioritize the machines to be worked on, based on the relative cost and difficulty of specific set-up reductions and the associated benefits to the system.

An application of the search simulation approach

The application chosen to examine the use of the search simulation approach is the problem of batch size determination. This is an example of stage (iii) of the phased implementation plan discussed above. The system parameters to be refined are the batch sizes.

It is generally recognized that the determination of appropriate batch sizes for the many products typically involved in a batch manufacturing system is a complex problem. Analytical methods are commonly used to solve this problem, the best known example of such an approach being the Economic Order Quantity (EOQ) model. The EOQ model and its derivatives have the advantage of simplicity, but are recognized as flawed in many ways when applied to a dynamic multi-item manufacturing environment. The problems of analytical batch sizing models are described and discussed in more detail in Byrne (1987) and O'Grady and Byrne (1987).

In this study the search simulation approach is applied to batch size determination in a theoretical job shop system. The approach avoids many of the problems of analytical approaches and appears to have a wide potential for use in practical systems.

Theoretical job shop system

The system used to examine the use of the approach was based on the model used by Wilbrecht and Prestcott (1969) to investigate the effects of set-up time on job shop performance. Certain modifications to this model were made to facilitate the representation of a repetitive batch manufacturing environment. The most significant characteristics of the system were as follows:

(1) The system contained nine machines, subject to no random breakdowns and allowing no job pre-emption.

(2) Twenty different products were modelled, with their routings, set-up and unit operation times determined by sampling from random distributions with parameters consistent with those used by Wilbrecht and Prestcott (1969).

(3) Demand levels for each product for a period of 48 weeks (12 months of 4 weeks) were randomly sampled from normal distributions truncated at zero. The means were selected such that the mean aggregate demand was within the capacity constraints of the system.

(4) A 48-week running-in period was simulated, starting from an empty shop, and the final state obtained from this run was stored and used as the initial state for subsequent experiments. This avoided the need for an extensive running-in period each time the model was run.

(5) A set of initial 'standard' batch sizes was devised on the basis of EOQ type calculations.

The AMASS simulation environment

The simulation model for these experiments was built and the experiments were run using a discrete event manufacturing system simulator written by the author. This simulator, known as AMASS (Adaptable MAnufacturing System Simulator), was developed as a tool to aid the rapid building and use of simulation models of a variety of different manufacturing systems. The simulator consists of a modular PASCAL program which makes use of many PASCAL's data structure and structured programming features to give a considerable amount of flexibility in adapting the simulator to model different systems while retaining a relatively efficient performance in running long simulations of complex systems.

AMASS allows the measurement of the performance of the simulated system by the collection of appropriate data as the model runs.

The search algorithm used was based directly on the adaptive pattern search of Taubert (1978).

Objective Functions

In order to compare the performance of this system using different batch sizes, two alternative objective functions were developed.

The first function was based on the variable production-related costs of the system. This total cost function included inventory holding costs for both WIP and finished stock (24% per annum), a backordering cost (48% p.a.) and a set-up cost based on the total time spent on set-ups for all machines. The unit of cost for the function was the standard hour (sth), with inventory values for each product based on the total operation times involved in their manufacture.

The second objective function was the average manufacturing lead time of batches completed by the system. This function has the advantage of being a directly measurable characteristic of a manufacturing system, as well as being an important determinant of the system's flexibility.

Experiments

The initial runs with this system were concerned with setting some standards for comparison with later results. Starting from the initial 'standard' batch sizes, simulations were performed with the actual lot sizes varied by factors between 2·0 and 0·25 times the initial values (in intervals of 0·25). The same factor was applied to the lot size of each product. The relevant parameters of these simulation runs are indicated, together with the results, in Table 1.

To consider a simple JIT type enhancement to the system, these initial runs were repeated with all set-up times reduced by 50%. The parameters and results of these runs are shown in Table 2.

The full search simulation approach was examined by using the search algorithm with the simulation model to vary the batch size factor to seek a lowest total cost, or lowest average lead time performance for the system. All the simulations were run for one simulated year (48-week) period.

Results

The results of the simulation runs described above are summarized in Tables 1–4 in terms of the components of the total cost function (average WIP level, average finished stock level, average backorder level and total set-up cost), the total cost and the measured average lead time. Figure 2 shows the variation of total cost with batch size factor for both the original and reduced set-up time runs.

Summaries of the initial runs are presented in Table 1 and Figure 2.

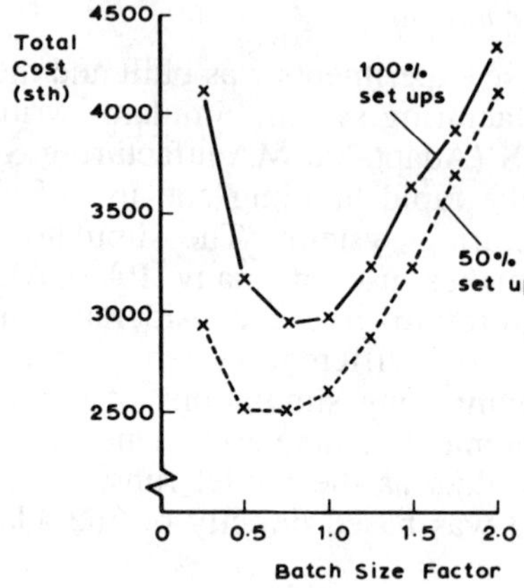

Figure 2. Variation of total cost with batch size.

Table 1. Results of simulation runs (original set-up times).

Batch size factor	Average WIP (sth)	Average F.S.[a] (sth)	Average backorder (sth)	Set-up cost (sth)	Total[b] cost (sth)	Average lead time (days)
2·00	11165	2268	1453	404	4325	105·4
1·75	9686	2482	1121	442	3900	91·4
1·50	8683	2672	761	525	3615	82·0
1·25	7176	2967	405	589	3218	67·7
1·00	6013	3201	89	715	2969	56·8
0·75	4787	3615	36	910	2944	45·2
0·50	3865	3877	12	1296	3160	36·5
0·25	2597	4440	10	2419	4112	24·5

[a] F.S., finished stock.
[b] Total cost = set-up cost + 24% × (average WIP + average F.S.) + 48% × average backorder.

The behaviour of the individual factors making up the total cost are each well characterized. The average WIP level varies approximately in proportion to the lot size factor, while the total set-up cost varies in approximately inverse proportion to the lot size factor. These effects are explained by the increase in the total number of lots which have to be processed when the lot size factor falls, and the associated reduction in total processing time for each lot. The WIP reduction is accompanied by a reduction in average lead time which in turn accounts for the increased finished stock level and reduced backorder level as the batch sizes are reduced.

The overall effect of these individual factors leads to a 'U-shaped' variation of total cost with batch size factor, with a minimum value at a batch size factor between 0·75 and 1·0 (see Figure 2). This relatively flat-bottomed curve is also characteristic of EOQ inventory theory, and a consequence of this shape is that the minimum total cost value (with a batch size factor of 0·75) is less than 1% below that for the run using the original batch sizes (i.e. batch size factor of 1·0).

The variation of average lead time with batch size factor is also straightforward in these runs, falling in direct proportion to the average WIP level. This effect is, however, not linear with batch size factor, as with the lower batch sizes the capacity of the machines is effectively reduced as more and more set-ups are required. There should therefore come a point at which no more set-ups can be accommodated, and hence the lead time will reach a minimum and then start to rise again. This point is clearly not reached within this range of batch size factors.

Table 2. Results of simulation runs (50% original set-up times).

Batch size factor	Average WIP (sth)	Average F.S.[a] (sth)	Average backorder (sth)	Set-up cost (sth)	Total[b] cost (sth)	Average lead time (days)
2·00	11041	2328	1410	204	4089	104·2
1·75	9596	2539	1109	229	3673	90·6
1·50	8312	2864	569	262	3217	78·5
1·25	6938	3121	320	299	2866	65·5
1·00	5680	3514	48	366	2595	53·6
0·75	4701	3719	23	467	2499	44·4
0·50	3800	3843	12	646	2510	35·9
0·25	2389	4668	10	1235	2933	22·6

[a] F.S., finished stock.
[b] Total cost = set-up cost + 24% × (average WIP + average F.S.) + 48% × average backorder.

Table 3. Results of minimum cost simulation run.

Batch size factor	Average WIP (sth)	Average F.S.[a] (sth)	Average backorder (sth)	Set-up cost (sth)	Total[b] cost (sth)	Average lead time (days)
0.90	5180	3561	31	783	2896	48·9

[a] F.S., finished stock.
[b] Total cost = set-up cost + 24% × (average WIP + average F.S.) + 48% × average backorder.

Results of the runs with set-up times reduced by 50% are shown in Table 2. The behaviour of the total cost function is similar to that with the original set-up times. The only significant change is in the total set-up cost component, which is approximately halved as a direct result of the set-up time reduction. As a result of this change the minimum total cost batch size factor is lowered slightly, as can be seen in Figure 2, although the lowest cost individual simulation run is again at a batch size factor of 0·75, with a total cost 15% lower than that with the original set-up times. (This is in accord with EOQ theory which would predict that the optimum batch size should be reduced to about 70% of its original value if the set-up time is halved.) Small reductions in average lead time are also observed.

In a system running at high machine utilization levels it is likely that this effect of reduced minimum cost batch size as set-up times are reduced would be magnified because of the extra capacity released by the reduced set-up times. This would result in less use of machine capacity for set-ups, and consequent reduced lead times and WIP. However the capacity levels in this particular system mean that the utilization (varying between 72% and 79%) is not high enough for this effect to be apparent.

The results obtained using the search algorithm to minimize total cost are presented in Table 3. This Table shows the cost function components and average lead time, as in Table 1, for the lowest cost batch size factor found by the search process (0·9). This result is consistent with the total costs found using predetermined lot size factors as shown in Table 1. The search algorithm appears to have successfully located the minimum cost operating point for the system.

The equivalent results with the search algorithm seeking lowest average lead time are shown in Table 4. The batch size factor giving the lowest average lead time was found to be 0·05. The total cost corresponding to this batch size factor is very much higher than the lowest cost value, but this is primarily due to the increase in total set-up cost, which is a somewhat artificial cost as long as both machine and setter capacity are available.

Table 4. Results of minimum lead time simulation run.

Batch size factor	Average WIP (sth)	Average F.S.[a] (sth)	Average backorder (sth)	Set-up cost (sth)	Total[b] cost (sth)	Average lead time (days)
0·05	2184	4353	0	4862	6431	20·6

[a] F.S., finished stock.
[b] Total cost = set-up cost + 24% × (average WIP + average F.S.) + 48% × average backorder.

Summary

The results of these experiments with the theoretical job shop system have illustrated the application of the proposed search simulation approach to the problem of batch size determination. It appears that in this system the 'optimum' batch sizes are not far from the original values chosen on the basis of EOQ theory.

The experiments with a global set-up time reduction show the expected consequent reduction in 'optimum' batch sizes. A more practical procedure would be to investigate the effects of specific set-up reductions on key machines to assess their relative cost-effectiveness.

The author has also investigated the use of the search simulation approach for batch sizing with multiple products or product groups, and for adjustment of other system parameters (see Byrne, 1987, 1990).

Conclusion

This chapter has presented an approach to aiding the implementation of Just-in-Time manufacturing. Using a simulation model of the system, changes in operating conditions can be investigated without the risk of disruption in the real system. Hence many of the benefits of the JIT approach may be achieved without some of its difficulties.

A possible phased implementation scheme for the approach was presented. This aims to avoid possible problems of using a simulation model which is not fully validated, by gradually refining and building up confidence in the model before relying too heavily on its predictions.

The search simulation approach was illustrated by its application to the determination of batch sizes in a theoretical job shop system. It was found to be effective for this particular problem.

It is suggested that the approach is potentially useful for a wide range of companies, due to its conceptual simplicity and the possibility of a gradual, confidence-building introduction.

References

Byrne, M.D., 1987, A search simulation approach to manufacturing planning and control. Ph.D. Thesis, University of Nottingham.

Byrne, M.D., 1990, Multi-item production lot sizing using a search simulation approach. *Engineering Costs and Production Economics*, **19**, 307–311.

Grasso, E.T. and Taylor, B.W., 1984, A study of supply/timing uncertainty in MRP systems. *International Journal of Production Research*, **22**, 3, 485–497.

Herschauer, J.L. and Ebert, R.J., 1975, Search simulation selection of a job-shop sequencing rule. *Management Science*, **21**, 7, 833–843.

Hooke, R. and Jeeves, T.A., 1961, Direct search solution of numerical and statistical problems. *Journal of the Association of Computer Machinery*, **8**, 2, 212–229.

Lee, W.B. and Khumawala, B.M., 1974, Simulation testing of aggregate production planning models in an implementation methodology. *Management Science*, **20**, 6, 903–912.

O'Grady, P.J. and Byrne, M.D., 1987, Lot sizing in a practical manufacturing system. *Engineering Costs and Production Economics*, **13**, 13–26.

Schonberger, R.J., 1982, *Japanese Manufacturing Techniques: Nine Hidden Lessons in Simplicity*. New York: Free Press.

Taubert, W.H., 1978, A search decision rule for the aggregate scheduling problem. *Management Science*, **14**, 16, B343–B359.

Voss, C.A., (Ed.), 1987, *Just in Time Manufacture*. (Bedford: IFS Publications).

Wilbrecht, J.K. and Prestcott, W.B., 1969, The influence of setup time on job shop performance. *Management Science*, **16**, 4, B274–B280.

Simulation of the production disturbances involving human intervention at danger zones

Kari K. Häkkinen,[1] Richard Booth,[2] Seppo Väyrynen,[3] Jari Haijanen,[4] Mikko Hirvonen[4]

[1]*Industrial Mutual, Risk Management Division, PO Box 12, SF-00210 Helsinki, Finland,* [2]*Aston University, Mechanical and Production Engineering, Health and Safety Unit, UK,* [3]*Oulu University, Institute of Work Science, Oulu, Finland and* [4]*Institute of Occupational Health, Vantaa, Finland*

Abstract

New technology and improved safety standards have brought about a decrease of serious machinery accidents. However, serious accidents still occur particularly in connection with the disturbances of automatic production. The increasing complexity of new production systems together with the great number and variation of hazard sources make safety assessment difficult. Simulation methods which today are much used in the optimum design of production systems may also be applied to a detailed a priori safety analysis. The objective of the project was to develop a simulation method by which knowledge can be acquired of human actions in disturbance situations and thereby to develop design criteria for the safety design of man–machine interfaces in computer-controlled production lines. The first experiments were conducted in order to analyse human actions and hand motion patterns when removing a faulty product from line.

Introduction

New technology combined with improved safety standards and better knowledge of the hazards of machinery has brought about a marked decrease of serious machinery accidents in many areas of industry. However, occasionally fatal accidents occur with automatic systems such as industrial robots. An example of a disturbance-induced fatality involving human intervention at the danger zone of a palletizing machine is given in Figure 1. The pallet fork struck a repairman on the head while he waited nearby for restart to check if the system would run correctly. These accidents represent a new generation of machinery accidents—the first generation appeared at the very beginning of mechanization when machines were running in routine production without guarding and protection. Now when the machines are guarded, or they function automatically during the course of production, the accidents take place during disturbances and trouble-shooting.

It is widely known that the potential for high severity injuries exists in many automatic production systems, but there are only a few incidents simply because the populations of those systems are still small and their operation time has been short. It has been estimated on the basis of the incident records available now, that

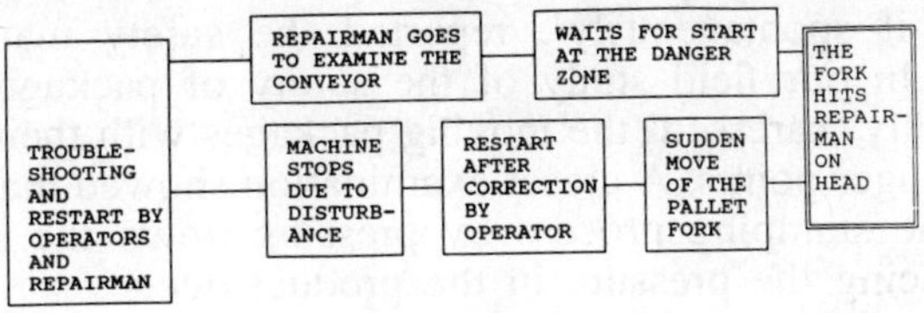

Figure 1. A fatal accident involving a disturbance in an automatic palletizing line (from Finnish Federation of Accident Insurance Companies, 1987).

the risk of industrial robots is at the same level as that of power presses (Helander and Karwan, 1988). In Japan, at least ten robot-related fatalities have been reported (Nagamachi, 1988). The increasing complexity of new production systems together with the great number and variation of hazard sources make safety management difficult. More information must be collected and analysed during the design and assessment of these systems and in evaluating their safety requirements.

Disturbances as a major source of production accidents

It has been noted, particularly in connection with accidents occurring with automatic machines, that in the design process sufficient attention has not been paid to disturbance situations. These are the situations in which the operator must enter the machine's hazard zone to carry out manual operations (Harms-Ringdahl, 1986). The machines are designed only for normal and automatic running; human intervention in disturbance situations has not been taken into account.

Accordingly, many of the solutions used to prevent accidents in automatic production lines are incompatible with respect to, for example, the demands of disturbance elimination and cleaning or maintenance operations. Therefore guards are often omitted, defeated or misused (Booth, 1979). Furthermore, there is a lack of specific procedures to maintain safe systems of work in disturbance situations. An analysis of the robot safety strategies in six British companies covering 84 robots showed that in 50% of the cases the hazard control devices were systematically circumvented by workers (Jones and Dawson, 1986).

Poor management of machine disturbances is not a problem of the new technology machines. For example, the so-called after-reach hazard in press operations has been much studied during the last few decades. After-reach is the incident which happens when the worker reaches spontaneously into the point of operation after the workcycle has been initiated in order to correct the misaligned workpiece. In spite of the several simulation studies, there is no firm consensus as to the hand speed value of human beings to be used as a design criterion. However, there is substantial evidence that in situations where the reflex action of humans are foreseeable in the correction of disturbances, the common standard value of 1·6 m/s is usually far too low (Winsemius, 1965; Van Ballegooijen, 1979; Pizatella and Moll, 1987).

The case of the after-reach hazard demonstrates that the human performance criteria for disturbances may not be equal to those in normal and smooth production. The variations seem to be greater and the extreme values may fall far beyond any expected distribution. This is especially important when human actions are considered in connection with automated systems and robots, where the configurations of motions for both humans and machines are far more complex than those with presses and other traditional machines.

Human actions and behaviour in accidents are to a large extent governed by various situational factors (Winsemius, 1965, 1969; Corlett and Gilbank, 1978). 'A plywood sheet rattling between conveyor belts can annoy a worker enough to make him grasp it spontaneously', reported the safety manager of a Finnish plywood factory. In our field study of the safety of packaging machines, some workers occasionally 'caressed' the moving packages with their hands and thereby reached near to danger points. A closer examination showed that they tried to avoid disturbances in the stamping process by pressing down the slack covers of the boxes, or by reducing the pressure in the product queue.

More than 50% of the accidents in connection with packaging machinery occur when people are trying to correct or to cope with disturbance situations (Rieppo and Häkkinen, 1985; Defren, 1988). Missing and deficient machine guards are typical contributing factors. The packaging machines involve several types of danger. The safety devices and guards must also meet many production requirements, e.g. they must allow easy cleaning at the points of operation and they must provide visibility. Thus the guarding problems are often difficult to resolve completely.

The role of simulation in production safety

The traditional method of collecting data for industrial safety studies has been to analyse actual accidents and to try to find typical accident scenarios on which to base the prevention measures. Apart from accident analysis, hazard analysis methods are also being increasingly used to ferret out the risky properties of workplaces and tasks before accidents have actually occurred. Systematic hazard analysis is usually the only method for novel automatic installations, because no accident experience is available and because there is a demand for safety from the very beginning of the new production technology. Moreover, the price of learning through accidents is now considered too high from both the human and the economic viewpoints.

According to Shannon (1975), simulation is 'the process of designing a model of a real system and conducting experiments with this model for the purpose of either understanding the behaviour of the system or of evaluating various strategies for the operation of the system'. A simulation model can be an exact physical model of the system or a scaled one presented on a computer screen or on a piece of paper. In simulation, the disturbance events can be repeated several times to provide enough material for a statistical analysis, and accurate measurements can be conducted during the simulated events. When the simulation results are combined and validated with the results of the accident and workplace studies, the production and safety experts can devise a profound safety picture of the system under development.

The situational effects in disturbances are very difficult to simulate and in many studies they are considered exceptional and totally ignored. Moreover, most simulation models have failed to demonstrate the trade-off situation between safety and productivity corresponding to that in the real production environment. Our view is that incorporating the situational variables into the simulation models yields a better understanding for the safety management of disturbances.

Simulation method and the study design

The objective of this study was to develop a simulation method by which knowledge can be acquired of human actions in disturbance situations and, on the basis of this, to develop criteria for the safety design of computer-controlled man–machine systems. The method is essentially based on a physical simulation model. Along with the physical model, however, a computer simulation model for the same process is developed simultaneously. These two approaches in simulation can be used to support each other.

The simulation model is developed so as to meet the following requirements:

— the test persons are not subjected to any danger or harm by the method,
— the method is realistic reflecting a packaging process or a part of it as used in production,
— the procedure is equally applicable to study disturbances induced by machine, product or human action,
— several alternative guarding solutions are available for studying the ranking order of preference as to various safety and efficiency criteria,
— the group of test persons are real operators of packaging machines,
— the method produces disturbance situations randomly; how the operator copes with them depends on his/her own 'cost–benefit' risk evaluation in that situation, which determines whether a safe or a hazardous practice is chosen,
— human operating modes and man–machine interactions in disturbance situations are measured quantitatively using motion analysis techniques.

The study is based on the presumption that the probability of the man–machine system's safe operation depends on (a) the necessity of entering the machine's hazardous zone (frequency, urgency, distance covered), and (b) the ease of hazardous access and the degree of difficulty of the safe operation.

Consequently, the simulation system is designed to manipulate the following variables quantitatively:

— frequency and urgency of the need to enter the hazardous zone,
— distance of the hazardous zone from the worker,
— degree of difficulty of entering the hazardous point, accomplished with machine guards and safety devices,
— degree of difficulty in defeating the machine's safety devices.

Preliminary experiments and results

For the development of the simulation model, the man–machine system of packaging work was selected. In earlier safety simulation studies, production-line work has seldom been considered. The experimental production system was constructed by using four belt conveyors (Figure 2). The conveyors were installed in the laboratory so that they constitute a rectangle and are connected with each other by slides and a carousel. During the experiments, empty boxes are continuously moving around. The conveyor system is controlled by a programmable logic unit. Various control devices and interlocking guards are coupled with the machine operation by the computer. Moreover, the logic system counts the number of packages 'produced'. Some disturbances can be programmed such as sudden stoppages of conveyors.

The first experiments were conducted in order to analyse human actions and hand motion patterns when removing a faulty package from the line. At one conveyor, there are two tunnel guards about 1 m apart. The package entering from the first tunnel may be faulty. In that case the operator should take the product from the line before the next production step in order to avoid major damage and production stop. When the action is delayed the operator may reach into the danger zone of the second tunnel in order to save production. As the speed of the conveyor is 0·35 m/s, the operator has nearly 3 s to remove the faulty box.

The hand motions were analysed by the Selspot computerized measurement system with two cameras and the infrared LEDs fastened to the hand. The measurement was triggered automatically when the photocell indicator at the end of the first tunnel received a special reflection from the faulty product. Three test persons participated. Their hand motion patterns were recorded, calculated and reported. A hand motion analysis program was developed for this experiment using the MultiLab command language provided by the manufacturer of the Selspot system.

The preliminary motion analysis results demonstrate the influence of the position

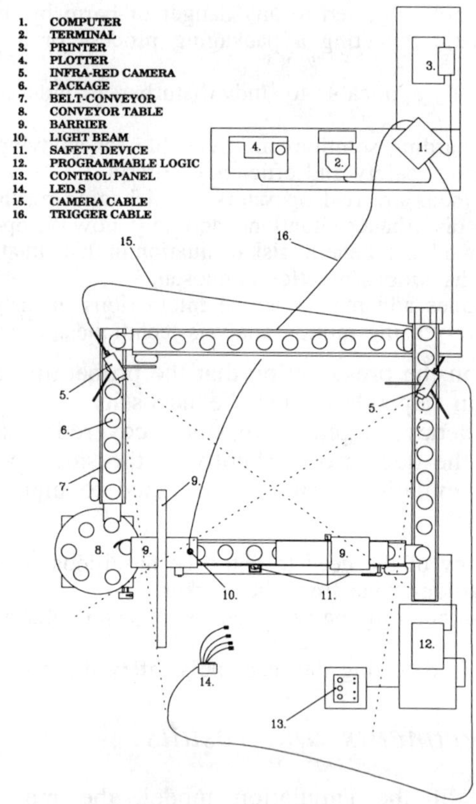

Figure 2. The laboratory installation for physical simulation experiments.

and the posture of the operator during the initiation of the disturbance. Figure 3 illustrates typical hand speed and motion path curves when the distance between the operator and the faulty product is only 30 cm. The maximum hand speeds with varying distances from 30 to 100 cm are presented in Figure 4.

The hand motion usually started within 0·5 s after the faulty product came from the tunnel. The rapid reaction in the experiments is natural because the persons were actually waiting for the broken package to come unlike in real life, where people have simultaneously many goals and tasks to co-ordinate. In all motions paths there was some irregularity and fuzziness, which tended to increase when the distance was lengthened. Also the speed values were widely scattered, even when the same person repeated the test at the same distance. One of the test persons maintained regularity in the motion patterns while the two other persons showed considerable irregularity with vague motion paths and high variations of speed. The maximum speeds at short distances (30 cm) were smaller than those for longer distances (60 and 100 cm). The greatest hand speeds were over 8 m/s.

Discussion and directions for future work

Simulation of production disturbances realistically in a laboratory is a controversial task, and therefore difficult. In the laboratory setting the variables should be strictly controlled and all the 'noise' factors eliminated. At workplaces, where disturbance-induced accidents happen, a complex and dynamic interaction of events and man–machine–environment elements takes place. Even in a research laboratory there are

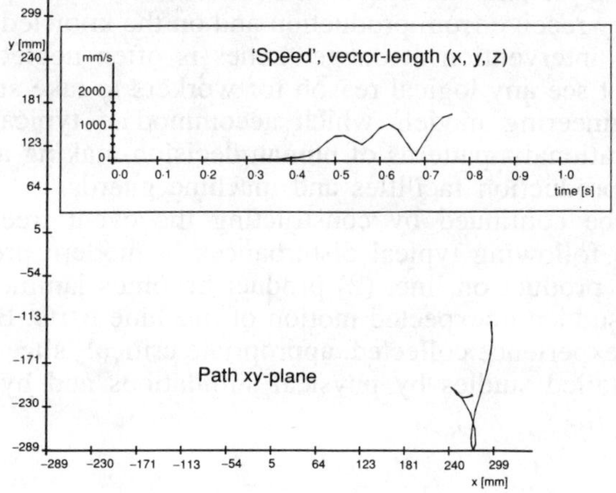

Figure 3. An example of some hand motion characteristics when a person is taking a faulty product from the experimental conveyor line starting at a distance of 30 cm from the line: speed at xyz-level and motion path at xy-level.

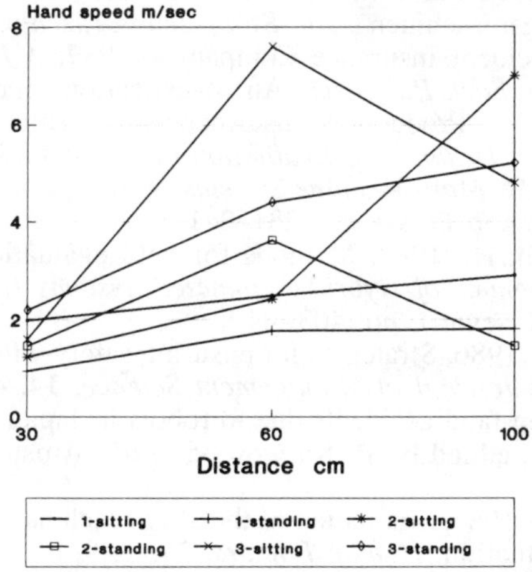

Figure 4. Maximum hand speeds measured while three persons were grasping the faulty product in sitting and standing positions and their distance from the line varied from 30 to 100 cm. Average values of three measurements are presented.

disturbing factors like the telephone ringing and persons walking through, looking at experiments or speaking. All those seem to cause changes in behaviour and actions, at least for some people. On the other hand, a highly controlled laboratory experiment may be unrealistic. Hence physical simulation in a laboratory is only one instrument in the safety study and incomplete by itself.

Simulation conducted in parallel with appropriate field studies may often be necessary to produce relevant new knowledge. The combination of computer simulation with the physical one is a promising new approach. For example, first the extremes of human actions in disturbances can be demonstrated by laboratory measurements, and then these extreme values can be transferred to a computer which shows by dynamic man–machine models if there are dangerous interactions.

The extreme values and vague patterns of human motion in disturbances are only one side of the coin. Actions are based on decisions which people make based on the information they receive from production and on the knowledge and experience they have. Human intervention at danger zones is often neglected by designers, because they do not see any logical reason for workers to take such actions. There is a need for engineering models which accommodate typical—and especially exceptional and irrational—patterns of human decision-making and risk behaviour for the design of production facilities and machine guards.

The study will be continued by constructing the event trees of the possible occurrences in the following typical disturbances in modern production systems: (1) faulty or loose product on line, (2) product becomes jammed and causes line blockage, and (3) sudden unexpected motion of machine parts. Based on the event trees and the field experience collected, appropriate critical 'slices' are cut from the event trees for detailed studies by physical simulations and by a dynamic CAD simulation model.

References

Booth, R., 1979, The design of effective machinery guards. *Design Studies*, **1**, 1, 12–14.

Corlett, E. and Gilbank, G., 1978, A systemic technique for accident analysis. *Journal of Occupational Accidents*, **2**, 25–28.

Defren, W., 1988, Methoden zur Ermittlung der Ursachen von Störungen dargestellt am Beispiel 'Verpackungsmaschinen'. *Die Berufsgenossenschaft*, **11**, 722–727.

Finnish Federation of Accident Insurance Companies, 1987, *A Fatality of a Maintenance Man at an Automatic Sack Palletizer*. An investigation report (in Finnish).

Harms-Ringdahl, L., 1986, Production disturbances—a safety problem in automatic machines. In *Analysis, Design and Evaluation of Man–Machine Systems*. Proceedings of the Conference *IFAC Man–Machine Systems*, Varese, Italy, 1985, edited by Mancini *et al.* (Oxford: Pergamon Press), pp. 231–241.

Helander, M. and Karwan, H., 1988, Methods for field evaluation of safety in a robotics workplace. In *Ergonomics of Hybrid Automated Systems I*, edited by B. Karwowski *et al.* (Amsterdam: Elsevier), pp. 403–410.

Jones, R. and Dawson, S., 1986, Strategies for ensuring safety with industrial robot systems. *Omega International Journal of Management Science*, **14**, 4, 287–297.

Nagamachi, M., 1988, Ten fatal accidents due to robots in Japan. In *Ergonomics of Hybrid Automated Systems I*, edited by B. Karwowski *et al.* (Amsterdam: Elsevier), pp. 391–396.

Pizatella, T. and Moll, M., 1987, Simulation of the after-reach hazard on power presses using dual palm button actuation. *Human Factors*, **29**, 1, 9–18.

Rieppo, K. and Häkkinen, K., 1985, Pakkauskoneiden työturvallisuus. [Safety of packaging machines, in Finnish]. *Pakkaus*, **7–8**, 49–52.

Shannon, R., 1975, *Systems Simulation—The Art and Science*. (Englewood Cliffs, NJ: Prentice-Hall).

Van Ballegooijen, A., 1979, Messungen an Mensch–Maschine–Systemen unter Berücksichtigung der Nachgreifgefahr (Mannheim: Berufsgenossenschaft für Nahrungsmittel und Gaststätten), *Symposium*, **6**, 79–89.

Winsemius, W., 1965, Some ergonomic aspects of safety. *Ergonomics*, **8**, 2, 151–162.

Winsemius, W., 1969, *On Single and Multipersonal Accidents* (Leyden: Netherlands Institute for Preventive Medicine).

87

The use of simulation for short-term decision making in production—a case study at a fabric dyeworks

B.L. MacCarthy

Department Production Engineering and Production Management, University of Nottingham, Nottingham, UK

Abstract

Successful simulation models in production have frequently been associated with high-level and long-term decision making. In recent years a large amount of simulation software has been aimed at the manufacturing sector, often with the claim that simulation can help short-term decision making. This chapter examines the use of simulation for short-term decision making in the context of a case study at a fabric dyeworks. Production management is complex in this environment because of the number of classes of input, large variations in batch size, the number of colour combinations and the number of routes through the plant. The construction of a simulation model for short-term production management is described. The difficulties encountered are highlighted. The wider issues and implications are discussed fully and a note of caution is sounded.

Introduction

Simulation, in particular discrete event simulation, is a powerful implement in the tool-kit of management scientists and operational researchers. In a survey of non-academic users representing a broad range of companies in the USA, Christy and Watson (1983) found that production was the most frequently quoted functional area in which simulation was applied. Nearly 60% of respondents in this survey identified production as an application area although no information is given on how simulation had been applied.

Simulation has frequently been employed successfully in the design and analysis of production systems. Case studies discussed in Hurrion (1986) demonstrate that simulation has often been associated with the successful introduction of new technology in the manufacturing sector. The power of simulation in new systems design, facilities layout and planning and in developing near-optimal, top-level management strategies, is generally accepted. A recent study by Brown (1988) at IBM illustrates the current state of the art. These application areas are normally associated with decision making in the context of long-term planning horizons.

In recent years the Information Technology (IT) revolution has generated much simulation software aimed at the manufacturing sector. Haider and Banks (1986) provide a good review. The trend in simulation software has been to move away from general purpose simulation languages towards special purpose simulators with more accessible user interfaces. The user may be sheltered, to some extent, from the underlying mechanics of the simulation. This trend raises key questions about how we use simulation now and will use simulation in the future in production environments. In particular, the possibility arises of simulation being of more immediate use in production. Can we envisage simulation being used frequently as a decision aid for short-term planning and ultimately for real-time decision making in production management? Simulation software vendors tend to support this view. However, because of the diversity of production environments, caution is necessary in considering this question.

Much of the literature from software vendors views the manufacturing sector with 'automation-tinted spectacles', where products take clear routes through perfect processes with well-defined scheduling rules always applying (and always applied). Although advanced automated manufacturing systems may be a reality in certain sectors and a justifiable goal in others, the manufacturing sector covers a 'multitude of sins' where this technology may be inappropriate.

The textile and clothing industries in the UK, for example, are very large employers. However, because of the volatile nature of the market and the inherent variability of raw materials, fully automated facilities are rare and these are likely to remain labour intensive industries. Can we envisage simulation models helping production management for short-term decisions in this wider context, where management decision making may be far more complex than in the highly automated sectors?

This chapter addresses these issues in the context of a case study at a fabric dyeworks and draws conclusions relevant in the wider context.

A case study at a fabric dyeworks

Textile fabrics are produced most frequently by weaving or knitting processes, but usually require further processing for specific end uses. The most common is dyeing to a particular shade. Fabrics may also be scoured or heat-set, or less frequently they may undergo surface treatments such as brushing or calendering. Because of the technical skill involved in all these processes, they are often performed by specialist companies.

The company in this case study is a fabric dyeworks which performs a variety of dyeing and finishing operations on a range of weft-knit and woven fabrics. It is a medium-sized company operating in this field and works on a commission basis. This means that it accepts fabric from any of its recognized customers at any time and charges them for storage. Fabric is processed when an instruction is received from the customer. The quantity requested may be less than the quantity stored. Fabric is never processed unless such an instruction is received. Indeed the customer is at liberty to send the fabric elsewhere for processing if it so desires.

Production control is complex in this environment because of the number of classes of input, large variations in batch size, the number of colour combinations and the number of routes through the plant. Management sought a model which might aid their day-to-day decision making.

Overview of the production system

The production system is indicated schematically in Figure 1. All fabric must be prepared before entering into process. This entails lapping fabric in a suitable manner and slitting tubular fabric so that it is a single thickness. Most fabric proceeds directly to dyeing which may be done on three classes of machine: JET,

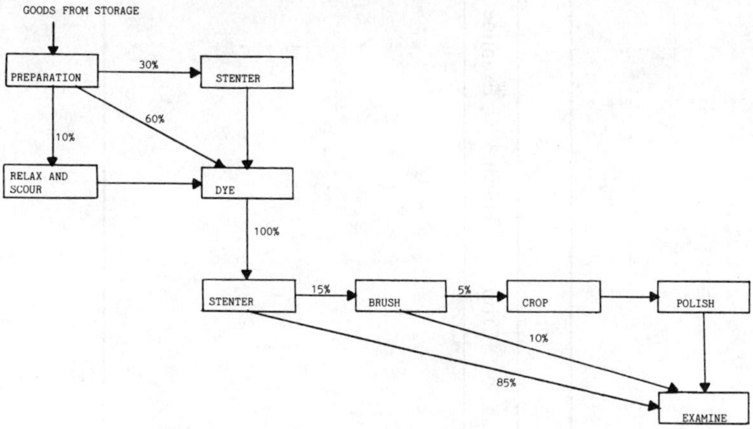

Figure 1. Fabric routes through the plant.

BEAM or WINCH. The details are unimportant here except to say that this implies known constraints on what can be dyed where. Once dyed, all fabric is dried and heat-set on a machine known as a stenter. Some fabric is relaxed and scoured prior to dyeing. However, this must be dried and heat-set again after dyeing. Once dyed and heat-set most fabric is examined and dispatched to the customer. However some classes of fabric may be cropped and polished before being examined. Figure 1 gives some indication of the volume of production flowing through each process.

The company processes 11 different classes of fabric as indicated in Table 1. These vary in their routes through the plant, their processing characteristics, handling methods, chemicals used, processing times, etc. A broader classification, useful for some purposes, is dress lace (1–3), curtain lace (4) and fabrics (5–11).

Prior to each week, the production control department produces a programme of jobs which should begin processing during the week. This is dictated in the main by due-dates, but consideration is also given to other priorities (based on customer and level of profit), current loading and available resources. This determines what is prepared. The dyeing controller has freedom to load the dyeing machines from what is prepared taking into account plant efficiency and quality. Similarly the stenter controller has the freedom and responsibility to operate in an efficient and quality-oriented manner. The loading of the dyeing machines in particular requires skill and judgement because of many constraints.

Objectives of the simulation and choice of software

From initial discussions with senior management it was felt that a simulation model might be developed to assist in:

(i) achieving production targets and meeting due-dates,
(ii) efficiently utilizing plant resources,
(iii) showing the effect of anticipated changes in production targets,
(iv) showing the effect of changes in plant or manpower resources.

Because of the complexity of the operation and time constraints the study was undertaken as a feasibility exercise. Initially a simplified view of the company's operations was considered. It was decided to attempt to model the preparation, dyeing, drying and heat-setting (one operation) operations only as all fabric went

Table 1. Classes of input.

Product		Operations								
		Preparation	Relax and scour	Heat-set	Dye	Dry	Brush	Crop	Polish	Examine
1	Helenca lace (Dress lace)	*	*		*	*				*
2	Leavers lace (Dress lace)	*	*		*	*				*
3	Raschel lace (Dress lace)	*	*		*	*				*
4	Curtain lace	*	*		*	*				*
Fabrics										
5	Warp knit crush panne	*		*	*	*	*	*	*	*
6	Warp knit smooth panne	*		*	*	*				*
7	Flat warp	*		*	*	*				*
8	Brush nylon	*		*	*	*	*			*
9	Weft knits and Tricelon	*			*	*				*
10	Woven polyesters type 1	*	*		*	*				*
11	Woven cottons	*			*	*				*

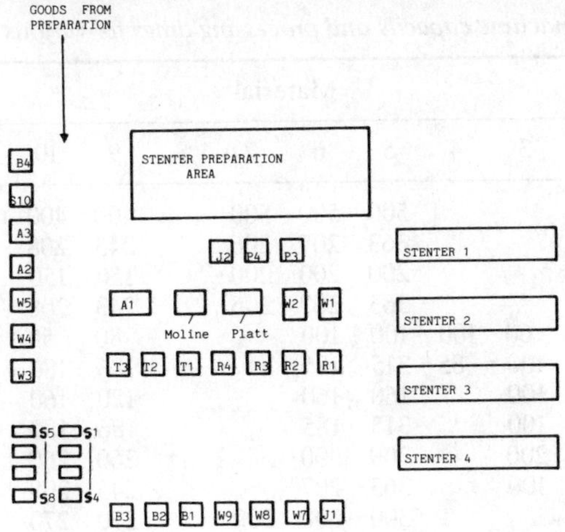

Figure 2. Schematic layout of the dyeing and stenter plant.

through these processes. The most complex aspect was dyeing machine loading. It was felt that a more detailed model could be developed incorporating all possible routes, if this aspect was adequately modelled. The initial study of the system showed that preparation was not a constrained or limited resource and was handled easily by the warehouse staff. Thus it was not necessary to include preparation explicitly in the initial model.

An analysis of the production system showed many aspects to be deterministic. However the arrival of jobs and some of the process times varied in a manner which could best be modelled stochastically. The simulation language SIMAN (Pegden, 1987) was used to model the system. SIMAN is an enhanced process-interaction simulation language, incorporating a network representation of the system and includes many features specifically designed for manufacturing systems (e.g. conveyor blocks).

As well as the power of a full, general purpose simulation language it has two facilities which make it particularly suitable for complex systems. First it has an easily accessible event orientation which can be used for complex logic not easily modelled by the blocks provided. Secondly, the model file is separate from the experimental file, allowing relatively easy modification of experimental conditions for different runs. Both of these features were useful in this study. SIMAN can be used for combined discrete/continuous models, although this facility was not found to be useful.

The simulation model

A plant layout of the dyeing and stentering facilities is shown in Figure 2. There are 37 dyeing machines and 4 stenters. Prepared fabric of the 11 classes of input arrrives as jobs for dyeing. Each job has four attributes: type of material, weight, width and shade. The width is relevant to stentering only and is discussed later.

Table 2 shows the maximum capacity of each dyeing machine and which classes of input can be processed. The dyehouse controller tries to load jobs on their closest capacity machines. However, constraints are introduced by the shade specified, the available machines, weight and the existing queues of jobs at each machine.

Table 2. Dyeing machine capacity and processing times for various classes of input.

Dyeing machine	Material 1	2	3	4	5	6	7	8	9	10	11	12	Dedication
R1					500	500	500		400	400			
					363	207	206		243	208		106	C
R2 R3 R4					200	200	200		150	150			
					363	207	206		243	208			C
T1	100	160	60	100	100	100			80	50			C
	182	100	100	85	315	185			186	180		85	C
T2	160	160	100		160	160			120	160			
	200	200	100		315	185			186	180		85	C
T3	300	300	200		300	300			260	260			
	260	260	100		363	207			243	208		106	C
JI J2					300	300	300		270	270			
					363	207	206		243	208		106	B/C
P3 P4									500	500			
									200	200		100	C
A1	270	270	200	400		400					400		
	200	400	200	280		210					352	120	W
A2	150	150	100	150		150					200		
	267	267	267	267		195					339	95	W
A3	150	150		150		150					200		
	267	267		267		195					339	95	
PLATT	250	250		250							280		
	280	280		280							352	120	C
MOLINE	150	150		150							100		
	280	280		280							352	120	C
B1 B2 B4	100	100	500	100									
	267	350	267	267								95	B/C/W
B3	30	30	20	20									
	267	350	267	267								95	C
W1 W2 W5	300	300	300	300				275			270		
	320	320	320	320				552			393	140	B/C/C
W3	60	60	60	60				45			60		
	267	350	267	267				497			337	95	C
W4	200	200	200	200				180			180		
	207	267	267	267				497			337	95	B/C/C
W7 W8 W9	120	120	120	120				90			120		
	267	267	267	267				497			337	95	B/C/C/B
S1 S2 S3 S4	30	30	15	30							25		
	245	245	245	245							296	85	C
S5 S6 S7	20	20	15	20							20		
	245	245	245	245							296	85	W
S8	35	35	20	35							35		
	245	245	245	245							296	85	W
S10	20	20	15	20							20		
	245	245	245	245							296	85	W

The top figures indicate maximum capacity (in kilograms) of each machine for each material. The bottom figure indicates dyeing time for any size of load or any shade. Column 12 indicates boil-out time and the dedication column indicates colour dedication (B, black; W, white, C; colour).

Shade specification

To achieve the specified shade a unique formulation of chemicals and processing conditions, known as the dye recipe, is required. This is specified by the dyeing laboratory before preparation. The range of shades dyed is very large. However, in a season (about six months) some shades are more prevalent. Laboratory files contain many thousands of recipes. The model tried to mimic the dye recipe process in determining the shade attribute.

Each shade is made up of three 'primary' colours—red, blue and yellow. Each of these 'primary' colours could be applied in a range from 0 to 10 parts. The shade white is made up of zero parts of each primary colour, whilst black is made up of 10 parts of each colour. Thus, excluding black and white, there are 1329 possible shades in the model which was roughly equivalent to the number of shades dyed in a season. The darkness of each shade is determined by the sum of the number of parts of each 'primary' colour used.

Loading the dyeing machines

This scheduling problem is complex as it entails many constraints. However management feel, it has a considerable effect on the efficiency of the plant, achievement of delivery dates and, not least, quality. From an analysis of the system, including discussions with key personnel, a number of general principles became apparent.

(1) Some machines are dedicated to white and some to black (see Table 2). This is justified both on the grounds of quality and efficiency. White is likely to be easily contaminated by other shades whilst black is likely to contaminate other shades. In addition white and black represent a significant proportion of dyeing production.
(2) Machines which are not dedicated to black or white, process jobs in order of increasing darkness, i.e. go from light shades to dark shades. This is to avoid colour contamination.
(3) The preferred light to dark sequence may be reversed provided the movement is relatively small (see below).
(4) Any machine can violate the light to dark sequence provided it goes through a process called boiling out. This may be costly, but is used frequently for versatile machines which are not dedicated to black or white.

When a job arrives the subset of dyeing machines which can process that job, in terms of machine compatibility and capacity are identified. The subset of machines is then checked for dedication to black or white.

If the specified job shade is black or white and a single machine is available, of sufficient capacity, then the job is either processed or queued if preceding jobs are incomplete. If more than one machine is available then the closest capacity machine is chosen. Further ties are resolved by queue length and finally by a random choice.

If the specified job shade is not black or white then the subset of machines is reduced further by the darkness level of the last job. Only those machines which are lighter or are darker within 1/5 of the level of the shade to be loaded, are included.

Ties are resolved by capacity first, then queue length and finally by a random choice. If no machine is available a boil-out is scheduled on the machine with the darkest shade and nearest in capacity.

All queues are processed using a First in First Out (FIFO) discipline.

Process times in dyeing

For dyeing, set-up times were negligible as chemicals were prepared in advance and loading and setting of machines were mostly automatic. The process time for each machine on each class of fabric is given in Table 2 (the bottom figure). The 12th column indicates the time taken to boil out the machine. These times are expected times and are independent of shade.

In practice, however, particularly for less frequent shades and for specific customers, the job is not completed in dyeing until it has been checked with standards. This often entails addition or subtractions of colour. The result is that process times may be extended considerably. However, in the initial model this stochastic element in process times was not included because the company did not maintain any past records on the frequency of these occurrences. With SIMAN this could easily be included if sufficient data were available.

Set-up and process times and job loading on stenters

There are four stenters, one restricted to curtain lace, one for dress lace and two for fabrics. For stentering there are three critical variables: temperature, width and chemicals to be applied by pad rollers. Temperature is determined primarily by the type of material and does not affect either process time or set-up time. Both other variables affect the set-up time for each job. Process time is determined by the length of fabric in a job.

The stenter room controller paid little attention to the sequencing of jobs, except to ensure that stenters were kept busy. The rationale for this was that stentering time was small in comparison to dyeing time (30 min against 3 h, for example). Set-up times were smaller still and frequently zero if no width changes were required. Stenter drivers were allowed almost complete control over sequencing and this was usually based on minimizing the frequency of width changes. A remarkable consequence of this was that process times for actual jobs were impossible to determine accurately as records (from bonus payments) were based on finished meterage which is not directly related to the unfinished weight of input jobs. Finished length is affected by processing conditions, shrinkage, shade and fabric construction. In addition no information whatsoever was available from historical records on set-up times.

As a first approximation in the initial model it was necessary to estimate an average process time per kilogram for each broad class. Set-up times were ignored. Again, if accurate historical data could be generated then the SIMAN model would require only minor modifications to incorporate these.

Data and validation

Much of the deterministic data has been given in the preceding sections. The business tends to have a strong seasonal element particularly with respect to certain classes of input and certain shades. The volume of curtain lace, for example, tends to be fairly constant whereas the volume of woven cottons shows far more variability. For this reason it was felt that collecting data by studying the system over a short period would be inappropriate. Historical records provided a more reliable source for input distributions of the stochastic elements in the system.

Problems arose with the mass of existing data because of the amount and level of detail provided for each class of input. The production records for a four-month period were analysed. Basically attributes of type, shade, weight and width had to be assigned to each job in a manner which accurately reflected past records.

First, frequency distributions were obtained for the three broad classes of input: dress lace, curtain lace and fabrics. This was relevant to the company's view of the business and the markets which they served. For dress lace three further distributions were obtained (classes 1–3) and for fabrics seven further distributions were obtained (classes 5–11). For each of these classes, distributions were formed reflecting the proportion of black, white or colour. In a similar manner the shade attribute indicating a colour was determined from a distribution reflecting the proportion of the 'primary' colours. Similarly the attributes of weight and width were assigned from distributions reflecting these characteristics in each class.

It was not possible to specify the arrival distribution of jobs to dyeing accurately. A negative exponential arrival distribution was assumed. The method of setting the parameter is discussed below. Specifying the initial state caused some problems. In our experience the real system was never observed in a completely unloaded state. Simulating with typical initial conditions was ruled out both on theoretical grounds (Pidd, 1988) and on practical grounds, as it was difficult to establish what a typical initial state might be. After some experimentation a run-in period of 200 jobs was used. During this period statistical output was not gathered.

For validation of the initial model the following aspects were considered:

(1) job throughput,
(2) job time in the system,
(3) behaviour of the queues in dyeing,
(4) utilization of the dyeing machines.

An appealing aspect of SIMAN is the ease with which modifications can be carried out and statistical output gathered. The arrival rate of jobs to dyeing was used as a 'tweaking' parameter. Numerous experiments were made to determine a mean inter-arrival interval which gave a throughput of 240 jobs per week, the average weekly throughput noted in the analysis of the historical records. It was found that with an 8-h per shift, 2 shift per day, 5 day per week system, a mean interval of 20 min was adequate. This is not claimed as hard validation but it was pleasing to note that this also gave a maximum and minimum job time in the system of approximately 4 days and 1 day, respectively. This agreed well with experience. The figures quoted relate to 100 replications of the system simulated over a week.

The dyeing controller's stated aim was to keep queues of less than 4 days work in front of dyeing machines. A study of the real system showed that the number of jobs in front of machines rarely exceeded eight and was rarely less than two. In simulation runs it was found that dyeing machines had maximum and minimum queue lengths of seven and zero jobs, respectively. This seemed reasonably close to reality.

The overall dyeing machine utilization observed in simulation was approximately 67%; this seemed low. The real system was believed to be closer to 80%. A possible explanation for this was the underestimate of process times, as discussed earlier. In addition, re-dyes (rejects from examination on the grounds of shade) were not included. These could be included in a more detailed model.

Uses, limitations and further development of the model

Modifying the input distributions for shade had interesting effects in simulation. A comparison was made of the control system (with the input data described above) with a system where all dress lace was black, all curtain lace was white and all fabrics were of one colour (all component parts set at 5). In the modified system throughput time reduced by about 10%. This is clearly due to less time being devoted to the non-productive boiling-out process. Overall machine utilization reduced very marginally (67·4% to 67·1%). The rather surprising result, that colour dedication had only marginal effects on the behaviour of the system, was the subject of some debate. This indicated the need for further experimentation with different shade profiles.

In contrast, varying the proportions of the main input classes had a very significant effect on throughput time, queue length and overall and individual machine utilization. In particular, the relative proportion of fabrics had a major effect. Increasing this proportion affected the utilization of some versatile machines with average queue lengths increasing rapidly. Machine utilization of high temperature JETS in particular approached 100%. These effects were clear and dramatic and provoked discussion. In particular, a factor which had not been included in the model—splitting large jobs to avail of low capacity machines—was discussed.

Further interesting experimentation on changes in plant resources showed how robust the company was. Prolonged breakdown of some versatile machinery could be tolerated with some increase in throughput time and in queue time. This accorded well with experience.

The current model reflects the real system moderately well. Further modifications and improvement might include:

(1) stenter set-up and process times,
(2) all routes through the plant,
(3) the effect on dyeing process times of chemical additions,
(4) the effect of re-working,
(5) an analysis of historical records over a longer period,
(6) more experimentation.

However, it is as well to bear in mind the more general points noted in the next section.

The model in its present form is useful to the perceptive analyst who is endeavouring to understand the system and make recommendations for changes in operating policy. Like many models, however, one suspects its major benefit is the knowledge gained about the real system by both the analyst and personnel involved. Is this the real value of simulation?

Wider implications

Simulation has wide appeal and a great expectancy of success amongst users, at the outset at least. Its generality of approach and application area, is responsible both for its appeal and its limitations. In the introduction, the usefulness of simulation in long-term, high-level planning has been noted. This may be because the longer term and higher level implies less complex decision making with fewer alternatives and clearer measures of effectiveness and performance.

This contrasts with the day-to-day production environment where fuzziness, fluidity and complexity abound. The primary objective in this study was to develop a model in just such an environment for the day-to-day management of the dyeworks. Although the basic model could be enhanced along the lines suggested in the previous section, it is unlikely that a generally useful model for short-term decision making is possible in this case. However, the case-study does raise important issues which may be of more general applicability in this type of model.

(1) *Difficulties in modelling may indicate difficulties in implementing and interpreting the results of a simulation. The more severe the difficulties the more likely the lack of success*—A number of examples of this were evident in the case study. The dyeing machine loading policy needed much clarification from more than one source. The implication is that actual policy may be fluid. The stenter loading policy was unknown to the nearest manager. Although the actual system may work, perhaps even optimally,

the simulation model cannot be expected to be illuminating where the system rules are fluid.

(2) *Lack of data may be symptomatic of more fundamental problems*—In our case, management could not provide information on the frequency of additions to dyeing. In this situation the analyst has identified logic which requires data which does not exist. Simulation may be too sophisticated a tool, in this case, to help. It is unlikely to illuminate further. More fundamental management practice must first be encouraged—basic production control, basic management information gathering and reporting.

(3) *How the user might interact with the model must be considered*—For a model of this type it is unlikely that the analyst will be the user. The user will not usually be skilled in programming, simulation or concepts of statistical experimentation. One cannot expect the user to program at any level. For instance, in the dyeworks model, a change in queue discipline from FIFO to priority would require a change to a model block in the SIMAN code. A more complex change might require programming in FORTRAN. This would clearly be inappropriate.

The model must be menu driven. The design, specification and implementation of such a system will be inherently costly except for the simplest systems. However, for the simplest systems, simulation may not be necessary.

(4) *Graphics may help but are not a panacea*—The use of graphics in simulation has rekindled its appeal and is being aggressively promoted by the software vendor. It may be used either as an aid in modelling or in interpreting output (Pidd, 1989). However graphics cannot clarify an inherently fuzzy, fluid system.

(5) *The more limited the objectives the more likely the success*—There are two ways in which this model might be developed to help management with this consideration in mind. Both would involve the restriction of objectives:

   (i) a model to evaluate dyeing machine scheduling policy. This would need critical measures of performance to be devised, different scheduling rules to be implemented and a simpler user interface. The model might be made to search for a good scheduling policy. There are a number of examples of good scheduling models using simulation in a creative fashion (Dale and Dewhurst, 1983; Udo Graefe and Chan, 1986; Siegel, 1987).

   (ii) a simplified, top-level model with the sole objective of determining throughput time. This would handle the major classes of input only and would be designed to give an estimate of throughput time under various conditions. This would help management with a current problem, giving realistic delivery dates to customers, particularly for smaller batches and unusual shades.

   Both of these models would benefit from the separation of the model file from the experimental file in SIMAN. This also allows the possibility of using real and up to date production control information to improve model accuracy.

Conclusions

Simulation studies usually provide some understanding of the complexity of underlying systems. However, if this is their only benefit they will turn out to be expensive and often inappropriate exercises. The thrust of this chapter has been to discuss, in the context of a specific case study, where and how simulation may be applied effectively in production management. Specifically, aspects such as difficulty in modelling, lack of data, the user interface and the level of detail need to be addressed. Problems encountered may indicate more fundamental difficulties in the real system and simulation may be ill-advised.

The specific case study highlighted many 'grey' areas. These were neither criticisms of management nor an indication that more could be achieved. They were simply an acknowledgement of the reality of many production environments which cannot be ignored.

It is incumbent on those who promote quantitative approaches in management decision making to consider how, with the proliferation of software, the tools may be profitably used. It is hoped that this chapter moves some way towards this in the context of production management.

References

Brown, E., 1988, Using different types of simulation to design an IBM facility. *Industrial Engineering*, **20**, 6, 22–26.

Christy, D.P. and Watson, H.J., 1983, The application of simulation: a survey of industry practice. *Interfaces*, **13**, 5, 46–52.

Dale, B.G. and Dewhurst, F., 1983, A microcomputer simulation of an automated non-ferrous sand casting foundry. *OMEGA*, **11**, 6, 599–606.

Haider, S.W. and Banks, J., 1986, Simulation software products for analysing manufacturing systems. *Industrial Engineering*, **18**, 7, 98–104.

Hurrion, R.D. (Ed.), 1986, *Simulation, Applications in Manufacturing* (Berlin: Springer).

Pegden, C.D., 1987, *Introduction to SIMAN* (State College, PA: Systems Modelling Corporation).

Pidd, M., 1988, *Computer Simulation in Management Science* (New York: Wiley).

Pidd, M., (Ed.), 1989, *Computer Modelling for Discrete Simulation* (New York: Wiley).

Siegel, S.L., 1987, Simulation of scheduling rules helps decision-making on various objectives in a manufacturing plant. *Industrial Engineering*, **19**, 10, 40–47.

Udo Graefe, P.W. and Chan, A.W., 1986, A production control aid for managers of manufacturing plants. In *Simulation, Applications in Manufacturing*, edited by R.D. Hurrion (Berlin: Springer), pp. 213–226.

PART VIII
COMPUTER-AIDED MANUFACTURE

Computer-aided manufacture

The papers in this section consider the techniques, applications and developments in fields such as computer-aided design (CAD), computer-aided process planning (CAPP) and computer-integrated manufacture (CIM). The use of computers throughout manufacturing systems has grown enormously, and new areas and applications are constantly being exploited.

In the CAD/CAM area Devgun describes a hybrid approach to feature representation and extraction in CAD models. Applications considered include design and manufacture of jewellery, Lim *et al.* and Grieve *et al.* describe a system for the application of two part adhesives using robots.

CIM implementation is undertaken as organizations pursue improved productivity and efficiency in their manufacturing systems. Vernadat reports on a methodology which covers the analysis of an organization, the modelling of its information system, and the design of a data base to support the operation of the organization's activities. Lung and Love investigate the benefits of a distributed planning and control concept, as opposed to the traditional centralized control, approach to CIM systems.

SECTION VIII.1. CAD/CAM

88

Vehicle scheduling models in automated production systems

J. Pemberton Cyrus

Department of Industrial Engineering, Technical University of Nova Scotia, PO Box 1000, Halifax, Nova Scotia, Canada B3J 2X4

Abstract

The objective of this chapter is to describe three automated manufacturing problems using vehicle scheduling models. The assignment of jobs to machines is modelled as a vehicle scheduling problem with time window and precedence constraints. The scheduling of tools, palettes and fixtures is modelled as a vehicle scheduling problem with time windows. The automated guided vehicle scheduling problem (VSP) is modelled similarly. These models are all solvable with vehicle scheduling algorithms. Some restrictive assumptions must be placed on the models in order to fit the VSP mould, but overall the solutions will be sufficiently useful to make decisions in an automated production environment.

Introduction

In an automated manufacturing system, the production schedule may be constructed in a hierarchical manner; typical approaches are described by Buzzacott and Yao (1986) and Jackson (1987). The solution to each problem in the hierarchy provides inputs to the problems at lower levels. The following hierarchy of problems is convenient for vehicle scheduling models. At the highest level, customer demands are exploded into discrete part requirements with due-dates and machine requirements. The first problem solved is the multi-processor scheduling problem (MPSP), where each job is assigned a start time on a specific machine. This schedule is used to determine a range of start times (optimal to the MPSP) which are used in the next level. A schedule is then determined for fixtures, tools and pallets. This schedule minimizes the number of each of these resources in use, such that all requirements of the MPSP solution are satisfied. The output of this level is an assignment of tools, fixtures and pallets to job-machine pairs in the production schedule from the previous level. At the lowest level, a schedule for the automated guided vehicle system (AGVS) is determined. This problem seeks to find the minimum number of AGVs which are required to meet the materials handling requirements of the jobs, pallets, fixtures and tools.

The objective of this chapter is to describe each of these manufacturing problems using a vehicle scheduling model. Vehicle scheduling problems have been the subject of an intense search for solutions, because of the immense potential savings in distribution costs which small improvements in solutions can achieve. The result of this research is a large number of effective solutions to large vehicle scheduling problems (Assad, 1988). By modelling manufacturing problems as vehicle scheduling problems, this work becomes immediately accessible.

Vehicle scheduling problems

The vehicle scheduling problem (VSP) can be stated as follows. Given is a set of jobs, each with an origin, a destination, an earliest start time, a latest start time and a duration; also given are empty travel times from the destination of one job to the origin of another. The objective is to find a set of schedules for the jobs such that the number of vehicles required is minimal and the total cost of empty trips is minimized. A job is usually a trip between two locations, but the job definition may be varied to model production jobs in a manufacturing system, delivery trips, crew schedules, airline flights and school bus tasks. In addition to the time windows on jobs, additional constraints allowing for some lateness and limiting the waiting time between jobs may be added. The objective function may also be changed to include waiting time costs and penalties for vehicles which do not return to their origins.

The VSP has been applied in a number of areas, including the scheduling of cargo ships and tankers (Dantzig and Fulkerson, 1954), airliners (Etschmaier and Mathaisel, 1985), AGVs (Kusiak and Cyrus, 1984) and school buses (Desrosiers *et al.*, 1984); applications have also been made in the postal service for the routing and scheduling of letter carriers and in the soft drink industry.

It is well known that vehicle routing and scheduling problems are very difficult to solve, and much work has been done in establishing their computational complexities. The most used approach to the VSP has been the solution of a restricted or relaxed problem by linear and integer programming methods such as column generation, branch-and-bound and lagrangean relaxation. Recently, the first optimal solutions to large problems were obtained by the column generation methods of Desrosiers *et al.* (1984). Heuristic methods have been proposed for the VSP with various constraints (see e.g. Cyrus, 1988; Solomon *et al.*, 1988).

The multi-processor scheduling problem (MPSP)

Consider a problem at the first optimization level: multi-processor scheduling. There is a set of machine clusters, referred to as flexible manufacturing cells in the context of Flexible Manufacturing Systems (FMSs); each cluster contains a number of machines of the same type and a list of jobs is available. Each job has an assigned machine type, processing time, earliest available time and deadline. There are change-over and materials handling transfer times between jobs. There are costs of change-over and transfer between jobs. There is an earliest available time and latest available time for the facility as a whole (the time window of the work shift or time horizon). There is a precedence matrix relating all jobs. The objective is to schedule all jobs on the machines in order to minimize the number of machines of each type and (secondarily) the total costs of change-over and materials handling transfer.

The costs of job lateness are assumed to be infinite. The transfer time between jobs in the same cluster is assumed to be much smaller than any transfer time between clusters. The time windows and precedence matrix are consistent, i.e. any precedence restriction between two jobs is implementable without violation of the time window constraints on the jobs.

Formulation

This problem can be formulated as an extension to the VSP with time windows. Machines and jobs in the MPSP correspond to vehicles and jobs respectively in the VSP. The VSP travel cost is the MPSP change-over cost. An additional set of scheduling constraints must be added to cover the precedence constraints. The parameters are defined as follows: J = the set of jobs; d_i = the processing time of job i, $i \in J$; b_i = earliest start time of job i, $i \in J$; e_i = latest start time of job i, $i \in J$; M = the set of machines; $[T_b, T_e]$ = the horizon of availability of the machines; K = the set of clusters, and the number of machine types = $|K|$; $\mu(k)$ = the cluster to which machine k belongs, $\mu(k) \in K$; k_i = the cluster identifier to job i, $i \in J$, $k_i \in K$;

$$u_{ij} = \begin{cases} \text{the change-over time from job } i \text{ to job } j & \text{if } k_i = k_j \\ \infty & \text{otherwise,} \end{cases}$$

$$t_{ij} = \begin{cases} \text{the transfer time from job } i \text{ to job } j & \text{if } k_i \neq k_j \\ \text{the material handling transfer time within cluster } k_i & \text{otherwise,} \end{cases}$$

$$r_{ij} = \begin{cases} 1 & \text{if job } i \text{ must precede job } j \\ 0 & \text{otherwise.} \end{cases}$$

The costs are defined as: $\alpha(k)$ = the unit cost of change-over time in cluster k; $\beta(k)$ = the unit cost of transfer time in cluster k; ζ_k = the fixed cost of machine k over the horizon; and

$$c_{ij} = \begin{cases} \alpha(k_i) \cdot u_{ij} + \beta(k_i) \cdot t_{ij} & \text{if } i \in J \text{ and } j \in J,\ i \neq j \\ \zeta_{\mu(i)} & \text{if } i \in M \text{ and } j \in J \text{ and } k_j = \mu(i),\ i \neq j \\ 0 & \text{if } j \in M \text{ and } i \in J \text{ and } k_i = \mu(j),\ i \neq j \\ 0 & \text{if } i \in M \text{ and } j \in M,\ i \neq j \\ \infty & \text{if } i = j \\ \infty & \text{otherwise.} \end{cases}$$

The variables are: s_i = the start time of job i, $i \in J$; and

$$x_{ij} = \begin{cases} 1 & \text{if job } i \text{ precedes job } j \text{ in the schedule, } i \in j \text{ and } j \in J \\ 1 & \text{if job } j \text{ is the first job in the schedule of machine } i,\ i \in M \text{ and } j \in J \\ 1 & \text{if job } i \text{ is the last job in a machine schedule, } i \in J \text{ and } j \in M \\ 0 & \text{otherwise.} \end{cases}$$

An integer programming formulation follows. Constraints (2) and (3) are flow conservation restrictions. Constraint (4) restricts links between jobs in the same cluster (N is some suitably large number), constraint (5) implements precedence requirements on travel between clusters, and (6) gives the processing start time windows for the jobs. It is assumed that b_i and e_i are chosen such that $T_b \leq b_i \leq e_i \leq T_e - d_i$ $\forall i \in J$, in order to respect the machine time horizon.

$$\text{(MPSP): min} \quad \sum_{i \in J \cup M} \sum_{j \in J \cup M} c_{ij} x_{ij} \tag{1}$$

$$\text{s.t.} \quad \sum_{i \in J \cup M} x_{ij} = 1 \qquad \forall j \in J \tag{2}$$

$$\sum_{j \in J \cup M} x_{ij} = 1 \qquad \forall i \in J \tag{3}$$

$$s_i - s_j + d_i + t_{ij} + u_{ij} \leq N \cdot (1 - x_{ij}) \qquad \forall i \in J, j \in J \tag{4}$$

$$(s_i - s_j + d_i + t_{ij}) \cdot r_{ij} \leq 0 \qquad \forall i \in J, j \in J \tag{5}$$

$$b_i \leq s_i \leq e_i \qquad \forall i \in J \tag{6}$$

$$x_{ij} \in \{0, 1\} \qquad \forall i \in J \cup M, \; j \in J \cup M \tag{7}$$

Solution methods

The formulation (MPSP) gives a solution with an underlying assignment problem structure. MPSP can also be formulated with an underlying network flow problem. These two structures suggest approaches to a solution.

Desrosiers *et al.* (1984) formulate the VSP with time windows as a network flow problem with time window constraints on the nodes. They relaxed one of the flow constraints to give a set partitioning master problem and a shortest path with time windows (SPTW) subproblem. The SPTW was solved by dynamic programming, making use of dominance rules to increase the solution speed by orders of magnitude. The VSP was then solved by a column generation algorithm. This algorithm was able to solve problems optimally with more than 250 jobs. The MPSP can be similarly formulated, with time window and precedence constraints on nodes. This problem can then be solved by a similar column generation algorithm, but with a different subproblem: the shortest path problem with time windows and precedence constraints. The addition of precedence constraints will have the effect of increasing the amount of dominance in the solution to the shortest path problem, thereby making a practical modification of the SPTW algorithm and dominance theorems possible.

In Cyrus (1988), the VSP is formulated with an underlying assignment structure as in (MPSP). This problem is solved approximately by a number of heuristics. The practical upper limit on problem size is similar to that of the assignment problem, i.e. thousands of jobs. A modification of the *Build-Graph* heuristic to solve the MPSP is as follows.

Define $f(\mathbf{S})$ as the solution to MPSP, given a feasible set of start times $\mathbf{S} = \{s_i : b_i \leq s_i \leq e_i, (s_i - s_j + d_i + t_{ij}) \cdot r_{ij} \leq 0, i \in J, j \in J\}$. The function $f(\mathbf{S})$ is given by the solution to an assignment problem with cost matrix C defined as

$$c_{ij} = \begin{cases} \alpha(k_i) \cdot u_{ij} + \beta(k_i) \cdot t_{ij} & \text{if } i \in J, j \in J, i \neq j \text{ and } s_i - s_j + d_i + t_{ij} + u_{ij} \leq 0 \\ \zeta_{\mu(i)} & \text{if } i \in M \text{ and } j \in J \text{ and } k_j = \mu(i), i \neq j \\ 0 & \text{if } j \in M \text{ and } i \in J \text{ and } k_i = \mu(j), i \neq j \\ 0 & \text{if } i \in M \text{ and } j \in M, i \neq j \\ \infty & \text{if } i = j \\ \infty & \text{otherwise.} \end{cases}$$

Define $E = \{(i, j) : b_i - e_j + d_i + t_{ij} + u_{ij} \leq 0, (b_i - e_j + d_i + t_{ij}) \cdot r_{ij} \leq 0, i \in J, j \in J\}$.

The algorithm proceeds as follows:

Algorithm: Build-Graph

Objective: improve an MPSP solution by adjusting **S** to include new (i, j) candidates

Input: feasible start times **S** Output: final start times **S**

(0) let $\mathbf{Z} = \mathbf{S}$;

(1) find $f(\mathbf{Z})$; let $Q = \varnothing$; let $P = \{(i, j) : s_i - s_j + d_i + t_{ij} + u_{ij} \leq 0\}$;

(2) find $(i^*, j^*) = \arg\min \{\alpha(k_i) \cdot u_{ij} + \beta(k_i) \cdot t_{ij} - u_i - v_j : \alpha(k_i) \cdot u_{ij} + \beta(k_i) \cdot t_{ij} - u_i - v_j < 0, (i, j) \in E, (i, j) \notin Q\}$, where u and v are the dual row and column solutions to the assignment problem; if there is no candidate (i^*, j^*), then STOP with solution **S**;

(3) let $R = \{(i, j) : (i, j) \in P \text{ and } x_{ij} = 1\}$;

(4) Let $j' = j$ such that $(i^*, j) \in R$, (j' is the follower of i^*); let $i' = i$ such that $(i, j^*) \in R$, (i' is the predecessor of j^*); $\forall k$ such that $(k, j') \in P$, let $R = R + (k, j')$; $\forall k$ such that $(i', k) \in P$, let $R = R + (i', k)$; let $R = R - (i', j^*) - (i^*, j')$; let $R = R + (i^*, j^*)$;

(5) try to find **Z** such that all elements of R are feasible:

(5·1) Topologically sort the graph $G(J, R)$; if G is not acyclic then go to (6): *there is no feasible* **Z** *for G*, otherwise let the topologically sorted list be Π;

(5·2) let $z_i = b_i \ \forall i \in J$; $k = 1$;

(5·3) $j = \pi_k$;

(5·4) $\forall i$ such that $(i, j) \in R$, if $z_i - e_j + d_i + t_{ij} + u_{ij} > 0$ or $(z_i - e_j + d_i + t_{ij}) \cdot r_{ij} > 0$ then go to (6): *there is no feasible* **Z** *for G*, otherwise let $z_j = \max\{z_i + d_i + t_{ij} + u_{ij}, z_j\}$;

(5·5) if $k = |J|$ then go to (6): **Z** *is a set of feasible start times for G*, otherwise let $k = k + 1$ and go to (5·3);

(6) if a feasible **Z** is found, let $\mathbf{S} = \mathbf{Z}$ and go to (1), otherwise let $Q = Q + (i^*, j^*)$ and go to (2).

Finding an initial feasible solution can be achieved by an algorithm similar to step (5) above. The algorithm Build-Graph has complexity $O(n^5)$ in the worst case, because the assignment problem can be solved in $O(n^3)$ time and $O(n^2)$ (i, j) pairs must be examined. In actual testing, the algorithm terminated within $|J|$ assignment problem evaluations with a solution within 10% of the optimum for typical vehicle scheduling problems with time windows, with sizes in the range of 8 to 223 jobs (Cyrus, 1988).

Fixture, pallet and tool scheduling

The VSP can also be applied to the scheduling of tools, pallets and fixtures for a group of FMCs, using the production schedule for the machines determined by the MPSP. The tools, pallets and fixtures can be thought of as resources required by the machines in the MPSP. Each resource requirement is modelled as a job with a time window when the resource is required. The duration of the job is the length of time for which the resource is used. The travel time is the time taken to transfer the resource from one demand point to another. There are fixed costs for resources over the horizon, and costs of transferring resources from one machine to another. The objective is to minimize the number of each resource required to meet the MPSP schedule and to minimize the total travel time. This problem is a VSP with time windows. It is an assignment problem if start times are given (equivalent to $f(\mathbf{S})$ above) and in this case the problem is separable into one assignment problem for each resource.

Automated guided vehicle scheduling

In an AGV system with a production schedule defined from the fixture, pallet, tool and job schedules above, a set of transportation requirements can be determined. A job involves loading a pallet, travel from one machine to another, then an unloading operation. The time window for a job is defined by the flexibility or slack of the production schedule. Travel times are those of empty trips from one workstation to another. The objective is to minimize the number of automated guided vehicles required and, secondarily, the total empty trip time. In some cases the fleet size may be fixed. If the assumption of no blocking on AGV paths is made, then this problem is a VSP with time windows, or an assignment problem if start times are given. If blocking is allowed, then a much more complex problem must be solved.

Conclusions

Three models of automated manufacturing scheduling problems were presented. These models are all solvable with vehicle scheduling algorithms. Some restrictive assumptions must be placed on the models in order to fit the VSP mould, but overall the solutions will be sufficiently useful to make decisions in an automated production environment. The models can also be used in a dynamic setting by incremental updating as the information changes. In a dynamic environment, the primary concern is solution speed, therefore the fast heuristics available for vehicle routing and scheduling are appropriate.

Acknowledgement

This research was partially supported by the Natural Sciences and Engineering Research Council of Canada, grant number OGP0036442.

References

Assad, A.A., 1988, Modelling and implementation issues in vehicle routing. In *Vehicle Routing: Methods and Studies*, edited by B.L. Golden and A.A. Assad (Amsterdam: Elsevier) pp. 7–45.

Buzacott, J.A. and Yao, D.D., 1986, Flexible manufacturing systems: a review of analytical models. *Management Science*, **32**, 7, 890–905.

Cyrus, J.P., 1988, The vehicle scheduling problem: models complexity and algorithms. Ph.D. Dissertation, Technical University of Nova Scotia, Halifax, Nova Scotia, Canada.

Dantzig, G.B. and Fulkerson, D.R., 1954, Minimizing the number of tankers to meet a fixed schedule. *Naval Research Logistics Quarterly*, **1**, 217–222.

Desrosiers, J., Soumis, F. and Desrochers, M., 1984, Routing with time windows by column generation. *Networks*, **14**, 545–565.

Etschmaier, M.M. and Mathaisel, D.F.X., 1985, Airline crew scheduling: an overview. *Transportation Science*, **19**, 2.

Jackson, H.F. and Jones, A.W.T., 1987, An architecture for decision making in the factory of the future. *Interfaces*, **17**, 6, 15–28.

Kusiak, A. and Cyrus, J.P., 1985, Routing and scheduling of automated guided vehicles.

Proceedings of the Eighth International Conference on Production Research, edited by H.-J. Bullinger and H.J. Warnecke (Berlin: Springer).

Solomon, M.M., Baker, E.K. and Schaffer, J.R., 1988, Vehicle routing and scheduling problems with time window constraints: efficient implementations of solution improvement procedures. In *Vehicle Routing: Methods and Studies*, edited by B.L. Golden and A.A. Assad (Amsterdam: Elsevier), pp. 85–105.

Form feature recognition for manufacturing

C.S. Chen[†] **and S.P. Owusu-Ofori**

Department of Industrial Engineering, North Carolina A&T State University, Greensboro, NC 27411, USA

Abstract

A methodology and a set of algorithms for the recognition of form features from a computer-aided design (CAD) file of a turned part are presented. The data originate from an engineering drawing. After digitization, the raster image is vectorized and converted into a wire frame data using IGES format. Algorithms have been developed to relate the geometries, extract the shapes, and to determine the features represented by the shapes. At this stage, the system can recognize form features which require a single view for interpretation.

Introduction

Engineering drawings have been used by designers as a major tool to document an engineering design accurately and communicate with the manufacturing personnel. Numerous engineering drawings were created and kept as hard copies before computer systems were introduced to the industry. The transfer of a drawing into a computer system has been a tedious process. Recently, technologies became available to convert these drawings into computer data files efficiently. These converted data are typically condensed and stored according to the Initial Graphics Exchange Specification (IGES) format for easy data storage and transfer between various CAD systems.

The automatic interpretation of such an IGES file for a three-dimensional (3-D) graphic presentation and computer-aided manufacture (CAM) applications is still a great research challenge. A typical engineering drawing contains geometric elements, notes and dimensions. Collectively, these geometric elements define the boundary of a part, from different viewpoints. In general, a single two-dimensional (2-D) view does not provide sufficient geometric information for a 3-D representation of the part. Interpretation of multiple views to synthesize the part geometry is needed. As geometric elements may contain ambiguous or inaccurate information engineers also have to rely on notes and dimensions to extract additional information to clear out ambiguities and inconsistency caused by omissions, abbreviations and inaccuracies in the geometric data. All these human expertises and drawing conventions are accumulated over a long period of time and cannot be readily programmed into a computer system.

The inclusion of drawing conventions and human experiences in drawing

---

[†] Present address: Department of Industrial Engineering, Florida International University, University Park, ECS 416, Miami, FL 33199, USA.

recognition is extremely critical to any computer system for engineering drawing recognition. The system, like a blueprint reader, must be able to interpret the individual views of projection and understand text and notes. It also has to analyse the relationships among the 2-D geometric elements defined in each of the views and extract its 3-D geometry and manufacturing features. To develop such a system, an understanding of descriptive geometry, drawing conventions, computer programming techniques, and CAD/CAM technologies such as part definition data formats is mandatory.

Related work

Recognition of line drawing representation of a scene has become the most recent target of computer vision research. An extensive review of this subject can be found in Ballard and Brown (1982) and Sugihara (1986). In these studies, a single view line drawing, which may not follow engineering drawing conventions, was typically used to represent a polyhedral or simply curved object in a wire frame format. The goal of these researches was to make a computer mimic human intelligence in picture perception.

In recognition of a mechanical part from a CAD data file for constructing a solid geometry and extracting machining information, several research efforts have been reported in the literature. Yoshiura *et al.* (1984) developed a natural language processor for the recognition and representation of notes in an engineering drawing. A pattern-matching approach was used to parse engineering notes. Each recognized note was stored in a frame structure of data file. The extracted knowledge in the frames was to be used to fill in missing (or omitted) geometric information regarding the part and to eliminate ambiguities in the drawing. Henderson (1984) reported the development of a set of algorithms to extract machining information for a part defined in a boundary representation (B-REP) data file. The extracted information was organized in a feature graph containing a set of connected form features. Each feature was considered as a cavity in the material, created by a specific machining operation. Template matching approach was used to recognize cavities. Matching a template was implemented in a set of pattern rules. The algorithms were aimed at prismatic parts. Nnaji and Vishnu (1986) developed a procedure for the conversion of a wire frame data file to a solid geometry. Using the geometric elements of points, lines, and arcs from a 3-D wire frame CAD system, it extracted surfaces, recognized solids and finally constructed a part geometry. Wang and Wysk (1988) proposed a methodology for identifying surface types of a symmetric rotational part defined in a single view wire frame graphic data file. Each surface type was meant to match one machining operation.

Presented in this chapter is a methodology and a set of algorithms for recognition of a machined part from an engineering drawing. The recognition is limited to a symmetric rotational part defined in a single view engineering drawing with no internal features. The general approach is described in the next section, followed by the implementation of the proposed algorithms.

Methodology

An ideal recognition process of an engineering drawing is outlined in Figure 1. Initially an engineering drawing is converted to an IGES file. Textual and dimension data are extracted from the IGES file and passed on to a text recognition system. The meanings of the recognized texts are used to assist in the process of geometry extraction. After the geometry is identified, the part is redefined in a feature file containing information for a complete part definition. The focus of this chapter is

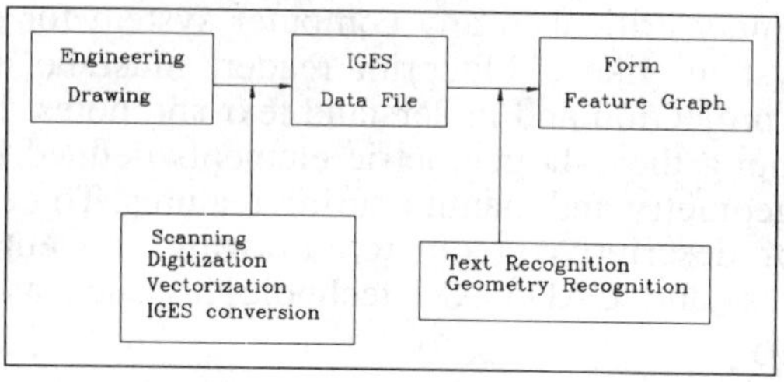

Figure 1. An engineering drawing recognition process.

on the geometry extraction process, aiming at geometric (form) feature recognition and part definition for manufacturing planning. A part drawing candidate is assumed accurate and complete so that it does not rely on textual data.

For recognition of a symmetric rotational part, one view drawing is sufficient. Actually a half drawing is all it needs to define the part geometry. The following facts regarding a symmetric rotational part are also observed.

(1) All the geometric elements are connected to one another, forming closed loops.
(2) There is no geometric element crossing another.
(3) Each vertical line connects to exactly two non-vertical geometric elements on each side. Therefore both the rightmost and leftmost vertical lines have two connections from only one side; and the connections occur at its end points.
(4) There is no loop nested in other loops; and there are no intersecting loops.
(5) All the loops are next to one another; together they represent the boundary of the part.
(6) Each loop contains an even number of geometric elements, due to its symmetric property.
(7) Loops are separated by vertical lines, due to their rotational properties. (Therefore there is a loop between two adjacent vertical lines.)
(8) Each loop indicates a 3-D shape on the part and also represents a form feature segment having some manufacturing implication.
(9) The loops, shapes, and therefore form feature segments, are symmetric about the centre line. Their adjacencies can be sufficiently defined according to their relative x-coordinates.

Based on these observations, a geometry recognition process for a rotational symmetrical part is proposed as in Figure 2. It is a bottom-up approach, with five stages of data conversion. The process first takes 2-D geometric elements from an IGES file, eliminates noises, mends broken lines, and then converts them into wire frame data. Noise detection and elimination is an essential operation in geometric element recognition. They are undesirable and unnecessary geometric elements in the IGES file.

In stage 2, these geometric data are classified into one of three data sets, namely, the upper set, the lower set, and the vertical line set. The classified data are then labelled in terms of their geometric element types. To do so, the centre line is first identified (or calculated, if it does not exist). All vertical lines across the centre line go to the vertical line set. All geometric elements above the centre line are in the upper set. The remaining are in the lower set. There are 15 geometric types identified in this study, including vertical line (V), slanted line (S), horizontal line (H), and 12 different arcs.

Connectivities between these elements are identified and used for loop search in stage 3. Connectivities are checked for between a vertical line and a non-vertical element and between two non-vertical elements within the same set. To search for loops, two adjacent vertical lines are picked and all the non-vertical lines which fall in between the two lines are selected from the upper and the lower sets,

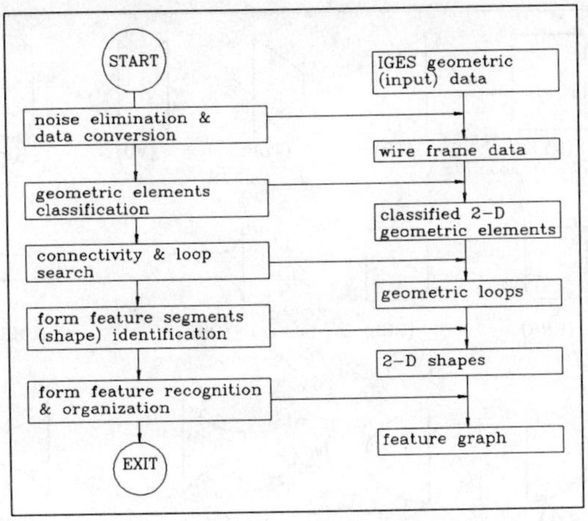

Figure 2. The geometry recognition procedure.

respectively. If they all connect to one another, including the two vertical lines, a closed loop is found. Otherwise an open loop is encountered. The closed loops are saved for shape recognition.

Commonly known shapes are identified at the fourth stage and used as templates for 2-D shape recognition. Each shape is represented by a pattern in terms of its boundary geometric elements. A shape is said to be recognized when a loop matches a pattern. Each time a new shape is encountered, the user is prompted for a name which is added to the known template list.

Templates are also used in stage 5 to identify form features. A common turned form feature may contain up to three shapes. Form feature templates are in terms of shape patterns. However, since each form feature implies certain machining requirements, the recognition also depends on the detail relations between the shapes in the pattern and the relationships with their adjacent shapes. Therefore, when a sequence of shapes fits a certain form feature pattern, additional rules are checked to verify its validity.

Feature extraction algorithm and implementation

A list of commonly known shapes for symmetric rotational parts are defined in Figure 3 and Table 1, containing 13 shapes. Each shape has a unique linkage pattern except the two trapezoids. A rectangle (R), for example, has a linkage pattern of V–H–V–H, meaning the boundary of the shape is a vertical line, a horizontal line, a vertical line, and then a horizontal line. A rectangle shape is said to be found if a loop matches this pattern.

Feature recognition

A list of form features is given in Table 2. Each form feature is defined by its shape pattern and a set of rules. Note that a shape pattern may be classified into several different form features due to its different geometric surroundings as specified in the rules. For example, the fourth form feature is called turned recess. It contains three shapes: a right trapezoid, a rectangle, and a left trapezoid, in this order. The recognition rule is: the diameter of the rectangle is equal to the smaller diameters of both trapezoids. As any form feature implies certain manufacturing

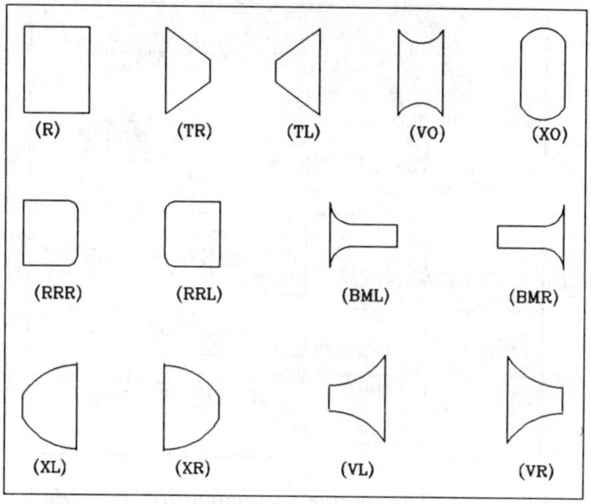

Figure 3. Basic shape on turned parts.

Table 1. Shapes definition.

Shape	Symbol	Loop
Rectangle	(R)	V-H-V-H
Trapezoid-right	(TR)	V-S-V-S
Trapezoid-left	(TL)	V-S-V-S
Concave-ortho.	(VO)	V-URVH-V-LLVH
Convex-ortho.	(XO)	V-ULXH-V-LRXH
Rd.-rectangle-right	(RRR)	V-H-ULXU-V-LRXU-H
Rd.-rectangle-left	(RRL)	V-ULXD-H-V-H-LRXD
Bell-mouth-right	(BMR)	V-URVD-H-V-H-LLVD
Bell-mouth-left	(BML)	V-H-URVH-V-H-LLVD-H
Convex-left	(XL)	V-ULXD-V-LRXD
Convex-right	(XR)	V-ULXU-V-LRXD
Concave-left	(VL)	V-URVU-V-LLVU
Concave-right	(VR)	V-URVD-V-LLVD

process, available machining technologies will affect the shape patterns and recognition rules. This list is not meant to be directly usable for any manufacturing planning system. A more comprehensive list containing 36 form features is given in Chen and Owusu-Ofori (1989).

Briefly, the algorithm sorts all the shapes in an ascending x-coordinate order and puts them in a shape list. It picks three adjacent shapes, starting from the one with the smallest x-coordinate. If the three shapes together do not match any three-shape pattern, the third shape is dropped from consideration. The remaining two are used to match two-shape patterns. If there is no match, the second shape is dropped. The first one alone is then used to match one-shape patterns. Whenever there is a match, the matched shape(s) is removed from the shape list. The procedure repeats until all the shapes are removed.

The recognized form features are stored in a feature graph. The relationships among the form features are defined by their relative x-coordinates. Attributes are also extracted sufficiently to define the geometry of each form feature.

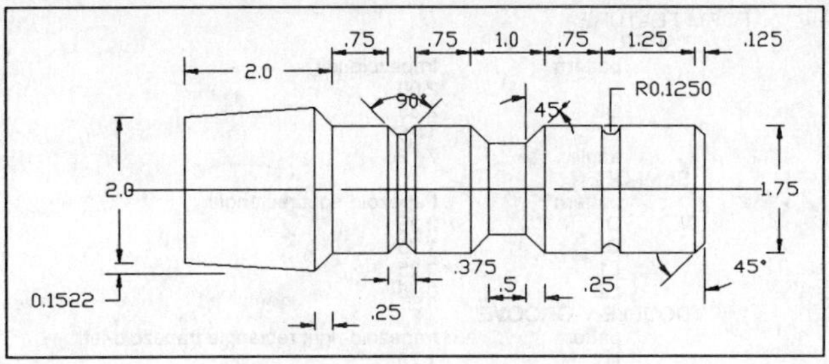

Figure 4. An example part.

Table 2. A list of form features.

No.	Name	Shape pattern(s)	Recognition rules
1.	Open	R	1·1 The diameter of the rectangle > = the diameter of the adjacent shapes
2.	Semi-open	TR-R or R-TL	2·1 The diameter of the rectangle = the smaller diameter of the trapezoid.
			2·2 The difference in the diameters of the trapezoid < = 0·5 inches
3.	Transition	TR-R or R-TR	3·1 The same as 2·1
			3·2 The difference in the diameters of the trapezoid > 0·5 inches
4.	Recess	TR-R-TL	4·1 The diameter of the rectangle = the smaller diameter of both the trapezoids
5.	Double V-Groove	TR-R-TL	5·1 The same as 4·1
			5·2 The length of the R < = 0·25 inches
6.	U-Groove	VO	6·1 The radius < = 3/8 inches
			6·2 The depth of the groove < = the radius
7.	Taper	TR or TL	7·1 The L/D ratio > = 1
			7·2 The L > = 0·2 inches
8.	Chamfer	TR or TL	8·1 The L/D ratio < = 1
			8·2 The L < = 0·25 inches

The program

The above algorithms are implemented in Common Lisp on a VAX 780 computer. Each algorithm is implemented in one function. The IGES data files are supplied by a SCANTEK SK-2020 system. Several sample parts have been used to test the system. It typically takes a few minutes to identify a part.

An example

An example part is shown in Figure 4. The drawing is digitized, scanned, vectorized and edited before it is converted into an IGES file. The IGES file goes through noise and broken line elimination, classification, connectivity and loop search, shape

```
FORM-FEATURE
    TAPER
        pattern         trapezoid-left
        D1              2·00
        D2              2·35
        L               1·75
        angle           7·125
    SEMI-OPEN
        pattern         trapezoid-right rectangle
        D1              2·35
        D2              1·75
        L1              0·25
        L2              0·75
    DOUBLE-V-GROOVE
        pattern         trapezoid-right rectangle trapezoid-left
        D1              1·75
        D2              1·25
        W1              0·375
        W2              0·125
        angle           90
    OPEN
        pattern         rectangle
        D               1·75
        L               0·5
    RECESS
        D1              1·75
        D2              1·25
        D3              1·75
        L1              0·25
        L2              0·50
        L3              0·25
    OPEN
        D               1·75
        L               0·50
    U-GROOVE
        D               1·75
        L               0·25
        radius          0·125
    OPEN
        D               1·75
        L               1·00
    CHAMFER
        D               1·75
        L               0·125
        angle           45
```

Figure 5. A feature frame.

matching, and finally feature recognition. The form features are recognized and organized in a form-feature frame as listed in Figure 5. This frame contains nine form features with a number of feature attributes. From left to right, they are taper, semi-open, V-groove, open, recess, open, U-groove, open and chamfer.

Concluding remark

A form feature extraction system has been described. It converts a digitized engineering drawing data file into a form feature graph containing manufacturing information. It recognizes external form features of a symmetric rotational part originally documented in a single view engineering drawing. The input data are organized in an IGES file. Only geometric data are extracted and recognized by this system. All the textual data and dimension information are ignored. It recognizes 36 external features.

The input data are assumed accurate and complete. The alignment problem during the scanning and digitization process is important to the recognition of orientation of digitized geometric elements. Tolerances assigned to broken element and connectivity detections are critical to the loop search process. More shape and form feature types need to be added to the template files to increase its recognition capability. Furthermore, learning ability of this system is crucial to its future success, given the fact that it will constantly encounter various unknown shapes and features.

Feature extraction from a multi-view drawing with internal features is a natural extension of this research. Text recognition capability is also a candidate for future research. Recognizing text helps in filling in missing information and clearing out ambiguities. It also provides accurate dimensional information, which currently is calculated from the vectors. Extension to prismatic part recognition is another fruitful research direction.

Acknowledgement

The research reported here was supported in part by the United States Air Force Wright Research and Development Center at WPAFB, Ohio, contract number F33615-89-C-5216 and is gratefully acknowledged.

References

Ballard, D.H. and Brown, C.M., 1982, *Computer Vision* (Englewood Cliffs, NJ: Prentice-Hall), pp. 295–307.

Chen, C.S. and Owusu-Ofori, S.P., 1989, Identification and Definition of Turned Part Form Features. A working paper, Department of Industrial Engineering, NC A&T State University, Greensboro, North Carolina.

Henderson, M., 1984, Extraction of feature information from 3-D CAD data. Ph.D. Thesis, Purdue University.

Nnaji, B. and Vishnu, A.K., 1986, A generalized shape descriptor from a wire frame model on a CAD/CAM system. *Journal of Manufacturing Systems*, **5**, 4, 231–241.

Sugihara, K., 1986, *Machine Interpretation of Line Drawings* (Boston, MA: MIT Press).

Wang, H. and Wysk, R.A., 1988, AIMSI: a prelude to a generation of integrated CAD/CAM system. *International Journal of Production Research*, **26**, 1, 119–131.

Yoshiura, H., Fujimura, K. and Kunii, T., 1984, Top-down construction of 3-D mechanical object shapes from engineering drawings. *Computer*, 32–40.

90

Computer-aided component selection for precision assembling

H.S. Shan and Alok Satyawadi

Department of Mechanical and Industrial Engineering, University of Roorkee, Roorkee, India

Abstract

The dual goal of very close tolerances on the final product and economy in component manufacture can be met by the use of selective assembly. Although it may appear simple, the selection of components becomes extremely difficult when accuracy requirements are stringent, and the mating parameters and lot size both increase.

This chapter describes a procedure for one-to-one pairing of component parts to be assembled. The proposed procedure not only reduces the overall labour content of fitting work but ensures a required level of product quality.

Introduction

Selective assembly is a procedure by which high-precision assemblies can be fabricated from relatively low-precision components. One common application of selective assembly is in the production of piston-cylinder assemblies. The manufacturing process of piston/cylinder may be such that the variability of the cylinder bore or the piston size is too large to meet the clearance specifications in an unacceptably large fraction of assemblies.

One solution to overcome this problem is to reduce the variability of the component-producing manufacturing process, but that may not always be feasible or economical.

Another alternative is to sort each component lot into two or more groups and randomly select components from within corresponding groups for assembly. For instance, large pistons would be matched with large cylinder bores, so that clearance variability is reduced. Such a technique in which sorted component groups represent specific dimension ranges of the mating parameter, is termed selective assembly.

Selective assembly is becoming an increasingly popular method of assembling due to its various advantages and the recent trends in manufacturing. These include flexible manufacturing systems (FMS), where a single process is required to produce a variety of part types and thus may lead to higher component variability over those processes where a machine is dedicated to the manufacture of a single part type.

Manual sorting of components into groups become impractical if the component batch size is quite large and the components are to mate on several surfaces (parameters). This is because the probability of coincidence of dimensional groups for all the parameters becomes almost negligible and the selection of the best pair for assembly becomes too tedious.

Not much work appears to have been done in this field. The earlier work of Desmond and Setty (1961) and Rabinovich (1968) indicate good attempts to improve the selective assembly technique. Shan and colleagues (Shan and Gupta, 1977; Gupta and Shan, 1982; Alok Satyawadi and Shan, 1989) suggested the use of computers for parts sorting on the basis of their dimensional groups. This chapter describes a relatively simple procedure for computer-aided matching of component parts which can be easily used on a PC.

Component matching scheme

The computer software developed provides one-to-one matching of components to be assembled. For two parts going in assembly, one can be called 'male' and the other 'female'. The input data required to obtain the 'best' pairing of parts are the dimensions of the mating parameters on both the male and female parts.

In order that the assembly takes minimum fitting time and is of high quality grade, the pattern of dimensional deviation from the standard dimension of the mating parameters on the male and female parts should be same, i.e. if the shaft size is on the positive side of the tolerance from the standard dimension, then so should the hole size. The dimensions of each of the mating parameters is input to obtain a code for each component part. This code, which can be termed a 'bit code', defines the pattern of dimensional deviation. If a particular dimension exceeds its corresponding standard dimension it is assigned code '0', while if it is equal to or less than the standard dimension, it is assigned '1'. Thus if the number of mating parameters is *m*, an *m*-digit code is assigned to each male and female component. For example, a component may have a code 111000, indicating that out of the six mating parameters, the dimensions of first three are either equal to or less than their respective standard dimensions whereas the dimensions of the last three exceed their respective standard dimensions.

To take into account the extent by which the dimensions of mating parameters deviate from their standard dimensions, the following procedure is adopted.

The total of process tolerance of all the mating parameters is obtained separately for male and female components (Figure 1). These cumulative figures are divided into a certain number of zones depending on the quality requirements of the lot, and the deviation limits for each zone or group are established.

From the dimension data of mating parameters of each component already in the computer the factor 'deviation sum' (DS) is obtained for each component. DS is the sum of the deviations of all the dimensions of mating parameters from their respective standard dimensions. By comparing DS with the zone limits already established, each component is allotted a zone number and is called the 'group number' of that component.

Pairing of male with female components is accomplished in a three-level program. In the first level, pairing is made if and only if the mating components have the same group number and the same bit code. If more than one female part qualifies for pairing with a male part, the condition of minimum deviation between respective parameter dimensions is used to make a one-to-one pairing. The components which cannot be paired by the above conditions or restrictions, i.e. in the first level, are then considered for pairing in the second level where the conditions are somewhat relaxed. Here, the condition of 'same group number' need only to be satisfied. Pairing of male and female components is done on the criterion of 'least bit code difference'.

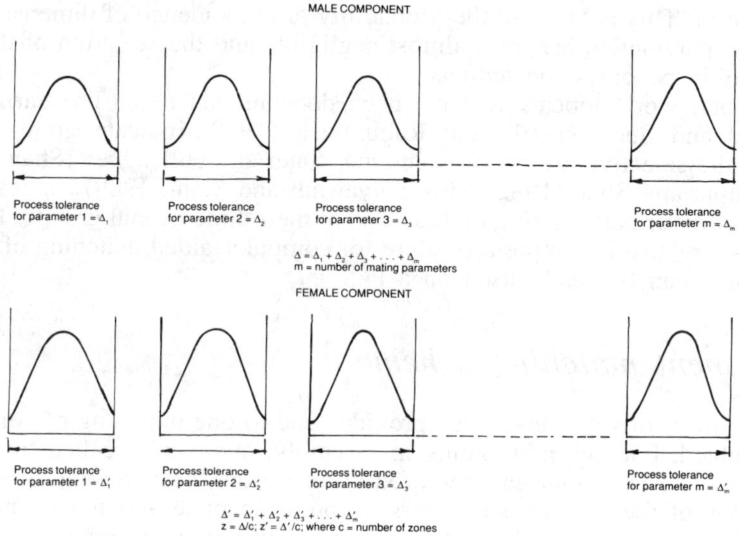

Figure 1. Process tolerances for male and female components.

The unqualified component parts at the end of second-level matching are considered for third-level matching. Here, pairing is done on the basis of 'minimum deviation between respective parameter dimensions'.

The component parts which could not qualify pairing even at the third-level matching are termed 'rejects'. These component parts can, however, take part in pairing of similar parts of the subsequent lots.

The flow chart of the complete matching scheme is shown in Figure 2.

Results and discussion

In the example that follows, two components of lot size 100 each were paired. These components were to mate on six mating parameters. The manufacturing process being under control, the dimensions of each mating parameter fell within ±3 sigma limits and followed a normal distribution pattern, as shown in Figure 3. The number of categories taken was 10. The limits of deviation of each category are shown in Table 1.

By inputting the dimension data of each mating parameter of all the components of the lot, the computer program was run on IBM PC AT. After the matching of parts was over, a list of matched pairs of component numbers of male and female parts was obtained. This list showed which pairs matched at the end of first-second- and third-level matching. It also gave the total of the cumulative deviation between the corresponding dimensions on the paired parts. In the above example a total of 88 pairs were made. Out of 88 pairs, 40 were made in first-level matching. The number of pairings accomplished in first-level matching is of interest as these could be considered to be the best pairs, the assembly of which would require the minimum fitting time and form what could be termed relatively 'Grade 1' assemblies.

The cumulative deviation figure is important because it gives a general idea about the quality grade of the whole lot of matched components. The smaller this figure,

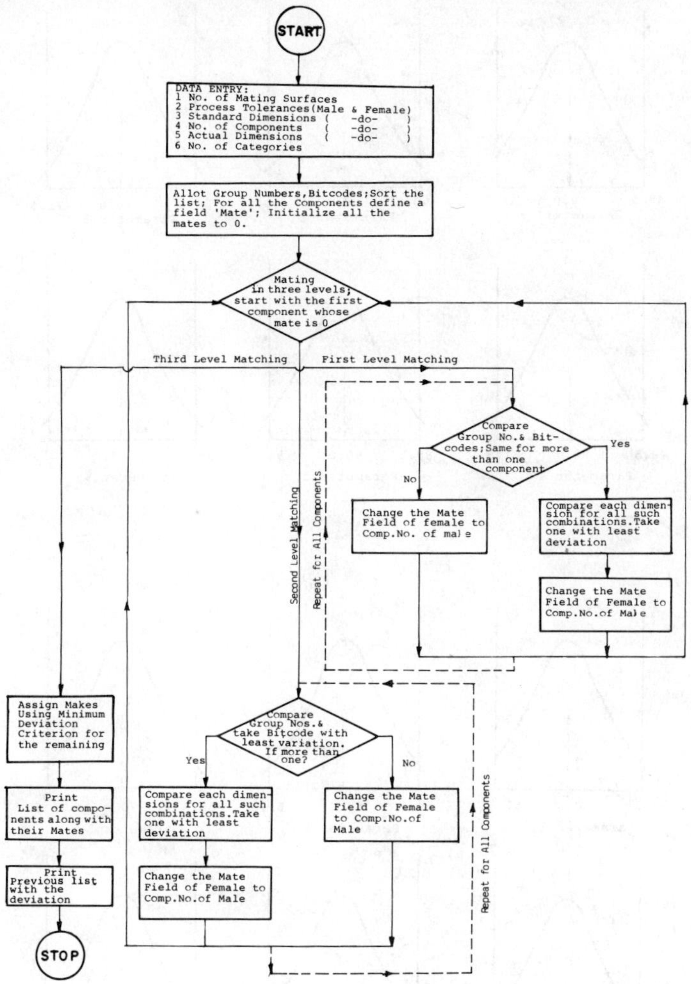

Figure 2. Flow chart of the component matching scheme.

the better the quality grade. In the example taken above, this cumulative figure was 250·99.

In order to examine the effect of changing the number of categories on the quality grade index and related factors, the computer program was run with the same component dimension data but with different number of categories. The results are shown in Figure 4. It can be seen that:

(i) Increase in number of categories deteriorates the quality level of pairs matched as indicated by the increase in the sum total of deviations. This does not seem logical. Actually, an increase in number of categories should improve the quality level of pairs matched. A reason for the above result obtained could be the small size of the component lot taken.
(ii) With an increase in the number of categories the number of pairs made in first-level matching decreases initially but then becomes nearly the same or asymptotic. This is because the conditions imposed for matching in the first level are stringent (same group number as well as bit code). As the number of categories is increased the number of components falling into each category decreases.

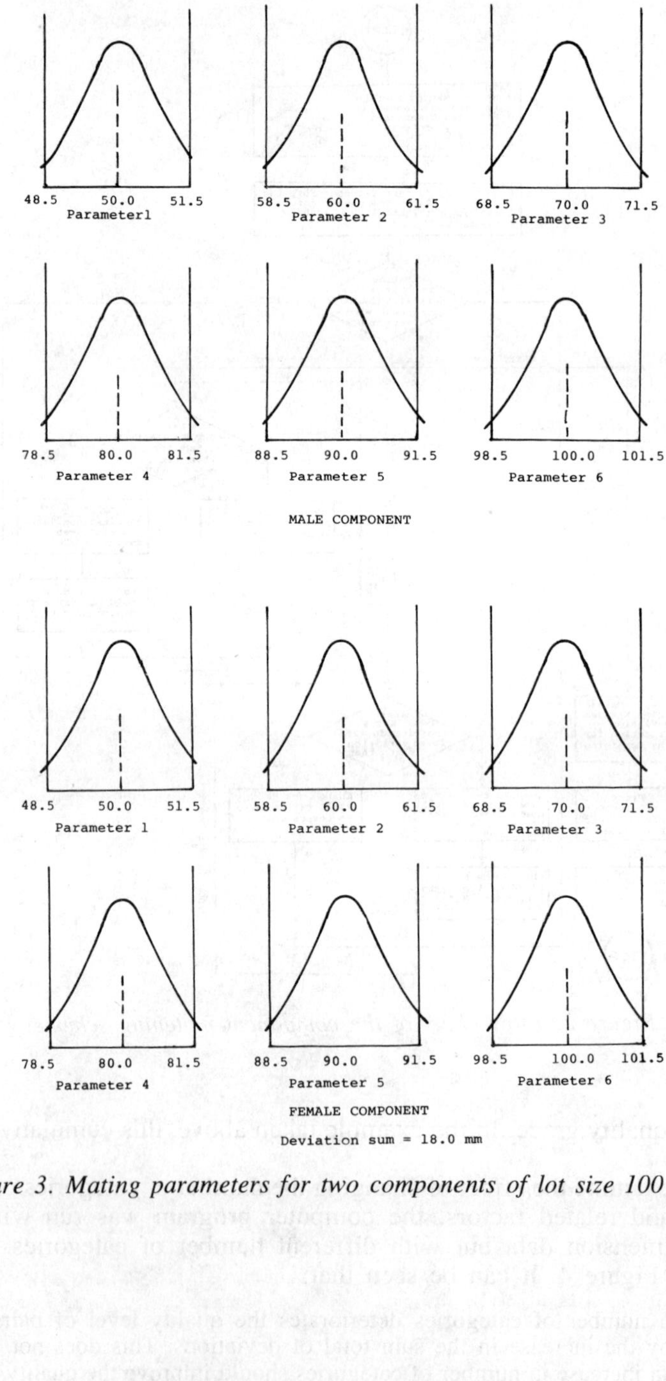

Figure 3. Mating parameters for two components of lot size 100 each.

Table 1. Limits of deviation of each category.

Category No.	Category limits
1	0·01– 1·80
2	1·81– 3·60
3	3·61– 5·40
4	5·41– 7·20
5	7·21– 9·00
6	9·01–10·80
7	10·81–12·60
8	12·61–14·40
9	14·41–16·20
10	16·21–18·00

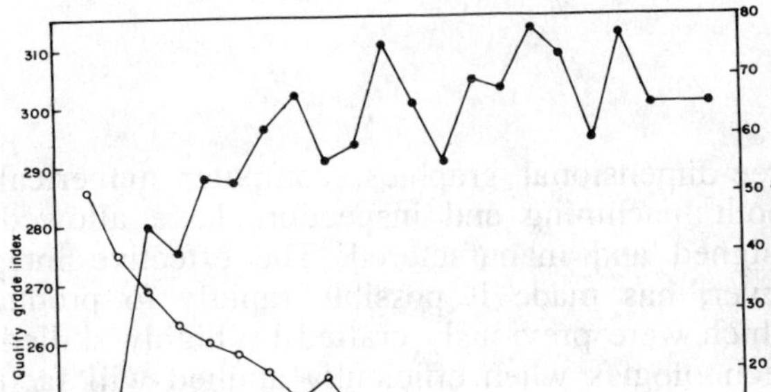

Figure 4. Effect of changing the number of categories on the quality grade index (●) and related factors (○).

References

Desmond, D.J. and Setty, C.A., 1962, Simplification of selective assembly. *International Journal of Production Research*, **1**(3), 3–18.

Gupta, A.K. and Shan, H.S., 1982, Computer aided component selection for assembly. *Mechanical Engineering Bulletin*, **13**, 4, 139–142.

Rabinovich, L.A., 1968, Improving the characteristics in relative assembly. *Russian Engineering Journal,* **48**, 54–59.

Satyawadi Alok and Shan, H.S., 1989, *Design and Development of a Software Package for Selective Assembly.* Project Report, Department of Mechanical and Industrial Engineering, University of Roorkee.

Shan, H.S. and Gupta, A.K., 1977, Computerized component matching for assembly. *Proceedings 3rd International Conference on Production Research*, Kyoto, Japan, 14–17 July.

91

Design and manufacture of jewellery using CAD/CAM/CAI techniques

B.S. Lim, C.K. Chua and T.T. Chong

Grumman International Nanyang Technological Institute CAD/CAM Centre, GINTIC, Nanyang Technological Institute, Nanyang Avenue, Singapore 2263

Abstract

The advent of three-dimensional graphics, computer numerical control (CNC) technologies for both machining and inspection, have allowed many different artifacts to be designed and manufactured. The effective integration of these technologies, however, has made it possible rapidly to produce accurate and complex objects which were previously crafted by highly skilled pattern makers. These integrated technologies when efficiently applied will facilitate many high value-added industries to develop appealing designs and manufacture cost-effective products rapidly to cater for fashionable trends.

The effective integration and application of this suite of technologies for the design and manufacture of jewellery for the fashion industry is one such contribution. This chapter describes an integrated suite of computer-assisted technologies together with the methodologies and techniques applied towards the following processes in the jewellery manufacturing industry:

(i) acquisition of geometrical features using a coordinate measuring machine;
(ii) reconstruction of digitized data to develop the master patterns on a CAD system;
(iii) generation of a two-dimensional profile from a three-dimensional model;
(iv) production of engineering drawings for human verification;
(v) parametric library of precious stones;
(vi) programming of computer numerical controlled (CNC) machining centre;
(vii) generation of cutting tool path;
(viii) machining operation of ring templates; and
(ix) inspection process.

Objectives

The principal aim of the research described in this chapter is focused at achieving the following objectives:

(1) Facilitate the function of the sales department by the introduction of advanced graphical and artistic representation media.
(2) Facilitate the development of a new ring design process based on established marketable products.
(3) Reduce the lead time associated with the manufacture of ring patterns.
(4) Relieve the jewellery master craftsmen from routine work so that they can concentrate on more creative tasks.

Figure 1. Photograph of rings from which variants are to be created.

(5) Enhance the productivity of a jewellery manufacturing organization to allow it to react rapidly and easily to fashionable trends.

Nature of the problems

The design and manufacture of jewellery is an artistic, iterative and time-consuming process which has to satisfy many criteria. Some of the main criteria to be resolved include the following requirements:

accurate anticipation of potential clients' preferences, tastes and income brackets,
selection of appropriate precious stones, metals and gems together with their attributes including sizes, values, shapes, cuts and colour,
configuration and design of both the styling and settings,
attractive presentations and market appeal,
competitiveness and market survival,
ease and flexibility of manufacture.

Acquisition of 3-D geometrical data

The prerequisite of the research described in this chapter was to acquire the geometrical dimensions of the actual rings as shown in Figure 1. A CMM was selected for the data acquisition process as it can provide a fast and accurate means of capturing the three-dimensional (3-D) geometrical data.

A CMM allows an operator to digitize the 3-D data using a force-sensitive probe. It is capable of digitizing a series of points along and across a non-mathematical contour. The CMM controller also allows a user to reconstruct complex curves by concatenating these digitized points together. Using these available facilities, the non-mathematical contours of the three rings can thus be effectively developed as shown in Figure 2.

Transfer and processing of the acquired data

The geometrical data acquired using a CMM can be transferred to a CAD system for the manual reconstruction and development of the design to ensure that most of the major objectives as described above can be simultaneously satisfied. Special

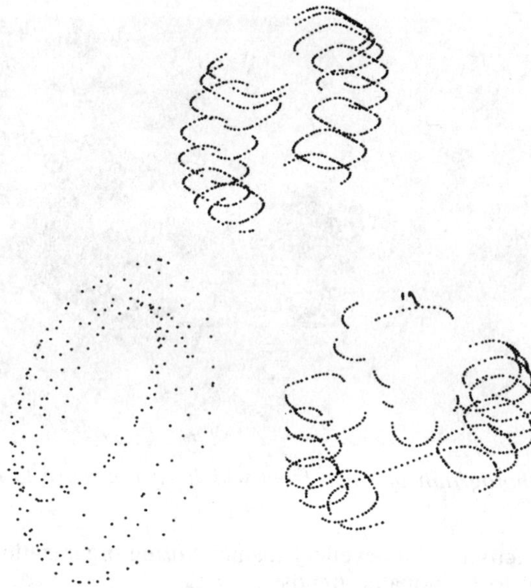

Figure 2. Digitized points of the three rings.

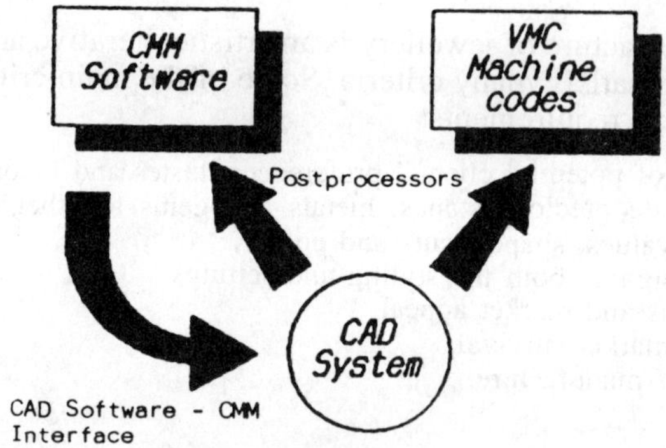

Figure 3. Post-processors between CMM, CAD system and CNC machine tool controller.

post-processors, pre-processors, interfaces and communication protocols were developed within GINTIC to facilitate the automatic transfer of data between the various computer-aided design, manufacturing and inspection facilities as shown in Figure 3. A communication link was also developed between the CMM controller and the IBM PC/AT to facilitate the transfer of data to the mainframe computer.

The coordinates from the CMM are translated into a set of geometrical data by an in-house specialized translator, developed using a geometry interface module available within a CAD system (Khoo *et al.*, 1987).

Generation and validation of the design

When the data has been processed by the above translator, a specific design of a ring can be further developed or enhanced to meet the functional, manufacturing and aesthetic requirements. This developed model serves as a common data base

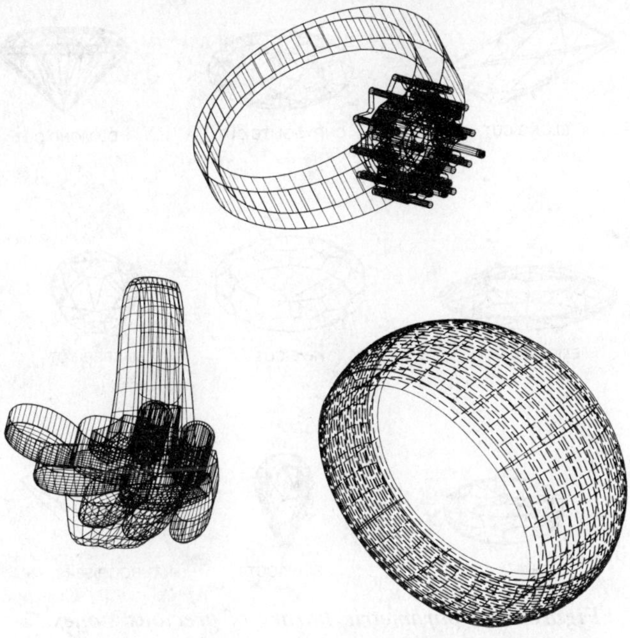

Figure 4. Wireframe representations of the three rings.

for all subsequent processes including CNC programming, cutting tool path generation and CMM probe path definition. This particular CAD system uses entities such as curves and splines as basic building blocks for the reconstruction and development of sculptured surfaces.

The initial design work involved the generation of the appropriate curves using the digitized data. The digitized points were manually concatenated together to form the required curves. The maximum deviation of the curves from the points was confined to be within 0·010 mm to allow for an accurate representation.

This specific CAD system allows a user to create an artifact from many different categories of curves. In the research described in this chapter, polynomial curves were chosen since the data points were digitized in a set of two-dimensional sections and these curves fitted adequately across the different sections. Polynomial curves are composite curves with successive low-degree polynomials fitted to a specific group of data points. Figure 4 shows the wireframe representations of the three rings.

Parametric library of precious stones

Following the reconstruction of a computer model of a specific ring, a new design can thus be easily developed, analysed and manufactured. A library of precious gems including stones such as diamond, ruby and sapphire was also developed based on parametric modelling techniques. This parametric facility allows a designer to develop a new ring based on existing patterns, choose and scale an appropriate stone to match a particular configuration together with its other attributes such as cuts, sizes and shapes. Figure 5 shows a portion of a parametric library of precious stones together with the reconstructed rings.

Shaded models of two additional rings developed from existing patterns are shown in Figure 6. Such a facility will assist a sales engineer in the proposals of new products, prior to the actual production stage.

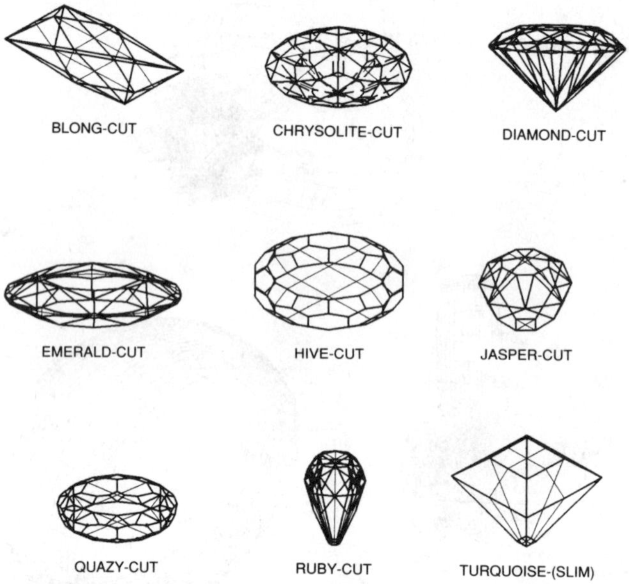

Figure 5. A parametric library of precious stones.

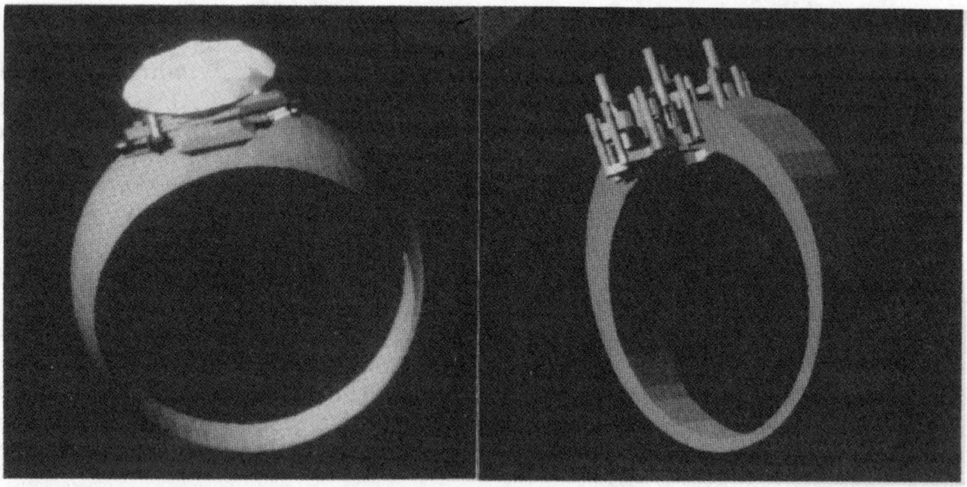

Figure 6. Shaded models of new design based on existing patterns.

Generation of two-dimensional models and engineering drawings

Specialized algorithms were developed to transform a 3-D model into a 2-D blank for the machining of the template. Equations from established sheet metal flow theory including the neutral axis axiom and bending allowance calculations theory were adopted for the development of these algorithms. Figure 7 shows a part of an engineering drawing of a specific ring generated primarily to facilitate communication between people.

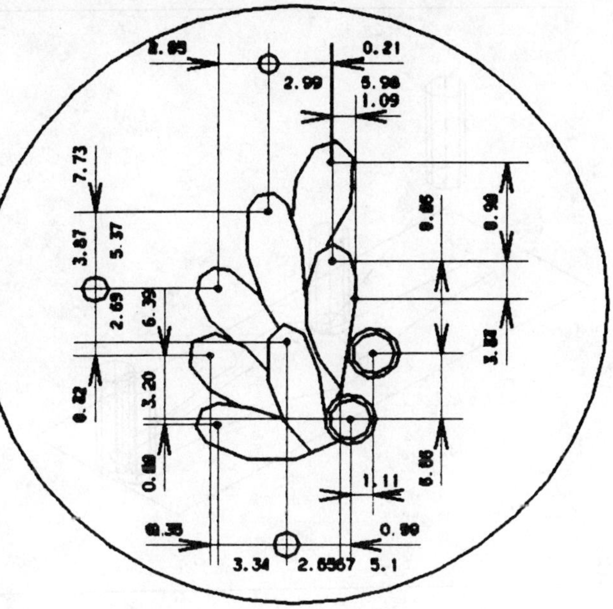

Figure 7. Part of an engineering drawing of a specific ring.

Generation of numerical control (NC) tool path

When a specific manufacturing order has been awarded, a process planner can proceed to develop the cutting tool path on a CAD model. The machining set-up requirements, potential interferences, number and depth of cuts, cutting speed, feed rate and NC programs can be generated within the same environment. Computer simulation of the cutting tool paths on a new design is shown in Figure 8.

Machining and verification of templates

A three-axis vertical machining centre was used to machine the plastic or silver patterns required for the manufacture of the mould cavities. It allows complex surfaces to be machined. A completed component can then be inspected using the same CMM to establish the differences between the actual and desired dimensions. Any specific deviation of the required dimensions can be easily adjusted by the machine operator. The mode of programming the CMM is similar to the method of programming the machining centre. A ring template being inspected by a CMM is shown in Figure 9.

Facilities used in this research

The facilities used in this research include the following equipment:

(1) Carl Zeiss Co-ordinate Measuring Machine Model UMC 550,
(2) IBM mainframe computer Model 4381 P13,
(3) CATIA software with sculptured surface, NC programming and parametric modelling facilities,
(4) Mitsui Seiki three-axis Vertical Machining Centre (VMC),

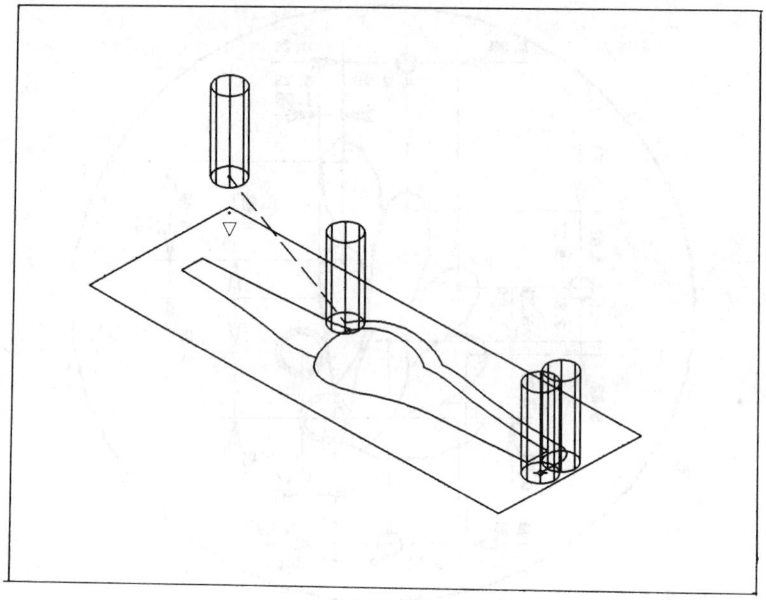

Figure 8. Simulation of cutting tools on a template.

Figure 9. Inspection of a template using a CMM.

(5) interfaces, post processors and translators developed within GINTIC for integrating the CAD and CAM systems together with the VMC and CMM.

Figure 10 shows the connections between the equipment. These computer-aided tools together with an automated storage and retrieval system, a fleet of three automated guided vehicles and a robotics modular fixturing workstation (Lam and Lim, 1989) will be integrated to form an intelligent flexible manufacturing system (Lim *et al.*, 1989) using artificial intelligence techniques for component design,

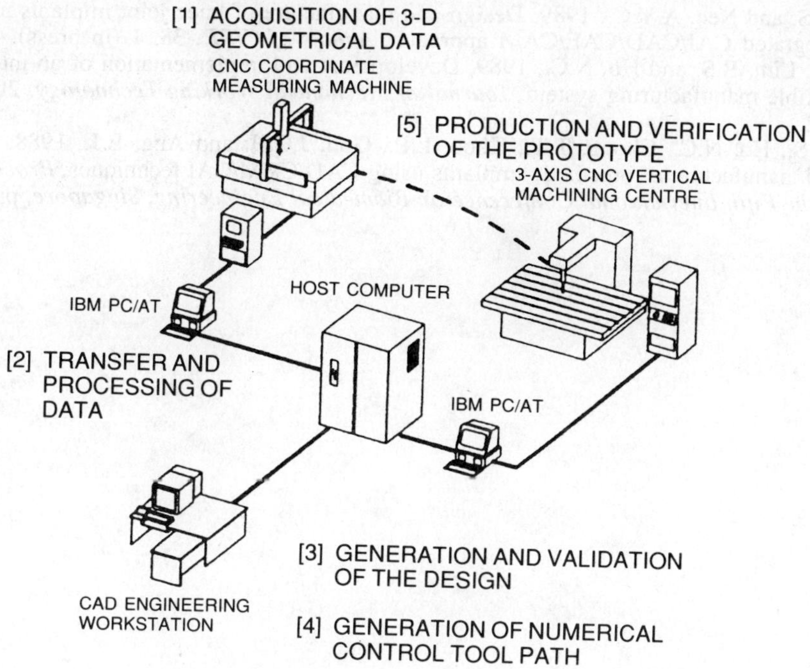

Figure 10. Flow of information between data acquisition, design, machining and inspection processes.

modular fixture design, process planning and production scheduling (Lim, 1987; Lim and Knight, 1987).

Conclusions

The research described in this chapter has shown that by combining the expertise of ring designers, production engineers, software engineers and tool makers together with CAD/CAM/CAI techniques and advanced production technology, many complex artifacts can be effectively manufactured. This set of CAD/CAM/CAI techniques has also been applied for the design and manufacture of knee joint prosthetic implants (Chong *et al.*, 1988; Lim *et al.*, 1989).

References

Chong, T.T., Lim, B.S. and Ho, N.C., 1988, Reverse engineering methodology using CAI/CAD/CAM techniques. *Automation in Manufacturing '88 Conference Proceedings*, Singapore, pp. 7·76–7·84.

Khoo, L.P., Yap, W.K., Wong, W.Y., Yeo, S.K. and Chong, T.T., 1987, GINTIC In-house documentation: Interfaces and Translators for integrating CMM with CAD system (Grumman International/Nanyang Technological Institute CAD/CAM Centre, Singapore).

Lam, S.L. and Lim, B.S., 1989, An integrated robotics off-line programming system for modular fixture assembly, *Journal of Mechanical Working Technology*, **20**, 261–271.

Lim, B.S., 1987, An IKBS for integrating component design with tool engineering. *International Journal of Knowledge Engineering*, **4**, 4, 252–291.

Lim, B.S. and Knight, J.A.G., 1987, A foundation for a knowledge based CIM system. *International Journal of Artificial Intelligence in Engineering*, **2**, 1, 11–22.

Lim, B.S. and Nee, A.Y.C., 1989, Design and manufacture of knee joint implants using an integrated CAI/CAD/CAE/CAM approach. *Annals of CIRP*, **38**, 1 (in press).

Lim, R., Lim, B.S. and Ho, N.C., 1989, Development and implementation of an intelligent flexible manufacturing system, *Journal of Mechanical Working Technology*, **20**, 389–402.

Lim, B.S., Ho, N.C., Chong, T.T., Khoo, L.P., Goh, J.C.H. and Ang, E.J., 1988, Design and manufacture of prosthetic implants using CAD/CAM/CAI techniques, *Proceedings of the Fifth International Conference on Biomedical Engineering, Singapore*, pp. 215–218.

92

Feature representation and extraction in CAD models for manufacturing: methods and issues

Mohan S. Devgun and Srikanth Padmanabhan

Department of Mechanical Engineering, Engineering Research Institute, Iowa State University, Ames, IA 50011, USA

Abstract

This chapter describes an effort to establish a perspective on the status of current research related to feature representation and extraction in geometric models for manufacturing applications. The first part of the chapter introduces the framework involved in characterizing physical objects in modelling systems and outlines the fundamental issues involved in the interplay of design in manufacturing. This framework is then used to describe and compare existing extraction schemes for manufacturing application mapping from a CAD data base, and the need for these techniques is questioned in the second section. The final section concentrates on a paradigm, 'feature-driven process-based design,' proposed by the authors in representing information on a CAD data base to support automatic programming of welding robots. The authors conclude that existing means are sufficient to represent information and that future generations of modelling systems can be created on the basis of general system techniques that are less analytical and depend more on heuristic decisions than on developing sophisticated extraction techniques for manufacturing automation.

Introduction

Contemporary geometric modelling systems are most useful for driving a growing array of computer-aided manufacturing (CAM) modules. However, the major problem in the integration of computer-aided design (CAD) and CAM is the representation and transformation of information among design and manufacturing subdivisions. In spite of all the claims, we do not see that CAD/CAM technology has provided anything more than a means of documentation, drafting, and specification of nominal geometry for numerical control (NC) applications (Shah and Wilson, 1988). This assessment has resulted in the redefinition of existing representation schemes to facilitate applications in manufacturing. This non-standardization in representation is attributable to our lack of knowledge of the interplay between design and manufacturing (Voelcker, 1988) and the mismatch of available tools to the tasks to be performed. The principal focus of this chapter is to describe the status of modelling systems research for manufacturing applications and to trace the efforts in representation and extraction of geometric features. This chapter summarizes the mathematical framework involved in characterizing physical objects in modelling systems. In an attempt to resolve difficulties such as specifying the characteristics of future geometric modelling systems, we hope to provide herein a clear and coherent view of what is currently known about modelling systems. Because representation of parts is informationally complete but lacks mathematical tasks and process models to drive application programs, we present the current trends toward feature and process-based design.

Extraction of geometric features from a CAD data base has received considerable focus. In the past decade, a variety of extraction schemes have been postulated on the basis of the premise that given a part detail, the process planning system should be able to obtain information automatically and generate the needed process plans. We enumerate the different schemes that have been developed and describe limitations in current systems, and we hypothesize the need to approach these problems differently. Two (of the many) issues, namely, the need for feature extraction schemes and how far we can go with feature-based design systems, are addressed in order to provide a background in creating the next generation of geometric modelling systems that can interface with manufacturing elements directly.

Existing computer-aided geometric modelling systems do not provide a structure for complete, correct and explicit product and process definitions; this is particularly true of geometric information on the weld features to support directly the automated creation of programs for welding robots. The final section presents a new paradigm, based on object-oriented programming techniques, that provides a structure in geometric modelling systems for supporting complete welding information and generating robot welding programs automatically. The requirements for developing the environment to support weld geometry details in conjunction with process knowledge are described; the merits of using a solid modelling system to support automatic programming of welding robots are particularly emphasized. We conclude that current modelling systems lack qualities for conceptual design, but are suited for manufacturing domain applications if additions are made to the existing structure. Geometric models should be complete not only in terms of geometrical information, but also in terms of the vocabulary of the tasks to be automated. Easy mechanisms for interpreting this information can be devised on the basis of sets of action-pair rules to permit mapping of different applications.

Modelling systems revisited

Computer-aided modelling systems have evolved over the past four decades, with early research propelled by advances in CAM. Figure 1 traces the evolution of this technology up to its current status.

A complete discussion of all the available modelling methods is beyond the scope of this chapter; the reader can refer to good surveys (Requicha, 1980; Requicha and Voelcker, 1982). Wireframe modelling systems supplied the base for all commercial CAD/CAM systems until the advent of solid modelling. Solid modelling emerged as the most unambiguous method of representing solids. Six unambiguous schemes are known (Requicha, 1980) but boundary representation (B-Rep) and constructive solid geometry (CSG) are the two most predominant techniques used in current modelling systems.

However, certain serious limitations exist in these modelling systems for supporting applications in the manufacturing domain. Current modelling methodologies do not provide means to model the effects of a process in the design stage that can also be used in the automatic planning of machine operations. As pointed out by Shah and Wilson (1988), the data base containing the geometric model is not complete. Incomplete representations occur because of the heavy reliance of the

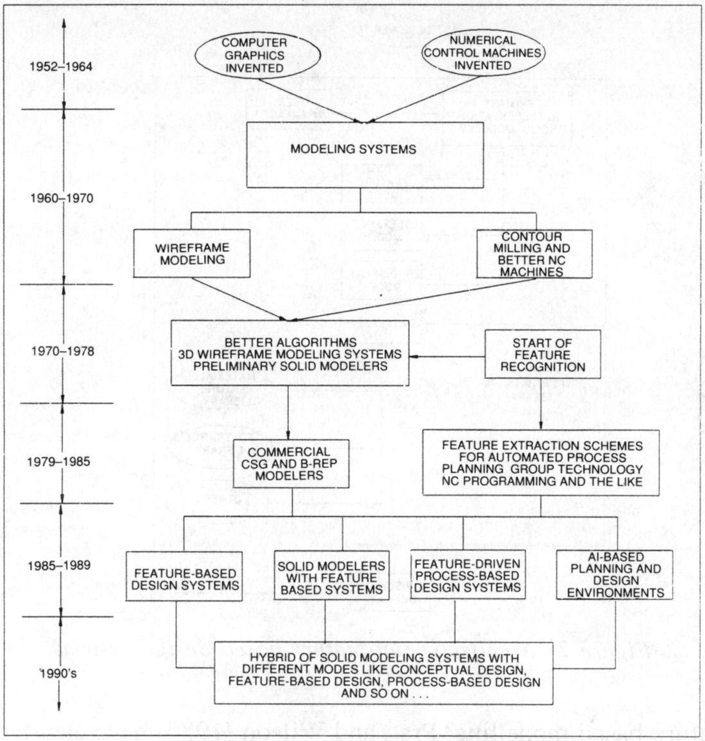

Figure 1. Evolution of geometric modelling technology.

model on the geometry rather than on the application. Also, the manufacturing information that needs to be incorporated is wide ranging, from simple materials specifications to representation of tolerances and surface finish. Furthermore, storage and retrieval of this information requires considerable effort. Tolerancing, surface finish, welding and other similar data have always depended on the visualization of the user; no information was held in the CAD data base. This gap resulted in an inability to support the automatic querying and retrieval schemes, which in turn inhibited the development of automatic planning/programming in the machining, assembly, and welding disciplines of manufacturing. This inability resulted in the criticism that solid modelling systems are mere documentation tools with limited applications to mass property calculations and interference checking. The link among the part model containing the geometry, topology, their relationships, and the application-specific spaces (like process planning and robot programming) can be thought of as a set of features.

With a view toward fulfilling the goal of automatic planning and noting the fact that certain dominant constituents model a solid object, certain researchers are developing sets of additions to existing solid modelling systems called 'feature-based design' systems (Pratt, 1988; Cunningham and Dixon, 1988; Shah and Rogers, 1988; Cutkosky *et al.*, 1988; Schulte *et al.*, 1990). A 'feature' is any geometric form or entity whose presence or dimensions in a domain are germane to manufacturing evaluation or planning, or to automation of functional analyses (Dixon, 1986). In this approach, the designer is constrained to work with a set of features that have significance for either design, analysis, or manufacturing. Instead of using a model consisting of graphic primitives (lines, points), the designer uses a set of features (holes, slots, pockets) from which manufacturing operations can be derived (Unger and Ray, 1988). Figure 2 gives an architecture that is currently

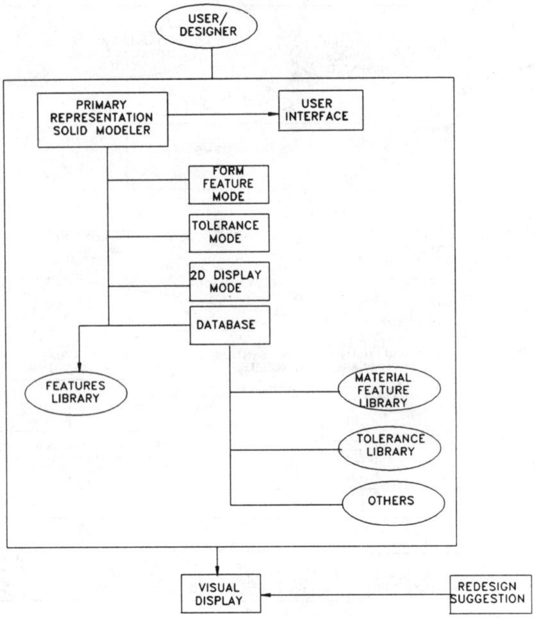

Figure 2. Architecture of feature-based design systems.

used for feature-based modelling. Pratt and Wilson (1985) have clearly outlined the requirements of such a feature-based modelling system for form features. This stream provides a means for building a complete data base at multiple abstraction levels right from the start of product development, and it allows users to build models by using features stored in libraries and networking them when and if needed. However, very little work has been done in modelling based on features other than form features modelling. This type of modelling is claimed to help the reasoning process of intelligent systems. Shah (1989) provides an excellent summary of the status of feature-based design systems.

In the following, we outline the different extraction schemes developed and provide a discussion on both lines of research (extraction of implicit information versus explicit representation) and propose our intermediate approach of 'feature-driven, process-based design' to solve the problem of automatic programming of welding robots. Table 1 lists all publications relevant to feature representation and extraction for manufacturing automation along with the researchers and their affiliations for further reference.

Feature extraction schemes

The principle behind the various extraction schemes is the development of a methodology for interpreting a CAD data base containing points, lines and surface information into a higher-level geometric feature useful for process applications. In this approach, a program examines the data base and makes deductions about the types of features present; this process amounts to making explicit what is implicit in the solid model. The motivation behind all research activities in this field was the need for automated process planning with available modelling systems. The available schemes can be classified as:

(1) volume decomposition,
(2) syntactic pattern recognition approach,

Table 1. Feature representation and extraction research summary.

Topic	Researcher and year	Affiliation
Solid modelling	Requicha, 1980	Initially as production automation group.
	Requicha and Voelcker, 1982	Now at University of Southern California
	Voelcker, 1988	and Cornell, respectively
Design by features	Dixon, 1986; Dixon, 1988	University of Massachussets at Amherst
	Cunningham and Dixon, 1988	
	Hummel and Brooks, 1986	Allied Space
	Pratt, 1988	Cranfield Institute of Technology, England
	Pratt and Wilson, 1985	Wilson at GE Corporate
	Shah and Rogers, 1988	Arizona State University
	Ostrowski, 1988	GE Corporate Research
	Unger and Ray, 1988	NIST
	Miner, 1985	MIT Master's work
Feature-driven, process-based design	Cutkosky *et al.*, 1988	Stanford University
	Schulte *et al.*, 1990	Iowa State University
Feature Extraction	Volume decomposition	University of Michigan
	Woo 1975, 1977, 1982, 1984	
	Syntactic pattern recognition	
	Kyprianou, 1980	Cambridge, England
	Jakubowski, 1982	Poland
	Choi *et al.*, 1984	Purdue University
	Srinivasan and Liu, 1985	Liu at Purdue and Srinivasan at IBM
	Staley *et al.*, 1983	Purdue University
	Li 1988	Arizona State University
	General Techniques	
	Grayer, 1977	Shape Data
	Chan, 1982	Then at Purdue
	Lee and Fu, 1987	University of Michigan
	Woodwark, 1988	
	Wang and Wysk, 1988	Wysk at Penn State
	Rule-based approach	
	Henderson, 1984, 1986	Arizona State University
	Joshi *et al.*, 1988	Chang at Purdue
	Kung, 1984	Then at Oklahoma State

(3) general techniques based on part data structure format,
(4) production rule-based systems,
(5) higher-level extraction from feature-based design systems and inverted forms of process models.

Extraction by volume decomposition

Research in feature extraction started more than 15 years ago when Woo (1975) described his work as 'the research on not how to manipulate data structures to enhance human visualization, but how to make computers visualize at all' for subsequent automation. This work was a precursor to much of the research that has followed since then.

Woo (1977) proposed a methodology for extracting geometric feature information from a CSG model for 2-1/2D workpieces. He focused on cavity recognition to transform volumetric designs of parts into descriptions for NC machining. The cavity recognition process is model driven in that a cavity is recognized only if it can be represented as a volume swept by a cylindrical cutter in 3-D space. Cavities studied in that research included counter-sunk holes and slots. In this method, the information provided at the design stage is pre-processed to obtain a concise algebraic description of the given volumetric design. The interpreter examines one construct at a time starting with the lowest levels of parentheses looking for cavities and working upward until all constructs are exhausted. If part of a cavity is found in one of the expressions, as set of 'hooks' is created to be interpreted later with newly formed 'hooks.' After the cavity is recognized, a program is called to take the vertices to form a new plane surface from them, and the source code for NC machining is synthesized. This work broke fertile ground for the next decade of research.

To improve the global understanding of the part, Woo (1982) proposed a novel technique for extracting features from a B-Rep file to yield a series of volumes with alternating signs for volume addition and subtraction called the Alternating Sum and Volume (ASV) expression. The ASV procedure basically resulted in a conversion from a B-Rep structure to a CSG-like tree. The algorithm takes a boundary representation of an object Ω as input and produces an expression of convex components having alternating signs. The algorithm is recursive and employs two operations, the convex hull operation H, which is actually the minimal volume that encases Ω, and the regularized set of difference operations that removes the difference between H and Ω.

Woo (1984) extended the ASV concept to provide a data structure conversion between two modelling interfaces (CSG and B-Rep). Symmetry was exploited in the part description format, and algorithms were presented to convert a graph to a tree and a graph to a different graph.

Syntactic pattern recognition approach

Numerous researchers have worked on applying syntactic pattern recognition for extracting features from edge list or line drawings (Jakubowski, 1982; Staley *et al.*, 1983; Choi *et al.*, 1984; Liu *et al.*, 1984; Srinivasan *et al.*, 1985; Li, 1988).

Syntactic pattern recognition is analytically analogous to formal language theory. In this technique, the structural information (like the topological information) is analogous to the syntax and grammar of the natural language. This information is represented as production rules. The language describes a class of patterns that have similar syntactic or structural relationships among the language's pattern primitives. A grammar G is formally defined (Fu, 1978) as a four-tuple:

$$G = (V_t, V_n, P, S)$$

where V_t is a finite set of terminals or primitives; V_n is a finite set of non-terminals; P is a set of production rules; $S \in V_n$ is the start symbol.

All the valid strings or patterns can be derived from the starting symbol by applying the production rules from a language or pattern class $L(G)$.

The typical application of syntactic pattern recognition for feature extraction starts with the selection of pattern primitives. These pattern primitives are then represented by the letters (generally the lower case) of the alphabet. A set of production rules (that describe the adjacency or the topology) are written for a class of parts that form a class on the basis of the features present. The given part is coded in terms of the primitives and is represented as an alphabet string. The recognition procedure consists of parsing the input string for acceptance by the grammar. This determines whether the part falls into the particular class represented by the grammar. Several parsing algorithms (Earley's algorithm, HYK algorithm, or the standard automaton) are used in practice.

Jakubowski (1982) used a syntactic pattern recognition approach to describe 2-D profile information on 2-D machined parts. He characterized all segments that are possible in machined parts to be 32 and distinguished them by length and orientation. Based on these primitives, the part is classified into similar shapes. The structure is then defined in a syntactic manner, by means of formal language theory as explained earlier, and produces an output in a chain format. This output is fed as an input to the parsing algorithm. A bottom-up parser is written for the chain structure and another parser (*LR*(*k*) method) is adapted to be applied to the modified chain structure called the regular right-part grammar. The output of the parser results in a set of patterns that have significance in manufacturing.

In a similar application Staley *et al.* (1983) applied syntactic pattern recognition to the classification of holes from a 2-D cross-sectional description extracted from a 3-D data base. The recognition process has been divided into two stages: extracting a 2-D edge list of a hole and performing parsing and classification. Rules have been devised to check the validity of the representation of the string. These rules form the production sets of the grammar. The process of classifying the strings that represent hole cross-sections has been done by the Earley's parser (Earley, 1977). It is an efficient technique applicable to general context-free grammars. Given an input string x and a context-free grammar G, the algorithm accepts or rejects x depending on whether x is derivable from G. The approach was used strictly for holes and the algorithm was coded in LISP.

Yet another group (Choi *et al.*, 1984; Choi and Barash, 1985), worked on applying the syntactic pattern recognition approach to identify machined surfaces from a 3-D B-Rep model. The geometry definition based on ROMULUS is given in a particular data structure to define seven types of machined surfaces: hole, slot, pocket, step, 2-D contour, plane, and sculptured surfaces. The recognition of the machined surfaces from 3-D data is done by a pattern-matching procedure. If a hole is recognized it is removed from the boundary representation file and defined in terms of a special data structure. Likewise, routines have been written for slots, pockets, and the like. A corresponding process cycle is attached to each of the elementary machined surfaces (EMS), appropriate tools are classified, and a process plan is generated.

Srinivasan *et al.* (1985) have also applied the same technique to extract details from a CAD data base for group technology and process planning. They describe efforts in addressing the problems of part family classification and generation of cutter paths for 3-D surfaces. Four separate grammars were constructed for representing four part families. A set of primitives was identified for representing axisymmetric components that were machined from the ends or from the centre, and strings developed were parsed by Earley's algorithm to classify the part. Also, 3-D parts were represented by a tree structure, extensions to the string structure were defined, and multi-dimensional pattern recognition procedures based on grammatical inference procedures were used to obtain process plans.

In one of the recent works, Li (1988) proposed the same approach to rotational parts without deviation. This approach consists of the primary pattern primitives and use of a character string to describe the part. The string matching proceeds from the left to the right, each time picking an unevaluated pattern string. The difference in style is in the set of pre-defined features that is stored in the data base. Further, a systematic check is done for all possible cases and the output results in a set of pre-defined feature strings whose grammar could be used to generate the process plan. Some refinements in terms of the raw material shape identification and tolerances are added interactively, and this format is suited for group technology coding and NC machine programming.

General extraction techniques from part data structure format

During the same time that other extraction techniques were being developed, some researchers presented general schemes that were not based on any particular theory. Grayer (1977) was one of the first to use a geometric description to produce machined components automatically. The input to the scheme is a B-Rep description of the geometry and the output is a collection of contours representing laminae to be machined, together with sequences incorporating minimal machining information. The B-Rep of the part is compared to the initial workpiece, and a set of disjointed laminae is separated to be machined after an overall sequencing operation. Each lamina is represented by one or more 2-D contours whose shape is given by its cross-section at different heights. Chan (1982) developed an interface called ROMAPT between ROMULUS, a solid modeller, and APT, a high level NC processor. This program scans through the data structure starting from the body, through the faces, to the points and edges. The output is then converted to standard APT geometry format and NC code is generated.

A theoretical approach to unify features in a CSG tree was developed by Lee and Fu (1987). They devised algorithms for extracting manufacturing features from a CSG tree automatically. The non-uniqueness of the CSG tree is presented with data structures and algorithms for supporting the extraction. The scheme that has been proposed first generates the principal axes of the participating primitives; they are used to locate the features. Once the features are identified, the CSG tree is unified by special algorithms. This brings together two nodes that are far away from each other in the original tree by way of their commonality. This is probably the only work other than Woo's (1977) that used a CSG tree as the basic input format.

In another work, Wang and Wysk (1988) report a system that identifies machined surfaces on symmetric rotational parts from a 2-D representation. This process basically extracts lines and curve entities from a wireframe modelling system. The whole system is divided into three stages: a pre-processor that extracts all the drawing entities from a CAD file, a main processor that performs the search and identification of all machined surfaces from a given workpiece in terms of rules, and a post-processor that writes information related to the machined surfaces in an output file, in a relational format, for subsequent applications.

Production rule-based systems

The use of artificial intelligence techniques in manufacturing led to the application of these tools for extracting form features from a solid modelling system for group technology, process planning, and other applications (Henderson, 1984, 1986a,b; Henderson and Anderson, 1984; Kung, 1984; Hirschtick and Gossard, 1986; Joshi *et al.*, 1988). It is based on the premise that feature recognition follows certain rules of logic and that a human normally examines a 3-D part and classifies features by matching the faces, edges, and vertices of the part with a pre-defined mental rule. This rule, if interpreted properly, would result in the successful recognition of a pattern and features can be identified.

Henderson (1984) used a production rule approach to recognize and extract features from a CAD model. One interesting approach taken by Henderson is to use cavity volumes to recognize the features rather than the part definition data itself. Cavity volumes are found by subtracting part definition from the raw stock definition. This operation leaves only that material that needs to be manufactured. The remaining part definition is then examined for features. When a feature is found, the respective cavity volume is subtracted by using standard techniques and what is left is searched. The search is limited only to swept features because the majority of the machining features can be classified to be 2-1/2-D volumes created by sweeping a tool profile through space. Prolog was the language chosen for encoding the production rules.

An extrusion advisor was developed by Hirschtick and Gossard (1986) to advise a user in the domain of aluminium extrusions. The system examines a geometric model, extracts features suitable for extrusion, and identifies features that will cause problems. Typical patterns are searched for and these are combined to form a characteristic pattern set such as parallel lines, a triangle, or the like. A production rule-based system is used and the system can recognize walls, hollows, knife edges and other specifics. The output is a list of manufacturing features and reasons for difficulty with possible remedies.

Instead of using the standard backward chaining strategy as proposed by Henderson (1984) and Kung (1984), Joshi *et al.* (1988) used a data-driven or forward-chaining recognition system. They contend that in order to recognize features, the adjacency relationship between faces, a qualifier for the type of adjacency, is needed. This relationship is represented by an attributed adjacency graph. Once this graph is created, recognition rules written for generic features like step, slot and pocket, based on this feature graph, can be attempted. Because searching the entire graph may be an exhaustive procedure, the feature sub-graphs are isolated before recognition starts. The observation that nodes—adjacent to all their neighbouring faces having convex angles—do not form parts of features is used to delete all those nodes. The remaining graph is searched to identify the sequence of features.

Extraction from feature-based design systems

The concept of feature-based design systems (FBDS) has proceeded by integrating the feature model with the solid model either by building subsidiary data structures with pointers back to the entities in a B-Rep model or by using features as the primary element in the solid modeller. However, the consensus on FBDS has been to integrate both models (Pratt, 1988). This is because in most cases the features created at the design stage are used for downstream applications; in these cases it is wasteful to extract and store them separately for each application. Also, by pointing the feature constituents in the solid model, we can resolve inconsistencies and perform updating at any stage. Interrogation, or feature extraction, is simple in such a model because high-level information is explicit, and, for each application, the features can be readily identified and used.

In another new approach called the inverted form of process models, low-level models can associate processes with features of a solid; that is, to establish a surface feature to be produced by machining, a set of options is available that can be narrowed down by optimizing and ranking candidate sets. In this case, the feature is less important than the process. Theories for achieving this are still evolving (Voelcker, 1988).

Discussion

The preceding section identified the available means for representation and extraction of features from geometric models for manufacturing applications.

Integration of CAD/CAM is often assumed to be the creation of the link between a CAD system and a process planning system suitable for machining. Hence, form features have always been identified by researchers as the key to integration. This has resulted in a wide variety of feature extraction schemes that can recognize form features from a geometric model. Feature extraction had to be attempted because information implicit in the data base had to be made explicit with the solid modellers then available. The extraction schemes were primarily designed for B-Rep modellers. Only two techniques have been proposed for CSG extraction. Though volume decomposition proposed by Woo (1975) and the work of Lee and Fu (1987) provided two excellent sets of algorithms for the manipulation of geometry from CSG modellers on a rigorous mathematical basis, they failed to include the process semantics in the geometry such that when features are identified the geometry can be used for manufacturing. Even so, the identification of cavities and the definition of features by Woo (1975) contributed significantly to further research in this area.

Recognition methods like syntactic pattern recognition have involved the use of separate grammars for each class of parts that consisted of certain features like stepped shaft, grooved shaft, and the like. This type of classification can be made more specific by including separate grammars for the left stepped shaft, right stepped shaft, and so on. However, the classification does not give information regarding the syntactic arrangement of features in the classified part, which is essential for generating process plans. This information is not explicit in the rule of the grammar, though it is used implicitly for writing the grammar. Furthermore, writing grammar for each possible case is not possible. We believe that writing rules with domain-specific knowledge embedded in the production rule helps in the easy creation of process plans and will result in a compact hierarchical representation of the parse as a result of parsing the part for acceptance by the grammar. Also, extraction schemes like production rule-based systems provided a good means for high-level interaction. The hypothesis that human mental rules can indeed be coded for computer understanding is an excellent concept. However, their use for composite and complicated features is still questionable.

It is clear that recognizing features in a CAD model is one of the most difficult problems in integration. The reason for this difficulty is that CAD models are incomplete and when human manufacturing experts interpret geometry, they apply a great deal of ancillary information, analyse the intentions of the designers, and so on. However, no general technique has evolved from the various schemes proposed, and the research has been limited to universities with no major application vendors actually implementing these concepts. The new generation of feature-based design systems promises to close the gap sufficiently. It represents the logical next step from existing solid modellers and captures the manufacturers' intent at the design stage. These systems provide explicit representation and reasoning for the part to be processed but have certain limitations. FBDS still needs some form of recognition mechanism when used with application shells. However, features are attractive for CAD and Computer-Aided Process Planning (CAPP) because they raise the levels of abstraction and provide more complete information to the planner. Features alone are not sufficient for CAD/CAM integration, and mapping shells are needed for all these systems. This in essence boils down to the problem of writing application programs to facilitate easy integration, though the mapping is done at a higher level. We do agree that features are important from a manufacturing standpoint, but an integrated approach toward process-driven design seems the best. This is very clearly echoed in the work of Cutkosky and Tannenbaum (1987), Cutkosky *et al.* (1988), and Schulte *et al.* (1990).

A closed set of features is always incomplete. Also, agreement on what sets or classes of features can represent a modelling system requires much compromise. Feature recognition and feature-based design systems have been primarily used for form features with minor exceptions for casting and extrusion. It is the authors' contention that there exists a whole branch of manufacturing disciplines that is totally neglected. The scope of the CAD/CAM link should include other manufacturing technologies such as metal forming, unconventional metal removal processes, metal joining processes, and others that would enable horizontal integration in the overall context (Srinivasan and Liu, 1987).

We conclude that automatic planning and programming are required in other manufacturing processes as in welding, a joining process, and are even more complicated with the use of robots. By adopting a mixture of process and geometry in design, we can emulate the present ways of programming the robots and provide a paradigm for automatic generation of robot programs. Also, planning at the design stage is better than doing it after the fact from a completed set of features in drawings. The latter results in suboptimal programming; thus, it is always better to have reasoning done at the beginning. The 'programming-as-design' concept best suits our needs, and this incremental mode derives benefits from feature-based systems and process-based planning systems. For these reasons, we argue that a feature-driven, process-based incremental design is most suitable for manufacturing automation.

A paradigm for automatic programming of welding robots

The main objective of our work, described in the following paragraphs, is to propose a paradigm, based on object-oriented programming techniques, to provide a structure in geometric modelling systems for supporting complete welding information. The process and geometry of welding features are thereby coupled tightly to automatically generated robot welding programs. This research is the first step in the automated planning and programming for robots in the area of welding and will eventually provide the means for an integrated, automatic programming system for welding robots that would derive benefits from the graphical manipulation capabilities of a feature-based CAD system and the inferencing capabilities of a process-based planning and reasoning system. Complete details and status of the project are available in Padmanabhan and Devgun (in preparation).

The goal is to develop a system that can produce an output that is a directly usable code and that can be tested on a target machine, namely, an arc welding robot, to perform the desired welding. The major research is being devoted to these areas:

(1) studying the requirements of incorporating welding feature information in a geometric modelling system,
(2) designing data abstractions in which all welding information can be encapsulated and providing means to extend these abstractions,
(3) developing a reasoning system in the feature-based modeller to select proper orientation methods and welding parameters for the robot program. The system uses expert information and captures weld feature information in a single structure,
(4) developing a user interface and a post processor to use this information to generate a robot program and execute the program on a GE-P50 robot (because of its availability in our welding lab).

A prototype system is being developed to implement the integrated modelling environment. The system architecture is shown in Figure 3. At any instant, the user interface will provide the user a menu of legal operations for the current mode. Menus will appear that show default parameters, with graphic tools for specifying the geometry and drawing. The process details will be obtained by checking with the reasoning system, and feasibility of orientations and sequence will be determined. The information will then be compiled to provide the output program that can execute the motion in a welding robot. The system is being developed in an object-oriented programming environment, C++, and the solid modelling system, IDEAS, runs as a child process.

The geometry information that will be captured in the modelling system is based on the current, standard, unambiguous methodology adopted by the American Welding Society to represent any weld, its type and its properties. This system for supporting welding information in a CAD system that can drive the robot

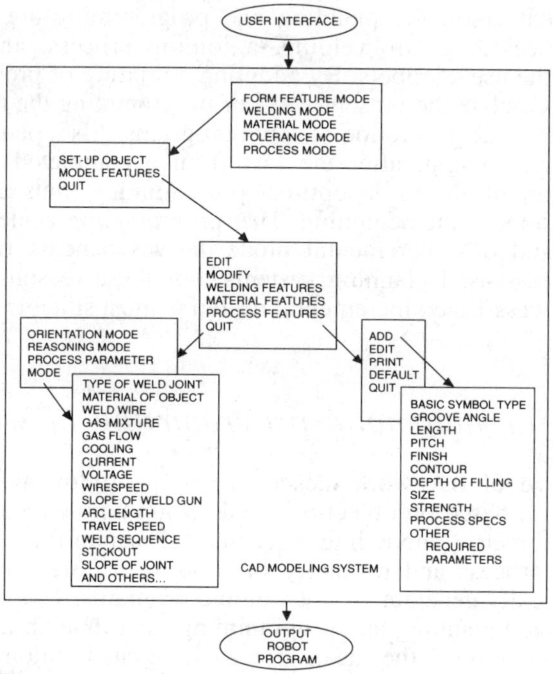

Figure 3. System architecture of the automatic programming system.

programming system is based on satisfying these requirements: (1) it can capture welding semantics, (2) it supports standards, (3) it is adaptable, (4) it can network with geometric entities, (5) it does validity checking of welds with geometric entities, (6) it does legality checking and default setting, and (7) it displays weld symbols. Current plans are under way to create this system simultaneously with the process reasoning system.

Considerable knowledge of the welding process is required to program a robot. The process parameters that we are considering are (1) type of weld joint, (2) material of object, (3) welding wire, (4) gas mixture, (5) gas flow, (6) cooling, (7) current, (8) voltage, (9) wire feed rate, (10) dwell, (11) torch angle, (12) displacement from the weld for multipass welds, (13) slope of the workpiece, (14) weld travel speed, (15) stick-out, and (16) weld sequence. The parameters 7–16 are interrelated. Current and wire speed are related in that a chosen current gives a particular wire speed or vice versa. This information can be categorized by data, facts and action-pair statements, and can be accessed by the modelling system as the design progresses. The remaining parameters are determined by manipulator constraints. The selection of these and the weld sequence is not algorithm-bound because of the freedom of the different objects and problems due to various factors including stresses and deformations. Though the interrelationships are not known, 'rules of thumb' always exist, and the practical approach of defining geometrical parameters first and then dictating the technological parameters is attempted in our incremental design. As the design progresses, with appropriate validity checks, the system is ready to write a neutral output file that can be directly used to generate the robot program in the high-level language required. The knowledge required for the welding process has been obtained through survey/interviews. Statistical analyses show certain major trends that have helped in the easy construction of the knowledge base. Proper interfaces and data structures are being developed to create this new generation of feature-driven, process-based systems.

Conclusions

The need for solid modelling systems to support tasks in automated manufacturing is clear. A consensus is evolving that a mixture of explicit feature representation and implicit feature description and extraction will be required in future commercial CAD systems that have capabilities for feature-oriented design. We believe that instead of using complicated analytical tools for feature extraction, we could raise the level of abstraction in geometric modelling systems and still provide comprehensive input to application tasks like process planning, group technology, and others. It is also clear that design is evolving in such a manner that future systems will allow designers to choose from different design modes such as conceptual, feature-oriented, process-oriented, hybrid and others. Also, the need for the extension of the scope of the CAD/CAM link to other manufacturing technologies is visible, and by our work we are attempting to close this gap in the area of welding with specific applications to automatic programming of welding robots. We hope that our intermediate, incremental approach of 'feature-driven, process-based' design will pave the way for automated planning and programming in welding and that it can be adopted in other areas of manufacturing as well because it provides an intelligent CAD environment that combines high-level planning and low-level operation details. Our system will grow over time into an integrated system that is competent and versatile in the automatic programming domain of robotic welding.

References

Chan, B.T.F., 1982, ROMAPT: a new link between CAD and CAM. *Computer Aided Design*, **14**, 5, 261–266.

Choi, B.K. and Barash, M.M., 1985, STOPP: an approach to CAD/CAM integration. *Computer Aided Design*, **17**, 4, 162–168.

Choi, B.K., Barash, M.M. and Anderson, D.C., 1984, Automatic recognition of machined surfaces from a 3-D solid model. *Computer Aided Design*, **16**, 2, 81–86.

Cunningham, J.J. and Dixon, J.R., 1988, Designing with features: the origin of features. *Proceedings of the ASME International Computers in Engineering Conference*, San Francisco, CA, August, pp. 237–243.

Cutkosky, M.R. and Tannenbaum, J.M., 1987, CAD/CAM integration through concurrent process and product design. *Proceedings of the ASME Winter Annual Meeting*, Boston, MA, December, pp. 1–10.

Cutkosky, M.R., Tannenbaum, J.M. and Muller, D., 1988, Features in process-based design. *Proceedings of the ASME International Computers in Engineering Conference*, San Francisco, CA, August, pp. 1–13.

Dixon, J.R., 1986, Artificial intelligence and design: a mechanical engineering view. *Proceedings of AAAI-86 Fifth National Conference on Artificial Intelligence*, Vol. 2, Philadelphia, PA, August, pp. 872–877.

Dixon, J.R., 1988, Designing with features: building manufacturing knowledge into more intelligent CAD systems. *Proceedings, Manufacturing International 88,* Vol. 1, *Product and Process Design*, Atlanta, GA, April, pp. 51–57.

Earley, J., 1977, An efficient context-free parsing algorithm. *Communications of the Association of Computing Machinery*, **13**, 2, 94–102.

Fu, K.S., 1978, *Syntactic Pattern Recognition, An Introduction.* (Reading, MA: Addison-Wesley).

Grayer, A.R., 1977, The automatic production of machined components starting from a stored geometric description. In *Advances in Computer-Aided Manufacture*, edited by D. McPherson (Amsterdam: North Holland), pp. 137–151.

Henderson, M.R., 1984, *Extraction of feature information from three dimensional CAD data.* Ph.D. Thesis, Purdue University, West Lafayette, Indiana.

Henderson, M.R., 1986a, Feature recognition in geometric modeling. *Proceedings of CAM-I's 13th Annual Meeting and Technical Conference*, Clearwater Beach, FL, November, pp. 5-1–5-12.

Henderson, M.R., 1986b, Automated group technology part coding from a three-dimensional CAD database. *Proceedings of ASME Winter Annual Meeting, Knowledge-Based Expert Systems in Manufacturing*, December, Anaheim, CA.

Henderson, M.R., and Anderson, D.C., 1984, Computer recognition and extraction of form features: a CAD/CAM link. *Computers in Industry*, **5**, 329–339.

Hirschtick, J.K. and Gossard, D.C., 1986, Geometric Reasoning For Design Advisory Systems. *Proceedings of the ASME International Computers in Engineering Conference*, Chicago, IL, July.

Hummel, K.E. and Brooks, S.L., 1986, Symbolic representation of manufacturing features for an automated process planning system. *Proceedings of ASME Winter Annual Meeting, Knowledge-Based Expert Systems in Manufacturing*, December, Anaheim, CA.

Jakubowski, R., 1982, Syntactic characterization of machine part shapes. *Cybernetics and Systems: An International Journal*, **13**, 1–24.

Jared, G.E.M., 1984, Shape features in geometric modeling. In *Solid Modeling By Computers: From Theory to Applications*, edited by M.S. Pickett, and J.W. Boyse, pp. 121–133.

Joshi, S., Vissa, N.N. and Chang, T.C., 1988, Expert process planning system with solid model interface. *International Journal of Production Research*, **26**, 5, 863–885.

Kung, H., 1984, An investigation into development of process plans from solid geometric modeling representation. Ph.D. Thesis, Oklahoma State University.

Kyprianou, L.K., 1980, Shape classification in computer-aided design. Ph.D. Thesis, University of Cambridge, Cambridge, UK.

Lee, Y.C. and Fu, K.S., 1987, Machine understanding of CSG: extraction and unification of manufacturing features. *IEEE Computer Graphics and Applications*, January, 20–32.

Li, R.K., 1988, A part-feature recognition system for rotational parts. *International Journal of Production Research*, **26**, 9, 1451–1475.

Liu, C.R. *et al.*, 1984, Generative process planning using syntactic pattern recognition. *Computers In Mechanical Engineering*, 63–66.

Miner, R.H., 1985, A method for the representation and manipulation of geometric features in a solid model. M.S. Thesis, Massachusetts Institute of Technology.

Ostrowski, M.C., 1988, A prototype feature modeler using constructive solid geometry (in press).

Pratt, M.J., 1988, Synthesis of an optimal approach to form feature modeling. *Proceedings of ASME International Computers In Engineering Conference*, San Francisco, CA, August.

Pratt, M.J. and Wilson, P.R., 1985, *Requirements for the Support of Form Features in a Solid Modeling System*. Report No. R-85-ASPP-01 (Arlington, TX: CAM-I Inc.).

Requicha, A.A.G., 1980, Representations for rigid solids: theory, methods, and systems. *Computing Surveys*, **12**, 4, 437–464.

Requicha, A.G.G. and Voelcker, H.B., 1982, Solid modeling: a historical summary and contemporary assessment. *IEEE Computer Graphics and Applications*, March, 9–24.

Schulte, R., Padmanabhan, S. and Devgun, M.S., 1990, A feature-driven, process-based approach to the integration of CAD/CAM in wireframe models. *International Journal of Production Research* (in press).

Shah, J.J., 1989, *Current Status of Features Technology*. Report No. R-88-GM-04·1 (Arlington, TX: CAM-1 Inc.).

Shah, J.J. and Rogers, M.T., 1988, Functional requirements and conceptual design of the feature-based modeling system. *Computer-Aided Engineering Journal*, February, 9–15.

Shah, J.J. and Wilson, P.R., 1988, Analysis of knowledge abstraction, representation and interaction requirements for computer aided engineering. *Proceedings of ASME International Computers In Engineering Conference*, San Francisco, CA, August.

Srinivasan, R. and Liu, C.R., 1987, On some important geometric issues in generative process planning. *Proceedings of ASME Winter Annual Meeting*, Boston, MA, December.

Srinivasan, R., Liu, C.R. and Fu, K.S., 1985, Extraction of manufacturing details from geometric models. *Computers In Industrial Engineering*, **9**, 2, 125–133.

Staley, S.M., Henderson, M.R. and Anderson, D.C., 1983, Using syntactic pattern recognition to extract features from a solid geometric data base. *Computers in Mechanical Engineering*, September, 61–66.

Unger, M.B. and Ray, S.R., 1988, Feature-based process planning in the AMRF.

Proceedings of ASME International Computers In Engineering Conference, San Francisco, CA, August.

Voelcker, H.B., 1988, Modeling in the design process. In *Design and Analysis of Integrated Manufacturing Systems*, edited by W.D. Compton (National Academy Press), pp. 167–199.

Wang, H.P. and Wysk, R.A., 1988, AIMSI: A prelude to a new generation of integrated CAD/CAM systems. *International Journal of Production Research*, **26**, 1, 119–131.

Woo, T.C., 1975, Computer understanding of designs. Ph.D. Thesis, University of Illinois, Urbana-Champaign.

Woo, T.C., 1977, Computer aided recognition of volumetric design. In *Advances in Computer Aided Manufacture*, edited by D. McPherson (Amsterdam: North Holland).

Woo, T.C., 1982, Feature extraction by volume decomposition. *Conference on CAD/CAM Technology in Mechanical Engineering*, MIT, MA, March.

Woo, T.C., 1984, Interfacing solid modeling to CAD and CAM: data structures and algorithms for decomposing a solid. *IEEE Computer*, December, 44–49.

Woodwark, J.R., 1988, Some speculation of feature recognition. *Computer-Aided Design*, **20**, 4, 189–196.

A CAD/CAM system for the application of two-part adhesives using robots

R.J. Grieve, R. Herfie and C. Hitchen

Department of Manufacturing and Engineering Systems, Brunel University, Uxbridge, Middlesex UB8 3PH, UK

Abstract

The use of adhesives in precision assembly tasks is growing; however, the software and automation hardware fully to take advantage are limited. The work described encompasses two areas: the development of a unit for dispensing fine epoxide beads and its integration with an Off-Line Programming Aid for an industrial robot to enable rapid application programming.

The Two-Part Adhesive Dispensing System is of small, compact design and can be manipulated by a robot over complex paths. The unit features dual component metering, mixing and variable flow rate dispensing controlled by stepping motors. Cleaning of the device is achieved by solvent flushing.

The CAD Robot Interface provides graphical representation of the component to be bonded and simulation of the adhesive path to be applied. The necessary robot code to achieve this is then generated by the menu-driven package. Reference points for fixturing devices can be added and the application parameters selected.

Introduction

The development of modern bonding materials has led to their increasing use as a medium for assembly. A number of dispensing and assembly cells have been set up but few have been incorporated into a computer-aided design/computer-aided manufacture (CAD/CAM) system. In addition these applications have centred on the dispensing of single-part adhesives or sealants. Typical examples are in the areas of printed circuit board (PCB) solder masking, surface-mount pre-bonding on PCBs, or low precision sealant and gasket applications for which *X-Y* cartesian manipulators are often used. Syringe-based dispensers are most frequently employed in these instances whereas the more complex robot processes for applying sealant to car windscreens incorporate bulk reservoirs with mechanical pumping units and variable-orifice, flow control valves. There are, however, many problems that need to be overcome and especially with two-part adhesive compounds, where accurately metering the correct ratio, thoroughly mixing and then being able to cope with a self-curing system are essential. With automatic cells effectively only able to apply single-part adhesives, the designer's choice of compound is limited.

Much of the commercial equipment on the market is associated with encapsulating or single-shot applications, although some robotic systems for composite panel bonding in the US automotive industry have been developed. However, again these are low-precision, high-volume applications. There are many applications for the use of two-part epoxies where precise deposition of small volumes is required which may be suitable for automation by reason of quantity and quality requirements. These applications exist in the automotive and electronics industries but as yet there is no suitable dispensing system available. There is a need therefore to develop dispensing heads capable of applying two-part adhesives.

This chapter describes the integration of two problem areas when applying adhesives using industrial robots, the manual input required when programming the robot, and the design of a suitable metering and dispensing head, which although independent in terms of automatic control, is attached to the robot arm. An assembly cell has been set up which incorporates an IBM 7547 industrial robot with a working area laid out to accommodate small components (typically flat and 20 x 20 cm in area). A two-part epoxy is used as the adhesive and fine, high-precision dispensing attempted. The CAD/CAM system consists of an off-line programming aid capable of producing robot program files in a suitable format for acceptance by the robot controller and an automatic dispensing system that allows the fully automated metering, mixing and flow-rate-controlled adhesive application.

The off-line programming aid

CAD systems for use with numerical control (NC) machines have been available for a number of years. These provide off-line programming capabilities and are recognized and proven techniques. Industrial robots on the other hand have lagged behind in this regard. There are some graphical robot modellers available allowing off-line programming, but these tend to be expensive and beyond the purchasing power of small companies. Many advantages could be gained if reasonably priced CAD packages could be modified so as to allow off-line programming of robots. The requirement would be that the component on which adhesive is to be dispensed is 'designed' on the CAD package. In addition the designer or manufacturing engineer could also show the track along which the adhesive bead is to be laid. The component with bead could be represented as a 3-D model or in orthogonal views. The user could then use the programming aid to interrogate the drawing and generate the high-level robot program including the data points. With the introduction of graphics standards the CAD system on which the original design was generated is irrelevant providing a system such as Initial Graphics Exchange Specification (IGES) is available.

In the present work a microcomputer-based CAD system is used (Micro CADDS Geometric Construction and Detailing—GCD) together with an accompanying high-level language User Programming Language (UPL) which is specifically designed by the suppliers to interact directly with the CAD data base. Using the on-screen menu facility it is possible to reconfigure the system to perform operations other than those supplied by the original package by writing functions in UPL that may be accessed by using modified icon-driven menus. UPL strongly supports geometrical programming since it allows variables of type COORD to be defined for example which contain x–y–z coordinates of a point or vector. Intrinsic functions that calculate the vector dot product, the vector cross product, the length of a vector or the coordinates of a unit vector encourage the use of vector analysis for calculations. Functions also allow data to be obtained from the data base, usually in an interactive mode, and stored in variables for direct access. GETDIG, for example, stores the x–y–z coordinates of a point that is digitized in a COORD variable. At present the system is configured so as to allow off-line programming of an IBM 7547 industrial robot. The high-level language used to drive this robot is A Manufacturing Language (AML). The structure of AML allows the user to define global points and constants at the beginning of a program. This is a useful facility when performing many similar applications using a robot differing only in

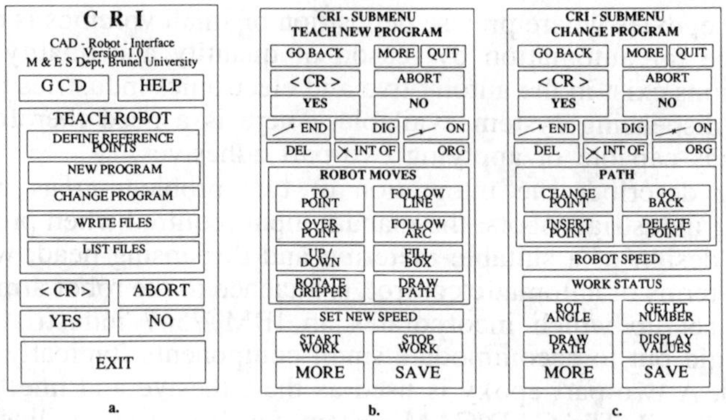

Figure 1. Typical on-screen menus.

the number of points and value of coordinates. An example of this is the laying of adhesive beads which simply involves moving the robot to a starting point, switching on a pump and moving through a series of points followed by switching off the pump and retraction of the head. If the point coordinate data could be generated off-line and the AML program generated transparently to the operator in a text file a reduction in robot programming time will result. This would also greatly assist the precise definition of the adhesive path on the component which may be difficult to teach manually. The AML program may then be compiled and run. In order for the system to operate it is necessary that three coordinate points on the component in the robot working environment be known and compared with the corresponding points in the programming system. This will enable all the necessary transformations from the programming aid to robot coordinates to be made.

CAD/robot interface (CRI)

CRI consists of two parts that may be described as an inner and outer package. The outer package is in the form of a menu that allows the user to select various options. The options are: load the AML system, load GCD, copy files to disc, list files, print files or robot programs. Help information is also available in the package.

The inner package consists of a set of UPL programs and operates within the GCD package. The UPL programs allow the user to create a robot program in AML with only the CAD drawing at hand.

Typical on-screen menus are shown in Figure 1. Sub-menus [Figure 1(b), (c)] may be called from the main menu shown in Figure 1(a). There are various options and procedures available (selected by mouse and cursor), a number of the most important are as follows:

— TO POINT When selected, a point must be digitized. The robot program will then cause the robot to move to that point.
— UP/DOWN Changes the height of the robot over a point.
— FOLLOW LINE This causes the robot to move in a straight line between two digitized points using linear interpolation.
— FILL BOX This is a function to allow the filling of areas on plane surfaces. Three or four digitized points are required to represent the corners of the area. This is useful for adhesive laying when a defined surface must be covered. The entry and exit points of the area can be defined together with the pitch of the bead track.
— DRAW PATH The path programmed by the user is superimposed on the drawing.
— SET NEW SPEED Allows robot speed to be changed.

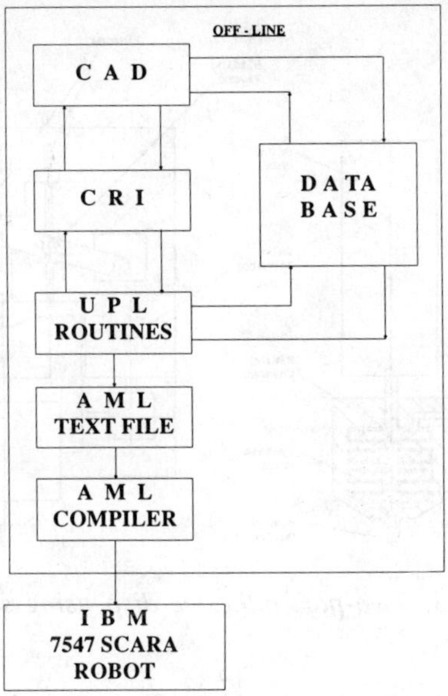

Figure 2. Schematic interaction of CAD/robot interface elements.

The system is configured in such way as to allow use by the designer or a person with minimal knowledge of robot programming. Component drawings held in the CAD data base may be accessed and robot motions defined by using the menus together with cursor movement over the display of the component.

After storing the data the AML program can be generated after responding to a number of prompts, the main one being defining three points on the CAD coordinate system in robot world coordinates. This is a relatively simple task if the fixturing arrangement is known. This allows the data generated from the CAD system to be transformed into robot working space. A schematic representation of the interaction of the various elements is shown in Figure 2.

Robot dispensing system

In the automated production environment of a robotic cell it is necessary for all the elements to perform their tasks repeatedly, reliably and with minimum manual input. For a two-part adhesive dispensing system to meet these requirements it should: meter the correct ratios of the adhesive components; thoroughly mix these together; and enable precise, programmable dispensing. To ensure continued trouble-free operation it should also be able to purge itself of potentially curing adhesive and completely clean out all residue on a regular basis.

The direct displacement Two-Part epoxide Adhesive Dispensing System (TP-ADS) has an overall size of 30 × 20 × 13 cm and weighs approximately 4 kg. The dispenser (Figure 3), is mounted on the robot tooling point with remote reservoirs for the adhesive components connected by flexible hoses. The TP-ADS features several elements to provide the facilities required for the automated application of fine beads of adhesive. Direct displacement metering pumps use two micro-bore pistons that act within a dual cylinder block, one for each adhesive component. These give a positive displacement, either drawing material into the chambers or pumping it out. The base resin piston is directly linked to a sliding beam. An adjustable ratio lever connects the hardener piston. As the pistons are retracted the material loaded into the chambers is thus metered to the mixing ratio. Reversing

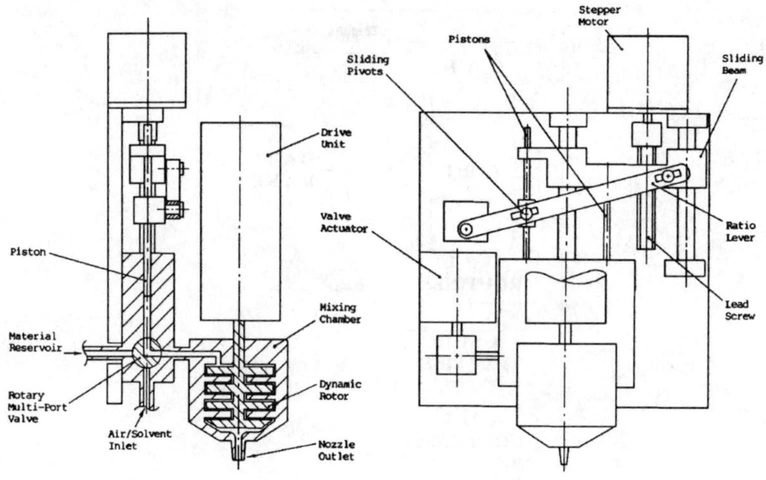

Figure 3. Two-part adhesive dispensing system.

the motion produces a direct pumping action. The central feature of the dispensing system is the double-banked, four-ported valve unit. By rotating a central shaft any adjacent pair of ports can be aligned. The two banks connect the reservoir inlet ports to the pump cylinders enabling reloading. Indexing the shaft through 90 degrees opens the mixer manifold ports and closes the inlets. Material displaced by the pumps is thus brought together on exiting the unit. The fourth port on each bank is a solvent inlet. These bypass the reservoir inlets and pumping cylinders allowing the mixer manifold and chamber to be flushed out with solvent. This is finally dried with filtered air before re-priming the mixing chamber. The reservoir and cleaning system is controlled by a separate PLC which monitors the material usage and adhesive working life status.

The mixing chamber is immediately prior to the outlet nozzle, hence providing a minimal mixed volume. This is of the dynamic rotor type, driven by a pneumatic motor and shears the adhesive components between the stator pins ensuring a thorough mix. The stepper motor provides the linear motion of the piston arrangement via a lead screw and is controlled by a programmable computer interface. Hence the flow rate can be varied and by using auxiliary inputs/outputs (I/Os) provide automatic or manual operation.

The stepper motor chopper driver oscillator which sets the base step pulse may be adjusted by varying an in-line resistor, thus providing a means of changing the dispensing rate. A suitable bank of parallel resistors in line with the driver oscillator may be configured in a number of combinations by using the binary outputs on the robot controller. The outputs are simple normally open circuit switches that may be operated from the AML program. Whenever there is a call for a change of speed from the AML program the status of the outputs are set to select the appropriate resistors and hence adjust the flow rate. It is possible to set up to eight discrete dispensing rates by this means involving the use of three outputs. More settings may be provided by expanding the output and resistor combinations used.

The TP-ADS in its present form is only capable of varying the dispensing rate by using the binary outputs from the robot controller. With many robots the status of the outputs can only be changed when a programmed point has been reached. It takes no account therefore of any change in velocity that may occur between points. The robot controller drives the manipulator arm radial shoulder and elbow sections through d.c. servo motors and monitors the optical encoder feedback loops. The resultant linear velocity of the tool point is thus dependent on the position of the arm and its movement direction. As the robot follows a path it accelerates and decelerates between intermediate target points and the distance separating these affects the extent to which it reaches its programmed speed. The net effect of this is a wide range of tooling point velocities which, with a constant dispensing flow rate, produces a non-uniform bead. If the change in velocity is to be taken into account it must be monitored on a continuous basis and either used to provide a direct feedback signal continuously to change the dispensing rate during program execution (with the attendant control problems), or be used to modify the adhesive application system before dispensing begins. Whichever of the options are used they will require a means of monitoring velocity which could require an investment in a relatively expensive robot where the velocity data may be extracted from the controller or use a quite separate measuring device. In the vast majority of instances such tight control over dispensing rate will not be required and the relatively simple solution of providing discrete dispensing rates will suffice.

SECTION VIII.2. CAPP

94

Computer-aided process planning in plastics manufacture

William E. Biles

Department of Industrial Engineering, University of Louisville, Louisville, KY 40292, USA

Abstract

This chapter describes a computer-aided process planning (CAPP) system for plastic parts fabrication. This system relies on computer-aided design/computer-aided manufacture (CAD/CAM) technology and a comprehensive data base relating the engineering properties of a plastic material to its compounding formulation and processing conditions. This system enables the design engineer to describe not only the part geometry, but also its method of fabrication, composition and processing conditions. The objective of this CAPP system is greatly to shrink the lead time in producing new plastic parts, from receipt of design requirements to delivery of manufactured items.

Introduction

Although increasing use is being made of plastic parts in manufactured products, scant attention has been given to the problem of automating plastics fabrication processes. Great strides have been made in CAD of plastics, but progress is much slower in CAM of plastics. This is particularly true in CAPP.

An area in which plastics technology differs markedly from metals processing is in selecting the compounding formulation of the plastic material in order to have the product meet selected physical, thermal, mechanical, chemical and electrical properties. In general, a plastic compound consists of a resin, one or more fillers to enhance its engineering properties, and selected chemical additives to aid processing. Choosing the type and proportion of each of these constituents in the compounding formulation is an important determinant in how the plastic material performs in service. Likewise, selecting the means of fabrication (injection moulding, extrusion, blow moulding, etc.), as well as the processing conditions employed in fabricating the part is crucially important to both product performance and manufacturability of the plastic part. Compared to metals processing, the range of choices in plastics fabrication is extremely broad. The approach to CAPP for plastics is, therefore, necessarily quite different from that for metals.

Computer-aided process planning

CAPP involves the automatic generation of a process plan to manufacture a part. This process plan is developed by recognizing the specific attributes of the part in question, including both design and manufacturing attributes, and relating these attributes to the corresponding manufacturing operations. Design attributes include geometric shape and size, as well as specifications on engineering properties (physical, thermal, mechanical, electrical and chemical). Manufacturing attributes include method of fabrication (injection moulding, blow moulding, extrusion, etc.), processing conditions (temperature, flow rate, hold time in the mould, etc.), and post-processing treatment (machining, spot welding, ultrasonic treatment, etc.).

CAPP in plastics parts manufacturing involves three primary functions:

(1) *Compounding*—Selection of the resin or polymer, the filler material(s), and any additives such as plasticizers, colourants and flame retardants. Crucial to compounding is the choice of the proportions in which the constituents are combined.
(2) *Forming process*—Selection of the basic method of forming the plastic to the desired size and shape. Common forming processes include injection moulding, extrusion, blow moulding, and compression moulding.
(3) *Processing conditions*—Selection of the mould and plastic temperatures, flow rates, holding times in the mould, moulding tolerances, etc.

Engineering design data base

An engineering design data base has been developed to relate the engineering properties of selected plastics materials to their compounding formulations and processing conditions. The purpose of this computerized data base is to allow the design engineer ready access to information on resins that have engineering properties closely matching those needed for a proposed new commercial application. If a plastic material can be found in the data base that provides the requisite properties, that resin is combined with selected fillers to give it any needed property enhancements and other additives to aid processing. Typical of the type of engineering data base described here is the GE Plastics Engineering Design Database developed by the GE Plastics Division of the General Electric Company (GE Plastics, 1989). This data base is offered as a service to GE Plastics' customers to aid in designing their products using GE resins.

The property data contained in the GE Plastics' Engineering Design Database are based on mean values achieved at several strain rates, stresses, temperatures and times for a particular resin. The approach proposed here is to interrogate the data base seeking a resin possessing mean values for its engineering properties that are within a target range of values specified for the new product. For each engineering property, a candidate set of resins meeting the stated specifications is listed. After all important design properties have been entered, a final list of candidate resins is produced which meets all engineering property specifications for the new plastic product, providing that at least one such resin exists.

An experiment file is then accessed for each of the candidate resins. This file contains experimental data for each experimental design point that has been evaluated in prior process development efforts. This file contains mean values for each engineering property, as measured through experimentation, along with the proportion of each constituent used in the compounding formulation, the preferred fabrication process, and the processing conditions employed to produce the plastic test specimens which provided the sample data in the experiment. The design engineer seeks a combination of compounding formulation and processing conditions which will produce the desired properties for a given fabrication method. If such a combination is already recorded in the data base, a production run is then conducted to manufacture the needed parts employing the selected composition and processing conditions. If such a compounding formulation does not exist, however, a structured approach must then be taken to isolate the appropriate formulation and processing conditions experimentally. This experimentation must be undertaken expeditiously in order to obtain the correct formulation in time for a scheduled production run. One might refer to this procedure as Just-in-Time process experimentation, resembling as it does the manufacturing philosophy that delivers parts, materials, and subassemblies to the production line 'just-in-time' for scheduled operations.

Compounding new plastic materials

When the data base fails to produce a compounding formulation providing the desired engineering properties, it becomes necessary experimentally to determine a new compounding formulation that does yield those properties. This 'compounding' issue is discussed by Wigotsky (1989). This is managed by considering selected existing entries in the data base as design points in a statistical experimental design, and augmenting these with new design points in promising, but as yet unexplored, regions in the experimental factor space.

This involves two types of design or 'independent' variables:

— composition variables, which derive from the level of each constituent in the mixture,
— process variables, which refer to the conditions under which the mixture is formed into a part.

Unlike the situation in which a commercially available plastic pellet is employed in the fabrication of a new plastic product, the need for a new compounding formulation requires process experimentation 'on the fly'. In the CAPP methodology proposed here, this experimental effort is based on the concept of the 'mixture problem', which was originally posed by Scheffe' (1958). Cornell (1981) provides an excellent, up-to-date treatment of the 'mixture problem'.

Biles (1971) developed an optimization-based experimental approach to determining the compounding formulation and processing conditions for a polymer-based engineering material. Biles combined experimental designs predicated on the 'mixture problem', together with response surface fitting and multivariable optimization. This three-phased approach involved the following steps:

(1) Process experimentation—Obtaining data through designed experiments.
(2) Model building—Fitting linear and/or quadratic models relating the engineering properties (i.e. 'responses') of the material to a set of composition variables and a set of process variables.
(3) Mathematical optimization—Using the fitted models in a mathematical search for the optimum compounding formulation and processing conditions.

The simplex and simplex lattice experimental designs shown in Figure 1 are based on the work by Scheffe' (1958) in which the vertices of a k-dimensional polyhedron represent a compounding formulation involving 100% of a given constituent. If x_i is the weight fraction of the ith constituent in the compounding formulation, it is clear that for a k-constituent mixture

$$0 \leq x_i \leq 1, \qquad i = 1, \ldots, k \tag{1}$$

$$\sum_{i=1}^{k} x_i = 1. \tag{2}$$

It is unlikely that compounding formulations involving pure constituents, i.e. 100% weight fraction of a given ingredient, have any usefulness in plastics technology; hence, formulations involving upper and/or lower bounds on constituent levels are more appropriate. Relation (1) can, therefore, be rewritten

$$0 \leq a_i \leq x_i \leq c_i \leq 1, \qquad i = 1, \ldots, k. \tag{1a}$$

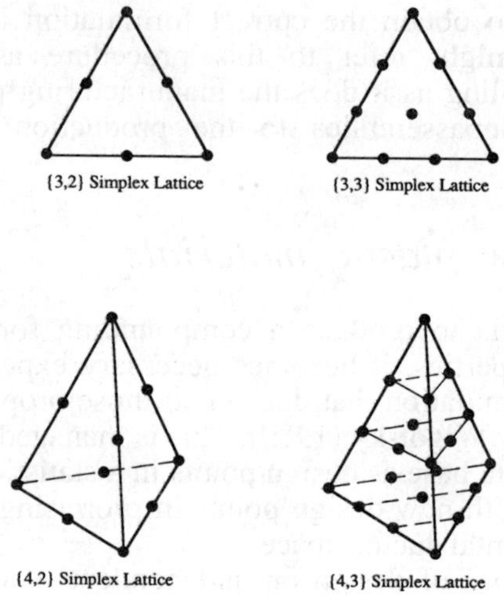

Figure 1. {k, r} *Simplex lattice designs for three- and four-component systems.*

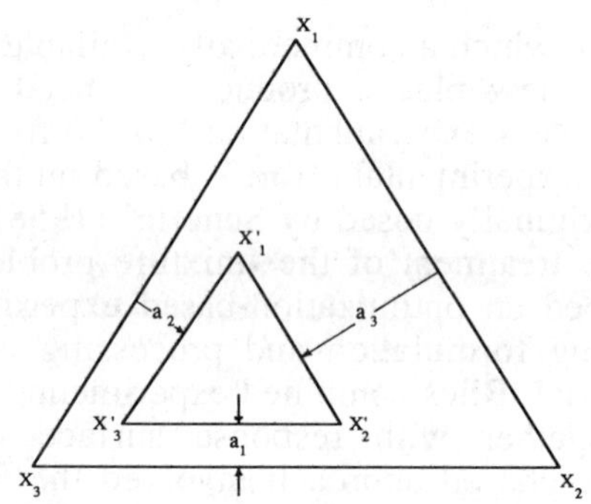

Figure 2. Triangular subspace representing a three-component system with lower bounds on all three components.

Figure 2 illustrates this relation for a three-component mixture.

It is also true that if any $k - 1$ constituent levels are fixed, then

$$x_k = 1 - \sum_{i=1}^{k-1} x_i . \tag{2a}$$

If we say that there are n total independent variables affecting each of the m engineering properties y_j, $j = 1, \ldots, m$, then if $k - 1$ of these are *composition* variables, the remaining $n - (k - 1)$ independent variables are *process* variables, such as plastic temperature, flow rate, mould temperature, etc. Hence, the total set of independent variables for the mixture problem becomes

$$\underset{\text{(composition variables)}}{x_1, x_2, \ldots, x_{k-1}} \qquad \underset{\text{(process variables)}}{x_k, x_{k+1}, \ldots, x_n .} \tag{3}$$

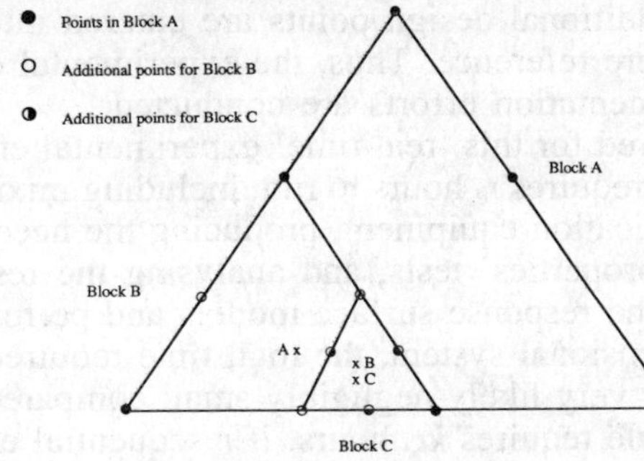

Figure 3. Sequential experimental blocks in a simplex lattice design for a three-component system.

The optimization problem in designing a new plastic material then becomes one of selecting the compounding formulation $(x_1, \ldots, x_{k-1})$ and processing conditions $(x_k, \ldots, x_n)$ in order to optimize the performance of the engineering properties $(y_1, \ldots, y_m)$. One mathematical statement of this problem (but by no means the only one or even the most appropriate one) is the 'constrained optimization' representation

$$\max(\min)\ y_j = g_j(x_1, \ldots, x_{k-1}, x_k, \ldots, x_n), \qquad j = 1, \ldots, m \tag{4}$$

subject to restrictions that

$$0 \le a_i \le x_i \le c_i \le 1, \qquad i = 1, \ldots, k-1 \tag{5}$$

$$a_i \le x_i \le c_i \qquad i = k, \ldots, n \tag{6}$$

where $g_j(x)$ is the real, but unknown, function relating the jth engineering property of the plastic material to its composition and to the processing conditions under which it is manufactured. The restrictions in relation (5) pertain to the k-dimensional composition factor space, while relation (6) says that processing conditions must remain within equipment capabilities and other known bounds.

To illustrate this three-phase approach to process experimentation, suppose that the existing experimental design points contained in the experimental data base form the {3, 2} simplex lattice shown as Block A in Figure 3 (solid circles). None of these available formulations produces the desired engineering properties, but a mathematical search employing fitting response surface models in the optimization problem (4)–(6) predicts that a suitable formulation might be found at point A. Since point A lies within a {3, 1} simplex subset of the original experimental region, three augmenting design points (open circles) are conducted and new response surface models fitted using the {3, 2} simplex lattice shown as Block B. A mathematical search of this region predicts an optimal solution at point B. Repeating this process with three more augmenting design points yields a solution at point C, which lies sufficiently close to point B to be able to accept point C as an 'optimal' compounding formulation.

Observe that this experimental procedure utilized six experimental design points that had already existed in the experimental data base. Six additional experimental design points were conducted in two segments of three points each. The response surface models fitted for Blocks B and C each utilized three 'old' design points and three 'new' design points. When this experimental process has been completed the data from these six additional design points are entered into the experimental data base for possible future reference. Thus, the experimental data base grows as additional process experimentation efforts are conducted.

Consider the time required for this 'real-time' experimental effort. Suppose each experimental design point requires t_e hours to run, including mixing the ingredients, loading them into the fabrication equipment, producing the needed test specimens, performing the requisite properties' tests, and analysing the results. Suppose that t_f hours are needed to fit the response surface models and perform a mathematical search. Then, for a k-dimensional system, the total time required per experimental cycle is $kt_e + t_f$. Since t_f is very likely negligibly small compared to t_e, we can say that each experimental cycle requires kt_e hours. If r sequential experimental cycles are needed to produce a formulation yielding the requisite engineering properties, the total time required is rkt_e hours. Thus, any savings in the number of experimental design points in each block translates directly to time (and cost) savings in the experimental programme.

By performing this three-phased approach to process optimization, the desired new compound can be determined 'just-in-time' to employ it in a production run. It may have been necessary, however, simultaneously to engage in the design of any new mould or die that might be needed to manufacture this newly designed plastic part. The next section briefly describes this procedure.

Computer-aided design and machining of moulds and dies

The time available for performing any needed process experimentation is a function of engineering design activities going on simultaneously with the process development effort. In general, these activities are as follows:

(1) CAD of the plastic part using 3-D graphical modelling techniques.
(2) CAD of the mould or die needed to fabricate the plastic part designed in step 1. This will entail design of gates and runners to allow flow of the molten plastic into the mould cavity, as well as coolant passages to allow flow of chill water through the mould.
(3) Simulation modelling of plastic flow and part cooling, using computer-based process simulation tools.
(4) Final selection of a mould or die design, and development of a computer program for computer numerical control (CNC) machining of the mould or die. This step is managed automatically by having the computer translate the 3-D model of the mould or die into a set of machining instructions.
(5) Downloading the CNC program to the most appropriate CNC equipment for machining of the mould or die.

Once the new mould or die is available, the process equipment is fully configured to produce the needed plastic parts. After the process experimentation phase has provided the desired compounding formulation and processing conditions, a pilot production run can get underway (it is usually desirable to perform a pilot production run to verify all process parameters before initiating the actual production run). The actual production run follows a successful pilot run. Rosato and Rosato (1986) provide a detailed treatment of the use of CAD/CAM technology in plastics design.

Conclusions

This chapter has described a workable methodology for CAPP in plastics manufacture. This approach differs markedly from the traditional approaches employed in metals fabrication processes, owing to the different processing methods employed in plastics manufacture. The most important departure, however, lies in the need to seek new and improved compounding formulations for plastic materials 'on the fly'. The method proposed here incorporates process experimentation based on the 'mixture problem', and utilizes as many prior experimental results as possible. Therefore, a computerized data base relating the engineering properties of plastics to their compounding formulation, method of fabrication, and processing conditions is essential to the successful implementation of a CAPP system for plastics. A second crucial component of the CAPP methodology is an algorithmic approach to process experimentation in a 'real-time' operational environment. This algorithmic approach is provided by a three-phased experimental optimization approach based on the well-known 'mixture problem'.

Acknowledgement

The assistance of Mr Anthony Springate in conducting this research is gratefully acknowledged.

References

Biles, W.E., 1971, A constrained optimization methodology for composite materials process development. Unpublished Ph.D. Dissertation, Virginia Polytechnic Institute and State University, Blacksburg, Virginia, USA.

Cornell, J.A., 1981, *Experiments with Mixtures* (New York: Wiley).

GE Plastics, 1989, *Product Information and Resources* (Pittsfield, MA: General Electric Company).

Rosato, D.V. and Rosato, D.V., 1986, *Injection Molding Handbook* (New York: Van Nostrand Reinhold).

Scheffe', H., 1958, Experiments with mixtures. *Journal of the Royal Statistical Society, Series B*, **20**, 344–360.

Wigotsky, V., 1989, Compounding. *Plastics Engineering*, February, 27–34.

95

Computer-aided process planning for electrochemical machining

Gursaran, J.L. Batra and V.K. Jain

Department of Mechanical Engineering, Indian Institute of Technology, Kanpur, India

Abstract

In this chapter a systematic methodology is presented for the generation of computerized process plans for electrochemically machined components produced on standard sinking machines. The proposed system is divided into two modules: the system input module and the planning module. The input module takes information on available machine tools, anode material, electrolyte properties and features to be machined on the workpiece. The planning module has an hierarchical structure. It embodies procedures for combining features for simultaneous machining, sequencing them and determining the optimal machining parameters. These procedures are repeated for each machine tool and the most economical machine tool for performing all the operations is selected. A fully interactive program, which ensures that the experience of the planner is not missed out at any stage, has been developed. The whole system has been coded in Turbo Pascal (version 3·0) and can be implemented on any IBM PC XT/AT compatible.

Introduction

Electrochemical machining (ECM) made its industrial debut in 1959–60. With the development of ECM as a qualified production process, the applications of this process have increased especially in the area of aerospace, gas turbine and nuclear industries. ECM represents a revolution in machining as metal is dissolved away atom by atom rather than torn away mechanically. Engineers in this field of manufacturing technology rely heavily on experience and empirical relations for the design of both the tool and the process.

Molloy (1965) has suggested an approach for the calculation of machining parameters and operating costs for the ECM operation by taking the example of a turbine blade. However, this planning procedure is incomplete, non-optimal and cannot be generalized for problems with constant feed rate. Calculation of optimal machining parameters has been attempted by Bhattacharya (1977) and more recently by Acharya *et al.* (1986). A review of the literature, however, suggests that a systematic procedure for process planning of components machined exclusively by ECM does not exist.

Chang and Wysk (1985) have discussed a large number of automated process planning techniques for conventional machining processes. Most of these systems

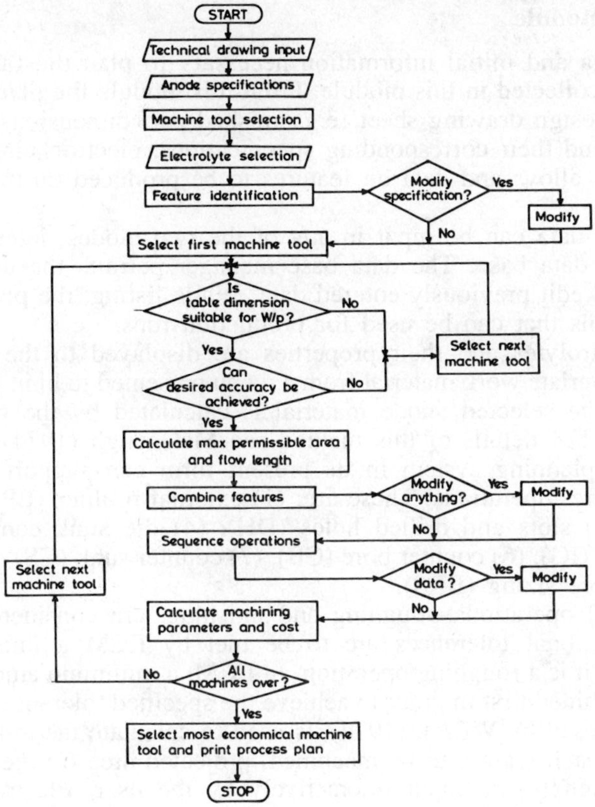

Figure 1. The CAPP system for ECM.

are far from universal in their applicability. Moreover, modifications on such systems to suit the ECM environment is not at all possible. Process planning in ECM does possess the basic structure of any conventional system, but the approach to each of the phases is radically different.

The intention of the present system is to systematize the process planning procedure for ECM based on logical theory and practice. A fully interactive program, which ensures that the experience of the planner is not missed out at any stage, has been developed. The whole system has been coded in Turbo Pascal (version 3·0) and can be implemented on any IBM PC XT/AT compatible. The system develops process plans for operations carried out on standard sinking machines. The application of the present methodology will significantly reduce the time, effort and costs involved in the development of process plans for electrochemically machined parts.

System description

The entire system is divided into two modules: (i) the system input module, (ii) the planning module. The planning module receives relevant information from the input module and all decision making is confined to this module. There is a provision to edit the generated data at more than one point. A description of the entire system is given in the form of a flow chart in Figure 1.

System input module

All relevant data and initial information necessary to plan the fabrication of the component are collected in this module. From this module the planner has general data from the design drawing sheet (e.g. finished part dimensions etc.), available machine tools and their corresponding specifications, electrochemical equivalents and densities of alloys and data on features to be produced on the workpiece by ECM.

Machine tool data can be input in any of the two modes, interactively by the user or from a data base. The data base manager permits the user to add new records, list and edit previously entered data. While listing, the planner can select the machine tools that can be used for production runs.

Suitable electrolytes and their properties are displayed to the planner, as he selects the appropriate work material from a menu presented to him. Electrochemical equivalent for the selected anode material is calculated by the superposition of charge method. For details of this method see McGeough (1974).

The process planning system in its present form can support only a limited number of features/operations. These are: (1) external profiles (EP), (2) trepanned shapes (TS), (3) slots and drilled holes (DH), (4) die sunk contours (DS), (5) internal grooves (IG), (6) counter bore (CB), (7) counter sunk (CS), (8) block sizing (BS) and (9) bore sizing (BRS).

Two principal operations, roughing and finishing, are considered. In case the specified dimensional tolerances are to be met by ECM a finish operation is designated; else it is a roughing operation. In ECM, a minimum amount of material needs to be machined first in order to achieve the specified tolerances. An analytical procedure, proposed by Wilson (1977), is used to calculate the minimum material allowance. For each feature to be machined, projected area (in the feed direction) and flow path length are input interactively by the user. Electrolyte flow path lengths are entered based on flow arrangements suggested by the system. Based on the surface finish required, the system may suggest more than one operation to be performed.

The planning module

The planning module has an hierarchical structure. First, available machine tools are screened and selected. Each machine tool is checked in terms of its capability to meet the accuracy and clamping requirements of the workpiece. Process plans for the selected machine tools are generated next.

During the electrochemical machining of metals, surfaces that are angled to the feed direction are subjected to a lower current density as compared to the areas square to feed direction. Hence, a minimum current density (0·8 A/mm^2) is used to calculate the maximum machinable area and corresponding electrolyte flow path length.

In ECM, it is possible to combine features for simultaneous machining by a single tool. However, selection of such features depends entirely on the machine capacity and the feasibility of designing and operating such tools. The information on maximum machinable area is used to obtain logical combinations of features/operations. Combinations are attempted using a sequential procedure. The procedure involves consideration of (i) embedded features, (ii) features using common natural points for electrolyte entry, (iii) similar features, (iv) block sizing operations and features produced subsequently on the same surface and (v) any other features that can be combined to utilize the machine capacity. The various combinations are illustrated in Figure 2. Large machining areas are, however, split into smaller areas

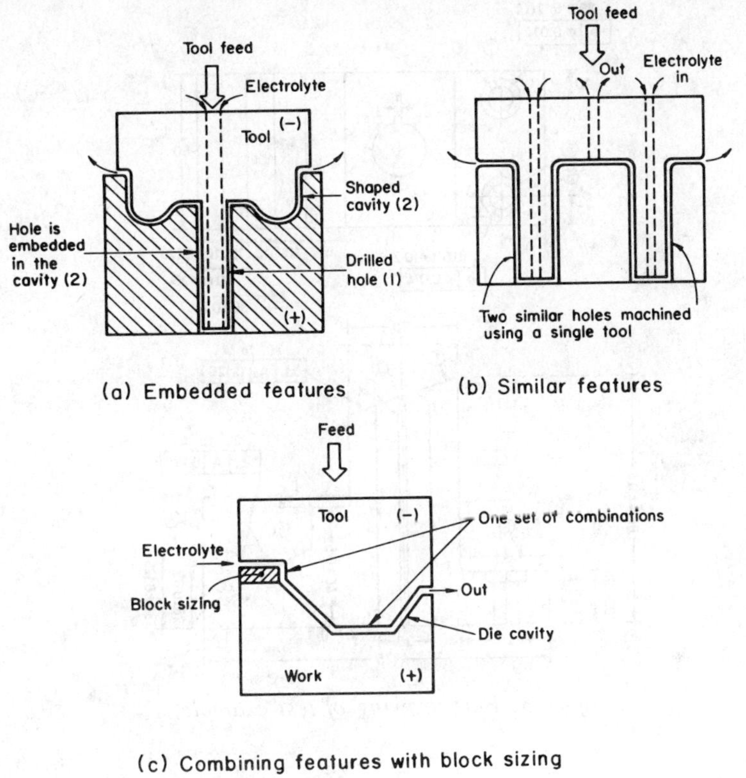

Figure 2. Illustrations of feature combinations for ECM.

and machined using a succession of tools. This is necessary in order to keep the current density above the minimum value.

Sequencing of the selected machining areas can be done both through manual and semi-automated procedures. The latter adopts the method of anteriorities, suggested by Eversheim *et al.* (1982), to establish the correct sequence. The various constraints considered are: technological, dimensional, geometrical, stray corrosion (corrosion by impinging electrolyte), component distortion (by passage of large currents through thin sections) and electrolyte flow arrangement (use of natural points of electrolyte entry). Preparation of the table of anteriorities requires a human interface. This table is then solved using a computerized procedure to establish the correct sequence. In the manual procedure, the user specifies the machining sequence.

Next, machining parameters are determined for the various operations. Optimal electrolyte flow velocity, voltage and tool feed rate are calculated through suitable optimization routines. A multiple objective optimization model for the above has been developed on the lines suggested by Acharya *et al.* (1986). The maximization of (i) material removal rate, (ii) geometrical accuracy and (iii) tool life are considered as the optimization objectives. These are subjected to various constraints: (i) electrolyte temperature at outlet is less than the boiling point, (ii) passivation of the work surface does not occur and (iii) chocking of electrolyte flow is avoided.

The model has been developed using realistic assumptions, i.e. (i) electrolyte pressure varies along the length of the gap and (ii) void fraction is not constant. The optimization problem has been solved using a non-linear goal programming

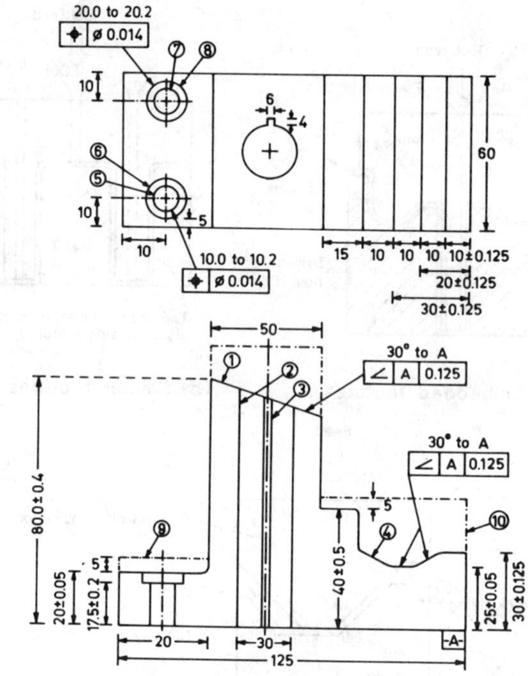

Figure 3. Part drawing of test example.

technique suggested by Rao *et al.* (1986). Priorities are attached to each of the objective functions depending on the machining cut (roughing or finishing) to be performed. In case of roughing operations, metal removal rate is given the first and geometrical accuracy the second priority. These priorities are reversed for finishing operations. Tool life is always assigned the last priority.

The operating cost of each of the machine tools, for all the operations, is calculated at the end of the planning procedure. The methodology suggested by Wilson (1971) is used for this purpose. The various costs considered are machine tool depreciation, tool costs, labour charges and cost of electricity.

Machine tool and process plan selection

The operating costs of the various machine tools are compared and the most economical machine tool is selected to carry out all the operations. Process plan for the selected machine tool is displayed to the user. Process plans for all the machine tools are also recorded on separate disk files.

Test example

A process plan has been prepared and presented for the part shown in Figure 3. The features to be produced on the workpiece using ECM are identified and numbered on the workpiece drawing. The specifications for each numbered surface are generated interactively through the feature selection module. Table 1 gives the details of these specifications. For each feature the flow path length is selected with the suggested electrolyte flow arrangement.

Table 1. Features and cutting data for test example.

Operation specifications	Feature Code									
	1F[a]	2R[b]	3F	4F	5R	6F	7R	8F	9F	10F
Projected area (mm^2)	2270·0	706·9	24·0	2400·0	78·5	235·6	78·5	235·6	1200·0	900·0
Flow length (mm)	20·00	15·00	6·00	60·00	5·00	5·00	5·00	5·00	30·00	30·00
Electrolyte flow arrangement	Cross	Forward	Cross	Cross	Forward	Straight	Forward	Straight	Straight	Straight
Slot location	Side	Central	Side	Side	Central	Central	Central	Central	Central	Central
Operation type	EP	DH	IG	DS	DH	CB	DH	CB	BS	BS
Inter-electrode gap (mm)	0·250	0·250	0·250	0·250	0·250	0·250	0·250	0·250	0·127	0·127
Min. materials allowance (mm)	1·13	0·00	0·00	1·57	0·00	0·00	0·00	0·00	0·00	0·00

[a] Finish cut; [b] Rough cut.

Part name: Test 1
Part number: 001
Drawing no·: DRG 1
Finished part length (mm): 120·00
Width (mm): 60·00
Height (mm): 80·00

Material specification: MS 1
Material name: Mild Steel
Batch size: 10

The data on work material composition and electrolyte are obtained interactively through the input module. The machine tools are identified from the data base and the planning procedure is carried out for each of them. In this chapter the planning procedure for only one of the machine tools is presented as an illustration.

For the machine tool (MC1), the first step is the selection of machining areas by combining the features to be produced. The resulting combinations are listed under the 'Operation code' column in Table 2. The features (2) and (3) are embedded in feature (1). Operations 2R and 3F were first combined. However, on further combining with 1F it was found that the combined area of the features exceeded the maximum permissible area limit. Hence, operations (2R + 3F) and 1F are carried out separately. All the other features are examined using similar logic. The number of stages required by each operation is listed in Table 2.

Selected machining operations are sequenced using the table of anteriorities, which is presented in Table 3. The sequence obtained is also shown in Table 3. It can be seen that the machining of features (1) to (3) can cause stray corrosion on the other surfaces, i.e. (4) to (10), as the electrolyte exits from the top and falls on other surfaces. Hence, operations (2R + 3F) and 1F are carried out first. Features (2) and (3) can also be used as an electrolyte supply port for operation 1F. It may be noted that the corrosion of the hole surface will not be significant, as it is not exposed to the electrolyte for prolonged periods of time. The above logic is used to determine the sequence of machining operations.

The next step is to calculate the machining parameters for each operation. The cost of operating the machine tool is calculated using the various cost parameters.

Process planning for a similar machine tool (MC2) of increased capacity i.e. higher voltage, rectified current and tool feed rate, is taken up next. With the increased capacity the number of operation steps is reduced as compared to machine one (MC1 has six steps and MC2, four steps). Further, in the case of machine two, the number of required stages is lower for some of the operations. This results in considerable savings in tooling costs. Considering the total cost of production of the batch, machine two is found to be more economical than machine one. Therefore, machine two is chosen to carry out all the operations. For complete details of the values used for various cost parameters, the reader may refer to Gursaran (1989).

Conclusion

The CAPP package developed in this work provides an important step towards a structured approach to automated process planning in ECM. It comprises interactive modules which decompose the various tasks and make generative process planning achievable.

The system helps the planner to generate logical and optimal process plans. It is also capable of generating optimal machining parameters. However, it is the users' responsibility to select and input the correct data and take important manufacturing decisions. This will help reduce any inconsistencies that may arise in the course of the planning.

Table 2. Process plan for MC1 of test example.

Machining parameters	Operation code					
	2R+3F	5R+6F+7R+8F	1F	4F	9F	10F
Projected area (mm²)	730·9	628·2	2270·0	2400·0	1200·0	900·0
Flow length (mm)	18·00	10·00	6·67	20·00	30·00	30·00
Machined depth (mm)	81·00	20·00	5·00	20·00	5·00	5·00
Optimum						
Velocity (m/s)	3·04	2·53	2·36	2·72	2·99	2·99
Voltage (V)	13·98	12·50	11·75	12·11	11·25	11·25
Feed (mm/s)	0·0321	0·0283	0·0280	0·0240	0·0246	0·0246
No. of stages	1	1	3	3	1	1
Inter-electrode gap (mm)	0·250	0·250	0·250	0·250	0·127	0·127
Pressure required kN/m²	5·863	3·665	3·086	4·645	6·394	6·394
Electrolyte						
Major constituent	$NaNO_3$	$NaNO_3$	$NaNO_3$	$NaNO_3$	$NaNO_3$	$NaNO_3$
Flow arrangement	Straight	Forward	Cross	Cross	Straight	Straight
Cathode material	Brass	Brass	Brass	Brass	Brass	Brass
Set-up time (s)	1009·346	282·286	71·428	333·333	81·300	81·300
Machining time (s)	2523·364	706·714	178·571	833·333	203·250	203·250
Handling time (s)	200·000	150·000	100·000	150·000	150·000	120·000

Machine tool code: MC 1
Maximum rectified current: 1000 A
voltage: 15 V
feed: 0·06 mm/s

Table 3. Anteriority table for sequencing of operations on MC 1.

Anteriorities	Operation code					
	2R+3F	5R+6F+7R+8F	1F	4F	9F	10F
Dimensional	—	(2R+3F)	—	—	—	—
Geometrical	—	—	—	—	—	—
Technological	—	—	—	—	—	—
Stray corrosion	—	(2R+3F),1F	(2R+3F)	(2R+3F),1F,9F	(2R+3F),1F	(2R+3F),1F
Component distortion	—	—	—	—	—	—
Economics of tooling	—	—	(2R+3F)	—	—	(5R+6F+7R+8F)

SEQUENCE OF OPERATIONS
2R+3F, 1F, 5R+6F+7R+8F, 10F, 9F, 4F

References

Acharya, B.G., Batra, J.L. and Jain, V.K., 1986, Multi-objective optimization of the ECM process. *Precision Engineering*, **8**, 2, 88–96.

Bhattacharya, A., 1977, *New Technology*, (India: The Institution of Engineers).

Chang, T.C. and Wysk, R.A., 1985, *An Introduction to Automated Process Planning Systems* (Englewood Cliffs, NJ: Prentice-Hall).

Eversheim, W., Spur, G. and Weill, R., 1982, Survey of computer aided process planning systems. *Annals of CIRP*, **31**, 2, 539–551.

Gursaran, 1989, Computer aided process planning for electrochemically machined components. Unpublished M. Tech. Thesis, Indian Institute of Technology, Kanpur.

McGeough, J.A., 1974, *Principles of Electrochemical Machining* (London: Chapman and Hall).

Molloy, J., 1965, Electrochemical machining of a turbine blade. *Machinery and Production Engineering*, **1**, 1198–1203.

Rao, P.N., Satyanarayana, B. and Tewari, N.K., 1986, Application of nonlinear goal programming technique in metal cutting. *AIMTDR*, 483–486.

Wilson, J.F., 1971, *Practice and Theory of Electrochemical Machining* (New York: Wiley Interscience).

SECTION VIII.3. CIM

96

Discrete cell control: framework and architecture

J. English, I. Franks, M. Loftus, G. Williams and N. Wood

Department of Engineering Production, University of Birmingham, PO Box 363, Birmingham B15 2TT, UK

Abstract

This chapter considers the approach taken by the Cell Control Group at the University of Birmingham, in collaboration with four industrial partners, to define the generic elements of a discrete cell controller. The research programme has determined the functions which should be performed at cell level and established how the cell controller should communicate both upwards to manufacturing information systems and downwards to the machine controllers on the shop floor.

Introduction

For many years industry has been faced with the problem of controlling diverse and often complex manufacturing systems. The earlier concepts of centralized control have been superseded by distributed processing power, which allows more flexibility and the focus of control at a local level. However, the implementation of the distributed philosophy has not been entirely without its problems, e.g. the lack of comprehensive and enforceable standards in a muti-vendor world.

The Open Systems Interconnect (OSI) communication standards are seen as the key to enable the implementation of the functions within a manufacturing cell controller. The replacement of custom-engineering with generic solutions will assist manufacturing industry to improve productivity and encourage the less inclined to adopt an Open Systems approach to integration.

Cell control environment

In the context of this chapter, a manufacturing cell is defined as a group of manufacturing resources which process the material input into value-added products. Cells may be automated, semi-automated, manual or combinations of all three. The Discrete Cell Controller (DCC) is seen to operate at the level between shop floor devices and the plant area control system. The function of the controller is to co-ordinate and manage the resources in the cell in an effective manner. This is a complex task because the scale of the problem is large; cells are dissimilar in character and within a cell there are major dynamics in operation. This scenario differs from the relatively stable process industry sector in which the cell control concept was developed and has now matured into robust products.

Current solutions

The discrete cell control market is fragmented with no specific vendor able to satisfy user requirements without some measure of customization. The extent of this customization may differ among applications and it introduces a major element of risk and uncertainty. The way in which the major suppliers are tackling this problem reflects their product backgrounds and experience (Loftus and Robson, 1989).

Honeywell's (MASC) and Texas Instruments' (TISTAR) credentials are based upon their considerable successes in the process area. IBM's solution (GEARBOX) is driven from higher level manufacturing information systems, whereas DEC's products (BASEWAY) are firmly routed in the manufacturing engineering area. Neither IBM nor DEC have been able to satisfy the control requirements at the machine tool level. This cannot be said about the two remaining major suppliers, namely Siemens (SIMATIC) and Allen Bradley (VISTA 2000). These companies are synonymous with programmable control at the shop floor level, but they have found it hard to satisfy the needs at higher levels. An interesting development is the recently announced Pyramid Integrator from the combined stables of DEC and Allen Bradley. Lastly, there are the solutions being offered by companies such as BAeCAM. Their CIMITAR concept is based upon the major developments within the Bristish Aerospace Group.

In conclusion, there are significant differences in the marketing strategies of the main suppliers. All appear to appreciate the need for comprehensive coverage of the cell control function, but as yet the general solution is not available. The MIRC study (1988), into the cell controller market in the USA, substantiates this finding.

Discrete cell control project

The objective of the project was to investigate the requirements for a generic cell control system which should operate in an Open Systems Environment. This was a major task requiring the combined efforts of Texas Instruments, GKN Technology, AFE Displays, OD Systems and the Cell Control Group at the University of Birmingham. The first phase of the work has been partially funded by the DTI's Innovation Support for Business initiative.

A team approach was adopted to merge the expertise within the consortium and direct the work on several fronts. Team One had to establish the user needs, and pass them to Team Two to address the technical requirements to satisfy these needs. Four major studies were conducted at GKN plants to assess their control requirements, and four brief studies in other manufacturing companies were conducted in parallel to reflect a more general view of the subject.

Team One activities

Preliminary activities were aimed at drawing upon the experience of team members to determine an initial draft model for a DCC. After a few iterations this model was used as a basis for determining a strategy, *inter alia*, to produce a more appropriate modelling technique and to design a user needs interview kit. Part of the model identified the principal functions to be discussed during the interview sessions (Table 1).

A standard interview kit was necessary to ensure a compatibility of approach to different types of companies and to facilitate access, by Team Two, to the

Table 1. Function modules identified in initial model.

Function modules	
Scheduling/dispatching	Quality
Engineering change	Process plan
Method analysis	Maintenance
Tool management	Event reporting
Time and attendance	Machine monitoring
Estimating	Simulation
Downlink control	Diagnostics
Operator interfaces	General communications

information gathered. The foundation of the kit was a matrix of those functions specified in the initial model against a number of general cell control elements e.g. inputs/outputs, failure/recovery, factory-wide considerations. Detailed check-lists for each function were then derived from the matrix. Each item in the check-list was treated solely as a 'discussion driver'; detailed questionnaires being very difficult to construct with a view to extracting a variety of good information from experts (Marca and McGowan, 1988) and not appropriate at this stage of dvelopment. Useful information was generated and recorded during the study sessions, particularly data lists for each functions, q.v. later. At the conclusion of each study a verbal presentation on the findings was made to Team Two, sometimes with company personnel present, and a report produced.

Concurrent with the user needs survey, a start was made on the construction of an IDEFo model of a generic cell controller. IDEFo is one of the IDEF suite of modelling methodologies used for CIM applications. It is particularly useful for undertaking requirements studies for the purpose of developing a system specification. The basis of IDEFo is the SADT methodology developed by SofTech Inc. which itself was the result of prolonged research by Ross (1985). The use of the function box, associated data arrows and decomposition, together with well-documented notes and procedures, make it well suited to a role as a foundation modelling methodology, for use in discussion and analysis sessions with manufacturing personnel and hence for system specification.

Initial model development had to be based on the existing function list, as sufficiently detailed data lists were not available until the industrial studies were completed. It is accepted that this order of analysis is probably not the best, but a start on the model had the advantage of making all project participants thoroughly familiar with IDEFo and identified other needs, e.g. computer-aided analysis. Later stages of the modelling process were influenced by the requirement to incorporate information emerging from the industrial studies into the generic model. As each study was completed a meeting was held between an IDEFo author(s) and the study team members, to examine the fit of that company's requirements to the generic model generated to date. Monthly meetings of all participants were also held to evaluate the current development of the generic model. It was in these later stages that analysis according to data lists rather than function lists demonstrated its superiority, making the resultant IDEFo diagrams more readily acceptable as a generic model. Another important contributory factor to a reduction in contentious argument was the compilation, from the beginning, of a Glossary of Terms with a well-structured data entry system.

As a result of Team One's activities a three-level IDEFo model of the fundamental functions and data flows of a generic DCC was produced. Further decomposition was not considered to be appropriate in this initial phase of the project. This model, together with other information derived from the studies were input to regular meetings of Team Two and to a rigorous three-day Joint Application Design session (IBM) with an independent consultant as Chairman.

Team Two activities

The definition of technical requirements for a cell controller is conditioned by the needs of the user, the application environment and the overall information structure in which it must operate. Team Two addressed the fundamental technical feasibility and established which standards or practices exist in this field today.

The project started from the premise that the Open System approach was necessary. The cost of custom software is one of the inhibiting factors in the adoption of computer solutions in manufacturing. The integration of islands of software was seen as a major objective. This leads to the two pivotal issues, connectivity and selection of computer architecture.

In order to establish the current position, visits were made to users in the UK and Europe and experts in several research institutes were consulted for comment. A remarkably consistent view has emerged.

The solution offered by OSI to the issue of connectivity is MAP (Pye, 1988). The major development for MAP in 1989 has been the emergence of MAP 3·0 with the valuable Manufacturing Message Service (MMS) interface. MMS programming will, in due course, prove a powerful tool for the Manufacturing Systems Engineer. There is now a set of standardized utilities which will permit the applications modules to be mounted on new platforms and reconfigured to support the manufacturing devices which are on line in any particular cell.

There has been much written about the underlying layers of MAP but it has emerged, during this study, that several types of network will have to co-exist and that they will have to inter-operate. OSI does not preclude such network architectures and it is a clear finding that these will have to be supported if a generic solution is to be achieved.

The services which are available at the applications interface (level 7) are the major attraction of this Standard and it is through the use of these that the islands can be integrated.

Computer architectures

In applying computers to the management of the cell there are a number of structural issues which have to be clarified. First, the main function of the cell controller is the manipulation and transfer of data. This suggests that a Relational Data base would be the most useful medium for primary operation. After a little more consideration it emerges that there are also a number of simple file transactions. The computer platform should be designed to carry out both of these types of operation in a timely manner. This demands real-time, on-line services. In the management of machine tools, real-time implies very rapid response (10–100 ms) and this defines the system performance.

Implicit within this architecture is the need for distributed services and some degree of fault tolerance. Work has been done in this field by Honeywell with their MASC system (Currie and Craigmyle, 1988). It is likely that this avenue will be pursued by a number of system developers.

The UNIX operating system was chosen to ease transportability, but it is acknowledged that there are well-established platforms, with their own operating systems, which must not be ignored (e.g. DEC with VMS).

Upward link with manufacturing information systems (MIS)

The top level IDEFo diagram identified controls as: the company manufacturing strategy, available resources, production and industrial engineering information and process planning. Inputs were identified as production plans or schedules from customers, emergency orders and cell loading and status information, data for use in other systems, and feedback information. Current computer-aided MIS include subsets of these elements, e.g. material requirements planning (MRP) computer-aided process planning (CAPP), etc. There is, therefore, an implication for integration with other facets of the manufacturing information system.

Manufacturing information systems have been defined as

> 'an arrangement of computer hardware and software, which includes an assembly of inter-related modules or elements directed towards providing manufacturing information, plans and schedules, in order to effect some control over the manufacturing operations of an organisation such that some benefit can be achieved.' (Williams, 1987).

There is a trend towards integration of manufacturing information systems. This has been instigated by the introduction of MRP packages which purported to link capacity planning, shop floor control, purchasing, etc. However, the problem of achieving comprehensive integration and communication across the vast number of proprietary software applications cannot be underestimated (Mantz *et al.*, 1983).

The studies revealed MRP to be a driving force for manufacturing information. The question is, which higher level systems to interface to the DCC. There is no general standard for MRP-type systems. White and Landvater (1978) attempted to resolve this issue with limited success. Sepehri (1985) and Williams *et al.* (1985) have offered some criteria to resolve this issue based upon hierarchical interfaces to suit a comprehensive manufacturing information system.

Scheduling and sequencing

In the majority of the studies the issues of scheduling and sequencing were identified as a major requirement. Within the various companies different strategies were in current use. In one case the company had formed production cells dedicated to the manufacture of a limited range of components or products. Another used an MRP system to generate lists of jobs based upon due-date priority. In a further case a simulation model was used to generate the required load on a particular process. In most cases the scheduling process was subject to some form of validation against available capacity. This scheduling function tends to be performed at a level above the cell and provides a work-to-list which may have prioritization other than a planned due-date.

With regard to the actual sequence, there was considerable reliance upon the experience and expertise of line supervisors and/or operators in selecting the next job from the list or queue. To an extent this was caused by the lack of systems for co-ordinating the arrival of items such as tools, fixtures, materials and other supporting resources.

The studies identified common factors affecting the process of scheduling at the planning level, but detailed scheduling and sequencing at the cell level is very specific. For example, a process within the production route may require the batching together of different orders to achieve a given loading on a machine. Alternatively, the allocation of labour from a crew may be required where manufacture is organized on a cellular basis. Consequently, in the derivation of the IDEFo model it was necessary to recognize the different levels of scheduling involved and the need for specific rules or algorithms as controls appropriate to these levels.

A further consideration is how the detailed decision on which job is to be performed next in a given cell is to be utilized when communicated back through the up-link. For example, this could allow the effect of the current decision upon future choices and upon the remaining cells to be assessed in terms of the overall performance of the manufacturing system. This requirement might be met by some form of simulation model at the higher level which could be used to assess the proposed decision and give a 'go' or 'no-go' signal back to the cell. The speed of response of the simulation model would be a critical factor in such a system but would be less critical if decisions at the cell level could be made sufficiently far in advance, for example, 30 min ahead of implementation.

Overall the model anticipates the need for generic requirements for scheduling at different levels with specific controls in the form of rules, whether these be heuristics or algorithms, for the detailed sequencing of work-to-lists at the cell level.

References

Currie, G.A.C. and Craigmyle, M., 1988, A system designed for manufacturing automation. *Proceedings of Factory 2000 Conference*, Cambridge, UK.

IBM, *Joint Application Design Brochure, G520–0008* (IBM).

Loftus, M. and Robson, A., 1989, The current status of cell controller. *5th National Conference on Production Research*, Huddersfield Polytechnic (London: Kogan Page).

Mantz, R.K., Merton, A.G. and Severance, D.G., 1983, *Senior Management Control of Computer Based Information Systems* (Financial Executive Research Foundation).

Marca, D.A. and McGowan, C.L., 1988, *SADT* (New York: McGraw-Hill).

MIRC, 1988, *Mainline Cell Controllers and Others* (Mount View, CA: MRC).

Pye, C., 1988, *What is MAP?* (Manchester: NCC Publications).

Ross, D.T., 1985, Doug Ross talks about Structured Analysis. *IEEE Computer*, July.

Sepehri, M., 1985, Newest manufacturing software packages offer modules to meet specialist needs. *Industrial Engineering*, **17**, 28–43.

White, O. and Landvater, D., 1978, *MRP Software on the Standard System* (Manufacturing Systems Inc. USA).

Williams, G.B., 1987, The evaluation and selection of manufacturing information systems. Ph.D. Thesis, University of Birmingham.

Williams, G.B., English, J.E., Fielden, M., Page, I.M. and Markoyiannopoulos, T., 1985, Material requirement planning packages: an evaluation. In *Advances in Manufacturing Technology, 1st National Conference on Production Research*, University of Nottingham (London: Kogan Page), pp. 21–24.

97

Strategies for factory automation in garment production

M. Sarhadi, P.R. Nicholson and S.J.L. Marshall

Department of Manufacturing and Engineering Systems, Brunel University, Uxbridge, Middlesex UB8 3PH, UK

Abstract

Improvements in shop floor automation in garment manufacture are now necessary on grounds of economics, quality control and speed of response. Automated assembly cells must be developed to replace operations which normally require skilled handling in the industry. Automated handling of non-rigid material is a difficult task involving long research and development phases to achieve reliable operation. Consequently, the initial capital investment costs in the more sophisticated and expensive machines present a barrier to progress. This chapter describes strategies applicable to assembly automation on the factory floor in garment manufacturing. Supported work in this field is outlined and recent work into the automation of a typical assembly process, by a fully integrated assembly cell, is reported. Performance results are given.

Introduction

The scope of this chapter concerns assembly in the factory floor within the garment assembly industry, where real-time control of manufacturing systems is critical to production. Introduction of assembly automation is desirable for three reasons:

(1) *Economics*—Assembly represents 20–30% of total manufacturing costs. To reduce this, manufacturers are basing assembly operations overseas; those who wish to support a home manufacturing base must compete with lower wage rates and narrower profit margins.

(2) *Product quality*—In assembly, quality is defined as conformance to a standard. Automated assembly is suited to consistent production and automated quality control assures standards are met. Adoption of BS 5750 (BSI, 1981–7) aids the integration of quality control and has already been adopted by garment manufacturers such as Jockey (UK) Ltd.

(3) *Speed of Response*—Design changes with small production runs or staple high volume products with variable size requirements would be better accommodated with flexible automation methods. In the first stage of manufacture, the cutting room, faster design is possible with the recent availability of computer based lay planning and numerical control (NC) lay cutting equipment, for example the Gerber Cutter. NC lay cutting has been relatively simple to implement as individual workpieces are not involved. Subsequent assembly stages have proved more resistant to automation due to the difficulties in limp fabric handling (AAMA, 1987). Following assembly, packaging has currently been automated only by customized machinery. Assembly operations are thus

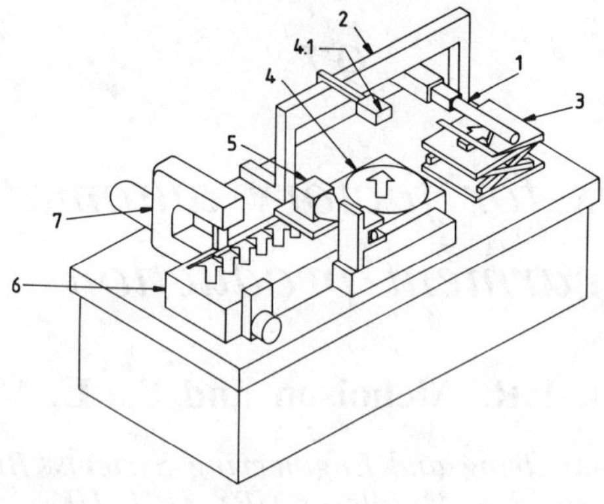

Figure 1. Diagram of the assembly cell: 1, gripper; 2, linear transporter; 3, elevating feed table; 4, alignment table; 4·1, alignment photosensors; 5, vacuum-based transfer mechanism; 6, conveyor; 7, automated sewing machine.

a good target for automation research. Initially the most suitable area of assembly comprised staple high-volume products as less flexibility in automation was required.

Early assembly research includes the US Transfer Street and German Rink projects. Both covered considerable basic research and improved assembly technology, producing aids to seam, cut, fold and restack. Later research has been supported in the UK by the ACME directorate of the SERC. In Europe the EC BRITE initiative supports numerous assembly projects. Major Japanese programmes are Toyota Sewing management System (TSS) and Japanese Automated Sewing System (ASS) (Disher, 1987). The US Textile/Clothing Technology Corporation [(TC) 2] and Manufacturing Applied Robotics System (MARS) developments are presently active.

The assembly cell

Conventional volume garment production lines are based upon batch processing using single operator workstations. To address their automation it was intended to automate a single two-dimensional process, reducing costs and improving quality. A production line automating men's briefs was examined and a specific operation, edge binding, identified for automation. This operation comprised the attachment of edge binding tape to one edge of a fabric component. It was the first activity in the production line and comprised 2 out of 14 activities within that line. Edge binding comprises destacking from a component stack, edge alignment, feeding into a sewing machine attaching edge binding tape, cutting and restacking.

Technical description of the assembly cell

The assembly cell is shown in Figures 1 and 2. The cell configuration used four electromechanical modules, controlled from a central cell controller, a Compaq 386s IMB PC compatible. The electromechanics for each module were mounted on a support frame which accommodated the cell wiring harness and pneumatic distribution system.

The first module implemented destacking. Batches of (typically 50) fabric plies were received from the cutting room and the module repeatedly destacked the topmost ply for subsequent onward processing in the cell. However, plies are

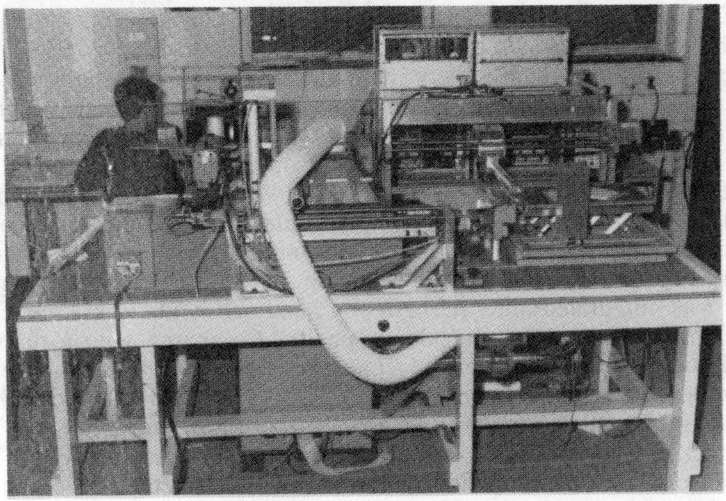

Figure 2. The assembly cell.

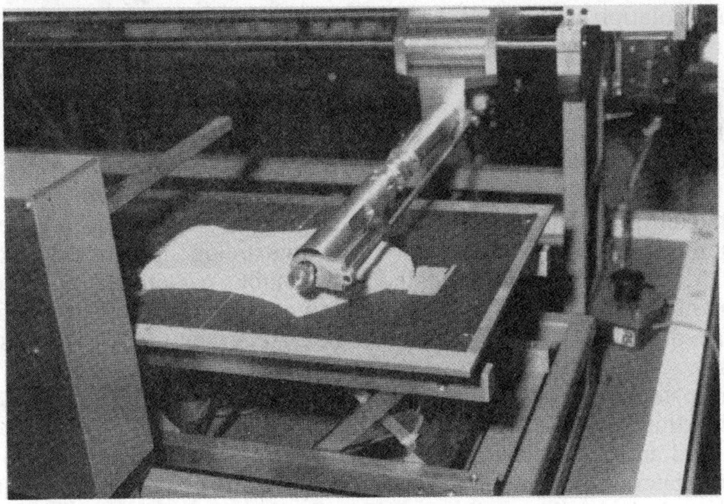

Figure 3. The destacking gripper.

normally edge fused due to fibre entanglement and welding in synthetic materials by heat generated in the cutting process. For this reason, destacking is a general problem and is difficult to automate reliably.

The destacking task has been approached by employing a roller type gripper (Figure 3). The gripper used an array of oblique pins arranged around its circumference to achieve surface engagement. Destacking procedure commenced by presenting the grip area to an edge of the topmost ply of the stack. Simultaneously, an articulated clamp bar was moved to hold the opposing side of the ply under controlled pressure. A partial rotation of the gripper then consolidated surface engagement and stretched the topmost ply, breaking edge fusing locally with the underlying plies. Having achieved local separation, the gripper rolled across the surface of the stack driven by a linear transporter, removing the ply. The clamp was retracted, allowing total separation with the ply rolled on the gripper. The gripper was then moved to a release position and an unrolling operation deposited the ply. The technique is described more fully by Sarhadi *et al.* (1986).

The gripper electromechanics were implemented with an aluminium cylinder (2) of 30 mm diameter by 100 mm, mounted on a linear transporter (3). In an open loop prototype version, a destacking reliability of 99% was achieved. Integration of sensory instrumentation into the destacking operation overcame unreliability to a large extent. Sensory feedback was provided with gripper force sensing. Strain gauge load cells instrumented gripper-to-ply contact force and engagement torque. These parameters could thus be controlled, improving destacking consistency. Furthermore, to detect failure of destacking, an infra-red emitter and sensor subsystem was integrated into the gripper head, allowing measurement of the destacked fabric infra-red transmission capacity. Compared with a reference cell integrated into the ply depth sensor system, using a sample fabric ply, determination could be made whether none, one or more than one ply had been destacked. Error correction could be carried out if other than one ply had been removed. Rejection of the destacked component or re-trial for separation was possible. The above analogue sensors were instrumented with an A to D converter card mounted in the cell controller.

The gripper was articulated with a proprietary motion control system comprising a three-axis stepper motor configuration. An accompanying indexer card allowed a preprogrammed sequence of movements to be carried out by the axes, including two-axis interpolation with the gripper and the linear transporter axes to effect the described rolling operation. A serial link connected the indexer and the cell controller, enabling integration with the cell control program. The third stepper motor axis was used to control an elevating table [Figure 1(3)]. This was necessary to adjust the stack height as plies were consumed during destacking. Auxiliary sensors such as a fabric presence detector and a table datum limit switch were interfaced via digital Input/Output provided by the indexer card.

The destacked plies were received at an alignment stage (4) to register one ply edge mechanically. This was necessary as downstream operations took place in open loop. Alignment was implemented with a table powered in X, Y and θ axes and instrumented with three overhead mounted photosensors. The axes and sensors were interfaced directly with the cell controller. Accuracy to ± 1 mm was achieved with the unit.

Having registered one edge, the ply was transferred onward by a pick and place unit (5). This third module employed an adaptive flat vacuum grip pad to grip the component and an X, Z cartesian articulation unit to effect raising and lowering of the pad and lateral movement. The module was provided with a stand-alone controller interfacing to the cell controller. Further description of the module is given by P.R. Nicholson and M. Sarhadi (this volume).

The final module attached edge binding tape to the transferred ply. An industrial sewing machine (7), adapted to attach tape with a dispenser and feed facility was used to join tape and ply. The machine was further modified with a d.c. drive controlled directly from the cell controller and instrumentation to monitor crankshaft angle. A synchronized stepper motor powered conveyor belt (6) was also provided to support sewing machine feed. The conveyor belt comprised several narrow individually tensioned strips driven from a main pulley to the stepper motor drive.

Operational results

A performance test using 500 components was conducted with the assembly cell. Cycle times and reliability were recorded and timing results are shown in Figure 4.

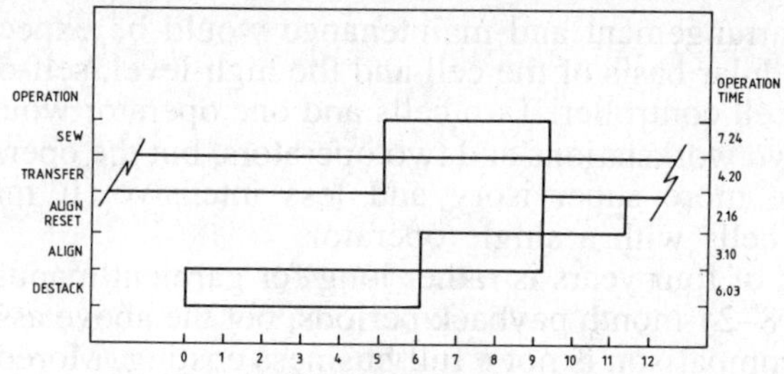

Figure 4. The edge binding process schedule.

Table 1. Costings for proposed assembly cell configuration.

Parameter	£
Operator salary	5000 p.a.
Operator overhead @ 25%	1250 p.a.
Assembly cell capital cost	25 000 p.a.
Assembly cell maintenance @ 5%	1250 p.a.
Conventional workstation cost	2500
Workstation maintenance @ 5%	125 p.a.

In the 11·29-s cycle time obtained for processing one ply, initial edge separation was achieved in 250 ms. To separate a fabric edge locally and commence rolling required 500 ms. It was not possible to reduce this time further as aerodynamic effects were found to decrease reliability. However, this term is a small component of the overall cycle time. With the concurrent schedule implemented in the cell controller, the remaining process times offer scope for further reduction but have not been optimized. The particular times returned from each module result from the specific implementation of their electromechanics: further industrial development may reduce the times substantially.

The test was run with error correction disabled in the destacking module. Two components were destacked incorrectly and no other failures were observed in the remaining modules. It was observed that plies delivered from the destacking module with minor creases were flattened by the adaptive vacuum transfer module as a result of its clamping method. An earlier ply destacking unit reported by Sarhadi *et al.* (1986) returned a 99% reliability in a test with 5000 components.

It is intended to conduct further scientific testing over a range of shapes and types of fabrics to determine the full performance of the cell.

Cost–performance evaluation

The commercial viability of the cell was assessed. The cell was evaluated by approximately 20 machine manufacturers and a commercial version of the prototype was estimated at £18 000–25 000.

A UK costing is given by taking a specific scenario and applying rates supplied by the industrial partner. If two edge binding operations (front and gusset panels) are automated using two assembly cells and one operator, this would replace two conventional workstations and two operators. Table 1 shows costings. Assuming that the cell productivity is equivalent to a conventional workstation, payback could be achieved in four years. Following payback, the cell will run at a lower cost that the conventional arrangement and maintenance would be expected to be reduced because of the modular basis of the cell and the high-level, self-diagnostic capacity possible with the cell controller. Two cells and one operator would take marginally more space than two workstations and two operators, but the operators' activity with the cell would be more supervisory and less intensive. It may be possible to incorporate more cells with a single operator.

A payback term of four years is rather long for garment manufacturers who base their costings on 18–24-month payback periods, but the above assumes a worst case situation and the comparison is not a full business costing. Moreover, with adaption of the cutting room procedures, it may be possible to integrate a buffered cassette loading system for parts which require less operator involvement.

Quality control integration aspects

The above presents an analysis on simple costing grounds only. In quality terms, the cell could be expected to deliver workpieces to a consistent standard. With the integration of supervisory and control systems presently under development within the research group, further quality improvement is possible. An example project is an active feed control system which inserts elastic feed at a constant tension into a seam. This uses force transducers and a closed loop step motor drive to facilitate feed (Marshall and Sarhadi, 1989). A second example is a yarn tension monitor which instruments yarn tension and can detect individual mis-stitches in seams or out of tolerance yarn tension in sewing machines (Catchpole and Sarhadi, 1989).

The cell may be programmed to produce reports on its performance, timings and quality results. These may be made available on a local area network (LAN) connection to Management Information Systems or Work-In-Progress Systems. With the PC controller unit presently used, a LAN may be implemented on the standard IBM token ring system or one of the available proprietary networks (Such a Novell or NineTiles SuperLink). Here, report files may be sent to remote nodes for analysis. Current work with an allied workstation has indicated the suitability of such interconnections. Likewise, maintenance requests and self-test procedures could be automatically generated. With integrated diagnostic facilities, identification of faults could be simplified and maintenance eased by the rationalized modular construction of the cell.

Assembly extensions

The cell provides an example of partial two-dimensional automation. Here its component modules are combined to produce a specific operation. To carry out other operations the modules may be used in differing configurations. By addition of further modules providing other operations, such as modules derived from (TC)2 technology, it may be possible to carry out full two-dimensional assembly. Thus in establishing practical generic sub-processes a more complicated assembly operation can be configured. If a central conveyor system were used, fabric subassemblies may be passed between modules until the required number of operations are completed.

Two-dimensional subassemblies will require three-dimensional manipulation to complete manufacture. In one of the few examples of such automation, Taylor and Taylor (1989) demonstrated a three-dimensional operation in the automated construction of a double concealed gusset where two manipulators are used to handle the gusset assembly.

Conclusions

Limp fabric handling has been identified as a critical area in implementing garment assembly automation. To address this, a two-dimensional garment assembly problem has been identified and is automated with an assembly cell, comprising a configuration of modules addressing individual handling and joining tasks. The cell operates with industrially acceptable reliability and has been costed with payback time in the order of commercial viability.

The overall cell cycle time is given in Figure 2 as 11·29 s. The timing schedule shows the destack, align and align reset stages on the critical path, so these are the targets for future cycle time reduction.

Acknowledgements

The authors express their acknowledgements to the ACME directorate of the SERC, to the BRITE programme of the European Community and to the project's industrial partner, Jockey (UK) Ltd, for support of this work.

References

AAMA (American Apparel Manufacturers Association), 1987, *Automated Limp Fabric Handling Devices*. Final Report, April 1987, The Limp Fabric Task Group of the Apparel Research Committee of the American Apparel Manufacturers Association.

BSI (British Standards Institute), 1981–7, British Standard Quality Systems, BS5750, Parts 0–6 (London: BSI).

Catchpole, J. and Sarhadi, M., 1989, Stitch quality monitoring using digital signal processing. *Proceedings SAMT89*, Sunderland Polytechnic.

Disher, M. 1987, Sewing techniques. *Manufacturing Clothier*, August, 8–11.

Marshall, S.J.L. and Sarhadi, M., 1989, In-process monitoring and control of elastic feed in sewing operations. Extended summary, presented at *The Xth International Conference on Production Research*, Nottingham, 14–17 August 1989.

Sarhadi, M., Nicholson, P.R. and Simmons, J.E., 1986, Advances in Gripper Technology for Apparel Manufacturing. *Proceedings of the Institute of Mechanical Engineering*, paper C372/86.

Taylor, P. and Taylor, G.E., 1989, Progress towards automated garment manufacture. In *Sensory Robotics For the Handling of Limp Materials (Proceedings Of a NATO Advanced Research Workshop*, Catelvechio Pascoli, Italy, October 1988), edited by P.M. Taylor (Berlin: Springer).

98

An algorithm for distributed planning and control in a computer-based manufacturing control system

A.W.M. Lung[1] and D.M. Love[2]

[1]*School of MAP Engineering, Kingston Polytechnic and* [2]*Department of Mechanical and Production Engineering, Aston University, UK*

Abstract

This chapter investigates the merit of using the concept of distributed planning and control instead of the more traditional and widely accepted centralized control approach in the design of Computer-Integrated Manufacturing (CIM) systems.

The distributed concept is founded on the principles of cellular manufacturing systems. Whilst the term cell is normally used to indicate a group of similar machines or processes, this chapter takes a broader view regarding a cell as a self-governing unit which has its own planning functions and exercises total control over its local resources. Material Requirements Planning (MRP) is likely to remain the core of many integrated systems for the foreseeable future. The idea of distributing the MRP system across a series of manufacturing cells will be explored and the potential benefits of such an approach discussed. A tentative algorithm for the operation of a distributed system will be discussed and a prototype computer system, developed to test these concepts, described.

Introduction

Computer-Integrated Manufacturing (CIM) is currently seen as a key mechanism to ensure the continued competitiveness of a manufacturing company. Indeed it has been suggested that survival depends on its successful implementation (Ingersoll Engineers, 1985). There exist many definitions of what constitutes, CIM, but however broad or narrow the definition it can be confidently assumed to represent a major undertaking for any manufacturing company. Considering the time it has taken for most companies to implement some of the individual elements [computer-aid design (CAD) and MRP for example] it is clear that the creation of a truly integrated system will prove to be beyond the capabilities of many companies.

Traditionally most CIM system design concentrates on data integration by adopting a centralized system built upon the use of common data base (Love *et al.*, 1986). It is clear from implementation experience of all types of computer system that as the size of the system grows it becomes more difficult to ensure design and operational integrity. These problems were evident in the development of early MRP systems and it should be noted that operational integrity remains a problem with many current MRP implementations. The application of these centralized principles to the design of even larger systems should at least be questioned.

Centralized systems in CIM design

According to Salzman and Little (1984) and Huges and Maull (1985), the different CIM system approaches can be generally grouped into two main global types. The essence of the first global approach is to build an integrated system from existing system elements, while the second approach is to build an integrated system from first principles.

The use of existing system elements

Within this category, there are three possible topologies (Lung, 1988): partial integration based on linking individual functions; the grouping of elements into large multi-functional systems which are themselves linked by simple data exchange paths; and total integration.

Single function integration

The 'islands of automation' present in many modern manufacturing companies are an example of this type of integration.

Most of these 'islands of automation' were derived originally from single functions or processes. These single functions have undergone rapid development, but have remained essentially isolated from each other. Examples include CAD, Computer-Aided Manufacturing (CAM), MRP [rather than Manufacturing Resources Planning (MRP2)], Computer Numerical Control (CNC) and Computer-Aided Process Planning (CAPP) (Gott, 1984; Groover, 1984; Willer, 1984; Blackstone and Cox, 1985; Callarman and Heyl, 1986). Although these islands may be isolated from one another in terms of hardware and software pedigree, a limited degree of integration may be achieved through dedicated links. An example of such a link would be the downloading of process plans from CAPP to the MRP system. They can be considered to represent the lowest degree of integration possible within a reasonable definition of CIM.

Multi-function integration—partial integration

The simple links typical of single function integration impose limitations in system performance and flexibility which may be alleviated by the grouping of the elements into major groupings, usually associated with traditional organizational boundaries. Examples of such integration include Flexible Manufacturing Systems (FMS), Engineering Database Systems [including CAD, Computer-Aided Engineering (CAE), CAPP, etc.] and Computer-Aided Production Management (CAPM) (Weatherall, 1984). They represent a middle stage in the achievement of total integration.

Total integration

Total integration may be regarded as the ultimate solution and a logical extension of the previous approaches (Lung, 1988). Figure 1 illustrates such a system formed by sub-group 1 (type 1), islands C, D and E (type 2), as well as some individual functions or processes. At this level of integration the system is at its most complex. Design of the shared data base will be a difficult if not impossible task. Inconsistencies between the sub-groups will generate difficulties and prevent the user achieving a seamless view of the integrated facility.

In all three approaches the integration is built around the functions performed by the various system elements. Like processes are grouped together and the group

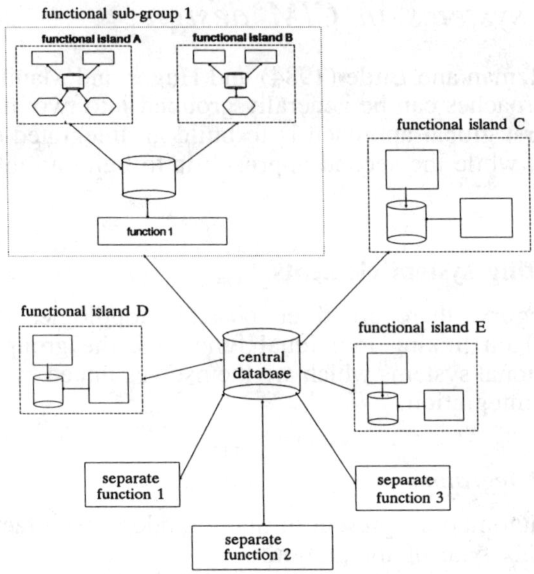

Figure 1. Layout of a total integrated system.

expands by incorporating the next set of related activities. This represents a centralized view of the organization; all CAD is provided by a single system; all production control by a single MRP system. As integration proceeds the degree of centralization increases to a point where the system will inevitably be huge, unwieldy and inflexible.

Design from first principles

The major alternative approach is to develop an integrated system from scratch using standard software only where it precisely meets the companies' needs (Salzman and Little, 1984). In addition a main characteristic of this approach is that the integration links should be designed and embedded into the essential system elements at conception. The lack of existing software limitations should permit consideration of more radical solutions.

In considering the possibilities it would appear sensible to review the suitability of the centralized methodology. As systems grow larger and more complex they become more difficult to comprehend and therefore to manage. A centralized approach to CIM in which a group of the major functional elements is expanded to absorb the others would inevitably exacerbate this problem. Functional centralization limits the degree to which the facilities offered can be tailored to meet local needs. In all but the smallest companies, needs vary across the organization. A single firm may produce a broad range of products which vary greatly in terms of design technology, market volumes and manufacturing processes. Thus conventional centralized systems tend to be developed to meet a variety of needs. As each new facility is added the system's complexity grows, as do the difficulties and costs of maintenance. In a CIM context these problems would be multiplied. Focused facilities can only be offered in a CIM environment if the system is designed to permit distribution of the various functions across a number of workstations or computers. For example it would be feasible to envisage different CAD or CAPP systems being used for different products, but the user still being permitted to move at will between them.

Centralization of a function in a single system is most apparent in production control where Manufacturing Resources Planning (MRP2) is seen as the obvious means to perform this function in CIM. Virtually all MRP2 systems run on a single computer. The performance limitations of the computer frequently limit the scope and speed with which manufacturing planning problems can be evaluated. Indeed without net-change facilities, what-if questions, which involve a regenerative MRP run, may be limited to no more than one per week. The dynamics of most manufacturing environments demand much faster responses.

The application of the concept of distribution to MRP2 systems appeared to offer some obvious advantages in improving the response of the system (by distributing the processing across several computers) and opening the way to 'bottom-up' definition of control system needs. To test the practical value of such a concept a prototype system has been developed by the authors.

The principle of the distributed planning and control

The roots of distributed planning and control for CIM are derived from the cellular manufacturing concepts associated with Group Technology (GT). Such systems are characterized by their ability completely to manufacture a 'family' of components, assemblies, or finished products. Thus each cell has a clearly defined product range. The cell manager may be regarded as the 'owner' of a manufacturing business unit and may be measured against broader performance objectives than would be common in conventional manufacturing plants. It is common GT practice for the cell to be responsible for local machine loading and sequencing, but all material planning decisions are normally performed by a centralized MRP system. In a distributed environment each cell is given a complete MRP system of its own. Thus the principle of treating each cell as a business is brought nearer fruition. Indeed the concept of distributed planning and control can be regarded as a means of achieving decentralization over a big organization, and yet the system interconnections would still permit the objective of total integration to be fulfilled. The distribution concept can be applied throughout so that the formation of integrated company cells is possible. This is illustrated in Figure 2. Each company cell is self-sufficient and has its own supporting data base. Data communication or information transfer is possible via the use of a Local Area Network (LAN) system.

Figure 3 outlines the principle of distributed planning and control applied across several cells. The central planning cell performs a special and supervisory role which is concerned with converting customer orders and market forecasts into a viable finished product manufacturing programme or Master Production Schedule (MPS). As such, this cell has no manufacturing facilities of its own but has to distribute product requirements to the final assembly cell(s). The viability of the overall MPS must be established through the use of rough-cut capacity planning techniques. The central planning cell must therefore hold bill-of-resource and family identification data for all the products manufactured by the company. The validated MPS is passed to the various assembly or, for spares, machining cells where it is treated as a 'Customer' schedule.

The individual cell plans its own local requirements using the same range of facilities which are normally to be found within a centralized system. The local MPS may be driven from customer requirements drawn from central planning or from orders generated by other manufacturing cells. The independence of the individual cell control and planning systems permits local management to re-plan their activities as and when required. Local policies and procedures may be implemented without the need extensively to update other parts of the system.

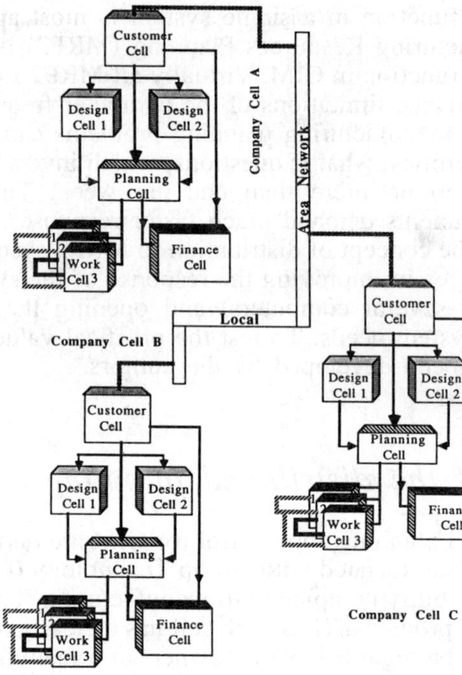

Figure 2. Formation of integrated company cells within a company.

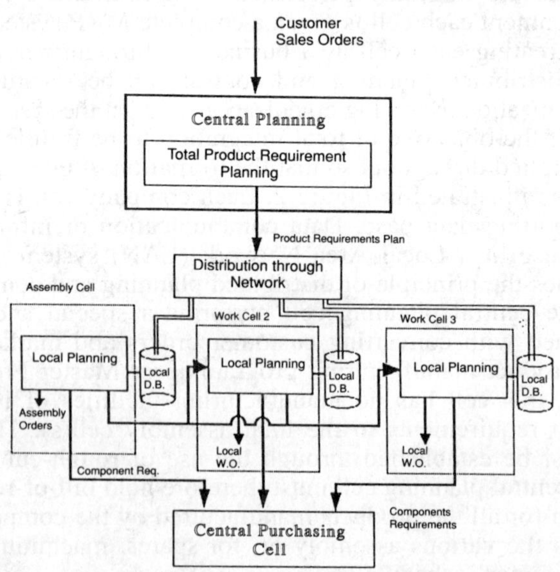

Figure 3. Principle of distributed planning and control.

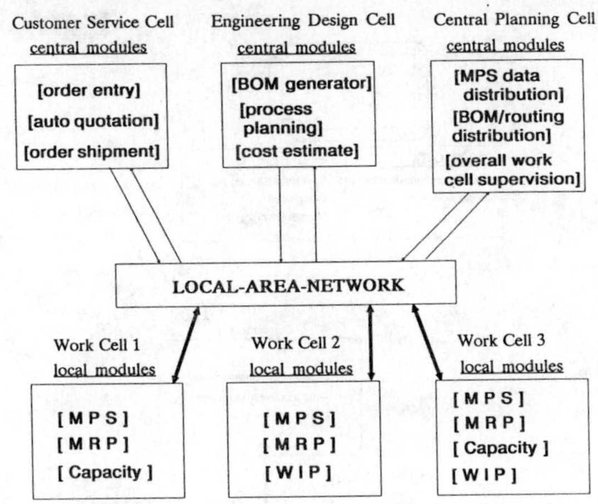

Figure 4. Six software sections within the embryo system.

The development of an embryo distributed CIM system

There are two main objectives of developing such an embryo system:

(1) to evaluate the benefits of the distribution concept,
(2) to develop design experience which could be used in the development of larger systems.

The system was first developed in Aston University, and is being continued as a joint project between Kingston Polytechnic and Aston University.

The system consists of six major sections of software modules distribution (Lung, 1988), as illustrated in Figure 4. Note that these software modules can be flexibly installed on any group of microcomputers that are networked together. The three work cells' configuration emulates a realistic workshop situation, each of them being assigned only the functions they need. For example only MPS, MRP and capacity planning are present in work cell 1, while work cell 3 is also assigned a Work-In-Progress (WIP) module.

The system was set up to represent two configurations. In the first experiment, only one work cell was used so as to represent a conventional centralized system in which all the planning is done by a single computer.

The second experiment contained three work cells: cells 1 and 2 were set up as machining cells and work cell 3 was configured as an assembly cell. All the MPS data were first processed by the assembly cell prior to subsequent distribution to the other cells. Figure 5 shows the set-up of the software modules and the hardware configurations used in the two experiments. Note that each work cell configuration would consist of the software modules capable of performing MPS, MRP, capacity planning and WIP monitoring.

The results obtained from both experiments were then broken down into different activities and analysed accordingly. A sample set of results is given in Table 1. As expected these simple tests showed that the multi-work cell configuration permits quicker evaluation of materials requirements in comparison with the conventional arrangement. It should be noted that the processing benefits (and consequent improvements in response times) are only one aspect of the benefits offered by this approach.

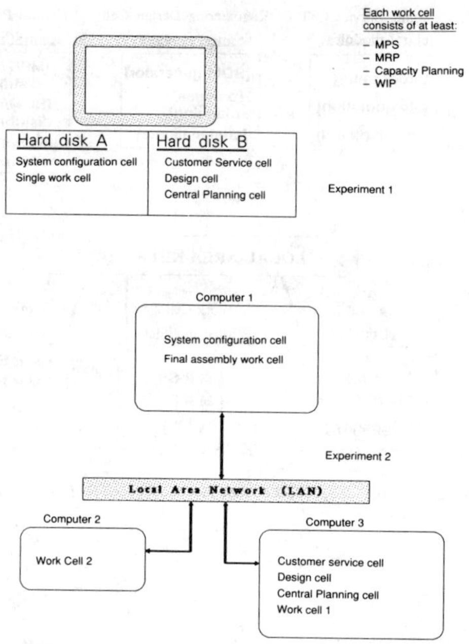

Figure 5. Hardware and software set-up for experiment 1 and experiment 2.

Table 1. Times recorded in the two test runs.

Activities	Times (s)			
	Experiment 1	Experiment 2		
	cell 3	cell 3	cell 2	cell 1
(1) Distribute BOM[a] routes	64		81 (total)	
(2) Distribute MPS	37	39		
(3) Local MRP	390	201	132	158
(4) Convert MRP to local work orders	665	442	302	875
(5) LSD calculations				
(a) Copy routing details	250	131	75	85
(b) Calculate LSD	480	270	159	151

[a] BOM, bills-of-materials.

Conclusion and further developments

This chapter has briefly reviewed the potential for the development of a distributed approach to the design of CIM systems. A specific application of the principle to requirements planning has been described. The potential of Distributed Material Requirements Planning (DMRP) appears to be considerable, particularly since subsequent work at Aston has shown that the system can be made capacity sensitive by the inclusion of feedback loops between the cells. It therefore offers an interesting alternative to techniques such as Optimized Production Technology (OPT) (Love and Barket, 1988).

However, for the system to provide a rapid response to customer enquiries or production problems it will be necessary to automate the local evaluation of capacity constraints and consequent revision of the cell's MPS. This problem is being examined and a solution anticipated based on the use of an expert system. The complete package (DMRP plus expert system) will be extensively tested using a factory simulator also developed at Aston (Love and Bridge, 1988).

References

Blackstone, J. and Cox, J.F., 1985, MRP design and implementation issue for small manufacturers. *Production and Inventory Management*, **3**, 65–76.

Callarman, T.E. and Heyl, J.E., 1986, A model for material requirements planning implementation. *Proceedings of the UK Operational Management Conference*, January Warwick, UK, pp. 31–43.

Gott, B., 1984, CAD and design options. *Cambridge Manufacturing Forum, Strategies for Survival and Success*, July, pp. 21.

Groover, M.P. and Zimmer E.W., 1984, CAD/CAM: *Computer-Aided Design and Manufacturing* (Englewood Cliffs, NJ: Prentice-Hall).

Huges, D.R. and Maull, R.S., 1985, Design of Computer Integrated Manufacturing systems. *8th International Conference on Production Research*, August, Stuttgart, pp. 372–379.

Ingersoll Engineers, 1985, It is not so much what the Japanese do it's what we don't do. Ingersoll Engineering Reports.

Love, D.M. and Barket, M., 1988, An introduction to the decentralized, distributive MRP philosophy; a solution to control problems in cellular manufacturing systems. *3rd International Conference on Computer-Aided Production Engineering*, University of Michigan, Ann Arbor, MI, 1–3 June, pp. 352–360.

Love, D.M. and Bridge, K., 1988, Specifications for a computer simulator to support the manufacturing system design process. *3rd International Conference on Computer-Aided Production Engineering*, University of Michigan, Ann Arbor, MI, 1–3 June, pp. 317–323.

Love, D.M., Lung, W.M. and Sawyer, J.H.F., 1986, Distributed planning and control in a cellular CIM system. *1st International Conference on Engineering Management: Theory and Application*, Swansea, pp. 215–224.

Lung, A.W.M., 1988, Computer-integrated manufacturing control. Ph.D. Thesis, University of Aston.

Salzman, R.M. and Little, A.D., 1984, The evolution from CAD/CAM to CIM, possibilities, problems and strategies for the future. *Proceedings of Computer Aided Manufacturing Process*, September, West Berlin, pp. 241–245.

Weatherall, A., 1984, The computer integrated business. *Cambridge Manufacturing Forum Paper* 264, 6/7.

Willer, D., 1984, Computer aided manufacturing. *Computers in Manufacturing Industry*, series in CME, 4, July/August, 22–24.

Information system design and analysis of CIM systems: the new M* approach

François Vernadat[1] Antonio Di Leva[2] and Piercarlo Giolito[2]

[1]*INRIA-Lorraine, BP 239, 54506 Vandoeuvre Cédex, France and* [2]*Università di Torino, Dipartimento di Informatica, Corso Svizzera 185, 10149 Torino, Italy*

Abstract

Information system design and analysis is an important issue in the design and implementation of CIM systems. This concerns enterprise analysis (in terms of functions and enterprise objects) and conceptual analysis (in terms of semantic data structures and data base transactions). Efficient modelling techniques and a sound methodology need to be used for this endeavour. In this chapter we report on M*, a methodology for information system modelling and analysis and data base design of manufacturing environments. It makes use of an integrated model for information representation based on high-level Petri nets and an Extended Entity-Relationship model. It allows modelling of both static and dynamic information properties, and it produces executable data base specifications.

Introduction

Computer-Integrated Manufacturing (CIM) is a new paradigm developed in the 1980s striving to improve productivity, efficiency or competitiveness of manufacturing systems (Groover and Zimmers, 1984; Ranky, 1986). Currently, it is well admitted that successful implementation of a CIM system relies on rational design of the data base system which is going to support the entire automated system.

To provide a contribution to this effort, the M* methodology has been developed for information system modelling and analysis and data base design of manufacturing environments. The M* methodology was initially developed by the authors as a joint effort between the Università di Torino in Italy and the National Research Council of Canada (NRCC) (Di Leva *et al.*, 1986, 1987).

A first originality of M* is to be one of the very few existing information system design and analysis methodologies adapted to industrial automation projects. A second originality concerns the fact that M* is based on an integrated model for information representation. This model is based on one hand on high-level Petri nets (Genrich and Lautenbach, 1981) and on the other hand on the Extended Entity-Relationship (EER) model (Bussolati *et al.*, 1983). It allows modelling of both static and dynamic information behaviour, and it allows the validation of the data schema by simulation of the execution of data base transactions expressed in a data base transaction language adapted to the model. A third originality of M* is to be an open methodology, i.e. it can be augmented with other modelling tools or methodologies (e.g. GRAI (Doumeingts, 1984) or CIM-OSA models (AMICE, 1989)) if needed.

The M* methodology

In CIM environments, it is now crucial that information system design and analysis, as well as data base design, must be performed in synergy with factory analysis. These design tasks can no longer be performed on the basis of more or less precise user requirements formulation, but must take into account the full reality of the manufacturing enterprise. The purpose is to build fully integrated systems.

The M* methodology, which is derived from the Italian DATAID methodology for general purpose data base design (Ceri, 1983; Albano *et al.*, 1985), has been designed to meet such requirements. It covers the analysis of an organization, the modelling of its information system, and the design of the data base environment which will have to support the operations of the organization activities.

The M* basic terminology

In M*, an object organization is a real-world production enterprise subject to be the focus of an analysis. An organization environment is a subset of the enterprise composed of users, objects and functions, and achieving a major goal of the enterprise. M* assumes that any manufacturing organization or enterprise can be decomposed into a set of functions and sub-functions. A function represents the task of processing information and/or materials using a set of resources (including people) under some constraints to achieve some enterprise objectives or sub-objectives. Within an organization environment, functions can be logically grouped into organization processes, which represent a group of inter-related functions sharing common resources and information. They interact by means of information exchange. Basic functions are called activities. They are described by their inputs, outputs, constraints or control conditions, and their supports. Activity executions are triggered by external happenings and/or information updates. Activity logic is composed of a sequence of operations described by process nets. Operations are the atomic units of decomposition.

The M* methodology

M* consists of three main design phases: (a) Enterprise Modelling and Analysis, (b) Conceptual Design, and (c) Implementation Design. Figure 1 provides a graphical illustration of the M* overall structure.

The Enterprise Modelling and Analysis phase provides managers, engineers and analysts with a set of tools for the analysis of the functional organization of the enterprise in terms of a hierarchically structured set of organization networks, which describe static aspects of the enterprise (i.e. the enterprise functional structure) and dynamic aspects (i.e. the enterprise functional behaviour) at different levels of detail. However, since one of the goals of the M* methodology is further to develop a global data base schema, the analysis process is also oriented towards the description of the information manipulated within the organization at a macro-level. The goal is here to provide managers with a clear and accurate view of the current status of the enterprise (AS-IS analysis), to detect organization anomalies, and to specify the structure of the new automated system (TO-BE analysis). This phase corresponds to the organization level of M*. It is supported by the M* organization model discussed later.

The Conceptual Design phase is used to analyse the enterprise functional description obtained previously in order to state a detailed conceptual specification of the organization in terms of a conceptual view. The conceptual view contains a unified description of all the different aspects of the organization reality, i.e. its static and dynamic features and its constraints. By static features, we mean the

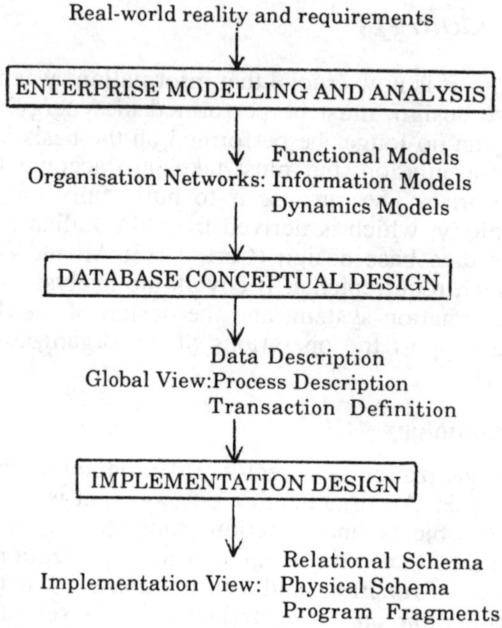

Figure 1. The M overall architecture.*

structure of the data involved and the integrity constraints existing on them (static features are expressed by means of a global data schema). Dynamic features concern both the operations on the data and the causal relationships existing between the organization operations (they are expressed in the form of process nets which represent the gross logic of application programs or transactions implementing enterprise activity execution). Manipulations on data are specified by means of a transaction definition language. This phase corresponds to the conceptual level of M* and is supported by the integrated Process and Data Net (PDN) model discussed later.

The Implementation Design phase translates the global view obtained at the conceptual level into a physical organization description (called implementation view) to be implemented on and processed by the target data base management system (DBMS) selected for physical implementation of the automated system. This phase corresponds to the implementation level of the M* methodology (not discussed in this chapter).

The most important feature of the M* models is their ability to describe, in an integrated way, both the organization behaviour and the manipulation actions on the data base which will have to support the information system activities. The new organization model presented in the chapter and based on high-level Petri nets (Predicate-Transition net models) is an alternative to the controversial SADT and IDEF formalisms (Bravoco and Yadav, 1985a, b).

High-level Petri nets

The main difference between the Predicate-Transition (PrT) net model (Genrich and Lautenbach, 1981; Jensen, 1981) and the basic Petri net model (Peterson, 1982) relies on the fact that tokens in the PrT net model can be structured objects and that the transition firing is controlled by imposing conditions on the token values or on the transition itself.

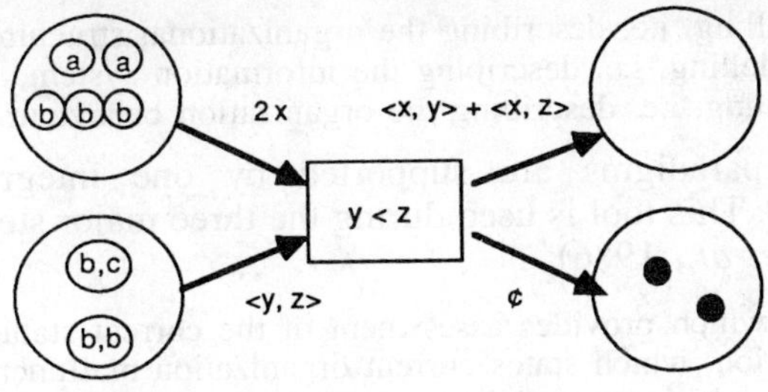

Figure 2. A PrT net example.

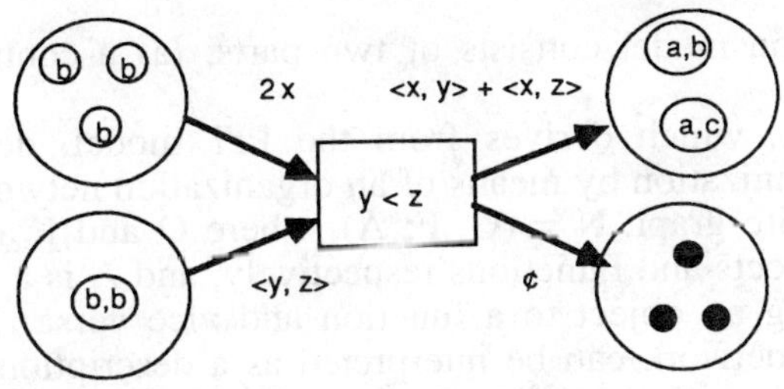

Figure 3. Firing of the transition.

A PrT net is a bipartite directed graph with two types of nodes, predicates (represented by circles), and transitions (represented by bars or boxes connected by arcs. Arcs can connect predicates to transitions and transitions to predicates. The set of tokens contained in a circle can be interpreted as the extension of the corresponding predicate, i.e. the set of objects for which the predicate is true. For example, a predicate M can represent a set of available machines and its extension may be $M = \{m1, m2, m3\}$ where m1, m2, and m3 are distinct occurrences of physical machines, the status of which is set to 'available'. A transition represents some processing of some input predicate occurrences which produces some output predicate occurrences under some conditions. For example, S may be a transition which selects a machine from M to allocate it to a job as an occurrence of predicate M' (the status of this machine then becomes 'on-use').

The extension of a predicate can be modified by transitions representing events occurring in the system being represented. In the PrT model, the set of objects affected by a transition (input/output sets) is indicated by means of variables and functions on the arcs connecting the input/output predicates to the transition. The firing of a transition consists of removing input objects from the input predicates and adding output objects to the output predicates. The same predicate can be an input and an output predicate of the same transition.

Figure 2 represents a simple PrT net with a condition on the execution of the transition. Figure 3 illustrates the firing of the transition.

Organization model and enterprise analysis

High-level dynamics description

Basically, the Enterprise Modelling and Analysis phase of M* is aimed at describing the static and dynamic features of the enterprise structure according to a top-down approach from three complementary viewpoints:

— functional modelling, i.e. describing the organizational structure,
— information modelling, i.e. describing the information system,
— dynamics modelling, i.e. describing the organization behaviour.

These modelling paradigms are supported by one integrated tool: the M* organization model. This tool is used during the three major steps of this M* phase, namely (Di Leva *et al.*, 1986):

— AS-IS analysis, which provides assessment of the current status of the enterprise,
— anomalies detection, which states current organization malfunctions,
— TO-BE analysis, which states the new organization structure to be installed.

The M* organization model

The M* organization model consists of two parts: (a) a control part, and (b) an information part.

The control part, which derives from the PrT model, describes the control structure of the organization by means of an organization network. An organization network is a bipartite graph N = (O, F; A), where O and F are two disjoint sets of nodes called Objects and Functions respectively, and A is a set of directed arcs, each one connecting an object to a function and vice versa.

An organization network can be interpreted as a description of an organization process which is oriented to specify the modifications produced in the context in which its component functions are executed, the relationships between its component functions, and its communications with other networks, i.e. other organization processes. An organization network is then a set of functions, object flows and precedence relationships among them in which the component functions (represented as boxes) can be activated only when their input objects (represented as circles) are present and its occurrence generates the output objects (consuming the input ones).

The information part of the organization model is used to record information obtained about the organization structure. It is based on a frame system (the frame idea was selected for flexibility in extending the model). A frame is associated to both object and function nodes of the network. The frame specifies the node characteristics and its relationships with other nodes.

An object frame contains:

— the object identifier, name, and description;
— the object type: an object node represents a set of object occurrences (object extension), and it may hold occurrences of a specified type;
— the object structure: it implies actions that can be applied to the object occurrences. Predefined object structures are: sets, ordered sets, stacks, and queues;
— the object state: it contains the description of the conditions that an object occurrence must satisfy in order to belong to the object extension;
— the object behaviour: it specifies what functions are connected to the object and how they are connected.

The set of primitive object types considered in the model concerns:

(1) *Data*—Data are information items relevant to the organization.
(2) *Files*—A data file is a structured set of computer records composed of data items.
(3) *Forms*—A form is a document which is used to collect, display, or communicate information within the organization.
(4) *Messages*—A message is a piece of information concerning the fact that something has happened, i.e. it represents an event of the real-world. If a message is an input to a function, its occurrence is a prerequisite to execute that function. If a message is an output of a function, it holds as soon as the function has been executed. Messages can be used to represent timed events scheduled at pre-defined instants.

(5) *Materials*—They include all the physical matters that enter the production process (raw materials, parts, assemblies, products, etc.).
(6) *Resources*—They include all the physical means used to carry out the manufacturing process (tools, fixtures, machines, people, etc.).

A function frame contains:

— the function identifier, name, and description;
— an input rule, which specifies the set of conditions that the input objects must satisfy to activate the function;
— a set (possibly empty) of control objects, which must be known to execute the function, but are not modified as a consequence of its execution;
— a set of actions: each action describes a change of the organization state, i.e. modification of the organization objects due to the function execution;
— a set of output rules: an output rule is a condition that can be associated to an output object of the function.

The state of an object and input/output rules of a function are usually described by means of natural language sentences or in pseudocode.

To specify the overall network behaviour, the state and behaviour of the objects and the input/output rules of the functions must be known. The causal model for the network simulation is based on the following function firing rule. A function F is *enabled* if the input objects I_1, I_2, ..., I_n and the control objects C_1, C_2, ..., C_i contain valid occurrences which satisfy the input rule, and if the input messages M_1, M_2, ..., M_k hold. After the execution of F:

— the input occurrences to be processed are removed from the input objects;
— a set of object occurrences is added to the output objects O_1, O_2, ..., O_m;
— the input messages M_1, M_2, ..., M_k cease to hold and the output messages N_1, N_2, ..., N_h begin to hold.

Organization analysis

The organization model provides suitable support for recursive refinement of the organization processes as discussed in the original description of structured system analysis techniques (Ross and Schoman, 1977). Process refinement consists in the identification of sub-processes (i.e. functions and sub-functions) and internal objects within the parent process. Sub-functions can in turn be further decomposed. This process results in a function refinement hierarchy. Both objects and functions can be refined and the information concerning the refinement step is stored in the frame system for further references, graphics representation and concept consistency checking. The refinement process ends when the component function granularity satisfies an organization specification consensus on which users can agree.

In the M* methodology, the behaviour of a function is simply described by means of a set of sentences in natural language, or pseudocode, and by arithmetic and logical expressions. These sentences specify the activities of the object function and will be processed in the following conceptual design phase to produce a formalized, computer-oriented specification.

The organization analysis starts with the AS-IS analysis task composed of two steps: top-down analysis and bottom-up analysis (Di Leva *et al.*, 1986). The aim is to find out the functional architecture of the enterprise and to describe its behaviour using the process refinement approach and the M* organization model. Then, the organization structure is analysed to verify its consistency in terms of functional architecture, information model and dynamics behaviour. Finally, if the system has to be partially restructured or if we are considering the design of a new system, the TO-BE analysis task is performed using again the organization model constructs to produce the new system specifications.

Conceptual model and conceptual design

The conceptual model is aimed at supporting the Conceptual Design phase of M*. This is a bottom-up approach for designing the enterprise data base system of the CIM system. This phase is composed of (Di Leva *et al.*, 1986):

— forms analysis, i.e. building data schemata from available formatted requirements identified in the previous phase (such as forms, documents, data file formats, etc.),
— user requirements analysis, i.e. building data schemata, operation schemata, process nets from unformatted requirements (user interviews, procedure manuals, etc.),
— view design, i.e. building a data and process description model for each data base view,
— view integration, i.e. merging all view descriptions in order to obtain a global view description for a given environment or for the whole organization.

These steps are supported by one integrated tool: the PDN (Process and Data Net) model in which the user requirements specifications are expressed in terms of a data description model (information part) and a process description model (control part). A transaction definition language is also used formally to express the data base transactions representing operation and activity executions on the data base.

The data description model

One form of the EER model (Bussolati *et al.*, 1983) is used in the M* methodology to describe the real-world objects of interest for the organization. This EER model provides the designer with four basic abstraction mechanisms for *static* data representation:

(1) *Entities*—An entity describes a set of objects (entity occurrences), which share common characteristics.
(2) *Relationships*—A relationship R among the entities E_1, E_2, ... describes the set of relations (relationship occurrences or links) relating the occurrences $e_1 \in E_1$, $e_2 \in E_2$, ..., and (possibly) sharing common characteristics. Mapping functionalities of a relationship are represented by means of two natural numbers: the min- and max-cardinalities. The min-cardinality is the minimum number of times (possibly 0) that each object can be involved in a link. The max-cardinality is the maximum number of times (possibly *n*) that each object can be involved in a link.
(3) *Attributes*—An attribute A describes a characteristic or property of an entity or a relationship, and takes its values from a value set called the domain of A.
(4) *Abstraction hierarchies*—A hierarchy represents entities in terms of more general entities. The first type of hierarchy is the subset hierarchy. An entity S is a subset of an entity E, E < = S (E is superset of S), if every occurrence of S is also an occurrence of E. The second type of hierarchy is the partition hierarchy. Entities P_1, P_2, ... are a partition of the entity E, E < = {P_1, P_2} (E is a generalization of P_1, P_2, ...) if each occurrence of E is also an occurrence of one (and only one) of the entities P_1, P_2, ... The partition or the subset structure on the occurrences of E can be explicitly represented by an attribute of E. In this case, it will be referred to as a discriminating attribute.

The graphical formalism used for these concepts is represented in Figure 4.

The process description model

In the M* methodology, the detailed, *dynamic* organization behaviour at the conceptual level is described by means of a set of process nets. Each one describes a set of operations on data base objects with their associated execution conditions, which must be executed to achieve some objectives of the object organization. Any event occurrence in the physical system is mirrored as a change in the data base state.

Execution conditions are usually related to messages coming from the external

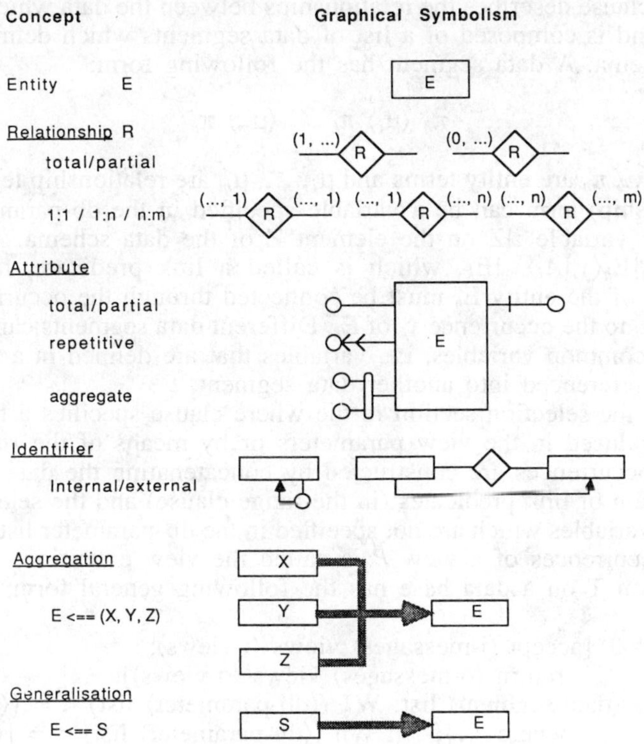

Figure 4. EER formalism.

world (with respect to the object function) and/or to the existence of some objects in a particular state (message and data-driven system).

The operation model

To specify data manipulations on a data base schema, the transaction definition language for the EER model is first presented. The language allows users to define views on the data base schema and to specify manipulations on it by means of transactions (i.e. units of activity upon the data base) which describe one or more operations on the data. Retrieve (get a piece of data), update (modify data), delete (erase data), and append (add data) are operations used to manage entity occurrences while connect (add a relationship occurrence) and disconnect (delete a relationship occurrence are operations used to manage relationship occurrences.

A view W on the data base has the following general form:
view W(⟨db-parameter⟩ list)
range ⟨data-segment⟩ list
where ⟨selection⟩

where W is the view name. A data base parameter (db-parameter) has the following form:

⟨variable⟩|⟨db-element⟩

where a db-element is an entity or a relationship, and the variable assumes, as values, an occurrence of the entity or the relationship, respectively.

The range clause describes the relationships between the data which are involved in the view and is composed of a list of data segments which define paths in the data base schema. A data segment has the following form:

$$\pi_h \langle \mu_i \rangle \pi_k \ldots \langle \mu_m \rangle \pi_n$$

where π_h, π_k ..., π_n are entity terms and μ_i, ..., μ_m are relationship terms. An entity (or a relationship) term can be a variable specified in the db-parameter list, or it can be a new variable z|Z on the element Z of the data schema. A sequence of terms like $x_h | E_h \langle y_i | A_i \rangle x_k | E_k$, which is called a link predicate, means that an occurrence x_h of the entity E_h must be connected through the occurrence y_i of the relationship A_i to the occurrence x_k of E_k. Different data segments can be connected by means of common variables, i.e. variables that are defined in a term of a data segment and referenced into another data segment.

Optionally, the selection section in the where clause specifies a formula on the variables introduced in the view parameters or by means of the *range* clause.

The view occurrences are constructed by concatenating the data objects which satisfy the chain of link predicates (in the range clause) and the selection formula. Then, all the variables which are not specified in the db-parameter list are discarded. The set of occurrences of a view *P* is called the view extension.

A transaction T on a data base has the following general form:

```
transaction  T [accept ⟨i-messages⟩ views ⟨i-views⟩;
                return ⟨o-messages⟩ views ⟨o-views⟩]:
range        ⟨data-segment⟩ list; W1 (⟨db-parameter⟩ list) < = {⟨data-segment⟩
                where ...}; ...; Wn (⟨db-parameter⟩ list) < = {⟨data-segment⟩
                where ...};
where        ⟨input-rule⟩
{⟨transaction-block}
```

where T is the transaction name. The accept clause describes the input messages ⟨i-messages⟩ and the input views ⟨i-views⟩ of T. The return clause describes the output messages ⟨o-messages⟩ and the output views ⟨o-views⟩ of T. A view is specified by a term of the form P(p_1, ..., p_n) where P is the view name and p_1, ..., p_n is a list of variables corresponding to the data base parameters of P as defined in the view definition.

Variables specified by the input/output (I/O) views are used in the transaction, together with the variables that may be introduced in the optional range clause, as in the view definition. The range clause specifies the relationships between the data which are involved in the transaction, and can be used to introduce local views (W_1, ..., W_n). The input rule in the optional where clause specifies a condition that must be satisfied by data which are involved in the transaction, and can be used to express conditions on the local views using a set of set-oriented operators (contained in, contains, difference, member of, and equals).

The ⟨transaction-block⟩ section consists of a sequence of (possibly one) data base operators on the data variables defined in the I/O views. Control structures (including sequence, iterative, and conditional structures) can be specified on the operators which manipulate data base objects and sets. Output rules of function descriptions at the organization level can be easily modelled at the conceptual level by means of transaction control structures.

The control model

At the conceptual level, the control structure of an organization is described by means of a set of process nets, which corresponds to the set of organization

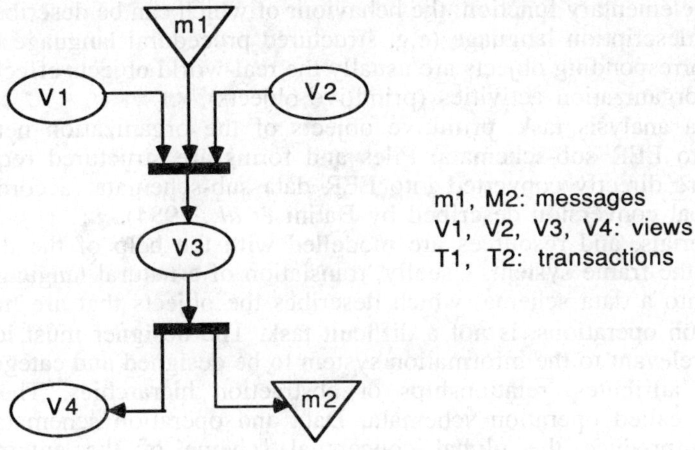

Figure 5. Example of a process net

networks defined at the organizational level. However, there can be several process nets associated to the same organization network.

The process net model used in M* is once again derived from the PrT net model. Each process net, related to a data base S, is defined by:

(1) A bipartite directed graph (P, T; F) where P = $\{p_i\}$ and T = $\{t_j\}$ are two disjoint sets of vertices called Places and data base Transactions, respectively, and F is a set of arcs, which connect a transition $t \in$ T to its input and output places p $\in$ P ($P \cap T \neq \emptyset$, $P \cup T \neq \emptyset$, $P \times T \cup T \times P \supseteq F$). The set of places P is composed of two disjoint subsets $P = V \cup M$, where $V = \{v_h\}$ is a set of data base views, and $M = \{m_k\}$ is a set of messages.
(2) Each transaction t_j represents possible indivisible changes of the adjacent view extensions.
(3) Given a transaction t of T, a substitution for t is a collection of occurrences extracted from the input views. A substitution for t is admissible if the range and where clauses of t are satisfied.

The transaction t is enabled whenever:

(1) an admissible substitution for t can be built,
(2) the input messages of t hold.

When a transaction is enabled, it can be fired. From the data base point of view, this means that:

(1) the occurrences involved in the admissible substitution are removed from the corresponding input views,
(2) the update actions specified in the transaction block are executed. Update actions generate new occurrences, which must be added to the output view extensions of t,
(3) input messages no longer hold and output messages begin to hold.

Process nets have a graphical representation in which bars represent transactions, circles represent views and triangles represent messages. Figure 5 provides an example of process net.

Conceptual design

The organization networks are usually further decomposed and analysed in the Conceptual Design phase because detailed information is needed at this stage to enable the design of the transactions which will support system operations. The refinement process ends at the elementary function level or activity level. An activity is an elementary function, the behaviour of which can be described in terms of a process description language (e.g. structured procedural language or pseudo-code). The corresponding objects are usually the real-world objects effectively used in the daily organization activities (primitive objects).

In the data analysis task, primitive objects of the organization networks are converted into EER sub-schemata. Files and forms are structured requirements. Thus, they are directly converted into EER data sub-schemata (according to the methodological conversion described by Batini *et al.*, 1984).

Data, materials, and resources are modelled with the help of the descriptions contained in the frame system. Usually, translation of a natural language function description into a data schema, which describes the objects that are manipulated by the function operations, is not a difficult task. The designer must identify the object items relevant to the information system to be designed and categorize them into entities, attributes, relationships or abstraction hierarchies. The resulting schemata are called operation schemata. Data and operation schemata are then integrated to produce the global conceptual schema of the enterprise. The methodological steps for the conceptual schema integration in M* are given by Di Leva *et al.* (1986) and issued from Batini *et al.* (1983).

In the transaction analysis task, object and function descriptions are analysed from a procedural viewpoint. The sub-schemata corresponding to the organization network objects are translated into views on the global data base schema. The object states must be expressed in the selection clause of the view specification. An elementary function or activity description is translated into a transaction specification. The input rule must be expressed in the where clause of the transaction. The output rules are expressed by the control structure of the transaction block.

The set of views and transactions specified on the data base schema forms the specification of the system. This specification can be directly prototyped and 'executed' to predict system behaviour.

Conclusion

An integrated approach, based on formal techniques from software engineering, for enterprise and modelling and information system modelling and analysis has been presented in this chapter. Integration of relevant concepts has been made possible by the use of high-level Petri nets. The formalism presented has proven to be adequate for modelling dynamic functional organizations as well as process and data representations in conjunction with the EER model. Each of the basic models of the M* approach is formally related to the others and can be computerized. This is one of the great advantages of M* as compared to other structured methodologies for manufacturing environment analysis and modelling. Computerization of the methodology is underway at the University of Turin.

References

Albano, A., DeAntonellis, V. and Di Leva, A. (Eds), 1985, *Computer-Aided Database Design* (Amsterdam: North-Holland).

AMICE, 1989, *Open System Architecture for CIM*, edited by ESPRIT Consortium AMICE. (Berlin: Springer).

Batini, C., Lenzerini, M. and Moscarini, M., 1983, Views integration. In *Methodology and Tools for Database Design*, edited by S. Ceri (Amsterdam: North Holland).

Batini, C., Demo, B. and Di Leva A., 1984, A methodology for conceptual design of office databases. *Information Systems*, **9**, 3/4, 251–263.

Bravoco, R.R. and Yadav, S.B., 1985a, A methodology to model the functional structure of an organization. *Computers in Industry*, **6**, 345–361.

Bravoco, R.R. and Yadav, S.B., 1985b, Requirements definition architecture—an overview. *Computers in Industry*, **6**, 237–251.

Bussolati, U., Ceri, S., DeAntonellis, V. and Zonta, B., 1983, Views conceptual design. In *Methodology and Tools for Database Design*, edited by S. Ceri, Amsterdam: North-Holland.

Ceri, S. (Ed.), 1983, *Methodology and Tools for Database Design* (Amsterdam: North-Holland).

Di Leva, A., Vernadat, F. and Bizier, D., 1986, M*: *a methodology for information system analysis and database design in production environments*. ERB-1000 (Ottawa, Canada: National Research Council).

Di Leva, A., Vernadat, F. and Bizier, D., 1987, Information system analysis and conceptual database design of production environments with M*. *Computers in Industry*, **9**, 183–217.

Doumeingts, G., 1984, Méthode GRAI: méthode de conception des systèmes de production. Thèse d'Etat, Bordeaux I.

Genrich, H.J. and Lautenbach, K., 1981, System modelling with high level Petri nets. *Theoretical Computer Science*, **13**, 109–136.

Groover, M.P. and Zimmers, E.W., 1984, *CAD/CAM: Computer-Aided Design and Manufacturing* (Englewood Cliffs, NJ: Prentice-Hall).

Jensen, K., 1981, Coloured Petri nets and the invariant-method. *Theoretical Computer Science*, **14**, 317–336.

Peterson, J.L., 1982, *Petri Net Theory and Modelling of Systems* (Englewood Cliffs, NJ: Prentice-Hall).

Ranky, P., 1986, *Computer-Integrated Manufacturing* (Englewood Cliffs, NJ: Prentice-Hall).

Ross, D.T. and Schoman, K.E., 1977, Structured analysis for requirements definition. *IEEE Transactions on Software Engineering*, **SE-3**, 16–34.

SECTION VIII.4. ORGANIZATION

100

Design justification in product development

James S. Noble and J.M.A. Tanchoco

School of Industrial Engineering, Purdue University, West Lafayette, IN 47907, USA

Abstract

The need to design products with maximized functionality and minimized costs is essential in today's globally competitive world. An environment that facilitates the interactions between design and justification would enhance such a decision process. The conceptual framework for such a design and justification system is presented in this chapter. The concept was implemented in the design of an electromagnetic/ radio frequency shield for electrical metering equipment. In this implementation the design justification environment enhanced the decision making process. It enabled the decision maker to take advantage of present knowledge and combine it with that of the environment better to analyse the design of the shield throughout the entire product development period.

Introduction

In today's economic and competitive environment a manufacturer must carefully weigh all decision trade-offs that it makes. The development of computer-aided manufacturing technology, combined with increased international competition, has raised the stakes of each decision. The realm of these decisions ranges from product features and manufacturing processes to where production should be located geographically.

In looking at the impact of design decisions, the product designer determines between 70 and 75% of the eventual cost of a product (Boothroyd, 1988; Wierda, 1988). This implies that each decision must be carefully justified before it can be implemented. Justification must be done with respect to the final cost and function of the product.

Typically, design and economic justification have been considered as two separate functions. Though they have a common goal of arriving at a competitive product, they have ended up making the competition between themselves. The goal of designing the best product possible is often contrary to the goal of cost minimization. It is necessary to bring these two separate, but related functions together in an environment where they can freely interact.

This chapter proposes an environment where the interaction between the design and economic justification functions is promoted. The purpose of creating such an environment is that better alternatives and solutions to the objective of maximizing the design function and economic feasibility can be found when allocating limited resources. We define the term *design justification* to denote the integration of the design and economic justification processes such that their interaction enhances the final design solution. This concept is explored in this research, and a prototype is developed to illustrate the concepts derived.

Until recently the issue of integrated product design and economic justification had not been dealt with much in the literature. An understanding of how these two areas can be integrated requires an understanding of them separately. Traditionally, justification has been performed using either the discounted cash flow, the internal rate of return or the payback period methodologies (Au and Au, 1983). These methodologies have been cited for having several shortcomings when applied to the current manufacturing environment and technologies: a short-term perspective on decisions (Hayes and Gavin, 1982), and consider each individual decision on equipment to be isolated from impacting the environment around it (Gold, 1982; Leung and Tanchoco, 1987).

Recent developments in justification methodologies have addressed the shortcomings of traditional justification procedures. Some of the new procedures that have been proposed are: weighted additive models (Sullivan, 1986), multi-attribute models (Randhawa and West, 1986), simulation in life cycle analysis (Haider and Blank, 1983), and expert systems (Sullivan and Leclair, 1985). All of the recent justification methodologies seek to bring more information into the decision process than has been traditionally considered. This information has primarily consisted of the strategic and integration effects associated with the decision.

The traditional design process has its roots in the scientific method. The general characteristics of the design process are: (1) problem formulation, (2) alternative generation (synthesis), (3) alternative evaluation (analysis), and (4) alternative selection (optimization) (Rouse, 1986).

Designers are in need of additional support in two areas: the conceptual development process for the design and the evaluation and generation of design alternatives. Both of these areas reflect a need for additional information. The characteristics of the information which would be of value to a designer have been proposed (Rouse, 1986). Therefore, the need for design support is to provide this relevant information (Sage, 1987).

The traditional design process has dealt with the issue of the economics of design in an implicit manner. This means that economics are always kept in mind, since minimization in the allocation of resources is required, but the actual evaluation is typically not done until after the design is complete. Early attempts at integrating the design and justification processes centred on developing design algorithms that were performance and cost driven (Boothroyd, 1982; Gustavson, 1984). This approach does not give additional information to aid the designer's decisions, rather the algorithms provide the design solution.

A different approach to integration has been to develop cost models for cost estimation to aid in the decision process. Daschbach and Apgar (1988) suggest that the parametric method for cost modelling should be used throughout the design process. Another approach to integration is to link economic measures to design data bases. Wierda (1988) reviewed the different types of cost estimation models that could be used to aid in product design and gave examples of how these have been linked with computer-aided design (CAD) systems to enable the designer to do cost estimation. Fleischer and Khoshnevis (1986) proposed a knowledge-based design system linking design, process planning and cost analysis within a CAD system. Boothroyd (1988) developed an actual implementation of a system to estimate the machining cost of a product.

Both of these last two approaches have provided some economic information to aid in the design process, but have not addressed the relationships between the engineering design factors and the economic ramifications of design decisions. The need for using economic justification measures during the design process has been proposed and discussed before (Boothroyd, 1988; Fleischer and Khoshnevis, 1986; Wierda, 1988). However, the question of how this should be accomplished has not been thoroughly examined.

Conceptual framework

Traditional economic justification models are used in the design process in one of two ways. One way is that a cost estimate is made from a budgetary limit, which is passed down to the designer. The designer can often become preoccupied with the cost constraint and becomes restricted in coming up with the best functional design.

The other way, as is most often the case, the designer is allowed to develop the product and then the economic justification is performed after the fact. In this case, little thought is initially made by the designer about cost since the functionality of the product is the primary concern. Therefore, when a product design is rejected based on cost considerations the whole design process must be reinvestigated to determine where cost might be reduced.

The basic concept of design justification is that there is a need to have meaningful interaction between design and economic justification at all major points of design trade-off. This will result in greater attention focused on the design factors which will enhance product performance with respect to the best cost efficiency. It also allows economic models to be developed incrementally with the design. Thus, it can be used as a decision tool throughout the entire design process. The added information within the design process enables the most economical and functionally acceptable design to be developed.

In the development of a system to aid design justification, it was found that the effectiveness of the system environment is determined by the way in which information is presented to the decision maker. The current trend in design and economic model methodologies is to work toward quantifying all information about a design in order to aid in the computer analysis of the design process. However, in doing this, the creativity, intuition and previous knowledge of the designer can be 'quantified out' of the process.

Therefore, the proposed framework for the development of economic models to aid in the design of a product, requires an environment consisting of the following:

(1) The ability to analyse and present quantifiable information. Quantifiable means that the information can be reduced to a numerical value.
(2) The presentation of uncertain or incomplete information in such a manner that the decision maker is able to process and integrate this information throughout the design process.
(3) The decision maker is the decision mechanism that drives the design process.

The solution approach for dealing with the design and justification interaction problem requires an environment that facilitates information interaction. An implementation of the environment makes several components necessary for design justification to occur. These components are: a knowledge base, an analysis procedure for design evaluation and economic model development, and a presentation medium that supports differing types of interaction. The goal of combining these components is to present information that will best aid the decision maker.

A knowledge base in necessary so that a body of design knowledge can be assessed throughout the design process. A knowledge base is a collection of data and component interactions within the design sphere (i.e. a manufacturing technique and the types of decisions which follow its selection). The knowledge base by its structure leads the designer down the feasible technical path of alternatives for the product design.

Figure 1. An EMI/RFI shield for electrical metering equipment.

The quantitative analysis component provides the decision maker with the opportunity to experiment with quantifiable design information and to develop the associated cost model for the design. The analysis relies upon information provided by both the knowledge base and the decision maker and consists of the ability to evaluate both the engineering and economic aspects of the design.

The final component of the environment is an engineering workstation that facilitates interaction between the knowledge base, quantitative analysis and decision maker. The workstation provides graphical capabilities to aid in assimilation of information by the decision maker. This platform makes the decision maker an essential component in the entire decision making process.

Prototype implementation

A prototype implementation of the design justification environment was developed. The product used is an electromagnetic interference/radio frequency interference (EMI/RFI) shield for an electrical metering device (Figure 1). The major function of the shield is to reduce EMI/RFI in the metering circuitry.

The design process required for an EMI/RFI shield consists of selecting the shield performance requirements, structural material and associated forming process, and the shielding material, with its associated application process. This implementation of design justification was conducted on the SUN Microsystem 3/50 workstation using the C programming language interfaced with the X-window windowing system.

The engineering knowledge base for EMI/RFI shielding design consists of information on plastic materials, plastic forming processes, conductive coating materials, coating application processes, and conductive fill materials. This information includes the current material and equipment costs, and provides the feasible technical path resulting from the selection of a given material or manufacturing technology.

The development of the engineering knowledge base can be affected by difficulty in collecting accurate engineering and economic information. It was found that for some information needed either the information did not exist in the necessary form or it was not available. This required the estimation of engineering and economic information from that which was available. Dealing with incomplete or estimated data requires different procedures than when accurate data is available. Information that was estimated or incomplete was presented as it is. The representation of known information was signified as discrete points within the estimated relationships. Therefore, the decision maker is able to deal with such information as is appropriate.

The exploration of the economic ramifications of the different design decisions is possible in two ways. First, a cost estimate is developed incrementally with the design. From this the decision maker is made aware of the unit cost and is able to design within the initial budget constraints for the project and the proposed unit price. Second, the decision maker can conduct an incremental payback analysis between two different design alternatives. Other economic analyses were explored such as an isoquant analysis, but due to the discrete nature of the design alternatives it was not deemed plausible.

The presentation of information to the decision maker takes several different forms. The decision maker determines the initial design criteria which are then posted on a bulletin board. The use of a bulletin board allows the design criteria to be visible to the decision maker and allows for changes to be made as the design process proceeds. The design bulletin board is also used to post the decision maker's decisions and associated economic information. As the design is developed the total fixed cost, total variable cost and variable unit cost are posted. This allows the decision maker to see the economic ramifications of a selection and to explore other possible solutions.

A graphical representation is another form in which information is presented to the decision maker. The design justification environment allows for the simultaneous analysis of engineering and economic information. Engineering information is presented in graphical form. The decision maker can analyse the graph and decide upon what level of a given material is needed. Economic information is also presented in graphical form to parallel the decision process of the engineering perspective. Graphs showing design criteria versus design criteria, design criteria versus cost, and incremental payback period versus production level are simultaneously presented.

Example—EMI/RFI shield design justification

A complete design justification session using the implemented environment consists of defining the design criteria, searching the engineering knowledge base to design each alternative, and comparing design alternatives based upon the available engineering and economic information. This product development process could require several iterations of design.

The following example is the result of one complete design iteration using the implemented design justification environment to design an EMI/RFI shield (see Figure 2).

The design criteria are specified: deflection temperature 250 F, part weight 0·3 lb, part surface area 35 in^2, cost estimate \$2, production range per year of 5000–30 000. Since there are several ways to measure shielding effectiveness, a graph showing the relationship between surface resistivity and attenuation is displayed.

The decision maker decides upon the type of shielding that will be the first design alternative, i.e. conductive coating v. conductive filler. In this example the first design alternative is selected to be a conductive coating. A search for engineering plastics which meet the temperature constraint is conducted. A polycarbon is

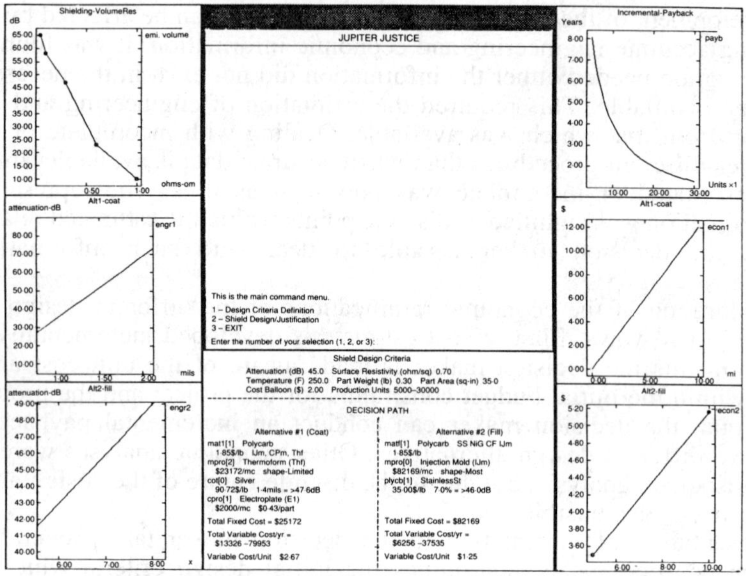

Figure 2. Example session.

selected and posted on the bulletin board, along with the associated plastic forming processes for polycarbon plastics. The current unit cost and the total variable cost range based upon the production range of 5000–30 000 units is posted.

The design process continues by searching the engineering knowledge base until a first alternative is developed. In the process of making design decisions the fixed and variable costs associated with each decision are displayed. The materials properties and cost trade-offs are posted so that the designer can consider the functional and economic factors concurrently.

The decision maker then develops a second alternative for comparison. A conductive filler is chosen as the second design alternative. A search of the engineering knowledge base continues until a second alternative is developed. Once the second alternative is developed an incremental payback analysis is performed to see the trade-off between the fixed cost and variable costs for each alternative. This process is continued until a design is developed which best satisfies the engineering and economic criteria.

The preceding example shows how the design justification environment can be used to aid the decision maker in the design process. The presentation of engineering and economic information in a form that can be utilized by the decision maker allows for consideration of design function, manufacturability, and cost in an integrated fashion.

Conclusion

It was discovered from the literature that there is a need to have new justification procedures to be able to consider the economic effect of design decisions. Several methodologies have been developed to deal with this problem. Few of these methodologies have considered integrating the product design process with that of economic justification in a way which reflects the designer's true information needs. A prototype integrated design and justification environment was developed and tested. It was found to provide the decision maker a large amount of interaction in both the engineering and economic aspects of the design process. The consideration of both engineering and economic factors concurrently allowed the decision maker to test the ramifications of different design components.

References

Au, T. and Au, Th., 1983, *Engineering Economics for Capital Investment Analysis* (Boston: Allyn and Bacon).

Boothroyd, G., 1982, Economics of assembly systems. *Journal of Manufacturing Systems*, **1**, 111–126.

Boothroyd, G., 1988, Estimate costs at an early stage. *American Machinist*, **132**, 54–57.

Daschbach, J.M. and Apgar, H., Design analysis through techniques of parametric cost estimation. *Engineering Costs and Production Economics*, **14**, 87–93.

Fleischer, G.A. and Khoshnevis, B., 1986, Incorporating economic impact assessment into computer-aided design. *Proceedings of the 1986 International Industrial Engineering Conference*, Dallas, TX, pp. 163–174.

Gold, B., 1982, CAM sets new rules for production. *Harvard Business Review*, **60**, 88–94.

Gustavson, R.E., 1984, Computer-aided synthesis of least cost assembly systems. *Proceedings of the 14th International Symposium on Industrial Robots*, Gothenburg, Sweden, pp. 95–106.

Haider, S. and Blank, L., 1983, A role for computer simulation in the economic analysis of manufacturing systems. *Proceedings of the 1983 Winter Simulation Conference*, Arlington, VA, pp. 199–206.

Hayes, R. and Garvin, D., 1982, Managing as if tomorrow mattered. *Harvard Business Review*, **60**, 70–79.

Leung, L. and Tanchoco, J.M.A., 1987, Multiple machine replacement within an integrated system framework. *The Engineering Economist*, **32**, 89–114.

Randhawa, S. and West, T., 1986, A multi-attribute methodology for the evaluation of automated assembly systems. *Proceedings of the 1986 International Industrial Engineering Conference*, Dallas, TX, pp. 526–533.

Rouse, W.B., 1986, On the value of information in system design: a framework for understanding and aiding designers. *Information Processing and Management*, **22**, 217–228.

Sage, A.P., 1987, Knowledge, skills, and information requirements for system design. In *System Design*, edited by W.B. Rouse and K.R. Boff (NY: North-Holland).

Sullivan, W., 1986, Models IE's can use to include strategic, non-monetary factors in automation decisions. *Industrial Engineering*, **18**, 42–50.

Sullivan, W. and Leclair, S., 1985, *Justification of Flexible Manufacturing Systems Using Expert System Technology*. SME Technical Paper MS85–1075 (Dearborn, MI: Society of Manufacturing Engineers).

Wierda, L.S., 1988, Product cost-estimation by the designer. *Engineering Costs and Production Economics*, **13**, 189–198.

Index

accident prevention, application of simulation methods 696–702
ACTS 551–7
adaptive scheduling, automatic 165–71
adhesives, 2-part, automatic dispensing 764–9
AGVs
  optimal flow path design 467–73
  routing control method 489–94
  scheduling 724
artificial intelligence (AI)
  cell management systems 578–87
  and manufacturing control 571–7
  vision systems 589–94
assembly
  application of AI-based vision system 589–94
  computer-aided component selection 734–9
  flexible
    demands on logistics and material supply 537–43
    versus manual 481–8
  paced, simulation of work-time distributions 653–9
  parallel systems, performance in 646–52
  precise, automated 431–5
assembly planning
  application of precedence and function diagrams 244–9
  model-mix balancing in 314–20
  using operation networks 495–503
automated guided vehicles *see* AGVs
automated manufacturing cells, cause-based failure recovery 504–9
automated production systems, vehicle scheduling models in 719–24
automatic cutting tool selection 551–7
automobile industry, Sweden, assembly systems evaluation 383–90

bandsaw teeth, computer-aided stress analysis 113–18
batch mode, relationship with bottle-neck machines and maximum throughput 283–9
batch supply systems 227
bills-of-material, generic 173–8
blackboard cell control system 571–7
BOM, generic 173–8
bottle-neck machines
  in cellular manufacturing systems 257–62
  generic scheduling 199–206
  relationship with batch mode and maximum throughput 283–9
broach design process, expert system for automation 39–44
broaching 39–40

C and P problems 46–53
cable harness manufacture, flexible loom board system 107–12
CAD
  in design and manufacture of jewellery 740–7
  feature representation and extraction in models 749–61
  two-part adhesive dispensing system 764–9
CAD/robot interface 766–7
CAI, in design and manufacture of jewellery 740–7
CAM
  in design and manufacture of jewellery 740–7
  two-part adhesive dispensing system 764–9
capital-back method 371–8
capital investments, flexible 371–8
CAPP
  in electrochemical machining 780–8
  in plastics manufacture 773–9
cause-based failure recovery 504–9
CC analysis 339–52
cell calibration, image-based 436–40
cell control, discrete 793–8

cell management systems 578–87
  use of blackboard architecture 571–7
cellular manufacturing systems
  bottle-neck machines in 257–62
  comprehensive group technology system 559–66
chatter
  cutting area change under 73–80
  predictive control in turning 65–72
China, production management practices 270–4
CIB
  development of a company-individual concept 21–5
  office automation as a component 19–21
CIM
  discrete cell control 793–8
  distributed planning and control algorithm 806–13
  flexible and porous materials handling 460–5
  garment production 799–805
  information system design and analysis 814–24
CINPRES gas-assisted moulding 98–103
classifier systems 611–12
combined manual and decision task (CMDT), relationship of learning and perceptual motor load in 392–9
component selection, computer-aided 734–9
computer-aided design *see* CAD
computer-aided manufacture *see* CAM
computer-aided process planning *see* CAPP
computer-integrated business *see* CIB
computer-integrated manufacturing *see* CIM
container loading 58
continuous supply systems 226
control selection and teach pendant design 413–19
controlled internal pressure process 98–103
cost management systems using MRP data base 353–9
creativity curve analysis 339–52
credit-period policies and inventory decisions 150–3
CRI 766–7
criterion index 32, 34–5
crystal glassware, robot cutting 132–5
cutting and packing problems 46–53
cutting area, change under chatter vibration 73–80
cutting tools
  automatic selection 551–7
  replacement in flexible manufacturing 617–23

decision-making, short-term 703–13
decision support system, knowledge-based 633–8
delivery system, flexible, simulation of 669–78
design justification in production development 829–35
DFA products, manual versus flexible assembly 481–8
discounts, effect on inventory policies 366–9
discrete cell control 793–8
disposals, effect on inventory policies 366–9
distributed information control structure for robotic systems 442–9
distributed planning and control in CIM 806–13
drilling small deep holes by electro-discharge machinery 92–7
DSS, knowledge-based 633–8
due-dates, optimal setting 297–303

electrochemical machining, CAPP 780–8
electro-discharge machining, in small deep hole drilling 92–7
engineering drawings, recognition 726–33
expert systems
  and the design of FMS 625–31
  tool replacement as 617–23

fabric dyeworks, use of simulation for short-term decision making 703–13
FACTOR 165, 166
factory co-ordination system 251–6
failure recovery, cause-based, in automated manufacturing cells 504–9
features
  extraction in CAD models 749–61
  recognition 726–33

representation in CAD models 749–61
fixed accelerator 142
flexibility 371, 372
flexible accelerator 142
flexible manufacturing systems *see* FMS
flexible materials handling 460–5, 799–805
flow path design, AGVs 467–73
FMS 513–14
in assembly lines 537–43
comparison with manual assembly 481–8
effective work organization 532–6
expert system approach to design 625–31
optimal part routing 544–50
part-sequencing problems in 305–13
performance analysis 525–30
performance evaluation of unreliable workstations 332–6
quality control problems 514–23
simulation for scheduling in 679–85
surface feet of metal cutting workstations 360–5
tool replacement as an expert system 617–23
forecasting methodology 271, 272
form feature recognition 726–33
framework systems 200
freight discounts, effect on inventory policies 366–9
function diagrams, application to assembly planning 244–9

garment production, factory automation strategies 799–805
gems, parametric library 743–4
glass, robot-cut, groove depth measurement 132–5
global scheduling 202
grinding, stability analysis 31–6
gripper, vacuum-based 460–5
groove depth measurement, robot-cut glass 132–5
group technology system for cellular manufacture 559–66
guide path layout design, AGVs 467–73

hierarchical production planning
integration of aggregate and detailed systems 290–6
in a two-stage production system 231–8
Hungary, production management practices 270–4
hybrid cell formation 260–1

industrial seating design 420–4
information manager, myths concerning 17–19
information system design and analysis 814–24
information technology 9–12
as a component of CIB 19–21
myths concerning 12–19
injection moulding, CINPRES process 98–103
inspection errors 179–88
inspection schedules, joint determination with production cycle 179–88
inventory control 141
effective 142–3, 148–9
historical context 141–2
inventory policies
and credit-period incentives 150–3
effect of freight discounts and disposals 366–9

jewellery design and manufacture using CAD/CAM/CAI 740–7
JIT 200–1
implementation planning 686–94
job-shop evaluation using Petri nets 189–95
jobbing industry, production scheduling 660–8
just-in-time 200–1
implementation planning 686–94

kanban system 353–4
KBS *see* knowledge-based systems
kitting systems 227
methods and techniques 229–30
knowledge-based systems (KBS)
in computer-assisted production 597–609
and manufacturing simulation 610–15
and production planning 633–8

learning, effect on perceptual motor load in a CMDT 392–9
logistics in flexible assembly 537–43
loom board assembly, flexible, in cable harness manufacture 107–12
lordosis, lumbar 422

M* methodology 814–24
machine cell formation 257
machine duplication 259–60
maintenance tasks, workplace design and posture 401–5
manual assembly
  comparison with flexible systems 481–8
  kitting systems for small parts 225–30
manufacturing simulation
  and analysis 202, 203–6
  and KBS 610–15
manufacturing systems
  defining 200–1
  intelligent control 571–7
MASI computer program 408–9
material flow, safety and ergonomic aspects 407–12
material requirements planning systems 251, 263–9, 353–9
metal removal rate in deep hole drilling 92–7
model-mix balancing 314–20
MPSP 720–3
MRP system 251
  in cost management 353–9
MRP II systems 263–9
MRP, in deep hole drilling 92–7
MSA generic modeller 202, 203–6
multi-processor scheduling problem 720–3

occupational safety, and material flow 407–12
off-line programming, and cell calibration 436–40
OFT, in surface texture assessment 123–30
OHMS 669–78
one-machine scheduling 208–14
operation networks 498–503
operational systems 200–1
OPNETs 498–503
OPT 201, 284
optical Fourier transform, in surface texture assessment 123–30
optimized production technology 201, 284
OR models of operations 496–7
order-handling manufacturing systems 669–78
organization of production, effective, in FMS 532–6
outsourcing 258–9

PAC system 251
paced assembly line, simulation of work-time distributions 653–9
packing problems
  interactive approaches 55–61
  solution approaches 46–53
pallet loading 56–7
parallel assembly systems, system performance in 646–52
parametric library of precious stones 743–4
part-sequencing, optimal, in FMS 305–13
parts feeder, programmable 453–9
pay-back period 372–4
penetration rate, effect in grinding process 32–3
perceptual motor load, relationship with learning in a CMDT 392–9
performance criteria 202–3
Petri nets, in job-shop evaluation 189–95
physical systems 200
plastic injection moulding, CINPRES process 98–103
plastics manufacture, CAPP 773–9
PML, relationship with learning in a CMDT 392–9
porous materials handling 460–5
posture
  monitoring 422
  in repair and maintenance tasks 401–5
  and seat design 422–3
precedence
  application to assembly planning 244–9
  graphs 497–8
pre-tinning process, evaluation 85–90
prismatic parts, quality control 513–23
product flow planning 251–6
production cells 275–82
production cycle, joint determination with inspection schedules 179–88
production development, design justification in 829–35
production disturbances, safety aspects 696–702
production management, comparisons 270–4
production planning
  hierarchical systems 231–8, 290–6
  international comparisons 271, 272
  knowledge-based decision support system 633–8

production research, current perspectives 3–8
production scheduling
  automatic 165–71
  microcomputer simulation software 660–8
  and optimal due-date setting 297–303
  simulation, in FMS 679–85
production structures, future-orientated 275–82
productivity 327–30
PROFEED 453–9
purchasing and materials management, international comparisons 273

Q-morphisms 611

rectangular parts, programmable feeder 453–9
reference measurement in precise automated assembly 431–5
regenerative chatter 31–2
repair tasks, workplace design and posture 401–5
research and development, and creativity measurements 340–4
resolution IV designs, using Taguchi's orthogonal arrays 641–4
robots
  comparison with manual assembly systems 481–8
  distributed information-control structure 442–9
  in precise assembly systems 431–5
  teach pendant design and control selection 413–19
routing
  control of AGVs, by shortest path 489–94
  modification, in bottle-neck machine 258
  optimal, in FMS 544–50

Saab, assembly systems, sociotechnical evaluation 388–9
safety analysis, application of simulation methods 696–702
seating design 420–4
selective assembly, computer-aided component selection 734–9
sensory feedback in precise automated assembly 431–5
sheet metal enclosures, flexible manufacture by task planning 474–80
shop floor control, international comparisons 272–3
short-term decision making, use of simulation 703–13
Silver-Meal heuristic, worst case bounds for finite horizons 239–42
simulation
  flexible delivery system 669–78
  and JIT implementation 686–94
  production disturbances at danger zones 696–702
  scheduling in FMS 679–85
  scheduling production in a dynamic and flexible jobbing environment 660–8
  and short-term decision making 703–13
  work-time distributions for a paced assembly line 653–9
sit-stand seat 423–4
small deep hole drilling by electro-discharge machining 92–7
South Korea, production management practices 270–4
stock holdings
  changing patterns 143–8
  historical context 141–2
stress analysis, computer-aided, in design of bandsaw teeth 113–18
stylus profilometry 123
substitutional sales promotion 155–61
supervisory control, modelling 610–15
supply systems
  categories 226–7
  factors affecting choice 227–9
surface texture assessment by OFT technique 123–30
Sweden, automobile assembly systems 383–90
synchronous components, sequence restrictions due to 646–52

Taguchi's orthogonal arrays, obtaining resolution IV designs using 641–4
task planning in flexible manufacture of sheet metal enclosures 474–80
teach pendant design and robot control selection 413–19
throughput maximization 283–9
total productivity 329
total-work-content due dates 297–303
trim loss 46, 47

turning, predictive control of
chatter 65–72
TWK due-dates 297–303
two-part adhesive dispensing system
764–9

United Kingdom, productivity practices
327–30

VCMS 217–24
vehicle scheduling models in automated
production systems 719–24
vehicle scheduling problems 720
virtual cell concept, expansions 217–24
virtual cellular manufacturing concept
217–24
vision system, AI-based 589–94
Volvo Car Corporation, assembly
systems, sociotechnical evaluation
385–6
Volvo Truck Corporation, assembly
systems, sociotechnical evaluation
386–8
VSP 720

wavesolder process, statistical quality
control design 85–90
Western Europe, production management
practices 270–4
workplace design in repair and
maintenance tasks 401–5
workstations
optimum surface feet in FMS 360–5
unreliable, performance evaluation
332–6

yield maps 115–18

zero-inventory manufacturing 141